Neurobiology of
Alzheimer's Disease

Molecular and Cellular Neurobiology Series

Series editors: G. L. Collingridge, S. P. Hunt, and N. J. Rothwell

Glial Cell Development 2e
Kristjan Jessen and William Richardson (eds)

Receptor and Ion-Channel Trafficking
Stephen Moss and Jeremy Henley (eds)

Immune and Inflammatory Responses in the Nervous System 2e
Nancy Rothwell and Sarah Loddick (eds)

The Molecular Biology of the Neuron 2e
Wayne Davies and Brian Morris (eds)

Neuroglycobiology
Minoru Fukuda, Urs Rutishauser, and Ronald Schnaar (eds)

The Neurobiology of Pain
Stephen Hunt and Martin Koltzenburg (eds)

Neurobiology of Alzheimer's Disease

(Molecular and Cellular Neurobiology)

THIRD EDITION

Edited by

David Dawbarn
Reader in Medicine, University of Bristol

Shelley J. Allen
Sigmund Gestetner Senior Research Fellow in Medicine,
University of Bristol

OXFORD
UNIVERSITY PRESS

OXFORD
UNIVERSITY PRESS

Great Clarendon Street, Oxford OX2 6DP

Oxford University Press is a department of the University of Oxford.
It furthers the University's objective of excellence in research, scholarship,
and education by publishing worldwide in

Oxford New York

Auckland Cape Town Dar es Salaam Hong Kong Karachi
Kuala Lumpur Madrid Melbourne Mexico City Nairobi
New Delhi Shanghai Taipei Toronto

With offices in

Argentina Austria Brazil Chile Czech Republic France Greece
Guatemala Hungary Italy Japan Poland Portugal Singapore
South Korea Switzerland Thailand Turkey Ukraine Vietnam

Oxford is a registered trade mark of Oxford University Press
in the UK and in certain other countries

Published in the United States
by Oxford University Press Inc., New York

British Library Cataloguing in Publication Data

Data available

Library of Congress Cataloguing in Publication Data

Neurobiology of Alzheimer's disease / edited by David Dawbarn, Shelley J. Allen. – 3rd ed.
    p. ; cm – (Molecular and cellular neurobiology series)
  Includes bibliographical references and index.
  1. Alzheimer's disease–Pathophysiology. 2. Molecular neurobiology. I. Dawbarn, David. II. Allen, Shelley J.
III. Series.
  [DNLM 1. Alzheimer Disease–physiopathology. 2. Neurobiology. WT 155 N494 2007]
  RC 523.N417 2007   616.8'3107—dc22   2006039204

ISBN 978-0-19-856661-8 (Hbk alk paper)

1 3 5 7 9 10 8 6 4 2

Typeset in Minion
by Cepha Imaging Pvt Ltd
Printed in Great Britain on acid-free paper by
Biddles Ltd., King's Lynn

Preface

It is now 100 years since Alois Alzheimer first reported his findings on his patient 'Auguste D' (1907). However, it is noteworthy that nearly half of the papers on Alzheimer's disease have been published since the year 2000.

By the beginning of this century we already understood much of the nature of the disease. The protein constituents forming the neuropathological hallmarks of the disease: plaques and tangles had been identified and the cDNAs had been cloned. Thus Aβ, a small fragment of the amyloid precursor protein (APP), was known to be located within plaques, and the tau protein was found to be within tangles. As the first edition of this book went to press in 1995, the discoveries were announced of the linkage of mutations in chromosomes 14 and 1 leading to early onset familial Alzheimer's disease; these were later shown to be due to mutations in the presenilin 1 and presenilin 2 genes, respectively. Two enzymes, β- and γ-secretase, were known to cleave APP for the release of Aβ. Later, γ-secretase was shown to be a complex of the proteins presenilin, nicastrin, APH1, and Pen2, with two presenilin aspartic acid residues forming the catalytic core of the complex. The identity of the enzyme β-secretase was revealed (coincidentally just as the second edition of this book went to press in late 1999) as BACE (β-APP site cleaving enzyme). The identification of these enzymes, together with the X-ray crystal structure of BACE, has led to the development of specific BACE and presenilin inhibitors with the aim of preventing amyloid plaque formation. As we go to press for the third edition many of these are now in clinical trials.

However, this is not the only credible therapeutic avenue. Very recently, evidence has lent weight to the theory that the cholinergic deficits seen early in the disease are important in molecular pathogenicity. In particular, specific muscarinic agonists have been shown to reverse Alzheimer pathology in transgenic animal models. In addition, although clinical trials have been temporarily halted because of side effects, the immunization of patients with Aβ still offers hope for a worldwide cost-effective therapeutic.

Thus, although Alzheimer's disease, a century after the first description, now presents a huge and complex puzzle, we can look forward with optimism to the availability of a variety of therapeutic solutions, probably within the next 5 years. However, it is most likely that finding a solution to the enigma of the molecular processes involved will take a good deal longer.

This book aims to present an accessible overview of the major areas of knowledge and controversy, each written by leaders in their area of expertise. Included are chapters on the protein hallmarks Aβ (how and why it is formed, the neuropathological consequences of its production, and how they may be halted) and tau, and its role in the disease process. The relevance of genetic mutations and polymorphisms are discussed, and disease-related neurochemical changes are described, in particular the molecular processes involved in cholinergic degeneration. There is emphasis on current and exciting new therapeutic avenues and possible future directions.

One hundred years on, the field of Alzheimer's disease is more exciting and contentious than it ever was. We trust that this new edition will be accessible to a range of readers, and that interested parties will enjoy the various perspectives offered, and hopefully will be filled with enthusiasm and optimism by the content.

David Dawbarn
Shelley J. Allen

Contents

List of Contributors

Allen, Shelley J.
Molecular Neurobiology Unit
University of Bristol
Dorothy Hodgkin Building
Whitson St
Bristol BS1 3NY, UK
shelley.allen@bristol.ac.uk

Alonso, Alejandra del C.
New York State Institute for Basic
Research in Developmental Disabilities
1050 Forest Hill Road
Staten Island NY 10314-6399, USA

Bhattacharyya, Sarmishtha
Wythenshawe Hospital
Manchester M23 9LT, UK

Blennow, Kaj
Institute of Clinical Neuroscience
Department of Experimental
Neuroscience
Sahlgrenska University Hospital
Göteborg University
Göteborg Sweden
kaj.blennow@neuro.gu.se

Borghgraef, Peter
Experimental Genetics Group
Department of Human Genetics,
KU Leuven B-3000
Leuven, Belgium

Burns, Alistair
Department of Old Age Psychiatry
Division of Psychiatry
University of Manchester Education and
Research Centre
School of Psychiatry and Behavioural
Sciences
Wythenshawe Hospital
Manchester M23 9LT, UK
Alistair.burns@manchester.ac.uk

Chen, Xi
Department of Neurology and VA
Medical Center
School of Medicine
Saint Louis University
St. Louis, MO 63106, USA

Chohan, M. Omar
New York State Institute for Basic
Research in Developmental Disabilities
1050 Forest Hill Road
Staten Island, NY 10314-6399, USA

Counts, Scott E.
Rush University Medical Center
Department of Neurological Sciences
Chicago, IL 60612, USA
scounts@rush.edu

Croes, Sophie
Experimental Genetics Group
Department of Human Genetics
KU Leuven, B-3000
Leuven, Belgium

Dawbarn, David
Molecular Neurobiology Unit
University of Bristol
Dorothy Hodgkin Building
Whitson Street
Bristol BS1 3NY, UK
dave.dawbarn@bristol.ac.uk

Dewachter, Ilse
Experimental Genetics Group
Department of Human Genetics,
KU Leuven, B-3000
Leuven, Belgium

El-Akkad, Ezzat
New York State Institute for Basic
Research in Developmental Disabilities
1050 Forest Hill Road
Staten Island, NY 10314-6399, USA

Esiri, Margaret M.
Neuropathology Department
Level 1, West Wing, John Radcliffe Hospital,
Headington, Oxford OX3 9DU
margaret.esiri@clneuro.ox.ac.uk

Fahnestock, Margaret
Department of Psychiatry and
Behavioural Neurosciences
McMaster University
Hamilton, Ontario L8N 3Z5, Canada

Francis, Paul T.
Wolfson Centre for Age-Related Diseases
GKT School of Biomedical Sciences,
King's College
London, UK
paul.francis@kcl.ac.uk

Ghiso, Jorge
Departments of Pathology and Psychiatry
NYU School of Medicine
TH–432, 550 First Avenue
New York, NY 10016, USA
ghisoj01@med.nyu.edu

Ginsberg, Stephen D.
Center for Dementia Research
Nathan Kline Institute
Department of Psychiatry and
Department of Physiology and
Neuroscience
New York University School of Medicine
Orangeburg, NY, USA
ginsberg@nki.rfmh.org

Goate, Alison M.
Department of Psychiatry
B8134, Washington University School of
Medicine
St. Louis, MO 63110, USA
goatea@icarus.wustl.edu

Gong, Cheng-Xin
New York State Institute for Basic
Research in Developmental Disabilities
1050 Forest Hill Road
Staten Island, NY 10314-6399, USA

Grundke-Iqbal, Inge
New York State Institute for Basic
Research in Developmental
Disabilities
1050 Forest Hill Road
Staten Island, NY 10314-6399, USA

Holtzman, David M.
Department of Neurology, Molecular
Biology and Pharmacology
Alzheimer's Disease Research Center and
the Hope Center for Neurological
Disorders
Washington University School
of Medicine
660 S. Euclid, Box 8111
St. Louis, MO 63110, USA
holtzman@neuro.wustl.edu

Iqbal, Khalid
New York State Institute for Basic
Research in Developmental Disabilities
1050 Forest Hill Road
Staten Island, NY 10314-6399, USA
iqbalk@worldnet.att.net

Kauwe, John S.K.
Department of Psychiatry
Washington University
660 St Euclid, Campus Box 8134
St Louis, MO 63110-1093, USA
keoni@icarus.wustl.edu

Khatoon, Sabiha
New York State Institute for Basic
Research in Developmental Disabilities
1050 Forest Hill Road
Staten Island, NY 10314-6399, USA

Lai, Mitchell K.P.
Wolfson Centre for Age-Related Diseases
GKT School of Biomedical Sciences
King's College
London, UK
and
Dementia Research Laboratory
Department of Clinical Research
Singapore General Hospital
Singapore

Lal, Ratnesh
Center for Nanomedicine
Department of Medicine
University of Chicago
MC 6076 Chicago, IL 60637, USA
rlal@uchicago.edu

Lee, Edward B.
The Center for Neurodegenerative Disease
Research
Department of Pathology and Laboratory
Medicine
HUP Maloney Building, 3rd Floor
University of Pennsylvania School of
Medicine
36th and Spruce Streets
Philadelphia, PA 19104, USA

Lee, Virginia M.-Y.
The Center for Neurodegenerative Disease
Research
Department of Pathology and Laboratory
Medicine
HUP Maloney Building 3rd Floor
University of Pennsylvania School of
Medicine
36th and Spruce Streets
Philadelphia, PA 19104, USA
vmylee@mail.med.upenn.edu

Liu, Fei
New York State Institute for Basic
Research in Developmental Disabilities
1050 Forest Hill Road
Staten Island, NY 10314-6399, USA

Masters, Colin L.
Department of Pathology and Mental
Health Research Institute
University of Melbourne
Parkville, Victoria 3010, Australia
c.masters@unimelb.edu.au

Mattson, Mark P.
Laboratory of Neurosciences
National Institute on Aging
Baltimore, MD 21224, USA
mattsonm@grc.nia.nih.gov

Mufson, Elliott J.
Rush University Medical Center
1735 W. Harrison Street, Suite 300
Chicago, IL 60612, USA
emufson@rush.edu

Muyllaert, David
Experimental Genetics Group
Department of Human Genetics,
KU Leuven, B-3000
Leuven, Belgium

Overshott, Ross
Trafford NHS Trust
Manchester, M41 5SL, UK

Ramírez, María J.
Laboratory of Neuropharmacology
Center for Applied Medical Research
(CIMA)
School of Medicine
University of Navarra
Pamplona, Spain

Rostagno, Agueda
Department of Pathology
NYU School of Medicine
TH 432, 550 First Avenue
New York, NY 10016, USA
rostaa02@med.nyu.edu

Stern, David
Dean's Office
University of Cincinnati College of
Medicine
Cincinnati, OH 45267. USA

Terwel, Dick
Experimental Genetics Group
Department of Human Genetics
KU Leuven, B-3000
Leuven, Belgium

Thinakaran, Gopal
Department of Neurobiology,
Pharmacology and Physiology
University of Chicago
Knapp R212
924 East 57th Street
Chicago, IL 60637, USA
gopal@uchicago.edu

Tsang, Shirley W.Y.
Dementia Research Laboratory
Department of Clinical Research
Singapore General Hospital
Singapore

Vandebroek, Tom
Experimental Genetics Group
Department of Human Genetics
KU Leuven, B-3000
Leuven, Belgium

Van Dooren, Tom
Experimental Genetics Group
Department of Human Genetics
KU Leuven, B-3000
Leuven, Belgium

Van Helmont, Thomas
Laboratory of Functional Biology
KU Leuven, B3001
Leuven, Belgium

Van Leuven, Fred
Experimental Genetics Group
Department of Human Genetics
Campus Gasthuisberg, O&N 1, niv6
KU Leuven, B-3000
Leuven, Belgium
Fred.VanLeuven@med.kuleuven.ac.be

Vetrivel, Kulandaivelu S.
Department of Neurobiology,
Pharmacology and Physiology
University of Chicago
Knapp R212
924 East 57th Street
Chicago, IL 60637, USA

Wahrle, Suzanne E.
Department of Neurology
Washington University School of
Medicine
660 S. Euclid, Box 8111
St. Louis, MO 63110, USA

Wilcock, Gordon K.
Nuffield Department of Medicine
University of Oxford
Level 7, John Radcliffe Hospital
Headington
Oxford, OX3 9DU, UK

Winderickx, Joris
Laboratory of Functional Biology
KU Leuven, B3001
Leuven, Belgium

Xia, Weiming
Center for Neurologic Diseases
Harvard Institute of Medicine
HIM 616, 77 Avenue Louis Pasteur
Boston, MA 02115, USA
wxia@rics.bwh.harvard.edu

Yan, Shi Du
Department of Pathology and Surgery
Taub Institute for Research on Alzheimer's
Disease and the Ageing Brain
College of Physicians and Surgeons
Columbia University
New York, NY 10032, USA
sdy1@columbia.edu

Zetterberg, Henrik
Institute of Clinical Neuroscience
Department of Experimental
Neuroscience
Sahlgrenska University Hospital
S-431 80 Göteborg
Sweden
Henrik.zetterberg@clinchem.gu.se

List of Abbreviations

Aβ	β amyloid peptide
ABAD	amyloid-β binding alcohol dehydrogenase (also called ERAB)
ABCA1	ATP-binding cassette transporter A1
ACAT	acyl-coenzyme A:cholesterol acyltransferase
ACE	angiotensin-converting enzyme
ACh	acetylcholine
AChE	acetylcholinesterase
AChEI	cholinesterase inhibitor
AChE-I	AChE inhibitor
ACT	a$_1$-antichymotrypsin
AD	Alzheimer's disease
ADAM	a disintegrin and metalloprotease family of proteases
ADAS	Alzheimer's disease assessment scale
ADDLs	Aβ derived diffusible ligands
ADFACS	Alzheimer's disease functional assessment and change scale
ADL	activities of daily living
ADRDA	Alzheimer's disease and related disorders association
AGECAT	computer-assisted algorithm
AICD	APP intracellular domain
AL	immunoglobulin light-chain systemic amyloidosis
AMPA	α-amino-3-hydroxy-5-methylisoxazole-4-propionic acid receptor
AMTS	abbreviated mental test score
Apaf	apoptotic protease-activating-factor
Aph	anterior pharynx defective
APLP	APP-like proteins
ApoE	apolipoprotein E
APP	amyloid precursor protein
ASPs	affected sibling pairs
β2m	β$_2$ microglobulin
BACE	β-site APP cleaving enzyme
BBB	blood–brain barrier
BDNF	brain-derived neurotrophic factor
BFC	basal forebrain cholinergic projection system
BPSD	behavioural and psychological symptoms of dementia
CAA	cerebral amyloid angiopathy
CAMDEX	Cambridge examination for mental disorders of the elderly (often referred to as the CAMCOG)
CAMKII	calcium/calmodulin-dependent protein kinase II
CAT	choline acetyl transferase
CBF	cholinergic basal forebrain
cdk	cyclin-dependent kinase
CEI	cholinesterase inhibitor
CH25H	cholesterol 25-hydroxylase
ChAT	choline acetyltransferase
CIE	Canberra interview for the elderly
CJD	Creutzfeldt–Jakob disease
CK	casein kinase
CNS	central nervous system
COX	cyclooxygenase
CREB	cyclic AMP response element binding protein
CSDD	Cornell scale for depression and dementia
CSF	cerebrospinal fluid
CTF	C-terminal fragment
CVD	cerebrovascular disease
DA	dopamine
DAD	disability assessment for dementia
DAPT	N-[N-(3,5-difluorophenacetyl)-L-alanyl]-S-phenylglycine t-butyl ester
DARPP-32	dopamine- and cAMP-regulated protein with molecular weight 32 kDa
DAT	DA transporter
DBB	diagonal band of Broca
DLB	dementia of Lewy bodies
DS	Down's syndrome
DSL	delta, serrate, and *Caenorhabditis elegans* lag-2 proteins family of ligands

DSM-IV	diagnostic and statistical manual (North American based)		ICAM	intercellular adhesion molecule
DZ	dizygotic twins		ICD 10	international classification of disease (European based)
EC	entorhinal cortex		ICV	intracerebroventricular
EEG	electroencephalography		IDDD	interview for deterioration in daily living activities in dementia
ELISA	enzyme-linked immunosorbent assay		IDE	insulin degrading enzyme
EPR	electron paramagnetic resonance spectroscopy		IGF	insulin-like growth factor
ER	endoplasmic reticulum		IP3	inositol triphosphate
ER/IC	endoplasmic reticulum/intermediate compartment		IQCODE	informant questionnaire on cognitive decline in the elderly
ERK	extracellular regulated kinase		JNK	N-terminal kinase
ERT	oestrogen replacement therapy		KA	kainate
FAD	familial AD		KPI	Kunitz-type serine protease inhibitor domain
FBD	familial British dementia		LBD	Lewy body dementia
FDD	familial Danish dementia		LDLR	low-density lipoprotein receptor
FDG-PET	2-[18F]fluoro-2-deoxy-D-glucose positron emission tomography		L-DOPA	3,4-dihydroxy-L-phenylalanine
FTD	frontotemporal dementia		LOAD	late-onset AD
FTDP-17	frontotemporal dementia with parkinsonism linked to chromosome-17		LOD	log of the likelihood odds ratio scores
FTIR	Fourier transform infrared spectroscopy		LPAC	lipid-protein attenuating compound
GABA	γ-aminobutyric acid		LRP	LDLR-related protein
GAL	neuropeptide galanin		LTD	long-term depression
GFAP	glial fibrillary acidic protein		LTP	long-term potentiation
GMSS	geriatric mental state schedule		LXR	liver X receptor
GnRH-1	gonadotropin-releasing hormone-1		MAO	monoamine oxidase
GSK-3	glycogen synthase kinase-3		MAP	microtubule-associated protein
HACU	high-affinity choline uptake system		MAPK	mitogen-activated protein kinase
HCHWA	hereditary cerebral haemorrhage with amyloidosis		MCI	mild cognitive impairment
			m-CPP	meta-chlorophenylpiperazine
HCHWA-D	hereditary cerebral haemorrhage with amyloidosis, Dutch type		M-CSF	macrophage colony-stimulating factor
HCHWA-I	hereditary cerebral haemorrhage with amyloidosis, Icelandic type		MEK	mitogen-activated or extracellular signal-regulated protein kinase
HDL	high-density lipoprotein		MHBD	methyl-3-hydroxybutyryl-CoA dehydrogenase
5-HIAA	5-hydroxyindoleacetic acid		MINT	munc18-interacting protein
HLA	human leucocyte antigen		MLS	multipoint LOD score
HMG	hydroxymethyl-CoA reductase,		MMSE	mini-mental state examination
HR	hydrophobic regions		MPAC	metal-protein attenuating compound
HRT	hormone replacement therapy		MPT	mitochondrial membrane permeability transition pore
5-HT	5-hydroxytryptamine (serotonin)		MPTP	1-methyl-4-phenyl-1,2,3,6-tetrahydropyridine
I-1	inhibitor-1			
IC	intermediate compartment		MRI	magnetic resonance imaging

MSR	macrophage scavenger receptor	PKA	protein kinase A
MTT	3-(4,5-dimethylthiazol-2-yl)-2,5-diphenyltetrazolium bromide	PKC	protein kinase C
		PLA2	phospholipase A2
MZ	monozygotic twins	PLC	phospholipase C
NA	noradrenaline	PLCγ	phospholipase C-γ
NACP	α-synuclein-derived non-amyloid component	PLD	phospholipase D
		PNS	peripheral nervous system
NBM	nucleus basalis of Meynert	PPAC	protein–protein attenuating compound
NADPH	nicotinamide adenine dinucleotide phosphate	PPARγ	peroxisome proliferators-activated receptor γ
NCAM	neural cell adhesion molecule		
NCI	no cognitive impairment	PS	presenilin
NEP	neutral endopeptidase, neprilysin	PSP	progressive supranuclear palsy
NET	NA transporter	P-tau	phosphorylated tau
NFκB	nuclear factor κB	QOL	quality of life
NFT	neurofibrillary tangles	RAGE	receptor for advanced glycation endproducts
NGF	nerve growth factor		
NICD	notch intracellular domain	RAP	receptor-associated protein
NICE	national institute for clinical excellence	RNAi	RNA interference silencing,
		ROCK	rho-associated kinase
NINCDS	national institute of neurological and communicative disorders and stroke	SAP	serum amyloid P component
		Sb	subiculum
		SDS–PAGE	sodium dodecyl sulfate polyacrylamide gel electrophoresis
NMDA	N-methyl-D-aspartate		
NMR	nuclear magnetic resonance	SELDITOF	surface-enhanced laser desorption/ionization time-of-flight mass spectrometry
NO	nitric oxide		
NPI	neuropsychiatric inventory		
NRADD	neurotrophin receptor like death domain protein	SF	straight filament
		SIDAM	structured interview for the diagnosis of dementia of the Alzheimer type, multi-infarct dementia and other dementias
NSAID	non-steroidal anti-inflammatory drug		
NT-3	neurotrophin-3		
NT-4	neurotrophin-4		
NTF	N-terminal fragment	SNP	single nucleotide polymorphism
p75NTR	p75 neurotrophin receptor	SOD	superoxide dismutase
PBE	present behavioural examination	SPECT	single-photon-emission computed tomography
PD	Parkinson's disease		
PDPK	proline-directed protein kinase	SSRI	selective serotonin-reuptake inhibitor
Pen	presenilin enhancer	TACE	TNF-a converting enzyme
PET	positron emission tomography	TCA	tricyclic antidepressant
PGE2	prostaglandin E prostanoid subtype 2 receptor	TGF-β	transforming growth factor-β
		TGN	trans-Golgi network
PHF	paired helical filament	THA	tacrine or tetrahydroaminoacridine
PI3K	phosphatidylinositol 3 kinase	TM	transmembrane
PIB	Pittsburgh compound B	TNF-α	tumour necrosis factor-a
PID	phosphotyrosine interaction domain	tPA	tissue type plasminogen activator
		Trk	tyrosine receptor kinase

T-tau	total tau (all tau isoforms irrespectively of phosphorylation status)
VAChT	vesicular ACh transporter
VAD	vascular dementia

VAMP	synaptobrevin/vesicle-associated membrane protein
VLDLR	very-low-density lipoprotein receptor
VMAT	vesicular monoamine transporter

Chapter 1

Alzheimer's disease: a hundred years of investigation

Shelley J. Allen

Alzheimer's disease (AD) is a condition with three major defining indicators: the symptoms of dementia and the occurrence of two neuropathological markers (in abundance), extracellular amyloid plaques and intracellular neurofibrillary tangles (referred to generally here as 'tangles'). We now know that a large number of influences, both genetic and environmental, are able to affect the likelihood and/or severity of AD. However, although the situation with regard to the molecular pathogenesis of AD appears extremely complex, this overview attempts to simplify matters and to present an introduction to the major ideas and arguments in the field. Different aspects are then considered in depth in the subsequent chapters.

1.1 Alzheimer: the beginnings

Alois Alzheimer (1864–1915) (Fig. 1.1) was a German physician at the Municipal Asylum for the Insane and Epileptic, Frankfurt, when in November 1901 he first interviewed a female patient with an early-onset form of dementia. Auguste D. was then 51 years old, with no family history of dementia. The onset of her dementia had been sudden, 8 months before her admission to hospital. She first suffered delusions, and then rapidly deteriorated, exhibiting signs of memory dysfunction, paranoia, and problems with language and behaviour. She died of septicaemia in 1906, apparently having spent most of the previous 2 years in a state of confusion and incomprehension. Meanwhile, Alzheimer had moved to Heidelberg to work with Emil Kraeplin, and then to Munich, again with Kraeplin, the following year. The autopsy samples from Auguste D. were sent to Alzheimer in Munich, and he presented the results of his neuropathological findings along with the case history to fellow psychiatrists in Tübingen 6 months later. The neuropathology described cerebral atrophy, intraneuronal fibrillar bundles (i.e. neurofibrillary tangles), and numerous extracellular deposits in the cerebral cortex (i.e. amyloid plaques), but also atherosclerotic changes. These findings were published in 1907 (Alzheimer 1907), and later Kraeplin referred to this and related cases as 'Alzheimer's disease' in his publication on dementia in 1910 (Kraeplin 1910).

Similar descriptions of neuropathology to that presented by Alzheimer had been reported previously by others, from 'senile' brains, albeit in less detail. However, despite

Fig. 1.1 Alois Alzheimer (1864–1915). In 1907 Alzheimer published his seminal paper: 'Uber eine eigenartige Erkrankung der Hirnrinde' [A novel disease of the cortex].

this, an artificial separation between 'senile' and 'pre-senile' became established and, because of the severity of the illness in younger patients, the term 'Alzheimer's disease' was associated with pre-senile cases. Thus for a long time it was assumed that the majority of senile dementias were due to atherosclerotic changes. It was not until the 1960s that it was realized that the majority of cases of dementia in the elderly were associated with AD neuropathology, and that the severity of this correlated with dementia scores (e.g. Roth *et al.* 1966). Eventually the pre-senile form of dementia and SDAT (senile dementia of the Alzheimer type) were seen to be essentially one syndrome and now the term 'Alzheimer's disease' refers to anyone with progressive dementia who presents with the requisite neuropathological indicators, retrospectively confirmed post-mortem.

1.2 **Today: the problem**

AD is the most common form of dementia in the elderly. Estimates of numbers affected worldwide are said to be approximately 18 million (statistics from http://www. alzheimers.org/). Looking at two Western countries as examples, the UK with a reported estimate of 420 000 AD sufferers in a population of approximately 60.6 million (0.7%) and the USA with reportedly 4.5 million sufferers in a total population of 298 million (1.5%), we have an insight into the possible number of cases in a worldwide population of 6.5 billion. The caveat to this is that there are extreme variations in life expectancy, and in some countries most of the populus do not reach their seventies, the age when signs of AD often appear. Conversely, in many societies in which support for the elderly is strong, changes in their behaviour may go unreported. The likelihood is that the prevalence of the disease often quoted as 1 in 10 of those over 65 years and 1 in 5 of those over 80 years is likely to be a conservative estimate.

As we live longer we become statistically more likely to suffer from AD; this is an increased burden on ourselves, our carers, and the health system (our taxes). Although the criteria of AD appear narrow, i.e. progressive dementia with the presence of plaques and tangles, it is perhaps better described as a set of symptoms or a syndrome since it is likely that the endpoint of the disease may be reached in a number of different ways. The situation appears extremely complex; it involves numerous possible interactions of molecules such as amyloid precursor protein (APP), tau, the constituents that comprise α-secretase (e.g. transforming growth factor-α (TNF-α) converting enzyme (TACE)), β-secretase (β-site APP cleaving enzyme (BACE-1 and BACE-2)), and γ-secretase (presenilin (PS), nicastrin, anterior pharynx defective (aph1), presenilin enhancer-2 (Pen2)), apolipoprotein E (apoE), a myriad of kinases (e.g. glycogen synthase kinase 3 (GSK-3)), neurotransmitters (e.g. acetylcholine (ACh)), numerous receptors (e.g. M1), the receptor for advanced glycation endproducts (RAGE), and so on apparently *ad infinitum*.

However, it may be possible to simplify the picture somewhat. The neuropathological defining features of AD, amyloid plaques and tangles, have as their major protein constituents amyloid/Aβ and tau, respectively. Aβ is formed from the larger protein APP. A mutation in the APP gene may result in early-onset AD, complete with dementia and plaques and tangles. Thus we may deduce that the presence (the action, production, deposition, lack of clearance) of Aβ precedes tangle formation and dementia. We also know that mutations in the tau gene (MAPT) do not lead to AD but result instead in other forms of dementia such as frontotemporal dementia of the Parkinson type associated with chromosome 17 (FTDP-17; the tau gene is on chromosome 17) or supranuclear palsy (PSP), also known as Steele–Richardson–Olszewski syndrome, where tangles are present but there are no amyloid plaques. Again, this seems to confirm that Aβ formation, in familial early-onset AD (FAD) and in sporadic AD, is 'upstream' of tangle formation, yet clearly the changes in tau must be instrumental in bringing about neurodegeneration and symptoms of dementia as seen in FTDP-17 and PSP. Thus we have a divide. Which is the greater contributor to the pathological process in AD and thus to the dementia?

1.3 Neuropathology and neurochemistry: selective vulnerability

1.3.1 Neuropathology

Post-mortem examination of the brain of a patient with severe AD immediately reveals an obvious reduction in size compared with a normal brain (the detailed neuropathology is described in Chapter 2). Atrophy particularly affects the medial temporal lobes and hippocampus, and presents as thinning of the cortical gyri, widening of the sulci, and enlargement of the lateral and third cerebral ventricles. There is loss of neurons, particularly in the neocortex and the hippocampus. Margaret Esiri (Chapter 2) refers to a loss, on average, of 40% of large pyramidal neurons in the neocortex, a 68% loss of hippocampal CA1 neurons, and also reductions of 40–70% in subcortical nuclei such as dorsal raphe, locus coeruleus, and basal nucleus.

With the aid of various staining techniques, amyloid plaques and tangles can be viewed microscopically. Plaques may have a dense 'core' of amyloid protein with amyloid aggregates forming β-pleated sheets, they may be 'neuritic' (surrounded and infiltrated by degenerating neurites), or they may be 'diffuse' in nature. Amyloid in the β-pleated sheet form can be visualized by staining with Congo red, which under polarized light produces a green birefringence, or by a stain such as thioflavin S which fluoresces. However, the 'diffuse' plaques, which are more abundant, are not in a β-pleated form and so cannot be visualized in this way. They require antibodies, usually to Aβ, in order to be made visible. About 80% of AD patients also have cerebrovascular amyloid in a β-pleated form deposited in the small vessels in the leptomeninges and cortices, i.e. congophilic angiopathy (Masters *et al.*1985).

The most severe neurofibrillary changes are generally found in the hippocampus, the entorhinal cortex, the amygdala, and the cortical association areas of the temporal, parietal and frontal cortex, with few or none in the motor area, primary visual areas, or the sensory regions (Pearson *et al.* 1985). Concomitant with this is the notion of selective vulnerability, i.e. certain areas are vulnerable to pathology while others are spared. The order of vulnerability of pyramidal neurons is discussed in relation to pathology in Chapter 2 as having an inverse relationship to the sequential anatomical connections linking the primary sensory cortex via the association cortex to the entorhinal cortex and hippocampus.

Wilcock and Esiri (1982) reported a strong correlation of clinical dementia with tangle formation and a much weaker association with amyloid plaque deposition, which supported the theory that tangles had a greater bearing on the ensuing dementia. However, it should also be considered that plaques, being more readily accessible to clearing mechanisms (see Chapters 4,17, and 18 for details) may therefore be less likely to act as reliable indicators of sequential time-dependent pathology. Such mechanisms include the action of the enzymes neutral endopeptidase (neprilysin), insulin-degrading enzyme (IDE), angiotensin-converting enzyme (ACE), and tissue type plasminogen activator (tPA). Nonetheless, the number of tangles was later shown to correlate with loss of hippocampal neurons (Bondareff *et al.* 1993) and loss of synapses (Terry *et al.* 1991). Furthermore, Braak and Braak (1991) showed that not only did neurofibrillary pathology correlate with severity of dementia, but it did so in an ordered fashion. Numbers and placement of tangles matched stages (I–VI) of dementia. Since, as mentioned above, mutations in APP which cause an increase in Aβ formation also lead to tangle formation, logically one assumes that there must be a connection between the two markers. This is discussed in terms of molecular pathology below; however, in terms of anatomical connection there is evidence to suggest that plaques are found at the axonal and dendritic terminations of tangle-bearing neurons (Chapter 2). This suggests a situation where axons of neurons are retrogradely affected by amyloid deposition or Aβ oligomers impinging on synapses, and would fit with a sequential pathway as suggested above.

1.3.2 Neurochemical changes

Selective vulnerability also occurs in terms of neurochemical markers (discussed in Chapter 12). In the late 1960s the finding of a loss of dopaminergic function in the substantia nigra in Parkinson's disease, and the subsequent success of the drug L-dopa, gave credence to the idea that there may be a corresponding change in a subset of neurons in AD. In the 1970s, investigation into the effects of scopolamine, an anaesthetic which induces memory loss (Drachman and Leavitt 1974), was rapidly followed by the discovery that cholinergic function was severely impaired in autopsy tisue (Bowen *et al.* 1976; Davies and Maloney 1976), and later in biopsy tissue (Sims *et al.* 1983), in AD. The 'cholinergic hypothesis' suggested that the loss of cholinergic function was pivotal to the neuropathology and dementia of AD, and so the search for cholinergic-based therapies ensued. It is intriguing that today, 30 years after the discovery of the cholinergic loss in AD, evidence is fast accumulating to show that cholinergic activity is not just important in a functional sense (i.e. cognitive processes, memory) but has fundamental significance in the molecular pathology of AD (Chapters 13 and 16).

In 1986 a trial of the drug tetrahydroaminoacridine (THA; tacrine) showed striking recovery in a small number of AD patients (Summers *et al.* 1986). This was the foundation for dozens of ensuing trials of this and other cholinesterase inhibitors (CEIs) (discussed in Chapter 16). The mode of action involved prevention of the degradation of ACh (by acetylcholinesterase), thus potentiating its response at the synapse. Although later trials did not reproduce the success of the initial findings, they still worked well enough for interest to continue. The side effects, which included hepatoxicity, lessened in second-generation CEIs such as rivastigmine (ENA 713, Exelon), donepezil (Aricept), and galantamine (Reminyl), and these drugs are now generally well tolerated. The latest ideas on the cholinergic hypothesis are discussed below (section 1.9), as they need to be placed in the wider context of amyloid formation and toxicity.

Other neuronal subsets, such as glutamatergic, noradrenergic, and serotonergic neurons, have subsequently been shown to be affected (e.g. Bowen *et al.* 1983). Loss of 5-HT, and losses of noradrenergic neurons in the locus coeruleus, may be associated with behavioural changes such as those seen in depression (see Chapter 12).

Even now, the major treatment for AD is cholinesterase inhibition. This is successful in over 50% of patients treated, but the effects are relatively short-lived probably because of the continuing failure of the cholinergic cells. In order to find new treatments it is necessary to understand the molecular mechanisms involved in the pathology of AD. The most obvious of these concerns the production of Aβ, the basic constituent of amyloid plaques.

1.4 Amyloid, amyloid precursor protein, and genetic clues

1.4.1 Amyloid precursor protein: chromosome 21

In the 1980s deposited amyloid (Aβ) was extracted and purified from human brain tissue and characterized, an arduous task because of the aggregative nature of the protein. The amino acid (aa) sequence of cerebrovascular amyloid was subsequently obtained by Glenner and Wong (1984). The plaque core of amyloid protein from the cerebral cortex of AD brain was also purified and characterized and found to be similar to that of amyloid of cerebrovascular origin and amyloid found in the brains of Down's syndrome patients (Masters *et al.* 1985). In all cases the protein consisted of multimeric aggregates of a 4 kDa polypeptide of about 40 residues. In 1987 this peptide was shown to be derived from a much larger precursor protein, APP (e.g. Goldgaber *et al.* 1987), and the gene was later localized to the long arm of chromosome 21. The APP gene contains 18 exons and there are seven known splice variants of the protein; the most common are 695, 751, and 770 aa in length. APP is expressed in all tissues, with the major neuronal transcript being 695 aa and the 770 aa transcript being expressed in most other cell types including glia. Down's syndrome sufferers usually develop AD symptoms and pathology by their forties, and it is usually thought that this is because of the presence of an extra whole or part of chromosome 21, and thus is a gene dosage effect. The two longer forms contain a 56 aa Kunitz protease inhibitor (KPI) domain near the N-terminus, are also known as protease nexin II, and are able to inhibit a number of serine proteases. APP has many putative functions including promoting neurite outgrowth, cell adhesion molecule, synaptogenesis, and cell survival (see Chapter 4 for more details).

In 1990 patients with amyloidosis (hereditary cerebral haemorrhage with amyloidosis-Dutch type (HCHWA-D)) were shown to have a mutation in the *APP* gene resulting in a protein mutation at Aβ22 (E693Q; the notation for *APP* is given according to the 770 splice variant) (Levy *et al.* 1990). In 1991 a mutation in the *APP* gene (V717I; the London mutation) was found in a family of FAD patients (Goate *et al.* 1991). There are now, at the latest count, 21 pathogenic mutations in *APP* which cause AD or a cerebral amyloid angiopathy CAA-associated dementia (see Appendix AI). These mutations are situated at three main sites which appear to correspond roughly to sites of cleavage of β- or γ-secretase or in the middle of the Aβ peptide (a detailed review of the genetic aspects of AD is given in Chapter 3). Aβ is formed by the cutting of *APP* by β-secretase and subsequently by γ-secretase (Fig. 1.2). This is predominantly the Aβ40 form, which is 40 aa long. However, a small amount of Aβ42 is also formed.

Mutations at the site of β-secretase cleavage, such as the Swedish double mutation (see Appendix AI), cause an increase in both forms of Aβ because of an increased affinity for the enzyme β-secretase or β-site APP cleaving enzyme (BACE) (Vassar *et al.* 1999). Mutations around the γ-secretase site increase the likelihood that γ-secretase will cut at a site 42 aa from the beginning of Aβ rather than the more common 40 aa form. The significance of this is thought to be that Aβ42 is more fibrillogenic and more toxic than Aβ40. Mutations in the middle of the Aβ sequence give rise to deposition of Aβ40 in the cerebral vessel walls and result in CAA.

1.4.2 Presenilin mutations: chromosomes 1 and 14

In 1995 two new AD genetic linkages were discovered. Mutations within the gene *PSEN1* on chromosome 14 (14q24.2) (Sherrington *et al.* 1995) and subsequently *PSEN2* on chromosome 1 (1q42.13) were found in a number of early-onset FAD kindreds (Levy-Lahad *et al.* 1995). These genes encode the multitransmembrane proteins presenilin 1 (PS1) and presenilin 2 (PS2), respectively, and have a greater than 65% sequence homology.

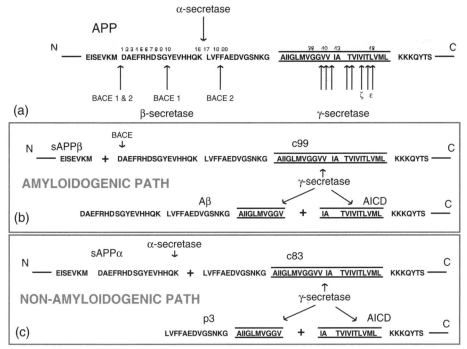

Fig. 1.2 (a) APP is a single transmembrane protein (putative membrane sequence shown between two lines) cleaved by enzyme activity from three main sources: α-, β-, and γ-secretase. This schematic diagram shows the principal cleavage sites. (b) Amyloidogenic pathway: BACE1 and BACE2 constitute β-secretase cleavage at M671/D672 (i.e. Aβ0 and Aβ1) to form sAPPβ and C99. BACE1 also cleaves APP at G680/Y681 (i.e. Aβ10 and Aβ11) to form sAPPβ' and C89. BACE2 cleaves APP at F690/F691 (i.e. Aβ19 and Aβ20) to form sAPPβ2 and C80. (c) Non-amyloidogenic pathway: α-secretase activity at APP comprises the activities of ADAM 17 (TACE), ADAM 10, and ADAM 9. α-Secretase cleaves between K687 and L688 (i.e. Aβ16/17) splitting APP to form the N-terminal sAPPα and C83 (83 aa). This cleavage precludes the production of the Aβ domain. (b and c) After α- and β-secretase cleavage, c83, c89, or c99 are cleaved by γ-secretase within the transmembrane domain (Aβ40/41 cleavage shown here) to form p3 or Aβ. γ-Secretase can also cleave C-terminal stubs to form Aβ38, Aβ39, Aβ42, or Aβ43. The APP intracellular domain (AICD) is formed simultaneously from these γ-secretase cleavages. C83 cleavage results in the production of the p3 peptide (Aβ17–40/42), C80 cleavage results in the production of Aβ20–40/42, and C89 cleavage results in the production of Aβ11–40/42. There is also another cleavage site called the ε (epsilon) cleavage site which occurs between Aβ49 and Aβ50, and a further site, the ζ (zeta) site, which occurs between Aβ46 and Aβ47. This diagram is not to scale.

Both genes have 12 exons. Mutations in *PSEN1* and *PSEN2*, like *APP* mutations, are autosomal dominant. However, mutations in PS1 give rise to extremely aggressive forms of dementia and can present symptoms in the twenties and thirties. There are now over 150 separate mutations at approximately 100 different sites on *PSEN1*. These occur throughout the protein and yet all seem to result in similar effects, i.e. increasing the ratio of Aβ42:Aβ40 levels (e.g. Citron *et al.* 1992; Borchelt *et al.* 1996). There are currently 10 known pathogenic mutations on the *PSEN2* gene (see Appendix AIV). They generally present later than *PSEN1* cases, between 40 and 75 years of age.

APP and *PS* mutations have been used as a basis for the generation of transgenic mouse models of AD (Chapter 9). Mice with *APP* FAD mutations can form diffuse and Congo red positive plaques; deposition of Aβ is vastly accelerated by crossing with transgenic *PS1* FAD mutant mice. *APP* mutations within the Aβ peptide, when used in transgenic mice, form the basis for models of CAA; increased Aβ40 formation results in deposition in the vasculature. It is possible that Aβ40, being less fibrillogenic than Aβ42, can be cleared to the blood vessels before forming deposits, and mutated forms may be more difficult to degrade.

Thus three genes are associated with FAD, *APP*, *PSEN1*, and *PSEN2*, and the majority of FAD cases are due to mutations in *PSEN1*. However, early-onset FAD represents less than 1% of all AD cases. These three genes can be used to supply clues to the possible pathogenic processes occurring in sporadic AD, and to isolate those proteins which may be interacting with APP downstream. In this way new candidate genes can be identified.

1.4.3 Apolipoprotein E: chromosome 19

In 1991 the protein apoE was shown by immunohistochemistry to be present in both amyloid plaques and tangles (Namba *et al.* 1991). In 1994 a late-onset AD family was shown to be associated with the presence of a polymorphism of apoE (apoE4). The *APOE* gene is located at chromosome 19q13.2 and has four exons. There are three common isoforms of *APOE* (see section 1.8 and Chapter 7 for further details); these are ε2 (8%), ε3 (78%), and ε4 (14%). Percentages of alleles in the population are given in parentheses but are conditional on the population. In all populations ε3 is the most common allele. The three haplotypes give rise to six genotypes: ε2ε2 (1%), ε2ε3 (15%) ε2ε4 (3%), ε3ε3 (55%), ε3ε4 (23%), and ε4ε4 (3%). These code for apoE proteins that differ by one or two amino acids out of 317: apoE3 has a cysteine at position 112 and arginine at 158, and apoE2 has cysteine residues and apoE4 has arginine residues at both positions (see Appendix AV for sequence and structure).

In the same year as its first association with late-onset AD, the ε4 allele was also found to be associated with sporadic AD (e.g. Chartier-Harlin *et al.* 1994). The presence of an ε4 allele appeared to be a risk factor rather than a direct causative factor; in other words, inheritance is not sufficient to cause the disease. However, the increased frequency of the ε4 allele is clear in late-onset families (~52%) and in sporadic AD (~40%) compared with age-matched normal subjects (~16%) (Saunders *et al.* 1993), whereas the ε2 allele has been suggested as having a protective effect, as evidenced by a decreased frequency of this allele in AD (Corder *et al.* 1994).

APOE status is the only robust genetic associative factor found so far for sporadic AD. Linkage analysis on large numbers of late-onset families has produced associations for a number of chromosomes, particularly chromosomes 9, 10, and 12. However, at least 12 chromosomes have some element of linkage, which suggests there are still many risk genes to be discovered. Candidate genes include ubiquilin and *ABCA1*. It seems likely that new genetic associations with AD will involve a multifaceted pattern of inheritance and, unlike the autosomal dominant forms discovered so far, they may be more easily modified by environmental influence. Therefore large numbers will need to be screened or new strategies developed in order to ascertain these genetic risk factors.

1.5 Processing of amyloid precursor protein

1.5.1 α-Secretase

APP is cleaved by enzymes known as α-secretase (Chapters 4 and 9), β-secretase (Chapters 4 and 17), and γ-secretase (Chapters 4, 8, and 17). The results of cleavage of APP by these enzymes are best shown in diagrammatic form (Fig. 1.2). Three members of the ADAMs (A disintegrin and metalloprotease) family are the most likely candidates for α-secretase activity: ADAM 17 (TACE), ADAM 10, and ADAM 9. There is some controversy as to which of these has constitutive/basal release and which has stimulated release, although it is most likely that TACE is activated by stimulation and ADAM 10 is constitutive. Some of the evidence for this is that TACE knockout cells lack inducible α-secretase activity (Buxbaum *et al.* 1998) and M1 muscarinic agonists, which stimulate α-secretase via protein kinase C (PKC), activate TACE and not ADAM 10 (Caccamo *et al.* 2006). α-Secretase cleaves between K687 and L688 (i.e. between positions Aβ16 and 17), splitting APP to form the N-terminal secreted sAPPα and a C-terminal fragment (CTF) C83 (83 aa). This cleavage precludes the production of the Aβ domain.

1.5.2 β-Secretase

β-Secretase cleavage is known to be due to two related aspartyl proteases: BACE1 (Vassar *et al.* 1999) and BACE2 (Acquati *et al.* 2000). The cleavage of APP by β-secretase determines the amyloidogenic pathway (Fig. 1.2b). BACE1 and BACE2 both cleave APP at M671/D672 (i.e. between Aβ0 and Aβ1) to form sAPPβ and C99. Other cleavage sites are shown in Fig. 1.2. Interestingly, the Flemish mutation (close to the BACE2 site of cleavage) significantly increases Aβ production generated by BACE2 but not by BACE1, whereas the Swedish mutation increases Aβ production by both BACE1 and BACE2 (Chapter 17; see also Fig.1.2).

1.5.3 γ-Secretase

Following α- and β-secretase cleavage, the C-terminal fragments can be cleaved by γ-secretase (Figs 1.2b and 1.2c), within the transmembrane domain, to form p3 or Aβ (Chapters 4, 8 and 17), commonly between Aβ40 and 41 or between Aβ42 and 43. The APP intracellular domain (AICD) is formed simultaneously from these γ-secretase cleavages; this is then involved in signalling pathways (see Chapter 4 for details). γ-Secretase cleavage of C83 results in the production of the p3 peptide (Aβ17–40/42). Significantly, α- and β-secretase compete for the APP substrate. When BACE is overexpressed, cleavage of APP to form sAPPβ and C99 is increased, and cleavage by α-secretase to form C83 is decreased.

γ-Secretase comprises a complex of protein components, the first of which to be discovered was PS1, with PS2 later being shown to act in a similar capacity. De Strooper *et al.* (1998), using neurons from a PS1 knockout mouse, reported a reduction in Aβ production and concomitant accumulation of APP CTF. In the light of the similarity between the Notch knockout mouse and the PS1 knockout (Shen *et al.* 1997), i.e. severe developmental abnormalities, a parallel was drawn between PS-mediated cleavage of APP and that of the Notch receptor (e.g. De Strooper *et al.* 1999). Prerequisites for γ-secretase cleavage are apparently sequence independent, and cleavage is possible in a variety of transmembrane proteins (Struhl and Adachi 2000) including Notch1, APP, APLP1 and APLP2, ErbB-4, CD44, CD43, LRP, N- and E-cadherin, p75, etc. (see Chapter 8).

PS1 and PS2 are multi-transmembrane spanning proteins; the number of transmembrane domains has generally been accepted to be eight with N- and C-termini, and a hydrophilic loop in the cytosol (between transmembrane regions 6 and 7) (Doan *et al.* 1996); however, new evidence suggests a nine-transmembrane model with the PS1 C-terminus in the lumen (e.g. Laudon *et al.* 2005; Kornilova *et al.* 2006).

PS1 holoprotein (approximately 43 kDa) is processed within its large cytoplasmic loop to produce an N-terminal fragment (NTF) (~27 kDa) and a C-terminal fragment (CTF) (~16 kDa) (Thinakaran *et al.* 1996). These fragments are stable (half-life 24 h), whereas the holoprotein has a short half-life (1–2 h). PS1 endoproteolysis is closely regulated, as in transfected cells only a small amount of overexpressed PS1 is converted to NTF and CTF; these accumulate in a 1:1 stoichiometry, suggesting that exogenous and endogenous PS1 compete for limiting factors (e.g.Thinakaran *et al.* 1997). PS1 is an aspartyl protease, and aspartate residues D275 (in transmembrane region 6) and D385 (in membrane region 7) (Wolfe *et al.* 1999), together with a region in the first membrane spanning domain (Brunkan *et al.* 2005), are critical for endoproteolysis and γ-secretase function, although these functions are probably separate (Beher *et al.* 2001).

Yeast two-hybrid assays and candidate approaches have been used to identify proteins that interact with PS, and numerous interacting proteins have been identified in this way (see Chapter 8 for details) including the enzyme GSK3β, tau, adaptor proteins X11a and X11b, BACE1, calsenilin, and proteins involved in vesicular transport. FAD-linked PS1 mutants or PS1 knockout cause an increase in GSK3β activity, increased phosphorylation in kinesin light chain, and reduction in kinesin-based axonal transport (Pigino *et al.* 2003). The presenilins are thought to have a role in protein trafficking, probably through binding partners. Absence or loss of function of PS1 results in an increase in maturation of APP, appearance at the cell surface, and half-life (e.g. Kaether *et al.* 2002; Cai *et al.* 2003). FAD-linked PS1 mutants produce the opposite effect (Cai *et al.* 2003), which suggests that mutants may increase the time that APP stays in the TGN so that more processing by β- and γ-secretases can occur. There is some evidence that other proteins may be similarly affected (Chapter 8).

In 2000 the type I transmembrane protein nicastrin (the mammalian homologue of aph-2) was identified as a PS1 binding protein (Goutte *et al.* 2000; Yu *et al.* 2000). A *Caenorhabditis elegans* screen also identified the genes *APH-1* and *PEN-2* as being able to augment PS function and being required for γ-secretase activity (Francis *et al.* 2002; Goutte *et al.* 2002). Aph-1 is a seven-transmembrane protein, and Pen-2 has two transmembrane domains. Association of the PS NTF and CTF with the proteins nicastrin, Aph-1a or Aph-1b, and Pen-2 forms the γ-secretase enzyme complex (Iwatsubo 2004), and their coexpression is adequate to reconstitute γ-secretase activity in yeast (Edbauer *et al.* 2003). In addition, the expression of these four proteins overcomes the 'limiting factor' problem mentioned earlier which prevented PS1 holoprotein being converted to NTF and CTF (e.g. Takasugi *et al.* 2003). Downregulation of any one of these proteins intrudes on the maturation and stability of the others (e.g. Li *et al.* 2003). Assembly of the four components is regulated, and it is likely that an intermediate complex comprising Aph-1 and nicastrin is formed first (LaVoie *et al.* 2003) and the C-terminus of the PS1 holoprotein binds to this at the nicastrin transmembrane domain (Kaether *et al.* 2002). Then Pen-2 associates with the complex via the fourth transmembrane region of PS1 (Kim and Sisodia 2005; Watanabe *et al.* 2005) and, simultaneously with PS1 endo-proteolysis, APP interacts with the extracellular domain of nicastrin (Shah *et al.* 2005). The whole complex probably acts in a dimeric formation (e.g. Schroeter *et al.* 2003).

Thus γ-secretase requires the assembly of these four components in order to initiate the cleavage of its substrates, with PS as the active aspartyl protease catalytic core of the complex.

1.5.4 Amyloid precursor protein and interacting proteins

APP is synthesized in the endoplasmic reticulum (ER) and glycosylated within the ER and Golgi apparatus (Chapters 4 and 17). APP transport in the cell is via the fast vesicular anterograde mechanism to the presynaptic terminal. The hypothesis that APP is a kinesin receptor, and that APP, β-secretase, and γ-secretase co-localize within a single axonal/synaptic vesicle, is discussed in Chapter 4, section 4.3, although it seems likely that APP and kinesin may interact via adaptor proteins. At the cell surface APP is cleaved by α-secretase or may undergo re-internalization. Under normal conditions very small amounts of Aβ1–40 and Aβ1–42 are formed in the cell; in both CSF and cell culture Aβ1–40 is the major component (50–70%), with Aβ1–42 as the minor component (5–20%) (Murphy *et al.* 1999). Neuronal and non-neuronal cells process APP differently. The α-secretase pathway predominates in non-neuronal cells and little Aβ is produced. However, in neuronal cells extracellular Aβ can be generated in the secretory pathway and also in endosomes after APP endocytosis. Most of the intracellular Aβ40 and Aβ42 are generated in the secretory pathway, and intracellular Aβ42 has been shown in the ER and intermediate compartment (IC) vesicles. This is important because the production and accumulation of intracellular Aβ may result in neuronal damage from within, rather than requiring its secretion and aggregation in plaques to initiate damage.

APP interacts with adaptor proteins including Fe65 (Fiore *et al.* 1995) and X11 (also called Mint or Munc18-interacting protein-1 because it binds Munc-18 which is important in synaptic vesicle docking and exocytosis) (see Chapter 4 for details). Interestingly, both Fe65 and X11/Mint bind competitively to the same 'GYENPTY' motif (a motif which interacts with phosphotyrosine-interaction domains (PIDs)) near the C-terminal of APP (Borg *et al.* 1996), and they have opposed effects on APP trafficking and proteolysis. Fe65 overexpression increases APP levels and turnover and increases Aβ formation (e.g. Sabo *et al.* 1999), whereas X11 overexpression decreases APP turnover which leads to a reduction in Aβ production and amyloid plaque deposition in a transgenic APP mouse model (Lee *et al.* 2003). Fe65 interacts, via its PID1 domain (it binds to APP with its PID2 domain), with LRP which mediates the endocytosis and degradation of ligands such as apoE (see section 1.8) (Trommsdorff *et al.* 1999), and a complex of Fe65, APP, and LRP has been isolated (Pietrzik *et al.* 2004). This is important as LRP alters the trafficking and proteolysis of APP. Also, part of the 'GYENPTY' motif, 'NPTY', is a consensus sequence for clathrin-coated pit-mediated internalization for endosomal/lysosomal sorting of APP.

In summary, it is likely that even small shifts of balance of interactions between APP and its processing enzymes and other interacting proteins may result in pathogenic outcomes over a long period of time.

1.6 Amyloid fibrillization and toxicity

1.6.1 Amyloidoses

A recurring theme, common to many neurodegenerative diseases including AD, is protein misfolding, aggregation, and the formation of fibrils (see Chapter 6 for details). Amyloid or amyloidosis is, of course, a general term relating not just to Aβ but to many insoluble proteins which form β-pleated sheet aggregates and therefore can be identified using such stains as Congo red. The molecular mechanisms involved in the fibrillization process appear to have common elements in this set of 'disorders of protein folding' and much may be learned by examining the process as a whole. To date twenty six unrelated proteins are so far known to produce amyloid diseases in humans (Ghiso and Frangione 2002) including polyglutamine-repeat disorders (e.g. Huntington's disease), cataracts, amyotropic lateral sclerosis, Parkinson's disease, prion diseases, type II diabetes, and AD. In these disorders, because of conditions of genetics and/or environmental factors, soluble proteins change their conformation and form insoluble structures which aggregate or form fibrillar lesions which may trigger inflammatory response and cell damage. Characteristics of these proteins are that they are poorly soluble, fibrillar, and weakly antigenic, and have a β-pleated sheet rich structure. Normally proteins are folded with the help of other proteins such as chaperones (Dobson 2001) which allow hydrophobic patches to be buried within the protein and hydrophilic patches to have access to the surrounding water-based environment. Chaperones may also assist in refolding misfolded proteins or target them for degradation by the proteasome.

It appears that self-assembly of amyloid into dimers, oligomers, and polymers proceeds in a concentration- and nucleation-dependent manner. Seeding with preformed fibrils can accelerate this process; the initial slow phase of nucleation is followed by a rapid aggregation step (e.g. McLaurin *et al.* 2000). One of the factors governing misfolding of a protein is the presence of mutations, or perhaps even polymorphisms, which increase propensity to aggregation. Mutations in APP seem to have three different types of effect: mutations at the β-secretase cutting site increase affinity for β-secretase cleavage and so increase production of all Aβ forms; mutations at the γ-secretase site increase the production of the more fibrillogenic form of Aβ, i.e. Aβ42; mutations within the Aβ sequence may affect the properties of the protein and its solubility. Examples of these internal mutations are seen in Flemish dementia associated with cerebrovascular pathology (mutation AβA21G), HCHWA-D which is accompanied by recurrent strokes and CAA without plaques or tangles (AβE22Q), the Arctic mutation producing an early-onset dementia with vascular pathology (AβE22G), the Italian I mutation causing early-onset dementia with cerebral haemorrhage (AβE22K), the Iowa aphasic dementia with severe CAA and leucoencephalopathy (AβD23N), and the Italian II mutation which, like the Dutch mutation, presents with severe CAA and recurrent haemorrhages without amyloid plaques or tangles (AβL34V). Common to all of these is the presence of vascular pathology.

Factors affecting the aggregation of these fibril-forming proteins are considered in detail in Chapter 6. These include concentration (as in Down's syndrome with an extra copy of chromosome 21 and thus an extra copy of APP), pH (microenvironment, for instance formation in lysosomes versus cytosol), ions (Zn^{2+}, Fe^{3+}, and Cu^{2+} which probably induce a β-structural transition (see Chapter 18)), the presence of nucleation factors (e.g. serum amyloid P component (SAP), α_1-antichymotrypsin (ACT), apoE, and glycosaminoglycans), and, interestingly, post-translational modifications such as phosphorylation (e.g. tau hyperphosphorylation), isomerization (as for Aβ, i.e. aspartate residues become L-isoAsp, aspartates isomerized at position 7 or 23 are differentially deposited in senile plaques, and vascular amyloid) and N-terminal pyroglutamate (confers resistance against N-terminal peptidases). This seemingly unrelated group of amyloid diseases has certain similarities, and comparing and contrasting similarities and differences may shed light on molecular mechanisms in AD.

1.6.2 Aβ: toxicity by stress and inflammatory processes

Aβ is known to have a toxic gain of function at the cellular level although there is still debate as to the process by which this occurs. It is apparent that certain neuronal populations are more at risk than others in AD. This selective vulnerability may be partly to do with their susceptibility to increased oxidative stress and impaired neuronal energy metabolism, leading to oxidative damage to proteins, DNA, and lipids (Chapter 10). Evidence suggests that oligomeric forms of Aβ may progress neuronal dysfunction by inducing oxidative stress and disrupting cellular calcium homeostasis, and synapses may be particularly vulnerable because they are subjected to extreme changes in levels of ions and oxidative stress (Mattson 2004). Post-translational modification of proteins which may result from such oxidative stress includes lipid peroxidation (Bruce-Keller *et al.* 1998) and glycation (Yan *et al.* 2000).

Inflammatory processes are concurrent with AD neuropathology; activated microglia and astrocytes associate with neuritic plaques, and there is evidence for an increase in inflammatory mediators including proteins involved in the complement cascade, particularly C1q. Aβ interacts with C1q to initiate the complement cascade and C1q enhances the aggregation of Aβ (see Chapter 10 for details). Microglia are activated by Aβ and increase their production of pro-inflammatory cytokines such as TNF-α and interferon-γ. Other cytokines, including interleukin 1β (IL-1β) and transforming growth factor-β_1 (TGF-β_1)

are elevated in cells associated with amyloid plaques, and IL-6, α_2-macroglobulin, cyclooxygenase-2 (COX-2), proteases, and protease inhibitors, such as α_1-antichymotrypsin and thrombin, are all increased in AD.

1.6.3 The RAGE receptor

In 1996 Shi Du Yan reported the discovery that the receptor for advanced glycation endproducts (RAGE), a member of the immunoglobulin superfamily of cell surface molecules, is able to bind Aβ (Yan *et al.* 1996). Advanced glycation endproducts (AGEs) form as a result of glycoxidation of free amino groups on proteins and lipids and are present in areas rich in reducing sugars and oxidant stress. RAGE is a cell surface multi-ligand signalling receptor which modulates cellular functions subsequent to ligand binding. It has a single transmembrane spanning domain and a short highly charged cytosolic tail which mediates intracellular signalling (Yan *et al.* 2000). The N-terminal V-domain appears to be responsible for ligand binding. In normal adulthood, expression of RAGE is low (Yan *et al.* 1996); however, in the AD brain, RAGE is present in high levels in neurons, microglia, and astrocytes, especially close to amyloid plaques and in tangle-bearing neurons. Exposure of cells to Aβ results in RAGE-dependent nuclear translocation of NF$\kappa\beta$ (Yan *et al.* 1996, 2000). NF$\kappa\beta$ then upregulates RAGE (Yan *et al.* 1997a) and turns on expression of cytokines, such as TNF-α and macrophage colony-stimulating factor (M-CSF) (Yan *et al.* 2000).

There is evidence that Aβ interacts with RAGE on the surface of microglia, neurons, and cells in the vasculature, and there is enough evidence to show that RAGE exacerbates AD pathology through neuronal stress. In one study transgenic mice with mutant APP (Tg mAPP) crossed with mice overexpressing RAGE (Tg mAPP/RAGE) displayed impaired learning at an earlier age (3–4 months instead of 6 months) (Arancio *et al.* 2004).

1.6.4 Aβ binding alcohol dehydrogenase

Aβ binding alcohol dehydrogenase (ABAD), also known as endoplasmic reticulum Aβ binding protein (ERAB) or 3-hydroxyacyl CoA dehydrogenase, is a mitochondrial enzyme which was discovered as an Aβ binding protein by yeast two-hybrid screening (Yan *et al.* 1997b). It has a broad substrate specificity and catalyses reversible NAD/NADH-dependent oxidation/reduction (Yan *et al.* 2000).

In post-mortem AD brain ABAD is dramatically increased in the cortex and hippocampus (Yan *et al.* 2000). Surface plasmon resonance studies revealed that Aβ40, Aβ42, and Aβ1–20 bind to ABAD in a dose-dependent manner with nanomolar affinity. Structural studies of ABAD show an NAD binding pocket (see Chapter 11 and Fig. 11.4); however, in the presence of Aβ, distortion prevents NAD binding. *In vitro* studies indicate that Aβ inhibits ABAD enzymatic activity towards its substrates (Yan *et al.* 1999b). However, despite the fact that nanomolar concentrations are able to bind ABAD, low micromolar concentrations of Aβ are required to inhibit activity. One hypothesis is that the N-terminus of ABAD serves to localize Aβ at critical sites within the cell while the C-terminal portion of the peptide is free to interact with additional molecules of Aβ, forming oligomers that further deform protein structure. Therefore while binding of the first Aβ molecule to ABAD would not affect the activity, oligomers of Aβ would eventually distort the enzyme and reduce its function. The downstream effect of suppression of ABAD activity would be to reduce the ability of the cell/animal to resist the effects of stress. Studies show that overexpression of ABAD reduces effects of stroke (Yan *et al.* 2000) and the toxic effects of 1-methyl-4-phenyl-1,2,3,6-tetrahydropyridine (MPTP) (Tieu *et al.* 2004) in mouse models.

There is much evidence to show that Aβ is able to affect cells adversely by a number of mechanisms and that it is likely that a build-up of this molecule will result in cell damage. Next, the link between the effects of Aβ and tangle formation is explored further.

1.7 Tau: hyperphosphorylation and pathogenicity

Although mutations in tau do not lead to AD, they produce dementias such as FTDP-17. Furthermore, the numbers of tangles correlate significantly with degree of dementia, more so than amyloid plaque numbers. Therefore the link between Aβ formation and tangle formation needs to be elucidated. Chapter 5 deals with this matter in depth. Evidence is presented to show that it is the state of abnormal hyperphosphorylation of tau which leads to its loss of function, and also a gain of toxic function whereby phosphorylated tau sequesters normal tau and other microtubule-associated proteins (MAPs), leading to disruption of microtubules.

In AD and other tauopathies, the MAP tau is abnormally hyperphosphorylated and is accumulated as intraneuronal tangles of paired helical filaments (PHF), twisted ribbons, and/or straight filaments (SFs) (e.g. Grundke-Iqbal *et al.* 1986; Iqbal *et al.*1989) (see Chapter 5). In 1988, by digestion of PHFs with proteases, a core of protein was obtained which produced a 12 kDa protein on acid treatment (Wischik *et al.* 1988). Sequence analysis showed that it comprised three of a possible four tandem repeat regions of the tau protein (Jakes *et al.* 1991). A cDNA library from AD cerebral cortex was screened and the cDNA for tau was isolated (Goedert *et al.* 1988). Tau exists in six isoforms (352–441 amino acids) by alternate splicing, with isoforms containing three or four tandem microtubule-binding domain repeats. In addition, there are two possible inserts near the amino terminus (Fig. 1.3). All six tau isoforms are expressed in adult human brain (Goedert 1993). The gene is located at chromosome 17q21–22.

Tau is one of a series of MAPs which maintain the microtubule network in neurons by facilitating assembly and stabilization of microtubules; the others are MAP1a or b and MAP2.

Tau function is affected by its degree of phosphorylation. Tau is both hyperphosphorylated and abnormally phosphorylated in AD (e.g. Morishima-Kawashima *et al.* 1995). Preparations of PHFs from AD and Down's syndrome brains reveal only three principal bands (corresponding to abnormally phosphorylated tau) on gels. This compares with the six bands seen in normal adults. By dephosphorylation it is possible to shift the three abnormal bands to align with the six non-phosphorylated tau isoforms (Goedert *et al.* 1992). Under normal conditions cytosolic tau has 2–3 moles of phosphate/mole protein, whereas in AD all six isoforms are hyperphosphorylated and the ratio appears to be 5–9 moles of

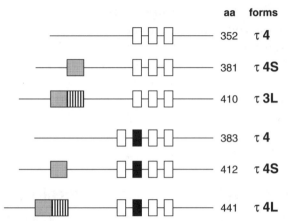

Fig. 1.3 Schematic representation of the six human brain tau isoforms. Proteins differ in whether they contain three or four tubulin binding domains (white or black) and no, one, or two inserts near the N-terminus. Protein inserts from exons 2 (shaded), 3 (vertical lines) (both 29aa), and 10 (solid black) (31 or 32 aa) are shown. The nomenclature is in accord with that used in Chapter 5 (t3/4 S/L, tau short/long forms).

phosphate/mole protein (Köpke *et al.* 1993). This hyperphosphorylation of tau depresses its binding to microtubules and therefore inhibits its ability to promote microtubule assembly (e.g. Alonso *et al.* 1994). The role of phosphorylation in tau is discussed in Chapter 5 and also in Chapter 9, particularly with regard to transgenic animals. Results from various studies suggest that the abnormal hyperphosphorylation of tau precedes its accumulation into tangles (Köpke *et al.* 1993). Hyperphosphorylated tau in AD brain is present in cytosol as well as in tangles (Iqbal *et al.* 1986); *in vitro* hyperphosphorylation promotes tau assembly into PHFs and SFs (Alonso *et al.* 2001a).

Classic APP transgenic models show little tau-associated pathology, and the fact that there is little neuronal death in most models lends weight to the importance of tau accumulation in the onset of symptoms of dementia. The reason mice are less likely to form tangles may lie in the intrinsic stability of the phosphorylating system within these animals, as hyperphosphorylated mouse tau assembles into tangles as readily as human tau (discussed in Chapters 5 and 9). However, transgenic models which have mutations in human *APP*, *PS1*, and *MAPT* genes (3xTg) now exist, and they show accelerated signs of AD pathology including tangles comprising abnormally hyperphosphorylated tau (e.g. Lewis *et al.* 2000; Oddo *et al.* 2004).

Various treatments of tau *in vitro* have been used for its artificial assembly into SFs and PHFs. One study used heparan sulphate and other negatively charged sugar-containing glycosaminoglycans to promote self-assembly of tau and destabilization of tubulin. This occurred equally in the presence or absence of phosphorylation, suggesting that phosphorylation of tau was not required for PHF formation (Goedert *et al.* 1996). However, abnormally hyperphosphorylated tau from AD brain is able to polymerize into tangles of PHFs/SFs *in vitro* (Alonso *et al.* 2001a) without addition of cofactors, and conversely dephosphorylation inhibits this (Iqbal *et al.* 1994). Tau from tangles is able to promote microtubule assembly in the same way as normal tau does after it has been dephosphory-lated *in vitro* (e.g. Iqbal *et al.* 1994). Hyperphosphorylated tau in the cytosol does not interact with microtubules but instead sequesters normal tau, MAP1a/MAP1b, and MAP2, causing disassembly of microtubules *in vitro* (e.g. Alonso *et al.* 1997).

Mutant forms of tau (FTDP-17) such as G272V, P301L, V337M, and R406W, can become hyperphosphorylated at a faster rate than normal tau and self-aggregate into filaments more readily, probably because of alteration in conformation of the protein so that it becomes more favourable as a substrate to brain protein kinases (Alonso *et al.* 2004). These mutant proteins are phosphorylated faster than wild-type tau and self-assemble at lower levels of phosphorylation than the wild-type protein. The six tau isoforms are differentially sequestered by phosphorylated tau *in vitro* (Alonso *et al.* 2001b), with t4RL being most and t3 least ready to be sequestered (see Fig. 1.3), making faster tangle formation more likely in FTDP-17 where there is an increase in the four repeat:three repeat ratio.

Tau is phosphorylated at over 30 serine/threonine residues in AD; most sites flank the microtubule binding domain. Many sites are proline-directed, i.e. serine/threonine followed by proline. A number of kinases, including proline-directed kinases (PDPKs), may be involved in this (see Chapter 5 for details). GSK-3β and cyclin-dependent protein kinase-5 (cdk5) are two PDPKs which have have been shown to be associated with all stages of neurofibrillary pathology in AD (Pei *et al.* 1998). Non-PDPKs such as protein kinase A (PKA) phosphorylate tau at a number of sites (e.g. Wang *et al.* 1998) and can cause a subsequent increase in phosphorylation by GSK-3β (e.g. Cho and Johnson 2003).

The balance between phosphorylation by kinase activity and dephosphorylation by phosphatase activity is crucial, and there is a reduction in phosphatase activity in AD brain. Studies suggest that, in particular, the phosphatase PP-2A is important in tau phosphorylation associated with inhibition of microtubule binding (e.g. Bennecib *et al.* 2001) and that this is reduced by 20–30% in AD brain (Gong *et al.* 1995).

In summary, there is a strong case for tangle formation being directly due to hyper-phosphorylation and abnormal phosphorylation of tau, which may be downstream of activation of kinases presumably as a consequence of the action of Aβ.

1.8 The role of apolipoprotein E

Transgenic models with FAD mutations have allowed dissection and exploration of the molecular processes governing amyloid plaque formation and, with the introduction of tau mutations, have enabled the examination of some of the processes involved in tangle formation. They also make useful models for the testing of candidate drugs. The down-side is that the models are not of the more common sporadic form of AD but of early-onset FAD, based on increased Aβ production and/or an increased ratio of Aβ42:Aβ40. Thus environmental factors, which may contribute to sporadic AD, and the presence of numerous other putative human proteins, which may interact, are not easily factored in.

Since the sporadic form is by far the most prevalent (99% of cases), we must consider what is likely to influence the onset of this form of the disease. Linkage studies tantalizingly hint at the influence of other genes, but the only strong and reproducible risk factor for sporadic AD is the presence of an ε4 allele of apoE (see Chapter 7).

1.8.1 The functions of apolipoprotein E

ApoE is a lipid transport molecule; it is a glycoprotein of 299 aa with isoform specificity through aa at position 112 and 158 (structural review by Mahley and Rall 2000). The lipid binding region near the carboxy terminus is critical for binding triglyceride-rich lipoproteins. ApoE2 and apoE3 preferentially bind to the smaller phospholipids rich in high-density lipoprotein (HDL), whereas apoE4 prefers larger triglyceride-rich very-low-density lipoproteins (VLDL). This is because of the conformational change that the one aa difference from apoE3 (R112C) has on the isoform. One of the results is that apoE4 carriers tend to have higher cholesterol levels (Mahley and Rall 2000). Although apoE is primarily produced by the liver in the periphery, apoE in the brain is separate and is produced mainly by astrocytes, although some is also secreted by microglia and neurons. Its function is to transport cholesterol and phospholipids as HDL particles and to mediate cellular uptake through binding to LDLR (the principal cholesterol-ester transporter in plasma) and to LRP. LRP binds to all isoforms of apoE equally (Takahashi *et al.* 1996) in both lipidated and poorly lipidated forms (e.g. Ruiz *et al.* 2005), but LDLR binds and effects endocytosis of apoE2 to a much lesser extent (approximately 2%) than apoE3 and apoE4 isoforms (Rall *et al.* 1982). Transgenic mice knockouts of the LDLR suggest that this receptor is important in regulating apoE3 and apoE4 levels in the brain.

1.8.2 The binding of apolipoprotein E to Aβ

ApoE has been suggested to facilitate aggregation of Aβ by converting it from an α-helix to a β-sheet conformation (Wisniewski *et al.* 1997). Studies of the effect of apoE status on Aβ fibrillogenesis has produced varied results, due partly to the fact that apoE can be lipidated or non-lipidated, and Aβ can be in different states of aggregation. However, de-lipidated apoE4 is reported to bind to Aβ more readily than de-lipidated apoE3 (Strittmatter *et al.* 1993), whereas lipidated apoE2 and apoE3 are reported as forming complexes with Aβ more easily than lipidated apoE4 (e.g. Tokuda *et al.* 2000). This is in accord with the finding that the ATP-binding cassette transporter A1 (ABCA1) is required for normal lipidation; ABCA1 knockout mice have little apoE, and that is poorly lipidated (e.g. Hirsch-Reinshagen *et al.* 2004) and when crossed with AD mouse models they have either no effect or increase Aβ deposition (e.g. Wahrle *et al.* 2005). It is possible that poorly lipidated apoE, especially apoE4, increases Aβ fibrillogenesis but richly lipidated apoE, in particular apoE2 and 3, are able to keep Aβ soluble and to clear it.

Crossing transgenic AD mouse models with apoE knockout mice results in markedly less amyloid plaque deposition (thioflavine S positive) (e.g. Bales *et al.* 1997). Surprisingly, replacement with any of the human isoforms of apoE delays Aβ deposition; with the caveat that those expressing apoE4 developed the most plaques and those with apoE2 developed the least (Fagan *et al.* 2002).

1.8.3 The effect of apolipoprotein E on synaptic plasticity

ApoE has a role in lipid redistribution during regeneration and thus is likely to influence the way in which neurons are able to cope with insults. There is evidence that synapse formation is related to apoE status. In normal brain the presence of an ε4 allele is associated with reduction in presynaptic markers (Love *et al.* 2006). This makes it likely that neuronal injury or insult will have more manifest effects in the presence of an ε4 allele. In accord with this the presence of an ε4 allele is detrimental to recovery from head trauma (Samatovicz 2000). Likewise, intracerebral haemorrhage patients with ε4 allele status have a much higher mortality and slower recovery rate than those with ε3, and those with dementia pugilistica suffer more severely with the ε4 allele (Jordan *et al.* 1997). Head injury alone does not increase risk for AD and so apoE is probably important during recovery from trauma. In addition, after excitotoxic insult apoE3 is neuroprotective whereas apoE4 is not (Buttini *et al.* 1999).

1.8.4 Apolipoprotein E binds to tau

ApoE4 has been linked to several other neurodegenerative disorders, including different tauopathies (Josephs *et al.* 2004), and has been associated with tau hyperphosphorylation in a range of animal models (e.g. Harris *et al.* 2004). ApoE is able to bind to tangles as well as amyloid plaques in AD brain and therefore it is possible that the 'risk' involved with apoE4 and AD involves an interaction with tau.

ApoE3 forms more stable complexes with tau *in vitro* than apoE4 does (Strittmatter *et al.* 1994), whereas phosphorylation of tau within the microtubule binding domain prevents apoE3 binding to tau (Huang *et al.* 1995). *In vitro*, apoE3 is more effective than apoE4 in promoting neurite sprouting and extension in neurons and neuronal cell lines (e.g. Narita *et al.* 1997); obviously tau function is vital to this. Age-related differences have been shown between apoE knockout and wild-type mice including lower levels of various neuronal markers (e.g. synaptophysin, MAP-2) in the neocortex. When either human apoE3 or apoE4 was re-introduced into the embryos of these null mice, those expressing human apoE3 were similar to the wild type whereas those with apoE4 were similar to null mice (Buttini *et al.* 1999). This accords with the idea that, under certain circumstances, apoE4 behaves like an apoE knockout in its involvement in synaptic plasticity.

The apoE receptors are involved in phosphorylation of tau via the glycoprotein reelin. Both the VLDL receptor and the apoE type 2 receptor (apoER2 or reelin receptor), which are members of the LDLR family of receptors, bind reelin; this protein is important in synaptic plasticity and is secreted by GABAergic inhibitory interneurons in the adult brain. Binding of reelin to its receptors promotes association of the NPxY sequence in the intracellular domain of the receptor with cytoplasmic mammalian disabled-1 (Dab 1) adaptor protein. This results in activation of non-receptor tyrosine kinases and a cascade of phosphorylation (Trommsdorff *et al.* 1999) which leads to, amongst other things, inhibition of tau phosphorylation (Cooper and Howell 1999). Dab 1 also binds to the NPxY motif on APP, as do Fe65 and X11/Mint (section 1.6). *In vitro* binding studies show a differential effect of apoE on reelin binding to its receptors (apoE2 inhibits the least, and apoE4 inhibits the most), and studies in knockout mice (apoE and reelin) also indicate a differential, albeit subtle (Deutsch *et al.* 2006).

Clearly there are a number of ways in which apoE4 may impinge on pathological processes and the relative importance of these is yet to be ascertained.

1.9 Cholinergic function and the molecular pathology of Alzheimer's disease

The vulnerability of cholinergic cells in Alzheimer's disease is discussed in detail in Chapter 13. Cholinergic cells of the nbM are positive for the receptors for nerve growth factor (NGF) p75NTR and tyrosine receptor kinase A (TrkA). p75NTR appears unchanged in AD (Allen *et al.* 1990; Treanor *et al.* 1991); in comparison, the number of TrkA positively staining cells has been shown to be reduced (Salehi *et al.* 1996; Mufson *et al.* 1997). NGF is produced in target tissue of the nbM, i.e. the cortex and hippocampus. This target tissue was found not to have reduced levels (Allen *et al.* 1991) although NGF levels in the AD basal forebrain are decreased (Mufson *et al.* 1999). Recently it has been shown that the predominant form of NGF in human brain is in the pro form, i.e. pro-NGF, and is increased in AD (Fahnestock *et al.* 2001). Pro-NGF appears to bind preferentially to p75NTR. Unfortunately, previous measurements of NGF, by means of ELISA, would not have been able to distinguish between mature NGF and the uncleaved proform. Evidence suggests that pro-NGF binding to p75NTR, in conjunction with binding to the sortilin receptor, may be associated with apoptosis (Beattie *et al.* 2002). The fact that cholinergic cells are almost unique in the forebrain in expressing p75NTR and the requirement of NGF for maintenance of these cells makes this an issue of selective vulnerability. p75NTR knockout mice have an increase in cholinergic neurons in the basal forebrain which fits with their role in apoptosis (Nykjaer *et al.* 2005).

In 2000 a transgenic mouse (AD11) which produced antibodies to NGF was shown to have loss of cholinergic cells in the basal forebrain (Capsoni *et al.* 2000). However, less predictably, there were reports of the presence of amyloid plaques and even hyperphosphorylated tau deposits, and more surprisingly cell loss as evidenced by enlarged lateral ventricles. Behavioural impairment was also noted, yet intranasal administration of NGF reversed symptoms and pathology. In addition, intraperitoneal injection of the CEI galantamine reversed amyloid pathology and restored cholinergic function (Capsoni *et al.* 2002). More recently the AD11 mouse has been crossed with a p75NTR knockout (now AD12) and much of the pathology is reduced. Characterization of the antibody produced by the mouse shows it to be specific for mature and not pro-NGF. Thus the antibody is able to sequester only mature NGF, leaving the pro-form untouched. One may hypothesize that the pro-NGF is able, unhindered, to bind at p75NTR and sortilin and to cause cell death, with no compensatory beneficial effects of the mature NGF acting at TrkA. One may enquire why there is an increase in cortical pro-NGF in AD. Obviously, the cellular location of mature and pro-NGF will determine availability at the TrkA and p75NTR receptors, and western blotting can only reflect pooled levels. It is possible that the retrograde transport of NGF from the cortex and hippocampus to the cholinergic cell bodies may be affected, causing accumulation of the pro-form in the target tissue. Retrograde transport of NGF from the hippocampus to the septum is dramatically reduced in aged rats (Cooper *et al.* 1994) and also in a mouse model of Down's syndrome (trisomy 16) (Cooper *et al.* 2001). Neither of these is necessarily a good model of AD; however, it would explain how target levels of pro-NGF could be increased yet basal fore-brain levels of NGF decreased in AD. However, the accumulation of Aβ in this model of NGF knockdown needs further explanation. One clue as to how this may happen can be seen in the latest work on M1 receptor agonists.

It has been known for some time that M1 and M3 muscarinic receptor stimulation activates protein kinase C (PKC). This results in activation of α-secretase and increased cleavage of APP along the non-amyloidogenic pathway. Recently, animal data from an M1 agonist, AF267B (NGX267), showed that it is able to inhibit Aβ and tau pathology by increasing α-secretase activity and reducing β-secretase and GSK-3β activity (Caccamo *et al.* 2006).

Therefore cholinergic function can be seen not only to be important in cognitive processes but also to be involved at the level of molecular pathology.

1.10 Diagnosis: the selection process

Before examining possible therapeutic solutions we need to look at who these therapies are aimed at. Diagnosing sufferers correctly is vital before we can assess the success or failure of our drugs. Since it is only at post-mortem examination that we know for sure whether the patient had AD, clinical diagnosis must proceed without knowing whether AD pathology is present. As new drugs begin to appear on the market, it will become more important that we find biochemical markers and new imaging techniques to facilitate diagnosis. Chapters 14 and 15 look at clinical assessment and biochemical markers, respectively.

The clinical presentation of AD (Chapter 14) is that of a syndrome with a triad of symptoms comprising neuropsychological and neuropsychiatric components and an inability to perform activities of daily living. The neuropsychological or cognitive element consists mainly of short-term memory loss (amnesia) with relative preservation of more distant events, aphasia (deficits in language), apraxia (inability to carry out tasks), and agnosia (inability to recognize objects). The neuropsychiatric component consists of associated psychiatric symptoms and behavioural disturbances, often present in a large proportion of patients. This includes depression, which is present at some point in up to 66% of patients, and also aggression (20%). The inability to perform activities of daily living includes self-care, and the patient may require help with such basic activities as feeding, dressing, and toileting. These features are common to all types of dementia; differentiation is based on clinical presentation and the presence of other features. Despite a paucity of good biochemical markers, accuracy of diagnosis is high (usually above 85%) and now is not just a diagnosis of exclusion.

However, biochemical markers (see Chapter 15) would assist diagnosis since it is still difficult to detect AD accurately in its presymptomatic stage. For obvious reasons the most likely candidates are Aβ1–42 and tau, since markers should mirror primary changes in the disease state. Unfortunately, it has not been possible to measure tau in blood, and although Aβ40 and Aβ42 can be measured, the results do not seem to be able to adequately distinguish AD patients from controls (e.g. Fukumoto *et al.* 2003). Thus it appears that CSF may be a better source of markers. Measurement of total tau (T-tau), i.e. all tau isoforms irrespective of phosphorylation status, in a number of studies is able to discriminate sporadic AD from controls (e.g. Blennow 2004). However, T-tau is not specific for AD and is found in CNS disorders with significant neuronal degeneration, including stroke. CSF ELISA measurements of phosphorylated tau, which are increased in AD (Blennow 2004) with no change after acute stroke or CJD even though there is a marked increase in T-tau in both, may be more useful.

Initial studies of total Aβ in the CSF using ELISA found a slight decrease or no change in AD, with a large overlap between AD patients and controls. No change in Aβ40 was seen in AD; however, Aβ42 measurement in CSF gave a reduction to about 50% of control levels in AD patients (Blennow 2004) with normal levels in psychiatric disorders such as depression and in chronic neurological disorders such as PD and PSP (Blennow 2004). Thus CSF measurements of phosphorylated tau and Aβ42 seem most promising as biomarkers, singly or in combination, and in general will give 80–90% differentiation.

In addition, there are new candidate CSF biomarkers including ubiquitin, neurofilament proteins, GAP43, isoprostanes, and several cytokines involved in microglia and astrocyte activation. Most recently two-dimensional gel electrophoresis coupled with mass spectrometry has been used successfully to identify AD-specific biomarkers in blood (Hue et al 2006). Proteomics analysis was carried out in blood samples from 50 AD and 50 normal elderly and validated in over 500 subjects with AD and other neurodegenerative diseases and in normal elderly. A number of proteins previously implicated in the disease were identified including complement factor H (CFH) precursor and alpha-2-macroglobulin (alpha-2M). These are now being further explored as specific identifiers of AD. Ideally, it should be possible to provide early diagnosis by combining blood and CSF biomarkers with neuroimaging techniques, such as 2-[^{18}F]fluoro-2-deoxy-D-glucose positron emission tomography (FDG-PET), and amyloid visualization using new molecular tracers in combination with PET or SPECT. Certainly in this climate of novel therapeutic possibilities, indicators of early changes such as blood alpha-2-macroglobulin CSF Aβ42 and p-tau levels should be used to assist appropriate identification of patients for clinical trials.

1.11 Possible therapeutic solutions

Although three chapters in this book are dedicated to current and future treatments (Chapters 16, 17 and 18), most of the authors in this collection see at least some aspects of their work as translational and have given clues as to how this may be implemented. The scope of the work is heroic and only a brief glimpse from a selection of potential therapeutics is given here.

1.11.1 Current available therapies

With the advent of the second-generation cholinesterase inhibitors donepezil (Aricept) and rivastigmine (ENA 713, Exelon), we now have well-tolerated drugs which are able to provide improvement or stabilization in cognitive function in over half the patients receiving them. Galanthamine is a reversible inhibitor of brain acetylcholinesterase but is also able to modulate nicotinic receptors, improving cognition, global function, and behavioural disturbance. The greatest turn-around seems to be in the area of muscarinic agonists (see sections 1.9 and 1.11.6).

Prevention should also be considered seriously, and there is enough evidence to suggest that diet and exercise are important factors in the process of amyloid formation. Five months of voluntary exercise in a transgenic mouse model of AD resulted in a decrease of up to a half of the number of plaques in the cortex and hippocampus as a result of alterations in the processing of APP (Adlard et al. 2005). Interestingly, BDNF mRNA levels in the hippocampus are significantly increased after physical exercise (Russo-Neustadt et al. 1999). There has now been reported a strong link between AD and reduction in BDNF, possibly because of its importance as a neurotransmitter and in neuronal plasticity (Lessmann et al. 1994). However, a more attractive alternative therapeutic may be the idea of a regular curry, particularly containing omega-3-rich oily fish. Curcumin, an active ingredient of turmeric, is able to prevent Aβ formation (section 1.11.5).

1.11.2 Modifiers of β-secretase activity

BACE-1 and BACE-2 are aspartyl proteases. BACE-1 is expressed at high levels in the brain, but at low levels in peripheral tissues; BACE-2 is expressed to a high level in the heart and kidney, but at a low level in the brain. A number of studies have now shown that there is an increase in activity of BACE in sporadic AD (Holsinger et al. 2002; Tyler et al. 2002; Yang et al. 2003) which lends weight to the hope that a BACE inhibitor will be of therapeutic use. As well as APP, additional BACE-1 substrates have been identified, i.e. LRP (von Arnim et al. 2005), P-selectin glycoprotein ligand-1 (Lichtenthaler et al. 2003), and the sialyl-transferase ST6Gal I (Kitazume et al. 2001), although studies of BACE knockouts suggest that there is no obvious resultant abnormality from a loss of β-secretase activity (e.g. Roberds et al. 2001). However, BACE has a large catalytic domain and this makes it hard to identify small drugs which will inhibit it. Nevertheless a number of BACE-1 inhibitors are approaching clinical trials (See Chapters 17 and 18).

1.11.3 Inhibitors and modulators of γ-secretase activity

Inhibition of γ-secretase results in a reduction of Aβ formation and a build-up of C-terminal fragments of APP. It also prevents the formation of AICD and subsequent signalling events. Thus on the surface the inhibition of this enzyme activity has more complex implications than inhibition of β-secretase, which results in less Aβ production with apparently no reduction in AICD formation. One compound in trial at the moment is LY-450139 which seems to cause lowering of plasma Aβ levels, although it appears not to lower Aβ levels in CSF (e.g. Siemers et al. 2006).

γ-Secretase is also more complicated than β-secretase as it is a composite of four separate components. In addition, it has a large number of substrates including APP, APLP1 and APLP2, Notch1, ErbB-4, nectin-1α, netrin receptor DCC, CD44, CD43, LRP, N- and E-cadherin, and p75. One can imagine that it will be difficult to avoid side effects from the inhibition of processing of these and other such substrates. However, there do appear to be subtle differences between processing of at least some of these substrates, and it may be possible to inhibit with some degree of selectivity. Clearly, factors do impinge on and modify the actions of γ-secretase. For instance, γ-secretase modulation by phospholipase D1 (PLD1) is discussed in Chapter 17. This catalyses the hydrolysis of phosphatidylcholine and is able to regulate membrane trafficking. It is able to rescue impaired neurite outgrowth in cells expressing FAD-linked mutant PS1 and also to inhibit γ-secretase activity. This enzyme binds PS1 and when overexpressed will cause dissociation of the γ-secretase complex, resulting in a reduction of γ-secretase cleavage of both APP and Notch.

Rather than inhibition of γ-secretase activity, the possibility of modifying the cleavage of APP may offer a better way forward. Epidemiological studies show that chronic users of NSAIDs have a lower incidence of AD. This was taken to be associated with a reduction in brain inflammation. However, the COX inhibitors rofecoxib and naproxen have not shown benefit in trials in AD (Aisen *et al.* 2003), although it has been shown that a subset of the NSAIDs are able to reduce Aβ production by modulating γ-secretase cleavage. Ibuprofen, indomethacin, and sulindac sulphide reduce Aβ42 production in cultured cells by shifting the cleavage site to produce Aβ38 (Weggen *et al.* 2001). By binding to a modulating site they are able to change the conformation of PS.

1.11.4 Clearing Aβ: a vaccine

Preclinical studies on transgenic mice overexpressing mutant human APP (V717F) showed that injecting with an Aβ1–42 preparation (AN1792) resulted in plaque clearance (Schenk *et al.* 1999). Animals immunized before the onset of AD-type neuropathology (at 6 weeks) or later (11 months) when amyloid deposition and neuropathology was well established showed that plaque formation, neuritic dystrophy, and astrogliosis were prevented in the young animals, and the extent and progression of the neuropathology was reduced in the older animals. Subsequent animal studies have shown improvement in memory impairment (e.g. Janus *et al.* 2000).

Clinical trials were commenced to test the efficacy of AN1792 in human volunteers. Unfortunately, the trial had to be halted as several patients developed aseptic meningoencephalitis (see Chapters 17 and 18 for details). Overall, 6% of patients were affected. Because the trial was stopped before the patients had received their planned booster injections it is difficult to see if real benefits were gained. However, from the results it seems unlikely that a reaction was provoked by formation of antibodies to endogenous APP or other cellular proteins (Hock *et al.* 2002) but was most likely due to a T-cell-mediated autoimmune response (Monsonego *et al.* 2006). The work is continuing with a new vaccine comprising the Aβ1–15 peptide, which is sufficient for B-cell but not T-cell response (Agadjanyan *et al.* 2005). Investigation of brain tissue from patients who took part in the trial and have subsequently died provides some interesting insights into the action of the vaccine (Preston *et al.* 2003). Areas of cortex appeared to have been cleared of plaques and there was some evidence that Aβ42 was present in vasculature, suggesting that it had been cleared and had moved to blood vessels. The alternative to active immunization is passive administration of antibodies. This approach has been shown to reduce Aβ plaque load in transgenic mice (e.g. Bard *et al.* 2000) and a humanized antibody is now in trials.

The mode of action of the vaccine approach is still being debated. One idea is that there is a 'peripheral sink' effect in which removal of Aβ from the periphery creates a concentration gradient between the CSF and plasma, resulting in clearance from the brain. In addition, examination of the autopsy brain tissue from patients who took part in the vaccine trial showed increased microglial activity, and active ingestion of Aβ by the microglia could be seen (Bard *et al.* 2000)

1.11.5 Degrading and clearing Aβ

One of the probable contributing factors to sporadic AD is a reduction in effectiveness of clearance mechanisms. The removal and clearance of Aβ is carried out in normal brain partly with the aid of the proteases neprilysin and IDE and a number of other proteins such as plasmin, ACE, and HrtA1. Interestingly, use of the ACE inhibitor captopril results in Aβ accumulation in cell culture, and perhaps this should be further investigated in patients (Chapter 17).

It seems likely that the small Aβ oligomers are damaging to neurons, rather than plaques and large aggregates of Aβ, and larger Aβ-derived diffusible ligands (ADDLs) are also toxic (see Chapters 4 and 17 for full details). Peptide and non-peptide therapeutics designed to inhibit the formation of these molecules, either by binding Aβ or by inhibiting the seeding effect of other proteins, are actively being pursued (Chapters 17 and 18). One of these molecules, curcumin, derived from the plant *Curcuma longa*, has long been described in traditional Chinese medicine as a treatment for inflammation. However, although it reduces inflammation in a mouse model of AD, it is also able to prevent Aβ42 oligomer formation *in vitro* and reduce plaque burden (Lim *et al.* 2001).

An interesting novel therapeutic principal is discussed in Chapter 18. Briefly, it makes use of the discovery of a metal binding domain near the N-terminus of Aβ which is capable of binding Zn^{2+} or redox-active Cu^{2+}. Binding of Aβ with these metal ions increases aggregation and makes clearance more difficult. However, it also results in the production of hydrogen peroxide and hydroxyl radicals which are able to cause oxidative damage to proteins, lipids, etc. A metal–protein attenuating compound (MPAC) has been devised to obviate this interaction. This binds weakly to metals and competes for the metal ion, thus removing it from interaction with Aβ. The antibiotic clioquinol (an 8-OH quinoline, which is a copper and zinc chelator) is an example of this; it has been found to reduce Aβ accumulation in mouse brain (Cherny *et al.* 2001; Raman *et al.* 2005) and has undergone phase II clinical trials (Ritchie *et al.* 2003; Ibach *et al.* 2005) (see Chapter 19 for details). Another such 8-OH quinoline derivative (PBT2, Prana Biotechnology) is now commencing phase II clinical trials.

1.11.6 Increasing α-secretase activity: cholinergic function

Although important in their ability to improve memory, CEIs do not appear to convey neuroprotection upon the cholinergic neurons and eventually therapeutic effects diminish with the failure of the cholinergic cells. Numerous studies show that NGF is able to increase cholinergic function and reverse memory deficits. In a Swedish trial of three patients mouse NGF was infused into the cerebral ventricles (Eriksdotter *et al.* 1998). Some improvement was seen in cognition and cerebral blood flow, but this was outweighed by back pain. This was later seen to be due to the effect of NGF on the p75NTR receptors, causing hypertrophism in Schwann cells. More recently, a phase I trial of eight AD patients, in which fibroblasts engineered to produce NGF were implanted into the nbM, showed slowing down of decline in cognitive scores, and marked increases in glucose utilization were seen in four patients (Tuszynski *et al.* 2005). This describes a good proof of principle and further trials are now being carried out with viral vectors able to secrete NGF.

The interrelationship between cholinergic function and the process of neuropathology in AD (see section 1.9) has opened up promising therapeutic avenues. By preserving the function of cholinergic cells or perhaps even by potentiating or increasing the functions of ACh it may be possible to halt the progression of the pathology. Under certain circumstances stimulation of PKC by muscarinic or nicotinic activation can result in increase in α-secretase cleavage at the expense of Aβ production. Use of the M1 agonist AF102B resulted in reduction of Aβ levels in CSF from AD patients (Nitsch *et al.* 2000), and subsequently a modification of this, AF267B (NX267), has entered clinical trials. Preclinical results with this drug have been impressive, showing recovery from cognitive impairment and reduction in plaques and hyperphosphorylated tau/tangle deposits in 3xTg-AD mice apparently directly due to an increase in TACE activity and reduction of GSK3β (Caccamo *et al.* 2006).

A phase I trial, completed at the end of 2005, reported the safety of single doses of the M1 agonist NGX267 in healthy adult volunteers, and a second phase I trial will measure the safety and tolerability of single and multiple doses in 65 healthy elderly subjects. The suggested explanation as to why this M1 agonist should work where others did not was that previous M1 agonists either were not specific or were not given at doses which are pharmacodynamically suitable, but that dosing was assessed for pharmacokinetic optimum (Fisher 2006). In other words the dose required for a downstream effect on APP processing is therapeutically feasible, whereas the dose required to keep the drug on the receptor for a prolonged period is not therapeutically beneficial.

1.11.7 Tau

Because evidence suggests that the formation and action of Aβ is responsible for abnormal phosphorylation of tau and subsequent aggregation, it seems reasonable that removal of Aβ will eventually result in a reduction in phosphorylation of tau (see Chapter 19 for a full discussion). However, it is also possible that removal of Aβ or prevention of its formation may do nothing to clear hyperphosphorylated tau, as was seen after Aβ vaccination (e.g. Oddo *et al.* 2004), and it is perhaps unlikely that aggregated tau would de-aggregate and be cleared without assistance. Thus inhibition of kinases has a clear therapeutic potential (Chapter 5). The problem associated with tau as a therapeutic target is that, unlike Aβ, it is critical for neuronal integrity and function and therefore the consequences of side effects are potentially greater. In addition, kinases and phosphatases have a number of substrates. Nevertheless, targeting of specific kinases implicated in the molecular pathology may prove extremely beneficial, particularly in treating the later stages of dementia; in addition, small molecules for inhibition of tau aggregation are being designed (e.g. Necula *et al.* 2005; Taniguchi *et al.* 2005).

This is an exciting era in which we see many possible avenues of therapeutic intervention. What is incredible is that in 10 years time there is the possibility that AD will no longer be the disease of hopelessness it is perceived as now. Of course, we will not know this for sure until many of these therapies have been tested thoroughly in the clinic, but now there is real hope.

1.12 Concluding remarks

A century after the initial observations by Alzheimer, the enigma of AD, despite its complexity, is beginning to clarify. We anticipate the eventual development of several innovative potential therapies, and despite drugs inevitably having side effects, it is necessary to bring their potential benefits to those who would otherwise go untreated. In conjunction with this goal-directed endeavour there needs to be continuing investment in basic science in order to achieve an increasing understanding of the aetiology of this disease.

References

Acquati F, M Accarino C Nucci P, *et al.* (2000). The gene encoding DRAP (BACE2), a glycosylated transmembrane protein of the aspartic protease family, maps to the down critical region: *FEBS Letters* **468**, 59–64.

Adlard PA, Perreau VM, Pop V, and Cotman CW. (2005). Voluntary exercise decreases amyloid load in a transgenic model of Alzheimer's disease. *Journal of Neuroscience* **25**, 4217–21.

Agadjanyan MG, Ghochikyan, A Petrushina I, *et al.* (2005). Prototype Alzheimer's disease vaccine using the immunodominant B cell epitope from beta-amyloid and promiscuous T cell epitope pan HLA DR-binding peptide: *Journal of Immunology* **174**, 1580–6.

Aisen PS, Schafer K A, Grundman M, *et al.* (2003). Effects of rofecoxib or naproxen vs placebo on Alzheimer disease progression: a randomized controlled trial: *Journal of the American Medical Association* **289**, 2819–26.

Allen SJ, Dawbarn D, MacGowan SH, Wilcock GK, Treanor JJS, and Moss TH (1990) A quantitative morphometric analysis of basal forebrain neurons expressing β-NGF receptors in normal and Alzheimer's disease brains. *Dementia* **1**, 125–137.

Allen SJ, MacGowan SH, Treanor JJ, Feeney R, Wilcock GK, Dawbarn D (1991). Normal beta-NGF content in Alzheimer's disease cerebral cortex and hippocampus, *Neuroscience Letters* **131**, 135–139

Alonso A del C, Zaidi T, Grundke-Iqbal I, and Iqbal K (1994). Role of abnormally phosphorylated tau in the breakdown of microtubules in Alzheimer disease. *Proceedings of the National Academy of Sciences of the USA* **91**, 5562–66.

Alonso A, Grundke-Iqbal I, Barra HS, and Iqbal K (1997). Abnormal phosphorylation of tau and the mechanism of Alzheimer neurofibrillary degeneration: sequestration of MAP1 and MAP2 and the disassembly of microtubules by the abnormal tau. *Proceedings of the National Academy of Sciences of the USA* **94**, 298–303.

Alonso A del C, Zaidi T, Novak M, Grundke-Iqbal I, and Iqbal K (2001a). Hyperphosphorylation induces self-assembly of tau into tangles of paired helical filaments/straight filaments. *Proceedings of the National Academy of Sciences of the USA* **98**, 6923–8.

Alonso A, Zaidi T, Wu Q, Novak M, Barra HS, Grundke-Iqbal I, and Iqbal K (2001b). Interaction of tau isoforms with Alzheimer's disease abnormally hyperphosphorylated tau and *in vitro* phosphorylation into the disease-like protein. *Journal of Biological Chemistry* **276**, 37967–73.

Alonso A del C, Mederlyova A, Novak M, Grundke-Iqbal I, and Iqbal K (2004). Promotion of hyperphosphorylation by frontotemporal dementia tau mutations. *Journal of Biological Chemistry* **279**, 34878–81.

Alzheimer, A. (1907). Über eine eigenartige Erkrankung der Hirnrinde [Concerning a novel disease of the cortex]. *Allgemeine Zeitschrift für Psychiatrie Psychisch-Gerichtlich Medizine* **64**, 146–48.

Arancio O, Zhang H P, Chen X, *et al.* (2004). RAGE potentiates Abeta-induced perturbation of neuronal function in transgenic mice. *EMBO Journal* **23**, 4096–105.

Bales KR, Verina T, Dodel RC, *et al.* (1997). Lack of apolipoprotein E dramatically reduces amyloid beta-peptide deposition. *Nature Genetics* **17**, 263–4.

Bard F, Cannon C, Barbour R, *et al.* (2000). Peripherally administered antibodies against amyloid beta-peptide enter the central nervous system and reduce pathology in a mouse model of Alzheimer disease. *Nature Medicine* **6**, 916–19.

Beattie MS, Harrington AW, Lee R, *et al.* (2002). ProNGF induces p75-mediated death of oligodendrocytes following spinal cord injury. *Neuron* **36**, 375–86.

Beher D, Wrigley JD, Nadin A, *et al.* (2001). Pharmacological knock-down of the presenilin 1 heterodimer by a novel γ-secretase inhibitor: implications for presenilin biology. *Journal of Biological Chemistry* **276**, 45394–402.

Bennecib M, Gong CX, Grundke-Iqbal I, and Iqbal K (2001). Inhibition of PP-2A upregulates CaMKII in rat forebrain and induces hyperphosphorylation of tau at Ser 262/356. *FEBS Letters* **490**, 15–22.

Blennow K (2004). Cerebrospinal fluid protein biomarkers for Alzheimer's disease. *NeuroRx* **1**, 213–25.

Bondareff W, Mountjoy CQ, Wischik CM, Hauser DL, LaBree LD, and Roth M, (1993). Evidence of subtypes of Alzheimer's disease and implications for etiology. *Archives of Genetics and Psychiatry* **50**, 350–6.

Borchelt DR, Thinakaran G, Eckman CB, *et al.* (1996). Familial Alzheimer's disease-linked presenilin 1 variants elevate Abeta1–42/1–40 ratio *in vitro* and *in vivo*. *Neuron* **5**, 1005–13.

Borg JP, Ooi J, Levy E, and Margolis B. (1996). The phosphotyrosine interaction domains of X11 and FE65 bind to distinct sites on the YENPTY motif of amyloid precursor protein. *Molecular Cell Biology* **16**, 6229–41.

Bowen DM, Smith CB, White P, and Davison AN. (1976). Neurotransmitter-related enzymes and indices of hypoxia in senile dementia and other abiotrophies. *Brain* **99**, 459–96.

Bowen DM, Allen SJ, Benton JS, *et al.* (1983). Biochemical assessment of serotonergic and cholinergic dysfunction and cerebral atrophy in Alzheimer's disease. *Journal of Neurochemistry* **41**, 266–72.

Braak H and Braak E (1991). Neuropathological staging of Alzheimer-related changes. *Acta Neuropathology* **82**, 239–59.

Bruce-Keller AJ, Li YJ, Lovell MA, *et al.* (1998). 4-Hydroxynonenal, a product of lipid peroxidation, damages cholinergic neurons and impairs visuospatial memory in rats. *Journal of Neuropathology and Experimental Neurology* **57**, 257–67.

Brunkan AL, Martinez M, Wang J, *et al.* (2005). Two domains within the first putative transmembrane domain of presenilin 1 differentially influence presenilinase and gamma-secretase activity. *Journal of Neurochemistry* **94**, 1315–28.

Buttini M, Orth M, Bellosta S, Akeefe H, *et al.* (1999). Expression of human apolipoprotein E3 or E4 in the brains of Apoe(-/-) mice: Isoform-specific effects on neurodegeneration. *Journal of Neuroscience* **19**, 4867–80.

Buxbaum JD, Liu KN, Luo YX, *et al.* (1998). Evidence that tumor necrosis factor alpha converting enzyme is involved in regulated alpha-secretase cleavage of the Alzheimer amyloid protein precursor. *Journal of Biological Chemistry* **273**, 27765–7.

Caccamo A, Oddo S, Billings LM, *et al* (2006). M1 receptors play a central role in modulating AD-like pathology in transgenic mice. *Neuron* **49**, 671–82.

Cai D, Leem JY, Greenfield JP, *et al.* (2003). Presenilin-1 regulates intracellular trafficking and cell surface delivery of beta-amyloid precursor protein. *Journal of Biological Chemistry* **278**, 3446–54.

Capsoni S, Ugolini G, Comparini A, Ruberti F, Berardi N, and Cattaneo A (2000). Alzheimer-like neurodegeneration in aged antinerve growth factor transgenic mice. *Proceedings of the National Academy of Sciences of the USA* **97**, 6826–31.

Capsoni S, Giannotta S, and Cattaneo A (2002). Nerve growth factor and galantamine ameliorate early signs of neurodegeneration in anti-nerve growth factor mice: *Proceedings of the National Academy of Sciences of the USA* **99**,12432–7.

Chartier-Harlin MC, Parfitt M, Legrain S, et al. (1994). Apolipoprotein E, epsilon 4 allele as a major risk factor for sporadic early and late-onset forms of Alzheimer's disease: analysis of the 19q13.2 chromosomal region. *Human.Molecular Genetics* **3**, 569–574.

Cherny RA, Atwood CS, Xilinas ME, et al (2001). Treatment with a copper–zinc chelator markedly and rapidly inhibits β-amyloid accumulation in Alzheimer's disease transgenic mice. *Neuron* **30**, 665–76.

Cho JH and Johnson GV (2003). Glycogen synthase kinase 3beta phosphorylates tau at both primed and unprimed sites: differential impact on microtubule binding. *Journal of Biological Chemistry* **278**, 187–93.

Citron M, Oltersdorf T, Haass C, et al. (1992). Mutation of the β-amyloid precursor protein in familial Alzheimer's disease increases β-protein production. *Nature* **360**, 672–674.

Cooper JA and Howell BW (1999). Lipoprotein receptors: signaling functions in the brain? *Cell* **97**, 671–4.

Cooper JD, Lindholm D, and Sofroniew MV (1994). Reduced transport of [125I]nerve growth factor by cholinergic neurons and down-regulated TrkA expression in the medial septum of aged rats. *Neuroscience* **62**, 625–9

Cooper JD, Salehi A, Delcroix JD, et al. (2001). Failed retrograde transport of NGF in a mouse model of Down's syndrome: reversal of cholinergic neurodegenerative phenotypes following NGF infusion. *Proceedings of the National Academy of Sciences of the USA* **98**, 10439–44

Corder EH, Saunders AM, Risch NJ, et al. (1994). Protective effect of apolipoprotein E type 2 allele for late onset Alzheimer disease. *Nature Genetics* **7**, 180–4.

Davies P and Maloney AJF (1976). Selective loss of central cholinergic neurons in Alzheimer's disease. *Lancet* **ii**, 1403.

De Strooper B, Saftig P, Craessaerts K, et al. (1998). Deficiency of presenilin-1 inhibits the normal cleavage of amyloid precursor protein. *Nature* **391**, 387–90.

De Strooper B, Annaert W, Cupers P, et al. (1999). A presenilin-1-dependent gamma-secretase-like protease mediates release of Notch intracellular domain. *Nature* **398**, 518–22.

Deutsch SI, Rosse RB, and Lakshman RM. (2006). Dysregulation of tau phosphorylation is a hypothe-sized point of convergence in the pathogenesis of Alzheimer's disease, frontotemporal dementia and schizophrenia with therapeutic implications. *Progress in Neuro-psychopharmacology and Biological Psychiatry* **30**, 1369–80.

Doan A, Thinakaran G, Borchelt DR, et al. (1996). Protein topology of presenilin 1. *Neuron* **17**, 1023–30

Dobson CM (2001). The structural basis of protein folding and its links with human disease. *Philosophical Transactions of the Royal Society of London. Series B: Biological Sciences*, **356**, 133–45.

Drachman DA and Leavitt J (1974). Human memory and the cholinergic system. A relationship to aging? *Archives of Neurology* **30**, 113–21

Edbauer D, Winkler E, Regula JT, Pesold B, Steiner H, and Haass C (2003). Reconstitution of γ-secretase activity. *Nature Cell Biology* **5**, 486–8.

Eriksdotter JM, Nordberg A, Amberla K, et al. (1998). Intracerebroventricular infusion of nerve growth factor in three patients with Alzheimer's disease. *Dementia and Geriatric Cognitive Disorders* **9**, 246–57.

Fagan AM, Watson M, Parsadanian M, Bales KR, Paul SM, and Holtzman DM (2002). Human and murine ApoE markedly alters A beta metabolism before and after plaque formation in a mouse model of Alzheimer's disease. *Neurobiology of Disease* **9**, 305–18.

Fahnestock M, Michalski B, Xu B, and Coughlin MD. (2001). The precursor pro-nerve growth factor is the predominant form of nerve growth factor in brain and is increased in Alzheimer's disease. *Molecular Cell Neuroscience* **18**, 210–20.

Fiore F, Zambrano N, Minopoli G, Donini V, Duilio A, and Russo T (1995). The regions of the Fe65 protein homologous to the phosphotyrosine interaction/phosphotyrosine binding domain of

Shc bind the intracellular domain of the Alzheimer's amyloid precursor protein. *Journal of Biological Chemistry* **270**, 30853–6.

Fisher, A. (2006). M1 muscarinic agonists as a comprehensive therapy in Alzheimer's disease (S4–04–02). Presented at 10th ICAD Conference, Madrid.

Francis R, McGrath G, Zhang J, *et al.* (2002). aph-1 and pen-2 are required for Notch pathway signaling, gamma-secretase cleavage of betaAPP, and presenilin protein accumulation. *Developmental Cell*, **3**, 85–97.

Fukumoto H, Tennis M, Locascio JJ, Hyman BT, Growdon JH, and Irizarry MC (2003). Age but not diagnosis is the main predictor of plasma amyloid beta-protein levels. *Archives of Neurology* **60**, 958–64.

Ghiso J and Frangione B (2002). Amyloidosis and Alzheimer's disease. *Advanced Drug Delivery Reviews* **54**, 1539–51.

Glenner GG and Wong CW (1984). Alzheimer's disease: initial report of the purification and characterisation of a novel cerebrovascular amyloid protein. *Biochemical Biophysical Research Communication* **120**, 885–90.

Goate A, Chartier-Harlin MC, Mullan M, *et al.* (1991). Segregation of a missense mutation in the amyloid precursor protein gene with familial Alzheimer's disease. *Nature* **349**, 704–6.

Goedert M (1993). Tau protein and the neurofibrillary pathology of Alzheimer's disease. *Trends in Neuroscience* **16**, 460–5.

Goedert M, Wischik CM, Crowther RA, Walker JE, and Klug A. (1988). Cloning and sequencing of the cDNA encoding a core protein of the paired helical filament of Alzheimer's disease: identification as the microtubule-associated protein. *Proceedings of the National Academy of Sciences of the USA* **85**, 4051–5.

Goedert M, Spillantini MG, Cairns NJ, and Crowther RA (1992). Tau proteins of Alzheimer paired helical filaments: abnormal phosphorylation of all six brain isoforms. *Neuron* **8**, 159–68.

Goedert M, Jakes R, Spillantini MG, Hasegawa M, Smith MJ, and Crowther RA. (1996). Assembly of microtubule-associated protein tau into Alzheimer-lke filaments induced by sulphated glycosaminoglycans. *Nature* **383**, 550–3.

Goldgaber D, Lerman MI, McBride OW, Saffiotti U, and Gajdusek DC (1987). Characterization and chromosomal localization of a cDNA encoding brain amyloid of Alzheimer's disease. *Science* **235**, 877–80.

Gong CX, Shaikh S, Wang JZ, Zaidi T, Grundke-Iqbal I, and Iqbal K (1995). Phosphatase activity toward abnormally phosphorylated tau: decrease in Alzheimer disease brain. *Journal of Neurochemistry* **65**, 732–8.

Goutte C, Hepler W, Mickey KM, and Priess JR (2000). aph-2 encodes a novel extracellular protein required for GLP-1-mediated signaling. *Development* **127**, 2481–92.

Goutte C, Tsunozaki M, Hale VA and Priess JR (2002). APH-1 is a multipass membrane protein essential for the Notch signaling pathway in *Caenorhabditis elegans* embryos. *Proceedings of the National Academy of Sciences of the USA* **99**, 775–9.

Grundke-Iqbal I, Iqbal K, Quinlan M, Tung YC, Zaidi MS, and Wisniewski HM (1986). Microtubule-associated protein tau: a component of Alzheimer paired helical filaments, *Journal of Biological Chemistry* **261**, 6084–9.

Harris FM, Brecht WJ, Xu Q, Mahley RW, and Huang Y (2004). Increased tau phosphorylation in apolipoprotein E4 transgenic mice is associated with activation of extracellular signal-regulated kinase: modulation by zinc, *Journal of Biological Chemistry* **279**, 44795–801

Hirsch-Reinshagen V, Zhou S, Burgess BL, *et al.*. (2004). Deficiency of ABCA1 impairs apolipoprotein E metabolism in brain. *Journal of Biological Chemistry* **279**, 41197–207.

Hock C, Konietzko U, Papassotiropoulos A, *et al.* (2002). Generation of antibodies specific for beta-amyloid by vaccination of patients with Alzheimer disease. *Nature Medicine* **8**, 1270–5.

Holsinger RM, McLean CA, Beyreuther K, *et al.* (2002). Increased expression of the amyloid precursor beta-secretase in Alzheimer's disease. *Annals of Neurology* **51**, 783–6.

Huang DY, Weisgraber KH, Goedert M, Saunders AM, Roses AD, Strittmatter WJ. (1995). ApoE3 binding to tau tandem repeat 1 is abolished by tau serine 262 phosphorylation. *Neuroscience Letters* **192**, 209–12.

Hye A, Lynham S, Thambisetty M, *et al.* (2006). Proteome-based plasma biomarkers for Alzheimer's disease. *Brain* **129**, 3042–50

Ibach B, Haen E, Marienhagen J, *et al* (2005). Clioquinol treatment in familiar early onset of Alzheimer's disease. A case report. *Pharmacopsychiatry* **38**, 178–9.

Iqbal K, Grundke-Iqbal I, Zaidi T, *et al.* (1986). Defective brain microtubule assembly in Alzheimer's disease. *Lancet* **ii**, 421–6.

Iqbal K, Grundke-Iqbal I, Smith AJ, George L, Tung YC, and Zaidi T (1989). Identification and localization of a tau peptide to paired helical filaments of Alzheimer disease. *Proceedings of the National Academy of Sciences of the USA* **86**, 5646–50.

Iqbal K, Zaidi T, Bancher C, and Grundke-Iqbal I (1994). Alzheimer paired helical filaments: restoration of the biological activity by dephosphorylation. *FEBS Letters* **349**, 104–8.

Iwatsubo T (2004). The γ-secretase complex: machinery for intramembrane proteolysis. *Current Opinions in Neurobiology* **14**, 379–83.

Jakes R, Novak M, Davison M, and Wischik CM (1991). Identification of 3- and 4-repeat tau isoforms within the PHF in Alzheimer's disease, *EMBO Journal* **10**, 2725–9.

Janus C, Pearson J, McLaurin J, *et al.* (2000). A beta peptide immunization reduces behavioural impairment and plaques in a model of Alzheimer's disease. *Nature* **408**, 979–82.

Jordan BD, Relkin NR, Ravdin LD, Jacobs AR, Bennett A, and Gandy S (1997). Apolipoprotein E epsilon4 associated with chronic traumatic brain injury in boxing. *Journal of the American Medical Association* **278**, 136–140.

Josephs KA, Tsuboi Y, Cookson N, Watt H, and Dickson DW (2004). Apolipoprotein E epsilon 4 is a determinant for Alzheimer-type pathologic features in tauopathies, synucleinopathies, and frontotemporal degeneration. *Archives of Neurology* **61**, 1579–84.

Kaether C, Lammich S, Edbauer D, *et al.* (2002). Presenilin-1 affects trafficking and processing of betaAPP and is targeted in a complex with nicastrin to the plasma membrane. *Journal of Cell Biology* **158**, 551–61.

Kim SH and Sisodia SS (2005). Evidence that the 'NF' motif in transmembrane domain 4 of presenilin 1 is critical for binding with PEN-2. *Journal of Biological Chemistry* **280**, 41953–66.

Kitazume S, Tachida Y, Oka R, Shirotani K, Saido TC, and Hashimoto Y (2001). Alzheimer's beta-secretase, beta-site amyloid precursor protein-cleaving enzyme, is responsible for cleavage secretion of a Golgi-resident sialyltransferase. *Proceedings of the National Academy of Sciences of the USA* **98**, 13554–9

Köpke E, Tung YC, Shaikh S, Alonso A del C, Iqbal K, and Grundke-Iqbal I (1993). Microtubule associated protein tau: abnormal phosphorylation of a non-paired helical filament pool in Alzheimer disease. *Journal of Biological Chemistry* **268**, 24374–84.

Kornilova AY, Kim J, Laudon H, and Wolfe MS (2006). Deducing the transmembrane domain organization of presenilin-1 in gamma-secretase by cysteine disulfide cross-linking, *Biochemistry* **45**, 7598–604

Kraeplin, E. (1910). *Psychiatrie: Ein Lehrbuch für Studierende und Aerzte* [Textbook of Pyschiatry], 8th edition. Barth, Leipzig.

Laudon H, Hansson EM, Melen K, *et al.* (2005). A nine-transmembrane domain topology for presenilin 1, *Journal of Biological Chemistry* **280**, 35352–60.

LaVoie MJ, Fraering PC, Ostaszewski BL, *et al.* (2003). Assembly of the γ-secretase complex involves early formation of an intermediate sub-complex of Aph-1 and nicastrin. *Journal of Biological Chemistry* **278**, 37213–22.

Lee JH, Lau KF, Perkinton MS, *et al.* (2003). The neuronal adaptor protein X11alpha reduces Abeta levels in the brains of Alzheimer's APPswe Tg2576 transgenic mice. *Journal of Biological Chemistry* **278**, 47025–9.

Lessmann V, Gottmann K and Heumann R (1994). BDNF and NT-4/5 enhance glutamatergic synaptic transmission in cultured hippocampal neurones. *NeuroReport* **6**, 21–5

Levy E, Carman MD, Fernandez-Madrid IJ, *et al.* (1990). Mutation of the Alzheimer's disease amyloid gene in hereditary cerebral hemorrhage, Dutch type. *Science* **248**, 1124–6.

Levy-Lahad E, Wasco W, Poorkaj P, *et al.* (1995). Candidate gene for the chromosome 1 familial Alzheimer's disease locus, *Science* **269**, 973–7.

Lewis J, McGowan E, Rockwood J, *et al.* (2000). Neurofibrillary tangles, amyotrophy and progressive motor disturbance in mice expressing mutant (P301L) tau protein. *Nature Genetics* **25**, 402–5.

Li T, Ma G, Cai H, Price DL and Wong PC (2003). Nicastrin is required for assembly of presenilin/ γ-secretase complexes to mediate Notch signaling and for processing and trafficking of β-amyloid precursor protein in mammals. *Journal of Neuroscience* **23**, 3272–7.

Lichtenthaler SF, Dominguez DI, Westmeyer GG, *et al.* (2003). The cell adhesion protein P-selectin glycoprotein ligand-1 is a substrate for the aspartyl protease BACE1. *Journal of Biological Chemistry* **278**, 48713–19.

Lim GP, Chu T, Yang F, Beech W, Frautschy SA, and Cole GM (2001). The curry spice curcumin reduces oxidative damage and amyloid pathology in an Alzheimer transgenic mouse. *Journal of Neuroscience* **21**, 8370–7

Love S, Siew LK, Dawbarn D, Wilcock GK, Ben Shlomo Y, and Allen SJ (2006). Premorbid effects of APOE on synaptic proteins in human temporal neocortex. *Neurobiology of Aging* **27**, 797–803.

McLaurin J, Yang DS, Yip CM, and Fraser PE (2000). Review: modulating factors in amyloid-β fibril formation. *Journal of Structural Biology* **130**, 259–70.

Mahley RW and Rall SC, Jr (2000). Apolipoprotein E: far more than a lipid transport protein. *Annual Review of Genomics and Human Genetics* **1**, 507–37.

Masters CL, Simms G, Weinman NA, Multhaup G, McDonald BL, and Beyreuther K (1985). Amyloid plaque core protein in Alzheimer's disease and Down's syndrome. *Proceedings of the National Academy of Sciences of the USA* **82**, 4245–9.

Mattson MP (2004). Pathways towards and away from Alzheimer's disease. *Nature* **430**, 631–9.

Monsonego A, Imitola J, Petrovic S, *et al.* (2006). Abeta-induced meningoencephalitis is IFN-gamma-dependent and is associated with T cell-dependent clearance of Abeta in a mouse model of Alzheimer's disease. *Proceedings of the National Academy of Sciences of the USA* **103**, 5048–53.

Morishima-Kawashima M, Hasegawa M, Takio K, *et al.* (1995). Proline-directed and non-proline-directed phosphorylation of PHF-tau. *Journal of Biological Chemistry* **270**, 823–9.

Mufson EJ, Lavine N, Jaffar S, Kordower JH, Quirion R, and Saragovi HU (1997). Reduction in p140-TrkA receptor protein within the nucleus basalis and cortex in Alzheimer's disease. *Experimental Neurology* **146**, 91–103

Mufson EJ, Kroin JS, Sendera TJ, and Sobreviela T (1999). Distribution and retrograde transport of trophic factors in the central nervous system: functional implications for the treatment of neurodegenerative diseases. *Progress in Neurobiology* **57**, 451–84.

Murphy MP, Hickman LJ, Eckman CB, Uljohn SN, Wang R, and Golde TE (1999). Gamma secretase, evidence for multiple proteolytic activities and influence of membrane positioning of substrate on generation of amyloid beta peptides of varying length. *Journal of Biological Chemistry* **274**, 11914–23.

Namba Y, Tomonaga M, Kawasaki H, Otomo E, and Ikeda K (1991). Apolipoprotein E immunoreactivity in cerebral amyloid deposits and neurofibrillary tangles in Alzheimer's disease and kuru plaque amyloid in Creutzfeldt–Jakob disease. *Brain Research* **541**, 163–6.

Narita M, Bu G, Holtzman DM, and Schwartz AL (1997). The low-density lipoprotein receptor-related protein, a multifunctional apolipoprotein E receptor, modulates hippocampal neurite development. *Journal of Neurochemistry* **68**, 587–95.

Necula M, Chirita CN, andKuret J (2005). Cyanine dye N744 inhibits tau fibrillization by blocking filament extension: implications for the treatment of tauopathic neurodegenerative diseases. *Biochemistry* **44**, 10227–37.

Nitsch RM, Deng M, Tennis M, Schoenfeld D, and Growdon JH (2000). The selective muscarinic M1 agonist AF102B decreases levels of total Abeta in cerebrospinal fluid of patients with Alzheimer's disease. *Annals of Neurology* **48**, 913–18.

Nykjaer A, Willnow TE, and Petersen C M (2005). p75NTR: live or let die. *Current Opinions in Neurobiology* **15**, 49–57.

Oddo S, Billings L, Kesslak JP, Cribbs DH, and LaFerla FM (2004). Abeta immunotherapy leads to clearance of early, but not late, hyperphosphorylated tau aggregates via the proteasome. *Neuron* **43**, 321–32.

Pearson RCA, Esiri MM, Hiorns RW, *et al.* (1985). Anatomical correlates of the distribution of the pathological changes in the neocortex in Alzheimer's disease. *Proceedings of the National Academy of Sciences of the USA* **82**, 4531–4.

Pei JJ, Grundke-Iqbal I, Iqbal K, Bogdanovic N, Winblad B, and Cowburn R (1998). Accumulation of cyclin-dependent kinase-5 (cdk5) in neurons with early stages of Alzheimer's disease neurofibrillary degeneration. *Brain Research* **797**, 267–77.

Pietrzik CU, Yoon IS, Jaeger S, Busse T, Weggen S, and Koo EH (2004). FE65 constitutes the functional link between the low-density lipoprotein receptor-related protein and the amyloid precursor protein. *Journal of Neuroscience* **24**, 4259–4265.

Pigino G, Morfini G, Pelsman A, Mattson MP, Brady ST, and Busciglio J (2003). Alzheimer's presenilin 1 mutations impair kinesin-based axonal transport. *Journal of Neuroscience* **23**, 4499–508.

Preston SD, Steart PV, Wilkinson A, Nicoll JA, and Weller RO. (2003). Capillary and arterial cerebral amyloid angiopathy in Alzheimer's disease: defining the perivascular route for the elimination of amyloid beta from the human brain. *Neuropathololology and Applied Neurobiology* **29**, 106–17.

Rall SC, Jr, Weisgraber KH, Innerarity TL, and Mahley RW (1982). Structural basis for receptor binding heterogeneity of apolipoprotein E from type III hyperlipoproteinemic subjects. *Proceedings of the National Academy of Sciences of the USA* **79**, 4696–700.

Raman B, Ban T, Yamaguchi K, *et al* (2005). Metal ion-dependent effects of clioquinol on the fibril growth of an amyloid β peptide. *Journal of Biological Chemistry* **280**, 16157–62.

Ritchie CW, Bush AI, Mackinnon A, *et al* (2003). Metal-protein attenuation with iodochlorydroxyquin (clioquinol) targeting Aβ amyloid deposition and toxicity in Alzheimer disease: a pilot phase 2 clinical trial. *Archives of Neurology* **60**, 1685–91.

Roberds SL, Anderson J, Basi G, *et al.* (2001). BACE knockout mice are healthy despite lacking the primary beta-secretase activity in brain: implications for Alzheimer's disease therapeutics *Human Molecular Genetics* **10**, 1317–24.

Roth M, Tomlinson BE, and Blessed G (1966). Correlation between scores for dementia and counts of 'senile plaques' in cerebral grey matter of elderly subjects. *Nature* **209**, 109–10.

Ruiz J, Kouiavskaia D, Migliorini M, *et al.* (2005). The apoE isoform binding properties of the VLDL receptor reveal marked differences from LRP and the LDL receptor. *Journal of Lipid Research* **46**, 1721–31.

Russo-Neustadt A, Beard RC, and Cotman CW (1999). Exercise, antidepressant medications, and enhanced brain derived neurotrophic factor expression. *Neuropsychopharmacology* **21**, 679–82.

Sabo SL, Lanier LM, Ikin AF, *et al.* (1999). Regulation of beta-amyloid secretion by FE65, an amyloid protein precursor-binding protein. *Journal of Biological Chemistry* **274**, 7952–7

Salehi A, Verhaagen J. Dijkhuizen PA, and Swaab DF (1996). Co-localization of high-affinity neurotrophin receptors in nucleus basalis of Meynert neurons and their differential reduction in Alzheimer's disease. *Neuroscience* **75**, 373–87

Samatovicz RA (2000). Genetics and brain injury: apolipoprotein E. *Journal of Head Trauma Rehabilitation* **15**, 869–74.

Saunders AM, Schmader K, Breitner JC, *et al.* (1993). Apolipoprotein E epsilon4 allele distributions in late-onset Alzheimer's disease and in other amyloid-forming disease. *Lancet* **342**, 710–11.

Schenk D, Barbour R, Dunn W, *et al.* (1999). Immunization with amyloid-β attenuates Alzheimer-disease-like pathology in the PDAPP mouse. *Nature* **400**, 173–7.

Schroeter EH, Ilagan MX, Brunkan AL, *et al.* (2003). A presenilin dimer at the core of the gamma-secretase enzyme: insights from parallel analysis of Notch 1 and APP proteolysis. *Proceedings of the National Academy of Sciences of the USA* **100**, 13075–80.

Shah S, Lee SF, Tabuchi K, Hao YH, *et al.* (2005). Nicastrin functions as a gamma-secretase-substrate receptor. *Cell* **122**, 435–47.

Shen J, Bronson RT, Chen DF, Xia W, Selkoe DJ, and Tonegawa S (1997). Skeletal and CNS defects in presenilin-1-deficient mice. *Cell* **89**, 629–39.

Sherrington R, Rogaev EI, Liang Y, *et al.* (1995). Cloning of a gene bearing missense mutations in early-onset familial Alzheimer's disease. *Nature* **375**, 754–60.

Siemers ER, Quinn JF, Kaye J, *et al.* (2006). Effects of a gamma-secretase inhibitor in a randomized study of patients with Alzheimer disease. *Neurology* **28**, 602–4.

Sims NR, Bowen DM, Allen SJ, Smith CCT, Neary D, Thomas DJ, and Davison AN (1983). Presynaptic cholinergic dysfunction in patients with dementia. *Journal of Neurochemistry* **40**, 503–9.

Strittmatter WJ, Weisgraber KH, Huang DY, *et al.* (1993). Binding of human apolipoprotein E to synthetic amyloid beta peptide: isoform-specific effects and implications for late-onset Alzheimer disease. *Proceedings of the National Academy of Sciences of the USA*, **90**, 8098–102.

Strittmatter WJ, Saunders AM, Goedert M, *et al.* (1994). Isoform-specific interactions of apolipoprotein E with microtubule- associated protein tau: implications for Alzheimer disease. *Proceedings of the National Academy of Sciences of the USA* **91**, 11183–6.

Struhl G and Adachi A (2000). Requirements for presenilin-dependent cleavage of notch and other transmembrane proteins. *Molecular Cell* **6**, 625–36

Summers WK, Majovski LV, Marsh GM, Tachiki K, and Kling A (1986). Oral tetrahydroaminoacridine in long-term treatment of senile dementia, Alzheimer type. *New England Journal of Medicine* **315**, 1241–5.

Takahashi S, Oida K, Ookubo M, *et al.* (1996). Very low density lipoprotein receptor binds apolipoprotein E2/2 as well as apolipoprotein E3/3. *FEBS Letters* **386**, 197–200.

Takasugi N, Tomita T, Hayashi I, *et al.* (2003). The role of presenilin cofactors in the γ-secretase complex. *Nature* **422**, 438–41.

Taniguchi S, Suzuki N, Masuda M, *et al.* (2005). Inhibition of heparin-induced tau filament formation by phenothiazines, polyphenols, and porphyrins. *Journal of Biological Chemistry* **280**, 7614–23.

Terry RD, Masliah E, Salmon DP, *et al.* (1991). Physical basis of cognitive alterations in Alzheimer's disease: synapse loss is the major correlate of cognitive impairment. *Annals of Neurology* **30**, 572–80.

Thinakaran G, Borchelt DR, Lee MK, *et al.* (1996). Endoproteolysis of presenilin 1 and accumulation of processed derivatives *in vivo*. *Neuron* **17**, 181–90.

Thinakaran G, Harris CL, Ratovitski T, *et al.* (1997). Evidence that levels of presenilins (PS1 and PS2) are coordinately regulated by competition for limiting cellular factors, *Journal of Biological Chemistry* **272**, 28415–22

Tieu K, Perier C, Vila M, *et al.* (2004). L-3-hydroxyacyl-CoA dehydrogenase II protects in a model of Parkinson's disease. *Annals of Neurology* **56**, 51–60.

Tokuda T, Calero M, Matsubara E, *et al.* (2000). Lipidation of apolipoprotein E influences its isoform-specific interaction with Alzheimer's amyloid beta peptides. *Biochemical Journal* **348**, 359–65.

Treanor JJ, Dawbarn D, Allen SJ, MacGowan SH, and Wilcock GK (1991). Low affinity nerve growth factor receptor binding in normal and Alzheimer's disease basal forebrain. *Neuroscience Letters* **121**, 73–6

Trommsdorff M, Gotthardt M, Hiesberger T, *et al.* (1999). Reeler/disabled-like disruption of neuronal migration in knockout mice lacking the VLDL receptor and apoE receptor. *Cell* **97**, 689–701.

Tuszynski MH, Thal L, Pay M, *et al.* (2005). A phase 1 clinical trial of nerve growth factor gene therapy for Alzheimer disease. *Nature Medicine* **11**, 551–5.

Tyler SJ, Dawbarn D, Wilcock GK, Allen SJ (2002). Alpha- and beta-secretase: profound changes in Alzheimer's disease. *Biochemical and Biophysical Research Communications* **299**, 373–6.

Vassar R, Bennett BD, BabuKhan S, *et al.* (1999). Beta-secretase cleavage of Alzheimer's amyloid precursor protein by the transmembrane aspartic protease BACE. *Science* **286**, 735–41.

von Arnim CA, Kinoshita A, Peltan ID, *et al.* (2005). The low density lipoprotein receptor-related protein (LRP) is a novel beta-secretase (BACE1) substrate. *Journal of Biological Chemistry* **280**, 17777–85.

Wahrle SE, Jiang H, Parsadanian M, *et al.* (2005). Deletion of Abca1 increases Abeta deposition in the PDAPP transgenic mouse model of Alzheimer disease. *Journal of Biological Chemistry* **280**, 43236–42.

Wang JZ, Wu Q, Smith A, Grundke-Iqbal I, and Iqbal K (1998). t is phosphorylated by GSK-3 at several sites found in Alzheimer disease and its biological activity markedly inhibited only after it is prephosphorylated by A-kinase. *FEBS Letters* **436**, 28–34.

Watanabe N, Tomita T, Sato C, Kitamura T, Morohashi Y, and Iwatsubo T (2005). Pen-2 is incorporated into the gamma-secretase complex through binding to transmembrane domain 4 of presenilin 1. *Journal of Biological Chemistry* **280**, 41967–75.

Weggen S, Eriksen J L, Das P, *et al.* (2001). A subset of NSAIDs lower amyloidogenic Abeta42 independently of cyclooxygenase activity. *Nature* **414**, 212–16.

Wilcock GK and Esiri MM (1982). Plaques, tangles and dementia. A quantitative study, *Journal of Neurological Science* **56**, 343–56.

Wischik CM, Novak M, Thogersen HC, *et al.* (1988). Isolation of a fragment of tau derived from the core of the paired helical filament of Alzheimer's disease. *Proceedings of the National Academy of Sciences of the USA* **85**, 4506–10.

Wisniewski T, Ghiso J, and Frangione B. (1997). Biology of A beta amyloid in Alzheimer's disease. *Neurobiology of Disease* **4**, 313–28.

Wolfe MS, Xia W, Ostaszewski BL, Diehl TS, Kimberly WT, and Selkoe DJ (1999). Two transmembrane aspartates in presenilin-1 required for presenilin endoproteolysis and γ-secretase activity. *Nature* **398**, 513–7.

Yan SD, Chen X, Fu J, *et al.* (1996). RAGE and amyloid-beta peptide neurotoxicity in Alzheimer's disease. *Nature* **382**, 685–91.

Yan SD, Zhu H, and Fu J (1997a). Amyloid-β peptide-receptor for advanced glycation end product interaction elicits neuronal expression of macrophage-colony stimulating factor. A pro-inflammatory pathway in Alzheimer's disease. *Proceedings of the National Academy of Sciences of the USA* **94**, 5296–301.

Yan SD, Fu J, Soto C, *et al.* (1997b). An intracellular protein that binds amyloid-beta peptide and mediates neurotoxicity in Alzheimer's disease. *Nature* **389**, 689–95.

Yan SD, Roher A, Chaney M, Zlokovic B, Schmidt AM, and Stern D (2000). Cellular cofactors potentiating induction of stress and cytotoxicity by amyloid beta-peptide. *Biochimica Biophysica Acta* **1502**, 145–57.

Yan SD, Shi Y, Zhu A, *et al.* (1999b). Role of ERAB/L-3-hydroxyacyl-coenzyme A dehydrogenase type II activity in Abeta-induced cytotoxicity. *Journal of Biological Chemistry* **274**, 2145–56.

Yang LB, Lindholm K, Yan R, *et al.* (2003). Elevated beta-secretase expression and enzymatic activity detected in sporadic Alzheimer disease. *Nature Medicine* **9**, 3–4.

Yu G, Nishimura M, Arawaka S, *et al.* (2000). Nicastrin modulates presenilin-mediated notch/glp-1 signal transduction and betaAPP processing. *Nature* **407**, 48–54.

Chapter 2

The neuropathology of Alzheimer's disease

Margaret M. Esiri

2.1 Introduction

The neuropathology of Alzheimer's disease (AD) is arguably its most distinguishing feature. Although its clinical features of progressive dementia with prominent and early memory failure are characteristic, there are other dementing disorders from which distinction can be difficult and the definite diagnosis of AD still requires neuropathological examination of the brain. The diagnostic pathological features are microscopic and were first described in the late nineteenth and early twentieth centuries (Beljahow 1889; Alzheimer 1907). Despite long familiarity with this pathology, it is only recently that understanding about its development is beginning to be achieved.

2.2 Gross neuropathology

The naked eye appearance of the brain in cases of AD varies from normal to grossly atrophic. Atrophy, when present, affects the cerebral hemispheres in a fairly generalized distribution, but in some cases the medial temporal lobes, hippocampus, and amygdala are relatively selectively picked out (Fig. 2.1). Severe generalized atrophy is more common in the minority of cases of early onset than in cases of AD developing very late in life. The early and selective medial temporal lobe atrophy has been detected using neuroimaging during life and found to be diagnostically useful. Some genetically affected members of families with a mutation that causes AD have had neuroimaging performed serially, commencing when they were asymptomatic. These studies have shown that in such cases cerebral atrophy can be detected before the onset of unequivocal psychological or clinical deterioration (Fox *et al.* 1996).

Externally, the brain usually shows some evidence of gyral narrowing and sulcal widening, although this may be no more obvious than is normally seen with increasing age. In slices through the cerebrum there may similarly be increased ventricular size ranging from well within the limits seen with normal ageing to very severe, involving the lateral and third ventricles but not the aqueduct and fourth ventricle. Cerebral white matter participates in the atrophic process as well as cortex (de la Monte 1989). The striatum may also show some atrophy. The only noteworthy feature in the brainstem is that the locus coeruleus is characteristically lacking some pigment. In contrast, unless

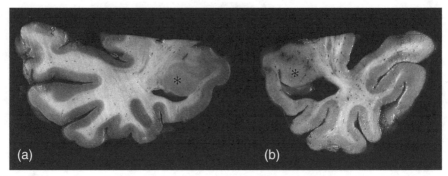

Fig. 2.1 Appearance of a coronal slice through the anterior temporal lobe and amygdala: (a) normal and (b) AD. Note reduction in size of both temporal lobe and amygdala (*) in AD with compensatory dilation of the inferior horn of the lateral ventricle.

Parkinson's disease pathology with Lewy bodies is also present, the substantia nigra appears normally pigmented in AD.

2.3 Microscopic neuropathology

There are several different components to the microscopic pathology of AD, the most distinctive and widespread of which are argyrophilic plaques and neurofibrillary tangles (NFTs). The other six components are neuron loss, glial cell reactions, neuropil threads, granulovacuolar degeneration, Hirano bodies, and amyloid angiopathy. These will be considered in turn.

2.3.1 Argyrophilic plaques

Argyrophilic plaques are complex extracellular foci best visualized with a variety of silver stains or with immunocytochemistry using antibodies to the chief protein constituent of plaques, β amyloid peptide (Aβ) (Fig. 2.2). They vary in size from about 5 to 200 mm across and can be divided on the basis of their structural appearance into two main types, diffuse and neuritic. Diffuse plaques consist of homogeneous deposits of fibrillary material unaccompanied by any local reactive glial cells or abnormal neuritic processes. They do not take up the Congo red stain for amyloid, and ultrastructurally contain no more than a few amyloid fibrils. An unusual variety of diffuse plaque, termed 'cotton wool plaque' is found mainly in forms of AD caused by a mutation in the presenilin 1 gene. Cotton wool plaques are larger than most diffuse plaques. In contrast, neuritic plaques have a more heterogeneous sculpted appearance with a central dense core which reacts with the Congo red stain. Around this core is clustered a peripheral, more or less circular, halo of similarly stained material. The halo contains additional elements in the form of glial and abnormal swollen neuritic processes (Fig. 2.2c, d). The processes of microglial cells are interposed between the core and the halo. Occasionally, plaque cores with no discernible halo can be seen in AD, but these make up only a very small proportion of total plaques even when present. Ultrastructurally the core contains a dense mass

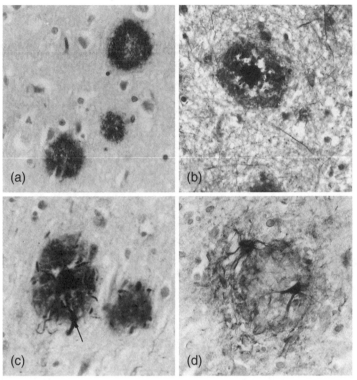

Fig. 2.2 Argyrophilic plaques in cerebral cortex: (a) diffuse plaques immunostained with an antibody to β amyloid protein; (b) plaque with a central core stained with silver; (c) plaque with a core and neuritic processes (arrow) around it stained with silver; (d) plaque immunostained for glial fibrillary acidic protein (GFAP) showing two astrocyte cell bodies near the margins of the plaque.

of extracellular amyloid fibrils and the neuritic processes contain collections of multi-vesicular bodies and mitochondria. Some are filled with abnormal bundles of helically wound paired filaments ('paired helical filaments' (PHFs)) (Kidd 1964; Terry *et al.* 1964). Some neuritic processes are axonal and others are dendritic. They display an abnormal deficiency of microtubules which causes their cytoskeleton to collapse (Gray *et al.* 1987). The glial components consist of the processes of astrocytes and of microglial cells, the latter displaying coated pits in intimate apposition to amyloid filaments.

The relationship between these two main types of plaques, diffuse and neuritic, is not certain. Diffuse plaques are the more abundant and in Down's syndrome, in which by late middle age the pathological features of AD have invariably already developed, they appear earlier than neuritic plaques (Iwatsubo *et al.* 1995; Lemere *et al.* 1996; Mann 1997). While both types of plaque contain Aβ, in diffuse plaques this consists of a polypeptide 42–43 amino acids (aa) in length (Aβ1–42), while in neuritic plaques much of it lacks the last one or two N-terminal amino acids (Aβ1–40). The two main possible interpretations are that diffuse plaques evolve into neuritic plaques with the passage of time and the interaction of the deposited protein with components of the neuropil, or that they have different origins and often occur together, but with a tendency for diffuse plaques to form more readily than neuritic plaques. At some sites in the brain, most notably the cerebellum, only diffuse plaques are found in AD, whereas in the other regions of the brain affected by plaque formation both diffuse and neuritic plaques are found (Joachim *et al.* 1989).

The exact nature and development of plaques is a subject of intense interest because it is known that mutations that cause familial AD either involve alterations to the structure of the precursor protein that gives rise on proteolytic processing to Aβ, or they influence the amount of this protein that is produced and skew its metabolism usually in such a way as to increase the amount of Aβ that is produced, particularly in its more fibrillo-genic 1–42 form.

Although Aβ is believed to be the main biochemical constituent of plaques, a host of other molecules are also concentrated in them. Many of these are proteins including enzymes such as acetylcholinesterase (Mesulam and Moran 1987) and α_1-antichy-motrypsin (Abraham *et al.* 1988), amyloid P component (Kalaria and Perry 1993), comple-ment components (Eikelenboom and Stam 1982; Eikelenboom *et al.* 1989), apolipoproteins E and J (clusterin) (Namba *et al.* 1991; McGeer *et al.* 2005), and growth factors and their receptors. In addition, there are other constituents such as glycosaminoglycans (Snow *et al.* 1988) and the ligands and receptor for advanced glycation endproducts (Yan *et al.* 1997). Yet further molecules, such as the amyloid precursor protein, neuropeptides, lysosomal enzymes, ubiquitin, α-synuclein-derived non-amyloid component (NACP), RNA, and markers of apoptosis, are found in the cellular processes in neuritic plaques. The extent to which these components are integral to the plaque and essential for its formation or evolu-tion, or are passively trapped in the fibrillary matrix, is unclear. The Congo red reactivity in the cores of neuritic plaques and the amyloid fibrils that can be found ultrastructurally are thought to represent an important difference from diffuse plaques; in neuritic plaques Aβ adopts the form of a twisted β-pleated sheet that is characteristic of all amyloids. Some studies of Aβ *in vitro* indicate that in this form it is more neurotoxic than when in a soluble form. However, others believe that soluble oligomers or membrane-trapped dimers of Aβ inside neurons or at the cell surface are the toxic species (Masters and Beyreuther 2006).

Argyrophilic plaques are not entirely specific for AD but also occur in normal ageing, Down's syndrome, and some other neurodegenerative conditions (dementia pugilistica, William's disease, progressive supranuclear palsy). The essential difference between AD and normal ageing with respect to plaques is that they are more numerous in AD and include, in particular, more neuritic elements than in normal ageing (see section 2.7). Their occurrence in Down's syndrome is considered to be a forerunner of AD. The most widely used pathological criteria for the diagnosis of AD are based on a semiquantitative assessment of the density of cortical neuritic plaques (CERAD) (Mirra *et al.* 1991, 1997).

2.3.2 **Neurofibrillary tangles (NFTs)**

NFTs are abnormal intraneuronal structures that are formed in the perikarya of neurons in AD. Like plaques, they are well visualized with silver stains or using immunocyto-chemistry with antibodies to their principal biochemical constituent, which in this case is

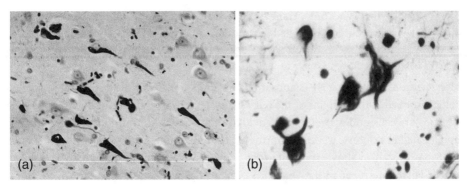

Fig. 2.3 Neurofibrillary tangles (a) in the hippocampus, immunostained with an antibody to hyperphosphorilated tau (AT8), and (b) in the entorhinal cortex stained with silver (at higher power). The fibrillary nature of these structures can be appreciated in (b).

hyperphosphorylated tau (Fig. 2.3). They also stain well with thioflavine S. They are generally smaller than plaques and their size relates to that of the neuron cell body containing them. Some of the largest are found in subcortical nuclei such as the locus coeruleus and raphe nuclei, where they are sometimes described as 'globose'. They appear as dense skein-like, looped, or flame-shaped fibrillary structures occupying the cell body and proximal apical dendrite of affected neurons. A nucleus can still be discerned in most NFT-bearing neurons, but in some cells the nucleus and cell outline may not be present and some NFTs are clearly extracellular and represent the insoluble contents of a neuron that has died ('ghost' tangle). Under these circumstances NFTs take on additional staining properties, presumably through accrual of additional molecules such as glial fibrillary acidic protein (GFAP) and Aβ.

The ultrastructure of NFTs is revealing (Terry 1963; Kidd 1964; Wisniewski et al. 1976). They consist of bundles of distinctive unbranched filaments measuring 20 nm across, pairs of which are wound helically around each other with a periodicity of 80 nm. They do not resemble any normal neuronal ultrastructural constituent. Mixed with these PHFs are a few straight unpaired filaments 5–20 nm across (Gibson et al. 1976; Yagishita et al. 1981). Some neurons containing NFTs also contain abnormal tubular or cylindrical cytoplasmic profiles of uncertain significance (Sloper et al. 1986; Ghatak 1992).

Despite the lack of resemblance of the PHFs forming NFTs to any normal neuronal cytoskeletal element, biochemical analysis of NFTs has shown that they are formed from a normal cytoskeletal component, the microtubule-associated protein tau (Grundke-Iqbal et al. 1986). However, compared with normal tau, the tau found in NFTs is abnormally highly phosphorylated.

Whilst being characteristic of AD, NFTs have also been described in human brain in a wide range of other diseases including rare inherited dementia associated with mutations in the tau gene on chromosome 17 (Poorkaj et al. 1998), progressive supranuclear palsy, dementia pugilistica, subacute sclerosing pancencephalitis, Guam disease, Niemann–Pick disease, lead poisoning, post-encephalitic parkinsonism, myotonic dystrophy, Kuf's disease, Cockayne's syndrome, Hallevorden–Spatz disease, and some varieties of prion disease. They also occur in a restricted distribution in normal ageing (see section 2.7 below).

2.3.3 Neuron loss

Many neurons containing NFTs appear healthy despite many of the perikaryal organelles being displaced to the margins of the cell. However, there are indications that NFT-bearing neurons are cells under stress. In particular, they express ubiquitin, a protein which is upregulated in response to various forms of stress including oxidative stress (Mayer *et al.* 1998). As described above, some neurons containing NFTs die, leaving behind a 'ghost' tangle. Furthermore, quantitative studies of neuronal populations have shown considerable neuron loss in AD (e.g. West *et al.* 1994). Drop-out of neurons occurs mainly in the types of neurons susceptible to NFT formation. For example, in the neocortex large pyramidal neuron density is reduced by 40%, hippocampal CA1 neurons by 68%, basal nucleus neurons by 40–70%, and dorsal raphe and locus coeruleus neurons by 40% and 55%, respectively (Terry *et al.* 1981; Arendt *et al.* 1983; Wilcock *et al.* 1988; Aletrino *et al.* 1992; West *et al.* 1994). Some shrinkage of remaining neurons is also seen in AD.

2.3.4 Glial cell reactions

As occurs with almost any type of neuropathology in the central nervous system (CNS), the glial cells show evidence of a reaction in AD. This is seen mainly in astrocytes and microglial cells. Astrocytes in grey and subcortical white matter appear enlarged with increased numbers and size of processes, and increased expression of GFAP. Many neuritic plaques are decorated with GFAP-positive astrocytic fibres (Fig. 2.2d).

Microglial cells are also increased in regions of grey matter containing neuritic plaques and NFTs in AD. The microglial cells are enlarged with increased numbers of processes and increased expression of MHC class II antigens as well as a moderately increased expression of lysosomal enzymes and complement receptors (McGeer *et al.* 2005; Itagaki *et al.* 1994). These are features of microglial cells displaying some evidence of activation. A particularly noteworthy involvement of microglial cells with the pathology of AD is an intimate relationship between these cells and the cores of neuritic plaques. This has led to the suggestion that microglial cells are either attempting to endocytose the amyloid fibrils or even that they contribute to amyloid formation (Gray *et al.* 1987; Wisniewski 1996). Microglial cells also express RAGE to which Aβ binds on their surface (Yan *et al.* 1997), and it is likely that interactions with these substances in plaques immobilizes microglial cells there and exposes components of plaques to neurotoxic influences such as nitric oxide and other free radicals as well as to lysosomal enzymes produced by microglial cells.

2.3.5 Neuropil threads

Neuropil threads are microscopic argyrophilic (straight or curved) linear structures resembling tiny threads that are present in large numbers in NFT-containing neuropil in AD (Braak *et al.* 1986) (Fig. 2.4). They are thought to consist predominantly of dendrites

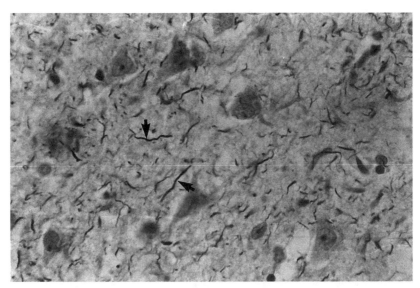

Fig. 2.4 Neuropil threads (arrows) in the neocortex from a case of AD, stained with silver.

of neurons that contain NFTs and they react in similar ways to NFTs with antibodies to hyperphosphorylated tau.

2.3.6 Granulovacuolar degeneration

Simchowicz (1911) was the first to describe granulovacuolar degeneration in cases of senile dementia. The change is largely confined to hippocampal pyramidal neurons which accumulate one or several cytoplasmic vacuoles in which dot-like granules are located (Fig. 2.5). The vacuoles measure 3–5 mm across and the granules are 1–2 mm across. Granulovacuolar degeneration is readily visible with routine stains and the granules are also prominently stained by silver. Ultrastructurally they consist of membrane-bound electron-dense granules (Hirano *et al.* 1968). Immunocytochemically the granules react with antibodies to phosphorylated neurofilaments, tubulin, tau, and ubiquitin. Granulovacuolar degeneration is thought to represent autophagic lysosomal degradation of cytoskeletal components. It occurs to a slight extent with normal ageing and can be found in other neurodegenerative conditions as well as in AD.

2.3.7 Hirano bodies

Hirano bodies were first described in cases of dementia–parkinsonism complex of Guam (Hirano *et al.* 1966), but they can be found in many cases of AD and to a much lesser extent with normal ageing. Like granulovacuolar degeneration, they are most prominent in the hippocampus. They are strongly eosinophilic rod- or carrot-shaped structures 10–30 mm in length and 6–8 mm in width. They are located adjacent to, or apparently within, pyramidal neurons. Ultrastructurally they consist of tightly aligned parallel filaments and they react immunocytochemically for actin and the actin-associated proteins

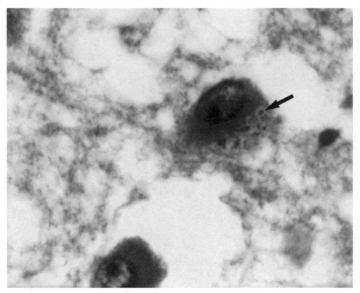

Fig. 2.5 Granulovacuolar degeneration in a hippocampal pyramidal neuron from a case of AD. One of the granules is arrowed.

α-actinin, tropomyosin and vinculin (Goldman 1983; Galloway *et al.* 1987). They are thought to represent an abnormal accumulation of cytoskeletal microfilaments.

2.3.8 Congophilic angiopathy

Congophilic angiopathy is a pathological change affecting small blood vessels in the leptomeninges and cerebral and cerebellar cortex in AD. Vessels deep to the cortex are hardly ever affected. The effect of the change is that the media of small arteries and arterioles is replaced by a homogeneous deposition of Aβ (Fig. 2.6). In fact, leptomeningeal vessels provided the original source from which Glenner and Wong (1984) isolated and characterized this protein. Both the 1–42 and 1–40 forms of the protein can be found in affected vessel walls, but the shorter 1–40 form predominates, particularly in severe cases (Prelli *et al.* 1988; Roher *et al.* 1993; Alonso *et al.* 1998). Incompletely affected vessels first show deposits of Aβ at the interface between the media and adventitia. The proportion of vessels affected varies greatly from case to case. Within a case there is a less marked variation in severity from one cerebral lobe to another. Most cases show one or a few vessels to be affected in a medium-sized microscopic section. The occipital lobe vessels are those most frequently affected, followed by the parietal and then the frontal and temporal lobes. The hippocampal vessels, in contrast, are rarely affected. Cerebellar leptomeningeal vessels are commonly mildly affected.

The vessels affected are predominantly small branches of leptomeningeal arteries or cortical arterioles. Their walls become converted to featureless, largely acellular, tubes of amyloid covered by a thin, sometimes apparently discontinuous, endothelium.

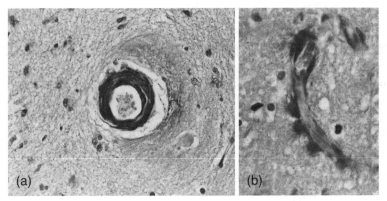

Fig. 2.6 (a) Small cortical arteriole stained with Congo red for amyloid from a case of AD. The media is replaced by a homogeneous deposit of amyloid. (b) Cortical capillary stained with Congo red from a case of AD. Subendothelial deposits of amyloid have fine spicules extending into the surrounding neuropil.

Smooth muscle cells eventually disappear. The adventitia is represented by a few collagen fibrils. In severe cases there may be secondary changes consisting of small perivascular haemorrhages, vascular occlusions or re-duplications, fibrinoid necrosis, or mild perivascular inflammation. Adjacent neuropil may show evidence of haemorrhage or infarction. Not infrequently a massive lobar and subarachnoid haemorrhage may occur from affected vessels and constitute a fatal terminal event. Occasionally the amyloid deposits in cortical vessels involve capillaries as well as arterioles, and spicules of amyloid are then seen radiating into the surrounding neuropil where perivascular plaques may be evident (Fig. 2.6b). In such cases a close association between plaques and congophilic angiopathy is clearly shown. More commonly the association is only qualitative, with no clear quantitative correlation between the extent of congophilic angiopathy and the density of plaques in a given region of cortex. Some genetically determined cases of AD have particularly severe congophilic angiopathy with much less neuritic pathology than in the typical case (Nochlin *et al.* 1998). Congophilic angiopathy also occurs in normal ageing, but it is then usually mild and is associated with some cortical plaque formation, although not necessarily enough to meet pathological criteria for AD. In one study amyloid angiopathy was detected in 30% of brains from elderly undemented subjects (Esiri and Wilcock 1986) and in another study it was detected in 36% (Vinters and Gilbert 1983). As with cases of AD with congophilic angiopathy, undemented elderly subjects with amyloid angiopathy may develop major lobar or subarachnoid haemorrhages. Among elderly subjects with such haemorrhages those with the *APOE ε2* (apolipoprotein E) genotype are over-represented (Nicoll *et al.* 1997). In contrast, among cases of AD with severe congophilic angiopathy those with the *APOE ε4* genotype are over-represented (Alonso *et al.* 1998). Congophilic angiopathy is more common among subjects with vascular dementia than in age-matched undemented subjects (Esiri *et al.* 1997).

There are other rare genetically determined diseases associated with congophilic angiopathy. One of these, in which recurrent cerebral haemorrhages occur, is due to a mutation in the *APP* gene close to sites at which other mutations cause AD (Levy *et al.* 1990; Van Broeckhoven *et al.* 1990) and Aβ is deposited, as in AD. There are other forms of inherited cerebral congophilic angiopathy in which the amyloid is not Aβ and combinations of stroke-like symptoms and dementia arise (reviewed by Plant *et al.* 2004).

The source of the Aβ that is deposited in vessel walls is not certain but smooth muscle is one possible source, the extracellular space another, and the nerve endings innervating the vessel walls a third.

2.4 Anatomy of plaque and tangle pathology

Having described the microscopic appearance of plaques and tangles we consider the relationships between these two hallmark lesions of AD and their distribution in the brain.

2.4.1 Relationship of plaques to tangles

Neuritic plaques have PHFs, as one of their many components, contained in abnormal neuritic processes which are identical ultrastructurally and immunocytochemically in PHFs, NFTs, and neuropil threads. Neurons containing NFTs probably have axonal processes that contribute to neuritic plaques and dendritic processes that form neuropil threads as well as contributing to neuritic plaques. These structural relationships are difficult to prove but they are supported by the common finding of a distinctive row of plaques in the molecular layer of the dentate gyrus of the hippocampus, at the known site of termination of axonal processes arising from entorhinal pyramidal neurons which are very prone to NFT formation in AD (Hyman *et al.* 1984), and by the location of neuropil threads in regions of the brain that are rich in NFTs.

Whether PHFs form first in neuronal processes or in the cell body, or at the two sites simultaneously, has not been determined. Evidence from studying Down's syndrome brains suggests that plaques make an appearance before NFTs and therefore that PHFs may form initially in plaques. What does seem clear is that only some of the brain's neurons are prone to NFT formation and that the distribution of plaques is consistent with these forming at sites to which NFT-prone neurons project. The idea that plaque formation and metabolic changes associated with this drives tangle formation is given strong support by work in transgenic mice in which APP mutations and tau mutations coexist. In this model, if Aβ loads are up- or downregulated there are secondary effects on tangle loads, whereas if tau loads are similarly manipulated no secondary changes in Aβ loads occur (Oddo *et al.* 2004; LaFerla and Oddo 2005). Another aspect of the plaque–tangle relationship that is worth bearing in mind is that neuritic processes in plaques display features suggestive of regenerative as well as degenerative activity (Hyman *et al.* 1987).

2.4.2 **Distribution of neurons that are prone to neurofibrillary tangle formation in Alzheimer's disease**

NFT-prone neurons are to be found in cerebral cortex, hippocampus, and certain subcortical nuclei. In the cortex it is the pyramidal neurons and some large stellate cells that are affected predominantly but their susceptibility to NFT formation varies markedly depending on where in the cortex they are positioned. Those that are most susceptible are situated in transentorhinal cortex occupying the anterior parahippocampal gyrus (Arnold *et al.* 1991; Braak and Braak 1991). Also highly susceptible are the large stellate cells of the adjacent entorhinal cortex followed in turn by the pyramidal neurons of the subiculum and CAI regions of the hippocampus, the pyramidal neurons of the corticomedial amygdaloid nuclei and periamygdaloid cortex, and the pyramidal neurons of the inferior and middle temporal gyrus, cingulate gyrus, insula, anterior olfactory nucleus, and association cortex of the frontal, parietal, and occipital lobes. Last in this susceptibility hierarchy of pyramidal neurons are those of the primary motor and sensory cortices which are notably spared from NFT formation (Pearson *et al.* 1985; Arnold *et al.* 1991; Esiri and Chance 2006). The ordering of this hierarchy shows a striking inverse resemblance to the sequential anatomical connections linking primary sensory cortex via association cortex to entorhinal cortex and hippocampus (Van Hoesen 1982). This resemblance has led to the suggestion that NFT pathology may spread from the transentorhinal region to both the hippocampus and the association cortex. In general, cortical brain regions to which the tangle pathology spreads account for the functional impairment in AD (Grabowski and Damasio 1997). Thus almost universal involvement of transentorhinal, entorhinal, and hippocampal regions underlies the characteristic early and severe memory deficits seen (Hyman *et al.* 1984; Van Hoesen and Damasio 1987; Nagy *et al.* 1996), whereas occasional cases with severe and early occipital and occipito-temporal pathology suffer from deficits in associative visual function (Balint's syndrome) (Hof *et al.* 1990).

The consistent hierarchical pattern of involvement of these different regions of cortex with NFT formation forms the basis of a practical staging scheme that has been developed by Braak and Braak (1991). The first two stages have NFTs confined to transentorhinal, entorhinal, and subicular/CAI regions of the hippocampus. At stages III and IV there are increased numbers of NFTs in these limbic regions. At stages V and VI isocortical areas are also involved. The first two or three stages are considered as largely subclinical and are exhibited commonly in the brains of undemented elderly patients, whereas most cases of pure AD reach stages V and VI. Stages III and IV are commonly found in cases of AD mixed with other pathology such as Parkinson's disease (Nagy *et al.* 1999).

If we accept the suggestion made in section 2.4.1 that plaques are to be found at the axonal and dendritic terminations of NFT-bearing neurons, we can predict where plaques may be expected to occur from knowledge based on animal, particularly primate, experimentation on cortical connectivity: in neuropil that is rich in NFTs (because many dendritic endings and some axonal endings project to local cortex), and in neuropil receiving a distant afferent supply from hippocampus and association cortex. This is, in general, the case. Plaques are particularly numerous in the outer three layers of the association cortex and in the molecular layer of the hippocampal dentate gyrus. They are less numerous, but nevertheless present, in sensory cortex where there is less sparing from plaques than from NFTs. If we extend consideration to the subcortical nuclei that are affected in AD, nuclei with rich projections to association cortex and hippocampus, such as the basal nucleus, dorsal raphe, locus coeruleus, some hypothalamic and thalamic nuclei (although not the sensory relay nuclei), and the amygdala develop NFTs, whereas those subcortical nuclei to which the cortex projects heavily, such as the striatum, develop plaques (Rogers and Morrison 1985; Lewis *et al.* 1987; Pearson and Powell 1989).

These anatomical projections between tangle-bearing neurons and plaques are liable to suffer from impaired anterograde and retrograde axonal transport as a consequence of NFT formation and depletion of normal tau. This has implications for transport of growth factors which depend on such mechanisms. In particular, delivery of nerve growth factor, which is synthesized in cortex and hippocampus and delivered to the cholinergic subcortical nuclei by retrograde transport, is likely to be impaired. Sofroniew and Mobley (1993) have drawn attention to the potential that such impairment might have to provoke a spiralling decline in which deprivation of a growth factor needed to support healthy cholinergic cells could lead to impaired cholinergic stimulation of the cortex and, in turn, to diminished growth factor production there.

2.5 Relationship between the neuropathology of Alzheimer's disease and dementia and behavioural scores

Recent longitudinal follow-up studies to autopsy of people with pathologically confirmed AD generally agree with Wilcock and Esiri (1982) that the severity of dementia correlates best with the density of NFTs in the neocortex. Likewise, NFT density in the hippocampus correlates closely and specifically with memory impairment (Nagy *et al.* 1996). Recent studies in transgenic mice suggest that soluble tau accumulation or cell cycle re-entry, leading to neuronal death, may be more critical for cognitive impairment than the NFTs themselves (Andorfer *et al.* 2005). In contrast, the total plaque density in neocortex is only weakly related to the severity of dementia (Arriagada *et al.* 1992; Berg *et al.* 1993; Bierer *et al.* 1995; Nagy *et al.* 1995). Neuritic plaque density in neocortex shows an intermediate level of correlation between that of total plaque and NFT densities with dementia severity. Other measures besides NFT density which show good correlations with dementia severity are measures of cortical synapse density (de Kosky and Scheff 1990) or indirect markers of this such as synaptophysin immunoreactivity (Terry *et al.* 1991), cortical neuron density (Neary *et al.* 1986; Hyman *et al.* 1994), and cortical neuritic threads (McKee *et al.* 1991). Biochemical markers which correlate well are the amount of hyperphosphorylated tau (Holzer *et al.* 1995) and markers of cholinergic enzyme activity (Francis *et al.* 1993).

Behavioural symptomatology, in contrast with cognitive decline, has been less studied in relation to neuropathology in AD. However, significant correlates have been found linking behavioural symptoms to subcortical pathology. Thus depressive symptoms have been linked to loss of neurons in the locus coeruleus (Zweig *et al.* 1988; Zubenko *et al.* 1990) and to loss of serotonergic endings in the cortex (Chen *et al.* 1996). Aggressive behaviour has also been correlated with reduced 5-hydroxytryptamine (5HT) function (Palmer *et al.* 1988) and preserved noradrenergic function (Russo-Neustadt *et al.* 1998), and anxiety to relative preservation of $5HT_{2A}$ receptors in cortex (Chen *et al.* 1994). Further work is needed before full understanding of the clinically important behavioural features of AD is reached. For example, the reason for the apparently beneficial effects of anticholinesterase therapy on the behavioural symptoms of AD is not readily apparent based on current understanding of the contribution of cholinergic deficits (Francis *et al.* 1999). These have been shown to correlate better with cognitive than behavioural scores in AD. However, the greater plasticity that is evident in the serotonergic and noradrenergic systems than in the cholinergic system renders brain functions that are dependent on these subcortical projections more readily supported by anticholinesterase therapy than cognitive function which depends heavily on an intact neocortex and hippocampus.

2.6 The heterogeneity of Alzheimer's disease and its complication of interpretation of clinico-pathological correlations

Clinico-pathological studies have shown that although most clinical AD sufferers have exclusively AD pathology at autopsy, a not unsubstantial minority show additional pathology as well. The best correlates between clinical dementia severity and AD pathology are, not surprisingly, obtained when cases of pure AD are considered (e.g. Nagy *et al.* 1995). The two most common additional pathologies found with AD are cerebrovascular disease and Parkinson's disease. Since these two pathological processes are known to be capable, on their own, of causing symptoms of dementia, it is logical to expect that when they are combined with AD the severity of AD pathology for a given severity of cognitive decline will tend to be less because the two types of pathology that are present can have additive effects. This has been shown to be the case in a study of AD combined with vascular disease in which the NFTs and plaque density were significantly lower than in a group of cases of pure AD with an equivalent severity of dementia (Nagy *et al.* 1997). In the presence of vascular disease it is of interest that the plaque load showed a more significant correlation with dementia severity than in pure AD. In the presence of Parkinson's disease pathology demented subjects have been described as having an abundance of plaques but little neuritic pathology—a finding that has led to the description of 'plaque-only' AD (Hansen *et al.* 1993). This finding is open to a similar interpretation to that suggested above with respect to vascular disease combined with AD, i.e. AD pathology when augmented by Parkinson's disease pathology becomes symptomatic at an earlier stage when plaques are present but few NFTs have developed. The finding generally of lower Braak stages in cases of mixed pathology also supports this view (Nagy *et al.* 1999).

Quite apart from the complications in clinicopathological correlation that are posed by cases of mixed pathology, there are also some subtypes of AD that are beginning to be discerned. For example, vascular complications occur with some forms of AD, particularly certain familial forms with severe amyloid angiopathy (Poorkaj *et al.* 1998). In such cases NFTs are not prominent and the vascular complications of the amyloid angiopathy may contribute more to dementia than neuritic pathology. Clearly, such cases do not conform to the usual finding of a close correlation between neuritic pathology and dementia severity.

2.7 The interface between Alzheimer's disease and normal ageing

It was mentioned in sections 2.2. and 2.3 that a number of the macroscopic and microscopic features of AD can be found, albeit in a less well developed form or in lower numbers, in the brains of elderly undemented subjects. Here we take a closer look at the overlap between ageing and AD.

The most conspicuous epidemiological characteristic of AD is its close relationship to ageing (Evans *et al.* 1989; Hofman *et al.* 1991; Ott *et al.* 1995), and some investigators view AD as an exaggerated form of brain ageing. Certainly, changes that occur with ageing present conditions that favour the development of AD. Age-related changes are diverse (reviewed by Esiri 1994; Mirra and Hyman 2002) and include slight reductions in the volume and weight of the cerebral hemispheres, enlargement of the ventricles and subarachnoid spaces, and mild fibrosis of the leptomeninges. The reduction in hemisphere weight and volume affects white as well as grey matter. At the microscopic level the most prominent changes are an increase in lipofuscin in neurons, the development of corpora amylacea, an increase in astrocytes, atrophy of neurons, and selective loss of neurons among some populations including cerebral cortical and hippocampal neurons, substantia nigra neurons, and cerebellar Purkinje cells. Also noted are a reduction in synaptic density, a change which accompanies reduction in number and length of dendrites and the number of dendritic spines, and a reduction in neuron nucleolar volume, reflecting a reduction of RNA transcription with an accompanying reduction in neuronal cytoplasmic RNA content. These changes do not affect all populations of neurons, but many of them affect neurons which are also vulnerable to drop-out or NFT formation in AD.

Added to these universal age-associated changes is the common appearance with ageing of argyrophilic plaques in neocortex and of NFTs in transentorhinal, entorhinal, and hippocampal neurons. The dorsal raphe is another site at which NFTs form in small numbers, even with normal ageing. Granulovacuolar degeneration and Hirano body formation also develop to a slight extent in the hippocampus, and amyloid angiopathy develops in leptomeningeal and cortical blood vessel walls. All these features tend to be displayed to a lesser extent in normal ageing than in AD, although there can be considerable overlap, particularly in the density of diffuse plaques and the extent of amyloid angiopathy. Neuritic pathology is much less in evidence in normal ageing. What determines its topographic restriction to transentorhinal and adjacent cortex (where it may be related to age-associated benign memory loss) in normal ageing and the loss of this restriction in AD is an important unanswered question.

2.8 The interface between Alzheimer's disease and dementia and Lewy bodies

Dementia with Lewy bodies is a term introduced relatively recently to describe a clinico-pathological syndrome in which on the clinical side there is fluctuating but progressive dementia, extrapyramidal motor symptoms, and visual hallucinations, and on the patho-logical side Lewy bodies in the substantia nigra and cerebral cortex, usually accompanied by considerable cortical AD pathology, particularly plaques (McKeith *et al.* 1996). Lewy bodies in the substantia nigra are, of course, a hallmark of Parkinson's disease, so that what, in pathological terms, is being described here is a combination of features of Parkinson's disease and AD. There may be a tendency for Parkinson's disease and AD pathology to occur together more frequently than would be expected by chance since in large pathological series of AD cases Lewy bodies are found consistently in the substantia nigra in about 20% of cases (Gearing *et al.* 1995), whereas they occur in only 1–2% of age-matched undemented controls (Smith *et al.* 1991). The extent of Lewy body forma-tion and neuron loss in Parkinson's disease is recognized to vary in subcortical nuclei such as the substantia nigra, nucleus basalis, locus coeruleus, hypothalamus, amygdala, and dorsal vagal nucleus, but all these nuclei are commonly affected (Braak *et al.* 2003). Improved methods of detection, using immunocytochemistry with antibodies to ubiqui-tin and α-synuclein (Lennox *et al.* 1989; Spillantini *et al.* 1998a), show that Lewy bodies also invariably occur in the neocortex, particularly the cortex of the parahippocampal gyrus, cingulate gyrus, and insula, in Parkinson's disease (Hughes *et al.* 1993). Therefore the pathological stage would seem to be set for the development of dementia with Lewy bodies when Parkinson's disease pathology bears particularly heavily on nuclei such as the nucleus basalis that project to the cortex as well as affecting the cortex itself quite severely, and when the cerebral cortex is simultaneously affected by considerable AD pathology, particularly argyrophilic plaque formation.

Clinical recognition of dementia with Lewy bodies is important because those suffering from it deteriorate significantly if treated with neuroleptic drugs (McShane *et al.* 1997). An explanation for the hallucinations that occur is not readily apparent. The severity of dementia has also been linked to Lewy body density (Lennox *et al.* 1989; Hurtig *et al.* 2000), and neurochemical studies have emphasized the pre-eminence of the severe cholinergic deficit in determining the severity of dementia (Perry *et al.* 1990).

2.9 A mention of frontal lobe dementia

Frontal lobe dementia is a clinical term applied to a subset of dementia subjects in whom symptoms of frontal lobe dysfunction, such as changes in personality, disinhibited behaviour, and impulsivity, predominate over those attributable to temporal lobe, pari-etal lobe, and hippocampal dysfunction, such as language disorders, memory deficits, and apraxia (Brun *et al.* 1994). Some cases of frontal lobe dementia show pathology of Pick's disease, but more than half of them show less distinctive pathology with less promi-nent frontal lobe atrophy than is typically seen in Pick's disease (Mirra and Hyman 2002). There are also relatively non-specific microscopic changes which are as characteristic for what they lack as for what they include. The features that are lacking are those of AD (in any more than a degree related to age alone), Lewy bodies, and Pick bodies or Pick cells. The features that are present are spongy change confined to lamina 2 of the cortex in frontal and anterior temporal lobes, neuron loss in the same distribution, and astrocytosis of affected cortex and subcortical white matter. Another noteworthy feature is the presence of ubiquitin-positive filamentous inclusions in neuron cytoplasm and processes in affected cortex. The nature of these inclusions has recently been clarified: at least in some cases they contain protein TDP-43, normally found in the nucleus (Neumann *et al.* 2006). Some subcortical sites, including substantia nigra, corpus striatum, and medial thalamus, may also show neuron loss and gliosis. Thus frontal lobe degeneration represents another form of partially selective neurodegeneration. It has a relatively high familial occurrence, and mutations in the progranulin gene account for most cases (Baker *et al.* 2006; Cruts *et al.* 2006). Rare familial forms of frontotemporal dementia with parkinsonism, associated with NFTs (but not plaques) in the brain, have already been linked to mutations in the gene coding for tau protein (Poorkaj *et al.* 1998; Spillantini *et al.* 1998b).

2.10 Conclusions and aetiological considerations

We now have a fairly clear picture of the pathology of AD and how this relates to symptomatology. There is also a remarkable amount of recently accumulated information about genetic influences. However, we still lack much understanding of pathogenesis and therefore of how best to intervene to slow the process down or prevent it. Other chapters in this volume address aetiology and pathogenesis more directly. The way in which ageing plays such a prominent part in most cases needs clarification. Likewise, the selectivity of the process for particular parts of the brain remains an enigma.

When the interconnectedness of the regions affected was first appreciated the question arose of whether a virus might be responsible (Esiri 1988). This was based on the known selectivity of some viruses for damaging specific subsets of anatomically connected neurons such as the predilection of herpes simplex virus, when causing acute encephalitis, to damage rather similar limbic regions of the brain to those damaged in AD (Esiri 1982). There has indeed been recent work purporting to implicate herpes simplex virus in the aetiology of AD, although the manner in which this ubiquitous virus might be involved is not at all clear and it will require more evidence to substantiate this suggestion (Lin *et al.* 1998). A more favoured possibility is that age-related free-radical damage to neurons has an important role and that apoptotic mechanisms of cell death play an important part. The selectivity of the neuroanatomical distribution of pathology may eventually find an explanation in a particular vulnerability of neurons that retain the long-term plasticity needed to enable lifelong learning to be maintained. Understanding the nature of this plasticity is likely to increase understanding of AD (Arendt *et al.* 1998; Arendt 2004; Esiri and Chance 2006).

References

Abraham CR, Selkoe DJ, and Potter H (1988). Immunochemical identification of the serine protease inhibitor a-1 antichymotrypsin in the brain amyloid deposits of Alzheimer's disease. *Cell* **52**, 487–501.

Aletrino MA, Vogels DJM, Van Doinburg PHMF, and Ten Donkelaar HJ (1992). Cell loss in the nucleus raphe dorsalis in Alzheimer's disease. *Neurobiology of Aging* **13**, 461–68.

Alonso NC, Hyman BT, Rebeck GW, and Greenberg SM (1998). Progression of cerebral amyloid angiopathy: accumulation of amyloid-beta 40 in affected vessels. *Journal of Neuropathology and Experimental Neurology* **57**, 353–9.

Alzheimer A (1907). Über eine eigen artige Erkrankung der Hirnrinde. *Allgemeine Zeitschrift für Psychiatrie Psychisch-Gerichtlich Medizine* **64**, 146–148.

Andorfer C, Acker CM, Kress Y, Hof PR, Duff K, and Davies P (2005). Cell-cycle reentry and cell death in transgenic mice expressing nonmutant human tau isoforms. *Journal of Neuroscience* **25**, 5446–54.

Arendt T (2004). Neurodegeneration and plasticity. *International Journal Developmental Neuroscience* **22**, 507–14.

Arendt T, Bigl V, Arendt A, and Tennstedt A (1983). Loss of neurons in the nucleus basalis of Meynert in Alzheimer's disease, paralysis agitans and Korsakoff's disease. *Acta Neuropathologica* **61**, 101–8.

Arendt T, Bruckner MK, Gertz HJ, Marcova L (1998). Cortical distribution of neurofibrillary tangles in Alzheimer's disease matches the pattern of neurons that retain their capacity of plastic remodelling in the adult brain. *Neuroscience* **83**, 991–1002.

Arnold SE, Hyman BT, Flory J, Damasio AR, and Van Hoesen GW (1991). The topographical and neuroanatomical distribution of neurofibrillary tangles and neuritic plaques in cerebral cortex of patients with Alzheimer's disease. *Cerebral Cortex* **1**,103–16.

Arriagada PV, Growdon JH, Jedley-Whyte ET, and Hyuman BT (1992). Neurofibrillary tangles but not senile plaques parallel duration and severity of Alzheimer's disease. *Neurology* **42**, 631–9.

Baker M, Mackenzie IR, Pickering-Brown SM, *et al.* (2006). Mutations in progranulin cause tau-negative frontotemporal dementia linked to chromosome 17. *Nature* **442**, 916–9.

Beljahow S (1889). Pathological changes in the brain in dementia senilis. *Journal of Mental Science* **35**, 261–2.

Berg L, McKeel DW, Miller JP, *et al.* (1993). Neuropathological indexes of Alzheimer's disease in demented and nondemented persons aged 80 and older. *Archives of Neurology* **50**, 349–58.

Bierer L, Hof P, Purohit D, *et al.* (1995). Neocortical neurofibrillary tangles correlate with dementia severity in Alzheimer's disease. *Archives of Neurology* **52**, 81–8.

Braak H and Braak E (1991). Neuropathological staging in Alzheimer-related changes. *Acta Neuropathologica* **82**, 239–59.

Braak H, Braak E, Grundke-Iqbal I, and Iqbal K (1986). Occurrence of neuropil threads in the senile human brain and in Alzheimer's disease: a third location of paired helical filaments outside of neurofibrillary tangles and neuritic plaques. *Neuroscience Letters* **65**, 351–5.

Braak H, Del Tredici K, Rub U, de Vos RA, Jansen Steur EN, and Braak E (2003). Staging of brain pathology related to sporadic Parkinson's disease. *Neurobiology of Aging* **24**, 197–211.

Brun A, England E, Gustafson L, *et al.* (1994). Clinical and neuropathological criteria for frontotemporal dementia. *Journal of Neurology, Neurosurgery and Psychiatry* **57**, 416–18.

Chen CPL-H, Hope RA, Alder JT, *et al.* (1994). Loss of $5HT_{2A}$ receptors in Alzheimer's disease neocortex is associated with cognitive decline whilst preservation of $5HT_{2A}$ receptors is associated with anxiety. *Annals of Neurology* **36**, 308–9.

Chen CPL-H, Alder JT, Bowen DM, *et al.* (1996). Presynaptic serotonergic markers in community recruited cases of Alzheimer's disease: correlations with depression and neuroleptic medication. *British Journal of Pharmacology* **66**, 1592–8.

Cruts M, Gijselinck I, van der Zee, *et al.* (2006). Null mutations in progranulin cause ubiquitin-positive frontotemporal dementia linked to chromosome 17q21. *Nature* **442**, 920–24.

de Kosky ST and Scheff SW (1990). Synapse loss in frontal cortex biopsies in Alzheimer's disease: correlation with cognitive severity. *Annals of Neurology* **27**, 457–64.

de la Monte SM (1989). Quantitation of cerebral atrophy in preclinical and end stage Alzheimer's disease. *Annals of Neurology* **25**, 450–9.

Eikelenboom P and Stam FC (1982). Immunogobulins and complement factors in senile plaques: an immunoperoxidase study. *Acta Neuropathologica* **57**, 239–42.

Eikelenboom P, Hack CE, Rozemuller JM, and Stam FC (1989). Complement activation in amyloid plaques in Alzheimer's dementia. *Virchows Archiv B Cell Pathology* **56**, 256–62.

Esiri MM (1982). Herpes simplex encephalitis. An immunohistological study of the distribution of viral antigen within the brain. *Journal of Neurological Science* **54**, 209–26.

Esiri MM (1988). Typical and atypical viruses in the aetiology of senile dementia of the Alzheimer type. In *Histology and histopathology of ageing brain. Interdisciplinary topics in gerontology* Vol. 25, pp. 119–39 (Karger, Basel).

Esiri MM (1994). Dementia and normal aging: neuropathology. In Huppert F, Brayne C, and O'Connor DW (eds), *Dementia and normal aging*, pp. 385–436 (Cambridge University Press, Cambridge).

Esiri MM and Chance SA (2006). Vulnerability to Alzheimer's pathology in neocortex: the roles of plasticity and columnar organisation. *Journal of Alzheimer's Disease* **9**, 79–89.

Esiri MM and Wilcock GK (1986). Cerebral amyloid angiopathy in dementia and old age. *Journal of Neurology, Neurosurgery and Psychiatry* **49**, 1221–6.

Esiri MM, Wilcock GK, and Morris JH (1997a). Neuropathological assessment of the lesions of significance in vascular dementia. *Journal of Neurology, Neurosurgery and Psychiatry* **63**, 749–53.

Evans DA, Funkenstein HH, Albert MS, *et al.* (1989). Prevalence of Alzheimer's disease in a community population of older persons. *Journal of the American Medical Association* **262**, 2551–6.

Fox NC, Warrington EK, Freeborough PA, *et al.* (1996). Presymptomatic hippocampal atrophy in Alzheimer's disease. *Brain*, 2001–7.

Francis P, Webster M, Chessell I, *et al.* (1993). Neurotransmitters and second messengers in aging and Alzheimer's disease. *Annals of the New York Academy of Sciences* **695**, 19–26.

Francis P, Palmer AM, Snape M, and Wilcock GK (1999). The cholinergic hypothesis of Alzheimer's disease: a review of progress. *Journal of Neurology, Neurosurgery and Psychiatry* **66**, 137–47.

Galloway PG, Perry G, and Gambetti P (1987). Hirano body filaments contain actin and actin-associated proteins. *Journal of Neuropathology and Experimental Neurology* **46**, 185–99.

Gearing M, Mirra S, Hedreen JC, *et al.* (1995). The Consortium to Establish a Registry for Alzheimer's Disease (CERAD). Part X: Neuropathology confirmation of the clinical diagnosis of Alzheimer's disease. *Neurology* **45**, 461–6.

Ghatak NR (1992). Intraneuronal cylindrical particles in Alzheimer's disease. *Acta Neuropathologica* **84**, 105–9.

Gibson PH, Stones M, and Tomlinson BE (1976). Senile changes in the human neocortex and hippocampus compared by the use of the electron and light microscopes. *Journal of Neurological Science* **27**, 389–405.

Glenner GG and Wong CW (1984). Alzheimer's disease: initial report of the purification and characterisation of a novel cerebrovascular amyloid protein. *Biochemical and Biophysical Research Communications* **120**, 885–90.

Goldman JE (1983). The association of actin with Hirano bodies. *Journal of Neuropathology and Experimental Neurology* **42**, 146–52.

Grabowski TJ and Damasio AR (1997). Definition, clinical features and neuroanatomical basis of dementia. In Esiri MM and Morris JH (eds), *The neuropathology of dementia*, pp 1–20 (Cambridge University Press, Cambridge).

Gray EG, Paula-Barbosa M, and Roher A (1987). Alzheimer's disease: paired helical filaments and cytomembranes. *Neuropathology and Applied Neurobiology* **13**, 91–110.

Grundke-Iqbal I, Iqbal K, Tung YCH, *et al.* (1986). Abnormal phosphorylation of the microtubule-associated protein tau in Alzheimer cytoskeletal pathology. *Proceedings of the National Academy of Sciences of the USA* **83**, 4913–17.

Hansen L, Masliah E, Galasko D, and Terry RD (1993). Plaque-only Alzheimer's disease is usually the Lewy body variant and vice versa. *Journal of Neuropathology and Experimental Neurology* **52**, 648–54.

Hirano A, Maland N, Elizan TS, and Kurland LT (1966). Amyotrophic lateral sclerosis and Parkinsonism–dementia complex of Guam. *Archives of Neurology* **15**, 35–51.

Hirano A, Dembitzer HM, Kurland LT, and Zimmerman HM (1968). The fine structure of some intraganglionic alterations. *Journal of Neuropathology and Experimental Neurology* **27**, 167–182.

Hof PR, Bouras C, Constantinidis J, and Morrison JH (1990). Balint's syndrome in Alzheimer's disease: specific disruption of the occipito-parietal visual pathway. *Brain Research* **493**, 368–75.

Hofman A, Rocca WA, Brayne C, *et al.* (1991). The prevalence of dementia in Europe: a collaborative study 1980–1990 findings. *International Journal of Epidemiology* **20**, 736–48.

Holzer M, Holzapfel HP, Zedlick D, Bruckner MK, and Arendt T (1995). Abnormally phosphorylated tau protein in Alzheimer's disease: heterogeneity of individual regional distribution and relationship to clinical severity. *Neuroscience* **63**, 499–516.

Hughes AJ, Daniel SE, Blankson S, and Lees AJ (1993). A clinico pathological study of 100 cases of Parkinson's disease. *Archives of Neurology* **50**, 140–80.

Hurtig HI, Trojanowski JQ, Galvin J, *et al.* (2000). Alpha-synuclein cortical Lewy bodies correlate with dementia in Parkinson's disease. *Neurology* **54**, 1916–21.

Hyman BT, Van Hoesen GW, Damasio AR, and Barnes CL (1984). Alzheimer's disease: cell specific pathology isolates the hippocampal formation in Alzheimer's disease. *Science* **225**, 1168–70.

Hyman BT, Kromer LJ, and Van Hoesen GW (1987). Reinnervation of the hippocampal perforant pathway zone in Alzheimer's disease. *Annals of Neurology* **21**, 259–67.

Hyman B, West H, Gomez-Isla T, and Mui S (1994). Quantitative neuropathology in Alzheimer's disease: neuronal loss in high order association cortex parallels dementia. In Iqbal K, Mortimer A, Winblad B, and Wisniewsk H (eds), *Research advances in Alzheimer's disease and related disorders*, pp. 363–70 (John Wiley, New York).

Itagaki S, Akiyama H, Saito H, and McGeer PL (1994). Ultrastructural localisation of complement membrane attack complex (MAC)-like immunoreactivity in brains of patients with Alzheimer's disease. *Brain Research* **645**, 78–84.

Iwatsubo T, Mann DMA, Odaka A, Suzuki N, and Ihara Y (1995). Amyloid β protein (Aβ) deposition: Aβ42(43) precedes Aβ40 in Down's syndrome. *Annals of Neurology* **37**, 294–9.

Joachim CL, Morris JHN, and Selkoe DJ (1989). Diffuse senile plaques occur commonly in the cerebellum in Alzheimer's disease. *American Journal of Pathology* **135**, 309–19.

Kalaria RN and Perry G (1993). Amyloid P component and other acute-phase proteins associated with cerebellar A-beta deposits in Alzheimer's disease. *Brain Research* **631**, 151–55.

Kidd M (1964). Alzheimer's disease: an electron microscopic study. *Brain* **87**, 307–20.

LaFerla FM and Oddo S (2005). Alzheimer's disease: Abeta, tau and synaptic dysfunction. *Trends in Molecular Medicine* **11**, 170–6.

Lemere CA, Blusztajn JK, Yamaguchi H, *et al.* (1996). Sequence of deposition of heterogeneous amyloid β-peptides and ApoE in Down syndrome: implications for initial events in amyloid plaque formation. *Neurobiology of Disease* **3**, 16–32.

Lennox G, Lowe J, Landon M, *et al.* (1989). Diffuse Lewy body disease: correlative neuropathology using anti-ubiquitin immunocytochemistry. *Journal of Neurology, Neurosurgery and Psychiatry* **52**, 1236–47.

Levy E, Carman MD, Fernandez-Madrid IJ, *et al.* (1990). Mutation of the Alzheimer's disease gene in hereditary cerebral haemorrhage, Dutch type. *Science* **248**, 1124–6.

Lewis DA, Campbell MJ, Terry RD, and Morrison JH (1987). Laminar and regional distributions of neurofibrillary tangles and neuritic plaques in Alzheimer's disease: a quantitative study of visual and auditory cortices. *Journal of Neuroscience* **7**, 1799–808.

Lin W-R, Graham J, MacGowan SM, Wilcock GK, and Itzhaki RF (1998). Alzheimer's disease, herpes virus in brain, apolipoprotein E_4 and herpes labialis. *Alzheimer's Reports* **1**, 173–8.

McGeer EG, Klegeris A, and McGeer PL (2005). Inflammation, the complement system and the diseases of aging. *Neurobiology Aging* **26**, Suppl 1, 94–97.

McKee A, Kosik K, and Kowall N (1991). Neuritic pathology and dementia in Alzheimer's disease. *Annals of Neurology* **30**, 156–65.

McKeith IG, Galasko D, Kosaka K, *et al.* (1996). Consensus guidelines for the clinical and pathological diagnosis of dementia with Lewy bodies. *Neurology* **47**, 1113–24.

McShane R, Keene J, Gedling K, *et al.* (1997). Do neuroleptic drugs hasten cognitive decline in dementia? Prospective study with necropsy follow-up. *British Medical Journal* **314**, 266–70.

Mann DMA (1997). Neuropathological changes of Alzheimer's disease in persons with Down's syndrome. In Esiri MM and Morris JH (eds), *The neuropathology of dementia*, pp. 122–36 (Cambridge University Press, Cambridge).

Masters CL and Beyreuther K (2006). Alzheimer's centennial legacy: prospects for rational therapeutic intervention targeting the Abeta amyloid pathway. *Brain* **129**, 2823–2839.

Mayer RJ, Landon M, and Lowe J (1998). Ubiquitin and the molecular pathology of human disease. In Peters J-M, Harris JR and Finley D (eds), *Ubiquitin and the biology of the cell*, pp. 429–62 (Plenum Press, New York).

Mesulam M-M and Moran AM (1987). Cholinesterases within neurofibrillary tangles related to age and Alzheimer's disease. *Annals of Neurology* **22**, 223–8.

Mirra SS (1997). Neuropathological assessment of Alzheimer's disease: the experience of the Consortium to Establish a Registry for Alzheimer's Disease. *International Psychogeriatrics* **1**, 263–72.

Mirra S and Hyman B (2002). Ageing and dementia. In Graham DI and Lantos PL (eds), *Greenfield's Neuropathology* (7th edn), pp. 195–272 (Arnold, London).

Mirra SS, Heyman A, McKeel D, *et al.* (1991). The Consortium to Establish a Registry of Alzheimer's Disease (CERAD). Part II: Standardisation of the neuron pathologic assessment of Alzheimer's disease. *Neurology* **41**, 479–86.

Nagy Zs, Esiri MM, Jobst KA, *et al.* (1995). Relative roles of plaques and tangles in the dementia of Alzheimer's disease: correlations using three sets of neuropathological criteria. *Dementia* **6**, 21–31.

Nagy Zs, Jobst KA, Esiri MM, *et al.* (1996). Hippocampal pathology reflects memory deficit and brain imaging measurements in Alzheimer's disease: clinicopathologic correlations using three sets of pathologic diagnostic criteria. *Dementia* **7**, 76–81.

Nagy Zs, Esiri MM, Jobst KA, *et al.* (1997). The effects of additional pathology on the cognitive deficit in Alzheimer's disease. *Journal of Neuropathology and Experimental Neurology* **56**, 165–70.

Nagy Zs, Hindley NJ, Braak H, *et al.* (1999). The progression of Alzheimer's disease from limbic regions to the neocortex: clinical, radiological and pathological relationships. *Dementia and Geriatric Cognitive Disorders* **10**, 115–20.

Namba Y, Tomonaga M, Kawaskai H, *et al.* (1991). Apolipoprotein E immuno-reacting in cerebral amyloid deposits and neurofibrillary tangles in Alzheimer's disease and Kuru plaque amyloid in Creutzfeldt–Jakob disease. *Brain Research* **541**, 163–66.

Neary D, Snowden J, Mann D, *et al.* (1986). Alzheimer's disease—a correlative study. *Journal of Neurology, Neurosurgery and Psychiatry* **49**, 229–37.

Neumann M, Sampathu DM, Kwong LK *et al.* (2006). Ubiquitinated TDP-43 in frontotemporal lobar degeneration and amyotrophic lateral sclerosis. *Science* **314**, 130–133.

Nicoll JA, Burnett C, Love S, *et al.* (1997). High frequency of apolipoprotein E epsilon 2 allele in haemorrhage due to cerebral amyloid angiopathy. *Annals of Neurology* **41**, 716–21.

Nochlin D, Bird TD, Nemens EJ, Ball MJ, and Sumi SM (1998). Amyloid angiopathy in a Volga German family with Alzheimer's disease and a presenilin-2 mutation (N141I). *Annals of Neurology* **43**, 131–5.

Oddo S, Billings L, Kesslak JP, Cribbs DH, and LaFerla FM (2004). Abeta immunotherapy leads to clearance of early, but not late, hyperphosphorylated tau aggregates via the proteasome. *Neuron* **43**, 321–32.

Ott A, Breteler MMB, Van Harskamp F, *et al.* (1995). Prevalence of Alzheimer's disease and vascular dementia: association with education. The Rotterdam study. *British Medical Journal* **310**, 970–3.

Palmer AM, Stratmann GC, Procter AW, and Bowen DM (1988). Possible neurotransmitter basis of behavioural changes in Alzheimer's disease. *Annals of Neurology* **23**, 616–20.

Pearson RCA and Powell TPS (1989). The neuroanatomy of Alzheimer's disease. *Reviews in Neuroscience* **2**, 101–21.

Pearson RCA, Esiri MM, Hiorns RW, *et al.* (1985). Anatomical correlates of the distribution of the pathological changes in the neocortex in Alzheimer's disease. *Proceedings of the National Academy of Sciences of the USA* **82**, 4531–4.

Perry EK, Marshall E, Perry RH, *et al.* (1990). Cholinergic and dopaminergic activities in senile dementia of Lewy body type. *Alzheimer Disease and Associated Disorders* **4**, 87–95.

Plant GT, Ghiso J, Holton JL, Frangione B, and Revesz T (2004). Familial and sporadic cerebral amyloid angiopathies associated with dementia and the BRI dementias. In Esiri MM, Lee VM-Y,and Trojanowski JQ (eds), *The neuropathology of dementia* (2nd edn), pp. 330–52 (Cambridge University Press, Cambridge).

Poorkaj P, Bird TD, Wijsman E, *et al.* (1998). Tau is a candidate gene for chromosome 17 frontotemporal dementia. *Annals of Neurology* **43**, 815–25.

Prelli F, Castano E, Glenner GG, and Frangione B (1988). Differences between vascular and plaque core amyloid in Alzheimer's disease. *Journal of Neurochemistry* **51**, 648–51.

Rogers J and Morrison JH (1985). Quantitative morphology and regional and laminar distributions of senile plaques in Alzheimer's disease. *Journal of Neuroscience* **5**, 2801–8.

Roher AE, Lowenson JD, Clark S, *et al.* (1993). β Amyloid (1–42) is a major component of cerebrovascular amyloid deposits: implications for the pathology of Alzheimer's disease. *Proceedings of the National Academy of Sciences of the USA* **90**, 10836–40.

Russo-Neustadt A, Zomorodian TJ, and Cotman CW (1998). Preserved cerebellar tyrosine hydroxylase-immunoreactive neuronal fibres in a behaviourally aggressive subgroup of Alzheimer's disease patients. *Neuroscience* **87**, 55–61.

Simchowicz T (1911). Histologische Studien ̦ber die senile Demenz. *Histologische und Histopathologische Arbeiten* **4**, 267–444.

Sloper JJ, Barnard RO, Eglin RP, and Powell TPS (1986). Abnormal tubular structures associated with the granular endoplasmic reticulum of the neocortical neurons in a biopsy from a patient with Alzheimer's disease. *Neuropathology and Applied Neurobiology* **12**, 491–501.

Smith PEM, Irving D, and Perry RH (1991). Density, distribution and prevalence of Lewy bodies in the elderly. *Neuroscience Research Communications* **8**, 127–35.

Snow AD, Mar H, Nochlin D, *et al.* (1988). The presence of heparan sulphate proteoglycans in the neuritic plaques and congophilic angiopathy in Alzheimer's disease. *American Journal of Pathology* **133**, 456–63.

Sofroniew MV and Mobley WC (1993). On the possibility of positive feedback in trophic interactions between afferent and target neurons. *Journal of Neuroscience* **5**, 309–12.

Spillantini MG, Crowther RA, Jakes R, Hasegawa M, and Goedert M (1998a). Alpha-synuclein in filamentous inclusions of Lewy bodies from Parkinson's disease and dementia with Lewy bodies. *Proceedings of the National Academy of Sciences of the USA* **95**, 6469–73.

Spillantini MG, Murrell JR, Goedert M, *et al.* (1998b). Mutation in the tau gene in familial multisystem tauopathy with presenile dementia. *Proceedings of the National Academy of Sciences of the USA*, **95**, 7737–41.

Terry RD (1963). The fine structure of neurofibrillary tangles in Alzheimer's disease. *Journal of Neuropathology and Experimental Neurology* **22**, 629–42

Terry RD, Gonatas NK, and Weiss M (1964). Ultrastructural studies in Alzheimer's presenile dementia. *American Journal of Pathology* **44**, 269–97.

Terry RD, Peck A, DeTeresa R, Schechter R, and Horoupian DS (1981). Some morphometric aspects of the brain in senile dementia of the Alzheimer type. *Annals of Neurology* **10**, 184–92.

Terry RD, Masliah E, Salmon DP, *et al.* (1991). Physical basis of cognitive alterations in Alzheimer's disease: synapse loss is the major correlate of cognitive impairment. *Annals of Neurology* **30**, 572–80.

Van Broeckhoven C, Haan J, Bakker F, *et al.* (1990). Amyloid β protein precursor gene and hereditary cerebral haemorrhage with amyloidosis (Dutch). *Science* **248**, 1120–22.

Van Hoesen GW (1982). The parahippocampal gyro: new observations regarding its cortical connections in the monkey. *Trends in Neuroscience* **5**, 345–50.

Van Hoesen G and Damasio A (1987). Neuronal correlates of cognitive impairment in Alzheimer's disease. In Plum F (ed.), *The handbook of physiology* Vol. 5, pp. 871–98 (American Physiological Society, Baltimore, MD).

Vinters HV and Gilbert JJ (1983). Cerebral amyloid angiopathy: incidence and complications in the ageing brain II. The distribution of amyloid vascular changes. *Stroke* **14**, 924–8.

West MJ, Coleman PD, Flood DG and Troncoso JC (1994). Differences in the pattern of hippocampal neuronal loss in normal ageing and Alzheimer's disease. *Lancet* **344**, 764–72.

Wilcock GK and Esiri MM (1982). Plaques, tangles and dementia: a quantitative study. *Journal of Neurological Science* **56**, 343–56.

Wilcock, GK, Esiri MM, Bowen DM and Hughes AO (1988). The differential involvement of subcortical nuclei in senile dementia of the Alzheimer type. *Journal of Neurology, Neurosurgery and Psychiatry* **51**, 842–9.

Wisniewski HM, Narang HK, Corsellis JAN, and Terry RD (1976). Ultrastructural studies of the neuropil and neurofibrillary tangles in Alzheimer's disease and post-traumatic dementia. *Journal of Neuropathology and Experimental Neurology* **35**, 367.

Wisniewski HM, Wegiel J, and Kotula L (1996). Some neuropathological aspects of Alzheimer's disease and its relevance to other disciplines. *Neuropathology and Applied Neurobiology* **22**, 3–11.

Yagishita ST, Itoh T, Nan W, and Amano N (1981). Reappraisal of the fine structure of Alzheimer's neurofibrillary tangles. *Acta Neuropathologica* **54**, 239–46.

Yan SD, Zhu H, and Fu J (1997). Amyloid-β peptide-receptor for advanced glycation end product interaction elicits neuronal expression of macrophage-colony stimulating factor: a pro-inflammatory pathway in Alzheimer's disease. *Proceedings of the National Academy of Sciences of the USA* **94**, 5296–301.

Zubenko GS, Moossy J, and Kopp U (1990). Major depression in primary dementia: clinical and neuropathologic correlates. *Archives of Neurology* **45**, 1182–6.

Zweig RM, Ross CA, Hedreen JC, *et al.* (1988). The neuropathology of aminergic nuclei in Alzheimer's disease. *Annals of Neurology* **24**, 233–42.

Chapter 3

Molecular genetics of Alzheimer's disease

John S.K. Kauwe and Alison M. Goate

3.1 Introduction

Alzheimer's disease (AD) is a complex neurodegenerative disorder characterized by gradual onset and progression of memory loss combined with deficits in executive functioning, language, visuospatial abilities, personality, behaviour and self-care. In the USA, more than 4.5 million people are estimated to have AD. Citing the rapid growth of the oldest age groups in the US population, recent studies predict a rapid increase in the prevalence of AD unless more effective treatments can be developed (Hebert *et al.* 2003). Individuals with AD live from eight to 20 or more years after the onset of symptoms, making this disease both emotionally and financially devastating. The national direct and indirect costs of caring for individuals with AD are in excess of 100 billion dollars (Ernst and Hay 1994). It has been suggested that a 5 year delay in the onset of disease would result in half as many cases in just one generation (Brookmeyer *et al.* 1998). In order to diagnose AD earlier and to treat it more efficiently, we must identify both the genetic and environmental factors which modulate risk for disease.

The National Institute of Neurological and Communicative Disorders and Stroke–Alzheimer's Disease and Related Disorders Association (NINCDS–ADRDA) has developed standardized criteria for the diagnosis of AD patients (McKhann *et al.* 1984). These criteria identify demented patients as definite AD, probable AD, or possible AD. A diagnosis of probable AD can be made following a clinical examination. It requires the identification of deficits in two or more areas of cognition and gradual progression of symptoms. It also requires that there be no disturbance of consciousness and an absence of other possible causes of the observed symptoms (such as medication). Cases in which dementia onset and progression are rapid, or cognitive deficits are specific to a single area of cognition, are diagnosed as possible AD. Cases in which another possible cause of dementia is present are also diagnosed as possible AD.

AD can be divided into two types based upon age of onset and familial aggregation: familial AD (FAD) and late-onset AD (LOAD). FAD is characterized by Mendelian inheritance (autosomal dominant) and early onset (<60 years). FAD represents less than 1% of all AD cases (Lopera *et al.* 1997; Crook *et al.* 1998; Ezquerra *et al.* 1999). LOAD is characterized by later onset (>60 years) and complex patterns of inheritance. Although they differ in age of onset, both forms of the disease are defined by the same pathological features: neuronal loss and the presence of β-amyloid (Aβ) plaques and neurofibrillary tangles. Plaques are extracellular deposits of insoluble proteins. These extracellular plaques mainly consist of Aβ peptides, which are derived from the β-amyloid precursor protein (APP) (Selkoe 1994). Tangles are intracellular deposits of hyperphosphorylated tau protein.

3.2 Genetic epidemiology

Familial aggregation of AD was first noted by Lowenberg and Waggoner (1934). Unfortunately, the late onset of AD adds some difficulty to family studies. For example, direct examination of parents in these families is often impossible. When parental information is available, it is usually inferred from medical records or from family history from living first-degree relatives. Siblings provide another set of challenges; older siblings may already be dead, while younger siblings and children may not have reached the risk period for AD. Despite these difficulties, risk for first-degree relatives has been reported to be 10–40% greater than in unrelated individuals (Mohs *et al.* 1987; Breitner *et al.* 1988; Huff *et al.* 1988; Mayeux *et al.* 1991). Sibling relative risk ratios (a comparison between the recurrence rate of AD in siblings of AD patients and the rate observed in the general population) have been reported in various studies (Breitner *et al.* 1988; Sadovnick *et al.* 1989; Hirst *et al.* 1994), and consistently estimated to be between 4 and 5. Studies using monozygotic (MZ) and dizygotic (DZ) twins also provide evidence that there is a genetic component to risk for AD. In these studies, the concordance of disease status is measured for both MZ and DZ twins. If there is a detectable genetic component to AD, MZ twins, who share 100% of their DNA, should have greater concordance than DZ twins, who share approximately half of their DNA. While the substantial variance in age of onset and the cross-sectional nature of most studies probably causes a reduction in concordance rates, MZ twin concordance (0.49) is generally higher than that observed in DZ twins (0.18) (Bergen 1994; Bergen *et al.* 1997). This indicates the presence of a genetic component, though the MZ concordance levels suggest that environmental factors also play an important role in AD.

Individuals with Down's syndrome (DS) who live beyond 30 years of age almost invariably develop the symptoms and pathological hallmarks of AD (Burger and Vogel 1973). In 1984, Glenner and Wong isolated and identified Aβ in the meningeal vessels of individuals with AD and DS. This discovery linked the two diseases and led to the proposal that the genetic defect in AD was located on chromosome 21 (Glenner and Wong 1984a,b). Several years later the gene encoding the Aβ protein precursor was cloned and mapped to chromosome 21 (Yoshikai *et al.* 1990).

3.3 Genetic risk factors for familial Alzheimer's disease

Mutations in three genes, *APP*, *PS1*, and *PS2*, have been shown to cause FAD by affecting the production of Aβ (Goate *et al.* 1991; Levy-Lahad *et al.* 1995; Sherrington *et al.* 1995). These genes were identified using positional cloning methods (*PS1*) or a combination of positional cloning and candidate gene methods (*PS2* and *APP*).

The *APP* gene spans 290 kb on chromosome 21q21.2 and contains 18 exons (Hattori *et al.* 2000). There are seven known alternative transcripts of *APP*; the longest transcript is 770 amino acids (aa) in length. Each of these transcripts encodes a multidomain protein with one transmembrane domain (Yoshikai *et al.* 1990). *APP* is expressed in all tissues, although the major transcript in neurons is shorter (*APP695*) than the major transcript in other cell types (*APP770*). *APP* undergoes proteolytic processing via two major pathways (Selkoe 1994). In most cell types, *APP* is cleaved first by α-secretase and then by γ-secretase. This pathway does not produce Aβ fragments. However, in neurons *APP* can be first cleaved by β-secretase and then by γ-secretase. γ-Secretase cleavage is promiscuous, and results in Aβ species varying in length from 37 to 43 aa. The most common fragment in bodily fluids is 40 aa in length (Aβ40), while the major component of amyloid plaques observed in AD is the 42 aa peptide (Aβ42). Since Goate *et al.* (1991) first reported that mutations in *APP* cause FAD, a total of 19 missense mutations in this gene have been shown to cause FAD, accounting for a small portion of FAD cases (about 10%). Most of these mutations occur at codons near the β- and γ-secretase cleavage sites in *APP*. The mutations near the γ-secretase cleavage site result in an increase in the ratio of Aβ42 to Aβ40 (Citron *et al.* 1992; De Strooper and Annaert 2000). In contrast, the so-called Swedish mutation at the β-secretase cleavage site results in an increase in total Aβ species without altering the ratio. Individuals with this mutation have both neuritic plaques and cerebral amyloid angiopathy (CAA). The first mutation to be described in the *APP* gene was observed in a large Dutch kindred with hereditary cerebral haemorrhage with amyloidosis (HCHWA), an autosomal dominant form of vascular amyloidosis in which Aβ is deposited in the cerebral blood vessels, leading to haemorrhagic strokes and dementia (Van Duinen *et al.* 1987; Levy *et al.* 1990). This disorder is caused by an amino acid substitution (E693) within the Aβ peptide that alters the physico-chemical properties of the peptide. Several other missense mutations within the Aβ sequence appear to lead to a mixed phenotype of dementia and haemorrhagic strokes. Duplications of *APP* have recently been shown to lead to a familial disorder characterized clinically by haemorrhagic strokes as well as dementia and pathologically by neuritic plaques and CAA (Rovelet-Lecrux *et al.* 2006).

The *PS1* gene contains 12 exons and spans 84 kb on chromosome 14q24.2. The full-length protein is composed of 467 aa, and includes eight, or possibly nine, transmembrane domains (Li and Greenwald 1996, 1997, 1998; Laudon *et al.* 2005). *PS2* is a homologous protein encoded by a gene on 1q42.13. Like *PS1* it has 12 exons, but it spans just 25 kb. A number of functions for the presenilins have been reported, but their role in γ-secretase activity is of particular interest to the molecular pathology of AD. *PS1* and *PS2* have distinct but overlapping γ-secretase activity (Lai *et al.* 2003). Anterior pharynx defective 1 homologue (APH1), nicastrin (NCT), presenilin 1 (PS1, or PS2) and presenilin enhancer 2 homologue (PEN2) associate with each other to create the stable proteolytically active γ-secretase complex (Kimberly *et al.* 2003). Several lines of evidence suggest that presenilin forms the catalytic core of this complex (Schroeter *et al.* 2003).

After linkage studies provided evidence for a locus on chromosome 14, Sherrington *et al* (1995) identified five missense mutations in *PS1* that segregated with FAD in their samples. Shortly thereafter, Levy-Lahad *et al.* (1995) identified a mutation in *PS2* that segregated with AD in the Volga German kindreds. Mutations in *PS1* increase the ratio of Aβ42 to Aβ40 and result in very early onset of disease (26–60 years) (Xia *et al.* 1997; Gustafson *et al.* 1998; Dermaut *et al.* 1999). Over 130 mutations in *PS1* have been identified, accounting for a large proportion of FAD cases. These mutations occur throughout the molecule but appear to have similar effects on Aβ levels. In contrast, just eight FAD mutations in *PS2* have been identified. The known mutations in *PS2* result in a later age of onset (40–75 years) than homologous mutations in *PS1* and may exhibit incomplete penetrance (Ezquerra *et al.* 2003). Mutations in *PS2* are the rarest of the known causes of FAD.

Several FAD mutations have been used to create transgenic mice that model age-dependent Aβ deposition. Interestingly, *PS* mutations that increase Aβ42/Aβ40 ratios result in neuritic plaques in transgenic mice, while mutations that increase total Aβ levels lead to both CAA and neuritic plaques. Transgenic mice overexpressing the HCHWA mutation in *APP* (E693Q) in neurons deposit Aβ in the cerebral vasculature, inducing haemorrhages (Herzig *et al.* 2004). However, when these mice are crossed with mice overexpressing an FAD mutation in *PS1*, the Aβ42/Aβ40 ratio is increased and the amyloid pathology redistributes to the parenchyma (Herzig *et al.* 2004). A second mutation in the Aβ sequence (the Arctic mutation E693G) also increases protofibril formation without altering Aβ ratios but leads to AD pathology (Cheng *et al* 2004). These results suggest that the differing pathologies are intrinsic to the mutations rather than resulting from genetic or environmental modifiers. Together, genetic studies in humans and transgenic mouse studies suggest that elevated levels of total Aβ or increased protofibril formation lead to both AD and CAA pathology, while elevation of the Aβ42/Aβ40 ratio leads to AD pathology. Thus it appears that mutations in APP can lead to a spectrum of clinical phenotypes including both dementia and haemorrhagic stroke.

3.4 Risk factors for late-onset Alzheimer's disease

While much is known about genetic risk factors for FAD, the vast majority of AD cases are late onset. The single most important known risk factor for LOAD is age (Jorm *et al.* 1987; Rocca *et al.* 1991; Ritchie and Kildea 1995). Both genetic linkage and association studies are being used to identify genetic risk factors for LOAD. To date, *APOE* is the only known genetic risk factor. The *APOE* gene is located in a region of chromosome 19 that has been identified as a risk region in genetic linkage studies in LOAD families (Strittmatter *et al.* 1993a). *APOE* has been consistently associated with risk for LOAD in many studies (Finckh 2003).

The *APOE* gene is located at 19q13.2. It spans less than 4 kb and has four exons. The protein has 317 residues and is produced in most organs. In the brain most APOE is synthesized by glial cells. There are three widely recognized isoforms of *APOE* (*APOE2*, *APOE3*, and *APOE4*). The *APOE3* allele is the most frequent isoform in all populations (Finckh 2003). *APOE2* and *APOE4* vary from *APOE3* at two amino acid residues (C112R and C158R). APOE mediates the binding, internalization, and catabolism of lipoprotein particles.

The *APOE* polymorphism has been genotyped in samples of many racial and ethnic origins, and consistently shows evidence for association with LOAD (Finckh 2003; Bertram *et al.* 2005b). The risk is dose dependent. European-Americans who are heterozygous for the *APOE4* allele exhibit a threefold increase in risk, while homozygotes exhibit an eightfold increase. Nearly all individuals homozygous for the *APOE4* allele will develop LOAD by 80 years of age (Corder *et al.* 1993). In addition, a recent study has shown that the *APOE* gene explains some of the variance in age of onset in families with a known FAD mutation (Pastor *et al.* 2003). *APOE4* is certainly a major risk factor for LOAD, however, there are likely to be other factors, because APOE4 shows only a modest effect on risk in Amish and Hispanic patients (Pericak-Vance *et al.* 1996; Tang *et al.* 1998). Furthermore, approximately 50% of Caucasian AD patients do not carry an *APOE4* allele.

A number of hypotheses of how APOE affects risk for AD have been proposed. Not surprisingly, many of them involve interactions with Aβ. Patients who carry at least one *APOE4* allele have a greater number of plaques than patients without an *APOE4* allele (Schmechel *et al.* 1993). *In vitro*, APOE4 binds to Aβ with higher affinity than it binds to APOE3 (Strittmatter *et al.* 1993b). There is also evidence that APOE and Aβ may compete for clearance through the same receptor (Kounnas *et al.* 1995). Mouse models that over-express *APP* carrying FAD mutations only show thioflavin-positive Aβ (amyloid) deposition when the *APOE* gene is expressed (Bales *et al.* 1997). A study by Holtzman *et al.* (2000) suggests that *APOE4* may influence fibril formation and clearance of Aβ, causing increased Aβ deposition. The same group also showed that mice with *APOE4* alleles developed earlier and more severe pathological phenotypes (Fryer *et al.* 2003). *In vitro* experiments show that *APOE3* binds to tau with a higher affinity than *APOE4*, suggesting that *APOE* may also have some effect on neurofibrillary tangles (Strittmatter *et al.* 1994). Some studies also suggest that promoter variants in *APOE* are associated with LOAD risk (Wang *et al.* 2000). *APOE2*, which is the rarest allele in most populations, is protective for AD, resulting in older ages of onset of disease. Also, mouse studies suggest that although Aβ is deposited it cannot form amyloid in the presence of *APOE2* (Fryer *et al.* 2003). Mayeux *et al.* (1995) showed that the biological effects of head injury may increase risk for AD through a synergistic relationship with the *APOE4* allele. The *APOE* genotype has also been implicated as a risk factor in a number of other diseases including coronary artery disease and CAA (Contois *et al.* 1996).

3.5 Novel risk loci

3.5.1 Linkage

Linkage studies identify regions of the genome that may have disease genes by examining the patterns of allele transmission or sharing between relatives. Linkage studies require familial samples, with multi-generational pedigrees providing the greatest statistical power. The characteristic late onset of LOAD makes it difficult to collect such samples. As a result, many linkage studies of LOAD are underpowered. These complications limit our ability to interpret the replication of results between studies. The lack of endophenotypes or quantitative traits for AD has also limited the statistical power of these studies. Despite these difficulties, several groups have identified linked regions by performing genome-wide genetic linkage studies in LOAD families or affected sibling pairs (Pericak-Vance et al. 1997; Kehoe et al. 1999; Myers et al. 2002; Blacker et al. 2003). Pericak-Vance et al. (1997) used a two-stage design for their linkage study. The first set of LOAD families was screened with markers at 10 cM intervals. Regions of the genome with LOD (log of the likelihood odds ratio) scores greater than 1 were investigated in an additional sample. The strongest evidence of linkage was found on chromosome 12: multipoint LOD score (MLS = 3.9). Evidence of linkage was also found on chromosomes 4, 6, and 20. In 2000, the same group performed a genome-wide screen on more than 450 families (Pericak-Vance et al. 2000). In this screen the strongest evidence for linkage was found on chromosome 9 (MLS = 4.3). Kehoe et al. (1999) reported the results of the first stage of their two-stage screen of LOAD-affected sibling pairs (ASPs). In the first stage, 292 ASPs were screened with markers at approximately 20 cM intervals. Regions with MLS > 1 were then screened with a denser set of markers in a total of 450 ASPs (Myers et al. 2002). The strongest evidence for linkage was found on chromosome 10 (MLS = 3.9). Scores reaching the level of 'suggestive' linkage were found on chromosomes 1 (MLS = 2.67), 9 (MLS = 2.38), 10 (MLS = 2.27), and 19 (MLS = 1.79). Blacker et al. (2003) performed a 9 cM screen of 437 families. They identified significant linkage on chromosome 19, and suggestive linkage in 12 regions on chromosomes 1, 3, 4, 5, 6, 9, 10, 11, 14, 15, and 21. Although there appears to be replication of several chromosomal regions in these studies, it should be noted that all the studies used overlapping datasets and thus the results are not truly independent.

A number of other studies have focused on the regions suggested in these earlier linkage results. Additional evidence for a LOAD locus on chromosome 10 has come from a number of sources. Ertekin-Taner et al. (2000) used a quantitative endophenotype to increase their power to detect susceptibility loci. They had previously shown that high plasma Aβ levels was a heritable trait and thus collected large families in which plasma Aβ levels were measured in all individuals. They followed up the linkage results reported by Kehoe et al. (1999) by genotyping markers on chromosomes 1, 5, 9, 10, and 19. The strongest evidence for linkage to high plasma Aβ levels was found on chromosome 10 (MLS = 3.9), a finding that is consistent with the results described by Myers et al. (2002).

A study by Bertram et al. (2000) identified another region of chromosome 10 which shows significant linkage to LOAD (MLS = 3.4). Li et al. (2002) performed a genome screen to identify regions influencing age of onset in LOAD and Parkinson's disease (PD). This study identified linkage around *APOE* on chromosome 19 and implicated a region of chromosome 10 near the one reported by Bertram et al. (2000).

The linkage on chromosome 12 reported by Pericak-Vance et al. (1997) has been confirmed by several other studies (Rogaeva et al. 1998; Wu et al. 1998). The original linkage report was refined when Scott et al. (2000) reported the results of a screen using samples from the Pericak-Vance et al. (1997) study and additional markers. A study using 35 markers in 79 Caribbean Hispanic families also found evidence for linkage on chromosome 12 (MLS = 3.15) (Mayeux et al. 2002).

Farrer *et al.* (2003), Sillen *et al.* (2006), and Hahs *et al.* (2006) recently reported the results of genome-wide linkage or association studies. Citing the possibility that genetic heterogeneity in earlier studies could have resulted in reduced power to detect disease loci, these studies focused on relatively homogeneous populations. Farrer *et al.* (2003) performed a 10 cM screen in an inbred Israeli Arab community with a high prevalence of dementia. They observed strong evidence for linkage with markers on chromosomes 9, 10, and 12, consistent with the results observed in outbred populations. A novel region of linkage on chromosome 2 was also observed. Sillen *et al.* (2006) performed a 10 cM screen in 71 Swedish families. This screen identified significant linkage on chromosome 19 (near *APOE*) and suggestive evidence consistent with reports by Kehoe *et al.* (1999) and Blacker *et al.* (2003) on chromosome 5. Hahs *et al.* (2006) performed an approximately 7 cM screen on 115 individuals from five Amish families. The strongest evidence for linkage in this study comes from chromosome 4 (MLS = 3.01). LOD scores >2.0 were observed on chromosomes 3, 4, 10, 11, and 19.

Rademakers *et al.* (2005) performed a whole-genome linkage scan in a single multiplex family with mean onset of 68 years and found strong evidence of linkage (LOD = 3.39) with markers at chromosome 7q36. They performed additional analyses that led to the identification of three additional families who shared haplotypes and narrowed the candidate region to 9.4 cM.

Genome screens consistently identify linkage to chromosome 19 near *APOE*. In addition, regions of chromosomes 9, 10, and 12 appear to be linked to LOAD or related phenotypes in a number of studies. Recent studies using dense screens and more homogeneous populations have not yet provided strong evidence for other risk loci. Table 3.1 summarizes the results of genome screens and genetic linkage studies for LOAD.

3.5.2 Candidate genes

The linkage studies presented above have led to the identification of many potential disease genes that are found in regions of linkage, or positional candidate genes. Many biological candidate genes for LOAD have also been identified. All known FAD mutations cause an increase in Aβ42 levels (Scheuner *et al.* 1996). It is also known that *APOE*, the other known risk factor for AD, influences the aggregation of Aβ. This suggests that Aβ plays a central role in AD pathology. Efforts to improve characterization of the role of Aβ have led to the identification of a large number of genes which are involved in Aβ production, aggregation, degradation, and clearance. Many of these genes are found in regions where linkage has been reported, making them both positional and biological candidate genes. Many of these strong candidate genes have been the subjects of multiple genetic association studies. Association studies have some advantages over linkage studies as they have greater power to detect disease loci. In addition, association studies can be performed using either family-based samples or case–control series. Case–control samples are easily collected and provide considerable statistical power, but may be subject to an increase in spurious associations due to population stratification if they are not carefully ascertained. Family-based samples, generally in the form of case–parent trios or ASPs, are not subject to the problem of stratification, but are difficult to collect given the

Table 3.1 Results of genome screens and linkage studies

Study	Chromosome[*]	Result[†]
Pericak-Vance et al. 1997[‡]	4, 6, **12**, 20	MLS:3.9
Wu et al. 1998	12	MLS:1.9
Rogaeva et al. 1998	12	NPLS:3.5
Zubenko et al. 1998[‡]	1, 10, 12, 19, ×	All P<0.05
Kehoe et al. 1999[‡]	**1**, 5, **9**, **10**, 12, 14, **19**, 21	All MLS~2
Pericak-Vance et al. 2000[‡]	9	MLS:4.3
Bertram et al. 2000	10	TLS:3.4
Ertekin-Taner et al. 2000	10	MLS:3.9
Mayeux et al. 2002	12	TLS:3.15
Myers et al. 2002[‡]	1, 5, 6, 9, **10**, 12, 19, 21, ×	MLS:3.9
Blacker et al. 2003[‡]	1, 3, 4, 5, 6, 9, 10, 11, 14, 15, 21	All MLS≥2
Farrer et al. 2003[‡]	2, 9, 10, 12	All P<0.05
Rademakers et al. 2005[‡]	7	MLS:3.39
Hahs et al. 2006[‡]	3, **4**, 10, 11, 19	MLS:3.01
Sillen et al. 2006[‡]	4, 5, **19**	MLS:2.99

MLS, multipoint LOD score; NPLS, non-parametric linkage score; TLS, two-point LOD score.
[*]The strongest finding is printed in bold.
[†]Result is for the strongest finding.
[‡]These studies are whole-genome screens

late onset of AD. A recent study comparing the power of the two methods showed that case–parent trios and unrelated cases and controls yielded higher odds ratios and more significant test values than ASPs when looking at known risk loci for type II diabetes. However, subsequent simulation studies suggest that no one strategy is ideal in every situation; the optimal ascertainment strategy depends on the underlying disease model, which is unknown (Howson et al. 2005). Both ubiquilin and ABCA1 (ATP-binding cassette subfamily A, member 1) are located in regions of linkage and are considered to be biological candidate genes. Association in families has been detected using single-nucleotide polymorphisms (SNPs) in ubiquilin (Bertram et al. 2005a) but subsequent studies in unrelated cases and controls seeking to validate those findings have found a similar effect in just two of the nine populations tested (Slifer et al. 2005; Brouwers et al. 2006; Kamboh et al. 2006; Smemo et al. 2006). Results for ABCA1 have also been inconsistent, with association being observed in some studies but not others (Wollmer et al. 2003; Katzov et al. 2004; Li et al. 2004b; Bertram et al. 2005; Shibata et al. 2006). Positional and biological candidate genes on chromosomes 10 and 12 exhibit the same trend (Table 3.2). A detailed summary of more than 1000 LOAD association studies can be found at http://www.alzgene.org (Bertram et al. 2005b). This website summarizes the results of studies of more than 800 polymorphisms in more than 300 genes. Alzgene also

Table 3.2 A summary of selected SNPs in positional and biological candidate genes on chromosomes 9, 10, and 12

Gene	Polymorphism	Chromosome/position	Association		Meta-analysis OR (95% CI)
			Positive populations	Negative populations	
UBQLN	rs12344615	9/83510749	7	7	1.08 (0.94, 1.22)
ABCA1	rs2066718*	9/104468810	2	3	1.16 (0.74, 1.83)
	rs2230806*	9/104700422	0	7	1.01 (0.88, 1.14)
	rs2230808*	9/104642359	0	6	1.04 (0.85, 1.26)
	rs4149313*	9/104666308	0	6	1.1 (0.94,1.29)
CHAT	7936C>T*	10/50487147	0	3	~1 (includes 1)
	rs180676	10/50494123	0	4	1.09 (0.83, 1.43)
	rs3810950 (A20T)*	10/50494625	2	4	1.19 (0.9,1.57)
	rs868750	10/50503845	2	2	1.21 (0.96, 1.52)
VR22	rs7070570	10/67534610	0	7	0.99 (0.89, 1.11)
	SNP1	See Busby et al. 2004	0	4	1.22 (0.96, 1.54)
PLAU	rs2227564 (P141L)*	10/75343107	2	12	0.96 (0.85, 1.08)
	rs2227566	10/75343737	1	6	1.14 (0.99, 1.32)
	rs2227568	10/75343885	0	7	0.97 (0.82, 1.15)
	rs2227571	10/75344746	1	6	1.13 (0.96, 1.32)
	rs2227580	10/75341362	0	3	1.12 (0.5,2.49)
	rs4065	10/75346470	1	6	1.14 (0.96, 1.34)
IDE	rs1832196	10/94258314	0	8	~1 (includes 1)
	rs1887922	10/94214145	0	3	~1 (includes 1)
	rs1999764	10/94310119	0	3	~1 (includes 1)

Continued

Table 3.2 A summary of selected SNPs in positional and biological candidate genes on chromosomes 9, 10, and 12—cont'd

Gene	Polymorphism	Chromosome/position	Association		Meta-analysis OR (95% CI)
			Positive populations	Negative populations	
	rs2251101	10/94201284	2	6	~1 (includes 1)
	rs3758505	10/94324758	0	7	1.11 (0.91,1.34)
	rs4646953	10/94323935	1	3	~1 (includes 1)
	rs4646954	10/94323807	0	4	~1 (includes 1)
	rs4646958	10/94204339	0	5	1.09 (0.88,1.36)
	rs551266	10/94212604	0	5	~1 (includes 1)
GAPDH	rs1060621	12/6514957	1	2	1.19 (0.98,1.44)
	rs3741916	12/6514252	1	2	1.15 (0.86,1.54)
A2M	5-bp in/del*	See Saunders *et al.* 2003	2	43	0.98 (0.91,1.05)
	V1000I*	See Saunders *et al.* 2003	3	22	0.98 (0.89,1.07)

*Putative functional polymorphism.

From www.alzgene.org

provides a meta-analysis of association for each SNP using these data. The website is updated frequently, and thus provides an interesting and dynamic view of LOAD association studies. With the exception of apolipoprotein E (*APOE*), association studies have been unsuccessful in identifying polymorphisms that provide consistent evidence for association with LOAD (Bertram and Tanzi 2004).

3.6 Future directions

3.6.1 Whole-genome association

Genotyping technology continues to advance, making it possible for many markers to be genotyped in a short amount of time and at a very low cost. As additional polymorphisms are genotyped in existing samples, we may gain insight into the association of positional and biological candidate genes.

These advances have also led to the development of large-scale association studies. These whole-genome studies face a number of difficulties. Care must be taken to design the study to deal with the statistical issues of multiple testing to identify true-positive results adequately while minimizing false-positive results. In addition, researchers must be sure to use carefully defined phenotypes, appropriately matched controls, and appropriate sample sizes. It is also important to account for other sources of error such as population stratification and genotyping error. Approaches have been developed to deal with most of these complicating factors (Ehm *et al.* 2005). Whole-genome studies provide increased power and efficiency compared with family-based linkage scans and are viewed by many scientists as a critical new tool for the identification of novel susceptibility loci for complex diseases (Risch and Merikangas 1996).

Recently, association studies for a number of diseases including PD and LOAD using hundreds or thousands of markers have been reported (Li *et al.* 2004a; Maraganore *et al.* 2005; Grupe *et al.* 2006). These studies generally take one of two approaches, attempting either to sample the linkage disequilibrium structure adequately or to screen the putative functional polymorphisms in the region.

Maraganore *et al.* (2005) used a two-stage approach to perform a whole-genome linkage disequilibrium screen for PD. They first screened 443 discordant sibling pairs with nearly 200 000 SNPs. The 1793 SNPs that were significant in that screen were then genotyped in 332 matched case–control pairs. The most significant association was detected for SNPs within the *PARK10* gene, a known PD susceptibility locus. The scan also detected evidence for association in the semaphorin 5A gene, which had previously not been implicated in PD. A similar study of LOAD is now underway.

While a genome-wide linkage disequilibrium screen of LOAD has not yet been reported, two studies focusing on the large-scale screening of putative functional polymorphisms have had some success. These studies performed association studies in three large case–control series using the same approach. First, a large number of markers were genotyped in a single sample set, which served as an exploratory set. Markers that showed association with the phenotype were genotyped in two validation sets. Fine mapping was then performed in the vicinity of markers that replicated. Li *et al.* (2004a) used this approach and detected association of multiple members of the *GAPD* (glyceraldehyde-3-phosphate-dehydrogenase) gene family with LOAD. The same group also reported association with a polymorphism on chromosome 10 (Grupe *et al.* 2006). These promising results suggest that large-scale association studies may indeed lead to the identification of novel susceptibility loci for LOAD.

3.6.2 Expression levels of familial Alzheimer's disease genes

Loci that cause autosomal dominant forms of many neurodegenerative diseases have been identified. In a number of cases, variation in the expression of these same genes is associated with increased risk for the sporadic forms of the disease (Hardy 2005). Mutations in *PRNP* (gene for prion protein) have been shown to segregate with Creutzfeldt–Jakob disease (CJD) (Owen *et al.* 1990). Further studies have shown that polymorphisms upstream of *PRNP* exon 1 and a missense polymorphism (M/V129) are associated with the sporadic form of CJD (Mead *et al.* 2001). Studies of frontal temporal dementia have identified missense and splicing mutations in the gene encoding the microtubule-associated protein tau that are associated with disease (Hutton *et al.* 1998). As the investigation of tau has continued, associations with the sporadic disorders progressive supranuclear palsy and corticobasal degeneration have also been identified, with the strongest evidence for association being observed with SNPs toward the 5' end of the gene (Baker *et al.* 1999; Houlden *et al.* 2001). Mutations in α-synuclein are associated with familial PD (Zarranz *et al.* 2004). In addition, polymorphisms that alter the expression of α-synuclein have been shown to contribute to risk for sporadic forms of PD (Chiba-Falek *et al.* 2003).

These studies suggest that in many neurodegenerative disorders familial early onset forms of the disease are caused by mutations in the gene encoding the protein that is deposited in the brain, while polymorphisms that lead to overexpression of the same gene may lead to sporadic or late-onset forms of the same disorder. It has been proposed that AD may have a similar aetiology (Hardy 2005). A comparison of known risk factors for AD and PD provides striking parallels. *APP*, like α-synuclein, codes for a deposited protein. Mutations in *APP* and α-synuclein have been shown to cause familial early-onset forms of AD and PD, respectively. Duplications and triplications of α-synuclein have been shown to cause familial PD (Singleton *et al.* 2003; Chartier-Harlin *et al.* 2004; Ibanez *et al.* 2004). Rovelet-Lecrux *et al.* (2006) documented several families carrying different duplications of *APP* that resulted in FAD and stroke. They identified duplications in five families and documented an approximately 50% increase in the expression of APP. This suggests that even modest increases in expression levels are sufficient to cause AD with an onset of clinical symptoms at approximately 50 years of age. Clinically, these duplication families would only meet diagnostic criteria for possible AD because of the co-occurrence of stroke, highlighting the fact that molecular diagnosis (β-amyloidopathy) and clinical diagnosis do not always agree. Similar observations have been made in other neurodegenerative disorders. For example, prion gene mutations are associated with CJD disease and familial fatal insomnia, and *MAPT* (microtubule-associated protein tau) mutations can be associated with a broad range of phenotypes from dementia to Parkinsonism and psychosis. These variable clinical phenotypes are also accompanied by great variation in pathological phenotypes.

Based on these similarities it seems likely that, like α-synuclein in PD, subtle changes in levels of expression of *APP* would affect risk for sporadic LOAD. Furthermore, based on the phenotype observed in the duplication families and in DS, overexpression of *APP* is likely to be associated with both neuritic plaques and CAA, and thus will be associated with both dementia and haemorrhagic strokes, while overexpression of the presenilins should lead to increased risk for neuritic plaques and dementia but not CAA or strokes. Only a handful of polymorphisms in the known risk factors for AD have been investigated in multiple studies (Table 3.3). To date, only two groups have reported the results of association studies in APP. The hypothesis that variation in the expression of the known familial risk factors can lead to pathology warrants further investigation of variation in *APP*, *PS1*, *PS2*, and *APOE* for association with LOAD.

3.6.3 Late-onset Alzheimer's disease biomarkers

The use of quantitative endophenotypes has greatly aided the search for genetic risk factors in complex disease. The use of cholesterol levels in coronary artery disease and various electrophysiological characteristics in alcoholism provide good examples (Sing and Moll 1989; Bierut *et al.* 2002). Few useful quantitative endophenotypes for AD are currently available. Phenotypes such as age at onset and various neuropsychiatric tests have been used as quantitative traits to identify less heterogeneous case samples, but they are not considered biomarkers for AD.

The hypothesis that AD is a result of an imbalance between Aβ production and clearance which results in the gradual accumulation of Aβ, ultimately leading to widespread neuronal loss, has been used to identify biological candidate genes for LOAD. It also suggests that Aβ levels may be good biomarkers for AD. Plasma Aβ levels have been shown to increase in individuals with FAD (Scheuner *et al.* 1996). Researchers have used plasma Aβ as an endophenotype for LOAD in both linkage and association studies (Ertekin-Taner *et al.* 2000, 2004, 2005). However, it does not appear that plasma Aβ levels increase in individuals with LOAD (DeMattos *et al.* 2002; Fukumoto *et al.* 2003). This makes it unclear how the results of these studies using plasma Aβ as a phenotype apply to genetic risk for LOAD. Furthermore, several studies have reported that BMI and creatinine levels correlate with plasma Aβ levels, suggesting that many factors unrelated to AD may influence plasma Aβ levels.

While plasma Aβ may be an interesting phenotype in its own right, it does not seem to be a useful biomarker for LOAD. Many other proteins and compounds are being considered (Frey *et al.* 2005). This is a considerable effort as a good quantitative endophenotype must have a number of characteristics. An ideal biomarker for AD should detect a fundamental feature of pathology and be validated in neuropathologically confirmed cases. It should have a diagnostic sensitivity and specificity of greater than 80%. Collection and measurement of samples should be reliable, reproducible, non-invasive, simple to perform, and inexpensive (Ronald and Nancy Reagan Research Institute 1998).

Table 3.3 A summary of selected SNPs in known risk factors for AD

Gene	Polymorphism	Chromosome/position	Association		Meta-analysis OR (95% CI)
			Positive populations	Negative populations	
APOE	E3 vs E4	19/50100904	34	2	3.68 (3.31,4.11)
	rs405509	19 50100676	5	10	0.79 (0.71,0.87)
	rs440446	19/50101007	3	1	0.58 (0.5,0.7)
	rs449647	19/50100404	14	20	0.72 (0.6,0.8)
	rs769446	See Lambert et al. 1998	3	9	0.85 (0.7,1)
PSEN1	−48 C/T	14/72672861	2	5	0.85 (0.67,1.08)
	rs165932	14/72734606	3	31	0.93 (0.87,0.98)
	rs362373 (E318G)†	See Mattila et al. 1998	1	4	1.37 (0.6,3.1)
PSEN2	5' in/del†	1/223364999	0	3	1.01 (0.82,1.25)
	rs8383	1/223390285	1	2	1.23 (0.88,1.74)

†Putative functional polymorphism

*Significant results from the meta-analysis are in bold

From www.alzgene.org

Levels of tau and Aβ42 in CSF have been shown to correlate with AD status (Sunderland *et al*. 2003). These traits are promising, although alone they lack the specificity necessary to be effective biomarkers. Many researchers are currently working on identifying other molecules or combinations of molecules in CSF that may serve as biomarkers for AD (Blasko *et al*. 2006).

Imaging of Aβ in the brain using Pittsburgh compound B (PIB) has shown that PIB retention is correlated with the presence of disease (Klunk *et al*. 2004). PIB imaging of amyloid plaques may soon be applied as a diagnostic tool (Mintun 2005), and pilot studies investigating the utility of this phenotype in genetic studies are underway. The development of valid biomarkers for AD will refine the phenotype, creating more homogenous samples and greatly improving our ability to detect disease loci.

3.7 **Conclusion**

Mutations in three genes have been shown to cause FAD by increasing the production of Aβ42. The only known genetic risk factor of LOAD is *APOE*, which has also been shown to affect Aβ fibril formation. In addition, *APOE* has been shown to affect the age at onset of AD in both FAD and sporadic forms of the disease.

Significant linkage has been observed in regions of chromosomes 9, 10, and 12 in a number of studies. Hundreds of positional and biological candidate genes have been tested for association with disease. To date, only the *APOE* gene has shown consistent association with LOAD. However, new technology and analytical tools for whole-genome association studies have recently been developed. These approaches have already resulted in the identification of putative risk factors for AD. In addition, the successful identification of risk loci in other neurodegenerative disorders has revealed promising new hypotheses for AD, suggesting that *APP*, *PS1*, *PS2*, and *APOE* should be subjected to further scrutiny in LOAD. Finally, the development of AD biomarkers such as CSF Aβ42 and PIB retention in the brain will refine the AD phenotype, providing additional power to detect disease loci. These new methods and hypotheses will accelerate our search for genes and variants that modify risk for AD.

References

Baker M, Litvan I, Houlden H, *et al*. (1999). Association of an extended haplotype in the tau gene with progressive supranuclear palsy. *Human Molecular Genetics* **8**, 711–15.

Bales KR, Verina T, Dodel RC, *et al*. (1997). Lack of apolipoprotein E dramatically reduces amyloid beta-peptide deposition. *Nature Genetics* **17**, 263–4.

Bergen A (1994). Hereditary dementia of the Alzheimer type. *Clinical Genetics* **46**, 144–9.

Bergen A, Engedal K, and Kringlen E (1997). The role of heredity in late-onset Alzheimer disase and vascular dementia: a twin study. *Archives of General Psychiatry* **54**, 264–70.

Bertram L and Tanzi RE (2004). Alzheimer's disease: one disorder, too many genes? *Human Molecular Genetics* **13**, R135–41.

Bertram L, Blacker D, Mullin K, *et al*. (2000). Evidence for genetic linkage of Alzheimer's disease to chromosome 10q. *Science* **290**, 2302–3.

Bertram L, Hiltunen M, Parkinson M, *et al*. (2005a). Family-based association between Alzheimer's disease and variants in UBQLN1. *New England Journal of Medicine* **352**, 884–94.

Bertram L, McQueen M, Mullin K, Blacker D, and Tanzi R (2005b). The AlzGene Database. Alzheimer Research Forum. Available online at: http://www.alzgene.org (accessed12 May 2005).

Bierut LJ, Saccone NL, Rice JP, *et al*. (2002). Defining alcohol-related phenotypes in humans. The Collaborative Study on the Genetics of Alcoholism. *Alcohol Research and Health* **26**, 208–13.

Blacker D, Bertram L, Saunders AJ, *et al*. (2003). Results of a high-resolution genome screen of 437 Alzheimer's disease families. *Human Molecular Genetics* **12**, 23–32.

Blasko I, Lederer W, Oberbauer H, *et al*. (2006). Measurement of thirteen biological markers in CSF of patients with Alzheimer's disease and other dementias. *Dementia and Geriatric Cognitive Disorders* **21**, 9–15.

Breitner JC, Murphy EA, Silverman JM, Mohs RC, and Davis KL (1988). Age-dependent expression of familial risk in Alzheimer's disease. *American Journal of Epidemiology* **128**, 536–48.

Brookmeyer R, Gray S, and Kawas C (1998). Projections of Alzheimer's disease in the United States and the public health impact of delaying disease onset. *American Journal of Public Health* **88**, 1337–42.

Brouwers N, Sleegers K, Engelborghs S, *et al*. (2006). The *UBQLN1* polymorphism, UBQ-8i, at 9q22 is not associated with Alzheimer's disease with onset before 70 years. *Neuroscience Letters* **392**, 72–4.

Burger PC and Vogel FS (1973). The development of the pathologic changes of Alzheimer's disease and senile dementia in patients with Down's syndrome. *American Journal of Pathology* **73**, 457–6.

Busby V, Goossens S, Nowotny P, *et al*. (2004). Alpha-T-catenin is expressed in human brain and interacts with the Wnt signaling pathway but is not responsible for linkage to chromosome 10 in Alzheimer's disease. *Neuromolecular Medicine* **5**, 133–46.

Chartier-Harlin MC, Kachergus J, Roumier C, *et al*. (2004). Alpha-synuclein locus duplication as a cause of familial Parkinson's disease. *Lancet* **364**, 1167–9.

Cheng IH, Palop JJ, Esposito LA, Bien-Ly N, Yan F, and Mucke L (2004). Aggressive amyloidosis in mice expressing human amyloid peptides with the Arctic mutation. *Nature Medicine* **10**, 1190–2.

Chiba-Falek O, Touchman JW, and Nussbaum RL (2003). Functional analysis of intra-allelic variation at NACP-Rep1 in the alpha-synuclein gene. *Human Genetics* **113**, 426–31.

Citron M, Oltersdorf T, Haass C, *et al*. (1992). Mutation of the beta-amyloid precursor protein in familial Alzheimer's disease increases beta-protein production. *Nature*. **360**, 672–4.

Contois JH, Anamani DE, and Tsongalis GJ (1996). The underlying molecular mechanism of apolipoprotein E polymorphism: relationships to lipid disorders, cardiovascular disease, and Alzheimer's disease. *Clinics in Laboratory Medicine* **16**, 105–23.

Corder EH, Saunders AM, Strittmatter WJ, *et al*. (1993). Gene dose of apolipoprotein E type 4 allele and the risk of Alzheimer's disease in late onset families. *Science* **261**, 921–3.

Crook R, Verkkoniemi A, Perez-Tur J, *et al*. (1998). A variant of Alzheimer's disease with spastic paraparesis and unusual plaques due to deletion of exon 9 of presenilin 1. *Nature Medicine* **4**, 452–5.

De Strooper B and Annaert W (2000). Proteolytic processing and cell biological functions of the amyloid precursor protein. *Journal of Cell Science* **113**, 1857–70.

DeMattos RB, Bales KR, Parsadanian M, *et al*. (2002). Plaque-associated disruption of CSF and plasma amyloid-beta (Abeta) equilibrium in a mouse model of Alzheimer's disease. *Journal of Neurochemistry* **81**, 229–36.

Dermaut B, Cruts M, Slooter AJ, *et al*. (1999). The Glu318Gly substitution in presenilin 1 is not causally related to Alzheimer disease. *American Journal of Human Genetics* **64**, 290–2.

Ehm MG, Nelson MR, and Spurr NK (2005). Guidelines for conducting and reporting whole genome/large-scale association studies. *Human Molecular Genetics* **14**, 2485–8.

Ernst RL and Hay JW (1994). The U.S. economic and social costs of Alzheimer's disease revisited. *American Journal of Public Health* **84**, 1261–4.

Ertekin-Taner N, Graff-Radford N, Younkin LH, *et al*. (2000). Linkage of plasma Abeta42 to a quantitative locus on chromosome 10 in late-onset Alzheimer's disease pedigrees. *Science* **290**, 2303–4.

Ertekin-Taner N, Allen M, Fadale D, *et al.* (2004). Genetic variants in a haplotype block spanning IDE are significantly associated with plasma Abeta42 levels and risk for Alzheimer disease. *Human Mutation* **23**, 334–42.

Ertekin-Taner N, Ronald J, Feuk L, *et al.* (2005). Elevated amyloid beta protein (Abeta42) and late onset Alzheimer's disease are associated with single nucleotide polymorphisms in the urokinase-type plasminogen activator gene. *Human Molecular Genetics* **14**, 447–60.

Ezquerra M, Carnero C, Blesa R, Gelpi JL, Ballesta F, and Oliva R (1999). A presenilin 1 mutation (Ser169Pro) associated with early-onset AD and myoclonic seizures. *Neurology* **52**, 566–70.

Ezquerra M, Lleo A, Castellvi M, *et al.* (2003). A novel mutation in the *PSEN2* gene (T430M) associated with variable expression in a family with early-onset Alzheimer disease. *Archives of Neurology* **60**, 1149–51.

Farrer LA, Bowirrat A, Friedland RP, Waraska K, Korczyn AD, and Baldwin CT (2003). Identification of multiple loci for Alzheimer disease in a consanguineous Israeli-Arab community. *Human Molecular Genetics* **12**, 415–22.

Finckh U (2003). The future of genetic association studies in Alzheimer disease. *Journal of Neural Transmission* **110**, 253–66.

Frey HJ, Mattila KM, Korolainen MA, and Pirttila T (2005). Problems associated with biological markers of Alzheimer's disease. *Neurochemical Research* **30**, 1501–10.

Fryer JD, Taylor JW, DeMattos RB, *et al.* (2003). Apolipoprotein E markedly facilitates age-dependent cerebral amyloid angiopathy and spontaneous hemorrhage in amyloid precursor protein transgenic mice. *Journal of Neuroscience* **23**, 7889–96.

Fukumoto H, Tennis M, Locascio JJ, Hyman BT, Growdon JH, and Irizarry MC (2003). Age but not diagnosis is the main predictor of plasma amyloid beta-protein levels. *Archives of Neurology* **60**, 958–64.

Glenner GG and Wong CW (1984a). Alzheimer's disease and Down's syndrome: sharing of a unique cerebrovascular amyloid fibril protein. *Biochemical and Biophysical Research Communications* **122**, 1131–5.

Glenner GG and Wong CW (1984b). Alzheimer's disease: initial report of the purification and characterization of a novel cerebrovascular amyloid protein. *Biochemical and Biophysical Research Communications* **120**, 885–90.

Goate A, Chartier-Harlin MC, Mullan M, *et al.* (1991). Segregation of a missense mutation in the amyloid precursor protein gene with familial Alzheimer's disease. *Nature* **349**, 704–6.

Grupe A, Li Y, Rowland C, *et al.* (2006). A scan of chromosome 10 identifies a novel locus showing strong association with late-onset Alzheimer disease. *American Journal of Human Genetics* **78**, 78–88.

Gustafson L, Brun A, Englund E, *et al.* (1998). A 50-year perspective of a family with chromosome-14-linked Alzheimer's disease. *Human Genetics* **102**, 253–7.

Hahs DW, McCauley JL, Crunk AE, *et al.* (2006). A genome-wide linkage analysis of dementia in the Amish. *American Journal of Medical Genetics. Part B, Neuropsychiatric Genetics* **141**, 160–6.

Hardy J (2005). Expression of normal sequence pathogenic proteins for neurodegenerative disease contributes to disease risk: 'permissive templating' as a general mechanism underlying neurodegeneration. *Biochemical Society Transactions* **33**, 578–81.

Hattori M, Fujiyama A, Taylor TD, *et al.* (2000). The DNA sequence of human chromosome 21. *Nature* **405**, 311–9.

Hebert LE, Scherr PA, Bienias JL, Bennett DA, and Evans DA (2003). Alzheimer disease in the US population: prevalence estimates using the 2000 census. *Archives of Neurology* **60**, 1119–22.

Herzig MC, Winkler DT, Burgermeister P, *et al.* (2004). Abeta is targeted to the vasculature in a mouse model of hereditary cerebral hemorrhage with amyloidosis. *Nature Neuroscience* **7**, 954–60.

Hirst C, Yee IM, and Sadovnick AD (1994). Familial risks for Alzheimer disease from a population-based series. *Genetic Epidemiology* **11**, 365–74.

Holtzman DM, Fagan AM, Mackey B, *et al.* (2000). Apolipoprotein E facilitates neuritic and cerebrovascular plaque formation in an Alzheimer's disease model. *Annals of Neurology* **47**, 739–47.

Houlden H, Baker M, Morris HR, *et al.* (2001). Corticobasal degeneration and progressive supranuclear palsy share a common tau haplotype. *Neurology* **56**, 1702–6.

Howson JM, Barratt BJ, Todd JA, and Cordell HJ (2005). Comparison of population- and family-based methods for genetic association analysis in the presence of interacting loci. *Genetic Epidemiology* **29**, 51–67.

Huff FJ, Auerbach J, Chakravarti A, and Boller F (1988). Risk of dementia in relatives of patients with Alzheimer's disease. *Neurology* **38**, 786–90.

Hutton M, Lendon CL, Rizzu P, *et al.* (1998). Association of missense and 5'-splice-site mutations in tau with the inherited dementia FTDP-17. *Nature* **393**, 702–5.

Ibanez P, Bonnet AM, Debarges B, *et al.* (2004). Causal relation between alpha-synuclein gene duplication and familial Parkinson's disease. *Lancet* **364**, 1169–71.

Jorm AF, Korten AE, and Henderson AS (1987). The prevalence of dementia: a quantitative integration of the literature. *Acta Psychiatrica Scand*inavica **76**, 465–79.

Kamboh MI, Minster RL, Feingold E, Dekosky ST (2006). Genetic association of ubiquilin with Alzheimer's disease and related quantitative measures. *Molecular Psychiatry* **11**, 273–9.

Katzov H, Chalmers K, Palmgren J, *et al.* (2004). Genetic variants of ABCA1 modify Alzheimer disease risk and quantitative traits related to beta-amyloid metabolism. *Human Mutation* **23**, 358–67.

Kehoe P, Wavrant-De Vrieze F, Crook R, *et al.* (1999). A full genome scan for late onset Alzheimer's disease. *Human Molecular Genetics* **8**, 237–45.

Kimberly WT, LaVoie MJ, Ostaszewski BL, Ye W, Wolfe MS, and Selkoe DJ (2003). Gamma-secretase is a membrane protein complex comprised of presenilin, nicastrin, Aph-1, and Pen-2. *Proceedings of the National Academy of Sciences of the USA* **100**, 6382–7.

Klunk WE, Engler H, Nordberg A, *et al.* (2004). Imaging brain amyloid in Alzheimer's disease with Pittsburgh compound-B. *Annals of Neurology* **55**, 306–19.

Kounnas MZ, Moir RD, Rebeck GW, *et al.* (1995). LDL receptor-related protein, a multifunctional ApoE receptor, binds secreted beta-amyloid precursor protein and mediates its degradation. *Cell* **82**, 331–40.

Lai MT, Chen E, Crouthamel MC, *et al.* (2003). Presenilin-1 and presenilin-2 exhibit distinct yet overlapping gamma-secretase activities. *Journal of Biological Chemistry* **278**, 22475–81.

Lambert JC, Pasquier F, Cottel D, Frigard B, Amouyel P, and Chartier-Harlin MC (1998). A new polymorphism in the APOE promoter associated with risk of developing Alzheimer's disease. *Human Molecular Genetics* **7**, 533–40.

Laudon H, Hansson EM, Melen K, *et al.* (2005). A nine-transmembrane domain topology for presenilin 1. *Journal of Biological Chemistry* **280**, 35352–60.

Levy E, Carman MD, Fernandez-Madrid IJ, *et al.* (1990). Mutation of the Alzheimer's disease amyloid gene in hereditary cerebral hemorrhage, Dutch type. *Science* **248**, 1124–6.

Levy-Lahad E, Wasco W, Poorkaj P, *et al.* (1995). Candidate gene for the chromosome 1 familial Alzheimer's disease locus. *Science* **269**, 973–7.

Li X and Greenwald I (1996). Membrane topology of the C. elegans SEL-12 presenilin. *Neuron* **17**, 1015–21.

Li X and Greenwald I (1997). HOP-1, a *Caenorhabditis elegans* presenilin, appears to be functionally redundant with SEL-12 presenilin and to facilitate LIN-12 and GLP-1 signaling. *Proceedings of the National Academy of Sciences of the USA* **94**, 12204–9.

Li X and Greenwald I (1998). Additional evidence for an eight-transmembrane-domain topology for Caenorhabditis elegans and human presenilins. *Proceedings of the National Academy of Sciences of the USA* **95**, 7109–14.

Li Y, Nowotny P, Holmans P, *et al*. (2004a). Association of late-onset Alzheimer's disease with genetic variation in multiple members of the *GAPD* gene family. *Proceedings of the National Academy of Sciences of the USA* **101**, 15688–93.

Li Y, Tacey K, Doil L, *et al*. (2004b). Association of *ABCA1* with late-onset Alzheimer's disease is not observed in a case–control study. *Neuroscience Letters* **366**, 268–71.

Li YJ, Scott WK, Hedges DJ, *et al*. (2002). Age at onset in two common neurodegenerative diseases is genetically controlled. *American Journal of Human Genetics* **70**, 985–93.

Lopera F, Ardilla A, Martinez A, *et al*. (1997). Clinical features of early-onset Alzheimer disease in a large kindred with an E280A presenilin-1 mutation. *Journal of the American Medical Association* **277**, 793–9.

Lowenberg K and Waggoner R (1934). Familial organic psychosis (Alzheimer's type). *Archives of Neurology and Psychiatry* **31**, 737.

McKhann G, Drachman D, Folstein M, Katzman R, Price D, and Stadlan EM (1984). Clinical diagnosis of Alzheimer's disease: report of the NINCDS-ADRDA Work Group under the auspices of Department of Health and Human Services Task Force on Alzheimer's Disease. *Neurology* **34**, 939–44.

Maraganore DM, de Andrade M, Lesnick TG, *et al*. (2005). High-resolution whole-genome association study of Parkinson disease. *American Journal of Human Genetics* **77**, 685–93.

Mattila KM, Forsell C, Pirttila T, *et al*. (1998). The Glu318Gly mutation of the presenilin-1 gene does not necessarily cause Alzheimer's disease. *Annals of Neurology* **44**, 965–7.

Mayeux R, Sano M, Chen J, Tatemichi T, Stern Y (1991). Risk of dementia in first-degree relatives of patients with Alzheimer's disease and related disorders. *Archives of Neurology* **48**, 269–73.

Mayeux R, Ottman R, Maestre G, *et al*. (1995). Synergistic effects of traumatic head injury and apolipoprotein-epsilon 4 in patients with Alzheimer's disease. *Neurology* **45**, 555–7.

Mayeux R, Lee JH, Romas SN, *et al*. (2002). Chromosome-12 mapping of late-onset Alzheimer disease among Caribbean Hispanics. *American Journal of Human Genetics* **70**, 237–43.

Mead S, Mahal SP, Beck J, *et al*. (2001). Sporadic—but not variant—Creutzfeldt–Jakob disease is associated with polymorphisms upstream of *PRNP* exon 1. *American Journal of Human Genetics* **69**, 1225–35.

Mintun MA (2005). Utilizing advanced imaging and surrogate markers across the spectrum of Alzheimer's disease. *CNS Spectrums* **10**, 13–16.

Mohs RC, Breitner JC, Silverman JM, and Davis KL (1987). Alzheimer's disease. Morbid risk among first-degree relatives approximates 50% by 90 years of age. *Archives of General Psychiatry* **44**, 405–8.

Myers A, Wavrant De-Vrieze F, Holmans P, *et al*. (2002). Full genome screen for Alzheimer disease: stage II analysis. *American Journal of Medical Genetics* **114**, 235–44.

Owen F, Poulter M, Shah T, *et al*. (1990). An in-frame insertion in the prion protein gene in familial Creutzfeldt–Jakob disease. *Brain Research: Molecular Brain Research* **7**, 273–6.

Pastor P, Roe CM, Villegas A, *et al*. (2003). Apolipoprotein Eepsilon4 modifies Alzheimer's disease onset in an E280A PS1 kindred. *Annals of Neurology* **54**, 163–9.

Pericak-Vance MA, Johnson CC, Rimmler JB, *et al*. (1996). Alzheimer's disease and apolipoprotein E-4 allele in an Amish population. *Annals of Neurology* **39**, 700–4.

Pericak-Vance MA, Bass MP, Yamaoka LH, *et al*. (1997). Complete genomic screen in late-onset familial Alzheimer disease. Evidence for a new locus on chromosome 12. *Journal of the American Medical Association* **278**, 1237–41.

Pericak-Vance MA, Grubber J, Bailey LR, *et al*. (2000). Identification of novel genes in late-onset Alzheimer's disease. *Experimental Gerontology* **35**, 1343–52.

Rademakers R, Cruts M, Sleegers K, *et al*. (2005). Linkage and association studies identify a novel locus for Alzheimer disease at 7q36 in a Dutch population-based sample. *American Journal of Human Genetics* **77**, 643–52.

Risch N and Merikangas K (1996). The future of genetic studies of complex human diseases. *Science* **273**, 1516–7.

Ritchie K and Kildea D (1995). Is senile dementia 'age-related' or 'ageing-related'? Evidence from meta-analysis of dementia prevalence in the oldest old. *Lancet* **346**, 931–4.

Rocca WA, Hofman A, Brayne C, *et al.* (1991). Frequency and distribution of Alzheimer's disease in Europe: a collaborative study of 1980–1990 prevalence findings. The EURODEM-Prevalence Research Group. *Annals of Neurology* **30**, 381–90.

Rogaeva E, Premkumar S, Song Y, *et al.* (1998). Evidence for an Alzheimer disease susceptibility locus on chromosome 12 and for further locus heterogeneity. *Journal of the American Medical Association* **280**, 614–8.

Ronald and Nancy Reagan Research Institute of the Alzheimer's Association and the National Institute on Aging Working Group (1998). Consensus report of the Working Group on: Molecular and Biochemical Markers of Alzheimer's Disease *Neurobiology of Aging* **19**, 109–16.

Rovelet-Lecrux A, Hannequin D, Raux G, *et al.* (2006). APP locus duplication causes autosomal dominant early-onset Alzheimer disease with cerebral amyloid angiopathy. *Nature Genetics* **38**, 24–6.

Sadovnick AD, Irwin ME, Baird PA, Beattie BL (1989). Genetic studies on an Alzheimer clinic population. *Genetic Epidemiology* **6**, 633–43.

Saunders AJ, Bertram L, Mullin K, *et al.* (2003). Genetic association of Alzheimer's disease with multiple polymorphisms in alpha-2-macroglobulin. *Human Molecular Genetics* **12**, 2765–76.

Scheuner D, Eckman C, Jensen M, *et al.* (1996). Secreted amyloid beta-protein similar to that in the senile plaques of Alzheimer's disease is increased *in vivo* by the presenilin 1 and 2 and APP mutations linked to familial Alzheimer's disease. *Nature Medicine* **2**, 864–70.

Schmechel DE, Saunders AM, Strittmatter WJ, *et al.* (1993). Increased amyloid beta-peptide deposition in cerebral cortex as a consequence of apolipoprotein E genotype in late-onset Alzheimer disease. *Proceedings of the National Academy of Sciences of the USA* **90**, 9649–53.

Schroeter EH, Ilagan MX, Brunkan AL, *et al.* (2003). A presenilin dimer at the core of the gamma-secretase enzyme: insights from parallel analysis of Notch 1 and APP proteolysis. *Proceedings of the National Academy of Sciences of the USA* **100**, 13075–80.

Scott WK, Grubber JM, Conneally PM, *et al.* (2000). Fine mapping of the chromosome 12 late-onset Alzheimer disease locus: potential genetic and phenotypic heterogeneity. *American Journal of Human Genetics* **66**, 922–32.

Selkoe DJ (1994). Normal and abnormal biology of the beta-amyloid precursor protein. *Annual Reviews of Neuroscience* **17**, 489–517.

Sherrington R, Rogaev EI, Liang Y, *et al.* (1995). Cloning of a gene bearing missense mutations in early-onset familial Alzheimer's disease. *Nature.* **375**, 754–60.

Shibata N, Kawarai T, Lee JH, *et al.* (2006). Association studies of cholesterol metabolism genes (*CH25H*, *ABCA1* and *CH24H*) in Alzheimer's disease. *Neuroscience Letters* **391**, 142–6.

Sillen A, Forsell C, Lilius L, *et al.* (2006). Genome scan on Swedish Alzheimer's disease families. *Molecular Psychiatry* **11**, 182–6.

Sing CF and Moll PP (1989). Genetics of variability of CHD risk. *International Journal of Epidemiology* **18**, S183–95.

Singleton AB, Farrer M, Johnson J, *et al.* (2003). Alpha-synuclein locus triplication causes Parkinson's disease. *Science* **302**, 841.

Slifer MA, Martin ER, Haines JL, Pericak-Vance MA (2005). The ubiquilin 1 gene and Alzheimer's disease. *New England Journal of Medicine* **352**, 2752–3; author reply 2752–3.

Smemo S, Nowotny P, Hinrichs AL, *et al.* (2006). Ubiquilin 1 polymorphisms are not associated with late-onset Alzheimer's disease. *Annals of Neurology* **59**, 21–6.

Strittmatter WJ, Saunders AM, Schmechel D, *et al*. (1993a). Apolipoprotein E: high-avidity binding to beta-amyloid and increased frequency of type 4 allele in late-onset familial Alzheimer disease. *Proceedings of the National Academy of Sciences of the USA* **90**, 1977–81.

Strittmatter WJ, Weisgraber KH, Huang DY, *et al*. (1993b). Binding of human apolipoprotein E to synthetic amyloid beta peptide: isoform-specific effects and implications for late-onset Alzheimer disease. *Proceedings of the National Academy of Sciences of the USA* **90**, 8098–102.

Strittmatter WJ, Saunders AM, Goedert M, *et al*. (1994). Isoform-specific interactions of apolipoprotein E with microtubule-associated protein tau: implications for Alzheimer disease. *Proceedings of the National Academy of Sciences of the USA* **91**, 11183–6.

Sunderland T, Linker G, Mirza N, *et al*. (2003). Decreased beta-amyloid1–42 and increased tau levels in cerebrospinal fluid of patients with Alzheimer disease. *Journal of the American Medical Association* **289**, 2094–103.

Tang MX, Stern Y, Marder K, *et al*. (1998). The APOE-epsilon4 allele and the risk of Alzheimer disease among African Americans, whites, and Hispanics. *Journal of the American Medical Association* **279**, 751–5.

Van Duinen SG, Castano EM, Prelli F, Bots GT, Luyendijk W, and Frangione B (1987). Hereditary cerebral hemorrhage with amyloidosis in patients of Dutch origin is related to Alzheimer disease. *Proceedings of the National Academy of Sciences of the USA* **84**, 5991–4.

Wang JC, Kwon JM, Shah P, Morris JC, Goate A (2000). Effect of APOE genotype and promoter polymorphism on risk of Alzheimer's disease. *Neurology* **55**, 1644–9.

Wollmer MA, Streffer JR, Lutjohann D, *et al*. (2003). *ABCA1* modulates CSF cholesterol levels and influences the age at onset of Alzheimer's disease. *Neurobiology of Aging* **24**, 421–6.

Wu WS, Holmans P, Wavrant-DeVrieze F, *et al*. (1998). Genetic studies on chromosome 12 in late-onset Alzheimer disease. *Journal of the American Medical Association* **280**, 619–22.

Xia W, Zhang J, Kholodenko D, *et al*. (1997). Enhanced production and oligomerization of the 42-residue amyloid beta-protein by Chinese hamster ovary cells stably expressing mutant presenilins. *Journal of Biological Chemistry* **272**, 7977–82.

Yoshikai S, Sasaki H, Doh-ura K, Furuya H, Sakaki Y (1990). Genomic organization of the human amyloid beta-protein precursor gene. *Gene* **87**, 257–63.

Zarranz JJ, Alegre J, Gomez-Esteban JC, *et al*. (2004). The new mutation, E46K, of alpha-synuclein causes Parkinson and Lewy body dementia. *Annals of Neurology* **55**, 164–73.

Zubenko GS, Hughes HB, Stiffler JS, Hurtt MR, and Kaplan BB (1998). A genome survey for novel Alzheimer disease risk loci: results at 10-cM resolution. *Genomics* **50**, 121–8.

Chapter 4

Biology and molecular neuropathology of β-amyloid protein

Edward B. Lee and Virginia M.-Y. Lee

4.1 Introduction: amyloid precursor protein and the amyloid hypothesis

A century ago, Alois Alzheimer described the post-mortem neuropathological findings from a woman who suffered from the disease which now bears his name (Alzheimer 1907). In addition to profound neuron loss, he found two major lesions in the cortex and hippocampus, now known as senile plaques and neurofibrillary tangles. Until the biochemical purification of the building blocks of these two pathological lesions, the mechanisms underlying the pathogenesis of Alzheimer's disease (AD) were difficult to ascertain. Eight decades after the initial description of AD, the molecular era of AD research was ushered in with the isolation of the Aβ peptide, initially from cerebrovascular amyloid (Glenner and Wong 1984) and subsequently from senile plaques (Masters *et al.* 1985). The elucidation of the 40–42 amino acid sequence of Aβ then led to the cloning of a gene on chromosome 21 for the amyloid precursor protein (APP), a 695–771 amino acid (aa) integral membrane protein which is subject to endoproteolysis to generate Aβ (Goldgaber *et al.* 1987; Kang *et al.* 1987; Tanzi *et al.* 1987). Neurofibrillary tangles were later shown to be comprised of insoluble hyperphosphory-lated forms of the microtubule-associated protein tau, which is discussed in greater detail in the next chapter.

Screening of several kindreds found linkage between mutations within the *APP* gene and autosomal dominant inheritance of diseases characterized by the accumulation of Aβ, including early-onset familial AD and hereditary cerebral haemorrhage with amyloidosis. More specifically, a double mutation within *APP*, termed the Swedish mutation, immediately adjacent to the N-terminus of Aβ results in the overproduction of Aβ1–40 and Aβ1–42 (Citron *et al.* 1992; Mullan *et al.* 1992; Cai *et al.* 1993). Furthermore, individuals with Down's syndrome harbouring an extra copy of chromosome 21 containing the *APP* gene invariably develop dementia and AD-like pathology. Other familial AD mutations, clustered around the C-terminus of Aβ, result in a shift towards the production of the longer Aβ42 variant (Haass *et al.* 1994a; Suzuki *et al.* 1994). Gene linkage analysis further identified mutations within two homologous genes encoding presenilin 1 and 2 (PS1, PS2) which had a similar effect on Aβ42 production (Levy-Lahad *et al.* 1995; Rogaev *et al.* 1995; Sherrington *et al.* 1995; Borchelt *et al.* 1996;

Citron *et al.* 1997; Tomita *et al.* 1997). Coupled with biophysical studies that showed that Aβ fibril formation was concentration dependent and that Aβ42 could seed amyloid fibril formation (Jarrett *et al.* 1993), the genetic linkage of APP mutations with AD remains the strongest argument for the amyloid hypothesis. Stated explicitly, the amyloid hypothesis states that the Aβ peptide is central to the pathogenesis of AD, and is the proximal cause of multiple effects including neurofibrillary tangle formation, synaptic dysfunction and loss, and neuronal death. The amyloid hypothesis was bolstered by the finding that Aβ fibrils are toxic to cultured neuronal cells (Yankner *et al.* 1990; Pike *et al.* 1991a,b). Additionally, increased production of Aβ or intracerebral injection of Aβ augments tau pathology in transgenic mice overexpressing mutant tau (Gotz *et al.* 2001; Lewis *et al.* 2001), suggesting that Aβ lies mechanistically upstream of tau.

However, the amyloid hypothesis remains a hypothesis because several findings do not fully support its claims. Indeed, Alois Alzheimer originally concluded that 'the plaques are not the cause of senile dementia but only an accompanying feature of senile involution of the central nervous system' because he observed regions of the brain containing neurofibrillary tangles in the absence of senile plaques (Alzheimer 1911), a finding which has been confirmed by more modern staging methods (Braak and Braak 1991). Furthermore, several transgenic mice overexpressing mutant forms of *APP* have been generated which accumulate high levels of Aβ with inflammatory changes, synaptic alterations, and cognitive deficits (Games *et al.* 1995; Hsiao *et al.* 1996; Sturchler-Pierrat *et al.* 1997). However, neurofibrillary tangle pathology is not induced in these models despite the abundance of cerebral Aβ. Furthermore, although Aβ fibrils may cause toxicity in cell culture models, neuronal viability is largely unaffected in most transgenic murine models. Finally, the extent of amyloid plaque pathology correlates poorly with clinical measures of dementia in AD patients, in contrast with other pathological features like neurofibrillary tau pathology and synaptic density (DeKosky and Scheff 1990; McKee *et al.* 1991; Terry *et al.* 1991; Berg *et al.* 1993; Scheff and Price 1993; Bierer *et al.* 1995; Nagy *et al.* 1995). These findings have called for further evaluation of the amyloid cascade hypothesis.

Despite the sometimes contradictory findings with regard to the amyloid hypothesis, the indisputable genetic evidence that APP is linked to the development of AD is difficult to counter. Therefore the ultimate hope is that continuing rational efforts aimed at understanding the biology and molecular neuropathology of APP will lead to further insights which will better elucidate the role of APP and Aβ in the development of AD. This chapter explores the advances made towards understanding APP biology with an emphasis on the findings and theories about the cell biology of Aβ generation and deposition.

4.2 Amyloid precursor protein structure and proteolysis

To better understand the functional and cellular aspects of APP biology, the current understanding of APP structure and proteolysis will be presented first. APP is a type I transmembrane glycoprotein with a relatively long N-terminal extracellular domain and a short C-terminal cytoplasmic domain (Fig. 4.1). The Aβ sequence begins in the extracellular domain of APP, close to the lumenal edge of the cellular membrane, and terminates within the transmembrane region. Of the 18 exons which encode for APP, exons 7, 8, and 15, which encode for regions in the extracellular domain, are subject to alternative splicing. Of the three major APP isoforms (695, 751, and 770 aa variants), neurons of the central and peripheral nervous system express almost exclusively the APP695 isoform. APP695 lacks the Kunitz-type serine protease inhibitor (KPI) domain encoded by exon 7, and a domain with homology to the MRC OX-2 antigen encoded by exon 8 (Kitaguchi *et al.* 1988; Ponte *et al.* 1988; Tanzi *et al.* 1988; Kang and Muller-Hill 1989). While exon 8 is of unknown significance owing to divergence from the MRC OX-2 antigen (Richards *et al.* 1995), the KPI domain functions in blood coagulation pathways (Oltersdorf *et al.* 1989; Van Nostrand *et al.* 1990a,b). However, since neurons which express APP695 lacking the KPI domain are the major source of APP and Aβ in the CNS (Tanzi *et al.* 1988; Weidemann *et al.* 1989), this functional domain does not appear to play a role in the development of AD. Furthermore, overexpressing APP695 results in Aβ production and amyloid plaque deposition, further indicating that the KPI domain is not intrinsic to AD pathology. Another non-neuronal isoform termed

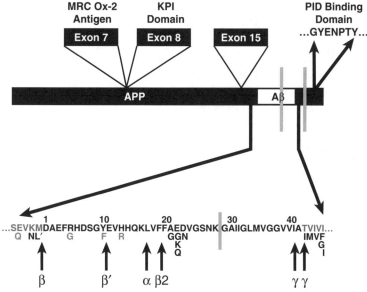

Fig. 4.1 Structure of APP. Schematic diagram of the structure of APP with the N-terminus to the left and the C-terminus to the right (not drawn to scale); the C-terminus of the Aβ domain lies within the membrane bilayer (grey lines). The major APP isoforms are formed via alternative splicing of exons 7, 8, and 15. The primary sequence of the region around Aβ (bottom) is also shown, with the major cleavage sites for α-, β-, and γ-secretases. The β2 cleavage site is due to the *BACE1* homologue, *BACE2*. The sequence corresponding to Aβ is in black, with autosomal dominant familial AD mutations shown underneath in black, and the murine polymorphisms shown underneath in grey.

L-APP, which results from the exclusion of exon 15, contains a recognition site for xylosyltransferase created by the conjoining sequences of exons 14 and 16, resulting in the post-translational modification by chondroitin sulphate glycosaminoglycans which are not seen in APP isoforms containing exon 15 (Thinakaran *et al.* 1995). Aside from subtle alterations in APP proteolysis and trafficking (Hartmann *et al.* 1996; Ho *et al.* 1996), alternative splicing of APP has not been shown to play a role in the development of AD pathology.

APP is subject to endoproteolytic cleavage by several proteases, named in order of their historical discovery as α-secretase, β-secretase, and γ-secretase. The existence of these proteases was surmised based on the presence of specific N- and C-terminal APP fragments in cell culture and *in vivo* models; the identification of the proteins responsible for these protease activities did not occur for many years. α-Secretase is the protease responsible for cleavage between lysine +16 and leucine +17 within the Aβ domain (VHHQK↓LVFFA) (Esch *et al.* 1990; Sisodia *et al.* 1990) and has been ascribed to members of the ADAM (a disintegrin and metalloprotease) family of proteases, namely TACE (tumour necrosis factor-α converting enzyme) or ADAM10 (Buxbaum *et al.* 1998a; Lammich *et al.* 1999). α-Secretase cleavage results in the secretion of an N-terminal ectodomain fragment called sAPPα, and the retention of an 83 aa C-terminal fragment called C83. α-Cleavage precludes the formation of intact Aβ, and therefore is non-amyloidogenic.

Although the majority of APP in the non-neuronal cells is cleaved by α-secretase, Aβ production was detected as a normal physiological APP processing event, attributed to proteases named β- and γ-secretase (Haass *et al.* 1992; Seubert *et al.* 1992; Shoji *et al.* 1992). β-Secretase cleaves APP between the methionine and aspartic acid immediately adjacent to the N-terminus of the first amino acid of Aβ (SEVKM↓DAEFR). The N-terminal ectodomain, termed sAPPβ, is no longer tethered to the membrane and is secreted, while the 99 aa C-terminus of APP, C99, remains associated with the membrane. The identity of β-secretase was determined by four independent groups, each finding the same protein, BACE (β-site APP cleaving enzyme) (Hussain *et al.* 1999; Vassar *et al.* 1999; Yan *et al.* 1999; Lin *et al.* 2000). BACE is a type I transmembrane pepstatin-insensitive aspartyl protease closely related to the pepsin family of proteases. Its expression and activity is highest in neurons. While cleavage at position +1 at the N-terminus of Aβ is its primary proteolytic activity, BACE also cuts APP between the tyrosine and glutamine at position +11 within Aβ (HDSGY↓EVHHQ), resulting in the formation of sAPPβ' and C89. Both cleavage events are increased upon BACE overexpression, although the latter β'-cleavage becomes even more prominent upon BACE overexpression which results in a higher proportion of APP cleavage at the alternative glutamine +11 site (Vassar *et al.* 1999; Huse *et al.* 2002; Liu *et al.* 2002; E.B. Lee *et al.* 2003, 2005). Both cleavages are completely inhibited upon genetic ablation of BACE by homologous recombination (Cai *et al.* 2001; Luo *et al.* 2001).

Since the C-terminus of Aβ lies within the transmembrane domain of C99, the final cleavage event towards the production of Aβ by γ-secretase occurs within the hydrophobic environment of the membrane's lipid bilayer. γ-Secretase cuts C99 within the transmembrane domain at either position 40 or 42 at the C-terminus of the Aβ domain (VGGVV↓IA↓TVIVI), thereby generating the lumenal/extracellular Aβ peptide and the cytoplasmic peptide AICD (APP intracellular domain). γ-Secretase is now believed to be a multimeric protein complex consisting of at least four proteins: PS, nicastrin, aph1 (anterior phanrynx defective), and pen2 (PS enhancer 2) (De Strooper et al. 1998; Herreman et al. 2000; Yu et al. 2000; Zhang et al. 2000; Francis et al. 2002; Goutte et al. 2002). Indeed, expression of all four proteins reconstitutes γ-secretase activity in yeast, an organism usually devoid of γ-secretase activity (Edbauer et al. 2003). Although all four proteins are necessary to form a stable γ-secretase complex and for catalytic activity, highly conserved aspartate residues within two of the trans-membrane domains of PS have been postulated to form the active-site of γ-secretase (Wolfe et al. 1999; Esler et al. 2000; Li et al. 2000). In support of this hypothesis, a novel family of proteases responsible for intramembranous proteolysis has been elucidated based on homology to PS (Grigorenko et al. 2002; Ponting et al. 2002; Weihofen et al. 2002). γ-Secretase is also able to cleave a variety of substrates including Notch, a signalling protein involved in lateral inhibition during development (De Strooper et al. 1999; Struhl and Greenwald 1999; Ye et al. 1999). Thus genetic ablation of members of the γ-secretase complex results in both an inhibition of γ-secretase activity and a lethal phenotype similar to the effects of blocking Notch activity (Shen et al. 1997; Wong et al. 1997). γ-Secretase also cuts the other C-terminal APP fragments generated by β'- or α-cleavage, with C89 cleavage resulting in the production of the N-terminally truncated Aβ11–40/42 and C83 cleavage resulting in the production of the p3 peptide (corresponding to Aβ17–40/42).

An understanding of APP proteolytic processing pathways is a necessary starting point towards elucidating the mechanisms which lead to Aβ production and deposition in the AD brain. However, a richer understanding of APP biology extends beyond structure to biological function and pathological effects of APP with regard to the development of AD.

4.3 Extracellular and intracellular amyloid precursor protein function

Despite intense efforts, the normal functions of APP and APP metabolites have been difficult to ascertain. Although APP-deficient mice are healthy and fertile, they develop locomotor defects, poor forelimb grip strength, reactive gliosis, decreased immunoreactivity for presynaptic terminals and dendrites, defects in long-term potentiation (LTP), and behavioural impairment in the Morris water maze (Zheng et al. 1995; Dawson et al. 1999). APP deficiency also exacerbates forebrain commissure defects in genetically susceptible strains of mice, and results in deficient neuronal survival and neurite outgrowth in cell culture (Perez et al. 1997; Magara et al. 1999). While single genetic deletion of homologous amyloid-precursor-like proteins (APLPs) are not lethal, the deletion of multiple members of the APP family is lethal, indicating that the function(s) of this family of proteins are vital (Heber et al. 2000). Given the complex and often subtle phenotype of APP knockout mice, the biological function of APP has been elusive.

The secretion of the relatively large N-terminal sAPP fragments suggested that APP might function in a paracrine or endocrine manner as an extracellular signalling molecule. One proposed role of sAPP is that of a neurotrophic factor promoting neuronal survival and neurite outgrowth (Mattson *et al.* 1993; Goodman and Mattson 1994; Small *et al.* 1994; Smith-Swintosky *et al.* 1994; Li *et al.* 1997; Perez *et al.* 1997), consistent with some of the effects of APP deficiency already mentioned. Furthermore, the neurotrophic effect of sAPP may manifest itself behaviourally, as intra-cerebroventricular administration of antibodies against sAPP impairs behaviour in passive avoidance learning tasks, while administration of sAPP significantly improves behaviour on a variety of long- and short-term memory tasks (Doyle *et al.* 1990; Huber *et al.* 1993; Meziane *et al.* 1998). Moreover, transgenic mice overexpressing wild-type APP show improved spatial learning and memory behaviour in the water maze relative to non-transgenic littermates (Westerman *et al.* 2002). At least one element responsible for sAPP activity may be a cell surface receptor activation via a conserved 'RERMS' sequence found in all sAPP isoforms (residues 328–332 of APP695) (Ninomiya *et al.* 1993; Jin *et al.* 1994). Alternatively, cell surface heparin sulphate proteoglycans have also been implicated in the binding of sAPP to cell surface membranes, indicating that two N-terminal heparin binding domains on sAPP may be involved in its activity (Multhaup 1994; Small *et al.* 1994; Snow *et al.* 1995; Williamson *et al.* 1996). Consistent with this hypothesis is the finding that sAPPα has been reported to be over 100-fold more potent than the sAPPβ, perhaps corresponding to the presence of an extra 'VHHQK' heparin binding domain (corresponding to Aβ residues 12–16) which is present on sAPPα but absent on sAPPβ (Furukawa *et al.* 1996). A receptor for sAPP has not been identified to date.

Again based on its structure, an alternative hypothesis has been generated in which APP acts a receptor triggering intracellular signal transduction pathways (Kang *et al.* 1987). To understand the intracellular function of APP better, adaptor proteins which interact with the cytoplasmic C-terminus of APP have been identified via yeast-two hybrid screens. One such protein is Fe65 (Fiore *et al.* 1995), a multifunctional protein with a proposed N-terminal transcriptional activator domain, a region displaying homology to retroviral integrases, two phosphotyrosine-interaction domains (PID), and a WW motif. Its multiple protein interaction domains indicate that Fe65 is an adaptor protein, proposed to mediate protein complex formation and play a role in transcription. PID domains (as originally described for the Shc protein), bind to NPXpY motifs (pY indicates a phosphorylated tyrosine residue, and X indicates any amino acid) (Blaikie *et al.* 1994). The first of two PID domains (PID1) interacts with the transcription factor CP2/LSF/LBP1 (Zambrano *et al.* 1998). PID1 also interacts with the low-density lipoprotein receptor-related protein (LRP), a 600 kDa transmembrane glycoprotein which mediates the endocytosis and degradation of extracellular ligands including apolipoprotein E and α_2-macroglobulin, both of which are implicated in either the clearance or the chaperoning of extracellular Aβ (Trommsdorff *et al.* 1998). LRP also mediates the endocytosis of KPI-containing isoforms of extracellular sAPP (Knauer *et al.* 1996; Kounnas *et al.* 1996).

PID2 binds to the 'GYENPTY' sequence in the C-terminus of APP (Fiore *et al.* 1995; Guenette *et al.* 1996; Zambrano *et al.* 1997). In contrast with the originally described Shc PID domain, interactions between PID2 and APP are independent of tyrosine phosphorylation, consistent with the significant divergence of the Fe65 PID domains from the Shc PID domain (Russo *et al.* 1998). However, phosphorylation of threonine-668 (numbering for APP695) which lies upstream of the Fe65 recognition site inhibits the APP-Fe65 interaction (Ando *et al.* 2001).

WW domains mediate interactions with proteins containing XPPXY or PPLP motifs (Chen and Sudol 1995; Sudol *et al.* 1995). The WW domain of Fe65 interacts with Mena, the mammalian homologue of *Drosophila* enabled, which interacts with actin and affects cytoskeleton dynamics, cell motility, and cell morphology, including axonal outgrowth and integrity (Ermekova *et al.* 1997). Thus *Drosophila* enabled mutants have defects in axonal structure (Gertler *et al.* 1995) and Mena-deficient mice have defects in neurulation and commisure formation (Lanier *et al.* 1999). Indeed, coexpression of APP and Fe65 enhances cell movement in a wound-healing assay, suggesting that one function of the APP-Fe65 complex is to regulate actin-based motility (Sabo *et al.* 2001).

Another PID-domain-containing adaptor protein, X11 (also known as mLin10 because of homology with the *Caenorhabditis elegans* Lin10) binds to the same 'GYENPTY' motif in the C-terminus of APP (Borg *et al.* 1996; McLoughlin and Miller 1996; Tanahashi and Tabira 1999; Tomita *et al.* 1999). In addition to the PID domain, the N-terminus of X11 contains a binding site for Munc-18, a protein essential for synaptic vesicle docking and exocytosis. Thus X11 has a third designation, MINT-1 (Munc18-interacting protein 1). X11 also interacts with CASK, a mammalian homologue of *C.elegans* Lin2, which in turn interacts with VELI (Lin7), forming a heterotrimeric complex localized to presynaptic terminals (Borg *et al.* 1998a; Butz *et al.* 1998). The homologous *C.elegans* tripartite complex comprising lin10, lin7, and lin2 is involved in the subcellular trafficking of an EGF receptor homologue in epithelial cells (Simske *et al.* 1996) and the glutamate receptor GLR-1 in neurons (Rongo *et al.* 1998). In addition to the Munc-18 binding site, X11 also has two PDZ domains which interact with a wide variety of proteins including spinophilin-neurabin II (Ide *et al.* 1998), a presynaptic voltage-gated calcium channel (Maximov *et al.* 1999), PS1 (Lau *et al.* 2000), the copper chaperone of SOD1 (McLoughlin *et al.* 2001), ADP-ribosylation factor GTPases (Hill *et al.* 2003), and KIF17, a member of the kinesin superfamily responsible for anterograde dendritic transport of NMDA-receptor-containing vesicles (Setou *et al.* 2000). X11 has also been proposed to form a homodimer via PDZ domain dimerization (Walhout *et al.* 2000). The promiscuity of the PDZ domain suggests that X11 is indeed an adapter protein providing a structural scaffold for protein complexes.

Consistent with the fact that Fe65 and X11 compete for the same C-terminal APP motif, they have been found to have antagonistic effects on APP trafficking and proteolysis. Through unknown mechanisms, Fe65 overexpression increases cell-surface APP levels and promotes β-cleavage of APP (Sabo *et al.* 1999). However, in addition to the described role in binding APP-associated proteins, the cytoplasmic 'NPTY' motif is also a consensus sequence for clathrin coated-pit mediated internalization and is required for endosomal/lysosomal sorting of APP (Chen *et al.* 1990; De Strooper *et al.* 1993; Lai *et al.* 1995), suggesting that interference with this signal may play a mechanistic role. A trimeric complex consisting of Fe65, APP, and LRP has been isolated (Pietrzik *et al.* 2004), which is significant because of the finding that LRP, via its cytoplasmic tail, alters APP trafficking and proteolysis by increasing the reinternalization of cell surface APP, the turnover of full-length APP and C-terminal APP fragments, the secretion of sAPP, and the secretion of Aβ (Ulery *et al.* 2000; Pietrzik *et al.* 2002). In contrast with Fe65, X11 overexpression decreases APP turnover, sAPP secretion, and Aβ generation in non-neuronal cell culture models (Borg *et al.* 1998b; Sastre *et al.* 1998; J.H. Lee *et al.* 2003), and inhibits Aβ levels and amyloid plaque deposition in a transgenic APP mouse model (J.H. Lee *et al.* 2003). While the mechanisms whereby Fe65 and X11 affect APP metabolism is unknown, Fe65 and X11 have been postulated to alter APP proteolysis via their effects on protein trafficking (King and Scott 2004; King *et al.* 2004).

Another alternative hypothesis for the function of APP has been extrapolated from the well-described function of Notch. Notch has been found to be involved in intracellular signalling controlling lateral inhibition during development. A parallel between Notch and APP can be made because, after binding to one of the DSL-family of ligands (Delta, Serrate, and *C.elegans* Lag-2 proteins), Notch is cleaved by α- and γ-secretase, releasing a cytoplasmic C-terminal domain NICD (Notch intracellular domain), which functions as a transcription factor after translocating into the nucleus (Brou *et al.* 2000; Mumm *et al.* 2000; Hartmann *et al.* 2002). By analogy, the AICD fragment has been proposed to translocate from the cytoplasm into the nucleus and affect gene transcription (Cao and Sudhof 2001; Gao and Pimplikar 2001; Kimberly *et al.* 2001). After its production, AICD is able to form a multimeric complex with Fe65 and the histone acetyltransferase Tip60, which becomes a potent transcriptional transactivator of a heterologous Ga14 or LexA expression cassettes (Cao and Sudhof 2001). Interestingly, the gene encoding the Aβ degrading enzyme neprilysin (NEP) has been suggested to be a genomic target of γ-secretase-dependent/AICD-mediated transcriptional activation (Pardossi-Piquard *et al.* 2005). However, in contrast with NICD which acts directly as a transcriptional transactivator within the nucleus, the actual translocation of AICD into the nucleus is not needed for transcriptional transactivation. Instead, AICD appears to activate Fe65, with in turn translocates into the nucleus and initiates transcription (Cao and Sudhof 2004). Also in contrast with Notch signalling, no ligands to the extracellular domain of APP have been found, and it is unclear whether ligand binding to APP is needed for the proposed AICD signalling cascade.

One final hypothesis related to kinesin-mediated intracellular transport has been proposed for the function of APP. The trafficking of APP within neurons is likely to be unique because of their polarized nature. Immunoctyochemical localization of APP in neurons indicates that APP is enriched in axonal and synaptic compartments (Schubert *et al.* 1991; Ferreira *et al.* 1993). Indeed, APP undergoes specialized sorting in polarized cell types, such as targeting to the basolateral compartment of Madin–Darby canine kidney epithelial cells (Haass *et al.* 1994b). *In vivo*, APP is subject to fast vesicular anterograde transport to the presynaptic terminal, where BACE-derived N- and C-terminal APP fragments can be detected (Koo *et al.* 1990; Morin *et al.* 1993; Amaratunga and Fine 1995; Buxbaum *et al.* 1998b; Kamal *et al.* 2000, 2001). Furthermore, evoked potentials increase the synaptic release of Aβ, which in turn results in synaptic depression (Kamenetz *et al.* 2003). Finally, lesioning the entorhinal cortex, thereby disrupting the perforant pathway, diminishes amyloid burden within the hippocampus, suggesting that synaptically released Aβ contributes to the development of Aβ amyloid pathology (Lazarov *et al.* 2002; Sheng *et al.* 2002).

These findings led to the hypothesis that APP acts as a kinesin receptor to tether vesicles subject to fast anterograde axonal transport to the kinesin. In addition to *in vitro* evidence for a direct interaction between APP and kinesin, Kamal and colleagues (Kamal *et al.* 2000, 2001) provide evidence that APP and the proteases responsible for Aβ generation are found within the same axonal vesicular compartment, and that Aβ can be generated within axonal vesicles. Surprisingly, this group found that a reduction in APP resulted in a decrease in fast anterograde transport. The same group also found that APP transgenic mice exhibit pathological axonal swellings consistent with an axonal transport deficit, and that reduction of kinesin in transgenic mice accelerates and enhances Aβ amyloid deposition (Stokin *et al.* 2005). However, several independent laboratories have failed to replicate these findings, casting doubt on whether there is a direct interaction between APP and kinesin, and whether the entire Aβ generating machinery is found in a single vesicular compartment within neurons (Cook *et al.* 2003; Inomata *et al.* 2003; Matsuda *et al.* 2003; Lazarov *et al.* 2005; E.B. Lee *et al.* 2005). More specifically, the direct binding of APP to kinesin is questionable as three reports have been unable to detect a direct kinesin–APP interaction, although indirect complexes which include APP and kinesin have been found (Inomata *et al.* 2003; Matsuda *et al.* 2003; Lazarov *et al.* 2005). The direct kinesin–APP interaction has been ascribed to artifactual non-specific binding kinetics (Matsuda *et al.* 2003). Furthermore, two reports do not find APP, BACE, and PS to be localized within the same vesicles (Cook *et al.* 2003; Lazarov *et al.* 2005). Additionally, the deletion of the APP gene does not affect fast anterograde transport (Lazarov *et al.* 2005). Finally, the depletion of axonal APP by BACE overexpression does not result in diminished steady-state levels of kinesin, indicating that APP is not a kinesin receptor mediating vesicular axonal transport (E.B. Lee *et al.* 2005). Therefore the hypothesis that APP is a kinesin receptor and that APP, β-secretase, and γ-secretase co-localize within a single axonal/synaptic vesicle remains controversial. In contrast, an interaction between APP and kinesin may be indirect via adaptor proteins such as JIP-1 (Inomata *et al.* 2003; Matsuda *et al.* 2003; Muresan and Muresan 2005).

4.4 Cell biology of amyloid precursor protein processing

While the normal function of APP is still unclear, more progress has been made in understanding the cell biology of APP trafficking and proteolysis of APP. In determining the trafficking and proteolysis of APP, careful regard towards the specialized sorting and metabolic pathways in neuronal cells is warranted. Seemingly conflicting data about the subcellular processing of APP may actually reflect the unique cell biology of neurons. APP is synthesized in the endoplasmic reticulum (ER) and is trafficked through the default secretory pathway. APP acquires N- and O-linked carbohydrates within the ER and Golgi apparatus, in addition to tyrosine sulphates and phosphates in the latter's secretory organelles (Schubert *et al.* 1989; Weidemann *et al.* 1989; Oltersdorf *et al.* 1990; Suzuki *et al.* 1992; Knops *et al.* 1993; Oishi *et al.* 1997). Cell-surface APP can then be re-internalized into endosomes and lysosomes where it may be degraded (Golde *et al.* 1992). As mentioned already, APP within neurons is also subject to fast anterograde axonal transport, although the specific signals for axonal targeting are unknown. However, the localization and abundance of phosphorylated APP within the axonal and synaptic compartments has led to the suggestion that phosphorylation may be a signal for axonal sorting of APP (Ando *et al.* 1999; Iijima *et al.* 2000; M.S. Lee *et al.* 2003; E.B. Lee *et al.* 2005; Muresan and Muresan 2005).

The subcellular localization of APP proteolysis has been the subject of intense investigation. In non-neuronal cells, the α-secretase pathway predominates and therefore non-neuronal cells do not generate high levels of Aβ under normal conditions. In contrast, α-secretase APP cleavage is a relatively minor component of APP metabolism in neuronal cells. Two α-secretase activities, constitutive and inducible, can be detected in non-neuronal and neuronal cells. Inducible α-secretase can be activated via pathways involving muscarinic/glutamate receptors and/or protein kinase C (Rossner *et al.* 1998). This inductible α-secretase activity can be attributed to TACE, since TACE knockout cells lack inducible α-secretase activity (Buxbaum *et al.* 1998a). However, these cells exhibit a basal level of α-secretase activity, indicating that at least one additional enzyme is capable of cleaving APP at the α-cleavage site in a non-inducible manner. The leading candidate for constitutive α-secretase is ADAM10. ADAM10 overexpression in cell cultures and transgenic mice increases α-cleavage (Lammich *et al.* 1999; Lopez-Perez *et al.* 2001; Postina *et al.* 2004), although ADAM10 knockout cells show no reduction in α-secretase activity in most isolated clones (Hartmann *et al.* 2002). Constitutive α-secretase is thought to occur at or close to the plasma membrane because cell-surface-labelled APP can be recovered rapidly in the extracellular medium as sAPPα (Sisodia 1992) and because the α-secretase-derived APP products, sAPPα and p3, cannot be detected intracellularly (Chyung *et al.* 1997; Forman *et al.* 1997; Parvathy *et al.* 1999; E.B. Lee *et al.* 2003). In contrast, inducible α-secretase activity occurs intracellularly within the trans-Golgi network (TGN), and therefore is able to compete with β-secretase for APP (Lammich *et al.* 1999; Skovronsky *et al.* 2000). Consistent with these findings is the presence of mature ADAM10 and TACE both on the cell surface and intracellularly (Black *et al.* 1997; Lammich *et al.* 1999; Skovronsky *et al.* 2000).

β-Cleavage and Aβ production have been detected throughout the secretory pathway, including endosomes/lysosomes, the TGN, and the endoplasmic reticulum/intermediate compartment (ER/IC). Unfortunately, the literature is somewhat confusing in that β-cleavage and Aβ production are often assumed to be synonymous, compounded by the use of both non-neuronal and neuronal cell culture models. For example, in non-neuronal cells, amyloidogenic C-terminal APP fragments are found within endosomes/lysosomes (Koo and Squazzo 1994), and agents which interfere with the pH of lysosomes, such as chlorquine or NH_4Cl, reduce Aβ secretion (Shoji et al. 1992; Haass et al. 1993). While these results suggest that γ-secretase activity is localized to endosomes/lysosomes, we have found these lysosomotropic agents also interfere with β-secretase because of the acidic pH optimum of BACE (unpublished observations). Regardless, in non-neuronal cells, Aβ secretion is reduced by blocking re-internalization of APP, and radiolabelled Aβ can be produced following re-internalization of radio-iodinated cell-surface APP (Koo and Squazzo 1994; Perez et al. 1996).

In stark contrast, inhibiting the endosomal/lysosomal system in neuronal cells does not reduce the secretion of Aβ (Tienari et al. 1997), indicating that the endosomal/ lysosomal pathway of Aβ production is negligible in cultured neuronal cells. Rather, in neuronal cells, a large proportion of secreted Aβ1–40 and Aβ1–42 is produced within the TGN (Forman et al. 1997; Hartmann et al. 1997; Xu et al. 1997; Greenfield et al. 1999). Another distinct pool of Aβ is also generated by neurons within the ER/IC (Chyung et al. 1997; Hartmann et al. 1997). Interestingly, Aβ produced within the ER/IC is almost exclusively Aβ1–42, and this pool of Aβ is not secreted but remains intracellularly and accumulates as an insoluble pool with time in culture (Skovronsky et al. 1998). Although ER/IC-derived Aβ has been detected in non-neuronal cells (Wild-Bode et al. 1997), it is more prominent in neurons and neuron-like cells (Cook et al. 1997; Hartmann et al. 1997; Greenfield et al. 1999; Grant et al. 2000).

The current understanding of the localization of BACE is consistent with its role in APP cleavage in Golgi and post-Golgi organelles. The majority of the published studies on BACE localization have been performed in non-neuronal cells. BACE is synthesized in the ER, and its propeptide domain is cleaved by furin or a related proprotein convertase within the Golgi, although propeptide cleavage does not appreciably alter catalytic activity (Bennett et al. 2000; Capell et al. 2000; Benjannet et al. 2001; Creemers et al. 2001). Consistent with an optimal activity at pH4.5, BACE is predominantly localized within acidic organelles in the secretory pathway including, but not limited to, the Golgi apparatus and endosomes of cultured non-neuronal cells (Vassar et al. 1999; Huse et al. 2000; Walter et al. 2001). A C-terminal dileucine motif promotes the targeting of BACE from the plasma membrane to endosomes (Huse et al. 2000). Furthermore, phosphory-lation of serine 498 by casein kinase I regulates trafficking between early endosomes and the TGN/late endosomes (Walter et al. 2001). However, these intrinsic trafficking signals do not appear to affect BACE activity or Aβ production. Studies investigating the subcellular localization of BACE in neuronal cells are sparse, but indicate that BACE protein or activity is enriched in autophagic vacuoles (Yu et al. 2004).

As for the localization of γ-secretase components and activity, original reports indicated that biochemically and immunocytochemically detectable PS was restricted to pre-Golgi organelles, mainly the ER and intermediate compartment (Annaert et al. 1999), and that co-localizing C99 with PS within the ER using a dileucine motif failed to generate secreted Aβ (Cupers et al. 2001). These results led to the 'spatial paradox', which stated that the lack of PS within Golgi and post-Golgi organelles cast doubt on the hypothesis that PS formed the active site of γ-secretase, given that Aβ generation and Notch cleavage are known to occur in these latter compartments. However, many groups, perhaps because of differentially sensitive experimental techniques, have found γ-secretase components throughout the secretory pathway, including the Golgi apparatus, endosomes, lysosomes, and the plasma membrane, apparently resolving the 'spatial paradox' (Takashima *et al.* 1996; Efthimiopoulos *et al.* 1998; Georgakopoulos *et al.* 1999; Ray *et al.* 1999; Schwarzman *et al.* 1999; Lah and Levey 2000; Singh *et al.* 2001; Kaether *et al.* 2002; Baulac *et al.* 2003; Rechards *et al.* 2003; Siman and Velji 2003; Tarassishin *et al.* 2004; Vetrivel *et al.* 2004; Wang *et al.* 2004; Hansson *et al.* 2005). With regard to the localization of γ-secretase and the subcellular site of Aβ generation, γ-cleavage of both APP and Notch does not occur until the precursor protein is cleaved by α- or β-secretase. Thus regulation of the site of Aβ production appears to be via pre-processing of the precursor protein, and not via subcellular localization of the γ-secretase components.

The mechanisms underlying the differences between the subcellular localization of APP cleavage between non-neuronal and neuronal cells have not been identified. The previously mentioned APP adaptor proteins, such as Fe65, are known to be preferentially expressed in neuronal cell types (Duilio *et al.* 1991; Duclos *et al.* 1993; Simeone *et al.* 1994). Certain post-translational modifications, specifically phosphorylation of Thr668 in the C-terminus of APP, are also neuron specific (Ando *et al.* 1999; Iijima *et al.* 2000; M.S. Lee *et al.* 2003; E.B. Lee *et al.* 2005; Muresan and Muresan 2005). However, as previously mentioned, the α-cleavage pathway is responsible for the most APP endoproteolysis in cultured non-neuronal cells (Esch *et al.* 1990; Sisodia *et al.* 1990). In contrast, BACE expression is high in neurons, and the β-cleavage pathway predominates in neuronal cell types (Wertkin *et al.* 1993; Turner *et al.* 1996; Chyung *et al.* 1997; Vassar *et al.* 1999; E.B. Lee *et al.* 2003). Therefore the differences between the subcellular localization may be due to differences in BACE expression, resulting in APP cleavage earlier in the secretory pathway within neurons. Several reports are consistent with the hypothesis that the level of β-secretase cleavage regulates the subcellular site of APP cleavage. For example, the relative amount of mature glycosylated APP is reduced in neuronal cells relative to non-neuronal cells, indicating that the higher basal level of BACE expression in neurons results in increased APP cleavage within the Golgi apparatus (E.B. Lee *et al.* 2003, 2005). Furthermore, in cell culture and transgenic mouse models, BACE overexpression results in a shift in APP proteolysis from synaptic terminals to neuronal perikarya (E.B. Lee *et al.* 2003, 2005). Finally, enhancing β-cleavage in non-neuronal cells by introducing the Swedish mutation within APP results in a shift in APP cleavage from endosomes/lysosomes to the TGN (Haass *et al.* 1995; Martin *et al.* 1995; Thinakaran *et al.* 1996). Therefore the level of β-cleavage occurring within a given cell is one factor which dictates the subcellular localization of β-cleavage, and therefore Aβ generation.

4.5 Aβ: extracellular and intracellular

The emphasis on the subcellular localization of Aβ generation bears relevance to the pathogenesis of AD. Increasingly the synaptic terminal is being recognized for its role in the amyloid cascade. The loss of synaptic markers correlates with the extent of cognitive dysfunction in AD, and is one of the strongest pathological correlates with disease severity (DeKosky and Scheff 1990; Terry *et al.* 1991; Scheff and Price 1993). Aβ is produced constitutively throughout life, with Aβ accumulation typically occurring only at advanced ages. This suggests that Aβ is usually non-pathogenic and does not result in toxicity until a threshold concentration is achieved. Transgenic murine models reiterate this finding in that, compared with non-transgenic littermates, certain spatial learning and memory behaviours become impaired only with increasing age, months after exogenous APP expression begins (Hsiao *et al.* 1996; Chapman *et al.* 1999; Chen *et al.* 2000; Westerman *et al.* 2002). The effects of Aβ on cognitive function are echoed by defects in long-term potentiation in transgenic mice (Chapman *et al.* 1999). Recently, both synthetic and cell-culture-derived Aβ has been shown to depress NMDA receptor levels on the cell surface (Snyder *et al.* 2005). Given that Aβ is capable of adopting multiple conformational states, the question remains as to which Aβ conformer is responsible for synaptic dysfunction and neuronal death in AD. *In vivo* microdialysis measurements using a 35 kDa molecular weight cut-off membrane demonstrated that concentrations of low molecular weight Aβ in APP transgenic mice do not increase with age (Cirrito *et al.* 2003). Therefore, although the long-term effects of exposure to monomeric Aβ are unknown, the alterations in cognitive behaviour in transgenic mice are likely to be independent of monomeric Aβ. In this section, the potential roles of insoluble extracellular Aβ, soluble extracellular Aβ and intracellular Aβ in the pathogenesis of AD will be framed in the context of synaptic function and dysfunction.

Growing evidence suggests that the synaptic terminal may be involved in the biogenesis of extracellular Aβ. Evoked potentials induced in slice cultures overexpressing APP results in the synaptic release of Aβ, which in turn depresses synaptic transmission (Kamenetz *et al.* 2003), suggesting indirectly that Aβ packaged within synaptic vesicles may be released upon excitation. Furthermore, synaptic zinc has been shown to contribute to amyloid plaque formation in transgenic mice (J.Y. Lee *et al.* 2002). Additionally, the disruption of the perforant pathway by ablating the entorhinal cortex of APP transgenic mice reduces Aβ deposition in the hippocampus, indicating that synaptically released Aβ may contribute to senile plaque formation (Lazarov *et al.* 2002; Sheng *et al.* 2002). However, ablation of the entorhinal cortex has profound effects on hippocampal physiology and therefore the decrease in hippocampal amyloid burden is difficult to assign solely to a decrease in synaptic Aβ. In a separate study, depletion of synaptic APP by BACE overexpression was also found to inhibit amyloid plaque deposition, lending further credence to the potential synaptic origin of amyloid plaques (E.B. Lee *et al.* 2005). While these studies are all highly suggestive of a link between synaptic Aβ and insoluble amyloid plaques, further investigations need to be performed to address this possibility.

Insoluble extracellular senile plaques have long been suggested to cause harm to surrounding neurons, either indirectly by inducing an inflammatory oxidative environment, or directly by altering neuronal structure and function. Consistent with the close spatial association with senile plaques and neuritic tau pathology, fibrillar Aβ enhances neurofibrillary tangle formation in tau transgenic mice (Gotz *et al.* 2001; Lewis *et al.* 2001).

Furthermore, the presence of senile plaques is closely related to microglia, suggesting that fibrillar Aβ triggers a chronic inflammatory response (Griffin *et al.* 1989; Itagaki *et al.* 1989; Rogers *et al.* 1992). The deposition of amyloid plaques within the extracellular milieu also distorts the local parenchyma and neurite geometry in both affected post-mortem human brains and transgenic mice, affecting neuronal function independent of neurofibrillary pathology (Knowles *et al.* 1998, 1999; Le *et al.* 2001; Brendza *et al.* 2003a,b; D'Amore *et al.* 2003; Tsai *et al.* 2004; Spires *et al.* 2005). These morphological effects of amyloid plaques can be reversed upon removing the amyloid via immunotherapy-based treatment of transgenic mice (Lombardo *et al.* 2003; Brendza *et al.* 2005). The presence of amyloid plaques also correlates with defects in synaptic signalling *in vivo* (Stern *et al.* 2004). Given the close link between neuronal structure and function as the cellular unit of a highly specialized geometric network, the presence of insoluble extracellular plaques undoubtedly has a negative impact on the neuronal geometry *in situ*, thereby affecting neuronal function.

Despite the evidence for impaired neuronal function due to the presence of insoluble amyloid plaques, soluble Aβ levels correlate better than the amount of fibrillar Aβ deposits with synaptic loss and/or cognitive dysfunction in both transgenic mice and humans with AD (Hsia *et al.* 1999; Lue *et al.* 1999; McLean *et al.* 1999; Mucke *et al.* 2000; Naslund *et al.* 2000). In transgenic mice, synaptic marker immunoreactivity and synaptic transmission defects have been found to precede the onset of microscopically detectable Aβ amyloid plaques (Hsia *et al.* 1999; Mucke *et al.* 2000). These findings have led to the hypothesis that a soluble Aβ oligomer is deleterious to neurons and contributes to the pathogenesis of AD.

In vitro, Aβ exists in a rapid equilibrium of small oligomeric assemblies (Bitan *et al.* 2001). Several non-fibrillar Aβ oligomer species have been isolated and have been shown to affect neuronal function. Soluble globular non-fibrillar Aβ aggregates, termed ADDLs (Aβ derived diffusible ligands), have been reported to be increased in AD compared with control subjects, are localized to dendritic post-synaptic sites on neurons, and negatively affect LTP in organotypic cultures (Lambert *et al.* 1998; Wang *et al.* 2002; Gong *et al.* 2003; Lacor *et al.* 2004). A second soluble non-fibrillar Aβ species was isolated from conditioned cell culture media from cells overexpressing APP harbouring the V717F familial AD mutation in APP_{751}. Conditioned media containing these soluble Aβ oligomers impair LTP *in vivo*, even when monomeric Aβ is removed by endoproteolytic digestion with insulin-degrading enzyme (IDE) (Walsh *et al.* 2002). Furthermore, this soluble Aβ oligomer preparation affects cognitive function in mice, which can be reversed by immunotherapy against the Aβ peptide (Cleary *et al.* 2005; Klyubin *et al.* 2005). Finally, a soluble Aβ 'globulomer', derived by incubating synthetic Aβ with fatty acids, which binds to dendritic neuronal processes has been isolated; it is found in AD brains and blocks LTP (Barghorn *et al.* 2005). Given the variety of isolation methods and the varied nature of these soluble Aβ oligomers, the convergence of effects on synaptic transmission is remarkable. The mechanisms whereby soluble Aβ oligomers affect neuronal function are as yet unknown. However, immunization of APP transgenic mice with an oligomer-selective monoclonal antibody (NAB61) improves cognitive function, adding evidence that targeting pathological conformers for the treatment of AD may be beneficial (E.B. Lee *et al.* 2006). Notably, the monoclonal antibody used in this final study recognizes an epitope on oligomeric Aβ which is found on Aβ fibrils. The soluble oligomers isolated by other groups are thought to result from a folding pathway independent of fibril formation, but the immunoreactivity of NAB61 against these oligomers is unknown.

Given that Aβ amyloid plaques are found extracellularly, much focus has been placed on the role of extracellular conformers of Aβ. However, intracellular pools of Aβ have been postulated to influence the development of AD. Intracellular Aβ was first detected in the NT2N human neuronal cell culture model (Wertkin *et al.* 1993). The level of intracellular Aβ was then found to increase with neuronal differentiation (Turner *et al.* 1996), and insoluble intracellular Aβ was found to accumulate with time in culture (Skovronsky *et al.* 1998). This insoluble pool of Aβ was derived from the endoplasmic reticulum, and comprised predominantly the longer and more amyloidogenic Aβ42 peptide. While intraneuronal Aβ42 production probably occurs within the ER/IC, intracellular Aβ42 accumulation may occur in the latter organelles, namely multivesicular bodies and endosomes within neuronal processes (Takahashi *et al.* 2002, 2004). It is unclear what forms the initial nidus for extracellular amyloid plaques, and the concentration of extracellular Aβ is below the threshold for Aβ fibril formation without the presence of mitigating factors. Therefore the presence of insoluble intracellular Aβ42 led to the hypothesis that the earliest steps towards Aβ aggregation occur intracellularly, resulting in Aβ aggregates which serve as seeds for fibril formation. These seeds are then released upon secretion or disruption of the cellular membrane secondary to neuronal insults. Consistent with this hypothesis is the finding that intracellular Aβ oligomers can be detected in cell culture and murine models (Walsh *et al.* 2000; Takahashi *et al.* 2004). Furthermore, amyloid plaques are found in close association with dead neurons (LaFerla *et al.* 1997), and intracellular debris is found within amyloid plaques (Ginsberg *et al.* 1999; D'Andrea *et al.* 2001).

While these studies opened a new field of intracellular amyloid biology, they also indicated that intracellular pools of Aβ were low and difficult to detect, especially given that the Aβ peptide shares the same sequence as APP, rendering most antibodies useless with regard to discerning Aβ from the full-length precursor. However, the development of antibodies which specifically recognize the Aβ peptide and not APP has led to the detection of intraneuronal Aβ42 in AD and Down's syndrome brains (Gouras *et al.* 2000; Gyure *et al.* 2001; Busciglio *et al.* 2002; Mori *et al.* 2002). Interestingly, intracellular Aβ42 immunoreactivity was prominent in individuals with mild cognitive impairment, or in individuals with Down's syndrome prior to the onset of extracellular Aβ amyloid pathology. Furthermore, Aβ42 immunoreactivity dissipated with disease progression, indicating that intracellular Aβ may play a role in the early pathogenesis of AD. Intraneuronal Aβ has also been detected in aged macaque monkeys (Martin *et al.* 1994) and transgenic mice (Wirths *et al.* 2000; Oddo *et al.* 2003). In the latter case, transgenic mice were generated harbouring $PS1_{M146V}$, APP_{Swe}, and tau_{P301L} transgenes, and intracellular Aβ immunoreactivity correlated with the onset of defects in LTP and cognition in the absence of extracellular amyloid plaques or intracellular tau pathology (Oddo *et al.* 2003; Billings *et al.* 2005). While these studies all suggest that intracellular Aβ may play an early role in the pathogenesis of AD, they rely heavily on the immunohistochemical specificities of anti-Aβ antibodies. Furthermore, intraneuronal Aβ is not uniformly detected in most transgenic models of AD, although the electrophysiological and behavioural manifestations of these murine models are remarkably similar. Thus further characterization of these models using a variety of methodologies is warranted. If validated, however, these studies indicate that intraneuronal Aβ has deleterious effects aside from its role in the formation of extracellular amyloid plaques. Indeed, akin to the potential pathogenic mechanism of other intracellular aggregates such as neurofibrillary tangles, intraneuronal Aβ may affect neuronal function and viability by disrupting intracellular processes such as trafficking and signalling pathways. Further investigation is also needed to elucidate better the relationship among intracellular pools of Aβ, soluble extracellular Aβ oligomers, and extracellular Aβ fibrils.

4.6 Summary and conclusions

The overwhelming breadth of literature on the biology and molecular neuropathology of APP and Aβ demonstrates the complexity inherent in the understanding of disease processes. Founded on the sequence of a small 4 kDa peptide and the strength of a genetic association between APP and AD, our understanding of APP and Aβ has grown enormously, with ever more questions to pursue. Remarkable progress has been made with regard to the elucidation of the proteolytic processing pathways of APP and the identification of the secretases responsible for Aβ production. While additional efforts are needed towards understanding the function of APP and the downstream effects of Aβ, promising efforts towards disrupting the pathological mechanisms leading to neurodegeneration are being pursued. The eventual validation of the amyloid cascade hypothesis may be the production of therapies aimed at disrupting the biochemical and cellular pathways detailed above, with the common goal of stopping, or at least mitigating, the disease process for those afflicted with AD.

References

Alzheimer A (1907). Über eine eigenartige Erkrankung der Hirnrinde. *Allgemeine Zeitschrift für Psychiatrie und psychisch-gerichtliche Medizin* **64**, 146–8.

Alzheimer A (1911). Über eigenartige Krankheitsfalle des späteren Alters. *Zeitschrift für die gesamte Neurologie und Psychiatrie* **4**, 356–85.

Amaratunga A and Fine RE (1995). Generation of amyloidogenic C-terminal fragments during rapid axonal transport *in vivo* of beta-amyloid precursor protein in the optic nerve. *Journal of Biological Chemistry* **270**, 17268–72.

Ando K, Oishi M, Takeda S, *et al.* (1999). Role of phosphorylation of Alzheimer's amyloid precursor protein during neuronal differentiation. *Journal of Neuroscience* **19**, 4421–7.

Ando K, Iijima KI, Elliott JI, Kirino Y, and Suzuki T (2001). Phosphorylation-dependent regulation of the interaction of amyloid precursor protein with Fe65 affects the production of beta-amyloid. *Journal of Biological Chemistry* **276**, 40353–61.

Annaert WG, Levesque L, Craessaerts K, *et al*. (1999). Presenilin 1 controls gamma-secretase processing of amyloid precursor protein in pre-Golgi compartments of hippocampal neurons. *Journal of Cell Biology* **147**, 277–294.

Barghorn S, Nimmrich V, Striebinger A, Krantz C, *et al*. (2005). Globular amyloid beta-peptide oligomer: a homogenous and stable neuropathological protein in Alzheimer's disease. *Journal of Neurochemistry* **95**, 834–47.

Baulac S, LaVoie MJ, Kimberly WT, *et al*. 2003. Functional gamma-secretase complex assembly in Golgi/trans-Golgi network: interactions among presenilin, nicastrin, Aph1, Pen-2, and gamma-secretase substrates. *Neurobiology of Disease* **14**, 194–204.

Benjannet S, Elagoz A, Wickham L, *et al*. (2001). Post-translational processing of beta-secretase (beta-amyloid-converting enzyme) and its ectodomain shedding. The pro- and transmembrane/cytosolic domains affect its cellular activity and amyloid-beta production. *Journal of Biological Chemistry* **276**, 10879–87.

Bennett BD, Denis P, Haniu M, *et al*. (2000). A furin-like convertase mediates propeptide cleavage of BACE, the Alzheimer's beta -secretase. *Journal of Biological Chemistry* **275**, 37712–17.

Berg L, McKeel DW, Jr, Miller JP, Baty J, and Morris JC (1993). Neuropathological indexes of Alzheimer's disease in demented and nondemented persons aged 80 years and older. *Archives of Neurology* **50**, 349–58.

Bierer LM, Hof PR, Purohit DP, *et al*. (1995). Neocortical neurofibrillary tangles correlate with dementia severity in Alzheimer's disease. *Archives of Neurology* **52**, 81–8.

Billings LM, Oddo S, Green KN, McGaugh JL, and LaFerla FM (2005). Intraneuronal Abeta causes the onset of early Alzheimer's disease-related cognitive deficits in transgenic mice. *Neuron* **45**, 675–88.

Bitan G, Lomakin A, and Teplow DB (2001). Amyloid beta-protein oligomerization: prenucleation interactions revealed by photo-induced cross-linking of unmodified proteins. *Journal of Biological Chemistry* **276**, 35176–84.

Black RA, Rauch CT, Kozlosky CJ, *et al*. (1997). A metalloproteinase disintegrin that releases tumour-necrosis factor-alpha from cells. *Nature* **385**, 729–33.

Blaikie P, Immanuel D, Wu J, Li N, Yajnik V, and Margolis B (1994). A region in Shc distinct from the SH2 domain can bind tyrosine-phosphorylated growth factor receptors. *Journal of Biological Chemistry* **269**, 32031–4.

Borchelt DR, Thinakaran G, Eckman CB, *et al*. (1996). Familial Alzheimer's disease-linked presenilin 1 variants elevate Abeta1–42/1–40 ratio *in vitro* and *in vivo. Neuron* **17**, 1005–13.

Borg JP, Ooi J, Levy E, and Margolis B (1996). The phosphotyrosine interaction domains of X11 and FE65 bind to distinct sites on the YENPTY motif of amyloid precursor protein. *Molecular and Cellular Biology* **16**, 6229–41.

Borg JP, Straight SW, Kaech SM, *et al*. (1998a). Identification of an evolutionarily conserved heterotrimeric protein complex involved in protein targeting. *Journal of Biological Chemistry* **273**, 31633–6.

Borg JP, Yang Y, Taddeo-Borg M, Margolis B, and Turner RS (1998b). The X11alpha protein slows cellular amyloid precursor protein processing and reduces Abeta40 and Abeta42 secretion. *Journal of Biological Chemistry* **273**, 14761–6.

Braak H and Braak E (1991). Neuropathological staging of Alzheimer-related changes. *Acta Neuropathologica* **82**, 239–59.

Brendza RP, O'Brien C, Simmons K, *et al*. (2003a). PDAPP; YFP double transgenic mice: a tool to study amyloid-beta associated changes in axonal, dendritic, and synaptic structures. *Journal of Comparative Neurology* **456**, 375–83.

Brendza RP, Simmons K, Bales KR, Paul SM, Goldberg MP, and Holtzman DM (2003b).Use of YFP to study amyloid-beta associated neurite alterations in live brain slices. *Neurobiology of Aging* **24**,1071–7.

Brendza RP, Bacskai BJ, Cirrito JR, *et al.* (2005). Anti-Abeta antibody treatment promotes the rapid recovery of amyloid-associated neuritic dystrophy in PDAPP transgenic mice. *Journal of Clinical Investigation* **115**, 428–33.

Brou C, Logeat F, Gupta N, *et al.* (2000). A novel proteolytic cleavage involved in Notch signaling: the role of the disintegrin-metalloprotease TACE. *Molecular Cell* **5**, 207–16.

Busciglio J, Pelsman A, Wong C, *et al.* (2002). Altered metabolism of the amyloid beta precursor protein is associated with mitochondrial dysfunction in Down's syndrome. *Neuron* **33**, 677–88.

Butz S, Okamoto M, and Sudhof TC (1998). A tripartite protein complex with the potential to couple synaptic vesicle exocytosis to cell adhesion in brain. *Cell* **94**, 773–82.

Buxbaum JD, Liu KN, Luo Y, *et al.* (1998a). Evidence that tumor necrosis factor alpha converting enzyme is involved in regulated alpha-secretase cleavage of the Alzheimer amyloid protein precursor. *Journal of Biological Chemistry* **273**, 27765–7.

Buxbaum JD, Thinakaran G, Koliatsos V, *et al.* (1998b). Alzheimer amyloid protein precursor in the rat hippocampus: transport and processing through the perforant path. *Journal of Neuroscience* **18**, 9629–37.

Cai XD, Golde TE, and Younkin SG (1993). Release of excess amyloid beta protein from a mutant amyloid beta protein precursor. *Science* **259**, 514–16.

Cai H, Wang Y, McCarthy D, *et al.* (2001). BACE1 is the major beta-secretase for generation of Abeta peptides by neurons. *Nature Neuroscience* **4**, 233–4.

Cao X and Sudhof TC (2001). A transcriptionally [correction of transcriptively] active complex of APP with Fe65 and histone acetyltransferase Tip60. *Science* **293**, 115–20.

Cao X and Sudhof TC (2004). Dissection of amyloid-beta precursor protein-dependent transcriptional transactivation. *Journal of Biological Chemistry* **279**, 24601–11.

Capell A, Steiner H, Willem M, *et al.* (2000). Maturation and pro-peptide cleavage of beta-secretase. *Journal of Biological Chemistry* **275**, 30849–54.

Chapman PF, White GL, Jones MW, *et al.* (1999). Impaired synaptic plasticity and learning in aged amyloid precursor protein transgenic mice. *Nature Neuroscience* **2**, 271–6.

Chen G, Chen KS, Knox J, *et al.* (2000). A learning deficit related to age and beta-amyloid plaques in a mouse model of Alzheimer's disease. *Nature* **408**, 975–9.

Chen HI and Sudol M (1995). The WW domain of Yes-associated protein binds a proline-rich ligand that differs from the consensus established for Src homology 3-binding modules. *Proceedings of the National Academy of Sciences of the USA* **92**, 7819–23.

Chen WJ, Goldstein JL, and Brown MS (1990). NPXY, a sequence often found in cytoplasmic tails, is required for coated pit-mediated internalization of the low density lipoprotein receptor. *Journal of Biological Chemistry* **265**, 3116–23.

Chyung AS, Greenberg BD, Cook DG, Doms RW, and Lee VMY (1997). Novel beta-secretase cleavage of beta-amyloid precursor protein in the endoplasmic reticulum/intermediate compartment of NT2N cells. *Journal of Cell Biology* **138**, 671–80.

Cirrito JR, May PC, O'Dell MA, *et al.* (2003). *In vivo* assessment of brain interstitial fluid with microdialysis reveals plaque-associated changes in amyloid-beta metabolism and half-life. *Journal of Neuroscience* **23**, 8844–53.

Citron M, Oltersdorf T, Haass C, *et al.* (1992). Mutation of the beta-amyloid precursor protein in familial Alzheimer's disease increases beta-protein production. *Nature* **360**, 672–4.

Citron M, Westaway D, Xia W, *et al.* (1997). Mutant presenilins of Alzheimer's disease increase production of 42-residue amyloid beta-protein in both transfected cells and transgenic mice. *Nature Medicine* **3**, 67–72.

Cleary JP, Walsh DM, Hofmeister JJ, *et al.* (2005). Natural oligomers of the amyloid-beta protein specifically disrupt cognitive function. *Nature Neuroscience* **8**, 79–84.

Cook DG, Forman MS, Sung JC, *et al.* (1997). Alzheimer's A beta(1–42) is generated in the endoplasmic reticulum/intermediate compartment of NT2N cells. *Nature Medicine* **3**, 1021–3.

Cook DG, Zhu P, Wang J, and Yang Y (2003). Beta-secretase (BACE) is highly enriched in neurotransmitter vesicles. *Society for Neuroscience 2003 Annual Meeting*, Abstract 445.2.

Creemers JW, Ines DD, Plets E, *et al.* (2001). Processing of beta-secretase by furin and other members of the proprotein convertase family. *Journal of Biological Chemistry* **276**, 4211–17.

Cupers P, Bentahir M, Craessaerts K, *et al.* (2001). The discrepancy between presenilin subcellular localization and gamma-secretase processing of amyloid precursor protein. *Journal of Cell Biology* **154**, 731–40.

D'Amore JD, Kajdasz ST, McLellan ME, *et al.* (2003). *In vivo* multiphoton imaging of a transgenic mouse model of Alzheimer's disease reveals marked thioflavine-S-associated alterations in neurite trajectories. *Journal of Neuropathology and Experimental Neurology* **62**, 137–45.

D'Andrea MR, Nagele RG, Wang HY, Peterson PA, and Lee DH (2001). Evidence that neurones accumulating amyloid can undergo lysis to form amyloid plaques in Alzheimer's disease. *Histopathology* **38**, 120–34.

Dawson GR, Seabrook GR, Zheng H, *et al.* (1999). Age-related cognitive deficits, impaired long-term potentiation and reduction in synaptic marker density in mice lacking the beta-amyloid precursor protein. *Neuroscience* **90**, 1–13.

DeKosky ST and Scheff SW (1990). Synapse loss in frontal cortex biopsies in Alzheimer's disease: correlation with cognitive severity. *Annals of Neurology* **27**, 457–64.

De Strooper B, Umans L, Van Leuven F, and Van Den BH (1993). Study of the synthesis and secretion of normal and artificial mutants of murine amyloid precursor protein (APP): cleavage of APP occurs in a late compartment of the default secretion pathway. *Journal of Cell Biology* **121**, 295–304.

De Strooper B, Saftig P, Craessaerts K, *et al.* (1998). Deficiency of presenilin-1 inhibits the normal cleavage of amyloid precursor protein. *Nature* **391**, 387–90.

De Strooper B, Annaert W, Cupers P, *et al.* (1999). A presenilin-1-dependent gamma-secretase-like protease mediates release of Notch intracellular domain. *Nature* **398**, 518–22.

Doyle E, Bruce MT, Breen KC, Smith DC, Anderton B, and Regan CM (1990). Intraventricular infusions of antibodies to amyloid-beta-protein precursor impair the acquisition of a passive avoidance response in the rat. *Neuroscience Letters* **115**, 97–102.

Duclos F, Boschert U, Sirugo G, Mandel JL, Hen R, and Koenig M (1993). Gene in the region of the Friedreich ataxia locus encodes a putative transmembrane protein expressed in the nervous system. *Proceedings of the National Academy of Sciences of the USA* **90**, 109–13.

Duilio A, Zambrano N, Mogavero AR, Ammendola R, Cimino F, and Russo T (1991). A rat brain mRNA encoding a transcriptional activator homologous to the DNA binding domain of retroviral integrases. *Nucleic Acids Research* **19**, 5269–74.

Edbauer D, Winkler E, Regula JT, Pesold B, Steiner H, and Haass C (2003). Reconstitution of gamma-secretase activity. *Nature Cell Biology* **5**, 486–8.

Efthimiopoulos S, Floor E, Georgakopoulos A, *et al.* (1998). Enrichment of presenilin 1 peptides in neuronal large dense-core and somatodendritic clathrin-coated vesicles. *Journal of Neurochemistry* **71**, 2365–72.

Ermekova KS, Zambrano N, Linn H, *et al.* (1997). The WW domain of neural protein FE65 interacts with proline-rich motifs in Mena, the mammalian homolog of *Drosophila* enabled. *Journal of Biological Chemistry* **272**, 32869–77.

Esch FS, Keim PS, Beattie EC, *et al.* (1990). Cleavage of amyloid beta peptide during constitutive processing of its precursor. *Science* **248**, 1122–4.

Esler WP, Kimberly WT, Ostaszewski BL, *et al.* (2000). Transition-state analogue inhibitors of gamma-secretase bind directly to presenilin-1. *Nature Cell Biology* **2**, 428–34.

Ferreira A, Caceres A, and Kosik KS (1993). Intraneuronal compartments of the amyloid precursor protein. *Journal of Neuroscience* **13**, 3112–23.

Fiore F, Zambrano N, Minopoli G, Donini V, Duilio A, and Russo T (1995). The regions of the Fe65 protein homologous to the phosphotyrosine interaction/phosphotyrosine binding domain of Shc bind the intracellular domain of the Alzheimer's amyloid precursor protein. *Journal of Biological Chemistry* **270**, 30853–6.

Forman MS., Cook DG, Leight S, Doms RW, and Lee VMY (1997). Differential effects of the Swedish mutant amyloid precursor protein on beta-amyloid accumulation and secretion in neurons and nonneuronal cells. *Journal of Biological Chemistry* **272**, 32247–53.

Francis R, McGrath G, Zhang J, *et al.* (2002). aph-1 and pen-2 are required for Notch pathway signaling, gamma-secretase cleavage of betaAPP, and presenilin protein accumulation. *Developmental Cell* **3**, 85–97.

Furukawa K, Sopher BL, Rydel RE, *et al.* (1996). Increased activity-regulating and neuroprotective efficacy of alpha-secretase-derived secreted amyloid precursor protein conferred by a C-terminal heparin-binding domain. *Journal of Neurochemistry* **67**, 1882–96.

Games D, Adams D, Alessandrini R, *et al.* (1995). Alzheimer-type neuropathology in transgenic mice overexpressing V717F beta-amyloid precursor protein. *Nature* **373**, 523–7.

Gao Y and Pimplikar SW (2001). The gamma -secretase-cleaved C-terminal fragment of amyloid precursor protein mediates signaling to the nucleus. *Proceedings of the National Academy of Sciences of the USA* **98**, 14979–84.

Gertler FB, Comer AR, Juang JL, *et al.* (1995). enabled, a dosage-sensitive suppressor of mutations in the *Drosophila* Abl tyrosine kinase, encodes an Abl substrate with SH3 domain-binding properties. *Genes and Development* **9**, 521–33.

Georgakopoulos A, Marambaud P, Efthimiopoulos S, *et al.* (1999). Presenilin-1 forms complexes with the cadherin/catenin cell-cell adhesion system and is recruited to intercellular and synaptic contacts. *Molecular Cell* **4**, 893–902.

Ginsberg SD, Crino PB, Hemby SE, *et al.* (1999). Predominance of neuronal mRNAs in individual Alzheimer's disease senile plaques. *Annals of Neurology* **45**, 174–81.

Glenner GG and Wong CW (1984). Alzheimer's disease and Down's syndrome: sharing of a unique cerebrovascular amyloid fibril protein. *Biochemical and Biophysical Research Communications* **122**, 1131–5.

Golde TE, Estus S, Younkin LH, Selkoe DJ, and Younkin SG (1992). Processing of the amyloid protein precursor to potentially amyloidogenic derivatives. *Science* **255**, 728–30.

Goldgaber D, Lerman MI, McBride OW, Saffiotti U, and Gajdusek DC (1987). Characterization and chromosomal localization of a cDNA encoding brain amyloid of Alzheimer's disease. *Science* **235**, 877–80.

Gong Y, Chang L, Viola KL *et al.* (2003). Alzheimer's disease-affected brain: presence of oligomeric A beta ligands (ADDLs) suggests a molecular basis for reversible memory loss. *Proceedings of the National Academy of Sciences of the USA* **100**, 10417–22.

Goodman Y. and Mattson MP (1994). Secreted forms of beta-amyloid precursor protein protect hippocampal neurons against amyloid beta-peptide-induced oxidative injury. *Experimental Neurology* **128**, 1–12.

Gotz J, Chen F, van Dorpe J, and Nitsch RM (2001). Formation of neurofibrillary tangles in P301l tau transgenic mice induced by Abeta 42 fibrils. *Science* **293**, 1491–5.

Gouras GK, Tsai J, Naslund J, *et al.* (2000). Intraneuronal Abeta42 accumulation in human brain. *American Journal of Pathology* **156**, 15–20.

Goutte C, Tsunozaki M, Hale VA, and Priess, JR (2002). APH-1 is a multipass membrane protein essential for the Notch signaling pathway in *Caenorhabditis elegans* embryos. *Proceedings of the National Academy of Sciences of the USA* **99**, 775–9.

Grant SM, Ducatenzeiler A, Szyf M, and Cuello AC (2000). Abeta immunoreactive material is present in several intracellular compartments in transfected, neuronally differentiated, P19 cells expressing the human amyloid beta-protein precursor. *Journal of Alzheimer's Disease* **2**, 207–22.

Greenfield JP, Tsai J, Gouras GK, *et al.* (1999). Endoplasmic reticulum and trans-Golgi network generate distinct populations of Alzheimer beta-amyloid peptides. *Proceedings of the National Academy of Sciences of the USA* **96**, 742–7.

Griffin WS, Stanley LC, Ling C, *et al*. (1989). Brain interleukin 1 and S-100 immunoreactivity are elevated in Down syndrome and Alzheimer disease. *Proceedings of the National Academy of Sciences of the USA* **86**, 7611–15.

Grigorenko AP, Moliaka YK, Korovaitseva GI, and Rogaev EI (2002). Novel class of polytopic proteins with domains associated with putative protease activity. *Biochemistry (Moscow)* **67**, 826–35.

Guenette SY, Chen J, Jondro PD, and Tanzi RE (1996). Association of a novel human FE65-like protein with the cytoplasmic domain of the beta-amyloid precursor protein. *Proceedings of the National Academy of Sciences of the USA* **93**, 10832–7.

Gyure KA, Durham R, Stewart WF, Smialek JE, and Troncoso JC (2001). Intraneuronal abeta-amyloid precedes development of amyloid plaques in Down syndrome. *Archives of Pathology and Laboratory Medicine* **125**, 489–92.

Haass C, Schlossmacher MG, Hung AY, *et al*. (1992). Amyloid beta-peptide is produced by cultured cells during normal metabolism. *Nature* **359**, 322–5.

Haass C, Hung AY, Schlossmacher MG, Teplow DB, and Selkoe DJ (1993). Beta-amyloid peptide and a 3-kDa fragment are derived by distinct cellular mechanisms. *Journal of Biological Chemistry* **268**, 3021–4.

Haass C, Hung AY, Selkoe DJ, and Teplow DB (1994a). Mutations associated with a locus for familial Alzheimer's disease result in alternative processing of amyloid beta-protein precursor. *Journal of Biological Chemistry* **269**, 17741–8.

Haass C, Koo EH, Teplow DB, and Selkoe DJ (1994b). Polarized secretion of beta-amyloid precursor protein and amyloid beta-peptide in MDCK cells. *Proceedings of the National Academy of Sciences of the USA* **91**, 1564–8.

Haass C, Lemere CA, Capell A, *et al*. (1995). The Swedish mutation causes early-onset Alzheimer's disease by beta-secretase cleavage within the secretory pathway. *Nature Medicine* **1**, 1291–6.

Hansson EM, Stromberg K, Bergstedt S, *et al*. (2005). Aph-1 interacts at the cell surface with proteins in the active gamma-secretase complex and membrane-tethered Notch. *Journal of Neurochemistry* **92**, 1010–20.

Hartmann D, De Strooper B, Serneels L, *et al*. (2002). The disintegrin/ metalloprotease ADAM 10 is essential for Notch signalling but not for alpha-secretase activity in fibroblasts. *Human Molecular Genetics* **11**, 2615–24.

Hartmann T, Bergsdorf C, Sandbrink R, *et al*. (1996). Alzheimer's disease betaA4 protein release and amyloid precursor protein sorting are regulated by alternative splicing. *Journal of Biological Chemistry* **271**, 13208–14.

Hartmann T, Bieger SC, Bruhl B, *et al*. (1997). Distinct sites of intracellular production for Alzheimer's disease A beta40/42 amyloid peptides. *Nature Medicine* **3**, 1016–20.

Herreman A, Serneels L, Annaert W, *et al*. (2000). Total inactivation of gamma-secretase activity in pre-senilin-deficient embryonic stem cells. *Nature Cell Biology* **2**, 461–2.

Heber S, Herms J, Gajic V, Hainfellner J, *et al*. (2000). Mice with combined gene knock-outs reveal essential and partially redundant functions of amyloid precursor protein family members. *Journal of Neuroscience* **20**, 7951–63.

Hill K, Li Y, Bennett M, McKay M, *et al*. (2003). Munc18 interacting proteins: ADP-ribosylation factor-dependent coat proteins that regulate the traffic of beta-Alzheimer's precursor protein. *Journal of Biological Chemistry* **278**, 36032–40.

Ho L, Fukuchi K, and Younkin SG (1996). The alternatively spliced Kunitz protease inhibitor domain alters amyloid beta protein precursor processing and amyloid beta protein production in cultured cells. *Journal of Biological Chemistry* **271**, 30929–34.

Hsia AY, Masliah E, McConlogue L, *et al*. (1999). Plaque-independent disruption of neural circuits in Alzheimer's disease mouse models. *Proceedings of the National Academy of Sciences of the USA* **96**, 3228–33.

Hsiao K, Chapman P, Nilsen S, *et al*. (1996). Correlative memory deficits, Abeta elevation, and amyloid plaques in transgenic mice. *Science* **274**, 99–102.

Huber G, Martin JR, Loffler J, and Moreau JL (1993). Involvement of amyloid precursor protein in memory formation in the rat: an indirect antibody approach. *Brain Research* **603**, 348–52.

Huse JT, Pijak DS, Leslie GJ, Lee VMY, and Doms RW (2000). Maturation and endosomal targeting of beta-site amyloid precursor protein-cleaving enzyme: the Alzheimer's disease beta-secretase. *Journal of Biological Chemistry* **275**, 33729–37.

Huse JT, Liu K, Pijak DS, Carlin, Lee VMY, and Doms RW (2002). Beta-secretase processing in the trans-Golgi network preferentially generates truncated amyloid species that accumulate in Alzheimer's disease brain. *Journal of Biological Chemistry* **277**, 16278–84.

Hussain I, Powell D, Howlett DR, *et al.* (1999). Identification of a novel aspartic protease (Asp 2) as beta-secretase. *Molecular and Cellular Neurosciences* **14**, 419–27.

Ide N, Hata Y, Hirao K, *et al.* (1998). Interaction of rat lin-10 with brain-enriched F-actin-binding protein, neurabin-II/spinophilin. *Biochemical and Biophysical Research Communications* **244**, 258–62.

Iijima K, Ando K, Takeda S, *et al.* (2000). Neuron-specific phosphorylation of Alzheimer's beta-amyloid precursor protein by cyclin-dependent kinase 5. *Journal of Neurochemistry* **75**, 1085–91.

Inomata H, Nakamura Y, Hayakawa A, *et al.* (2003). A scaffold protein JIP-1b enhances amyloid precursor protein phosphorylation by JNK and its association with kinesin light chain 1. *Journal of Biological Chemistry* **278**, 22946–55.

Itagaki S, McGeer PL, Akiyama H, Zhu S, and Selkoe D (1989). Relationship of microglia and astrocytes to amyloid deposits of Alzheimer disease. *Journal of Neuroimmunology* **24**, 173–82.

Jarrett JT, Berger EP, and Lansbury PT, Jr (1993). The carboxy terminus of the beta amyloid protein is critical for the seeding of amyloid formation: implications for the pathogenesis of Alzheimer's disease. *Biochemistry* **32**, 4693–7.

Jin LW, Ninomiya H, Roch JM, *et al.* (1994). Peptides containing the RERMS sequence of amyloid beta/A4 protein precursor bind cell surface and promote neurite extension. *Journal of Neuroscience* **14**, 5461–70.

Kaether C, Lammich S, Edbauer D, *et al.* (2002). Presenilin-1 affects trafficking and processing of betaAPP and is targeted in a complex with nicastrin to the plasma membrane. *Journal of Cell Biology* **158**, 551–61.

Kamal A, Stokin GB, Yang Z, Xia CH, and Goldstein LS (2000). Axonal transport of amyloid precursor protein is mediated by direct binding to the kinesin light chain subunit of kinesin-I. *Neuron* **28**, 449–59.

Kamal A, Almenar-Queralt A, LeBlanc JF, Roberts EA, and Goldstein LS (2001). Kinesin-mediated axonal transport of a membrane compartment containing beta-secretase and presenilin-1 requires APP. *Nature* **414**, 643–8.

Kamenetz F, Tomita T, Hsieh H, *et al.* (2003). APP processing and synaptic function. *Neuron* **37**, 925–37.

Kang J and Muller-Hill B (1989). The sequence of the two extra exons in rat preA4. *Nucleic Acids Research* **17**, 2130.

Kang J, Lemaire HG, Unterbeck A, *et al.* (1987). The precursor of Alzheimer's disease amyloid A4 protein resembles a cell-surface receptor. *Nature* **325**, 733–6.

Kimberly WT, Zheng JB, Guenette SY, and Selkoe DJ (2001). The intracellular domain of the beta-amyloid precursor protein is stabilized by Fe65 and translocates to the nucleus in a Notch-like manner. *Journal of Biological Chemistry* **276**, 40288–92.

King GD and Scott TR (2004). Adaptor protein interactions: modulators of amyloid precursor protein metabolism and Alzheimer's disease risk? *Experimental Neurology* **185**, 208–19.

King GD, Cherian K, and Turner RS (2004). X11alpha impairs gamma- but not beta-cleavage of amyloid precursor protein. *Journal of Neurochemistry* **88**, 971–82.

Kitaguchi N, Takahashi Y, Tokushima Y, Shiojiri S, and Ito H (1988). Novel precursor of Alzheimer's disease amyloid protein shows protease inhibitory activity. *Nature* **331**, 530–2.

Klyubin I, Walsh DM, Lemere CA *et al.* (2005). Amyloid beta protein immunotherapy neutralizes Abeta oligomers that disrupt synaptic plasticity *in vivo*. *Nature Medicine* **11**, 556–61.

Knauer MF, Orlando RA, and Glabe CG (1996). Cell surface APP751 forms complexes with protease nexin 2 ligands and is internalized via the low density lipoprotein receptor-related protein (LRP). *Brain Research* **740**, 6–14.

Knops J, Gandy S, Greengard P, Lieberburg I, and Sinha S (1993). Serine phosphorylation of the secreted extracellular domain of APP. *Biochemical and Biophysical Research Communications* **197**, 380–5.

Knowles RB, Gomez-Isla T, and Hyman BT (1998). Abeta associated neuropil changes: correlation with neuronal loss and dementia. *Journal of Neuropathology and Experimental Neurology* **57**,1122–30.

Knowles RB, Wyart C, Buldyrev SV, *et al.* (1999). Plaque-induced neurite abnormalities: implications for disruption of neural networks in Alzheimer's disease. *Proceedings of the National Academy of Sciences of the USA* **96**, 5274–9.

Koo EH and Squazzo SL (1994). Evidence that production and release of amyloid beta-protein involves the endocytic pathway. *Journal of Biological Chemistry* **269**, 17386–9.

Koo EH, Sisodia SS, Archer DR, *et al.* (1990). Precursor of amyloid protein in Alzheimer disease undergoes fast anterograde axonal transport. *Proceedings of the National Academy of Sciences of the USA* **87**, 1561–5.

Kounnas MZ, Church FC, Argraves WS, and Strickland DK (1996). Cellular internalization and degradation of antithrombin III-thrombin, heparin cofactor II-thrombin, and alpha 1-antitrypsin-trypsin complexes is mediated by the low density lipoprotein receptor-related protein. *Journal of Biological Chemistry* **271**, 6523–9.

Lacor PN, Buniel MC, Chang L, *et al.* (2004). Synaptic targeting by Alzheimer's-related amyloid beta oligomers. *Journal of Neuroscience* **24**, 10191–200.

LaFerla FM, Troncoso JC, Strickland DK, Kawas CH, and Jay G (1997). Neuronal cell death in Alzheimer's disease correlates with apoE uptake and intracellular Abeta stabilization. *Journal of Clinical Investigation* **100**, 310–20.

Lah JJ and Levey AI. (2000). Endogenous presenilin-1 targets to endocytic rather than biosynthetic compartments. *Molecular and Cellular Neuroscience* **16**, 111–26.

Lai A, Sisodia SS, and Trowbridge IS (1995). Characterization of sorting signals in the beta-amyloid precursor protein cytoplasmic domain. *Journal of Biological Chemistry* **270**, 3565–73.

Lambert MP, Barlow AK, Chromy BA, *et al.* (1998). Diffusible, nonfibrillar ligands derived from Abeta1–42 are potent central nervous system neurotoxins. *Proceedings of the National Academy of Sciences of the USA* **95**, 6448–53.

Lammich S, Kojro E, Postina R, *et al.* (1999). Constitutive and regulated alpha-secretase cleavage of Alzheimer's amyloid precursor protein by a disintegrin metalloprotease. *Proceedings of the National Academy of Sciences of the USA* **96**, 3922–7.

Lanier LM, Gates MA, Witke W, *et al.* (1999). Mena is required for neurulation and commissure formation. *Neuron* **22**, 313–25.

Lau KF, McLoughlin DM, Standen C, and Miller CC (2000). X11 alpha and X11 beta interact with presenilin-1 via their PDZ domains. *Molecular and Cellular Neurosciences* **16**, 557–65.

Lazarov O, Lee M, Peterson DA, and Sisodia SS (2002). Evidence that synaptically released beta-amyloid accumulates as extracellular deposits in the hippocampus of transgenic mice. *Journal of Neuroscience* **22**, 9785–93.

Lazarov O, Morfini GA, Lee EB, *et al.* (2005). Axonal transport, amyloid precursor protein, kinesin-1, and the processing apparatus: revisited. *Journal of Neuroscience* **25**, 2386–95.

Le R, Cruz L, Urbanc B, *et al.* (2001). Plaque-induced abnormalities in neurite geometry in transgenic models of Alzheimer disease: implications for neural system disruption. *Journal of Neuropathology and Experimental Neurology* **60**, 753–8.

Lee EB, Skovronsky DM, Abtahian F, Doms RW, and Lee VMY (2003). Secretion and intracellular generation of truncated Abeta in beta-site amyloid-beta precursor protein-cleaving enzyme expressing human neurons. *Journal of Biological Chemistry* **278**, 4458–66.

Lee EB, Zhang B, Liu K, *et al.* (2005). BACE overexpression alters the subcellular processing of APP and inhibits Abeta deposition *in vivo*. *Journal of Cell Biology* **168**, 291–302.

Lee EB, Leng LZ, Zhang B, *et al.* (2006). Targeting amyloid-beta peptide (Abeta) oligomers by passive immunization with a conformation-selective monoclonal antibody improves learning and memory in Abeta precursor protein (APP) transgenic mice. *Journal of Biological Chemistry* **281**, 4292–9.

Lee JH, Lau KF, Perkinton MS, *et al.* (2003). The neuronal adaptor protein X11alpha reduces Abeta levels in the brains of Alzheimer's APPswe Tg2576 transgenic mice. *Journal of Biological Chemistry* **278**, 47025–9.

Lee JY, Cole TB, Palmiter RD, Suh SW, and Koh JY (2002). Contribution by synaptic zinc to the gender-disparate plaque formation in human Swedish mutant APP transgenic mice. *Proceedings of the National Academy of Sciences of the USA* **99**, 7705–10.

Lee MS, Kao SC, Lemere CA, *et al.* (2003). APP processing is regulated by cytoplasmic phosphorylation. *Journal of Cell Biology* **163**, 83–95.

Levy-Lahad E, Wasco W, Poorkaj P, *et al.* (1995). Candidate gene for the chromosome 1 familial Alzheimer's disease locus. *Science* **269**, 973–7.

Lewis J, Dickson DW, Lin WL, *et al.* (2001). Enhanced neurofibrillary degeneration in transgenic mice expressing mutant tau and APP. *Science* **293**, 1487–91.

Li HL, Roch JM, Sundsmo M, *et al.* (1997). Defective neurite extension is caused by a mutation in amyloid beta/A4 (A beta) protein precursor found in familial Alzheimer's disease. *Journal of Neurobiology* **32**, 469–80.

Li YM, Xu M, Lai MT, *et al.* (2000). Photoactivated gamma-secretase inhibitors directed to the active site covalently label presenilin 1. *Nature* **405**, 689–94.

Lin X, Koelsch G, Wu S, Downs D, Dashti A, and Tang J. (2000). Human aspartic protease memapsin 2 cleaves the beta-secretase site of beta-amyloid precursor protein. *Proceedings of the National Academy of Sciences of the USA* **97**, 1456–60.

Liu K, Doms RW, and LeeVMY (2002). Glu11 site cleavage and N-terminally truncated A beta production upon BACE overexpression. *Biochemistry* **41**, 3128–36.

Lombardo JA, Stern EA, McLellan ME, *et al.* (2003). Amyloid-beta antibody treatment leads to rapid normalization of plaque-induced neuritic alterations. *Journal of Neuroscience* **23**, 10879–83.

Lopez-Perez E, Zhang Y, Frank SJ, Creemers J, Seidah N, and Checler F (2001). Constitutive alpha-secretase cleavage of the beta-amyloid precursor protein in the furin-deficient LoVo cell line: involvement of the pro-hormone convertase 7 and the disintegrin metalloprotease ADAM10. *Journal of Neurochemistry* **76**, 1532–9.

Lue LF, Kuo YM, Roher AE, *et al.* (1999). Soluble amyloid beta peptide concentration as a predictor of synaptic change in Alzheimer's disease. *American Journal of Pathology* **155**, 853–62.

Luo Y, Bolon B, Kahn S, *et al.* (2001). Mice deficient in BACE1, the Alzheimer's beta-secretase, have normal phenotype and abolished beta-amyloid generation. *Nature Neuroscience* **4**, 231–2.

McKee AC, Kosik KS, and Kowall NW (1991). Neuritic pathology and dementia in Alzheimer's disease. *Annals of Neurology* **30**, 156–65.

McLean CA, Cherny RA, Fraser FW, *et al.* (1999). Soluble pool of Abeta amyloid as a determinant of severity of neurodegeneration in Alzheimer's disease. *Annals of Neurology* **46**, 860–6.

McLoughlin DM and Miller CC (1996). The intracellular cytoplasmic domain of the Alzheimer's disease amyloid precursor protein interacts with phosphotyrosine-binding domain proteins in the yeast two-hybrid system. *FEBS Letters* **397**, 197–200.

McLoughlin DM, Standen CL, Lau KF, *et al.* (2001). The neuronal adaptor protein X11alpha interacts with the copper chaperone for SOD1 and regulates SOD1 activity. *Journal of Biological Chemistry* **276**, 9303–7.

Magara F, Muller U, Li ZW, *et al.* (1999). Genetic background changes the pattern of forebrain commissure defects in transgenic mice underexpressing the beta-amyloid-precursor protein. *Proceedings of the National Academy of Sciences of the USA* **96**, 4656–61.

Martin BL, Schrader-Fischer G, Busciglio J, Duke M, Paganetti P, and Yankner BA (1995). Intracellular accumulation of beta-amyloid in cells expressing the Swedish mutant amyloid precursor protein. *Journal of Biological Chemistry* **270**, 26727–30.

Martin LJ, Pardo CA, Cork LC, and Price DL (1994). Synaptic pathology and glial responses to neuronal injury precede the formation of senile plaques and amyloid deposits in the aging cerebral cortex. *American Journal of Pathology* **145**, 1358–81.

Masters CL, Simms G, Weinman NA, Multhaup G, McDonald BL, and Beyreuther K (1985). Amyloid plaque core protein in Alzheimer disease and Down syndrome. *Proceedings of the National Academy of Sciences of the USA* **82**, 4245–9.

Matsuda S, Matsuda Y, and D'Adamio L (2003). Amyloid beta protein precursor (AbetaPP), but not AbetaPP-like protein 2, is bridged to the kinesin light chain by the scaffold protein JNK-interacting protein 1. *Journal of Biological Chemistry* **278**, 38601–6.

Mattson MP, Cheng B, Culwell AR, Esch FS, Lieberburg I, and Rydel RE (1993). Evidence for excitoprotective and intraneuronal calcium-regulating roles for secreted forms of the beta-amyloid precursor protein. *Neuron* **10**, 243–54.

Maximov A, Sudhof TC, and Bezprozvanny I (1999). Association of neuronal calcium channels with modular adaptor proteins. *Journal of Biological Chemistry* **274**, 24453–6.

Meziane H, Dodart JC, Mathis C, *et al.* (1998). Memory-enhancing effects of secreted forms of the beta-amyloid precursor protein in normal and amnestic mice. *Proceedings of the National Academy of Sciences of the USA* **95**, 12683–8.

Mori C, Spooner ET, Wisniewsk KE, *et al.* (2002). Intraneuronal Abeta42 accumulation in Down syndrome brain. *Amyloid* **9**, 88–102.

Morin PJ, Abraham CR, Amaratunga A, *et al.* (1993). Amyloid precursor protein is synthesized by retinal ganglion cells, rapidly transported to the optic nerve plasma membrane and nerve terminals, and metabolized. *Journal of Neurochemistry* **61**, 464–73.

Mucke L, Masliah E, Yu GQ, *et al.* (2000). High-level neuronal expression of abeta 1–42 in wild-type human amyloid protein precursor transgenic mice: synaptotoxicity without plaque formation. *Journal of Neuroscience* **20**, 4050–8.

Mullan M, Crawford F, Axelman K, *et al.* (1992). A pathogenic mutation for probable Alzheimer's disease in the APP gene at the N-terminus of beta-amyloid. *Nature Genetics* **1**, 345–7.

Multhaup,G. (1994). Identification and regulation of the high affinity binding site of the Alzheimer's disease amyloid protein precursor (APP) to glycosaminoglycans. *Biochimie* **76**, 304–11.

Mumm JS, Schroeter EH, Saxena MT, *et al.* (2000). A ligand-induced extracellular cleavage regulates gamma-secretase-like proteolytic activation of Notch1. *Molecular Cell* **5**, 197–206.

Muresan Z and Muresan V (2005). Coordinated transport of phosphorylated amyloid-(beta) precursor protein and c-Jun NH2-terminal kinase-interacting protein-1. *Journal of Cell Biology* **171**, 615–25.

Nagy Z, Esiri MM, Jobst KA, *et al.* (1995). Relative roles of plaques and tangles in the dementia of Alzheimer's disease: correlations using three sets of neuropathological criteria. *Dementia* **6**, 21–31.

Naslund J, Haroutunian V, Mohs R, *et al.* (2000). Correlation between elevated levels of amyloid beta-peptide in the brain and cognitive decline. *Journal of the American Medical Association* **283**, 1571–7.

Ninomiya H, Roch JM, Sundsmo MP, Otero DA, and Saitoh T (1993). Amino acid sequence RERMS represents the active domain of amyloid beta/A4 protein precursor that promotes fibroblast growth. *Journal of Cell Biology* **121**, 879–86.

Oddo S, Caccamo A, Shepherd JD, *et al.* (2003). Triple-transgenic model of Alzheimer's disease with plaques and tangles: intracellular Abeta and synaptic dysfunction. *Neuron* **39**, 409–21.

Oishi M, Nairn AC, Czernik AJ, *et al.* (1997). The cytoplasmic domain of Alzheimer's amyloid precursor protein is phosphorylated at Thr654, Ser655, and Thr668 in adult rat brain and cultured cells. *Molecular Medicine* **3**, 111–23.

Oltersdorf T, Fritz LC, Schenk DB, *et al.* (1989). The secreted form of the Alzheimer's amyloid precursor protein with the Kunitz domain is protease nexin-II. *Nature* **341**, 144–7.

Oltersdorf T, Ward PJ, Henriksson T, *et al.* (1990). The Alzheimer amyloid precursor protein. Identification of a stable intermediate in the biosynthetic/degradative pathway. *Journal of Biological Chemistry* **265**, 4492–7.

Pardossi-Piquard R, Petit A, KawaraiT, *et al.* (2005). Presenilin-dependent transcriptional control of the Abeta-degrading enzyme neprilysin by intracellular domains of betaAPP and APLP. *Neuron* **46**, 541–54.

Parvathy S, Hussain I, Karran EH, Turner AJ, and Hooper NM (1999). Cleavage of Alzheimer's amyloid precursor protein by alpha-secretase occurs at the surface of neuronal cells. *Biochemistry* **38**, 9728–34.

Perez RG, Squazzo SL, and Koo EH (1996). Enhanced release of amyloid beta-protein from codon 670/671 'Swedish' mutant beta-amyloid precursor protein occurs in both secretory and endocytic pathways. *Journal of Biological Chemistry* **271**, 9100–7.

Perez RG, Zheng H, Van der Ploeg LH, and Koo EH (1997). The beta-amyloid precursor protein of Alzheimer's disease enhances neuron viability and modulates neuronal polarity. *Journal of Neuroscience* **17**, 9407–14.

Pietrzik CU, Busse T, Merriam DE, Weggen S, and Koo EH (2002). The cytoplasmic domain of the LDL receptor-related protein regulates multiple steps in APP processing. *EMBO Journal* **21**, 5691–700.

Pietrzik CU, Yoon IS, Jaeger S, Busse T, Weggen S, and Koo EH (2004). FE65 constitutes the functional link between the low-density lipoprotein receptor-related protein and the amyloid precursor protein. *Journal of Neuroscience* **24**, 4259–65.

Pike CJ, Walencewicz AJ, Glabe CG, and Cotman CW (1991a). Aggregation-related toxicity of synthetic beta-amyloid protein in hippocampal cultures. *European Journal of Pharmacology* **207**, 367–8.

Pike CJ, Walencewicz AJ, Glabe CG, and Cotman CW (1991b). *In vitro* aging of beta-amyloid protein causes peptide aggregation and neurotoxicity. *Brain Research* **563**, 311–14.

Ponte P, Gonzalez-DeWhitt P, Schilling J, *et al.* (1988). A new A4 amyloid mRNA contains a domain homologous to serine proteinase inhibitors. *Nature* **331**, 525–7.

Ponting CP, Hutton M, Nyborg A, Baker M, Jansen K, and Golde TE (2002). Identification of a novel family of presenilin homologues. *Human Molecular Genetics* **11**, 1037–44.

Postina R, Schroeder A, Dewachter I, *et al.* (2004). A disintegrin-metalloproteinase prevents amyloid plaque formation and hippocampal defects in an Alzheimer disease mouse model. *Journal of Clinical Investigation* **113**, 1456–64.

Ray WJ, Yao M, Mumm J, *et al.* (1999). Cell surface presenilin-1 participates in the gamma-secretase-like proteolysis of Notch. *Journal of Biological Chemistry* **274**, 36801–7.

Rechards M, Xia W, Oorschot VM, Selkoe DJ, and Klumperman J (2003). Presenilin-1 exists in both pre- and post-Golgi compartments and recycles via COPI-coated membranes. *Traffic* **4**, 553–565.

Richards SJ, Hodgman C, and Sharpe M (1995). Reported sequence homology between Alzheimer amyloid770 and the MRC OX-2 antigen does not predict function. *Brain Research Bulletin* **38**, 305–6.

Rogaev EI, Sherrington R, Rogaeva EA, *et al.* (1995). Familial Alzheimer's disease in kindreds with missense mutations in a gene on chromosome 1 related to the Alzheimer's disease type 3 gene. *Nature* **376**, 775–8.

Rogers J, Cooper NR, Webster S, *et al.* (1992). Complement activation by beta-amyloid in Alzheimer disease. *Proceedings of the National Academy of Sciences of the USA* **89**, 10016–20.

Rongo C, Whitfield CW, Rodal A, Kim SK, and Kaplan JM (1998). LIN-10 is a shared component of the polarized protein localization pathways in neurons and epithelia. *Cell* **94**, 751–9.

Rossner S, Ueberham U, Schliebs R, Perez-Polo JR, and Bigl V (1998). The regulation of amyloid precursor protein metabolism by cholinergic mechanisms and neurotrophin receptor signaling. *Progress in Neurobiology* **56**, 541–69.

Russo T, Faraonio R, Minopoli G, de Candia P, De Renzis S, and Zambrano N (1998). Fe65 and the protein network centered around the cytosolic domain of the Alzheimer's beta-amyloid precursor protein. *FEBS Letters* **434**, 1–7.

Sabo SL, Lanier LM, Ikin AF, *et al.* (1999). Regulation of beta-amyloid secretion by FE65, an amyloid protein precursor-binding protein. *Journal of Biological Chemistry* **274**, 7952–7.

Sabo SL, Ikin AF, Buxbaum JD, and Greengard P (2001). The Alzheimer amyloid precursor protein (APP) and FE65, an APP-binding protein, regulate cell movement. *Journal of Cell Biology* **153**, 1403–14.

Sastre M, Turner RS, and Levy E (1998). X11 interaction with beta-amyloid precursor protein modulates its cellular stabilization and reduces amyloid beta-protein secretion. *Journal of Biological Chemistry* **273**, 22351–7.

Scheff SW and Price DA (1993). Synapse loss in the temporal lobe in Alzheimer's disease. *Annals of Neurology* **33**, 190–9.

Schubert D, LaCorbiere M, Saitoh T, and Cole G. (1989). Characterization of an amyloid beta precursor protein that binds heparin and contains tyrosine sulfate. *Proceedings of the National Academy of Sciences of the USA* **86**, 2066–9.

Schubert W, Prior R, Weidemann A, *et al.* (1991). Localization of Alzheimer beta A4 amyloid precursor protein at central and peripheral synaptic sites. *Brain Research* **563**, 184–94.

Schwarzman AL, Singh N, Tsiper M, *et al.* (1999). Endogenous presenilin 1 redistributes to the surface of lamellipodia upon adhesion of Jurkat cells to a collagen matrix. *Proceedings of the National Academy of Sciences of the USA* **96**, 7932–7.

Setou M, Nakagawa T, Seog DH, and Hirokawa N (2000). Kinesin superfamily motor protein KIF17 and mLin-10 in NMDA receptor-containing vesicle transport. *Science* **288**, 1796–1802.

Seubert P, Vigo-Pelfrey C, Esch F, *et al.* (1992). Isolation and quantification of soluble Alzheimer's beta-peptide from biological fluids. *Nature* **359**, 325–7.

Shen J, Bronson RT, Chen DF, Xia W, Selkoe DJ, and Tonegawa S (1997). Skeletal and CNS defects in Presenilin-1-deficient mice. *Cell* **89**, 629–39.

Sheng JG, Price DL, and Koliatsos VE (2002). Disruption of corticocortical connections ameliorates amyloid burden in terminal fields in a transgenic model of Abeta amyloidosis. *Journal of Neuroscience* **22**, 9794–9.

Sherrington R, Rogaev EI, Liang Y, *et al.* (1995). Cloning of a gene bearing missense mutations in early-onset familial Alzheimer's disease. *Nature* **375**, 754–60.

Shoji M, Golde TE, Ghiso J, *et al.* (1992). Production of the Alzheimer amyloid beta protein by normal proteolytic processing. *Science* **258**, 126–9.

Siman R and Velji J (2003). Localization of presenilin–nicastrin complexes and gamma-secretase activity to the trans-Golgi network. *Journal of Neurochemistry* **84**, 1143–53.

Simeone A, Duilio A., Fiore F, *et al.* (1994). Expression of the neuron-specific FE65 gene marks the development of embryo ganglionic derivatives. *Developmental Neuroscience* **16**, 53–60.

Simske JS, Kaech SM, Harp SA, and Kim SK (1996). LET-23 receptor localization by the cell junction protein LIN-7 during *C.elegans* vulval induction. *Cell* **85**, 195–204.

Singh N, Talalayeva Y, Tsiper M, *et al.* 2001.The role of Alzheimer's disease-related presenilin 1 in intercellular adhesion. *Experimental Cell Research* **263**, 1–13,

Sisodia SS (1992). Beta-amyloid precursor protein cleavage by a membrane-bound protease. *Proceedings of the National Academy of Sciences of the USA* **89**, 6075–9.

Sisodia SS, Koo EH, Beyreuther K, Unterbeck A, and Price DL (1990). Evidence that beta-amyloid protein in Alzheimer's disease is not derived by normal processing. *Science* **248**, 492–5.

Skovronsky DM, Doms RW, and Lee VMY (1998). Detection of a novel intraneuronal pool of insoluble amyloid beta protein that accumulates with time in culture. *Journal of Cell Biology* **141**, 1031–9.

Skovronsky DM, Moore DB, Milla ME, Doms RW, and Lee VMY (2000). Protein kinase C-dependent alpha-secretase competes with beta-secretase for cleavage of amyloid-beta precursor protein in the trans-Golgi network. *Journal of Biological Chemistry* **275**, 2568–75.

Small DH, Nurcombe V, Reed G, *et al.* (1994). A heparin-binding domain in the amyloid protein precursor of Alzheimer's disease is involved in the regulation of neurite outgrowth. *Journal of Neuroscience* **14**, 2117–27.

Smith-Swintosky VL, Pettigrew LC, Craddock SD, Culwell AR, Rydel RE, and Mattson MP (1994). Secreted forms of beta-amyloid precursor protein protect against ischemic brain injury. *Journal of Neurochemistry* **63**, 781–4.

Snow AD, Kinsella MG, Parks E, *et al.* (1995). Differential binding of vascular cell-derived proteoglycans (perlecan, biglycan, decorin, and versican) to the beta-amyloid protein of Alzheimer's disease. *Archives of Biochemistry and Biophysics* **320**, 84–95.

Snyder EM, Nong Y, Almeida CG, *et al.* (2005). Regulation of NMDA receptor trafficking by amyloid-beta. *Nature Neuroscience* **8**, 1051–8.

Spires TL, Meyer-Luehmann M, Stern EA, *et al.* (2005). Dendritic spine abnormalities in amyloid precursor protein transgenic mice demonstrated by gene transfer and intravital multiphoton microscopy. *Journal of Neuroscience* **25**, 7278–87.

Stern EA, Bacskai BJ, Hickey GA, Attenello FJ, Lombardo JA, and Hyman BT (2004). Cortical synaptic integration *in vivo* is disrupted by amyloid-beta plaques. *Journal of Neuroscience* **24**, 4535–40.

Stokin GB, Lillo C, Falzone TL, *et al.* (2005). Axonopathy and transport deficits early in the pathogenesis of Alzheimer's disease. *Science* **307**, 1282–8.

Struhl G. and Greenwald I. (1999). Presenilin is required for activity and nuclear access of Notch in *Drosophila*. *Nature* **398**, 522–5.

Sturchler-Pierrat C, Abramowski D, Duke M, *et al.* (1997). Two amyloid precursor protein transgenic mouse models with Alzheimer disease-like pathology. *Proceedings of the National Academy of Sciences of the USA* **94**, 13287–92.

Sudol M, Chen HI, Bougeret C, Einbond A, and Bork P (1995). Characterization of a novel protein-binding module: the WW domain. *FEBS Letters* **369**, 67–71.

Suzuki N, Cheung TT, Cai XD, *et al.* (1994). An increased percentage of long amyloid beta protein secreted by familial amyloid beta protein precursor (beta APP717) mutants. *Science* **264**, 1336–40.

Suzuki T, Nairn AC, Gandy SE, and Greengard P (1992). Phosphorylation of Alzheimer amyloid precursor protein by protein kinase C. *Neuroscience* **48**, 755–61.

Takahashi RH, Milner TA, Li F, *et al.* (2002). Intraneuronal Alzheimer abeta42 accumulates in multivesicular bodies and is associated with synaptic pathology. *American Journal of Pathology* **161**, 1869–79.

Takahashi RH, Almeida CG, Kearney PF, *et al.* (2004). Oligomerization of Alzheimer's beta-amyloid within processes and synapses of cultured neurons and brain. *Journal of Neuroscience* **24**, 3592–9.

Takashima A, Sato M, Mercken M, *et al.* (1996). Localization of Alzheimer-associated presenilin 1 in transfected COS-7 cells. *Biochemical and Biophysical Research Communications* **227**, 423–6.

Tanahashi H and Tabira T. (1999). X11L2, a new member of the X11 protein family, interacts with Alzheimer's beta-amyloid precursor protein. *Biochemical and Biophysical Research Communications* **255**, 663–7.

Tanzi RE, Gusella JF, Watkins PC, *et al.* (1987). Amyloid beta protein gene: cDNA, mRNA distribution, and genetic linkage near the Alzheimer locus. *Science* **235**, 880–4.

Tanzi RE, McClatchey AI, Lamperti ED, Villa-Komaroff L, Gusella JF, and Neve RL (1988). Protease inhibitor domain encoded by an amyloid protein precursor mRNA associated with Alzheimer's disease. *Nature* **331**, 528–30.

Tarassishin L, Yin YI, Bassit B, and Li YM (2004). Processing of Notch and amyloid precursor protein by gamma-secretase is spatially distinct. *Proceedings of the National Academy of Sciences of the USA* **101**, 17050–5.

Terry RD, Masliah E, Salmon DP, *et al.* (1991). Physical basis of cognitive alterations in Alzheimer's disease: synapse loss is the major correlate of cognitive impairment. *Annals of Neurology* **30**, 572–80.

Thinakaran G, Slunt HH, and Sisodia SS (1995). Novel regulation of chondroitin sulfate glycosamino-glycan modification of amyloid precursor protein and its homologue, APLP2. *Journal of Biological Chemistry* **270**, 16522–5.

Thinakaran G, Teplow DB, Siman R, Greenberg B, and Sisodia SS (1996). Metabolism of the 'Swedish' amyloid precursor protein variant in neuro2a (N2a) cells. Evidence that cleavage at the 'beta-secretase' site occurs in the Golgi apparatus. *Journal of Biological Chemistry* **271**, 9390–9397.

Tienari PJ, Ida N, Ikonen E, *et al.* (1997). Intracellular and secreted Alzheimer beta-amyloid species are generated by distinct mechanisms in cultured hippocampal neurons. *Proceedings of the National Academy of Sciences of the USA* **94**, 4125–30.

Tomita S, Ozaki T, Taru H, *et al.* (1999). Interaction of a neuron-specific protein containing PDZ domains with Alzheimer's amyloid precursor protein. *Journal of Biological Chemistry* **274**, 2243–54.

Tomita T, Maruyama K, Saido TC, *et al.* (1997). The presenilin 2 mutation (N141I) linked to familial Alzheimer disease (Volga German families) increases the secretion of amyloid beta protein ending at the 42nd (or 43rd) residue. *Proceedings of the National Academy of Sciences of the USA* **94**, 2025–30.

Trommsdorff M, Borg JP, Margolis B, and Herz J (1998). Interaction of cytosolic adaptor proteins with neuronal apolipoprotein E receptors and the amyloid precursor protein. *Journal of Biological Chemistry* **273**, 33556–60.

Tsai J, Grutzendler J, Duff K and Gan WB. (2004). Fibrillar amyloid deposition leads to local synaptic abnormalities and breakage of neuronal branches. *Nature Neuroscience* **7**, 1181–3.

Turner RS, Suzuki N, Chyung AS, Younkin SG, and Lee VMY (1996). Amyloids beta40 and beta42 are generated intracellularly in cultured human neurons and their secretion increases with maturation. *Journal of Biological Chemistry* **271**, 8966–70.

Ulery PG, Beers J, Mikhailenko I, *et al.* (2000). Modulation of beta-amyloid precursor protein process-ing by the low density lipoprotein receptor-related protein (LRP). Evidence that LRP contributes to the pathogenesis of Alzheimer's disease. *Journal of Biological Chemistry* **275**, 7410–15.

Van Nostrand WE, Schmaier AH, Farrow JS, and Cunningham DD (1990a). Protease nexin-II (amyloid beta-protein precursor): a platelet alpha-granule protein. *Science* **248**, 745–8.

Van Nostrand WE, Wagner SL, Farrow JS, and Cunningham DD (1990b). Immunopurification and protease inhibitory properties of protease nexin-2/amyloid beta-protein precursor. *Journal of Biological Chemistry* **265**, 9591–4.

Vassar R, Bennett BD, Babu-Khan S, *et al.* (1999). Beta-secretase cleavage of Alzheimer's amyloid precursor protein by the transmembrane aspartic protease BACE. *Science* **286**, 735–41.

Vetrivel KS, Cheng H, Lin W, *et al.* (2004). Association of gamma-secretase with lipid rafts in post-Golgi and endosome membranes. *Journal of Biological Chemistry* **279**, 44945–54

Walhout AJ, Sordella R, Lu X, *et al.* (2000). Protein interaction mapping in *C.elegans* using proteins involved in vulval development. *Science* **287**, 116–22.

Walsh DM, Tseng BP, Rydel RE, Podlisny MB, and Selkoe DJ (2000). The oligomerization of amyloid beta-protein begins intracellularly in cells derived from human brain. *Biochemistry* **39**, 10831–9.

Walsh DM, Klyubin I, Fadeeva JV, *et al.* (2002). Naturally secreted oligomers of amyloid beta protein potently inhibit hippocampal long-term potentiation *in vivo*. *Nature* **416**, 535–9.

Walter J, Fluhrer R, Hartung B, *et al.* (2001). Phosphorylation regulates intracellular trafficking of beta-secretase. *Journal of Biological Chemistry* **276**, 14634–41.

Wang HW, Pasternak JF, Kuo H, *et al.* (2002). Soluble oligomers of beta amyloid (1–42) inhibit long-term potentiation but not long-term depression in rat dentate gyrus. *Brain Research* **924**, 133–40.

Wang H, Luo WJ, Zhang YW, *et al*. (2004). Presenilins and gamma-secretase inhibitors affect intracellular trafficking and cell surface localization of the gamma-secretase complex components. *Journal of Biological Chemistry* **279**, 40560–6.

Weidemann A, Konig G, Bunke D, *et al*. (1989). Identification, biogenesis, and localization of precursors of Alzheimer's disease A4 amyloid protein. *Cell* **57**, 115–26.

Weihofen A, Binns K, Lemberg MK, Ashman K, and Martoglio B (2002). Identification of signal peptide peptidase, a presenilin-type aspartic protease. *Science* **296**, 2215–18.

Wertkin AM, Turner RS, Pleasure SJ, *et al*. (1993). Human neurons derived from a teratocarcinoma cell line express solely the 695-amino acid amyloid precursor protein and produce intracellular beta-amyloid or A4 peptides. *Proceedings of the National Academy of Sciences of the USA* **90**, 9513–17.

Westerman MA, Cooper-Blacketer D, Mariash A, *et al*. (2002). The relationship between Abeta and memory in the Tg2576 mouse model of Alzheimer's disease. *Journal of Neuroscience* **22**, 1858–67.

Wild-Bode C, Yamazaki T, Capell A, *et al*. (1997). Intracellular generation and accumulation of amyloid beta-peptide terminating at amino acid 42. *Journal of Biological Chemistry* **272**, 16085–8.

Williamson TG, Mok SS, Henry A, *et al*. (1996). Secreted glypican binds to the amyloid precursor protein of Alzheimer's disease (APP) and inhibits APP-induced neurite outgrowth. *Journal of Biological Chemistry* **271**, 31215–21.

Wirths O, Weickert S, Majtenyi K, *et al*. (2000). Lewy body variant of Alzheimer's disease: alpha-synuclein in dystrophic neurites of A beta plaques. *Neuroreport* **11**, 3737–41.

Wolfe MS, Xia W, Ostaszewski BL, Diehl TS, Kimberly WT, and Selkoe DJ (1999). Two transmembrane aspartates in presenilin-1 required for presenilin endoproteolysis and gamma-secretase activity. *Nature* **398**, 513–17.

Wong PC, Zheng H, Chen H, *et al*. (1997). Presenilin 1 is required for Notch1 and DII1 expression in the paraxial mesoderm. *Nature* **387**, 288–292.

Xu H, Sweeney D, Wang R, *et al*. (1997). Generation of Alzheimer beta-amyloid protein in the trans-Golgi network in the apparent absence of vesicle formation. *Proceedings of the National Academy of Sciences of the USA* **94**, 3748–52.

Yan R, Bienkowski MJ, Shuck ME, *et al*. (1999). Membrane-anchored aspartyl protease with Alzheimer's disease beta-secretase activity. *Nature* **402** 533–7.

Yankner BA, Duffy LK, and Kirschner DA (1990). Neurotrophic and neurotoxic effects of amyloid beta protein: reversal by tachykinin neuropeptides. *Science* **250**, 279–82.

Ye Y, Lukinova N, and Fortini ME (1999). Neurogenic phenotypes and altered Notch processing in *Drosophila* Presenilin mutants. *Nature* **398**, 525–9.

Yu G, Nishimura M, Arawaka S, *et al*. (2000). Nicastrin modulates presenilin-mediated notch/glp-1 signal transduction and betaAPP processing. *Nature* **407**, 48–54.

Yu WH, Kumar A, Peterhoff C, *et al*. (2004). Autophagic vacuoles are enriched in amyloid precursor protein-secretase activities: implications for beta-amyloid peptide over-production and localization in Alzheimer's disease. *International Journal of Biochemistry and Cell Biology* **36**, 2531–40.

Zambrano N, Buxbaum JD, Minopoli G, *et al*. (1997). Interaction of the phosphotyrosine interaction/phosphotyrosine binding-related domains of Fe65 with wild-type and mutant Alzheimer's beta-amyloid precursor proteins. *Journal of Biological Chemistry* **272**, 6399–405.

Zambrano N, Minopoli G, de Candia P, and Russo T (1998). The Fe65 adaptor protein interacts through its PID1 domain with the transcription factor CP2/LSF/LBP1. *Journal of Biological Chemistry* **273**, 20128–33.

Zhang Z, Nadeau P, Song W, *et al*. (2000). Presenilins are required for gamma-secretase cleavage of beta-APP and transmembrane cleavage of Notch-1. *Nature Cell Biology* **2**, 463–5.

Zheng H, Jiang M, Trumbauer ME, *et al*. (1995). Beta-amyloid precursor protein-deficient mice show reactive gliosis and decreased locomotor activity. *Cell* **81**, 525–31.

Chapter 5

Molecular basis of tau protein pathology: role of abnormal hyperphosphorylation

Khalid Iqbal, Alejandra del C. Alonso,
M. Omar Chohan, Ezzat El-Akkad,
Cheng-Xin Gong, Sabiha Khatoon,
Fei Liu, and Inge Grundke-Iqbal

Tau protein pathology, seen as neurofibrillary degeneration, is a hallmark of Alzheimer's disease (AD) and related tauopathies. To date, the most established and the most compelling cause of dysfunctional tau in AD and related tauopathies is the abnormal hyperphosphorylation of tau. The abnormal hyperphosphorylation results not only in the loss of tau function of promoting assembly and stabilizing microtubules but also in a gain of a toxic function whereby the pathological tau sequesters not only normal tau, but also the other two neuronal microtubule-associated proteins (MAPs), MAP1A/MAP1B and MAP2, and causes inhibition and disruption of microtubules. This toxic gain of function of the pathological tau appears to be solely due to its abnormal hyperphosphorylation because dephosphorylation converts it functionally into a normal-like state. The sequestration of normal MAPs leads to a slow but progressive degeneration of the affected neurons. The hyperphosphorylation of tau results from both an imbalance between the activities of tau kinases and tau phosphatases and changes in the conformation of tau which affects its interaction with these enzymes. The affected neurons defend against the toxic tau by both continually synthesizing new normal tau and packaging the abnormally hyperphosphorylated tau into polymers, i.e. neurofibrillary tangles of paired helical filaments (PHFs), twisted ribbons, and straight filaments (SFs). The filamentous tau is inert; it neither interacts with tubulin and stimulates it assembly, nor binds to normal MAPs and causes disruption of microtubules.

5.1 Introduction

A neuron, with its long axonal and dendritic arborization, is strongly dependent on its microtubule network which, among other functions, is critical for axonal transport. In the brain, the microtubule network of neurons is maintained by three MAPs—tau, MAP1A/MAP1B, and MAP2—which apparently perform similar functions, i.e. promote assembly and stabilize microtubules. Tau is the major, the smallest in size, and apparently the most dynamic of the three MAPs. It is also apparently most vulnerable in a disease situation. The biological activity of tau, a phosphoprotein, in promoting the assembly and stability of microtubules is regulated by its degree of phosphorylation. Normal tau contains 2–3 moles of phosphate/mole of the protein (Köpke *et al.* 1993), the level of phosphorylation for its optimal activity. Hyperphosphorylation of tau depresses its microtubule assembly activity and its binding to microtubules (Lindwall and Cole 1984; Alonso *et al.* 1994).

Human brain tau is a family of six proteins derived from a single gene by alternative mRNA splicing (Goedert *et al.* 1989; Himmler *et al.* 1989). These proteins differ in whether they contain three (τ3L, τ3S, or τ3) or four (τ4L, τ4S, or τ4) tubulin binding domains (repeats, R) of 31 or 32 amino acids (aa) each near the C-terminal and two (τ3L, τ4L), one (τ3S, τ4S), or no (τ3, τ4) inserts of 29 aa each in the N-terminal portion of the molecule; the two amino-terminal inserts, 1 and 2, are coded by exon 2 and exon 3, respectively.

In AD and related disorders characterized by tau pathology, called tauopathies, tau is abnormally hyperphosphorylated and is accumulated as intraneuronal tangles of PHFs, twisted ribbons, and/or SFs (Grundke-Iqbal *et al.* 1986a,b; Iqbal *et al.* 1986, 1989; Lee *et al.* 1991). This hallmark brain lesion of these diseases directly correlates with dementia in these patients (Tomlinson *et al.* 1970; Alafuzoff *et al.* 1987; Arrigada *et al.* 1992). The aetiology and pathogenesis of neurofibrillary degeneration and therapeutic strategies to inhibit this lesion have been the subject of several recent reviews (Iqbal *et al.* 2000, 2003; Iqbal and Grundke-Iqbal 2004). In this chapter the molecular basis of tau pathology is described.

5.2 Dysregulation of tau phosphorylation

Since the discovery of tau as the major protein subunit of PHF in 1986 (Grundke-Iqbal *et al.* 1986a), a number of post-translational modifications, i.e. abnormal hyperphosphorylation, ubiquitination, glycation, N-glycosylation, O-GlcNAcylation, polyamination, nitration, and truncation, have been implicated in its pathology (reviewed by Gong *et al.* 2005). To date, the most established cause of dysfunction of tau in AD and related tauopathies is its abnormal hyperphosphorylation. Tau in AD brain is abnormally hyperphosphorylated and in this state is the major protein subunit of the PHFs/SFs which form neurofibrillary tangles, a hallmark lesion of this disease (Grundke-Iqbal *et al.* 1986a,b; Iqbal *et al.* 1986, 1989; Lee *et al.* 1991). Tau pathology, which is seen only as accumulation of abnormally hyperphosphorylated protein, is also seen in several other human neurodegenerative disorders, including frontotemporal dementia, Pick's disease, corticobasal degeneration, dementia pugilistica, and progressive supranuclear palsy. In every one of these disorders, called tauopathies, the accumulation of the abnormally hyperphosphorylated tau is associated with neurofibrillary degeneration and dementia. The discovery of mutations in the *tau* gene and their cosegregation with the disease in the inherited frontotemporal dementia with parkinsonism linked to chromosome 17 (FTDP-17) has established that abnormalities in tau protein as a primary event can lead to neurodegeneration and dementia (Hutton *et al.* 1998; Poorkaj *et al.* 1998; Spillantini *et al.* 1998).

The abnormal hyperphosphorylation of tau appears to precede its accumulation in the affected neurons in AD. *In vitro* hyperphosphorylation promotes tau's assembly into bundles of PHFs and SFs (Alonso *et al.* 2001a, 2004). Induction of hyperphosphorylation of tau in metabolically active rat brain slices by inhibition of protein phosphatase 2A (PP-2A) activity with okadaic acid, and in normal adult rats by activation of protein kinase-A, leads to accumulation of tau (Gong *et al.* 2000; Liu *et al.* 2004a; Sun *et al.* 2005). The abnormally hyperphosphorylated tau was discovered not only in neurofibrillary tangles (Grundke-Iqbal *et al.* 1986a) but also in cytosol from AD brains (Iqbal *et al.* 1986). Quantitative immunocytochemical studies with mAb Tau-1 have revealed deposits of only abnormally phosphorylated tau, but not normal tau, in neurons without tangles (stage 0 tangles) in both AD and normal aged hippocampi (Bancher *et al.* 1989, 1991). Tau in tangles, mostly ghost tangles, is known to be ubiquitinated (Mori *et al.* 1987; Perry *et al.* 1987; Grundke-Iqbal *et al.* 1988), whereas the abnormally hyperphosphorylated tau isolated from AD brain cytosol was found to have no ubiquitin reactivity. All these studies suggest that the abnormal hyperphosphorylation of tau precedes its accumulation into neurofibrillary tangles (Köpke *et al.* 1993).

One of the possibilities is that the abnormal hyperphosphorylation of tau might be due to conformational change(s) in tau in the diseased brain, which might make it a better substrate for phosphorylation and/or a worse substrate for dephosphorylation. Davies and his colleagues have developed a series of monoclonal antibodies to conformational alterations of tau, and, employing these antibodies, have shown that tau is conformationally altered in AD (Jicha *et al.* 1997, 1999a,b) and in transgenic mice overexpressing human tau (Duff *et al.* 2000). In inherited cases of FTDP-17, where the disease is caused by certain missense mutations in tau, these mutations make tau a more favourable substrate for hyperphosphorylation by brain protein kinases (Alonso *et al.* 2004; Bhaskar *et al.* 2005); such a scenario is less likely in AD because tau is not the only neuronal protein which is hyperphosphorylated in AD as a result of the protein phosphorylation–dephosphorylation imbalance. Biochemically, tubulin and neurofilaments (Vijayan *et al.* 2001; Wang *et al.* 2001) and immunohistochemically neurofilaments and MAP1B (Sternberger *et al.* 1985; Hasegawa *et al.* 1990; Ulloa *et al.* 1994) have been found to be hyperphosphorylated in AD brain. Furthermore, both the cytosolic and PHF abnormally hyperphosphorylated taus are readily dephosphorylated by phosphatases *in vitro* (Grundke-Iqbal *et al.* 1986b; Iqbal *et al.* 1989, 1994; Wang *et al.* 1995, 1996a; 1998).

Neurofibrillary degeneration of the Alzheimer type is seen only sparsely in aged animals and in experimentally induced conditions. To date, none of the mutations in β-amyloid precursor protein (β-APP), presenilin-1, or presenilin-2, which have been found to cause familial AD, have been shown to produce AD-like extensive tau pathology in transgenic mice overexpressing these human mutant proteins (Games *et al.* 1995; Hsiao *et al.* 1996). On the other hand, overexpression of FTDP-17 mutant tau, as well its coexpression with APP/PS1 mutations in transgenic mice, has been found to produce neurofibrillary tangles of SFs/PHFs of abnormally hyperphosphorylated tau (e.g. Lewis *et al.* 2000;

Götz *et al.* 2001; Tanemura *et al.* 2001; Oddo *et al.* 2004). A recent study has shown that, on hyperphosphorylation, murine tau self-assembles into tangles of filaments (PHFs/SFs) as readily as the corresponding human brain tau (Chohan *et al.* 2005), suggesting that the protein phosphorylation–dephosphorylation system is probably more stable and resistant to changes in lower-order species than in higher-order species, such as humans. Consistent with these suggestions, overexpression of p25, the activator of cdk5 in transgenic mice which promotes the hyperphosphorylation of tau, has been found to result in self-assembly of filaments, although sparsely (Cruz *et al.* 2003; Noble *et al.* 2003).

Tau in AD and other tauopathies appears to be mostly intact (Grundke-Iqbal *et al.* 1986a,b; Iqbal *et al.* 1986, 1989; Lee *et al.* 1991; Goedert *et al.* 1992). However, immuno-histochemically, tau in AD neurofibrillary tangles has been shown to be truncated at both Glu 391 and Ser-421 (Novak *et al.* 1991; Gamblin *et al.* 2003; Cotman *et al.* 2005). These truncated taus have been shown to be associated with apoptosis in cultured cells (Fasulo *et al.* 2000; Rissman *et al.* 2004). However, what percentage of tau is truncated at these sites at what stage of neurofibrillary pathology in AD brain has not been reported to date. Furthermore, unlike the monomeric truncated tau employed in the cell biological studies, this protein polymerized in neurofibrillary tangles/PHF might not have any biological activity. Since Alzheimer neurofibrillary degeneration takes place over a period of several months to years, it should not be surprising to have certain truncated taus in AD brain, particularly resulting from neurofibrillary tangles which are exposed to hydrolases, in both the affected neurons and the ghost tangles in extracellular space (Skrabana *et al.* 2004). Both N- and C-terminal regions flanking the microtubule binding domains of tau are inhibitory to its self-assembly into filaments (Alonso *et al.* 2001a). Thus neutralization of these inhibitory domains by abnormal hyperphosphorylation, a major mechanism probably involved in AD and related tauopathies, or partially by truncation might result in the formation of neurofibrillary tangles (see Alonso *et al.* 2001a). Consistent with this hypothesis, a transgenic rat model overexpressing truncated human tau has been shown to produce a significant number of neurofibrillary tangles and tau in these lesions is abnormally hyperphosphorylated (Hrnkova *et al.* 2004).

5.3 Assembly of tau into filaments

Normal tau is immunohistochemically negative under most standard conditions of tissue fixation, whereas the opposite is true for tau aggregates. The latter is the only state in which tau pathology has been found to date. In AD and other tauopathies, tau aggregates are seen as PHFs, twisted ribbons, or SFs of various diameters ranging from ~2.1 nm (protofilaments) to ~15 nm (Ruben *et al.* 1992). In both human diseases (AD and other tauopathies) and experimentally-induced animal conditions, the tau aggregates are invariably made up of abnormally hyperphosphorylated protein (reviewed by Iqbal *et al.* 2000, 2003; Iqbal and Grundke-Iqbal 2004). Thus, ever since the discovery of tau as the major protein subunit of PHFs/SFs (Grundke-Iqbal *et al.* 1986a,b), there has been considerable interest in understanding the aggregation of tau into filaments.

All six tau isoforms are present in a hyperphosphorylated state in PHFs from AD brain (Grundke-Iqbal *et al.* 1986a,b; Iqbal *et al.* 1986, 1989; Lee *et al.* 1991; Goedert et al. 1992).

In AD brain, abnormally hyperphosphorylated tau is present both as a cytosolic protein (Iqbal *et al.* 1986; Harada *et al.* 1994) and polymerized into PHFs (Grundke-Iqbal *et al.* 1986a,b; Iqbal *et al.* 1989; Lee *et al.* 1991). Unlike normal tau, which contains two or three phosphate groups, the cytosolic hyperphosphorylated tau from AD brain (AD P-tau) contains 5–9 moles of phosphate per mole of protein (Köpke *et al.* 1993).

In vitro assembly of tau into SF- and PHF-like structures has been achieved under different conditions, such as urea treatment for 60 hours, incubation with unsaturated free fatty acids, tRNA, heparin, or polyglutamic acid, employing a tau fragment, tau concentrations up to 12 mg/ml, and incubation times up to several days (Montejo de Garcini *et al.* 1986; Crowther *et al.* 1992; Wille *et al.* 1992; Schweers *et al.* 1995; Wilson and Binder 1995; Goedert *et al.* 1996; Kampers *et al.* 1996; Yanagawa *et al.* 1998; Friedhoff *et al.* 2000; Perez *et al.* 2000; von Bergen *et al.* 2000). However, none of these conditions used for tau assembly is consistent with the presence of all six tau isoforms in an abnormally hyperphosphorylated state in PHFs as entire or nearly entire protein molecules. Although heparan sulphate and hyperphosphorylated tau coexist in neurons in AD brain (Goedert *et al.* 1996), the disassembly of PHF/neurofibrillary tangles by dephosphorylation (Wang *et al.* 1995) and the self-assembly of recombinant tau by hyperphosphorylation alone suggest that sulphated glycosaminoglyans might not play a critical role in neurofibrillary degeneration.

Abnormally hyperphosphorylated tau (0.4 mg/ml) isolated from AD brain polymerizes into PHF/SF tangles *in vitro* at pH 6.9 under reducing conditions at 35°C over a period of 90 minutes (Alonso *et al.* 2001a), and these self-assembly conditions, which are consistent with the findings in AD and other tauopathies, do not require any cofactor. This self-assembly of tau requires hyperphosphorylation because dephosphorylation inhibits it. Unlike dephosphorylation, deglycosylation of AD tau does not inhibit its ability to self-assemble into filaments (Wang *et al.* 1996b). Furthermore, on *in vitro* hyperphosphorylation, each of the six recombinant human brain tau isoforms self-assemble into PHFs/SFs. All these findings taken together suggest that the abnormal hyperphosphorylation is probably required to cause the assembly of tau into filaments and might be the molecular mechanism involved in the formation of tau lesions in AD and other tauopathies.

Tau is an unusual protein which has long stretches of (positively and negatively) charged regions that are not conducive to intermolecular hydrophobic association (Ruben *et al.* 1991). Of the four microtubule binding repeats in tau, the predicted amino acids with β-structure are concentrated in R2 and R3 (Von Bergen *et al.* 2000) and can self-assemble into filaments *in vitro*; R2 and R3 have also been shown to co-assemble with heparin into PHFs (Arrasate *et al.* 1999). It is likely that, because of the way the charged regions are located, the rest of the molecule has an inhibitory effect upon the self-polymerization of tau. Of all of the tau isoforms, this inhibitory effect seems to be the least in τ4L. The N-terminal inserts are highly acidic, and their presence markedly neutralizes the basic charge of tau. For instance, the theoretical isoelectric points of τ4, τ4S, and τ4L are 9.46, 9.86, and 8.24, respectively. The presence of the extra repeat, the R2, and the two N-terminal inserts probably promotes the intermolecular hydrophobic interaction in τ4L sufficiently to result in its self-assembly into PHFs, and hyperphosphorylation further enhances this process. The abnormal hyperphosphorylation that occurs in AD and other tauopathies neutralizes the basic inhibitory charge of tau. Most of the sites at which tau is hyperphosphorylated flank the microtubule binding domains. Neutralization of the basic charge by hyperphosphorylation in these flanking regions probably neutralizes their inhibitory effect and allows tau to self-assemble into filaments. However, the nature of the neutralization by the two N-terminal inserts and by the abnormal hyperphosphorylation is most probably different, as evidenced by the formation of filaments with different morphologies.

Tau probably self-assembles by intermolecular hydrophobic interaction and through its microtubule binding repeat R3 (in the case of 3R taus), and R2 and R3 (in the case of 4R taus), but only when the rest of the molecule (i.e. the amino-terminal and carboxyl-terminal regions flanking the repeats, which are inhibitory) is neutralized. In AD and related tauopathies, these inhibitory regions are neutralized by abnormal hyperphosphorylation.

The FTDP-17 mutations appear to alter conformation of the protein such that it becomes a more favourable substrate for brain protein kinases (Alonso *et al.* 2004). The mutated taus are more rapidly hyperphosphorylated and can self-assemble at a lower level of hyperphosphorylation than the wild type tau.

5.4 Molecular basis of neurofibrillary degeneration

In AD brains the levels of tau, but not the mRNA for this protein, are four- to eightfold increased compared with age-matched control brains, and this increase is in the form of the abnormally hyperphosphorylated tau (Khatoon *et al.* 1992, 1994). The abnormally hyperphosphorylated tau is found in AD brain in two subcellular pools: (i) polymerized into neurofibrillary tangles of PHFs mixed with SFs; (ii) in a non-fibrillized form in the cytosol (Iqbal *et al.* 1986; Bancher *et al.* 1989; Köpke *et al.* 1993). The tau polymerized into neurofibrillary tangles is apparently inert and behaves like normal tau in promoting microtubule assembly only on enzymatic dephosphorylation *in vitro* when released from PHFs/tangles (Iqbal *et al.* 1994; Wang *et al.* 1995). In contrast, the cytosolic abnormally hyperphosphorylated tau (AD P-tau), which can be as much as ~ 40% of the total abnormal tau in AD brain (Köpke *et al.* 1993), does not interact with tubulin/microtubules but instead sequesters normal tau, MAP1A/MAP1B, and MAP2, causing inhibition and disassembly of microtubules *in vitro* (Alonso *et al.* 1994, 1996, 1997, 2006). The association between AD P-tau and normal tau is not saturable and *in vitro* results in the formation of tangles of ~ 2.1 mm filaments (Alonso *et al.* 1996). The association between AD P-tau and MAP1A/MAP1B or MAP2 is weaker than that between the AD P-tau and normal tau and does not result in the formation of filaments (Alonso *et al.* 1997). This toxic property of the AD P-tau appears to be solely due to its abnormal hyperphosphorylation because dephosphorylation by alkaline phosphatase, protein phosphatase (PP)-2A, PP-2B and to a lesser degree by PP-1 converts the abnormal tau into a normal-like protein in promoting the microtubule assembly *in vitro* (Alonso *et al.* 1994, 1996, 1997;

Iqbal *et al.* 1994; Wang *et al.* 1995, 1996a). The sequestration of functional tau by the abnormally hyperphosphorylated tau causes disruption of microtubule network and thereby leads to neurodegeneration.

Several missense mutations in tau cosegregate with the disease in FTDP-17 (Hutton *et al.* 1998; Poorkaj *et al.* 1998; Spillantini *et al.* 1998). Four of these missense mutations, G272V, P301L, V337M and R406W which have been studied to date make tau a more favourable substrate than the wild-type human tau for abnormal hyperphosphorylation by brain protein kinases *in vitro* (Alonso *et al.* 2006). These mutated taus become hyper-phosphorylated at a faster rate and self-aggregate into filaments more readily, i.e. at a phosphorylation stoichiometry of 4–6 as compared with 10 or more in the case of the wild-type protein. These faster kinetics of the hyperphosphorylation of the mutated tau might explain a relatively early onset, severity and autosomal dominance of the disease in the inherited FTDP-17 cases.

The six human tau isoforms, τ4RL (4R, 2N), τ4S (4R, 1N), τ4 (4R, 0N), τ3RL (3R, 2N), τ3RS (3R, 1N), and τ3 (3R, 0N), also called fetal tau, are differentially sequestered by AD P-tau, *in vitro* (Alonso *et al.* 2001b). The association of AD P-tau to normal human brain recombinant taus is τ4RL > τ4RS > τ4R and τ3RL > τ3RS > τ3, and τ4RL > τ3RL. AD P-tau also inhibits the assembly and disrupts microtubules pre-assembled with each tau isoform with an efficiency which corresponds directly to the degree of interaction with these isoforms. *In vitro* hyperphosphorylation of recombinant tau converts it into an AD-P-tau-like state in sequestering normal tau and inhibiting microtubule assembly. The preferential sequestration of 4R taus and taus with amino terminal inserts explains both (i) why fetal brain (fetal tau is with 3R and no N) is protected from Alzheimer neurofibrillary pathology and (ii) why intronic mutations seen in certain inherited cases of FTDP-17, which result in alternate splicing of tau mRNA and consequently an increase in 4R:3R ratio, lead to neurofibrillary degeneration and the disease. *In vitro*, hyperphosphorylated tau sequesters normal tau at a phosphorylation stoichiometry of ~4 and above, whereas it requires a stoichiometry of 10 or more to self-aggregate into filaments (Alonso *et al.* 2004). On aggregation into filaments tau loses its ability to sequester normal tau. Furthermore, AD P-tau, but not PHFs, inhibits regeneration of microtubule network in detergent-extracted PC12 cells, indicating that the formation of filaments might be initiated as a self-defence response by the affected neurons (see Alonso *et al.* 2004).

The abnormal hyperphosphorylation of tau makes it resistant to proteolysis by the calcium-activated neutral protease (Wang *et al.* 1995, 1996a); this is probably the reason why the levels of tau are several-fold increased in AD (Khatoon *et al.* 1992, 1994). Some increase in tau level in AD brain can also result from the activation of p70 S6 kinase which upregulates the translation of tau (An *et al.* 2003; Pei *et al.* 2006). It is likely that, in order to neutralize the AD P-tau's ability to sequester normal MAPs and cause disassembly of microtubules, the affected neurons promote the self-assembly of the abnormal tau into tangles of PHFs. The fact that the tangle-bearing neurons seem to survive for many years (Morsch *et al.* 1999) is consistent with such a self-defence role of the formation of tangles. The AD P-tau readily self-assembles into PHF/SF tangles *in vitro* under

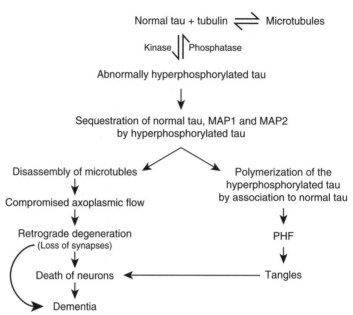

Fig. 5.1 Molecular mechanism of neurofibrillary degeneration. Normal tau interacts with tubulin, stimulating its assembly and stabilizing microtubules. In AD brain, because of an imbalance in tau kinase and phosphatase activities and a change in its conformation induced by other post-translational changes or mutations as in inherited cases of FTDP-17, tau becomes abnormally hyperphosphorylated. The abnormally hyperphosphorylated tau resulting from any one of the above causes behaves as an inhibitory/toxic protein; not only is it unable to stimulate micro-tubule assembly and bind to microtubules, but it also sequesters normal tau, MAP 1A/MAP1B, and MAP2, and leads to inhibition of assembly and disruption of microtubules. The breakdown of the microtubule network in the affected neurons compromises axonal transport, leading to retrograde degeneration which, in turn, results in dementia. The association between the AD P-tau and normal tau in the presence of glycosylation results in the formation of neurofibrillary tangles. The tangles are ubiquitinated for degradation by the non-lysosomal ubiquitin pathway, but apparently this degradation, if any, is minimal. Unlike the non-polymerized abnormally hyperphosphorylated tau, the neurofibrillary tangles are inert but, with disease progression, these lesions grow in size and eventually may physically choke the affected cells to death.

physiological conditions of protein concentration, pH, ionic strength, and reducing conditions (Alonso *et al.* 2001a). Furthermore, dephosphorylation inhibits the self-assembly of AD P-tau into PHFs/SFs, and the *in vitro* abnormal hyperphosphorylation of each of the six recombinant human brain tau isoforms promotes their assembly into PHF/SF tangles. Thus, all these studies taken together demonstrate the pivotal involvement of abnormal hyperphosphorylation in neurofibrillary degeneration (Fig. 5.1).

5.5 Factors involved in abnormal hyperphosphorylation of tau

A number of factors might be responsible for the abnormal hyperphosphorylation of tau. The most important of these factors are changes in activities of one or more of the tau kinases and/or phosphatases that regulate phosphorylation of tau under normal conditions, and conformational changes in tau produced either by a mutation or by one or more post-translational modifications.

5.5.1 Tau kinases

The state of phosphorylation of a phosphoprotein is a function of the balance between the activities of the protein kinases and the protein phosphatases that regulate its phosphorylation. Tau, which is phosphorylated at over 30 serine/threonine residues in AD (Morishima-Kawashima *et al.* 1995; Hanger *et al.* 1998), is a substrate for several protein kinases (Singh *et al.* 1994; Johnson and Hartigan 1998). Among these, glycogen synthase kinase-3 (GSK-3), cyclin-dependent protein kinase-5 (cdk5), protein kinase A (PKA), calcium- and calmodulin-dependent protein kinase-II (CaMKII), mitogen-activated protein (MAP) kinase ERK 1/2, and stress-activated protein kinases have been most implicated in the abnormal hyperphosphorylation of tau (Pei *et al.* 2003). MARK/Par-1 has been shown to phosphorylate tau *in vitro* and in transgenic mice and *Drosophila* at Ser-262/356, although the exact physiological role of this kinase in brain has not yet been established (Drewes *et al.* 1997; Nishimura *et al.* 2004; Matenia *et al.* 2005). A large number of the abnormally hyperphosphorylated sites in tau are proline directed, i.e. serine/threonine followed by proline which are canonical sites of proline-directed protein kinases (PDPKs). All three major PDPKs (GSK-3β, cdk5, and ERK 1/2) have been shown to phosphorylate tau at a large number of the same sites seen in AD.

GSK-3β and cdk5 phosphorylate tau at a large number of sites, most of which are common to the two enzymes (Wang *et al.* 1998; Anderton *et al.* 2001; Liu *et al.* 2002a). Expression of GSK-3β and cdk5 is high in the brain (Woodgett 1990; Tsai *et al.* 1993; Lew *et al.* 1994) and both enzymes have been shown to be associated with all stages of neurofibrillary pathology in AD (Pei *et al.* 1998, 1999). Overexpression of GSK-3β in cultured cells and in transgenic mice results in hyperphosphorylation of tau at several of the same sites seen in AD, and inhibition of this enzyme by lithium chloride attenuates phosphorylation in these models (Lovestone *et al.* 1996; Stambolic *et al.* 1996; Wagner *et al.* 1996; Hong *et al.* 1997; Spittaels *et al.* 2000; Lucas *et al.* 2001; Perez *et al.* 2003; Tatebayashi *et al.* 2004).

The enzyme cdk5 requires for its activity interaction with p39 or p35 or, better, their proteolytic products p29 or p25, respectively, which are generated in post-mitotic neurons by digestion with calpains (Kusakawa *et al.* 2000; Patzke and Tsai 2002). Overexpression of p25 in transgenic mice, which results in an increase in the activity of this enzyme, also produces hyperphosphorylation of tau (Cruz *et al.* 2003; Noble *et al.* 2003).

Members of the MAP kinase family, which includes ERK1, ERK2, p70S6 kinase, and the stress-activated kinases JNK and p38 kinase, have been shown to phosphorylate tau at several of the same sites as the abnormally hyperphosphorylated tau, indicating that these enzymes are associated with the progression of neurofibrillary degeneration in AD (Drewes *et al.* 1992; Ledesma *et al.* 1992; Lu and Wood 1993; Roder *et al.* 1993; Pei *et al.* 2001, 2003; An *et al.* 2003; Kins *et al.* 2003; Pearson *et al.* 2006).

Unlike the PDPKs, the non-PDPKs have been shown to phosphorylate tau at only a few of the sites. CaMKII phosphorylates tau at Ser-262/356 (Singh *et al.* 1996; Sironi *et al.* 1998; Bennecib *et al.* 2001) and at Ser-416 (Steiner *et al.* 1990). Both PKA and MARK kinase have also been shown to phosphorylate tau at Ser-262 (Scott *et al.* 1993; Brandt *et al.* 1994; Drewes *et al.* 1995, 1997; Sironi *et al.* 1998). PKA phosphorylates tau at Ser-214, Ser-217, Ser-262, Ser-324, Ser-356, Ser-396/404, and Ser-416 (Litersky *et al.* 1996; Wang *et al.* 1998). However, phosphorylation of tau by these non-PDPKs markedly increases its phosphorylation by GSK-3β (Singh *et al.* 1995; Wang *et al.* 1998; Cho and Johnson 2003). The priming of tau by PKA appears to be sufficient to enhance the abnormal hyperphosphorylation of tau by the basal level of GSK-3β activity in normal adult rat brain and leads to an impairment of spatial memory in these animals (Liu *et al.* 2004?a/b). Although, to date, none of the activities of these protein kinases have been reproducibly shown to be upregulated in AD brain, transient stimulation of these enzymes, especially the priming kinases such as PKA or CaMKII, might be sufficient to result in the abnormal hyperphosphorylation of tau. Recent studies have shown that tau is also phosphorylated at tyrosines by src family non-receptor tyrosine kinases (reviewed by Lee 2005), and that FTDP-17 mutations increase association of tau with Fyn (Bhaskar *et al.* 2005).

5.5.2 Tau phosphatases

The activities of protein phosphatases 2A and 1 (PP-2A and PP-1) are compromised by ~20–30% in AD brain (Gong *et al.* 1993, 1995), and the phosphorylation of tau that suppresses its microtubule binding and assembly activities in adult mammalian brain is regulated by PP-2A and not by PP-2B (Gong *et al.* 2000; Bennecib *et al.* 2001). PP-2A, which accounts for ~70% of the tau Ser/Thr phosphatase activity in human brain (Liu *et al.* 2005), also regulates the activities of several tau kinases in brain. Inhibition of PP-2A activity by okadaic acid in cultured cells and in metabolically active rat brain slices results in abnormal hyperphosphorylation of tau at several of the same sites as in AD, not only directly by a decrease in dephosphorylation but also indirectly by promoting the activities of CaMKII (Bennecib *et al.* 2001), PKA (Tanaka *et al.* 1998; Li *et al.* 2004), MAP kinase kinase (MEK1/2), extracellular regulated kinase (ERK 1/2) and P70S6 kinase (An *et al.* 2003; Pei *et al.* 2003). Thus, barring the fact that tau is not the only neuronal substrate of these protein kinases and phosphatases, it should be possible to inhibit the abnormal hyperphosphorylation of tau by inhibiting the activity of one or more tau kinases and/or restoring or upregulating the activity of PP-2A.

Although the brain has several tau phosphatase activities (Cheng *et al.* 2000, 2001; Rahman *et al.* 2005, 2006), PP-2A and PP-1 produce more than 90% of the serine/threonine protein phosphatase activity in mammalian cells (Oliver and Shenolikar 1998). The intracellular activities of these enzymes are regulated by endogenous inhibitors. PP-1 activity is regulated mainly by a 18.7 kDa heat-stable protein called inhibitor-1 (I-1) (Cohen *et al.* 1988; Cohen 1989). In addition, a structurally related protein, DARPP-32 (dopamine and cAMP-regulated phosphoprotein of apparent molecular weight 32 000), is expressed predominantly in the brain (Walaas and Greengard 1991).

I-1 and DARPP-32 are activated on phosphorylation by protein kinase A and inactivated at basal calcium level by PP-2A. Thus inhibition of PP-2A activity would keep I-1/DARPP-32 in active form and thereby result in a decrease in PP-1 activity. In AD brain a reduction in PP-2A activity might decrease the PP-1 activity by allowing the upregulation of the I-1/DARPP-32 activity.

The activity of PP-2A in mammalian tissue is regulated by two heat-stable proteins: I_1^{PP2A}, a 30 kDa cytosolic protein (Li *et al.* 1995) which inhibits PP-2A with a K_i of 30 nM, and I_2^{PP2A}, a 39 kDa nuclear protein which inhibits PP-2A with a K_i of 23 nM (Li *et al.* 1996a). Both I_1^{PP2A} and I_2^{PP2A} have been cloned from human kidney (Li *et al.* 1996a,b) and brain (Tsujio *et al.* 2005). I_1^{PP2A} has been found to be the same protein as the putative histocompatibility leucocyte antigen class II associated protein (PHAP-1). This protein, which has also been described as mapmodulin, pp32, and LANP (Ulitzur *et al.* 1997) is 249 amino acids long and has an apparent molecular weight of 30 kDa on SDS–PAGE. I_2^{PP2A}, which is the same as TAF-1b or PHAPII, is a nuclear protein that is a homologue of the human SETa protein (von Lindern *et al.* 1992). In AD brain, the neo-cortical levels of I_1^{PP2A} and I_2^{PP2A} are increased by ~20%, and there is a cleavage and shift from nuclear to cytoplasmic localization of I_2^{PP2A} (Tanimukai *et al.* 2005). These changes in I_1^{PP2A} and I_2^{PP2A} may be responsible for the downregulation of PP-2A and hyperphos-phorylation of tau in AD brain.

5.5.3 Changes in tau conformation

In addition to the activities of the tau kinases and phosphatases, the phosphorylation of tau is also regulated by its conformational state. Free tau is more readily hyperphospho-rylated than the microtubule-bound tau. The rate and extent of tau phosphorylation by PKA, CaMKII, C-kinase, casein kinase I (CK-I), cdk5, and GSK-3 are dependent on its initial phosphorylation state. For instance, when recombinant human brain tau is prephosphorylated by one of several non-PDPKs (i.e. PKA, CaMKII, or C-kinase) or by cdk5, a PDPK, then its subsequent phosphorylations catalysed by GSK-3 are stimulated several-fold (Singh *et al.* 1995; Sengupta *et al.* 1997). In addition, the rate and extent to which various tau isoforms are phosphorylated also depend on whether tau contains three repeats or four repeats and zero, one, or two N-terminal inserts (Singh *et al.* 1997).

In addition to abnormal hyperphosphorylation, tau is also abnormally glycosylated and the latter appears to precede the former in AD brain (Wang *et al.* 1996b; Liu *et al.* 2002a). *In vitro* studies indicate that the abnormal glycosylation promotes tau phospho-rylation with PKA, GSK-3β, and cdk5, and inhibits dephosphorylation of tau with PP2A and PP5 (Liu *et al.* 2002b,c). In addition, like some other neuronal phosphoproteins, tau is also O-GlcNAcylated (Arnold *et al.* 1996). In contrast with classical N- or O-glycosylation, O-GlcNAcylation, which involves the addition of a single sugar at serine/threonine residues of a protein, dynamically post-translationally modifies cytoplasmic and nuclear proteins in a manner analogous to protein phosphorylation (see Hart 1997). O-GlcNAcylation and phosphorylation reciprocally regulate each other. In AD the O-GlcNAcylation of tau is significantly reduced, probably because of impaired glucose uptake/metabolism, and decreased glucose metabolism in cultured cells and in mice, which decreases the O-GlcNAcylation of tau, produces abnormal hyperphosphorylation of this protein (Liu *et al.* 2004b).

In inherited frontotemporal dementia linked to chromosome 17 (FTDP-17), certain mutations in the tau gene co-segregate with the disease (Hutton *et al.* 1998; Poorkaj *et al.* 1998; Spillantini *et al.* 1998). The most studied of these mutations are the missense mutations G272V, P301L, V337M, and R406W. Tau with these mutations is a more favourable substrate for hyperphosphorylation than wild-type tau; the mutated taus are hyperphosphorylated much faster and polymerize into filaments at lower stoichiometry than identically treated wild-type tau (Alonso *et al.* 2004). Thus all the studies described in this section, taken together, suggest that, in addition to the levels of the activities of tau kinases and phosphatases, the phosphorylation of tau is regulated at the substrate (tau) level.

5.6 Critical abnormally phosphorylated sites

Tau is abnormally hyperphosphorylated at over 30 sites in AD. However, not all of these sites may be involved in converting normal tau into a toxic molecule. Identification of these critical sites has been most difficult. Phosphorylation of tau at Ser-262, Thr-231, and Ser-235 inhibits its binding to microtubules by ~35%, ~25%, and 10%, respectively (Sengupta *et al.* 1998). Hyperphosphorylation of tau at the level of 4–6 moles phosphate per mole of protein induces the toxic property where it sequesters normal tau (Alonso *et al.* 2004). Additional phosphorylation to a level of ~10 moles phosphate per mole of protein is required to induce its self-assembly into filaments. Taus with FTDP-17 mutations (G272V, P301L, V337M, and R406W) are phosphorylated much faster than wild-type tau and self-assemble at lower levels of phosphorylation than the wild-type protein. Time kinetics of phosphorylation of these mutated and wild-type taus at various abnormally phosphorylated sites and the ability of these proteins to bind normal tau suggest that Ser-199/202/205, Thr-212, Thr-231/Ser-235, Ser-262/356, and Ser-404 are among the critical sites that convert tau to a toxic-like protein. Further phosphorylation at Thr-231, Ser-396, and Ser-422 promotes self-assembly of tau into filaments (Alonso *et al.* 2004). These sites are known to be substrates of PKA, CaMKII, GSK-3β, and cdk5, among other protein kinases.

5.7 Conclusions

In conclusion, the abnormal hyperphosphorylation of tau is critically involved in its pathology which leads to neurofibrillary degeneration and dementia in AD and related tauopathies. The abnormal hyperphosphorylation of tau leads to neurodegeneration by sequestration of normal MAPs and disruption of the microtubule network. In contrast, tau filaments (PHFs/SFs) are apparently inert, and thus may represent a defence approach of affected neurons.

Acknowledgments

We are grateful to Janet Murphy and Sonia Warren for secretarial assistance. Studies in our laboratories were supported in part by the New York State Office of Mental

Retardation and Developmental Disabilities and NIH grants AG019158 and AG16760, Alzheimer's Association (Chicago, IL) grants IIRG-00–2002 and NIG-03–4721, and a grant from the Institute for the Study of Aging (ISOA), New York.

References

Alafuzoff I, Iqbal K, Friden H, Adolfsson R, and Winblad B (1987). Histopathological criteria for progressive dementia disorders: clinical-pathological correlation and classification by multivariate data analysis. *Acta Neuropathologica (Berlin)* **74**, 209–25.

Alonso A del C, Zaidi T, Grundke-Iqbal I, and Iqbal K (1994). Role of abnormally phosphorylated tau in the breakdown of microtubules in Alzheimer disease. *Proceedings of the National Academy of Sciences of the USA* **91**, 5562–6.

Alonso A del C, I. Grundke-Iqbal I, and Iqbal K (1996). Alzheimer's disease hyperphosphorylated tau sequesters normal tau into tangles of filaments and disassembles microtubules. *Nature Medicine* **2**, 783–7.

Alonso A, Grundke-Iqbal I, Barra HS, and Iqbal K (1997). Abnormal phosphorylation of tau and the mechanism of Alzheimer neurofibrillary degeneration: sequestration of MAP1 and MAP2 and the disassembly of microtubules by the abnormal tau. *Proceedings of the National Academy of Sciences of the USA* **94**, 298–303.

Alonso A del C, Zaidi T, Novak M, Grundke-Iqbal I, and Iqbal K (2001a). Hyperphosphorylation induces self-assembly of tau into tangles of paired helical filaments/straight filaments. *Proceedings of the National Academy of Sciences of the USA* **98**, 6923–8.

Alonso A, Zaidi T, Wu Q, *et al.* (2001b). Interaction of tau isoforms with Alzheimer's disease abnormally hyperphosphorylated tau and *in vitro* phosphorylation into the disease-like protein. *Journal of Biological Chemistry* **276**, 37967–73.

Alonso A del C, Mederlyova A, Novak M, Grundke-Iqbal I, and Iqbal K (2004). Promotion of hyperphos-phorylation by frontotemporal dementia tau mutations. *Journal of Biological Chemistry* **279**, 34878–81.

Alonso A del C, Li B, Grundke-Iqbal I, Iqbal K. (2006). Polymerization of hyperphosphorylated tau into filaments eliminates its inhibitory activity. *Proceedings of the National Academy of Sciences of the USA* **103**, 8864–9.

An WL, Cowburn RF, Li L, *et al.* (2003). Up-regulation of phosphorylated/activated p70 S6 kinase and its relationship to neurofibrillary pathology in Alzheimer's disease. *American Journal of Pathology* **163**, 591–607.

Anderton BH, Betts J, Blackstock WP, *et al.* (2001). Sites of phosphorylation in tau and factors affecting their regulation. *Biochemical Society Symposium* **67**, 73–80.

Arnold CS, Johnson GV, Cole RN, Dong DL, Lee M, and Hart GW (1996). The microtubule-associated protein tau is extensively modified with O-linked *N*-acetylglucosamine. *Journal of Biological Chemistry* **271**, 28741–4.

Arrasate M, Perez M, Armas-Portela R, and J Avila (1999). Polymerization of tau peptides into fibrillar structures. The effect of FTDP-17 mutations. *FEBS Letters* **446**, 199–202.

Arrigada PA, Growdon JH, Hedley-White ET, and Hyman BT (1992). Neurofibrillary tangles but not senile plaques parallel duration and severity of Alzheimer's disease. *Neurology* **42**, 631–9.

Bancher C, Brunner C, Lassmann H, *et al.* (1989). Accumulation of abnormally phosphorylated tau precedes the formation of neurofibrillary tangles in Alzheimer's disease. *Brain Research* **477**, 90–9.

Bancher C, Grundke-Iqbal I, Iqbal K, Fried VA, Smith HT, and Wisniewski HM (1991). Abnormal phosphorylation of tau precedes ubiquitination in neurofibrillary pathology of Alzheimer disease. *Brain Research* **539**, 11–8.

Bennecib M, Gong CX, Grundke-Iqbal I, and Iqbal K (2001). Inhibition of PP-2A upregulates CaMKII in rat forebrain and induces hyperphosphorylation of tau at Ser 262/356. *FEBS Letters* **490**, 15–22.

Bhaskar K, Yen SH, and Lee G (2005). Disease-related modifications in tau affect the interaction between Fyn and tau. *Journal of Biology Chemistry* **280**, 35119–25.

Brandt R, Lee G, Teplow DB, Shalloway D, and Abdel-Ghany M (1994). Differential effect of phosphorylation and substrate modulation on tau's ability to promote microtubule growth and nucleation. *Journal of Biological Chemistry* **269**, 11776–82.

Cheng LY, Wang JZ, Gong CX, *et al.* (2000). Multiple forms of phosphates from human brain: isolation and partial characterization of affi-gel blue binding phosphatases. *Neurochemical Research* **25**, 107–20.

Cheng LY, Wang JZ, Gong CX, *et al.* (2001). Multiple forms of phosphatase from human brain isolation and partial characterization of affi-gel blue nonbinding phosphatase activities. *Neurochemical Research* **26**, 425–38.

Cho JH and Johnson GV (2003). Glycogen synthase kinase 3beta phosphorylates tau at both primed and unprimed sites. Differential impact on microtubule binding. *Journal of Biological Chemistry* **278**, 187–93.

Chohan MO, Haque N, Alonso A, *et al.* (2005). Hyperphosphorylation-induced self assembly of murine tau: a comparison with human tau. *Journal of Neural Transmission* **112**, 1035–47.

Cohen P (1989). The structure and regulation of protein phosphatases. *Annual Review of Biochemistry* **58**, 453–508.

Cohen P, Alemany S, Hemmings BA, Resink TJ, Stralfors P, and Tung HYL (1988). Protein phosphatase-1 and protein phosphatase-2A from rabbit skeletal muscle. *Methods in Enzymology* **159**, 390–408.

Cotman CW, Poon WW, Rissman RA. and Blurton-Jones M (2005). The role of caspase cleavage of tau in Alzheimer disease neuropathology. *Journal of Neuropathology and Experimental Neurology* **64**, 104–12.

Crowther RA, Olesen OF, Jakes R, and Goedert M (1992). The microtubule binding repeats of tau protein assemble into filaments like those found in Alzheimer's disease. *FEBS Letters* **309**, 199–202.

Cruz JC, Tseng HC, Goldman JA, Shih H, and Tsai LH (2003). Aberrant cdk5 activation by p25 triggers pathological events leading to neurodegeneration and neurofibrillary tangles. *Neuron* **40**, 471–83.

Drewes G, Lichtenberg-Kraag B, Döring F, *et al.* (1992). Mitogen activated protein (MAP) kinase transforms tau protein into an Alzheimer-like state. *EMBO Journal* **11**, 2131–8.

Drewes G, Trinczek B, Illenberger, S *et al.* (1995). Microtubule-associated protein/microtubule affinity-regulating kinase (p110mark). A novel protein kinase that regulates tau-microtubule interactions and dynamic instability by phosphorylation at the Alzheimer-specific site serine 262. *Journal of Biological Chemistry* **270**, 7679–88.

Drewes G, Ebneth A, Preuss U, Mandelkow EM, and Mandelkow E (1997). MARK, a novel family of protein kinases that phosphorylate microtubule-associated proteins and trigger microtubule disruption. *Cell* **89**, 297–308.

Duff K, Knight H, Refolo LM, *et al.* (2000). Characterization of pathology in transgenic mice overexpressing human genomic and cDNA tau transgenes. *Neurobiology of Disease* **7**, 87–98.

Fasulo L, Ugolini G, Visintin M, *et al.* (2000). The neuronal microtubule-associated protein tau is a substrate for caspase-3 and an effector of apoptosis. *Journal of Neurochemistry* **75** 624–33.

Friedhoff P, von Bergen M, Mandelkow EM, and Mandelkow E (2000). Structure of tau protein and assembly into paired helical filaments. *Biochimica et Biophysica Acta* **1502**, 122–32.

Gamblin TC, Chen F, Zambrano A, *et al.* (2003). Caspase cleavage of tau: linking amyloid and neurofibrillary tangles in Alzheimer's disease. *Proceedings of the National Academy of Sciences of the USA* **100**, 10032–7.

Games D, Adams D, Alessandrini R, *et al.* (1995). Alzheimer-type neuropathology in transgenic mice overexpressing V717F beta-amyloid precursor protein. *Nature* **373**, 523–7.

Goedert M, Spillantini MG, Jakes R, Rutherford D, and Crowther RA (1989). Multiple isoforms of human microtubule-associated protein tau: sequences and localization in neurofibrillary tangles of Alzheimer's disease. *Neuron* **3**, 519–26.

Goedert M, Spillantini MG, Cairns NJ, and Crowther RA (1992). Tau proteins of Alzheimer paired helical filaments: abnormal phosphorylation of all six brain isoforms. *Neuron* **8**, 159–68.

Goedert M, Jakes R, Spillantini MG, Hasegawa M, Smith MJ, and Crowther RA (1996). Assembly of microtubule-associated protein tau into Alzheimer-like filaments induced by sulphated glycosaminoglycans. *Nature* **383**, 550–3.

Gong CX, Singh TJ, Grundke-Iqbal I, and Iqbal K (1993). Phosphoprotein phosphatase activities in Alzheimer disease brain. *Journal of Neurochemistry* **61**, 921–7.

Gong CX, Shaikh S, Wang JZ, Zaidi T, Grundke-Iqbal I, and Iqbal K (1995). Phosphatase activity towards abnormally phosphorylated t: decrease in Alzheimer disease brain. *Journal of Neurochemistry* **65**, 732–8.

Gong CX, Lidsky T, Wegiel J, Zuck L, Grundke-Iqbal I, and Iqbal K (2000). The phosphorylation state of microtubule-associated protein tau is regulated by protein phosphatase 2A in mammalian brain. *Journal of Biological Chemistry* **275**, 5534–44.

Gong CX, Liu F, Grundke-Iqbal I, and Iqbal K (2005). Post-translational modifications of tau protein in Alzheimer's Disease. *Journal of Neural Transmission* **112**, 813–38.

Götz J, Chen F, van Dorpe J, and Nitsch RM (2001). Formation of neurofibrillary tangles in P301l tau transgenic mice induced by Abeta 42 fibrils. *Science* **293**, 1491–5.

Grundke-Iqbal I, Iqbal K, Quinlan M, Tung YC, Zaidi MS, and Wisniewski HM (1986a). Microtubule-associated protein tau: a component of Alzheimer paired helical filaments, *Journal of Biological Chemistry* **261**, 6084–9.

Grundke-Iqbal I, Iqbal K, Tung YC, Quinlan M Wisniewski HM, and Binder LI (1986b). Abnormal phosphorylation of the microtubule-associated protein (tau) in Alzheimer cytoskeletal pathology. *Proceedings of the National Academy of Sciences of the USA* **93**, 4913–17.

Grundke-Iqbal I, Vorbrodt AW, Iqbal K, Tung YC, Wang GP, and Wisniewski HM (1988). Microtubule-associated polypeptides tau are altered in Alzheimer paired helical filaments. *Molecular Brain Research* **4**, 43–52.

Hanger DP, Betts JB, Loviny TL, Blackstock WP, and Anderton BH (1998). New phosphorylation sites identified in hyperphosphorylated tau (paired helical filament-tau) from Alzheimer's disease brain using nanoelectrospray mass spectrometry. *Journal of Neurochemistry* **71**, 2465–76.

Harada A, Oguchi K, Okabe S, *et al.* (1994). Altered microtubule organization in small-calibre axons of mice lacking tau protein. *Nature* **369**, 488–91.

Hart GW (1997). Dynamic O-linked glycosylation of nuclear and cytoskeletal proteins. *Annual Review of Biochemistry* **66**, 315–35.

Hasegawa M, Arai T, and Ihara Y (1990). Immunochemical evidence that fragments of phosphorylated MAP5 (MAP1B) are bound to neurofibrillary tangles in Alzheimer's disease. *Neuron* **4**, 909–18.

Himmler A, Drechsel D, Kirschner MW, and Martin DW, Jr (1989). Tau consists of a set of proteins with repeated C-terminal microtubule binding domains and variable N-terminal domains. *Molecular and Cellular Biology* **9**, 1381–8.

Hong M, Chen DC, Klein PS, and Lee VM (1997). Lithium reduces tau phosphorylation by inhibition of glycogen synthase kinase-3. *Journal of Biological Chemistry* **272**, 25326–32.

Hrnkova M, Zilka N, Filipcik P, and Novak M (2004). Cognitive deficit and progressive motor impairment in AD rat model. *Neurobiology of Aging* **25**, S233.

Hsiao K, Chapman P, Nilsen S, *et al.* (1996). Correlative memory deficits, Abeta elevation, and amyloid plaques in transgenic mice. *Science* **274**, 99–102.

Hutton M, Lendon CL, Rizsazu P, *et al.* (1998). Association of missense and 5'-splice-site mutations in tau with the inherited dementia FTDP-17. *Nature* **393**, 702–5.

Iqbal K and Grundke-Iqbal I (2004). Inhibition of neurofibrillary degeneration: A promising approach to Alzheimer's disease and other tauopathies. *Current Drug Targets* **5**, 495–502.

Iqbal K, Grundke-Iqbal I, Zaidi T, *et al.* (1986). Defective brain microtubule assembly in Alzheimer's disease. *Lancet* **ii**, 421–6.

Iqbal K, Grundke-Iqbal I, Smith AJ, George L, Tung YC, and Zaidi T (1989). Identification and localization of a tau peptide to paired helical filaments of Alzheimer disease. *Proceedings of the National Academy of Sciences of the USA* **86**, 5646–50.

Iqbal K, Zaidi T, Bancher C, and Grundke-Iqbal I (1994). Alzheimer paired helical filaments: Restoration of the biological activity by dephosphorylation. *FEBS Letters* **349**, 104–8.

Iqbal K, Alonso A del C, Gondal JA, *et al.* (2000). Mechanism of neurofibrillary degeneration and pharmacologic therapeutic approach. *Journal of Neural Transmission* **59**, 213–22.

Iqbal K, Alonso A del C, El-Akkad E, *et al.* (2003). Alzheimer neurofibrillary degeneration, Significance, mechanism and therapeutic targets. In Iqbal K and Winblad B (eds), *Alzheimer's Disease and Related Disorders: Research Advances*, pp. 277–292 (Ana Aslan Academy of Aging, Bucharest).

Jicha GA, Lane E, Vincent I, Otvos L, Jr, Hoffman R, and Davies P (1997). A conformation- and phosphorylation-dependent antibody recognizing the paired helical filaments of Alzheimer's disease. *Journal of Neurochemistry* **69**, 2087–95.

Jicha GA, Berenfeld B, and Davies P (1999a). Sequence requirements for formation of conformational variants of tau similar to those found in Alzheimer's disease. *Journal of Neuroscience Research* **55**, 713–23.

Jicha GA, Rockwood JM, Berenfeld B, Hutton M, and Davies P (1999b). Altered conformation of recombinant frontotemporal dementia-17 mutant tau proteins. *Journal of Neuroscience Letters* **260**, 153–6.

Johnson GVW and Hartigan JA (1998). Tau protein in normal and Alzheimer's disease brain: an update. *Alzheimer's Disease Reviews* **3**, 125–41.

Kampers T, Friedhoff P, Biernat J, Mandelkow EM, and Mandelkow E (1996). RNA stimulates aggregation of microtubule-associated protein tau into Alzheimer-like paired helical filaments. *FEBS Letters* **399**, 344–9.

Khatoon S, Grundke-Iqbal I, and Iqbal K (1992). Brain levels of microtubule associated protein tau are elevated in Alzheimer's disease brain: a radioimmunoslot-blot assay for nanograms of the protein. *Journal of Neurochemistry* **59**, 750–3.

Khatoon S, Grundke-Iqbal I, and Iqbal K (1994). Levels of normal and abnormally phosphorylated tau in different cellular and regional compartments of Alzheimer disease and control brains. *FEBS Letters* **351**, 80–4.

Kins S, Kurosinski P, Nitsch RM, and Gotz J (2003). Activation of the ERK and JNK signaling pathways caused by neuron-specific inhibition of PP-2A in transgenic mice. *American Journal of Pathology* **163**, 833–43.

Köpke E, Tung YC, Shaikh S, Alonso A del C, Iqbal K, and Grundke-Iqbal I (1993). Microtubule associated protein tau: abnormal phosphorylation of a non-paired helical filament pool in Alzheimer disease, *Journal of Biological Chemistry* **268**, 24374–84.

Kusakawa G, Saito T, Onuki R, Ishiguro K, Kishimoto T, and Hisanaga S (2000). Calpain-dependent proteolytic cleavage of the p35 cyclin-dependent kinase 5 activator to p25. *Journal of Biological Chemistry* **275**, 17166–72.

Ledesma MD, Correas I, Avila J, and Diaz-Nido J (1992). Implication of brain cdc2 and MAP2 kinases in the phosphorylation of tau protein in Alzheimer's disease. *FEBS Letters* **308**, 218–24.

Lee G (2005). Tau and src family tyrosine kinases. *Biochimica et Biophysica Acta* **1739**, 323–30.

Lee VMY, Balin BJ, Otvos L, Jr, and Trojanowski JQ (1991). A68: A major subunit of paired helical filaments and derivatized forms of normal tau. *Science* **251**, 675–8.

Lew J, Huang QQ, Qi Z, *et al.* (1994). A brain-specific activator of cyclin-dependent kinase-5. *Nature* **371**, 423–6.

Lewis J, McGowan E, Rockwood J, *et al.* (2000). Neurofibrillary tangles, amyotrophy and progressive motor disturbance in mice expressing mutant (P301L) tau protein. *Nature Genetics* **25**, 402–5.

Li L, Sengupta A, Haque N, Grundke-Iqbal I, and Iqbal K (2004). Memantine inhibits and reverses the Alzheimer type abnormal hyperphosphorylation of tau and associated neurodegeneration. *FEBS Letters* **566**, 261–9.

Li M, Guo H and Damuni Z (1995). Purification and characterization of two potent heat-stable protein inhibitors of protein phosphatase 2A from bovine kidney. *Biochemistry* **34**, 1988–96.

Li M, A. Makkinje A, and Damuni Z (1996a). Molecular identification of I1PP2A, a novel potent heat-stable inhibitor protein of protein phosphatase 2A. *Biochemistry* **35**, 6998–7002.

Li M, Makkinje A, and Damuni Z (1996b). The myeloid leukemia-associated protein SET is a potent inhibitor of protein phosphatase 2A. *Journal of Biological Chemistry* **271**, 11059–62.

Lindwall G and Cole R.D (1984). Phosphorylation affects the ability of tau protein to promote microtubule assembly. *Journal of Biological Chemistry* **259**, 5301–5.

Litersky JM, Johnson GV, Jakes R, Goedert M, Lee M, and Seubert P (1996). Tau protein is phosphory-lated by cyclic AMP-dependent protein kinase and calcium/calmodulin-dependent protein kinase II within its microtubule binding domains at Ser-262 and Ser-356. *Biochemical Journal* **316**, 655–60.

Liu F, Zaidi T, Iqbal K, Grundke-Iqbal I, Merkle RK, and Gong CX (2002a). Role of glycosylation in hyperphosphorylation of tau in Alzheimer's disease. *FEBS Letters* **512**, 101–6.

Liu F, Zaidi T, Grundke-Iqbal I, Iqbal K, and Gong CX (2002b). Aberrant glycosylation modulates phosphorylation of tau by protein kinase A and dephosphorylation of tau by protein phosphatase 2A and 5. *Neuroscience* **115**, 829–37.

Liu F, Grundke-Iqbal I, Iqbal K, and Gong CX (2002c). Involvement of aberrant glycosylation in phosphorylation of tau by cdk5 and GSK-3β. *FEBS Letters* **530**, 209–14.

Liu SJ, Zhang JY, Li HL, *et al.* (2004a). Tau becomes a more favorable substrate for GSK-3 when it is prephosphorylated by PKA in rat brain. *Journal of Biological Chemistry* **279**, 50078–88.

Liu F, Iqbal K, Grundke-Iqbal I, Hart GW, and CX Gong (2004b). O-GlcNAcylation regulates phospho-rylation of tau: a novel mechanism involved in Alzheimer's disease. *Proceedings of the National Academy of Science USA* **101**, 10804–9.

Liu F, Grundke-Iqbal I, Iqbal K and Gong, CX (2005). Contributions of protein phosphatases PP1, PP2A, PP2B and PP5 to the regulation of tau phosphorylation. *European Journal of Neuroscience* **22**, 1942–50.

Lovestone S, Hartley CL, Pearce J, and Anderton BH (1996). Phosphorylation of tau by glycogen synthase kinase-3 beta in intact mammalian cells: the effects on the organization and stability of microtubules. *Neuroscience* **73**, 1145–57.

Lu Q and Wood JG (1993). Functional studies of Alzheimer's disease tau protein. *Journal of Neuroscience* **13**, 508–15.

Lucas JJ, Hernandez F, Gomez-Ramos P, Moran MA, Hen R, and Avila J (2001). Decreased nuclear beta-catenin, tau hyperphosphorylation and neurodegeneration in GSK-3beta conditional transgenic mice. *EMBO Journal* **20**, 27–39.

Matenia D, Griesshaber B, Li XY, *et al.* (2005). PAK5 kinase is an inhibitor of MARK/Par-1, which leads to stable microtubules and dynamic actin. *Molecular and Cellular Biology* **16**, 4410–22.

Montejo de Garcini E, Serrano L, and Avila J (1986). Self assembly of microtubule associated protein tau into filaments resembling those found in Alzheimer disease. *Biochemical and Biophysical Research Communications* **141**, 790–6.

Mori H, Kondo J, and Ihara Y (1987). Ubiquitin is a component of paired helical filaments in Alzheimer's disease. *Science* **235**, 1641–4.

Morishima-Kawashima M, Hasegawa M, Takio K, *et al.* (1995). Proline-directed and non-proline-directed phosphorylation of PHF-tau. *Journal of Biological Chemistry* **270**, 823–9.

Morsch R, Simon W, and Coleman PD (1999). Neurons may live for decades with neurofibrillary tangles. *Journal of Neuropathology and Experimental Neurology* **58**, 188–97.

Nishimura I, Yang Y, and Lu B (2004). PAR-1 kinase plays an initiator role in a temporally ordered phosphorylation process that confers tau toxicity in *Drosophila*. *Cell* **116**, 671–82.

Noble W, Olm V, Takata K, *et al.* (2003). Cdk5 is a key factor in tau aggregation and tangle formation *in vivo*. *Neuron* **38**, 555–65.

Novak M, Jakes R, Edwards PC, Milstein C, and Wischik CM (1991). Difference between the tau protein of Alzheimer paired helical filament core and normal tau revealed by epitope analysis of monoclonal antibodies 423 and 7.51. *Proceedings of the National Academy of Sciences of the USA* **88**, 5837–41.

Oddo S, Billings L, Kesslak JP, Cribbs DH, and LaFerla FM (2004). Abeta immunotherapy leads to clearance of early, but not late, hyperphosphorylated tau aggregates via the proteasome. *Neuron* **43**, 321–32.

Oliver CJ and Shenolikar S (1998). Physiologic importance of protein phosphatase inhibitors. *Frontiers in Bioscience* **3**, 961–72.

Patzke H and Tsai LH (2002). Calpain-mediated cleavage of the cyclin-dependent kinase-5 activator p39 to p29. *Journal of Biological Chemistry* **277**, 8054–60.

Pearson AG, Byrne UT, MacGibbon GA, Faull RL, and Dragunow M (2006). Activated c-Jun is present in neurofibrillary tangles in Alzheimer's disease brains. *Neuroscience Letters* **398**, 246–50.

Pei JJ, Grundke-Iqbal I, Iqbal K, Bogdanovic N, Winblad B, and Cowburn R (1998). Accumulation of cyclin-dependent kinase-5 (cdk5) in neurons with early stages of Alzheimer's disease neurofibrillary degeneration. *Brain Research* **797**, 267–77.

Pei JJ, Braak E, Braak H, *et al.* (1999). Distribution of active glycogen synthase kinase 3β (GSK-3β) in brains staged for Alzheimer's disease neurofibrillary changes. *Journal of Neuropathology and Experimental Neurology* **58**, 1010–19.

Pei JJ, Braak E, Braak H, *et al.* (2001). Localization of active forms of C-jun kinase (JNK) and p38 kinase in Alzheimer's disease brains at different stages of neurofibrillary degeneration. *Journal of Alzheimer's Disease* **3**, 41–8.

Pei JJ, Gong CX, An WL, *et al.* (2003). Okadaic-acid-induced inhibition of protein phosphatase 2A produces activation of mitogen-activated protein kinases ERK 1/2, MEK 1/2, and p70 S6, similar to that in Alzheimer's disease. *American Journal of Pathology* **163**, 845–58.

Pei JJ, An WL, Zhou XW, *et al.* (2006). +/-P70 S6 kinase mediates tau phosphorylation and synthesis. *FEBS Letters* **580**, 107–14.

Perez M, Valpuesta JM, Medina M, Montejo de Garcini E, and Avila J (2000). Phosphorylated, but not native, tau protein assembles following reaction with the lipid peroxidation product, 4-hydroxy-2-nonenal. *FEBS Letters* **486**, 270–4.

Perez M, Hernandez F, Lim F, Diaz-Nido J, and Avila J (2003). Chronic lithium treatment decreases mutant tau protein aggregation in a transgenic mouse model. *Journal of Alzheimer's Disease* **5**, 301–8.

Perry G, Friedman R, Shaw G, and Chau V (1987). Ubiquitin is detected in neurofibrillary tangles and senile plaque neurites of Alzheimer disease brains *Proceedings of the National Academy of Sciences of the USA* **84**, 3033–6.

Poorkaj P, Bird TD, Wijsman E, Nemens E, *et al.* (1998). Tau is a candidate gene for chromosome 17 frontotemporal dementia, *Annals of Neurology* **43**, 815–25.

Rahman A, Grundke-Iqbal I, and Iqbal K (2005). Phosphothreonine-212 of Alzheimer abnormally hyperphosphorylated tau is a preferred substrate of protein phosphatase-1. *Neurochemical Research* **30**, 277–87.

Rahman A, Grundke-Iqbal I, and Iqbal K (2006). PP2B isolated from human brain preferentially dephosphorylates Ser-262 and Ser-396 of the Alzheimer Disease abnormally hyperphosphorylated tau. *Journal of Neural Transmission* **113**, 219–30.

Rissman RA, Poon WW, Blurton-Jones M, *et al.* (2004). Caspase-cleavage of tau is an early event in Alzheimer disease tangle pathology. *Journal of Clinical Investigation* **114**, 121–30.

Roder HM, Eden PA, and Ingram VM (1993). Brain protein kinase PK40erk converts tau into a PHF-like form as found in Alzheimer disease. *Biochemical and Biophysical Research Communications* **193**, 639–47.

Ruben GC, Iqbal K, Grundke-Iqbal I, Wisniewski HM, Ciardelli TL, and Johnson JE (1991). The microtubule associated protein tau forms a triple-stranded left-hand helical polymer. *Journal of Biological Chemistry* **266**, 22019–27.

Ruben GC, Iqbal K, Wisniewski HM, Johnson JE, Jr, and Grundke-Iqbal I (1992). Alzheimer neurofibrillary tangles contain 2.1 nm filaments structurally identical to the microtubule associated protein tau: a high resolution transmission electron microscope study of tangles and senile plaque core amyloid. *Brain Research* **590**, 164–79.

Schweers O, Mandelkow EM, Biernat J, and Mandelkow E (1995). Oxidation of cysteine-322 in the repeat domain of microtubule-associated protein t controls the *in vitro* assembly of paired helical filaments. *Proceedings of the National Academy of Sciences of the USA* **92**, 8463–7.

Scott CW, Spreen RC, Herman JL, *et al.* (1993). Phosphorylation of recombinant tau by cAMP-dependent protein kinase. Identification of phosphorylation sites and effect on microtubule assembly. *Journal of Biological Chemistry* **268**, 1166–73.

Sengupta A, Wu Q, Grundke-Iqbal I, Iqbal K, and Singh TJ (1997). Potentiation of GSK-3-catalyzed Alzheimer-like phosphorylation of human tau by cdk5. *Molecular and Cellular Biochemistry* **167**, 99–105.

Sengupta A, Kabat J, Novak M, Wu Q, Grundke-Iqbal I, and Iqbal K (1998). Phosphorylation of tau at both Thr 231 and Ser 262 is required for maximal inhibition of its binging to microtubules. *Archives of Biochemistry and Biophysics* **2**, 299–309.

Singh TJ, Grundke-Iqbal I, McDonald B, and Iqbal K (1994). Comparison of the phosphorylation of microtubule associated protein tau by non-proline dependent protein kinases. *Molecular and Cellular Biochemistry* **131**, 181–9.

Singh TJ, Zaidi T, Grundke-Iqbal I, and Iqbal K (1995). Modulation of GSK-3-catalyzed phosphorylation of microtubule-associated protein tau by non-proline-dependent protein kinases. *FEBS Letters* **358**, 4–8.

Singh TJ, Wang JZ, Novak M, Kontzekova E, Grundke-Iqbal I, and Iqbal K (1996). Calcium/calmodulin-dependent protein kinase II phosphorylates tau at Ser 262 but only partially inhibits its binding to microtubules. *FEBS Letters* **387**, 145–8.

Singh TJ, Grundke-Iqbal I, Wu Q, *et al.* (1997). Protein kinase C and calcium/calmodulin-dependent protein kinase II phosphorylate three-repeat and four-repeat tau isoforms at different rates. *Molecular and Cellular Biochemistry* **168**, 141–8.

Sironi JJ, Yen SH, Gondal JA, Wu Q, Grundke-Iqbal I, and Iqbal K (1998). Ser-262 in human recombinant tau protein is a markedly more favorable site for phosphorylation by CaMKII than PKA or PhK. *FEBS Letters* **436**, 471–5.

Skrabana R, Kontsek P, Mederlyova A, Iqbal K, and Novak M (2004). Folding of Alzheimer's core PHF subunit revealed by monoclonal antibody 423. *FEBS Letters* **568**, 178–82.

Spillantini MG, Murrell JR, Goedert M, Farlow MR, Klug A, and Ghetti B (1998). Mutation in the tau gene in familial multiple system tauopathy with presenile dementia. *Proceedings of the National Academy of Sciences of the USA* **95**, 7737–41.

Spittaels K, Vanden Haute C, Van Dorpe J, *et al.* (2000). Glycogen synthase kinase-3beta phosphorylates protein tau and rescues the axonopathy in the central nervous system of human four-repeat tau transgenic mice. *Journal of Biological Chemistry* **275**, 41340–9.

Stambolic V, Ruel L, and Woodgett JR (1996). Lithium inhibits glycogen synthase kinase-3 activity and mimics wingless signaling in intact cells. *Current Biology* **6**, 1664–8.

Steiner B, Mandelkow EM, Biernat J, *et al.* (1990). Phosphorylation of microtubule-associated protein tau: identification of the site for Ca2(+)-calmodulin dependent kinase and relationship with tau phosphorylation in Alzheimer tangles. *EMBO Journal* **9**, 3539–44.

Sternberger NH, Sternberger LA, and Ulrich J (1985). Aberrant neurofilament phosphorylation in Alzheimer disease. *Proceedings of the National Academy of Sciences of the USA* **82**, 4274–6.

Sun L, Wang X, Liu S, *et al.* (2005). Bilateral injection of isoproterenol into hippocampus induces Alzheimer-like hyperphosphorylation of tau and spatial memory deficit in rat. *FEBS Letters* **579**, 251–8.

Tanaka T, Zhong J, Iqbal K, Trenkner E, and Grundke-Iqbal I (1998). The regulation of phosphorylation of t in SY5Y neuroblastoma cells: the role of protein phosphatases. *FEBS Letters* **426**, 248–54.

Tanemura K, Akagi T, Murayama M, *et al.* (2001). Formation of filamentous tau aggregations in transgenic mice expressing V337M human tau. *Neurobiology of Disease* **8**, 1036–45.

Tanimukai H, Grundke-Iqbal I, and Iqbal K (2005). Up-regulation of inhibitors of protein phosphatase-2A in Alzheimer's disease. *American Journal of Pathology* **166**, 1761–71.

Tatebayashi Y, Haque N, Tung YC, Iqbal K, and Grundke-Iqbal I (2004). Role of tau phosphorylation by glycogen synthase kinase-3β in the regulation of organelle transport. *Journal of Cell Science* **117**, 1653–63.

Tomlinson BE, Blessed G, and Roth M (1970). Observations on the brains of demented old people. *Journal of Neuroscience* **11**, 205–42.

Tsai LH, Takahashi T, Caviness VS, Jr, and Harlow E (1993). Activity and expression pattern of cyclin-dependent kinase 5 in the embryonic mouse nervous system. *Development* **119**, 1029–40.

Tsujio I, Zaidi T, Xu J, Kotula L, Grundke-Iqbal I, and Iqbal K (2005). Inhibitors of protein phosphatase-2A from human brain structures, immunocytological localization and activities towards dephosphorylation of the Alzheimer type hyperphosphorylated tau. *FEBS Letters* **579**, 363–72.

Ulitzur N, Rancano C, and Pfeffer SR (1997). Biochemical characterization of mapmodulin, a protein that binds microtubule-associated proteins. *Journal of Biological Chemistry* **272**, 30577–82.

Ulloa L, de Garcini EM, GÚmez-Ramos P, Moran MA, and Avila J (1994). Microtubule-associated protein MAP1B showing a fetal phosphorylation pattern is present in sites of neurofibrillary degeneration in brains of Alzheimer's disease patients. *Molecular Brain Research* **26**, 113–22.

Vijayan S, El-Akkad E, Grundke-Iqbal I, and Iqbal K (2001). A pool of t-tubulin is hyperphosphorylated at serine residues in Alzheimer disease brain. *FEBS Letters* **509**, 375–81.

von Bergen M, Friedhoff P, Biernat J, Heberle J, Mandelkow EM, and Mandelkow E (2000). Assembly of tau protein into Alzheimer paired helical filaments depends on a local sequence motif ((306)VQIVYK(311)) forming beta structure. *Proceedings of the National Academy of Sciences of the USA* **97**, 5129–34.

von Lindern M, van Baal S, Wiegant J, Raap A, Hagemeijer A, and Grosveld G (1992). Can, a putative ongogene associated with Myeloid Leukomogenesis, may be activated by fusion of its 3' half to different genes: characterization of the set gene. *Molecular and Cellular Biology* **12**, 3346–3355.

Wagner U, Utton M, Gallo JM, and Miller CC (1996). Cellular phosphorylation of tau by GSK-3 beta influences tau binding to microtubules and microtubule organization. *Journal of Cell Science* **109**, 1537–43.

Walaas SI and Greengard P (1991). Protein phosphorylation and neuronal function. *Pharmacological Reviews* **43**, 299–349.

Wang JZ, Gong CX, Zaidi T, Grundke-Iqbal I, and Iqbal, K (1995). Dephosphorylation of Alzheimer paired helical filaments by protein phosphatase-2A and -2B. *Journal of Biological Chemistry* **270**, 4854–60.

Wang JZ, I. Grundke-Iqbal I, and Iqbal K (1996a). Restoration of biological activity of Alzheimer abnormally phosphorylated t by dephosphorylation with protein phosphatase-2A, -2B and -1. *Molecular Brain Research* **38**, 200–8.

Wang JZ, Grundke-Iqbal I, and Iqbal K (1996b). Glycosylation of microtubule-associated protein tau: An abnormal posttranslational modification in Alzheimer's disease. *Nature Medicine* **2**, 871–5.

Wang JZ, Wu Q, Smith A, Grundke-Iqbal I, and Iqbal K (1998). τ is phosphorylated by GSK-3 at several sites found in Alzheimer disease and its biological activity markedly inhibited only after it is prephosphorylated by A-kinase. *FEBS Letters* **436**, 28–34.

Wang J, Tung YC, Wang Y, Li XT, Iqbal K, and Grundke-Iqbal I (2001). Hyperphosphorylation and accumulation of neurofilament proteins in Alzheimer disease brain and in okadaic acid-treated SY5Y cells. *FEBS Letters* **507**, 81–7.

Wille H, Drewes G, Biernat J, Mandelkow EM, and Mandelkow E (1992). Alzheimer-like paired helical filaments and antiparallel dimers formed from microtubule-associated protein tau *in vitro*. *Journal of Cell Biology* **118**, 573–84.

Wilson DM and Binder LI (1995). Polymerization of microtubule-associated protein tau under near-physiological conditions. *Journal of Biological Chemistry* **270**, 24306–14.

Woodgett JR (1990). Molecular cloning and expression of glycogen synthase kinase-3/factor A. *EMBO Journal* **9**, 2431–8.

Yanagawa H, Chung SH, Ogawa Y, *et al.* (1998). Protein anatomy: C-tail region of human tau protein as a crucial structural element in Alzheimer's paired helical filament formation *in vitro*. *Biochemistry* **37**, 1979–88.

Chapter 6

Protein misfolding, aggregation, and fibril formation: common features of cerebral and non-cerebral amyloidosis

Agueda Rostagno, Ratnesh Lal, and Jorge Ghiso

Cerebral amyloidosis is primarily associated with neurodegeneration and dementia; in cases in which vessels are largely compromised, the vascular pathology translates to ischaemia and/or cerebral haemorrhage. The molecular mechanisms involved in the fibrillization process are slowly starting to emerge. Unexpected misfolding of normal proteins, a process shared by a variety of unrelated amyloid subunits including those producing systemic forms of amyloidosis, generates a diversity of intermediate assemblies in the form of oligomers and protofibrils with the capability to induce cell toxicity. Amyloidogenesis and cell toxicity are complex highly intermingled processes regulated by a variety of structural and environmental factors. Intermediate products of aggregation can assemble into ion-channel-like structures in the cell membrane, induce apoptotic/necrotic mechanisms, trigger inflammatory responses, activate the complement cascade, or generate oxidative stress, or a combination thereof. In turn, these pathways, separately or synergistically, are capable of producing different levels of cell damage and modulating cytotoxicity. The importance of amyloidosis in the mechanism of neurodegeneration is stressed by the association of unrelated amyloid subunits with almost identical neurofibrillary pathology and clinical dementia.

6.1 Introduction

Cerebral amyloid diseases are considered to be part of an emerging complex group of chronic and progressive entities collectively known as 'disorders of protein folding' which include, among many others, Alzheimer's disease, polyglutamine-repeat disorders, cataracts, amyotropic lateral sclerosis, Parkinson's disease and other synucleinopathies, tauopathies, a variety of systemic amyloidoses, prion diseases, cerebellar ataxias, and type II diabetes (Dobson 2001a; Lovestone and McLoughlin 2002; Taylor *et al.* 2002; Temussi *et al.* 2003). In these diseases, through mechanistic pathways poorly understood, soluble proteins normally found in biological fluids change their conformation and form either insoluble structures that accumulate in the form of intra- and extracellular aggregates or fibrillar lesions usually associated with local release of inflammatory mediators, oxidative stress, complement activation, cell toxicity, apoptosis, or a combination thereof resulting in cell damage, organ dysfunction, and eventually death.

Regardless of the organ targeted and the misfolded protein involved, all extracellular fibrillar deposits, generically referred to as amyloid, share common physical, structural, and tinctoral properties: (i) they are highly polymerized and poorly soluble assemblies, features that preclude their efficient physiological *in vivo* removal by macrophages and require the use of strong detergents, harsh acid conditions, or concentrated chaotropes to partially extract them *in vitro* from the tissue deposits; (ii) they are weakly antigenic, consistently failing to induce a high titre response or high affinity antibodies in different animal species; (iii) they are structurally rich in β-pleated sheet conformations, a property which accounts for the apple-green birefringence of the deposits when observed under polarized light after Congo red staining as well as for their yellow-green fluorescence after thioflavin S staining; (iv) they are fibrillar in shape when observed by electron microscopy and atomic force microscopy (ATM) (LeVine 1995a; Klunk *et al.* 1999; Ghiso and Frangione 2002; Srinivasan *et al.* 2003).

6.2 Normal and abnormal protein folding: the process of aggregation and fibril formation

It is well established that following synthesis on the ribosome, each of the proteins of a living organism must successfully fold into a specific conformational state in order to be able to carry out its biological function. Understanding the mechanism of this normally occurring phenomenon is one of the most fascinating and challenging problems of structural biology today. Within the cells, folding takes place in a complex and highly crowded environment with the aid of many families of auxiliary proteins, primarily molecular chaperones which either reduce the chance of aggregation, allow aggregated proteins to refold or target them for degradation (Dobson 2001b). A major driving force for protein folding is the internalization of hydrophobic stretches and the exposure of hydrophilic side chains to the solvent, resulting in a final structure usually stabilized by many weak non-covalent interactions (Ellis 2004). On the route to adopting their native structure during normal folding, proteins pass through partially folded intermediates of extremely short lifetime, features that reduce the probability of intermolecular interactions. During these normal physiological processes molecular chaperones bind to transiently exposed hydrophobic residues and protect the incompletely folded polypeptide from undesired interactions, particularly those that may result in aggregation. However, under certain circumstances, during both the synthesis of the peptide and the unfolding of proteins induced by stress, hydrophobic residues may become exposed and able to interact with neighboring chains to form non-functional aggregates. Here again, chaperone molecules contribute to protein solubilization by participating in the refolding processes of the aggregated proteins. Ultimately, certain chaperones also participate in the elimination of aggregated proteins by targeting them for degradation by the proteasome (Dobson 2001b).

The ability of soluble proteins or protein fragments to change their conformation under particular circumstances and convert spontaneously into amyloid fibrils is a highly challenging and puzzling biological phenomenon, taking into consideration that once native folding is achieved, proceeding to aggregation or fibril formation is thermodynamically unfavourable. However, despite this energy barrier, 26 unrelated proteins are so far known to produce amyloid diseases in humans (reviewed by Ghiso and Frangione 2002). Notably, proteins not associated with human disease have also been found to convert *in vitro* into aggregates with structural and cytotoxic properties almost indistinguishable from those exhibited by amyloid assemblies associated with pathological conditions (Guijarro *et al.* 1998; Litvinovich *et al.* 1998; Chiti *et al.* 1999; Ramirez-Alvarado *et al.* 2000; Bucciantini *et al.* 2002), indicating that fibrillization may be a more generic property of polypeptide chains rather than being restricted to a small number of sequences (Chiti *et al.* 1999; Takahashi *et al.* 2000; Dobson 2001b; Ventura *et al.* 2004) and in agreement with the notion that, under appropriate conditions, many proteins may undergo comparable abnormal folding (Villegas *et al.* 2000; Fandrich *et al.* 2001; Pertinhez *et al.* 2001; Lopez De La Paz *et al.* 2002). The formation of amyloid fibrils has been also successfully modelled *in vitro* using synthetic homologues of amyloid subunits (Castano *et al.* 1986; Wisniewski *et al.* 1991; Goldfarb *et al.* 1993). In most cases, significant secondary structure differences between the monomeric soluble peptides and the aggregated/fibrillar end-products have been documented (reviewed by Soto 2003), supporting the concept that conformational modifications are associated with the aggregation process.

Regardless of all the differences in primary structure and molecular size, this heterogeneous group of native and synthetic molecules share comparable endstage fibrillar structures which are morphologically indistinguishable from one another. X-ray fibre diffraction data indicate that these molecules share a characteristic cross-β structure (Dobson 2001b), organized as fibrils 7–10 nm in diameter, in which the long axis of the fibril is parallel to the helical axis and perpendicular to the β-strands (reviewed by Serpell 2000). Experiments with photoaffinity cross-linking, Fourier transform infrared spectroscopy (FTIR), and solid state nuclear magnetic resonance (NMR) demonstrated that both parallel and antiparallel β-sheet orientations are present in the amyloid fibrils (Serpell 2000). All these striking similarities among unrelated subunits suggest that at least part of the amyloid assembly processes may proceed through the formation of structurally related intermediates and involve similar mechanisms (Sunde *et al.* 1997; Dobson 1999).

There is considerable *in vitro* data indicating that the process of self-assembly towards the formation of amyloid fibrils proceeds in a nucleation-dependent polymerization mechanism. Above a critical concentration, below which polymerization does not occur, monomers self-assemble to form dimers, oligomers, and polymers, with kinetics that are characteristic of each amyloid subunit. In addition to differences in rate and degree of polymerization, all molecules exhibit a variable lag phase that can be minimized or even eliminated by the addition of preformed aggregates. In this sense, *in vitro* experiments indicate that the process of fibril formation is substantially accelerated when non-fibrillar solutions are seeded with preformed fibrils, a mechanism that has been correlated with the infectivity of prion diseases (Pan *et al.* 1993). From the thermodynamic point of view, this process occurs in a two-step reaction involving an initial slow lag period characteristic of each amyloid subunit which reflects the energy barrier necessary for the formation of a nucleation 'seed', followed by a rapid fibril propagation and aggregation state (reviewed by McLaurin *et al.* 2000; Serpell 2000). Using kinetic studies, the participation of seeding/nucleation mechanisms has been demonstrated in an array of different proteins, including PrP, huntingtin, α-synuclein, Aβ, and ABri (Scherzinger *et al.* 1999; Wood *et al.* 1999; Srinivasan *et al.* 2003).

Biochemical data from cerebral and non-cerebral amyloidosis indicate that all amyloid subunits have a high tendency to form these poorly soluble polymeric structures. Amyloid extracted from tissue deposits is always highly aggregated regardless of the organ studied, the amyloid subunits involved, or the methodology used to disrupt and partially solubilize the lesions (Fig. 6.1a). Amyloid extracts analysed by size exclusion chromatography (native or SDS–PAGE) consistently show heterogeneous multimeric species which are usually composed of a mixture of intact subunits at different stages of aggregation, (Figs 6.1a and 6.1b) N- and C-terminal degradation products of variable length, and often derivatives bearing a variety of post-translational modifications. Whether these multiple components reflect a clearance mechanism or play a role in the formation and stability of the amyloid lesions is still controversial.

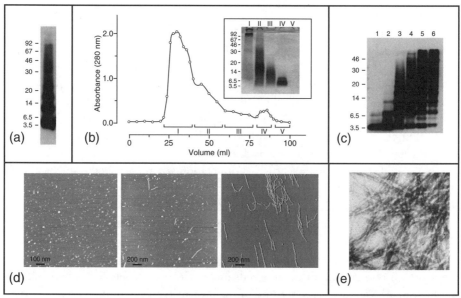

Fig. 6.1 *In vivo* and *in vitro* oligomerization. (a) Western blot after SDS–PAGE illustrating a crude formic acid extract of leptomeningeal amyloid ABri. (b) Size exclusion chromatography of a comparable sample showing oligomer subfractionation. The inset shows western blot of fractions I–V collected as specified in the graph. (c) The oligomerization process of monomeric ABri is mimicked by a synthetic peptide incubated in PBS at 37°C for different periods of time: lane 1, 0 h; lane 2, 2 h; lane 3, 6 h; lane 4, 12 h; lane 5, 24 h; lane 6, 48 h. (d) Atomic force microscopy images illustrating globular oligomers (left), protofibrils (centre), and fibrils (right). (e) Electron microscopy image of extracted amyloid fibrils negatively stained with uranyl acetate (magnification 120 000×).

The process of polymerization has been studied in detail in *in vitro* paradigms using synthetic and recombinant versions of different amyloid subunits (Fig. 6.1c). In the case of synthetic Aβ peptides, fibril polymerization occurs through the sequential formation of dimers, tetramers, and oligomers, in which the charged residues form ion pairs and the hydrophobic residues form a hydrophobic core (Harper *et al.* 1997; Walsh *et al.* 1999). Oligomers, which have been found in conditioned medium of cells constitutively secreting Aβ peptides as well as in cerebrospinal fluid (CSF) and human brain homogenates (LeVine 1995b; Kuo *et al.* 1996; Lambert *et al.* 1998; Walsh *et al.* 2002), appear to lead to the formation of protofibrils. These protofibrils (Fig 6.1d), believed to be the precursors of full-length fibrils, have a rod-like appearance and are shorter than fully formed fibrils (Fig 6.1e), generally consisting of unbranched structures 3–6 nm wide and up to 100 nm long (Serpell 2000, Lazo *et al.* 2005). Electron and atomic force microscopy have demonstrated the presence of these small ordered aggregates at early stages of fibril formation (Lin *et al.* 1999; Zhu *et al.* 2000); protofibrils subsequently elongate as a function of time with a rate that depends on many factors, among them the peptide concentration, temperature, ionic strength, and pH (Harper *et al.* 1999). Secondary structure analyses indicate that protofibrils have high β-sheet content comparable to that of the amyloid fibrils themselves and also bind Congo red and thioflavin S. These protofibrillar intermediates coexist with both non-aggregated forms of the Aβ peptide and fibrillar aggregates in a dynamic equilibrium (Teplow 1998), and have been claimed to be responsible for the detrimental effect on cell viability (Lambert *et al.* 1998; Walsh *et al.* 1999).

6.3 Modulation of fibril formation: lessons from cerebral and non-cerebral amyloidosis

An increasing number of genetic, structural, and environmental factors play a role in the modulation of the *in vitro* and *in vivo* misfolding and aggregation of amyloidogenic subunits (Howland *et al.* 1998; Kelly 1998; Chiti *et al.* 1999; Lazo *et al.* 2005) which, in turn, translate into a direct cytotoxic effect not only through caspase-dependent pathways or the formation of ion-channel structures but through the induction of oxidative stress and inflammation (Fig. 6.2).

6.3.1 Mutations

The most frequent forms of amyloidosis are those localized to the CNS. About one-third of the 26 different proteins so far known to produce amyloid diseases in humans cause fibrillar deposits in the CNS, resulting in cognitive deficits, dementia, stroke, cerebellar and extrapyramidal signs, or a combination thereof (Ghiso and Frangione 2002; Westermark *et al.* 2002, 2005). In terms of frequency, Alzheimer's Aβ is the most common amyloid subunit, and curiously the only one so far restricted to the brain. An important argument supporting the importance of amyloid in disease pathogenesis comes from rare familial disorders in which the presence of mutated amyloid subunits is consistently associated with early onset of the disease. In the *APP* gene on chromosome 21,

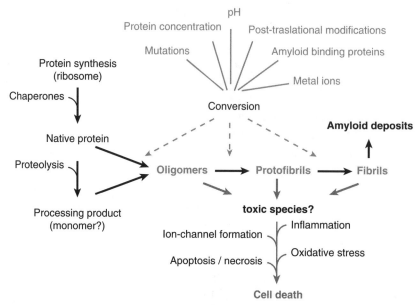

Fig. 6.2 Oligomerization, fibril formation, and cell death. Schematic representation of the factors involved in the modulation of fibril formation and influencing the various pathways leading to cell toxicity.

multiple mutation sites either within or immediately flanking the Aβ peptide coding sequence are clinically associated with dementia, ischaemic strokes, and/or cerebral haemorrhage phenotypes (Fig. 6.3). Notably, those mutations surrounding the β- and γ-secretase cleavage sites increase the production of Aβ (mainly Aβ42) and are linked to a dementia phenotype, whereas those mutations located within the Aβ peptide sequence are primarily associated with cerebral amyloid angiopathy (CAA), ischaemia, and/or cerebral haemorrhage. The first mutation in the *APP* gene was described in an autosomal dominant condition known as hereditary cerebral haemorrhage with amyloidosis, Dutch type (HCHWA-D) (Van Duinen *et al.* 1987), clinically defined by recurrent strokes, vascular dementia, and fatal cerebral bleeding in the fifth to sixth decades of life (Luyendijk *et al.* 1988). Parenchymal pre-amyloid deposits and massive cerebral amyloid angiopathy affecting the walls of leptomeningeal and cortical vessels are constant findings, whereas dense plaque cores and neurofibrillary tangles are consistently absent. A single nucleotide change (G for C) at codon 693 of APP (Levy *et al.* 1990) results in a single amino acid substitution (glutamine for glutamic acid) at position 22 of the Aβ peptide (Van Duinen *et al.* 1987; Prelli *et al.* 1990). As well as the Dutch mutation, other genetic variants are concentrated in the middle of the Aβ sequence (positions 21–23, corresponding to codons 692–694 of *APP*) and are invariably associated with extensive cerebrovascular pathology. The Flemish mutation, a C to G transversion at codon 692, is the only variant so far described at residue 21, resulting in an alanine to glycine substitution and a clinical phenotype of presenile dementia and cerebral haemorrhage (Hendriks

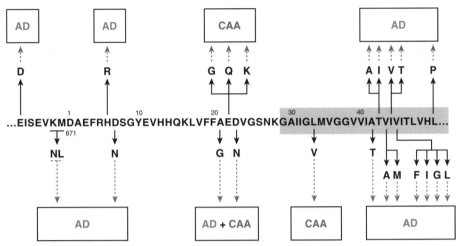

Fig. 6.3 APP genetic variants. Fragment 664–722 of APP illustrating the location of the known mutants producing distinct phenotypes in familial forms of AD. The shaded box denotes the intramembrane region of APP.

et al. 1992). At position 22 of Aβ, in addition to the Dutch mutant, two other genetic variants have been described: the Arctic mutation (A to G at codon 693, resulting in the replacement of glutamic acid for glycine), presenting as an early-onset AD with prominent vascular symptomathology (Kamino *et al.* 1992), and the Italian mutation (G to A at codon 693, resulting in lysine for glutamic acid substitution), presenting as a presenile dementia with cerebral haemorrhage (Miravalle *et al.* 2000). At position 23, the Iowa mutant is the result of a G to A transition at codon 694 which translates in the substitution of the normally occurring aspartic acid residue for asparagine and a clinical picture of aphasic dementia, severe CAA, and leucoencephalopathy (Grabowski *et al.* 2001). Recently, a new genetic variant associated with cerebral haemorrhages was reported outside the Aβ21–23 'hot spot'. The novel L705V mutation, also described in an Italian family (Piedmont), is due to a G to C transversion in the first nucleotide of codon 705 of APP, changing a leucine residue at position 34 in the Aβ sequence for valine (Obici *et al.* 2005). The AβV34 variant deposits primarily in cerebral vessels, causing severe CAA and recurrent haemorrhages without parenchymal amyloid plaques or neurofibrillary tangles, a phenotype very similar to that produced by the Dutch mutation.

Several other mutated non-Aβ amyloid subunits linked to early-onset cerebral amyloidosis have been reported. The first amyloid purified and characterized from the CNS (Cohen *et al.* 1983; Ghiso *et al.* 1986a,b) was a genetic variant of cystatin C (ACys-Q68), a ubiquitously expressed inhibitor of cysteine proteases codified by a single gene located on chromosome 20. It was found associated with cases of hereditary cerebral haemorrhage with amyloidosis, Icelandic type (HCHWA-I), an autosomal dominant disorder characterized by massive amyloid deposition in small arteries and arterioles of the leptomeninges, cerebral cortex, basal ganglia, brainstem, and cerebellum

(Gudmundsson *et al.* 1972) which clinically manifests with cerebral haemorrhages of fatal outcome in the third to fourth decade of life. The ACys-Q68 amyloid subunit is the result of a single nucleotide change, A for T at codon 68 of cystatin C, changing at the protein level the normally occurring leucine for glutamine (Ghiso *et al.* 1986a,b; Levy *et al.* 1989). Recently, further support for the importance of amyloid in disease pathogenesis came from the genetically altered type II transmembrane precursor protein BRI2 codified by the *BRI2* gene (also known as *ITM2B*) (Pittois *et al.* 1998) located on chromosome 13. A single nucleotide substitution (T for A at codon 267) results in the presence of an arginine residue replacing the normally occurring stop codon in the wild-type precursor molecule and a longer open-reading frame of 277 amino acids (aa) instead of the 266 residues of BRI2 (Vidal *et al.* 1999). As a result of a normal furin-like proteolytic processing (Kim *et al.* 1999), the C-terminal comprising 34 aa of the mutated precursor protein is released, generating the ABri amyloid peptide which forms amyloid deposits in patients with familial British dementia (FBD). In this early-onset autosomal dominant disorder, clinically characterized by progressive dementia, spastic tetraparesis, and cerebellar ataxia (Mead *et al.* 2000), the main neuropathological hallmarks are severe amyloid angiopathy of the brain and spinal cord with perivascular amyloid plaque formation, parenchymal plaques affecting limbic areas, cerebellum, and occasionally cerebral cortex, neurofibrillary degeneration of hippocampal neurons, and periventricular white matter changes. A different defect, namely the presence of a 10 nucleotide duplication (795–796insTTTAATTTGT) between codons 265 and 266, three nucleotides before the normal stop codon 267, also produces a frame-shift in the BRI2 sequence generating a larger than normal precursor protein, of which the last 34 C-terminal aa compose the ADan subunit found associated with amyloid deposits in familial Danish dementia (FDD) (Vidal *et al.* 2000). This early-onset autosomal dominant disorder clinically characterized by cataracts, deafness, progressive ataxia, and dementia (Strömgren *et al.* 1970, Bek 2000), exhibits widespread cerebral amyloid angiopathy, parenchymal diffuse plaques, and neurofibrillary tangles.

Despite the variety of amyloid subunits involved and the vast array of mutations reported, it is still unclear whether all of them have a similar effect on the fibrillization pathways. In general, mutations seem to destabilize native structures by increasing their β-sheet content and producing shorter lag phases and accelerated kinetic rates of aggregation and fibrillization, although in some cases the presence of a mutation affects the protein clearance through the blood–brain barrier (Wisniewski *et al.* 1991; Van Nostrand *et al.* 2001; Monro *et al.* 2002).

6.3.2 **Protein concentration**

There is a direct effect between high protein concentration and the tendency to form aggregates and/or fibrillar structures. The most common examples are found in dialysis-related amyloidosis and certain lymphoproliferative disorders, e.g. multiple myeloma. The former is related to the systemic deposition of β_2 microglobulin (β2m), the non-covalently bound light chain of the class I human leucocyte antigen (HLA class I). β2m is 99 residues long and has seven-stranded β-sandwich folds typical of the immunoglobulin superfamily (Saper *et al.* 1991). *In vivo*, β2m is continuously shed from the surface of cells displaying HLA I molecules into the serum, where it is transported to the kidneys for degradation and excretion. In cases of renal insufficiency, the concentration of β2m circulating in serum increases up to 60-fold (Floege and Ehlerding 1996) and free β2m chain then associates to form amyloid fibrils which typically accumulate in the musculoskeletal system and result in the development of dialysis-related amyloidosis and carpal tunnel syndrome. The majority of the amyloid deposits are composed of full-length wild-type protein with less than 30% of modified or truncated forms (Linke *et al.* 1987; Floege and Ehlerding 1996; Bellotti *et al.* 1998). Although protein concentration is an important factor in the formation of β2m amyloid, *in vitro* studies indicate that conformational rearrangements of the normally highly soluble native protein are also required to trigger fibril formation. In agreement with findings for other amyloid proteins, acidic pH, protein fragmentation, and the presence of Cu^{2+} ions are some of the interlinking mechanisms enhancing β2m aggregation (Trinh *et al.* 2002).

The role of enhanced protein production in aggregation/fibrillization is also highlighted in immunoglobulin light chain (AL) systemic amyloidosis related to B-cell lymphoproliferative diseases. In these disorders the clonal expansion of plasma cells results in the overproduction of the monoclonal protein leading to the tissue deposition of the light chains or their proteolytically derived fragments in various organs, particularly kidney and cardiac muscle (Gallo *et al.* 1989). As in other amyloidoses, additional factors, such as acidic micro-environments and proteolysis of the parent protein, enhance light chain aggregation and fibril formation. In particular, proteolytic degradation has been attributed a high relevance in the biology of AL amyloidosis; material extracted from tissue deposits is most frequently comprised of N-terminal proteolytic fragments of the light chain variable region. It is still debatable whether this proteolysis occurs before the formation of the deposits, thus probably causing an alteration of the native folding state favouring deposition or, on the contrary, enzymatic cleavage occurs after the tissue deposition and reflects a clearance effort (Bellotti *et al.* 2000; Buxbaum 2004).

Elevated concentration of amyloidogenic proteins also plays a role in cerebral amyloidosis and various examples exist in which the elevated levels of Aβ peptides have been linked to enhanced tissue deposition. One of them is Down's syndrome (DS) in which an extra copy of the chromosome 21 produces an overexpression of the precursor APP and a concomitant overproduction of the amyloidogenic Aβ peptides. Interestingly, along with the enhanced expression of APP, triplication of chromosome 21 also causes overexpression of BACE-2 which may contribute to the increased Aβ production (reviewed by Lott and Head 2005). In DS the characteristic cerebral lesions of AD appear at childhood with virtually all carriers of the trisomy 21 developing AD neuropathology by the age of 40 years (Mann and Esiri 1989). Not only do DS cases show significantly higher Aβ plasma levels than age-matched controls (Mehta *et al.* 1998; Schupf *et al.* 2002), but those individuals with more elevated plasma levels of both Aβ40 and Aβ42 have a higher incidence of dementia (Schupf *et al.* 2002), pointing to a pathogenic role of the enhanced protein production. Similar effect is seen in association with certain *APP* mutations resulting in early-onset FAD, as highlighted above (reviewed by Ghiso and Frangione 2002).

The role of elevated Aβ concentrations in the amyloidogenic pathway is also stressed by the successful construction of multiple lines of transgenic animals in which the overexpression of various mutated genes promotes the expression of increased quantity of Aβ peptides (reviewed by Sinha 2002; Phinney *et al.* 2003). Many comparable models have been reported using various mutant forms of both *APP* and *PS1*, both singly and in combination. Aβ deposits are achieved by increasing APP overexpression whereas the presence of *PS1* mutations not only contributes to the Aβ load but to the early onset of the deposits (Citron *et al.* 1997). Interestingly, the presence of a *PS1* mutant can also modulate the length of the Aβ peptides and shift the localization of the deposits from the vessels to the parenchyma (Herzig *et al.* 2004).

Further evidence for the important role of elevated protein concentration in the amyloidogenesis mechanism is provided by *in vitro* studies which showed unequivocally that amyloid peptides at their low physiological concentrations (pico- to nanomolar range) are unable to form either fibrillar or protofibrillar assemblies under physiological pH and ionic strength conditions.

6.3.3 pH

A growing number of dissimilar proteins associated with various systemic and cerebral amyloidosis form fibrils readily under mild acidic conditions (pH ~4.7–5.5), among them lysozyme variants (Morozova-Roche *et al.* 2000), immunoglobulin light chains (Rostagno *et al.* 1999), β_2 microglobulin (McParland *et al.* 2000; Smith and Radford 2001; Trinh *et al.* 2002), transthyretin (Damas and Saraiva 2005), ABri (Srinivasan *et al.* 2003), Aβ, and PrP. In the last two instances it was demonstrated that the acidic pH resulted in an increment of the β-sheet structure and an enhanced exposure of hydrophobic patches on the surface of the molecule which correlated with the tendency for fibrillization (Colon and Kelly 1992; Wood *et al.* 1996; Swietnicki *et al.* 1997). In the case of two lysozyme mutants studied in depth, the destabilization under mild acidic conditions was attributed to altered folding kinetics (reviewed by Dumoulin *et al.* 2005). Although extreme conditions of pH generally favour the formation of aggregates through denaturation mechanisms, in some cases the production of amyloid-like fibrils *in vitro* at low pH has been also reported, for example cystatin C variant (Jascolski and Grubb 2005).

The remarkable behaviour of ABri peptide at different pH conditions is a perfect example of the complexity of the fibrillization process and how it is highly influenced by the micro-environment. In the 3.1–4.3 pH range, ABri adopts almost exclusively a random structure and a predominantly monomeric aggregation state as visualized by analytical ultracentrifugation. At neutral pH (7.1–7.3), the peptide shows limited solubility and produces spherical and amorphous aggregates with predominantly β-sheet secondary structure, whereas at the slightly acidic pH of 4.9, spherical aggregates, intermediate sized protofibrils, and larger-sized mature amyloid fibrils are detected by atomic force microscopy. With ageing at pH 4.9, these protofibrils undergo further association and eventually elongate, forming mature fibrils (Srinivasan *et al.* 2003).

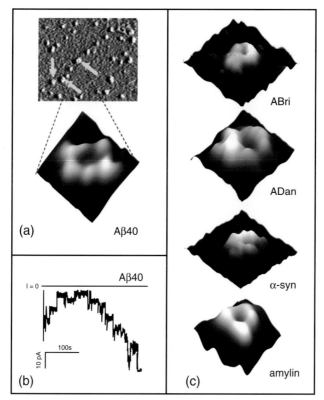

Plate 1 Formation of ion channels in lipid bilayers. (a) Atomic force microscopy image of Aβ1–40 reconstituted in membrane bilayers showing channel-like structures with a central pore (yellow arrows) at low and high resolution (image size, 25 nm). (b) Single-channel record for Aβ1–40 showing current traces as a function of time under voltage-clamp conditions. (c) Individual channel-like structures at high resolution formed by ABri, ADan, α-synuclein, and amylin (image size: 35 nm for ABri, 20 nm for ADan, and 25 nm for α-synuclein and amylin). See p. 147.

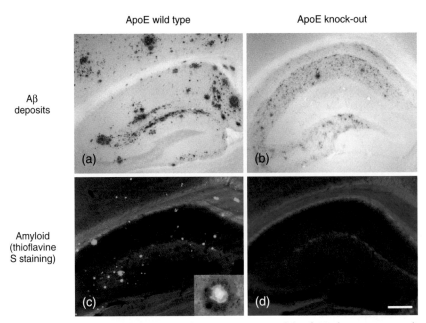

ApoE wild type   ApoE knock-out

Aβ
deposits

(a)   (b)

Amyloid
(thioflavine
S staining)

(c)   (d)

Plate 2 Effect of apoE on Aβ deposition. Relative to mouse models of AD that express murine apoE (a, c), mouse models of AD lacking apoE have substantially less Aβ deposition (b) and virtually no thioflavine S positive plaques (d) in the hippocampus. Reprinted by permission of Macmillan Publishers Ltd, *Molecular Psychiatry*, 2002. See p. 166.

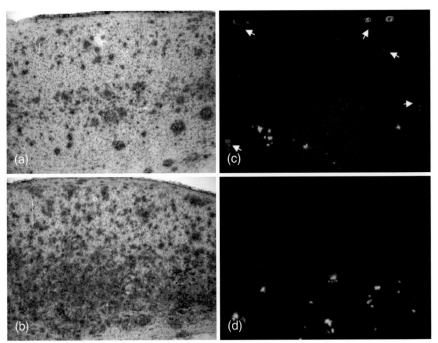

(a)   (c)

(b)   (d)

Plate 3 Amyloid pathology in brain parenchyma of APP-V717I × BACE1 double transgenic mice at age 22 months (b, d) compared with age-matched APP-V717I single transgenic mice (a,c). Amyloid deposits in subiculum immunostained with antibody 3D6 (a,b) and with thioflavin S (c, d). Note the complete absence of amyloid pathology in the vasculature of the double transgenic mice (d) as opposed to the CAA in the parental APP transgenic mice (c, arrows). (from Willem *et al.* 2004). See p. 200.

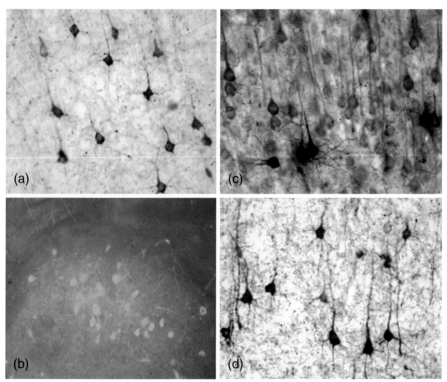

Plate 4 The unique tau-pathological changes in tau-P301L transgenic mice (age 9 months) are identified by immunostaining with (a) PG5 in cerebral cortex and (b) X-34 in medial fastigial nucleus of the cerebellum. (c) Tauopathy defined in the cortex of tau-P301L transgenic mouse by AT8 immunohistochemistry; (d) similar but after 30 minutes post-mortem delay to demonstrate that most normal tau phosphorylation recognized by AT8 has disappeared, while the pathological hyperphosphorylation remains unchanged, typical for the paired helical filaments in AD brain. See p. 205.

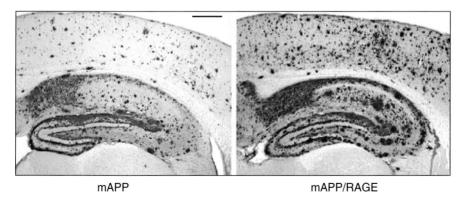

mAPP   mAPP/RAGE

Plate 5 Amyloid load in the cerebral cortex and hippocampus in Tg mAPP mice and Tg mAPP × RAGE mice after staining the section with 3D6 antibody. See p. 233.

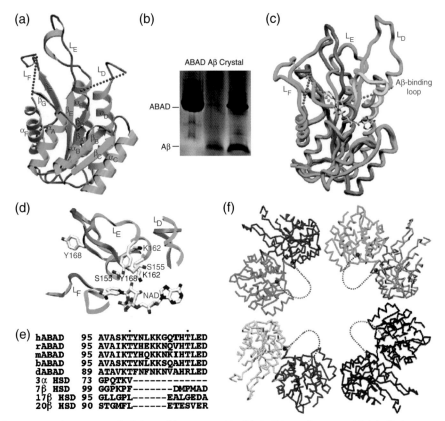

Plate 6 Crystal structure of Aβ-bound human ABAD. (a) A ribbon diagram with labelled secondary structures and the LB$_{DB}$, LB$_{EB}$, and LB$_{FB}$ loops. Helices are shown in green, β strands are shown in pink, and disordered regions are shown by dotted lines. (b) SDS–PAGE of washed and dissolved crystals of human ABAD and Aβ. Lanes from left to right: ABAD standard, Aβ(1–40) standard and dissolved crystals. (c) Superposition of Aβ-bound human ABAD (pink) and rat ABAD in complex with NAD (blue). The LB$_{DB}$ loop of 3β-hydroxysteroid dehydrogenase (3β-HSD) is shown in yellow. NAD is shown as a stick model with grey for carbon atoms, red for oxygen atoms, blue for nitrogen atoms, and yellow for phosphate atoms. The proposed Aβ-binding loop is indicated. (d) Superposition of the active sites of Aβ-bound human ABAD (pink) and rat ABAD (blue), showing distortion of the NAD binding site and the catalytic triad S^{155}, K^{162}, and Y^{168}. Colours are the same as in (c). (e) Sequence alignment of the disordered part of the LB$_{DB}$ loop (residues 95–113) among human, rat, mouse, bovine, and *Drosophila* ABAD and several hydroxysteroid dehydrogenases (HSDs), showing the insertion in ABAD relative to other HSDs. (f) Section of the crystal packing interactions, showing the large solvent channels. Each ABAD molecule is shown in a different colour. The ordered ends of the LB$_{DB}$ loop, residues 94 and 114, are marked as red and blue balls, respectively, and the hypothetical loops are shown as pink dotted lines. See p. 236.

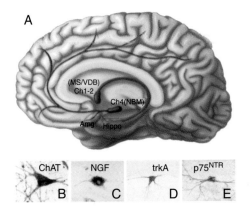
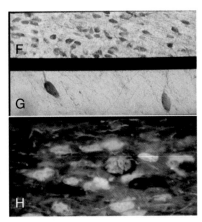

Plate 7 (a). Diagram showing the cholinergic projection neurons from Ch4 (NB) (blue) to the entire cortical mantle and the cholinergic projections from Ch2 (medial septal diagonal band complex (MS–VDB)) (red) to the hippocampus in a sagittal view of the human brain. Note that the Ch4 neurons also innervate the amygdala (Amg). The lower panels show that the cholinergic neurons are immunopositive for (b) ChAT, (c) NGF, (d) trkA, and (e) p75NTR (f) Low-power photomicrograph showing numerous ChAT-immunopositive cholinergic neurons within the NB from an aged control subject. (g) Higher-magnification photomicrograph showing the extensive reduction in ChAT-positive neurons within the NB from a patient with end stage AD (h) Section dual stained for p75NTR (dark blue) and thioflavin S (yellow) showing both tangle-bearing (yellow) and non-tangle-bearing (dark blue) neurons in endstage AD. See p. 285.

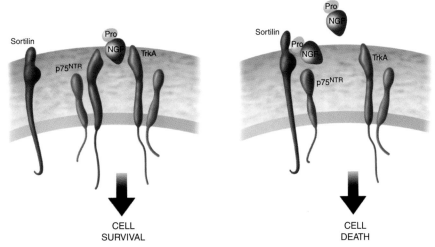

Plate 8 Schematic model illustrating putative scenarios for proNGF–receptor binding during the progression of AD. (a). ProNGF complexes with TrkA to activate cell survival mechanisms in the aged human healthy brain. This binding event is facilitated by the coexpression of p75NTR on the cell surface. (b). Elevated levels of cortical proNGF in the face of reduced cortical TrkA result in increased binding of proNGF to p75NTR–sortilin complexes, enhancing the activation of pro-apoptotic mechanisms during the transition from MCI to AD. (Reproduced with permission from Counts and Mufson 2005.) See p. 296.

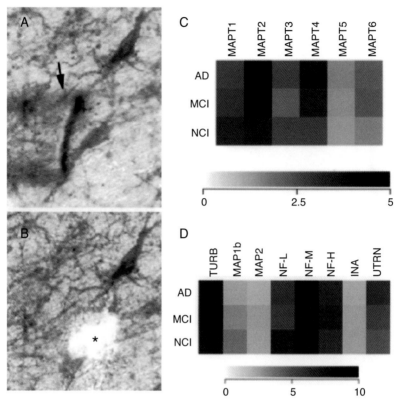

Plate 9 Single cholinergic NB neuron stained for p75^NTR (a) before and (b) after (asterisk). Microaspiration for gene array expression analysis. The arrow in (a). indicates the position of the glass pipette prior to aspiration. (c) Dendrogram illustrating the stability of relative tau gene expression levels across NCI, MCI, and AD. In contrast, significant differences were found when a 3Rtau/4Rtau ratio was calculated (see Table 13.1). (d) Dendrogram illustrating a non-significant difference in relative expression levels for β-tubulin (TuBB), microtubule-associated proteins MAP1b and MAP2, neurofilament subunits NF-L (light molecular weight). NF-M (medium), and NF-H (heavy), alpha internexin (INA), and utrophin (UTRN) across clinical groups. See p. 297.

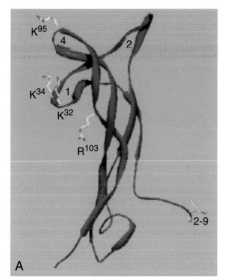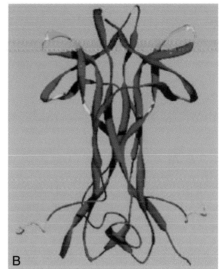

Plate 10 Structure of NGF. (a). NGF monomer structure, with the variable β-turn loops 1, 2, and 4 labelled. Further denoted are residues of particular importance for binding and/or biological activity to p75^{NTR} (lysines 32, 34, and 95). or TrkA (arginine 103, and residues 2–9 of the N-terminus of the mature form, shown in yellow). (b). NGF dimer structure with mapping of residues interacting with p75^{NTR} (red: W21, D30, I31, K32, K34, E35, K74, H75, K88, K95, R100, and R103), TrkA (green: S2, S3, H4, P5, I6, F7, H8, R9, I31, N45, G94, K95, Q96, A97, A98, and R103), or both (yellow: I31, K95, and R103). Note the overlap of the binding sites for the two receptors. (Reproduced with permission from Longo *et al.* 2005.) See p. 301.

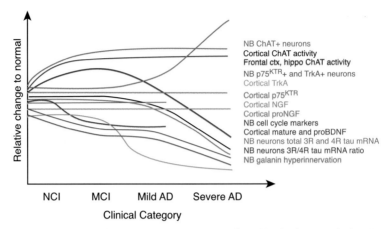

Plate 11 Summary diagram of cholinotrophic alterations found in the basocortical projection system during the progression of AD. Note that the number of ChAT-positive NB neurons is preserved while the number of NB neurons expressing the NGF receptor proteins TrkA and p75^{NTR} is reduced in MCI and early-stage AD. The frank loss of ChAT-positive NB neurons is associated with late-stage AD. In contrast, cortical levels of TrkA protein are stable in MCI but reduced in AD, while p75^{NTR} levels in the cortex are stable throughout the disease progression. ProNGF protein levels increase during MCI and AD. Cortical levels of both proBDNF and mature BDNF decrease during AD progression and may also impact cholinergic cell survival as well as many other neuronal populations subserving cognitive function. In addition to these cholinotrophic alterations, there is a *de novo* expression of cell cycle markers in NB neurons in MCI and early AD that may precede apoptotic signalling attributable to increased p75^{NTR} signalling. Moreover, the ratio of 3R to 4R tau transcripts is reduced in MCI and AD in NB neurons, potentially impacting neurofibrillary pathogenesis. Finally, levels of galanin hyperinnervation within the NB are increased only during the advanced stages of the disease but may have a beneficial impact on remaining neurons. The lines do not represent actual numerical values but a general trend for each factor depicted. See p. 308.

Based on the enhancement of fibrillization under acidic conditions a putative role in amyloidogenesis was suggested for acidic lysosomes (Durie *et al.* 1982; Colon and Kelly 1992), although it is not clear how intracellular processing in these subcellular compart ments may lead to extracellular amyloid deposition. Only recently have some investigations indicated that in the case of Aβ the initial accumulation may actually begin inside neurons, with the earliest detectable material located in membranous compartments identified as endosomes or lysosomes (D'Andrea *et al.* 2001, Kienlen-Campard *et al.* 2002, Takahashi R. H. *et al.* 2002), therefore providing evidence for a likely scenario in which acidic micro-environments may actively participate in the *in vivo* aggregation mechanisms.

6.3.4 Post-translational modifications

Amyloid subunits and their degradation products are well known to sustain a number of post-translational modifications which in turn largely contribute to the heterogeneity of the amyloid deposits. Among the many post-translational modifications identified in systemic and cerebral amyloidosis, phosphorylation, isomerization, racemization, oxidation, and cyclation occurring in conjunction with proteolytic fragmentation are the most relevant (Roher *et al.* 1993a,b; Saido *et al.* 1995; Bellotti *et al.* 2000; Ghiso *et al.* 2001a; Shimizu *et al.* 2002; Tomidokoro *et al.* 2005). Hyperphosphorylation is the common post-translational modification of intracellular tau composing the paired helical filaments accumulating as neurofibrillary tangles in several neurodegenerative disorders (Watanabe *et al.* 1999). For Aβ, heavy isomerization and racemization of the aspartyl residues at positions 1 and 7 have been reported, with more than half of the residues composed of L-isoAsp in some AD cases (Roher et al. 1993a,b); similar modifications were reported in the Iowa variant of FAD (Shin *et al.* 2003). In general these non-enzymatic chemical modifications of aspartyl residues may occur spontaneously, contribute to protein degradation, and lead to protein inactivation, loss of biological activity, and accelerated peptide aggregation and fibril formation (Geiger and Clarke 1987; Tomiyama *et al.* 1994). Perhaps one of the most frequent post-translational modifications is the formation of N-terminal pyroglutamate which is believed to confer resistance against N-terminal peptidases, thus delaying and/or preventing their removal from the lesions. This modification also makes their identification by N-terminal sequence analysis difficult, as no N-terminal amino group is available for Edman degradation. Interestingly, the formation of pyroglutamate seems to take place via two different molecular mechanisms according to the N-terminal amino acid involved. If the amino acid is glutamine, as observed with many immunoglobulin molecules, pyroglutamate arise through a process of deamination; however, if the N-terminus is aspartic acid, as seen in Aβ species truncated at position 3 and 11 as well as in the ABri and ADan molecules, pyroglutamate formation occurs through a process of dehydration (Ghiso *et al.* 2001a; Tomidokoro *et al.* 2005). In some reports, pyroglutamate-containing peptides account for more than 50% of the truncated Aβ species accumulated in plaques, being particularly enriched with Aβ3(pE)40/42 (Mori *et al.* 1992; Saido *et al.* 1995), a component present not only in senile plaques but also in pre-amyloid lesions (Lalowski *et al.* 1996). Aβ3(pE) appears more resistant than full-length peptides to proteolytic cleavage *in vitro*, confers enhanced tendency to oligomerization (Harigaya *et al.* 2000), and exerts higher toxicity for both neuronal and glial cell cultures (Russo *et al.* 2002). Pyroglutamate derivatives of Aβ are not major species in plasma or CSF, suggesting that pyroglutamate formation takes place at the site of deposition. In cases of familial British and Danish dementias, circulating ABri and ADan peptides feature only glutamate at their N-terminus, whereas their pyroglutamate counterparts are the heavily dominant species in the amyloid deposits. Since the conversion to pyroglutamate is chemically stable and poorly reversible, the presence of glutamate-only species in plasma is a clear indication that the circulating material does not represent a clearance mechanism for the cerebral deposits, but rather their immediate precursors (Rostagno *et al.* 2005). Oxidation of methionine (in the form of methionine sulphoxide) at position 35 of the Aβ peptide has been widely documented in AD (Naslund *et al.* 1994; Kuo *et al.* 2001; Barnham *et al.* 2003; Dong *et al.* 2003). Interestingly, Aβ peptides containing Met35 sulphoxide are able to induce oxidative stress and are more cytotoxic (Varadarajan *et al.* 2001).

6.3.5 Metal ions

The contribution of metal ions to the modulation of the aggregation/fibrillization process has been primarily studied in AD in which Zn^{2+}, Fe^{3+}, and Cu^{2+} are found associated with Aβ amyloid deposits (Lovell *et al.* 1998; McLaurin *et al.* 2000). The precise source of these ions is unclear; they may originate from the circulating pool and/or be released from their transport molecules and other metalloproteins under slightly acidic conditions during inflammatory processes (McLaurin *et al.* 2000). In the presence of metal chelators, extraction of Aβ deposits from the CNS seems more efficient, suggesting a role for metals in the aggregation/fibrillization process. In agreement with this notion it was reported that binding to Zn^{2+} ions is able to induce Aβ aggregation in a pH-dependent manner (Bush *et al.* 1994), with comparable effects also reported for Cu^{2+}, Ni^{2+}, and $Fe3^{+}$ (Atwood *et al.* 1998). Current lines of investigation suggest that the presence of Zn^{2+} affects both the nucleation and the aggregation stages of Aβ fibrillization, inducing an instantaneous β-structural transition and, in the case of preformed fibrils, causing their aggregation into large precipitating masses (McLaurin *et al.* 2000). Based on the current knowledge indicating that metal ions are powerful kinetic accelerators of fibril formation, they are actively being pursued as potential target for therapeutics aiming at modulating amyloid deposition (Barnham *et al.* 2004). In addition to their potential contribution to fibrillogenesis, metal ions are intricately linked to the detrimental effects induced by oxidative stress mechanisms discussed below, since the chemical origin of the majority of reactive oxygen species arise from the reaction with the redox-active metals copper and iron (Huang *et al.* 1999a/b).

6.3.6 Amyloid-associated proteins

Complex mixtures of unrelated molecules, collectively known as amyloid-associated proteins, co-localize with all amyloid deposits (including those of Aβ), but are not a structural part of the final fibril (reviewed by Ghiso and Frangione 2002). Serum amyloid P component (SAP), α_1-antichymotrypsin (ACT), apoE, apoJ, complement components, vitronectin, glycosaminoglycans, interleukins, and extracellular matrix proteins are among the many amyloid-associated proteins so far described in all forms of cerebral and systemic amyloidosis (Coria *et al.* 1988; Kalaria *et al.* 1991; Namba *et al.* 1991; Wisniewski and Frangione 1992; Yamada *et al.* 1994; Kindy *et al.* 1995; Rostagno *et al.* 2003, 2005; Ghiso *et al.* 2005; Veerhuis *et al.* 2005). At present, it is still unclear whether these molecules are innocent bystanders or whether their presence is related to the mechanism of amyloidogenesis. Several lines of investigation favour the latter notion, at least for some of them. For example, apoE and SAP have been found in all AL fibrillar deposits, whereas their presence cannot be demonstrated in the non-fibrillar Congo red negative immunoglobulin deposits in cases of light chain deposition disease (Gallo *et al.* 1988, 1994). Similar findings have been reported in Aβ and non-Aβ cerebral amyloidosis in which SAP and activation-derived products of the complement system are present in amyloid deposits but are consistently absent in non-fibrillar pre-amyloid lesions (Gallo *et al.* 1988; Rostagno *et al.* 2003). Crossings of APP transgenics with mice knockouts for either SAP or apoE exhibit fewer amyloid lesions and delay in their onset, although neither *SAP* nor *apoE* gene ablation prevents the formation of amyloid deposits (Botto *et al.* 1997; Holtzman *et al.* 1999). Studies mostly limited to Aβ indicate that many amyloid-associated proteins also have the ability to modulate the formation of amyloid fibrils *in vitro*. Some of them (e.g.C1q, apoE4, SAP, and ACT) enhance Aβ fibril formation (Wisniewski *et al.* 1993; LaDu *et al.* 1994; Ma *et al.* 1994; Webster *et al.* 1994; Tennent *et al.* 1995), while others (e.g. apoJ) contribute to peptide solubility precluding fibrillogenesis *in vitro* (Matsubara *et al.* 1996, Veerhuis *et al.* 2005). In the latter case, this protective effect has been proposed to contribute to the enhanced production of slowly sedimenting Aβ-derived diffusible ligands (ADDLs) which are highly toxic to neurons in culture at nanomolar concentrations (Lambert *et al.* 1998).

6.4 Potential mechanisms of cell death

It is clear that more than a single factor is involved in the mechanism of cell death, extensively studied for neurons and other brain cells in Aβ and non-Aβ cerebral amyloidosis, although poorly explored in systemic forms of the disease. Mounting evidence indicates that, in addition to a direct effect on cell viability related to the formation of oligomers/protofibrils and their ability to assemble into functional ion-channel-like structures in lipidic environments, amyloids are also able to induce apoptosis/necrosis, trigger oxidative stress, generate an inflammatory response, and/or activate the complement cascade. In turn, these pathways, separately or synergistically, are capable of producing different levels of cell damage and modulate cytotoxicity.

6.4.1 Nature of the toxic species

The amyloid cascade hypothesis proposes Aβ fibrillization as the probable cause of AD (Selkoe 1994). The original *in vitro* observations that fibrillar Aβ is neurotoxic (Pike *et al.* 1993;

Lorenzo and Yankner 1994) and that Aβ42 is more fibrillogenic than Aβ40 (Jarrett *et al.* 1993), together with the *in vivo* finding that patients with trisomy 21 develop the same amyloid lesions as in AD and patients with mutations in the *APP* gene have an early-onset of the disease (reviewed by Ghiso *et al.* 2001b; Ghiso and Frangione 2002) boosted the hypothesis. However, since amyloid plaques correlate poorly in number, time of appearance, or distribution with neurodegeneration or clinical dementia, the exact nature of the toxic elements was questioned (Lansbury 1999; Haass and Steiner 2001; Klein *et al.* 2001). A constant finding in cerebral amyloid disorders is the presence of non-fibrillar Congo-red-negative parenchymal lesions (referred to as pre-amyloid deposits) composed of similar species to those identified in fibrillar lesions, although less polymerized, circumventing the need for formic acid for their tissue extraction. Notably, these lesions coexist with neurofibrillary tangles identical to those found in AD even in the absence of mature cored plaques (e.g. in the Iowa variant of FAD or in the non-Aβ cerebral amyloidosis FBD and FDD, etc.). Since it is likely that the disease pathogenesis underlies in the folding and assembly of Aβ monomers into oligomers and higher-order multimers (Lazo *et al.* 2005), pre-amyloid deposits might be central in the mechanism and warrant further studies. Soluble oligomers have been implicated as the primary pathological species and are believed to be the ultimate cause of the synaptic loss and dementia associated with AD (Small 1998; Lansbury 1999; Haass and Steiner 2001; Klein *et al.* 2001). Stable oligomers of Aβ42 have been identified *in vivo* and isolated from brain, plasma, and CSF (Kuo *et al.* 1996; Roher *et al.* 1996) and their presence seems to correlate better with the severity of neurodegeneration (Lue *et al.* 1999; McLean *et al.* 1999). Furthermore, experiments with non-fibrillar structures including oligomers, amyloid-derived diffusible ligands (ADDLs) (Oda *et al.* 1995; Lambert *et al.* 1998, 2001; Walsh *et al.* 2002), and protofibrils (Volles and Lansbury 2003) demonstrated that these peptide assemblies are also neurotoxic.

Oligomers permeabilize cell membranes and assemble into lipid bilayers regardless of the sequence from which they are derived (Bucciantini *et al.* 2002, 2004; Stefani and Dobson 2003) and may share common structural motifs, as suggested by limited data on common immunoreactivity with antibodies recognizing oligomeric assemblies (Kayed *et al.* 2003). There is experimental data indicating that these oligomeric structures localize in dynamic membrane microdomains known as 'lipid rafts', rich in cholesterol and glycosphingolipids (Simons and Toomre 2000). There are reports indicating that in the human brain Aβ largely resides in lipid rafts (Lee *et al.* 1998; Morishima-Kawashima and Ihara 1998), and in the Tg2576 transgenic mouse model Aβ in a dimeric form was reported as highly concentrated in lipid rafts (Kawarabayashi *et al.* 2004). Lipid rafts also appear to foster the transformation of PrP[Res] into PrP[Scr], suggesting a role in the pathogenesis of prion diseases (reviewed by Hooper 2005). It is conceivable that these lipid structures provide the proper environment for oligomerization and the assembly of ion-channel-like structures.

6.4.2 Ion-channel formation

The cytotoxic action of amyloids might be initiated by interaction with cell membranes and the subsequent formation of ion channels (reviewed byArispe 2004). In fact it was the early demonstration that the formation of ion-channel-like assemblies was inhibited by Congo red, which provided one of the first indications that channels were formed by peptide aggregation in the lipid membranes (Hirakura *et al.* 1999). Atomic force microscopy revealed the formation of doughnut-shaped structures protruding from the membrane surface with a centralized pore-like depression presumably representing individual channels (Fig. 6.4 [**Plate 1**]) (Lin *et al.* 2001). Electrophysiological data corroborated the formation of ion-permeable channels and demonstrated their dependence on the aggregation state of the peptides with a shift to larger single-channel conductances with increased peptide aggregation (Hirakura *et al.* 1999). As illustrated in Fig. 6.4, the formation of ion channels has also been observed for many other amyloid peptides, either related to systemic and localized non-cerebral amyloidosis (including β2m, serum amyloid A, atrial natriuretic factor, calcitonin, lysozyme, transthyretin, calcitonin, lysozyme, and amylin) or associated with cerebral amyloidosis such as Aβ, α-synuclein,

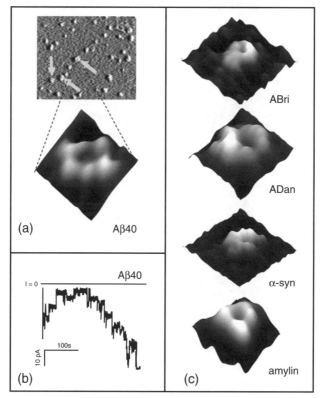

Fig. 6.4 Formation of ion channels in lipid bilayers. (a) Atomic force microscopy image of Aβ1–40 reconstituted in membrane bilayers showing channel-like structures with a central pore (yellow arrows) at low and high resolution (image size, 25 nm). (b) Single-channel record for Aβ1–40 showing current traces as a function of time under voltage-clamp conditions. (c) Individual channel-like structures at high resolution formed by ABri, ADan, α-synuclein, and amylin (image size: 35 nm for ABri, 20 nm for ADan, and 25 nm for α-synuclein and amylin). Please see the colour plate section for a colour version of this figure.

and the more recently described ABri and ADan peptides (Fig. 6.4c) (Lin *et al.* 2001; Kagan *et al.* 2004; Quist *et al.* 2005). Despite the common capacity to generate channels, structural, biochemical, and electrophysiological data demonstrated a certain heterogeneity in the different multimeric channels assembled by the different amyloid species. This heterogeneity could either reflect conformational changes in the different amyloid structures, a simple difference in the number of subunits that form a single channel (Rhee *et al.* 1998; Hirakura *et al.* 1999; Kawahara *et al.* 2000; Lin *et al.* 2001), or a varying channel-forming activity depending on the nature of the lipid mixtures (Kagan *et al.* 2002; Quist *et al.* 2005). Although the contribution of channel formation to disease pathogenesis remains to be further elucidated, many of the effects of amyloid *in vivo*, including Ca^{2+} dysregulation, membrane depolarization, mitochondrial dysfunction, inhibition of long-term potentiation, and cytotoxicity may be attributed to channel formation in both plasma and intracellular membranes.

6.4.3 **Necrosis/apoptosis induction**

The mechanisms leading to cell toxicity and underlying human amyloid disorders are the focus of intensive studies in neurodegenerative diseases, particularly in AD. Although the precise intracellular signalling triggered by the amyloid peptides is not fully understood, Aβ exerts a detrimental effect on neuronal cell viability, and both necrotic and apoptotic processes have been reported to occur *in vitro* and *in vivo* in AD brains (Yankner *et al.* 1990; Loo *et al.* 1993; Behl *et al.* 1994; Su *et al.* 1994). Unlike the multistep apoptotic process, necrosis represents the rapid collapse of internal cell homeostasis, exhibiting cell swelling, organelle damage, chromatin clumping, and breakdown of the plasma membrane. DNA degradation occurs late in the destruction of the cell and is random; it is detectable as a smear in DNA gels rather than the laddering found in cells undergoing apoptosis (Behl *et al.* 1994).

The apoptotic mechanisms are extremely complex, involving an increasing number of proteins, a regulatory balance between pro- and anti-apoptotic pathways, and different subcellular organelles (e.g. lysosomes, ER, and mitochondria). Although apoptosis was originally assumed to participate mainly in physiological cell death processes during brain development, new evidence seems to indicate a role in neurodegenerative disorders. The execution of neuronal apoptosis involves relatively few pathways that converge on the activation of the caspase family of cysteine proteases (Cribbs *et al.* 2004). Two principal pathways involving either cell death receptors (e.g. Fas and the p75 neurotrophin receptor) or the mitochondria, lead to the specific and sequential activation of a series of caspases (reviewed by Takuma *et al.* 2005). In the cell death receptor path, the critical event transmitting the death signal is the activation of caspase 8. However, in the mitochondrial pathway the mechanism is triggered by the formation of an apoptotic protease activating factor-1 (Apaf-1)–cytochome *c* complex which is fully functional in recruiting and activating pro-caspase 9 (Takuma *et al.* 2005) and subsequently cleaves and activates the downstream caspases 3, 6, and 7. Interestingly, a significant participation in mitochondrial dysfunction has been attributed to the formation of ion channels or non-selective pores on the organelle membrane with the resulting release of cytochrome c in a mechanism which mimics the formation of ion channels on lipid bilayers discussed above. Regardless of the mechanism, the activation of effector caspases leads to (i) DNAse activation with subsequent DNA cleavage into oligonucleosome-sized fragments and production of nuclear fragmentation, and (ii) the enzymatic degradation of a number of proteins that coordinate the cell death process (reviewed by Mattson 2000).

6.4.4 Oxidative stress

A common feature of neurodegenerative diseases is the extensive evidence of oxidative stress contributing to the dysfunction and/or death of neuronal cells. As a consequence of its high oxygen consumption, relatively low antioxidant levels, and low regenerative capacity, brain tissue is particularly susceptible to oxidative damage (Barnham *et al.* 2004) resulting from the unregulated production of reactive oxygen species, such as hydrogen peroxide, nitric oxide, superoxide, and the highly reactive hydroxyl radicals. Lipid peroxidation of unsaturated fatty acids together with oxidation of proteins and DNA result in impaired cellular functions, formation of toxic species (e.g. peroxides, alcohols, aldehydes, ketones, and cholesterol oxides), altered enzymatic activity, dysregulation of intracellular calcium signalling, and triggering of a cascade of events leading to cell death (Ferrari 2000; Mattson and Chan 2003; Tamagno *et al.* 2003; Barnham *et al.* 2004). In this sense, there is interesting data that links the formation of ion-channel structures with the production of oxidative stress. Assembling of Aβ ion-channel-like structures in astrocytic mitochondrial membranes mediates calcium-dependent activation of NADPH (nicotinamide adenine dinucleotide phosphate) oxidase, generating an oxidative stress response in the astrocytes which, in turn, contributes to neuronal cell death (Abramov and Duchen 2005). The mechanisms of oxidative stress are intrinsically linked to the detrimental effects of metal ions, since the chemical origin of the majority of reactive oxygen species arise from the Fentom reaction with the redox-active metals copper and iron (Huang *et al.* 1999a,b; Barnham *et al.* 2004).

6.4.5 Inflammation

Compelling evidence has accumulated during the last decade pointing to a significant role of local inflammatory processes in the progression of neurodegenerative disorders, including AD, Parkinson's disease, and prion diseases (reviewed by Minghetti 2005). These inflammation-related mechanisms have been widely studied in AD (Akiyama *et al.* 2000). In particular, complement activation and its pro-inflammatory consequences have been demonstrated to contribute extensively to disease pathogenesis (Tenner 2001), and inflammation-related cytokines are now considered to be a driving force in the neuropathological cascade associated with AD (reviewed by Griffin *et al.* 1998; Akiyama *et al.* 2000). Complement activation products co-localize with cerebral parenchymal and vascular deposits in AD and non-Aβ cerebral amyloidosis, thereby indicating that the chronic inflammatory response, most likely initiated by the deposits, is probably a general phenomenon (Ishii *et al.* 1984; Akiyama *et al.* 2000; Emmerling *et al.* 2000; Rostagno *et al.* 2003). These deposit-associated components originate from direct activation of the complement system by Aβ and non-Aβ amyloid peptides (e.g. ABri and ADan) and, once generated, seem to participate, at least in the case of Aβ, in several key steps of amyloidogenesis including aggregation, microglial activation, and phagocytosis (Rogers *et al.* 1992; Webster *et al.* 1994, 2004; Bradt *et al.* 1998; Emmerling *et al.* 2000; Rostagno *et al.* 2003). The presence of activated cytokine-expressing microglia co-localizing with amyloid deposits in affected brain areas is a common finding in AD and other non-Aβ neurodegenerative disorders (Griffin *et al.* 1998; Ghiso *et al.* 2001a,b). This, together with the presence of increased levels of complement activation products, cytokines, chemokines, and free radicals, leads to the concept of neuroinflammation as a self-propagating toxic cycle in which several factors—protein aggregates, abnormal cellular components, injured neurons, and abnormal synapses—activate microglia to release inflammatory mediators, which in turn exacerbate amyloid deposition and neuronal injury (Mrak and Griffin 2005).

6.5 Concluding remarks

The mechanism of amyloidogenesis is undoubtedly very complex and interlinks a diversity of mechanistic pathways leading to cell toxicity and death. Although AD is by far the most common and best studied form of amyloid-related disorder in humans, recent findings on the molecular mechanisms of brain degeneration have demonstrated common features among this heterogeneous group of disorders. Histopathological, genetic, biochemical, and physicochemical studies, together with the generation of transgenic animal models, strongly support the notion that the members of this diverse group of pathological entities are caused by the abnormal folding, aggregation/fibrilization, and subsequent tissue accumulation of particular proteins, specific for each disorder. The similarities in the physicochemical mechanisms ruling the aggregation/fibrillogenesis pathways and leading to similar endpoint structures, as well as the existence of common factors able to modulate the fibrillogenesis process, bridge together this wide range of dissimilar pathological entities and suggest unifying mechanisms of disease pathogenesis. The lessons learned from the study of systemic and CNS amyloid disorders may shed light on the field of neurodegeneration by providing alternative models to study the molecular basis of neuronal cell death.

Acknowledgments

This work was supported in part by the National Institutes of Health Grants AG08721 and NS38777, by the Alzheimer Association, and by the Alzheimer's Disease Research Program of the American Health Assistance Foundation.

References

Abramov AY and Duchen MR (2005). The role of an astrocytic NADPH oxidase in the neurotoxicity of amyloid beta peptides. *Philosophical Transactions of the Royal Society of London, Series B* **360**, 2309–14.

Akiyama H, Barger S, Barnum S, *et al.* (2000). Inflammation and Alzheimer's disease. *Neurobiology of Aging* **21**, 383–421.

Arispe N (2004). Architecture of Alzheimer's Aβ ion channel pore. *Journal of Membrane Biology* **197**, 33–48.

Atwood CS, Moir RD, Huang X, *et al.* (1998). Dramatic aggregation of Alzheimer Aβ by Cu(II) is induced by conditions representing physiological acidosis. *Journal of Biological Chemistry*, **273**, 12817–26.

Barnham KJ, Ciccotosto GD, Tickler AK, *et al.* (2003). Neurotoxic, redox-competent Alzheimer's β-amyloid is released from lipid membrane by methionine oxidation. *Journal of Biological Chemistry* **278**, 42959–65.

Barnham KJ, Masters CL, and Bush AI (2004). Neurodegenerative diseases and oxidative stress. *Nature Reviews*, **3**, 205–14.

Behl C, Davis JB, Klier FG, and Schubert D (1994). Amyloid β peptide induces necrosis rather than apoptosis. *Brain Research* **645**, 253–64.

Bek T (2000). Ocular changes in heredo-oto-ophtalmo-encephalopathy. *British Journal of Ophthalmology* **84**, 1298–302.

Bellotti V, Stoppini M, Mangione P, *et al.* (1998). Beta2-microglobulin can be refolded into a native state from *ex vivo* amyloid fibrils. *European Journal of Chemistry* **258**, 61–7.

Bellotti V, Mangione P, and Merlini G (2000). Review: immunoglobulin light chain amyloidosis–the archetype of structural and pathogenic variability. *Journal of Structural Biology* **130**, 280–9.

Botto M, Hawkins PN, Bickerstaff MCM, *et al.* (1997). Amyloid deposition is delayed in mice with targeted deletion of serum amyloid P component. *Nature Medicine* **3**, 855–9.

Bradt B, Kolb WP, and Cooper NR (1998). Complement-dependent proinflammatory properties of the Alzheimer's disease β-peptide. *Journal of Experimental Medicine* **188**, 431–8.

Bucciantini M, Giannoni E, Chiti F, *et al.* (2002). Inherent toxicity of aggregates implies a common mechanism for protein misfolding diseases. *Nature* **416**, 507–11.

Bucciantini M, Calloni G, Chiti F, *et al.* (2004). Prefibrillar amyloid protein aggregates share common features of cytotoxicity. *Journal of Biological Chemistry* **279**, 31374–82.

Bush AI, Pettingell WH, Multhaup G, *et al.* (1994). Rapid induction of Alzheimer Aβ amyloid formation by zinc. *Science* **265**, 1464–7.

Buxbaum JN (2004). The systemic amyloidoses. *Current Opinion in Rheumatology* **16**, 67–75.

Castano EM, Ghiso J, Prelli F, Gorevic PD, Migheli A, and Frangione B (1986). *In vitro* formation of amyloid fibrils from two synthetic peptides of different lengths homologous to Alzheimer's disease beta-protein. *Biochemical and Biophysical Research Communications* **141**, 782–9.

Chiti F, Webster P, Taddei N, *et al.* (1999). Designing conditions for *in vitro* formation of amyloid protofilaments and fibrils. *Proceedings of the National Academy of Sciences of the USA* **96**, 3590–4.

Citron M, Westaway D, Xia W, *et al.* (1997). Mutant presenilins of Alzheimer's disease increase production of 42-residue amyloid beta-protein in both transfected cells and transgenic mice. *Nature Medicine* **3**, 67–72.

Cohen DH, Feiner H, Jensson O, and Frangione B (1983). Amyloid fibril in hereditary cerebral hemorrhage with amyloidosis (HCHWA) is related to the gastroentero pancreatic neuroendocrine protein, gamma trace. *Journal of Experimental Medicine* **158**, 623–8.

Colon W and Kelly JW (1992). Partial denaturation of transthyretin is sufficient for amyloid fibril formation *in vitro*. *Biochemistry* **31**, 8654–60.

Coria F, Castaño EM, Prelli F, *et al.* (1988). Isolation and characterization of amyloid P component from Alzheimer's disease and other types of cerebral amyloidosis. *Laboratory Investigation* **58**, 454–7.

Cribbs DH, Poon WW, Rissman RA, and Blurton-Jones M (2004). Caspase-mediated degeneration in Alzheimer's disease. *American Journal of Pathology* **165**, 353–5.

Damas AM and Saraiva MJ (2005). Transthyretin. In Sipe JD (ed.), *Amyloid proteins. The beta sheet conformation and disease* Vol. 2, pp. 571–88 (Wiley–VCH Verlag, Weinheim, Germany).

D'Andrea MR, Nagele RG, Wang HY, Peterson PA, and Lee DH (2001). Evidence that neurones accumulating amyloid can undergo lysis to form amyloid plaques in Alzheimer's disease. *Histopathology* **38**, 120–34.

Dobson CM (1999). Protein misfolding, evolution and disease. *Trends in Biochemical Sciences* **24**, 329–32.

Dobson CM (2001a). Protein folding and its links with human disease. *Biochemical Society Symposium* **68**, 1–26.

Dobson CM (2001b). The structural basis of protein folding and its links with human disease. *Philosophical Transactions of the Royal Society of London, Series B* **356**, 133–45.

Dong J, Atwood CS, Anderson VE, *et al.* (2003). Metal binding and oxidation of amyloid-beta within isolated senile plaque cores: raman microscopic evidence. *BioChemistry* **42**, 2768–73.

Dumoulin M, Belloti V, and Dobson CM (2005). Lysozyme. In Sipe JD (ed.), *Amyloid proteins. The beta sheet conformation and disease*, Vol. 2, pp. 635–56 (Wiley–VCH Verlag, Weinheim, Germany).

Durie BGM, Persky B, Soehnlen BJ, Grogan TM, and Salmon SE (1982). Amyloid production in human myeloma stem-cell culture with morphologic evidence of amyloid secretion by associated macrophages. *New England Journal of Medicine* **307**, 1689–92.

Ellis RJ (2004). Editorial. *Seminars in Cell and Developmental Biology* **15**, 1–2.

Emmerling MR, Watson MD, Raby CA, and Spiegel K (2000). The role of complement in Alzheimer's disease pathology. *Biochimica Biophysica Acta* **1502**, 158–71.

Fandrich M, Forge V, Buder K, Kittler M, Dobson CM, and Diekmann S (2001). Myoglobin forms amyloid fibrils by association of unfolded polypeptide segments. *Proceedings of the National Academy of Sciences of the USA* **100**, 15463–8.

Ferrari CKB (2000). Free radicals, lipid peroxidation and antioxidants in apoptosis: implications in cancer, cardiovascular and neurological disease. *Biologia* **55**, 581–90.

Floege J and Ehlerding G (1996). Beta-2-microglobulin-associated amyloidosis. *Nephron* **72**, 9–26.

Gallo G, Picken M, Frangione B, and Buxbaum JN (1988). Nonamyloidotic monoclonal immunoglobulin deposits lack amyloid P component. *Modern Pathology* **1**, 453–6.

Gallo G, Picken M, Buxbaum J, and Frangione B (1989). The spectrum of monoclonal immunoglobulin deposition disease associated with immunocytic dyscrasias. *Seminars in Hematology* **26**, 234–45.

Gallo G, Wisniewski T, Choi-Miura NH, Ghiso J, and Frangione B (1994). Potential role of apolipoprotein-E in fibrillogenesis. *American Journal of Pathology* **145**, 526–30.

Geiger T and Clarke S (1987). Deamidation, isomerization, and racemization at asparaginyl and aspartyl residues in peptides. Succinimide-linked reactions that contribute to protein degradation. *Journal of Biological Chemistry* **262**, 785–94.

Ghiso J and Frangione B (2002). Amyloidosis and Alzheimer's disease. *Advanced Drug Delivery Reviews* **54**, 1539–51.

Ghiso J, Jensson O, and Frangione B (1986a). Amyloid fibrils in hereditary cerebral hemorrhage with amyloidosis of Icelandic type is a variant of gamma-trace basic protein (cystatin C). *Proceedings of the National Academy of Sciences of the USA* **83**, 2974–8.

Ghiso J, Pons-Estel B, and Frangione B (1986b). Hereditary cerebral amyloid angiopathy: the amyloid contain a protein which is a variant of cystatin C, an inhibitor of lysosomal cysteine proteases. *Biochemical and Biophysical Research Communications* **136**, 548–54.

Ghiso J, Holton J, Miravalle L, *et al.* (2001a). Systemic amyloid deposits in Familial British Dementia. *Journal of Biological Chemistry* **276**, 43909–14.

Ghiso J, Revesz T, Holton J, *et al.* (2001b). Chromosome 13 dementia syndromes as models of neurodegeneration. *Amyloid* **8**, 277–284.

Ghiso J, Rostagno A, Tomidokoro Y, *et al.* (2005). Familial British and Danish dementias. In Sipe JD (ed.), *Amyloid proteins. The beta sheet conformation and disease*, Vol. 2, pp. 515–26 (Wiley–VCH Verlag, Weinheim, Germany).

Goldfarb LG, Brown P, Haltia M, Ghiso J, Frangione B, and Gajdusek DC (1993). Synthetic peptides corresponding to different mutated regions of the amyloid gene in familial Creutzfeldt–Jakob disease show enhanced *in vitro* formation of morphologically different amyloid fibrils. *Proceedings of the National Academy of Sciences of the USA* **90**, 4451–4.

Grabowski TJ, Cho HS, Vonsattel JPG, Rebeck GW, and Greenberg SM (2001). A novel APP mutation in an Iowa family with dementia and severe cerebral amyloid angiopathy. *Annals of Neurology* **49**, 697–705.

Griffin WST, Sheng JG, Royston MC, *et al.* (1998). Glial-neuronal interactions in Alzheimer's disease: the potential role of a 'cytokine cycle' in disease progression. *Brain Pathology* **8**, 65–72.

Gudmundsson G, Hallgrimsson J, Jonasson T, and Bjarnason O (1972). Hereditary cerebral hemorrhage with amyloidosis. *Brain* **95**, 387–404.

Guijarro JI, Sunde M, Jones JA, Campbell ID, and Dobson CM (1998). Amyloid formation by an SH3 domain. *Proceedings of the National Academy of Sciences of the USA* **95**, 4224–8.

Haass C and Steiner H (2001). Protofibrils, the unifying toxic molecule of neurodegenerative disorders? *Nature Neuroscience* **4**, 859–60.

Harigaya Y, Saido TC, Eckman CB, Prada CM, Shoji M, and Younkin SG (2000). Amyloid β protein starting pyroglutamate at position 3 is a major component of the amyloid deposits in the Alzheimer's disease brain. *Biochemical and Biophysical Research Communications* **24**, 422–7.

Harper JD, Lieber CM, and Lansbury PT (1997). Observation of metastable Aβ amyloid protofibrils by atomic force microscopy. *Chemistry and Biology* **4**, 951–9.

Harper JD, Wong CW, Lieber CM, and Lansbury PT (1999). Assembly of Aβ amyloid protofibrils: an *in vitro* model for a possible early event in Alzheimer's disease. *BioChemistry* **38**, 8972–80.

Hendriks L, Van Duijn CM, Cras P, *et al.* (1992). Presenile dementia and cerebral haemorrhage linked to a mutation at codon 692 of the beta-amyloid precursor protein gene. *Nature Genetics* **1**, 218–21.

Herzig MC, Winkler DT, Burgermeister P, *et al.* (2004). Aβ is targeted to the vasculature in a mouse model of hereditary cerebral hemorrhage with amyloidosis. *Nature Neuroscience* **7**, 954–60.

Hirakura Y, Lin M-C, and Kagan BL (1999). Alzheimers amyloid Aβ 1–42 channels: Effect of solvent, pH and Congo red. *Journal of Neuroscience Research* **57**, 458–66.

Holtzman DM, Bales KR, Wu S, *et al.* (1999). Expression of human apolipoprotein E reduces amyloid-beta deposition in a mouse model of Alzheimer's disease. *Journal of Clinical Investigation* **103**, R15–21.

Hooper NM (2005). Roles of proteolysis and lipid rafts in the processing of the amyloid precursor protein and prion protein. *Biochemical Society Transactions* **33**, 335–8.

Howland DS, Trusko SP, Savage MJ, *et al.* (1998). Modulation of secreted beta-amyloid precursor protein and amyloid-beta peptide in brain by cholesterol. *Journal of Biological Chemistry* **273**, 16576–82.

Huang X, Atwood CS, Hartshorn MA, *et al.* (1999a). The Aβ peptide of Alzheimer's disease directly produces hydrogen peroxide through metal ion reduction. *BioChemistry* **38**, 7609–16.

Huang X, Cuajungco MP, Atwood CS, *et al.* (1999b). Cu(II) potentiation of Aβ Alzheimer Aβ neurotoxicity. Correlation with cell-free hydrogen peroxide production and metal reduction. *Journal of Biological Chemistry* **274**, 37111–16.

Ishii T, Haga S, Yagishita S and Tateishi J (1984). The presence of complement in amyloid plaques of Creutzfeldt–Jakob disease and Gerstmann–Straussler–Scheinker disease. *Applied Pathology* **2**, 370–9.

Jarrett JT, Berger EP, and Lansbury PT (1993). The C-terminus of the beta protein is critical in amyloidogenesis. *Annals of the New York Academy of Sciences* **695**, 144–8.

Jascolski M and Grubb A (2005). Cystatin C. In Sipe JD (ed.), *Amyloid proteins. The beta sheet conformatio and disease*, Vol. 2, pp. 697–722 (Wiley–VCH Verlag, Weinheim, Germany).

Kagan BL, Hirakura Y, Azimov R, Azimova R, and Lin MC (2002). The channel hypothesis of Alzheimer's disease: current status. *Peptides* **23**, 1311–5.

Kagan BL, Azimov R, and Azimova R (2004). Amyloid peptide channels. *Journal of Membrane Biology* **202**, 1–10.

Kalaria RN, Galloway PG, and Perry G (1991). Widespread serum amyloid P immunoreactivity in cortical amyloid deposits and the neurofibrillary pathology of Alzheimer's disease and other degenerative disorders. *Neuropathology and Applied Neurobiology* **17**, 189–201.

Kamino K, Orr HT, Payami H, *et al.* (1992). Linkage and mutational analysis of familial Alzheimer disease kindreds for the APP gene region. *American Journal of Human Genetics* **51**, 998–1014.

Kawahara M, Kuroda Y, Arispe N, and Rojas E (2000). Alzheimer's beta-amyloid, human islet amylin, and prion protein fragment evoke intracellular free calcium elevations by a common mechanism in a hypothalamic GnRH neuronal cell line. *Journal of Biological Chemistry* **275**, 14077–83.

Kawarabayashi T, Shoji M, Younkin LH, *et al.* (2004). Dimeric amyloid β protein rapidly accumulates in lipid rafts followed by apolipoprotein E and phosphorylated tau accumulation in the Tg2576 mouse model of Alzheimer's disease. *Journal of Neuroscience* **24**, 3801–9.

Kayed R, Head E, Thompson JL, *et al.* (2003). Common structure of soluble amyloid oligomers implies common mechanism of pathogenesis. *Science* **300**, 486–9.

Kelly JW (1998). The alternative conformations of amyloidogenic proteins and their multi-step assembly pathways. *Current Opinion in Structural Biology* **8**, 101–6.

Kienlen-Campard P, Miolet S, Tasiaux B, and Octave JN (2002). Intracellular amyloid-beta 1–42, but not extracellular soluble amyloid-beta peptides, induces neuronal apoptosis. *Journal of Biological Chemistry* **277**, 15666–70.

Kim S-H, Wang R, Gordon DJ, *et al.* (1999). Furin mediates enhanced production of fibrillogenic ABri peptides in familial British dementia. *Nature Neuroscience* **2**, 984–8.

Kindy MS, King AR, Perry G, de Beer MC, and de Beer FC (1995). Association of apolipoprotein E with murine amyloid A protein amyloid. *Laboratory Investigation* **73**, 469–75.

Klein WL, Krafft GA, and Finch C (2001). Targeting small Aβ oligomers: the solution to an Alzheimer's disease conundrum? *Trends in Neuroscience* **24**, 219–24.

Klunk WE, Jacob R,F and Mason RP (1999). Quantifying amyloid by Congo red spectral shift assay. *Methods in Enzymology* **309**, 285–305.

Kuo YM, Emmerling MR, Vigo-Pelfrey C, *et al.* (1996). Water-soluble Aβ (N-40, N-42) oligomers in normal and Alzheimer disease brains. *Journal of Biological Chemistry* **271**, 4077–81.

Kuo Y-M, Kokjohn TA, Beach T, *et al.* (2001). Comparative analysis of amyloid-β chemical structure and amyloid plaque morphology of transgenic mouse and Alzheimer's disease brains. *Journal of Biological Chemistry*, **276**, 12991–8.

LaDu MJ, Falduto MT, Manelli AM, Reardon CA, Getz GS, and Frail DE (1994). Isoform-specific binding of apolipoprotein E to beta-amyloid. *Journal of Biological Chemistry* **269**, 23403–6.

Lalowski M, Golabek A, Lemere CA, *et al.* (1996). The 'nonamyloidogenic' p3 fragment (amyloid beta17–42) is a major constituent of Down's syndrome cerebellar preamyloid. *Journal of Biological Chemistry* **271**, 33623–31.

Lambert MP, Barlow AK, Chromy BA, *et al.* (1998). Diffusible, nonfibrillar ligands derived from Aβ1–42 are potent central nervous system neurotoxins. *Proceedings of the National Academy of Sciences of the USA* **95**, 6448–53.

Lambert MP, Viola KL, Chromy BA, *et al.* (2001). Vaccination with soluble Aβ oligomers generates toxicity-neutralizing antibodies. *Journal of Neurochemistry* **79**, 595–605.

Lansbury PT (1999). Amyloid diseases: abnormal protein aggregation in neurodegeneration. *Proceedings of the National Academy of Sciences of the USA* **96**, 3342–4.

Lazo ND, Maji SK, Fradinger EA, Bitan G, and Teplow D (2005). The amyloid β protein. In Sipe JD (ed.), *Amyloid proteins. The beta sheet conformation and disease*, Vol. 2, pp. 385–491 (Wiley–VCH Verlag, Weinheim, Germany).

Lee S-J, Liyanage U, Bickel PE, Xia W, Lansbury PTJ, and Kosik KS (1998). A detergent-insoluble membrane compartment contains Aβ *in vivo*. *Nature Medicine* **4**, 730–4.

LeVine H (1995a). Thioflavine T interaction with amyloid β-sheet structures. *Amyloid* **2**, 1–6.

LeVine H (1995b). Soluble multimeric Alzheimer Aβ1–40 pre-amyloid complexes in dilute solution. *Neurobiology of Aging* **16**, 755–764.

Levy E, Lopez-Otin C, Ghiso J, Geltner D, and Frangione B (1989). Stroke in Icelandic patients with hereditary amyloid angiopathy is related to a mutation in the cystatin C gene, an inhibitor of cysteine proteases. *Journal of Experimental Medicine* **169**, 1771–8.

Levy E, Carman MD, Fernandez Madrid IJ, *et al.* (1990). Mutation of the Alzheimer's disease amyloid gene in hereditary cerebral hemorrhage, Dutch type. *Science* **248**, 1124–6.

Lin H, Zhu YJ, and Lal R (1999). Amyloid beta protein (1–40) forms calcium-permeable, Zn2+-sensitive channel in reconstituted lipid vesicles. *BioChemistry* **38**, 11189–96.

Lin H, Bhatia R, and Lal R (2001). Amyloid beta protein forms ion channels: implications for Alzheimer's disease pathophysiology. *Faseb Journal* **15**, 2433–44.

Linke RP, Hampl H, Bartel-Schwarze S, and Eulitz M (1987). Beta 2-microglobulin, different fragments and polymers thereof in synovial amyloid in long-term hemodialysis. *Biological Chemistry Hoppe Seyler* **368**, 137–44.

Litvinovich SV, Brew SA, Aota S, Akiyama SK, Haudenschild C, and Ingham KC (1998). Formation of amyloid-like fibrils by self-association of a partially unfolded fibronectin type III module. *Journal of Molecular Biology* **280**, 245–58.

Loo DT, Copani A, Pike CJ, Whittemore ER, Walencewicz AJ and Cotman CW (1993). Apoptosis is induced by β-amyloid in cultured central nervous system neurons. *Proceedings of the National Academy of Sciences of the USA* **90**, 7951–5.

Lopez De La Paz M, Goldie K, Zurdo J, *et al.* (2002). De novo designed peptide-based amyloid fibrils. *Proceedings of the National Academy of Sciences of the USA* **99**, 16052–7.

Lorenzo A and Yankner BA (1994). β-amyloid neurotoxicity requires fibril formation and is inhibited by Congo red. *Proceedings of the National Academy of Sciences of the USA* **91**, 12243–7.

Lott IT and Head E (2005). Alzheimer disease and Down syndrome: factors in pathogenesis. *Neurobiology of Aging* **26**, 383–9.

Lovell MA, Robertson JD, Teesdale WJ, Campbell JL and Markesbery W (1998). Copper, iron and zinc in Alzheimer's disease senile plaques. *J. Neurol. Sci.*, **158**, 47–52.

Lovestone S and McLoughlin DM (2002). Protein aggregates and dementia: is there a common toxicity? *J. Neurol. Neurosurg. Psychiatry*, **2002**, 152–61.

Lue L-F, Kuo Y-M, Roher A, *et al.* (1999). Soluble amyloid beta peptide concentration as a predictor of synaptic change in Alzheimer's disease. *American Journal of Pathology* **155**, 853–62.

Luyendijk W, Bofs GT, Vegter-van der Vlis M, *et al.* (1988). Heredital cerebral hemorrhage caused by cortical amyloid angiopathy. *Journal of Neurological Science* **85**, 267–80.

McLaurin J, Yang DS, Yip CM and Fraser PE (2000). Review: Modulating Factors in Amyloid-β fibril formation. *Journal of Structural Biology* **130**, 259–70.

McLean CA, Cherny RA, Fraser FW, *et al.* (1999). Soluble pool of Abeta amyloid as a determinant of severity of neurodegeneration in Alzheimer's disease. *Ann. Neurol.*, **46**, 860–6.

McParland VJ, Kad NM, Kalverda AP, *et al.* (2000). Partially unfolded states of beta(2)-microglobulin and amyloid formation *in vitro*. *Biochemistry* **39**, 8735–46.

Ma J, Yee A, Brewer HB, Jr., Das S and Potter H (1994). Amyloid-associated proteins alpha 1-antichymotrypsin and apolipoprotein E promote assembly of Alzheimer beta-protein into filaments. *Nature*, **392**, 92–4.

Mann DMA and Esiri MM (1989). The pattern of acquisition of plaques and tangles in the brains of patients under 50 years of age with Down's syndrome. *J. Neurol. Sci.*, **89**, 169–79.

Matsubara E, Soto C, Governale S, Frangione B and Ghiso J (1996). Apolipoprotein J and Alzheimer's amyloid beta solubility. *Biochem J.*, **316**, 671–9.

Mattson MP (2000). Apoptosis in neurodegenerative disorders. *Nat. Rev. Molec. Cell Biol.*, **1**, 120–9.

Mattson MP and Chan SL (2003). Neuronal and glial calcium signaling in Alzheimer's disease. *Cell Calcium*, **34**, 385–97.

Mead S, James Galton M, Revesz T, *et al.* (2000). Familial British dementia with amyloid angiopathy: Early clinical, neuropsychological and imaging findings. *Brain*, **123**, 975–86.

Mehta PD, Dalton AJ, Mehta SP, Kim KS, Sersen EA and Wisniewski HM (1998). Increased plasma amyloid beta protein 1–42 levels in Down syndrome. *Neurosc. Lett.*, **241**, 13–6.

Minghetti L (2005). Role of inflammation in neurodegenerative diseases. *Curr. Opin. Neurol.*, **18**, 315–321.

Miravalle L, Tokuda T, Chiarle R, *et al.* (2000). Substitution at codon 22 of Alzheimer's Aβ peptide induce diverse conformational changes and apoptotic effects in human cerebral endothelial cells. *Journal of Biological Chemistry* **275**, 27110–6.

Monro OR, Mackic JB, Yamada S, *et al.* (2002). Substitution at codon 22 reduces clearance of Alzheimer's amyloid-beta peptide from the cerebrospinal fluid and prevents its transport from the central nervous system into blood. *Neurobiol Aging.*, **23**, 405–12.

Mori H, Takio k, Ogawara M and Selkoe D (1992). Mass spectromety of purified amyloid beta protein in Alzheimer's disease. *Journal of Biological Chemistry* **267**, 17082–6.

Morishima-Kawashima M and Ihara Y (1998). The presence of amyloid β-protein in the detergent-insoluble membrane compartment of human neuroblastoma cells. *Biochemistry* **37**, 15427–53.

Morozova-Roche LA, Zurdo J, Spencer A, *et al.* (2000). Amyloid fibril formation and seeding by wild-type human lysozyme and its disease-related mutational variants. *Journal of Structural Biology* **130**, 339–51.

Mrak RE and Griffin WST (2005). Glia and their cytokines in progression of neurodegeneration. *Neurobiol Aging.*, **26**, 349–54.

Namba Y, Tomonaga M, Kawasaki H, Otomo E and Ikeda K (1991). Apolipoprotein E immunoreactivity in cerebral amyloid deposits and neurofibrillary tangles in Alzheimer's disease and kuru plaque amyloid in Creutzfeldt-Jakob disease. *Brain Research* **541**, 163–6.

Naslund J, Schierhorn A, Hellman U, *et al.* (1994). Relative abundance of Alzheimer A beta amyloid peptide variants in Alzheimer disease and normal aging. *Proceedings of the National Academy of Sciences of the USA* **91**, 8378–82.

Obici L, Demarchi A, de Rosa G, *et al.* (2005). A novel APP mutation exclusively associated with cerebral amyloid angiopathy. *Ann. Neurol.*, **58**, 639–44.

Oda T, Wals P, Osterburg HH, *et al.* (1995). Clusterin (apoJ) alters the aggregation of amyloid beta-peptide (A beta 1–42) and forms slowly sedimenting A beta complexes that cause oxidative stress. *Exp. Neurol.*, **136**, 22–31.

Pan KM, Baldwin M, Nguyen J, *et al.* (1993). Conversion of alpha-helices into beta-sheets features in the formation of the scrapie prion proteins. *Proceedings of the National Academy of Sciences of the USA* **90**, 10962–6.

Pertinhez TA, Bouchard M, Tomlinson EJ, *et al.* (2001). Amyloid fibril formation by a helical cytochrome. *FEBS Lett.*, **495**, 184–6.

Phinney AL, Horne P, Yang J, Janus C, Bergeron C and Westaway D (2003). Mouse models of Alzheimer's disease: the long and filamentous road. *Neurol. Res.*, **25**, 590–600.

Pike CJ, Burdick D, Walencewicz AJ, Glabe CG and Cotman CW (1993). Neurodegeneration induced by beta-amyloid peptides *in vitro*: the role of peptide assembly state. *Journal of Neuroscience* **13**, 1676–87.

Pittois K, Deleersnijder W and Merregaert J (1998). cDNA sequence analysis, chromosomal assignment and expression pattern of the gene coding for integral membrane protein 2B. *Gene*, **217**, 141–9.

Prelli F, Levy E, Van Duinen SG, Bots GT, Luyendijk W and Frangione B (1990). Expression of a normal and variant Alzheimer's β-protein gene in amyloid of hereditary cerebral hemorrhage, Dutch type: DNA and protein diagnostic assays. *Biochemical and Biophysical Research Communications* **170**, 301–7.

Quist A, Doudevski I, Lin H, *et al.* (2005). Amyloids form membrane pores: a common structural link for protein-misfolding disease. *Proceedings of the National Academy of Sciences of the USA* **102**, 10427–32.

Ramirez-Alvarado M, Merkel JS and Regan L (2000). A systematic exploration of the influence of the protein stability on amyloid fibril formation *in vitro*. *Proceedings of the National Academy of Sciences of the USA* **97**, 8979–84.

Rhee SK, Quist AP and Lal R (1998). Amyloid beta protein-(1–42) forms calcium-permeable, Zn2+-sensitive channel. *Journal of Biological Chemistry* **273**, 13379–82.

Rogers J, Cooper NR, Webster S, *et al.* (1992). Complement activation by β-amyloid in Alzheimer's disease. *Proceedings of the National Academy of Sciences of the USA* **89**, 10016.

Roher AE, Lowenson JD, Clarke S, *et al.* (1993a). Beta-Amyloid-(1–42) is a major component of cerebrovascular amyloid deposits: implications for the pathology of Alzheimer disease. *Proceedings of the National Academy of Sciences of the USA* **90**, 10836–40.

Roher AE, Lowenson JD, Clarke S, *et al.* (1993b). Structural alterations in the peptide backbone of beta-amyloid core protein may account for its deposition and stability in Alzheimer's disease. *Journal of Biological Chemistry* **268**, 3072–83.

Roher AE, Chaney MO, Kuo YM, *et al.* (1996). Morphology and toxicity of Aβ-(1–42) dimer derived from neuritic and vascular amyloid deposits of Alzheimer's disease. *Journal of Biological Chemistry* **271**, 20631–5.

Rostagno A and Ghiso J (2003). Amyloidosis. In M Aminoff and R Daroff, eds. *Encyclopedia of neurological sciences.* **1**, pp. 129–135. Academic Press, San Diego.

Rostagno A, Vidal R, Kaplan B, *et al.* (1999). pH-dependent fibrillogenesis of a VkIII Bence Jones protein. *Br.J. Haematology*, **107**, 835–43.

Rostagno A, Revesz T, Lashley T, *et al.* (2003). Complement activation in Chromosome 13 dementias: Similarities with Alzheimer's Disease. *Journal of Biological Chemistry* **277**, 49782–90.

Rostagno A, Tomidokoro Y, Lashley T, *et al.* (2005). Chromosome 13 dementias. *Cell. Mol. Life Sci.*, **62**, 1814–25.

Russo C, Violani E, Salis S, *et al.* (2002). Pyroglutamate-modified amyloid β-peptides- AβN3(pE)- strongly affect cultured neuron and astrocyte survival. *Journal of Neurochemistry* **82**, 1480–9.

Saido T, Iwatsubo T, Mann DM, Shimada H, Ihara Y and Kawashima S (1995). Dominant and differential deposition of distinct β-amyloid peptide species, $A\beta_{N3(pE)}$, in senile plaques. *Neuron*, **14**, 457–66.

Saper MA, Bjorkman PJ and Wiley DC (1991). Refined structure of the human histocompatibility antigen HLA-A2 at 2.6 A resolution. *Journal of Molecular Biology* **219**, 277–319.

Scherzinger E, Sittler A, Schweiger K, *et al.* (1999). Self-assembly of polyglutamine-containing huntingtin fragments into amyloid-like fibrils: implications for Huntington's disease pathology. *Proceedings of the National Academy of Sciences of the USA* **96**, 4604–9.

Schupf N, Patel B, Silverman W, *et al.* (2002). Elevated plasma amyloid beta-peptide 1–42 and onset of dementia in adults with Down syndrome. *Neurosc. Lett.*, **301**, 199–203.

Selkoe D (1994). Alzheimer's disease: a central role for amyloid. *J. Neuropathol. Exp. Neurol.*, **53**, 438–47.

Serpell L (2000). Alzheimer's amyloid fibrils: structure and assembly. *Biochimica Biophysica Acta* **1502**, 16–30.

Shimizu T, Fukuda H, Murayama S, Izumiyama N and Shirasawa T (2002). Isoaspartate formation at position 23 of amyloid beta peptide enhanced fibril formation and deposited onto senile plaques and vascular amyloids in Alzheimer's disease. *Journal of Neuroscience Research* **70**, 451–61.

Shin Y, Cho HS, Fukumoto H, *et al.* (2003). Abeta species, including IsoAsp23 Abeta, in Iowa-type familial cerebral amyloid angiopathy. *Acta neuropathol.*, **105**, 252–258.

Simons K and Toomre D (2000). Lipid rafts and signal transduction. *Nat. Rev. Molec. Cell Biol.*, **1**, 31–40.

Sinha S (2002). The role of beta-amyloid in Alzheimer's disease. *Med. clin. North Am.*, **86**, 629–39.

Small DH (1998). The amyloid cascade hypothesis debate: emerging consensus on the role of A beta and amyloid in Alzheimer's disease. *Amyloid: Intl. J. Exp. Clin. Invest.*, **5**, 301–4.

Smith DP and Radford SE (2001). Role of the single disulphide bond of beta(2)-microglobulin in amyloidosis *in vitro*. *Protein Sci.*, **10**, 1775–84.

Soto C (2003). Unfolding the role of protein misfolding in neurodegenerative diseases. *Nat. Rev. Neurosci*, **4**, 49–59.

Srinivasan R, Jones EM, Liu K, Ghiso J, Marchant RE and Zagorsky MG (2003). pH-dependent amyloid and protofibrils formation by the ABri peptide of familial British dementia. *Journal of Molecular Biology* **333**, 1003–23.

Stefani M and Dobson CM (2003). Protein aggregation and aggregate toxicity: new insights into protein folding, misfolding diseases and biological evolution. *Journal of Molecular Medicine* **81**, 678–99.

Strömgren E, Dalby A, Dalby MA, and Ranheim B (1970). Cataract, deafness, cerebellar ataxia, psychosis and dementia: a new syndrome. *Acta Neurologica Scandinavica* **46**, 261–2.

Su JH, Anderson AJ, Cummings BJ, and Cotman CW (1994). Immunohistochemical evidence for apoptosis in Alzheimer's disease. *Neuroreport* **5**, 2529–33.

Sunde M, Serpell LC, Bartlam M, Fraser PE, Pepys MB, and Blake CC (1997). Common core structure of amyloid fibrils by synchrotron X-ray diffraction. *Journal of Molecular Biology* **273**, 729–39.

Swietnicki W, Petersen RB, Gambetti P, and Surewicz WK (1997). pH-dependent stability and conformation of the recombinant human prion protein PrP. *Journal of Biological Chemistry* **272**, 27517–20.

Takahashi RH, Milner TA, Li F, *et al.* (2002). Intraneuronal Alzheimer Aβ42 accumulates in multivesicular bodies and is associated with synaptic pathology. *American Journal of Pathology* **161**, 1869–79.

Takahashi Y, Ueno A, and Mihara H (2000). Mutational analysis of designed peptides that undergo structural transition from alpha helix to beta sheet and amyloid fibril formation. *Structure* **8**, 915–25.

Takuma K, Yan SD, Stern D, and Yamada K (2005). Mitochondrial dysfunction, Endoplasmic reticulum stress, and apoptosis in Alzheimer's disease. *Journal of Pharmacological Sciences* **97**, 312–6.

Tamagno E, Robino G, Obbili A, *et al.* (2003). H2O2 and 4-hydroxynonenal mediate amyloid beta-induced neuronal apoptosis by activating JNKs and p38MAPK. *Experimental Neurology* **180**, 144–55.

Taylor JP, Hardy J, and Fischbeck KH (2002). Toxic proteins in neurodegenerative disease. *Science* **296**, 1991–5.

Temussi PA, Masino L, and Pastore A (2003). From Alzheimer to Huntington: why is a structural understanding so difficult? *EMBO Journal* **22**, 355–61.

Tennent GA, Lovat LB, and Pepys MB (1995). Serum amyloid P component prevents proteolysis of the amyloid fibrils of Alzheimer's disease and systemic amyloidosis. *Proceedings of the National Academy of Sciences of the USA* **92**, 4299–303.

Tenner AJ (2001). Complement in Alzheimer's disease: opportunities for modulating protective and pathogenic events. *Neurobiology of Aging* **22**, 849–61.

Teplow D (1998). Structural and kinetic features of amyloid β-protein fibrillogenesis. *Amyloid* **5**, 121–42.

Tomidokoro Y, Lashley T, Rostagno A, *et al.* (2005). Familial Danish dementia: Co-existence of ADan and Aβ amyloid subunits in the absence of compact plaques. *Journal of Biological Chemistry* **280**, 36883–94.

Tomiyama T, Asano S, Furiya Y, Shirasawa T, Endo N, and Mori H (1994). Racemization of Asp23 residue affects the aggregation properties of Alzheimer amyloid beta protein analogues. *Journal of Biological Chemistry* **269**, 10205–8.

Trinh CH, Smith DP, Kalverda AP, Phillips SE and Radford SE (2002). Crystal structure of monomeric human beta-2-microglobulin reveals clues to its amyloidogenic properties. *Proceedings of the National Academy of Sciences of the USA* **99**, 9771–6.

Van Duinen SG, Castaño EM, Prelli F, Bots GT, Luyendijk W, and Frangione B (1987). Hereditary cerebral hemorrhage with amyloidosis in patients of Dutch origin is related to Alzheimer disease. *Proceedings of the National Academy of Sciences of the USA* **84**, 5991–4.

Van Nostrand WE, Melchor JP, Cho HS, Greenberg SM, and Rebeck GW (2001). Pathogenic effects of D23N Iowa mutant amyloid b-protein. *Journal of Biological Chemistry* **276**, 32860–6.

Varadarajan S, Kanski J, Aksenova M, Lauderback C, and Butterfield DA (2001). Different mechanisms of oxidative stress and neurotoxicity for Alzheimer's Aβ1–42 Aβ25–35. *Journal of the American Chemical Society* **123**, 5625–31.

Veerhuis R, Boshuizen RS, and Familian A (2005). Amyloid associated proteins in Alzheimer's and prion disease. *Current Drug Targets. CNS and Neurological Disorders* **4**, 235–48.

Ventura S, Zurdo J, Narayanan S, *et al.* (2004). Short amino acid stretches can mediate amyloid formation in globular proteins: the Src homology 3 (SH3) case. *Proceedings of the National Academy of Sciences of the USA* **101**, 7258–7263.

Vidal R, Frangione B, Rostagno A, *et al.* (1999). A stop-codon mutation in the *BRI* gene associated with familial British dementia. *Nature* **399**, 776–81.

Vidal R, Ghiso J, Revesz T, *et al.* (2000). A decamer duplication in the 3' region of the *BRI* gene originates a new amyloid peptide that is associated with dementia in a Danish kindred. *Proceedings of the National Academy of Sciences of the USA* **97**, 4920–5.

Villegas V, Zurdo J, Filimonov VV, Aviles FX, Dobson CM, and Serrano L (2000). Protein engineering as a strategy to avoid formation of amyloid fibrils. *Protein Science* **9**, 1700–8.

Volles MJ and Lansbury PT (2003). Zeroing in on the pathogenic form of α-synuclein and its mechanism of neurotoxicity in Parkinson's disease. *Biochemistry* **42**, 7871–8.

Walsh DM, Hartley DM, Kusumoto Y, *et al.* (1999). Amyloid β-protein fibrillogenesis. Structure and biological activity of protofibrillar intermediates. *Journal of Biological Chemistry* **274**, 25945–52.

Walsh DM, Klyubin I, Fadeeva JV, *et al.* (2002). Naturally secreted oligomers of amyloid beta protein potently inhibit hippocampal long-term potentiation *in vivo*. *Nature* **416**, 535–9.

Watanabe A, Takio K, and Ihara Y (1999). Deamidation and isoaspartate formation in smeared tau in paired helical filaments. Unusual properties of the microtubule-binding domain of tau. *Journal of Biological Chemistry* **274**, 7368–78.

Webster S, O'Barr S, and Rogers J (1994). Enhanced aggregation and β structure of amyloid β peptide after coincubation with C1q. *Journal of Neuroscience Research* **39**, 448–56.

Webster S, Yang AJ, Margol L, Garzon-Rodriguez W, Glabe CG, and Tenner AJ (2000). Complement component C1q modulates the phagocytosis of Aβ by microglia. *Experimental Neurology* **161**, 127–38.

Westermark P, Benson MD, Buxbaum JN, *et al.* (2002). Amyloid fibril protein nomenclature. *Amyloid* **9**, 197–200.

Westermark P, Benson MD, Buxbaum JN, *et al.* (2005). Amyloid: toward terminology clarification. Report from the Nomenclature Committee of the International Society of Amyloidosis. *Amyloid* **12**, 1–4.

Wisniewski T and Frangione B (1992). Apolipoprotein E: a pathological chaperone protein in patients with cerebral and systemic amyloid. *Neuroscience Letters* **135**, 235–238.

Wisniewski T, Ghiso J and Frangione B (1991). Peptides homologous to the amyloid protein of Alzheimer's disease containing a glutamine for glutamic acid substitution have accelerated amyloid fibril formation. *Biochemical and Biophysical Research Communications* **179**, 1247–54.

Wisniewski T, Castaño EM, Golabek AA, Vogel T and Frangione B (1993). Acceleration of Alzheimer's fibril formation by apolipoprotein E *in vitro*. *American Journal of Pathology* **145**, 1030–5.

Wood SJ, Maleeff B, Hart T and Wetzel R (1996). Physical, morphological and functional differences between pH 5.8 and 7.4 aggregates of the Alzheimer's amyloid peptide Aβ. *Journal of Molecular Biology* **256**, 870–7.

Wood SJ, Wypych J, Steavenson S, Louis JC, Citron M, and Biere AL (1999). Alpha-synuclein fibrillogenesis is nucleation-dependent. Implications for the pathogenesis of Parkinson's disease. *Journal of Biological Chemistry* **274**, 19509–12.

Yamada T, Kakihara T, Gejyo F, and Okada M (1994). A monoclonal antibody recognizing apolipoprotein E peptides in systemic amyloid deposits. *Annals of Clinical Laboratory Science*, **24**, 243–249.

Yankner BA, Duffy LK, and Kirschner D (1990). Neurotrophic and neurotoxic effects of amyloid β protein: reversal by tachykinin neuropeptides. *Science* **250**, 279–82.

Zhu YJ, Lin H, and Lal R (2000). Fresh and nonfibrillar amyloid beta protein (1–40) induces rapid cellular degeneration in aged human fibroblasts: evidence for Aβ-channel-mediated cellular toxicity. *Faseb Journal* **14**, 1244–54.

Chapter 7

Apolipoprotein E, amyloid β peptide, and Alzheimer's disease

Suzanne E. Wahrle and David M. Holtzman

Alzheimer's disease (AD) is the most common cause of dementia. Pathological examination of brains from AD patients reveals two key features: extracellular deposits of amyloid β peptide (Aβ) and intracellular aggregates of hyperphosphorylated tau protein. It is hypothesized that Aβ aggregation and deposition is a critical early event in AD pathogenesis that initiates a cascade of events including neuronal dysfunction, tau aggregation and hyperphosphorylation, and neuronal death. These downstream events ultimately result in the cognitive impairments characteristic of AD (Hardy and Selkoe 2002). In a related condition seen in many individuals with AD, and sometimes independently, patients have accumulation of Aβ in blood vessels of the cortex and leptomeninges that is termed cerebral amyloid angiopathy (CAA). CAA may lead to spontaneous intracerebral haemorrhage and ischaemic changes (Castellani *et al.* 2004). Evidence from *in vitro* and *in vivo* studies suggests that the deposition and clearance of Aβ in the brain and blood vessels is modified by apolipoprotein E (apoE), which acts as a chaperone for Aβ.

7.1 Genetics of Alzheimer's disease

Some rare familial forms of AD are caused by alterations in Aβ production. Aβ and other products are formed when amyloid precursor protein (APP) is cleaved by the enzymatic activities known as β- and γ-secretase (Wilquet and De Strooper 2004). Presenilin-1 (PS1) and presenilin-2 (PS2) are essential components of the γ-secretase enzyme complex (Brunkan and Goate 2005). Rare families have early-onset AD (onset before age 60) with an autosomal dominant pattern of inheritance which can be caused by mutations in *APP*, *PS1*, and *PS2* (Selkoe 1997; Tanzi and Bertram 2005). Importantly, most mutations that cause early-onset familial AD either increase production of all Aβ species, including forms with lengths of 40 and 42 amino acids (Aβ40 and Aβ42), or selectively increase production of the more fibrillogenic Aβ42 relative to Aβ40 (Selkoe 1997). Additionally, the gene encoding APP is located on chromosome 21 and patients with Down's syndrome (trisomy 21) develop early-onset AD pathology (Mann 1989), probably as a result of increased Aβ production due to the presence of an extra copy of *APP*. There are also mutations in the Aβ region of *APP* that do not affect Aβ production but appear to affect the propensity of Aβ to aggregate (Remes *et al.* 2004). Most of the mutations in this region cause familial forms of CAA but at least one of the mutations (Arctic) causes deposition of Aβ in brain parenchyma and AD (Nilsberth *et al.* 2001). These rare genetic causes of AD have emphasized the importance of Aβ metabolism in AD pathogenesis.

Although increases in Aβ production, an elevated Aβ42:Aβ40 ratio, and Aβ mutations are the major causes of early-onset AD, they are probably not the main mechanisms underlying the common late-onset AD (onset after age 60) which accounts for more than 99% of AD cases. Late-onset AD has a complex pattern of inheritance, suggesting that multiple genes and environmental influences are involved in determining risk for disease (Kamboh 2004). *APOE* genotype is the best studied and most robust genetic risk factor for late-onset AD. The *APOE* gene is located on chromosome 19 and has three common alleles (e2, e3, and e4) which result in isoforms of apoE with differences at two amino acid positions (apoE2, cys^{112}, cys^{158}; apoE3, cys^{112}, arg^{158}; apoE4, arg^{112}, arg^{158}) (Mahley 1988). These alterations in only two of 299 amino acids lead to large differences in risk for AD and CAA. Compared with subjects homozygous for the e3 allele, subjects with one or more copies of the e4 allele are at higher risk for AD and CAA (Corder *et al.* 1993; Mayeux *et al.* 1993; Schmechel *et al.* 1993; Strittmatter *et al.* 1993a; Greenberg *et al.* 1995) and subjects with one or more copies of the e2 allele are at lower risk for AD (Fig. 7.1)

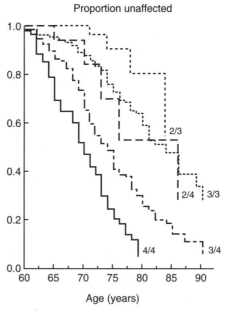

Fig. 7.1 Percentage population with AD: ApoE status and age. Compared with subjects homozygous for the apoE e3 allele (3/3), subjects with one or more copies of the apoE e4 allele (3/4, 4/4) have a higher rate of AD while subjects with the e2 allele have a lower rate of AD (2/3). Note that the plot depicts the proportion unaffected or 1 − (proportion affected). Reprinted, with permission, from the *Annual Review of Neuroscience*, Volume 19 © 1996 by Annual Reviews, www.annualreviews.org.

(Corder *et al.* 1993; Mayeux *et al.* 1993; Schmechel *et al.* 1993). The e4 allele is also a risk factor for diseases not involving amyloid, most notably cardiovascular disease (Smith 2000), and may predict a poor prognosis in multiple sclerosis (Chapman *et al.* 2001) and traumatic brain injury (Friedman *et al.* 1999). Some data suggest that in individuals with CAA, the e2 allele increases the susceptibility to CAA-related haemorrhage (Nicoll *et al.* 1997; Greenberg *et al.* 1998). Interestingly, apoE isoforms do not appear to affect APP processing or the production of Aβ (Biere *et al.* 1995). Therefore the importance of apoE to multiple diseases has prompted investigations of the normal function of apoE and of the mechanisms by which *APOE* genotype affects risk for AD.

7.2 Normal function and metabolism of apolipoprotein E

Much research has focused on understanding the normal function and metabolism of apoE in the blood, other organs, and the central nervous system (CNS). ApoE is a 34 kDa protein with a C-terminal lipid binding region and an N-terminal receptor binding domain (Wetterau *et al.* 1988; Morrow *et al.* 2000). Plasma apoE is primarily produced by the liver and is one of several apolipoproteins that carry cholesterol and phospholipids in high-density lipoprotein (HDL) particles as well as in large lipoprotein remnant particles (β-VLDL) (Mahley *et al.* 1984). HDL has a major role in regulating lipid homeostasis in body tissues and is protective against atherosclerosis (Linsel-Nitschke and Tall 2005). Several receptors mediate endocytosis of plasma apoE by the liver. The low-density lipoprotein receptor (LDLR) is the major receptor for plasma apoE (Brown and Goldstein 1983). LDLR binds and mediates endocytosis of apoE3 and apoE4 isoforms similarly, but binds poorly to apoE2 (Rall *et al.* 1982). The decreased binding of apoE2 to LDLR appears to play a role in the development of type III hyperlipidemia in some apoE2 homozygous patients (Mahley *et al.* 1999) and is the probable reason for plasma apoE levels being highest in carriers of the e2 allele (Schiele *et al.* 2000). In contrast, the LDLR-related protein (LRP) and very low density lipoprotein receptor (VLDLR) bind to all isoforms of apoE (Takahashi *et al.* 1996), and bind both lipidated and poorly lipidated forms of apoE (Narita *et al.* 2002; Ruiz *et al.* 2005). Following receptor-mediated endocytosis of apoE-containing lipoproteins by the liver, apoE may be recycled or degraded (Fazio *et al.* 1999). In addition to its well known function as a lipid carrier, recent work shows that apoE is involved in the presentation of lipid antigens to immune cells (van den Elzen *et al.* 2005). Substantial previous work also suggests a possible role of apoE in immune system function (Mahley 1988).

ApoE is the major extracellular lipid carrier in the CNS, and some studies suggest that it can play a role in neuronal plasticity under certain conditions. CNS apoE is not derived from liver (Linton *et al.* 1991), but is synthesized by brain cells, mainly by astrocytes (Boyles *et al.* 1985; Pitas *et al.* 1987a 1987b) and to some extent by microglia (Stone *et al.* 1997). In humans, neurons may also synthesize a small percentage of CNS apoE (Xu *et al.* 1999). In the CNS, apoE is secreted predominantly by glial cells and is associated with brain tissue; it is also is present as HDL-like particles in the cerebrospinal fluid (CSF) (Pitas *et al.* 1987b; Ladu *et al.* 1998; Koch *et al.* 2001). There is no correlation between apoE levels in the plasma and CSF, suggesting that the pools of plasma apoE and CNS apoE are distinct (Fagan *et al.* 2000). In addition, a study in a patient who received a liver transplant showed that apoE in the CSF is derived from within the CNS and not the periphery (Linton *et al.* 1991). The metabolism of the different apoE isoforms in the CNS has not been well characterized, but one study suggests that apoE4 may be catabolized more slowly than apoE3 in the CNS, which differs from the approximately equal rates of catabolism of the isoforms in plasma (Fukumoto *et al.* 2003). ApoE in the brain also undergoes endocytosis, which appears to be mediated by LDLR (Pitas *et al.* 1987b; Fryer *et al.* 2005) and other receptors including LRP (Kowal *et al.* 1989). Recent studies show that murine apoE is elevated ~1.5–2-fold in the CNS of $Ldlr^{-/-}$ mice (Fryer *et al.* 2005; Cao *et al.* 2006). Interestingly, apoE3 and apoE4 are elevated ~3–4-fold in the CNS of human apoE knockin mice on an $Ldlr^{-/-}$ background, while apoE2 is elevated to a similar extent in the presence and absence of LDLR (Fryer *et al.* 2005). These findings suggest that LDLR regulates the levels of apoE in the brain. In addition to functioning as a lipid transporter, apoE has been shown to play a role in neuronal plasticity in certain experimental paradigms. The apoE3 isoform enhances neurite outgrowth but, interestingly, the apoE4 isoform induces less or no sprouting, depending on the paradigm utilized (Teter 2004). For example, following an excitotoxic insult apoE3 expression results in neuroprotection whereas apoE4 does not (Buttini *et al.* 1999). This experiment and others have led to the hypothesis that patients with the e4 allele are less able to compensate for neurodegeneration and therefore develop dementia due to AD at an accelerated rate.

Recently, the ATP-binding cassette transporter A1 (ABCA1) was found to regulate levels and lipidation of apoE in the brain. It was previously known that loss-of-function mutations in *ABCA1* cause Tangier's disease (Bodzioch *et al.* 1999; Brooks-Wilson *et al.* 1999; Remaley *et al.* 1999; Rust *et al.* 1999), which is characterized by deposition of lipid in lymphatic tissue and increased catabolism of poorly lipidated lipoproteins, resulting in very low levels of plasma apoAI and HDL-C (Schaefer *et al.* 1978, 1981). From these and other studies it was determined that ABCA1 transports cholesterol and phospholipids onto plasma apolipoproteins, especially apoAI, to form pre-HDL. Mice lacking *Abca1* develop a similar phenotype to Tangier's disease patients in the periphery, with decreased plasma apoAI and HDL-C (McNeish *et al.* 2000). Interestingly, $Abca1^{-/-}$ mice have greatly reduced apoE not only in the plasma (4% of wild type) but also in the CSF (~2% of wild type) and cortex (~20% of wild type) (Hirsch-Reinshagen *et al.* 2004; Wahrle *et al.* 2004). Additionally, the CSF of $Abca1^{-/-}$ mice contains abnormally small apoE particles, which indicates that apoE is not being properly lipidated in the CNS (Wahrle *et al.* 2004). ApoE secreted by primary astrocyte cultures derived from $Abca1^{-/-}$ mice is also in smaller-sized HDL-like particles which contain less cholesterol (Wahrle *et al.* 2004). Futhermore, experiments have shown that in primary cultures of astrocytes and microglia derived from $Abca1^{-/-}$ mice, cholesterol is not as efficiently transported onto apolipoproteins and glia develop intracellular lipid accumulation (Hirsch-Reinshagen *et al.* 2004). These findings suggest that ABCA1 is required for normal lipidation of brain apoE and that ABCA1 can modulate levels of brain apoE.

7.3 **Apolipoprotein E–Aβ interactions**

In the CNS, in addition to its normal functions as a lipid carrier and modulator of neurite plasticity, apoE binds to and codeposits with amyloid-containing protein aggregates of all kinds (Wisniewski and Frangione 1992). β-Amyloid plaques isolated from the brains of AD patients contain apoE, as do extracellular deposits of many other amyloid-forming proteins (Wisniewski and Frangione 1992). The association of apoE with Aβ in plasma, CSF, and brain homogenates led to the hypothesis that apoE is a chaperone for Aβ and regulates its conversion from a mixed random coil/α-helix conformation to a β-sheet amyloid conformation (Wisniewski *et al.* 1997). Many studies have attempted to examine the interaction between apoE and Aβ *in vitro*. These studies have used different forms and preparations of apoE and Aβ, as well as different conditions for interaction. ApoE forms SDS-stable complexes with Aβ, and de-lipidated apoE4 binds to Aβ more readily than de-lipidated apoE3 (Strittmatter *et al.* 1993b). Lipidated apoE can also form SDS-stable complexes with Aβ, with lipidated apoE2 and apoE3 forming complexes to a much greater extent than lipidated apoE4 (Ladu *et al.* 1994; Tokuda *et al.* 2000). Studies have also examined whether apoE affects Aβ aggregation *in vitro*. The results are varied, with some studies showing that apoE promotes and others that apoE inhibits fibrillogenesis (Ma *et al.* 1994; Sanan *et al.* 1994; Castano *et al.* 1995; Evans *et al.* 1995; Wood *et al.* 1996). The conflicting results of the studies may be due to differences between the type of apoE used and the conditions of aggregation. Since the microenvironment in which Aβ aggregates *in vivo* is not known, and therefore cannot be replicated *in vitro*, studies of Aβ deposition *in vivo* using animal models has been very informative.

7.4 **Studies of apolipoprotein E–Aβ interactions in mice**

A useful method of studying the effects of specific molecules on human Aβ deposition and fibrillogenesis is with mouse models that overexpress human APP containing familial AD mutations in the brain. In these mice, Aβ aggregates, deposits, and forms amyloid fibrils in an age-related fashion in an environment that is probably similar to that found in human brain. Two of the best studied APP transgenic mouse models of AD are the PDAPP and APPsw/Tg2576 models, which overexpress APP with the V717F and the K670N, M671L mutations, respectively (Games *et al.* 1995; Hsiao *et al.* 1996). These mice have no Aβ deposition in the brain when young but, as they age, develop Aβ deposits in their brains which resemble the amyloid plaques found in AD patients. Some deposits are diffuse and contain Aβ that is not in a β-sheet conformation, whereas other deposits contain Aβ fibrils with a high β-sheet conformation that stain with the amyloid binding dyes Congo red and thioflavine S. These fibrillar deposits are surrounded by numerous large swollen dystrophic axons and dendrites (neuritic plaques) as well as by microglial and astrocytic hypertrophy and proliferation (Brendza *et al.* 2003). Additionally, the mice develop CAA to various degrees (Holtzman 2001). Breeding mouse models with AD-like pathology with mice that lack or overexpress other genes has contributed much to our understanding of Aβ metabolism and AD.

Mice lacking endogenous murine apoE, which is ~70% identical to human apoE, have been bred with the PDAPP and APPsw mouse models of AD. In both models, mice lacking apoE developed less plaque deposition and, interestingly, the Aβ that does deposit has little or no fibrillar (thioflavine S positive) amyloid (Bales *et al.* 1997, 1999; Holtzman *et al.* 2000a,b) (Fig. 7.2 [Plate 2]). PDAPP, *Apoe*[+/−] and APPsw, *Apoe*[+/−] mice have intermediate amounts of Aβ deposition and thioflavine-S-positive plaques. CAA and CAA-associated micro-haemorrhage is also virtually absent in mouse models of AD that lack apoE (Bales *et al.* 1997, 1999; Holtzman *et al.* 2000b; Fryer *et al.* 2003). These dramatic results demonstrate not only that apoE levels modulate the amount of Aβ deposition, but also that apoE is required for the formation of fibrillar Aβ, neuritic plaques, and CAA. Interestingly, prior to plaque deposition, one study has noted that apoE knockout mice have rather higher levels of soluble Aβ and altered Aβ half-life in brain interstitial fluid (Demattos *et al.* 2004). This suggests that, in addition to regulating the amount and conformation of Aβ deposits, apoE may facilitate transport and clearance of soluble Aβ.

Because of the large effects of *APOE* genotype on AD and CAA in humans, investigators were interested to determine whether the different isoforms of human apoE affected the deposition and conformation of Aβ *in vivo*. They found that while mice expressing murine apoE have early deposition of large amounts of fibrillar Aβ, expression of any of

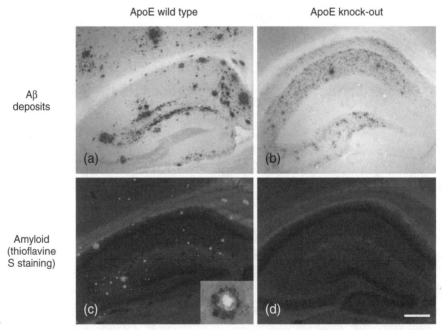

Fig. 7.2 Effect of apoE on Aβ deposition. Relative to mouse models of AD that express murine apoE (a, c), mouse models of AD lacking apoE have substantially less Aβ deposition (b) and virtually no thioflavine S positive plaques (d) in the hippocampus. Reprinted by permission of Macmillan Publishers Ltd, *Molecular Psychiatry*, 2002. Please see the colour plate section for a colour version of this figure.

the human apoE isoforms markedly delays the development of Aβ and amyloid deposition (Holtzman *et al.* 1999; Fagan *et al.* 2002). Importantly, among mice expressing the human apoE isoforms, those expressing apoE4 develop the most plaque deposition and those expressing apoE2 develop the least deposition (Fagan *et al.* 2002). These effects in mice mimic the findings in human AD and CAA. The reason for the delay in Aβ deposition in APP transgenic mice expressing human apoE isoforms compared with murine or no apoE may be due to enhanced Aβ clearance or to inhibition of Aβ fibrillogenesis or to both of these processes by the human apoE isoforms. Further studies are required to determine the mechanism of these effects.

As previously discussed, deletion of *Abca1* leads to poor lipidation of apoE and large reductions in apoE levels in the plasma, CSF, and brain parenchyma (Hirsch-Reinshagen *et al.* 2004; Wahrle *et al.* 2004). Since mouse models of AD that have reduced or no expression of mouse apoE develop significantly less Aβ deposition, it might be expected that the decreased levels of apoE present in *Abca1* knockout mice would lead to less Aβ-related pathology in *Abca1$^{-/-}$* mice bred with mouse models of AD. Contrary to this hypothesis, three laboratories found that deletion of *Abca1* either has no effect or even increases Aβ deposition and CAA in four different mouse models of AD (Hirsch-Reinshagen *et al.* 2005; Koldamova *et al.* 2005; Wahrle *et al.* 2005). These results suggest that the poorly lipidated apoE produced by *Abca1$^{-/-}$* mice increases Aβ fibrillogenesis. Additionally, apoE codeposited with Aβ in brain (Wahrle *et al.* 2005) and *Abca1$^{-/-}$* mice accumulated insoluble apoE to a greater extent than *Abca1$^{+/+}$* mice (Hirsch-Reinshagen *et al.* 2005; Wahrle *et al.* 2005), demonstrating that lipid-poor apoE binds to fibrillar Aβ *in vivo* and may be more amyloidogenic.

ApoE binds to Aβ, forming ApoE-Aβ complexes that could theoretically undergo receptor-mediated endocytosis. As discussed earlier, LDLR and LRP are two major receptors for apoE and are present in the CNS. To determine whether members of the LDLR family affect the level of Aβ in the brain, mice that lack receptor-associated protein (RAP), a chaperone for LDLR-family members, were bred with a mouse model of AD (Van Uden *et al.* 2002). The *Rap$^{-/-}$* mice had reduced levels of LDLR and LRP, and APP transgenic *Rap$^{-/-}$* mice developed more Aβ deposition than APP transgenic *Rap$^{+/+}$* mice (Van Uden *et al.* 2002), suggesting that receptor-mediated clearance of apoE may play some role in removing Aβ from the brain. When transgenic mice overexpressing a functional LRP mini-receptor predominantly by CNS neurons were crossed with an AD mouse model, soluble levels of Aβ in brain were higher in transgenic mice than in non-transgenic littermates but insoluble levels were not affected (Zerbinatti *et al.* 2004). This suggests that, somehow, LRP-mediated endocytosis of apoE–Aβ complexes or Aβ alone by neurons may not lead to effective Aβ clearance. Mice lacking LDLR were also bred with a mouse model of AD. In one study, murine apoE levels were increased 1.5-fold and Aβ deposition and pathology were higher, although the increase was not significant at the one age studied (Fryer *et al.* 2005). Another study found significant increases in both apoE and Aβ deposition (Cao *et al.* 2006). Collectively, these results suggest that receptor-mediated endocytosis of apoE–Aβ complexes may play a role in regulating CNS Aβ metabolism *in vivo*, but this hypothesis requires further testing.

7.5 **Summary**

Genetic evidence demonstrates that *APOE* genotype is one of the most important factors in determining risk for sporadic AD. To understand the mechanism(s) behind this effect, studies have examined the normal function of apoE and the effects of apoE and apoE-related molecules on Aβ metabolism. Whilst the primary role of apoE in the plasma and CNS is to transport lipids, it has additional functions in the nervous system, such as facilitating neural repair, that are relevant to understanding its role in CNS disease. Since Aβ plays a central role in AD pathogenesis, the effects of apoE isoform on Aβ metabolism are likely to be the primary reason that *APOE* genotype is a strong AD risk factor. Mouse models of AD have provided strong evidence that apoE affects the amount, type, and onset of Aβ deposition in the brain, which suggests that apoE influences the conformation and clearance of Aβ. Additional studies examining ABCA1, which influences apoE levels and lipidation, and the apoE receptors LDLR and LRP show that molecules which affect the metabolism of apoE also affect Aβ. The important role that apoE has on modifying Aβ levels, deposition, and clearance makes apoE a therapeutic target for AD. Modulation of apoE levels, lipidation, or isoforms in the brain may significantly alter the development and progression of AD.

References

Bales KR, Verina T, Dodel RC, *et al.* (1997). Lack of apolipoprotein E dramatically reduces amyloid beta-peptide deposition. *Nature Genetics* **17**, 263–4.

Bales KR, Verina T, Cummins DJ, *et al.* (1999). Apolipoprotein E is essential for amyloid deposition in the APP(V717F) transgenic mouse model of Alzheimer's disease. *Proceedings of the National Academy of Sciences of the USA* **96**, 15233–8.

Biere AL, Ostaszewski B, Zhao H, Gillespie S, Younkin SG, and Selkoe DJ (1995). Co-expression of beta-amyloid precursor protein (betaAPP) and apolipoprotein E in cell culture: analysis of betaAPP processing. *Neurobiology of Disease* **2**, 177–87.

Bodzioch M, Orso E, Klucken J, *et al.* (1999). The gene encoding ATP-binding cassette transporter 1 is mutated in Tangier disease. *Nature Genetics* **22**, 347–51.

Boyles JK, Pitas RE, Wilson E, Mahley RW, and Taylor JM (1985). Apolipoprotein E associated with astrocytic glia of the central nervous system and with nonmyelinating glia of the peripheral nervous system. *Journal of Clinical Investigation* **76**, 1501–13.

Brendza RP, O'Brien C, Simmons K, *et al.* (2003). PDAPP; YFP double transgenic mice: a tool to study amyloid-beta associated changes in axonal, dendritic, and synaptic structures. *Journal of Comparative Neurology* **456**, 375–83.

Brooks-Wilson A, Marcil M, Clee SM, *et al.* (1999). Mutations in ABC1 in Tangier disease and familial high-density lipoprotein deficiency. *Nature Genetics* **22**, 336–45.

Brown MS and Goldstein JL (1983). Lipoprotein receptors in the liver. Control signals for plasma cholesterol traffic. *Journal of Clinical Investigation* **72**, 743–7.

Brunkan AL and Goate AM (2005). Presenilin function and gamma-secretase activity. *Journal of Neurochemistry* **93**, 769–92.

Buttini M, Orth M, Bellosta S *et al.* (1999). Expression of human apolipoprotein E3 or E4 in the brains of Apoe–/– mice: isoform-specific effects on neurodegeneration. *Journal of Neuroscience* **19**, 4867–4880.

Cao D, Fukuchi KI, Wan H, Kim H, and Li L (2006). Lack of LDL receptor aggravates learning deficits and amyloid deposits in Alzheimer transgenic mice. *Neurobiology of Aging* **27**, 1632–43.

Castano EM, Prelli F, Wisniewski T, *et al.* (1995). Fibrillogenesis in Alzheimer's disease of amyloid beta peptides and apolipoprotein E. *Biochemical Journal* **306**, 599–604.

Castellani RJ, Smith MA, Perry G, and Friedland RP (2004). Cerebral amyloid angiopathy: major contributor or decorative response to Alzheimer's disease pathogenesis. *Neurobiology of Aging*, **25**, 599–602.

Chapman J, Vinokurov S, Achiron A, *et al.* (2001). APOE genotype is a major predictor of long-term progression of disability in MS. *Neurology* **56**, 312–16.

Corder EH, Saunders AM, Strittmatter WJ, *et al.* (1993). Gene dose of apolipoprotein E type 4 allele and the risk of Alzheimer's disease in late onset families. *Science* **261**, 921–3.

Demattos RB, Cirrito JR, Parsadanian M, *et al.* (2004). ApoE and clusterin cooperatively suppress Abeta levels and deposition: evidence that ApoE regulates extracellular Abeta metabolism *in vivo*. *Neuron* **41**, 193–202.

Evans KC, Berger EP, Cho CG, Weisgraber KH, and Lansbury PT, Jr (1995). Apolipoprotein E is a kinetic but not a thermodynamic inhibitor of amyloid formation: implications for the pathogenesis and treatment of Alzheimer disease. *Proceedings of the National Academy of Sciences of the USA* **92**, 763–7.

Fagan AM, Younkin LH, Morris JC, *et al.* (2000). Differences in the Abeta40/Abeta42 ratio associated with cerebrospinal fluid lipoproteins as a function of apolipoprotein E genotype. *Annals of Neurology* **48**, 201–10.

Fagan AM, Watson M, Parsadanian M, Bales KR, Paul SM, and Holtzman DM (2002). Human and murine ApoE markedly alters A beta metabolism before and after plaque formation in a mouse model of Alzheimer's disease. *Neurobiology of Disease* **9**, 305–18.

Fazio S, Linton MF, Hasty AH, and Swift LL (1999). Recycling of apolipoprotein E in mouse liver. *Journal of Biological Chemistry* **274**, 8247–53.

Friedman G, Froom P, Sazbon L, *et al.* (1999). Apolipoprotein E-epsilon4 genotype predicts a poor outcome in survivors of traumatic brain injury. *Neurology* **52**, 244–8.

Fryer JD, Taylor JW, Demattos RB, *et al.* (2003). Apolipoprotein E markedly facilitates age-dependent cerebral amyloid angiopathy and spontaneous hemorrhage in amyloid precursor protein transgenic mice. *Journal of Neuroscience* **23**, 7889–96.

Fryer JD, Demattos RB, McCormick LM, *et al.* (2005). The low density lipoprotein receptor regulates the level of central nervous system human and murine apolipoprotein E but does not modify amyloid plaque pathology in PDAPP mice. *Journal of Biological Chemistry* **280**, 25754–9.

Fukumoto H, Ingelsson M, Garevik N, *et al.* (2003). APOE epsilon 3/epsilon 4 heterozygotes have an elevated proportion of apolipoprotein E4 in cerebrospinal fluid relative to plasma, independent of Alzheimer's disease diagnosis. *Experimental Neurology* **183**, 249–53.

Games D, Adams D, Alessandrini R, *et al.* (1995). Alzheimer-type neuropathology in transgenic mice overexpressing V717F beta-amyloid precursor protein. *Nature* **373**, 523–7.

Greenberg SM, Rebeck GW, Vonsattel JP, Gomez-Isla T, and Hyman BT (1995). Apolipoprotein E epsilon 4 and cerebral hemorrhage associated with amyloid angiopathy. *Annals of Neurology* **38**, 254–59.

Greenberg SM, Vonsattel JP, Segal AZ, *et al.* (1998). Association of apolipoprotein E epsilon2 and vasculopathy in cerebral amyloid angiopathy. *Neurology* **50**, 961–5.

Hardy J and Selkoe DJ (2002). The amyloid hypothesis of Alzheimer's disease: progress and problems on the road to therapeutics. *Science* **297**, 353–6.

Hirsch-Reinshagen V, Zhou S, Burgess BL, *et al.* (2004). Deficiency of ABCA1 impairs apolipoprotein E metabolism in brain. *Journal of Biological Chemistry* **279**, 41197–207.

Hirsch-Reinshagen V, Maia LF, Burgess BL, *et al.* (2005). The absence of ABCA1 decreases soluble ApoE levels but does not diminish amyloid deposition in two murine models of Alzheimer disease. *Journal of Biological Chemistry* **280**, 43243–56.

Holtzman DM (2001). Role of apoe/Abeta interactions in the pathogenesis of Alzheimer's disease and cerebral amyloid angiopathy. *Journal of Molecular Neuroscience* **17**, 147–55.

Holtzman DM, Bales KR, Wu S, *et al.* (1999). Expression of human apolipoprotein E reduces amyloid-beta deposition in a mouse model of Alzheimer's disease. *Journal of Clinical Investigation* **103**, R15–21.

Holtzman DM, Bales KR, Tenkova T, *et al.* (2000a). Apolipoprotein E isoform-dependent amyloid deposition and neuritic degeneration in a mouse model of Alzheimer's disease. *Proceedings of the National Academy of Sciences of the USA* **97**, 2892–7.

Holtzman DM, Fagan AM, Mackey B, *et al.* (2000b). Apolipoprotein E facilitates neuritic and cerebrovascular plaque formation in an Alzheimer's disease model. *Annals of Neurology* **47**, 739–47.

Hsiao K, Chapman P, Nilsen S, *et al.* (1996). Correlative memory deficits, Abeta elevation, and amyloid plaques in transgenic mice. *Science* **274**, 99–102.

Kamboh MI (2004). Molecular genetics of late-onset Alzheimer's disease. *Annals of Human Genetics* **68**, 381–404.

Koch S, Donarski N, Goetze K, *et al.* (2001). Characterization of four lipoprotein classes in human cerebrospinal fluid. *Journal of Lipid Research* **42**, 1143–51.

Koldamova R, Staufenbiel M, and Lefterov I (2005). Lack of ABCA1 considerably decreases brain ApoE level and increases amyloid deposition in APP23 mice. *Journal of Biological Chemistry* **280**, 43224–35.

Kowal RC, Herz J, Goldstein JL, Esser V, and Brown MS (1989). Low density lipoprotein receptor-related protein mediates uptake of cholesteryl esters derived from apoprotein E-enriched lipoproteins. *Proceedings of the National Academy of Sciences of the USA* **86**, 5810–14.

Ladu MJ, Falduto MT, Manelli AM, Reardon CA, Getz GS, and Frail DE (1994). Isoform-specific binding of apolipoprotein E to beta-amyloid. *Journal of Biological Chemistry* **269**, 23403–6.

Ladu MJ, Gilligan SM, Lukens JR, *et al.* (1998). Nascent astrocyte particles differ from lipoproteins in CSF. *Journal of Neurochemistry* **70**, 2070–81.

Linsel-Nitschke P and Tall AR (2005). HDL as a target in the treatment of atherosclerotic cardiovascular disease. *Nature Reviews Drug Discovery* **4**, 193–205.

Linton MF, Gish R, Hubl ST, *et al.* (1991). Phenotypes of apolipoprotein B and apolipoprotein E after liver transplantation. *Journal of Clinical Investigation* **88**, 270–81.

McNeish J, Aiello RJ, Guyot D, *et al.* (2000). High density lipoprotein deficiency and foam cell accumulation in mice with targeted disruption of ATP-binding cassette transporter-1. *Proceedings of the National Academy of Sciences of the USA* **97**, 4245–50.

Ma J, Yee A, Brewer HB, Jr, Das S, and Potter H (1994). Amyloid-associated proteins alpha 1-antichymotrypsin and apolipoprotein E promote assembly of Alzheimer beta-protein into filaments. *Nature* **372**, 92–4.

Mahley RW (1988). Apolipoprotein E: cholesterol transport protein with expanding role in cell biology. *Science* **240**, 622–30.

Mahley RW, Innerarity TL, Rall SC, Jr, and Weisgraber KH (1984). Plasma lipoproteins: apolipoprotein structure and function. *Journal of Lipid Research* **25**, 1277–94.

Mahley RW, Huang Y, and Rall SC, Jr (1999). Pathogenesis of type III hyperlipoproteinemia (dysbetalipoproteinemia). Questions, quandaries, and paradoxes. *Journal of Lipid Research* **40**, 1933–49.

Mann D.M. (1989). Cerebral amyloidosis, ageing and Alzheimer's disease: a contribution from studies on Down's syndrome. *Neurobiology of Aging*, **10**, 397–9.

Mayeux R, Stern Y, Ottman R., *et al.* (1993). The apolipoprotein epsilon 4 allele in patients with Alzheimer's disease. *Annals of Neurology* **34**, 752–4.

Morrow JA, Segall ML, Lund-Katz S, *et al.* (2000). Differences in stability among the human apolipoprotein E isoforms determined by the amino-terminal domain. *Biochemistry*, **39**, 11657–66.

Narita M, Holtzman DM, Fagan AM, *et al.* (2002). Cellular catabolism of lipid poor apolipoprotein E via cell surface LDL receptor-related protein. *Journal of Biochemistry* **132**, 743–9.

Nicoll JA, Burnett C, Love S, *et al.* (1997). High frequency of apolipoprotein E epsilon 2 allele in hemorrhage due to cerebral amyloid angiopathy. *Annals of Neurology* **41**, 716–21.

Nilsberth C, Westlind-Danielsson A, Eckman CB, *et al.* (2001). The 'Arctic' APP mutation (E693G) causes Alzheimer's disease by enhanced Abeta protofibril formation. *Nature Neuroscience* **4**, 887–93.

Pitas RE, Boyles JK, Lee SH, Foss D, and Mahley RW (1987a). Astrocytes synthesize apolipoprotein E and metabolize apolipoprotein E-containing lipoproteins. *Biochimica Biophysica Acta* **917**, 148–61.

Pitas RE, Boyles JK, Lee SH, Hui D, and Weisgraber KH (1987b). Lipoproteins and their receptors in the central nervous system. Characterization of the lipoproteins in cerebrospinal fluid and identification of apolipoprotein B,E(LDL) receptors in the brain. *Journal of Biological Chemistry* **262**, 14352–0.

Rall SC, Jr, Weisgraber KH, Innerarity TL, and Mahley RW (1982). Structural basis for receptor binding heterogeneity of apolipoprotein E from type III hyperlipoproteinemic subjects. *Proceedings of the National Academy of Sciences of the USA* **79**, 4696–700.

Remaley AT, Rust S, Rosier M, *et al.* (1999). Human ATP-binding cassette transporter 1 (ABC1): genomic organization and identification of the genetic defect in the original Tangier disease kindred. *Proceedings of the National Academy of Sciences of the USA* **96**, 12685–90.

Remes AM, Finnila S, Mononen H, *et al.* (2004). Hereditary dementia with intracerebral hemorrhages and cerebral amyloid angiopathy. *Neurology* **63**, 234–40.

Ruiz J, Kouiavskaia D, Migliorini M, *et al.* (2005). The apoE isoform binding properties of the VLDL receptor reveal marked differences from LRP and the LDL receptor. *Journal of Lipid Research* **46**, 1721–31.

Rust S, Rosier M, Funke H, *et al.* (1999). Tangier disease is caused by mutations in the gene encoding ATP-binding cassette transporter 1. *Nature Genetics* **22**, 352–5.

Sanan DA, Weisgraber KH, Russell SJ, *et al.* (1994). Apolipoprotein E associates with beta amyloid peptide of Alzheimer's disease to form novel monofibrils. Isoform apoE4 associates more efficiently than apoE3. *Journal of Clinical Investigation* **94**, 860–9.

Schaefer EJ, Blum CB, Levy RI, *et al.* (1978). Metabolism of high-density lipoprotein apolipoproteins in Tangier disease. *New England Journal of Medicine* **299**, 905–10.

Schaefer EJ, Anderson DW, Zech LA, *et al.* (1981). Metabolism of high density lipoprotein subfractions and constituents in Tangier disease following the infusion of high density lipoproteins. *Journal of Lipid Research* **22**, 217–28.

Schiele F, De Bacquer D, Vincent-Viry M, *et al.* (2000). Apolipoprotein E serum concentration and polymorphism in six European countries: the ApoEurope Project. *Atherosclerosis* **152**, 475–88.

Schmechel DE, Saunders AM, Strittmatter WJ, *et al.* (1993). Increased amyloid beta-peptide deposition in cerebral cortex as a consequence of apolipoprotein E genotype in late-onset Alzheimer disease. *Proceedings of the National Academy of Sciences of the USA* **90**, 9649–53.

Selkoe DJ (1997). Alzheimer's disease: genotypes, phenotypes, and treatments. *Science* **275**, 630–1.

Smith JD (2000). Apolipoprotein E4: an allele associated with many diseases. *Annals of Medicine* **32**, 118–27.

Stone DJ, Rozovsky I, Morgan TE, Anderson CP, Hajian H, and Finch CE (1997). Astrocytes and microglia respond to estrogen with increased apoE mRNA *in vivo* and *in vitro*. *Experimental Neurology* **143**, 313–18.

Strittmatter WJ, Saunders AM, Schmechel D, *et al.* (1993a). Apolipoprotein E: high-avidity binding to beta-amyloid and increased frequency of type 4 allele in late-onset familial Alzheimer disease. *Proceedings of the National Academy of Sciences of the USA* **90**, 1977–81.

StrittmatterWJ, Weisgraber KH, Huang DY, *et al.* (1993b). Binding of human apolipoprotein E to synthetic amyloid beta peptide: isoform-specific effects and implications for late-onset Alzheimer disease. *Proceedings of the National Academy of Sciences of the USA* **90**, 8098–102.

Takahashi S, Oida K, Ookubo M, *et al.* (1996). Very low density lipoprotein receptor binds apolipoprotein E2/2 as well as apolipoprotein E3/3. *FEBS Letters* **386**, 197–200.

Tanzi RE and Bertram L (2005). Twenty years of the Alzheimer's disease amyloid hypothesis: a genetic perspective. *Cell* **120**, 545–55.

Teter B (2004). ApoE-dependent plasticity in Alzheimer's disease. *Journal of Molecular Neuroscience* **23**, 167–79.

Tokuda T, Calero M, Matsubara E, *et al.* (2000). Lipidation of apolipoprotein E influences its isoform-specific interaction with Alzheimer's amyloid beta peptides. *Biochemical Journal* **348**, 359–65.

van den Elzen P, Garg S, Leon L, *et al.* (2005). Apolipoprotein-mediated pathways of lipid antigen presentation. *Nature* **437**, 906–10.

Van Uden E, Mallory M, Veinbergs I, Alford M, Rockenstein E, and Masliah E (2002). Increased extracellular amyloid deposition and neurodegeneration in human amyloid precursor protein transgenic mice deficient in receptor-associated protein. *Journal of Neuroscience* **22**, 9298–304.

Wahrle SE, Jiang H, Parsadanian M, *et al.* (2004). ABCA1 is required for normal central nervous system ApoE levels and for lipidation of astrocyte-secreted apoE. *Journal of Biological Chemistry* **279**, 40987–93.

Wahrle SE, Jiang H, Parsadanian M, *et al.* (2005). Deletion of Abca1 increases Abeta deposition in the PDAPP transgenic mouse model of Alzheimer disease. *Journal of Biological Chemistry* **280**, 43236–42.

Wetterau JR, Aggerbeck LP, Rall SC, Jr, and Weisgraber KH (1988). Human apolipoprotein E3 in aqueous solution. I. Evidence for two structural domains. *Journal of Biological Chemistry* **263**, 6240–8.

Wilquet V and De Strooper B (2004). Amyloid-beta precursor protein processing in neurodegeneration. *Current Opinion in Neurobiology* **14**, 582–8.

Wisniewski T and Frangione B (1992). Apolipoprotein E: a pathological chaperone protein in patients with cerebral and systemic amyloid. *Neuroscience Letters* **135**, 235–8.

Wisniewski T, Ghiso J, and Frangione B (1997). Biology of A beta amyloid in Alzheimer's disease. *Neurobiology of Disease* **4**, 313–28.

Wood SJ, Chan W, and Wetzel R (1996). Seeding of A beta fibril formation is inhibited by all three isotypes of apolipoprotein E. *Biochemistry* **35**, 12623–8.

Xu PT, Schmechel D, Qiu HL, *et al.* (1999). Sialylated human apolipoprotein E (apoEs) is preferentially associated with neuron-enriched cultures from APOE transgenic mice. *Neurobiology of Disease* **6**, 63–75.

Zerbinatti CV, Wozniak DF, Cirrito J, *et al.* (2004). Increased soluble amyloid-beta peptide and memory deficits in amyloid model mice overexpressing the low-density lipoprotein receptor-related protein. *Proceedings of the National Academy of Sciences of the USA* **101**, 1075–80.

Chapter 8

Presenilins

Kulandaivelu S. Vetrivel and Gopal Thinakaran

Alzheimer's disease (AD) is the leading cause of dementia in the elderly. While the vast majority of AD occurs as an age-associated disorder, autosomal dominant inheritance of mutant genes *APP*, *PSEN1*, or *PSEN2*, which encode the amyloid precursor protein (APP), presenilin 1 (PS1), or presenilin 2 (PS2), respectively, cause early-onset AD in several families (familial Alzheimer's disease (FAD)). Mutations in *PSEN1* and *PSEN2* account for the majority of cases of FAD. Since the first discovery of a genetic link between *PSEN1* and *PSEN2* with AD, several research groups have sought to unravel how pathogenic mutations in *PSEN* genes cause dysfunction of select populations of neurons in the brain leading to AD pathogenesis. PS1 and PS2 are polytopic membrane proteins which function as the catalytic subunit of an enzyme complex, termed γ-secretase. β-Amyloid (Aβ) peptides, which accumulate in brain senile plaques of aged individuals and patients with AD, are generated by γ-secretase cleavage of APP. FAD-associated PS variants exert their pathogenic function by selectively elevating the levels of highly amyloidogenic Aβ42 peptides. In addition, γ-secretase cleaves a wide spectrum of type I membrane protein substrates, including Notch receptor, within their transmembrane domains. Here, we review the biology of PS1 and its role in γ-secretase activity, and discuss recent developments in the cell biology of PS1 with respect to AD pathogenesis.

8.1 Introduction

In 1995 independent groups identified genetic linkage and mutations within *PSEN1* (chromosome 14) and *PSEN2* (chromosome 1) genes in several early-onset FAD kindreds (Levy-Lahad *et al.* 1995; Rogaev *et al.* 1995; Sherrington *et al.* 1995). Since then, a number of research groups have focused on the biology of proteins encoded by these homologous genes, and how FAD-linked mutations lead to AD pathogenesis. *PSEN1* and *PSEN2* encode 467 and 448 amino acids (aa) long polytopic transmembrane proteins termed presenilin 1 (PS1) and presenilin 2 (PS2), respectively. The sequence identity between these two highly conserved proteins is greater than 65%. PS1 is expressed earlier than PS2 during mouse embryonic development (Lee *et al.* 1996). In rodents and humans both PS1 and PS2 are ubiquitously expressed in the brain and peripheral tissues; in general, PS1 is relatively expressed at higher levels than PS2. So far 140 and 10 mutations that co-segregate with FAD (in most cases before 60 years of age) have been identified in *PSEN1* and *PSEN2*, respectively. With the exception of two deletion mutants, all pathogenic mutations are point mutations found throughout the polypeptide, with a greater preponderance in hydrophobic stretches (Fig. 8.1 [**Plate 2**]).

Mutations in *PSEN1* and *PSEN2* are autosomal dominant and highly penetrant, and cause the most aggressive form of AD, in some cases with onset at less than 30 years of age. Even though FAD-linked mutations in amyloid precursor protein (*APP*) and *PSEN* genes account for less than 5% of total AD cases, the phenocopies of these FAD mutations are reminiscent of late-onset sporadic AD. As discussed below, PS1 and PS2 are subunits of a protein complex, termed γ-secretase, which cleaves several type I membrane proteins, including APP, within their transmembrane domain. In the case of APP, γ-secretase cleavage generates Aβ peptides 39–43 aa long, which accumulate in the brains of aged individuals and patients with AD. FAD-linked mutations in PS selectively elevate the levels of highly amyloidogenic Aβ42 peptides (Borchelt *et al.* 1996; Scheuner *et al.* 1996). Because of its central role in the pathogenesis of AD, PS has become the focus of much scrutiny and has been considered as a potential therapeutic target for the treatment of AD. Furthermore, certain *PSEN1* mutations as well as *PSEN1* splice variants have been genetically linked to familial frontotemporal dementia. In addition, several studies have identified physiological functions of PS1 beyond AD, including apoptosis, calcium homeostasis, neurite outgrowth, and synaptic plasticity (reviewed by

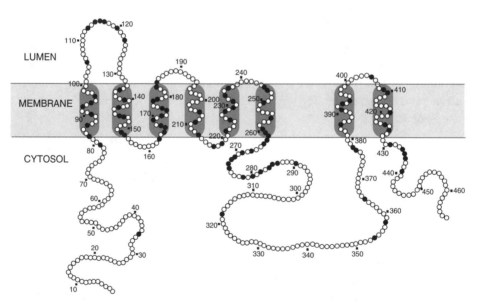

Fig. 8.1 Membrane organization of presenilin 1. In this model, PS1 is depicted traversing the membrane eight times. Each circle represents an amino acid, and black circles represent individual residues mutated in FAD cases. In addition to point mutations, two deletion mutants, DI83/M84 and DS290-S319, have been identified. The arrow marks the major site of endoproteolytic cleavage, which occurs when the nascent PS1 molecule assembles with nicastrin, APH-1, and PEN-2.

Thinakaran and Parent 2004). In this chapter, we mainly focus on recent advances in the cell biology of PS, and discuss the function of PS as it relates to AD.

8.2 Presenilin topology

Determination of PS protein topology is of particular interest because it may facilitate a better understanding of the structural and functional relationship of γ-secretase activity. The transmembrane (TM) topology of PS1 is still being debated, although several topology models have been proposed based on different experimental approaches. Most of these studies used antibodies, engineered N-glycosylation acceptor sites, protease digestion, and gene fusions with reporters to map the cytosolic or lumen regions of the protein. The amino acid sequence of PS contains 10 hydrophobic regions (HRs) that can potentially function as TM domains, leading to the proposal of several models for PS with six to nine TM segments. Nevertheless, an eight-TM topology model with N- and C-termini, and a hydrophilic loop domain between TM6 and TM7 facing the cytosol, has been widely accepted (Doan *et al.* 1996) (Fig. 8.1 [**Plate 2**]). While there is uniform agreement between various models in cytosolic orientation of the N-terminus and the TM assignments of the first six HRs, the models differ in the transmembrane segment assignment of HR7–HR10, and the orientation of the PS1 C-terminus. A recent study using glycosylation acceptor sequences also confirmed the luminal orientation of the first hydrophilic loop and cytosolic orientation of the N-terminus and a large hydrophilic loop (between TM6 and TM7), which is in agreement with the eight-TM model (Laudon *et al.* 2005). However, in contrast with the eight-TM model, the C-terminus was found oriented to the lumen, suggesting that an HR near the C-terminus functions as the ninth TM segment.

8.3 Endoproteolysis of presenilin

PS1 is synthesized as a 42–43 kDa polypeptide which undergoes highly regulated endoproteolytic processing within the large cytoplasmic loop domain connecting putative TM6 and TM7 to generate a stable 27–28 kDa N-terminal fragment (NTF) and a 16–17 kDa C-terminal fragment (CTF) (Thinakaran *et al.* 1996) by an uncharacterized proteolytic activity. Endoproteolytic processing of PS is a highly conserved, and perhaps critical, event that regulates the stability of PS1 and possibly the biological activity of PS. While full-length PS1 is relatively short-lived with a half-life of 1–2 h, the endoproteolytically processed derivatives (NTF and CTF) have a half-life of 24 h (Podlisny *et al.* 1997; Ratovitski *et al.* 1997). Moreover, the endoproteolytic event has been identified as the activation step in the process of PS1 maturation as it assembles with three other γ-secretase subunits, namely nicastrin, anterior pharynx defective (APH-1), and presenilin enhancer-2 (PEN-2) (described below) (Iwatsubo 2004). An exception to this initial activation event is the maturation of the FAD-associated deletion mutant PS1ΔE9, which lacks a region including the endoproteolytic cleavage site encoded by exon 9 (Fig. 8.1 [**Plate 2**]). PS1ΔE9 holoprotein is metabolically stable, forms a complex with other γ-secretase subunits, and generates functional γ-secretase activity.

In transfected cells the vast majority of nascent PS1 holoprotein is highly unstable and is rapidly degraded, and only a fraction of overexpressed PS1 is converted to stable NTF and CTF, which accumulate in 1:1 stoichiometry (Thinakaran *et al.* 1996, 1997). Thus PS1 endoproteolysis is a tightly regulated and saturable event. Transgenic overexpression of human PS1 in mice replaces endogenous mouse PS1 by a highly selective and compensatory mechanism, and the extent of replacement is proportional to the level of exogenous PS1, which suggests that exogenous and endogenous PS compete for limiting cellular factors (Thinakaran *et al.* 1996, 1997). In cells co-expressing PS1 and PS2, the assemblies consist of either PS1 derivatives or PS2 derivatives but not mixed assemblies (e.g. PS1NTF–PS2CTF). However, expression of chimeric PS1/PS2 holoprotein is endoproteolysed and forms heteromeric assemblies made of PS1NTF–PS2CTF (Saura *et al.* 1999). Furthermore, exogenous PS1 NTF does not co-assemble with endogenous PS1 CTF. Nevertheless, co-expression of exogenous NTFs and CTFs can reconstitute functional γ-secretase in PS-deficient cells (Laudon *et al.* 2004; Stromberg *et al.* 2005). These findings indicate that association between N- and C-terminal domains of PS1 holoprotein occurs prior to endoproteolysis, and, following assembly, PS-derived NTFs/CTFs do not exchange between γ-secretase complexes containing either PS1 or PS2.

Experimental deletion and replacement mutants and domain swap experiments have been useful in identifying amino acid residues that are critical for PS1 endoproteolysis as well as γ-secretase activity. Deletions and substitutions near the putative endoproteolytic cleavage sites between Thr291 and Ala299 (Podlisny *et al.* 1997) abolished PS1 endoproteolysis, but the resulting stable holoprotein still retained γ-secretase activity (Steiner *et al.* 1999). In contrast, aspartate residues at position D275 (in TM6) and D385 (in TM7) of PS1 are critical for both PS1 endoproteolysis and γ-secretase function (Wolfe *et al.* 1999). Furthermore, residues critical for both PS1 endoproteolysis and γ-secretase activity have been identified within TM1 of PS1 (Brunkan *et al.* 2005). These findings have been inconclusive with reference to the correspondence between PS endoproteolysis and γ-secretase activities. With a few exceptions, well-characterized and highly potent γ-secretase inhibitors do not affect PS1 endoproteolysis, providing more definite proof that PS endoproteolysis and γ-secretase activity are pharmacologically distinct (Beher *et al.* 2001). The enzyme activity responsible for PS1 endoproteolysis has not been identified. Based on the available data one cannot exclude the possibility that PS1 endoproteolysis may be an autocatalytic event which occurs during the maturation of unstable nascent PS1 holoprotein into stable derivatives. Clues from the deletion of domains located far from the endoproteolytic site are consistent with a potential conformational change associated with the process of PS1 endoproteolysis and maturation. For example, deletion of TM1 to TM2 resulted in an endoproteolysis-defective mutant which, unlike PS1 holoprotein, is extremely stable (Leem *et al.* 2002a). Interestingly, a large hydrophilic loop domain connecting TM6 and TM7 (residues 304–371) of PS1, which includes a caspase cleavage site (D345) and serves as the interaction domain for several PS-associated proteins (see below), is dispensable for PS1 endoproteolysis and γ-secretase activity (Saura *et al.* 2000).

8.4 Presenilin-dependent γ-secretase is involved in intramembranous proteolysis of type I membrane proteins

In the past few years there has been accumulating interest in understanding regulated cleavage of type I membrane proteins within their transmembrane domains. Aβ40 and Aβ42 peptides are released from APP by intramembranous cleavage at two major sites within the TM. In keeping with the nomenclature of previously described α- and β-secretase cleavage of APP within the extracellular domain, the intramembraneous cleavage of APP was termed 'γ-secretase' cleavage. The realization that cleavage at one of the sites that corresponds to +42 residue of Aβ is enhanced 2-fold in FAD cases (Scheuner et al. 1996) led to increased scrutiny of this cleavage process. De Strooper et al. (1998) first reported a direct role for PS1 in γ-secretase cleavage when they noted loss of Aβ secretion and accumulation of APP C-terminal fragments (CTF) in *PS1*$^{-/-}$ neurons. Soon after, a striking similarity between PS-mediated cleavage of APP and the proteolytic cleavage of transmembrane Notch receptor began to emerge (De Strooper et al. 1999; Struhl and Greenwald 1999; Mumm et al. 2000), although cleavage of Notch by γ-secretase occurred several residues C-terminal to the scissile bonds for Aβ production in APP. In both cases membrane-tethered CTFs, rather than the full-length proteins, serve as the substrates for γ-secretase cleavage. Struhl and Adachi (2000) systematically examined the requirements for PS1/γ-secretase cleavage and concluded that PS1 can mediate sequence-independent cleavage of a diverse set of transmembrane proteins with short extracellular domains. Consistent with APP and Notch processing, cleavage efficiency of experimental proteins was inversely proportional to the length of the extracellular domain of the proteins (Struhl and Adachi 2000). More recently, detailed characterization of γ-secretase cleavage of APP in transfected cells and *in vitro* assays led to the identification of additional cleavage sites termed ε (epsilon cleavage occurring between residues 49 and 50 relative to the N-terminus of Aβ, and corresponding to γ-secretase cleavage site in Notch) (Gu et al. 2001; Sastre et al. 2001; Yu et al. 2001), and ζ (zeta cleavage occurring between residues 46 and 47) (Zhao et al. 2004).

As expected from the relaxed sequence specificity of γ-secretase, a wide range of substrates has been described within the last few years, extending a physiological role for PS1 beyond the nervous system and AD. These include homologues of APP (APP-like proteins APLP1 and APLP2) Notch1 homologues, Notch ligands Delta and Jagged, the receptor tyrosine kinase ErbB-4, cell surface adhesion protein CD44, the mucin-type molecule CD43, low-density lipoprotein receptor-related protein, cell adhesion receptors, N- and E-cadherins, cadherin-related gamma-protocadherins, synaptic adhesion protein nectin-1α, netrin receptor DCC, cell surface heparin sulphate proteoglycan syndecan-1, p75 neurotrophin receptor and its homologue NRADD, the voltage-gated sodium channel β$_2$ subunit, etc. As with the case of APP and Notch1, γ-secretase cleavage of these additional substrates is preceded by cleavage(s) within their extracellular domain. In several cases PS1/γ-secretase cleavage releases an intracellular domain analogous to the Notch intracellular domain that can translocate into the nucleus. It remains to be determined whether the intracellular domains of γ-secretase substrates other than Notch homologues engage in signal transduction upon gaining entry into the nucleus. Interestingly, γ-secretase cleavage of DCC terminates the cAMP/PKA signalling cascade and potentially modulates glutamatergic synaptic transmission in neurons (Parent *et al.* 2005).

8.5 Presenilin 1 stability and functions in γ-secretase require other components

As described above, overexpression of PS1 alone neither substantially increases the levels of mature PS1 derivatives nor elevates Aβ generation in cultured cell lines and transgenic mice. Based on these findings, a model was proposed wherein the abundance of mature PS1 fragments is regulated by interaction with limiting cellular factor(s) (Thinakaran *et al.* 1997). Genetic and biochemical approaches have identified three additional proteins that are crucial for the stability of PS1 as well as γ-secretase activity. Goutte *et al.* (2000, 2002) used genetic screens to identify two genes, named *aph-1* and *aph-2*, which act in the GLP-1/Notch pathway during the early stages of *Caenorhabditis elegans* embryogenesis. A type I transmembrane protein, termed Nicastrin, which is the mammalian homologue of *aph-2*, was identified simultaneously by biochemical methods as a protein bound to PS1 (Yu *et al.* 2000). An independent *C.elegans* genetic screen identified *aph-1* and *pen-2* as enhancers of sel-12/Presenilin function, and showed that *aph-2*, *aph-1*, and *pen-2* are required for Notch signalling, PS1 accumulation, and γ-secretase activity (Francis *et al.* 2002). APH-1 is a multipass transmembrane protein encoded by two (human) or three (rodent) genes that can be alternatively spliced. PEN-2 is a protein with two transmembrane domains encoded by one gene. It is now clear that PS, nicastrin, APH-1, and PEN-2 are stoichiometric components of the high molecular weight γ-secretase complex. Co-expression of these four transmembrane proteins is sufficient to reconstitute γ-secretase activity in yeast, which lacks these mammalian orthologues (Edbauer *et al.* 2003). Furthermore, co-expression of all four proteins is sufficient to overcome the limitation in generating excess PS-derived NTF and CTF in transfected cells (Kim *et al.* 2003; Kimberly *et al.* 2003; Takasugi *et al.* 2003). Thus PS1 endoproteolysis, stability, and accumulation of PS1 NTF/CTF are regulated by the availability of stoichiometric levels of nicastrin, APH-1, and PEN-2. Most interestingly, biogenesis, maturation, stability, and the steady-state levels of γ-secretase components are co-dependent. Downregulation or targeted gene disruption of any one of these components affects maturation and stability of other subunits, indicating that their assembly into a high molecular weight complex is a highly regulated process that occurs during biosynthesis of these polypeptides. For example, the heavily glycosylated type I membrane protein nicastrin does not mature and exit the endoplasmic reticulum (ER) in cells lacking PS1 expression (Leem *et al.* 2002b). On the other hand, PS1 fails to undergo endoproteolysis to generate stable NTFs and CTFs in nicastrin$^{-/-}$ cells (Li *et al.* 2003).

The events in the assembly of the γ-secretase complex are only beginning to unravel at this time. Several pharmacological inhibitors, selected for their ability to inhibit Aβ production, do not seem to have a marked effect on PS1 complex formation or localization (Esler *et al.* 2002; Beher *et al.* 2003), thus posing difficulty in exploring the details of γ-secretase complex assembly process. Available evidence supports the formation of an early intermediate subcomplex made of APH-1 and nicastrin (LaVoie *et al.* 2003). The proximal C-terminus region PS1 holoprotein then binds to this subcomplex by interacting with nicastrin TM domain (Kaether *et al.* 2004). During the final step in the assembly, PEN-2 associates with the complex by interaction with TM4 of PS1 2 (Kim and Sisodia 2005; Watanabe *et al.* 2005), concurrent with PS1 endoproteolysis. Despite the identification of the core components, the precise stoichiometry of the active γ-secretase complex remains obscure. First, the existence of distinct subsets of γ-secretase complexes consisting of different PS or APH-1 subunits have been reported from studies conducted in cell culture and animal models (Shirotani *et al.* 2004; Mastrangelo *et al.* 2005). Secondly, a recent study demonstrated the existence of PS dimer at the core of the γ-secretase complex with the substrates being processed between the PS monomers (Schroeter *et al.* 2003). In addition, the use of different biochemical purification methods and detergents has led to an apparent size discrepancy of active γ-secretase complexes, with estimates ranging from 250 kDa (Kimberly *et al.* 2003) to 2 MDa (Li *et al.* 2000a).

8.6 Presenilin 1 functions as the catalytic centre of γ-secretase

The exact functional contribution of each γ-secretase subunit to enzyme activity still remains elusive. Two lines of investigation strongly support the notion that PS1 may form the catalytic centre of γ-secretase. First, Wolfe *et al.* (1999) identified the presence of two highly conserved aspartate residues in TM domains 6 (D257 in PS1 and D263 in PS2) and 7 (D385 in PS1 and D366 in PS2) that are indispensable for γ-secretase activity, and suggested that γ-secretase is a transmembrane aspartyl protease. Secondly, it was found that highly specific transition-state analogue inhibitors of γ-secretase specifically bound to PS1 NTF/CTF heterodimer (Li *et al.* 2000b). A recent study suggested that PS exists as a dimer within the γ-secretase and proposed a model where the substrates are processed between the two PS molecules (Schroeter *et al.* 2003). Interestingly, APP and Notch can interact with PS1 in the presence of active site γ-secretase inhibitors, revealing the presence of a docking site within PS1 that is distinct from the catalytic site (Esler *et al.* 2002; Berezovska *et al.* 2003). In a recent study, Yu and colleagues reported that the extracellular domain of nicastrin interacts with ectodomain cleaved APP and Notch, and thus functions as the γ-secretase substrate receptor (Shah *et al.* 2005). A three-dimensional electron microscopic structure of purified γ-secretase complex has been generated very recently, and shows a proteinaceous catalytic site that is occluded from the hydrophobic environment of the lipid bilayer and two ~20 Å pores which could potentially allow cleavage products to exit an interior chamber ~20–40 Å long (Lazarov *et al.* 2006; Ogura *et al.* 2006). Despite these recent advances, it is still intriguing how FAD-associated point mutations that are widely scattered throughout the protein, many of which are located far away from the putative catalytic site, selectively alter the cleavage site preference in APP in a manner that increases Aβ42 production.

In addition to the core subunits, it is likely that γ-secretase activity is modulated by interaction with regulatory subunits or accessory molecules. A recent study suggested that the widely expressed cell surface type I transmembrane glycoprotein CD147 (also called basigin or EMMPRIN) may function as a regulatory subunit of the γ-secretase complex (Zhou *et al.* 2005). Downregulation of CD147 expression caused a modest increase in Aβ production. It is likely that additional proteins which modulate overall enzymatic activity as well as cleavage site and substrate specificity will be identified.

8.7 Subcellular localization of presenilin 1

PS1 has been localized to multiple organelles including the ER, ER/Golgi intermediate compartments, Golgi apparatus, endosomes, lysosomes, phagosomes, plasma membrane, and mitochondria. Quantitative immunoelectron microscopy estimates of endogenous PS1 in CHO cells showed that the vast majority (52%) of PS1 is present in pre-Golgi membranes, including the nuclear envelope, ER, and vesicular–tubular clusters that are positive for COP1, whereas only about 1% of total label was localized in the Golgi complex (Rechards *et al.* 2003). A significant amount of PS1 (25%) was localized in the plasma membrane and 13% was localized in endocytic compartments. However, a recent biochemical study estimated that only 6% of PS1 and γ-secretase activity exists at the cell surface (Chyung *et al.* 2005). Consistent with the latter estimates, cell fractionation studies show that the majority of the mature components of endogenous γ-secretase complex are present in intracellular organelles (Vetrivel *et al.* 2004). Furthermore, non-ionic detergent extraction revealed the presence of PS1 and other γ-secretase subunits in cholesterol- and sphingolipid-rich detergent-resistant membrane microdomains of post-Golgi, TGN, and endosome membranes. Association of γ-secretase components with detergent-resistant membranes is sensitive to cholesterol depletion, fulfilling a stringent criterion expected of bona fide lipid-raft-associated proteins. Using magnetic immunoisolation, active and mature components of γ-secretase complex were found to co-reside in lipid raft microdomains with VAMP-4 (TGN), syntaxin 6 (TGN vesicles), and syntaxin 13 (late endosomes). Interestingly, the cell surface raft-associated protein SNAP-23 does not co-reside with γ-secretase subunits, further validating intracellular compartmentalization of the γ-secretase complex (Vetrivel *et al.* 2004).

8.8 Differential functions of wild-type and FAD-associated presenilins 1 and 2

Over 130 mutations (>258 families) in *PSEN1* and nine mutations (15 families) in *PSEN2* have been identified as autosomal dominant fully penetrant mutations which cause early-onset AD. In general, mutations in *PSEN2* are associated with a later age of onset compared with *PSEN1* (mean familial age of onset 57.1 years and 44.1 years, respectively) and slower disease progression (Bertram and Tanzi 2004). Intriguingly FAD-linked mutations in PS, by and large considered *gain-of-function* mutations, influence γ-secretase cleavage by an elusive mechanism that modulates the proteolysis of APP to selectively enhance the generation of highly amyloidogenic Aβ42 peptides

(Borchelt *et al.* 1996; Scheuner *et al.* 1996). However, the high frequency of FAD-associated mutations in *PSEN1* compared with *PSEN2* may be an indicator of differential functions of γ-secretase containing PS1 or PS2. Indeed, *in vivo* complementation of PS1 deficiency with PS2 in transgenic mice indicates functionally different roles for PS1 and PS2 (Mastrangelo *et al.* 2005). In the absence of PS1, transgenic expression of wild-type PS2 neither increases Aβ40/42 nor rescues Notch-associated skeletal defects in embryos. However, complementation of FAD-associated variants of PS2 selectively elevated Aβ42 levels and rescued Notch-associated defects. These studies provide compelling support for the existence of functionally distinct γ-secretase complexes, and further support the idea of gain of function by FAD-linked PS variants.

Precisely how FAD-linked PS mutations influence Aβ42 production is still not understood. Altered conformation has been suggested as a mechanism by which FAD-associated PS variants may influence specificity of cleavage site within APP, thus elevating Aβ42 production. For example, it has been shown that FAD-associated mutations change the proximity of the N- and C-termini of PS1 (Berezovska *et al.* 2005). On the other hand, non-steroidal anti-inflammatory drugs such as sulindac sulphide, ibuprofen, indomethacin, and flurbiprofen, which specifically lower Aβ42 levels without affecting Aβ40 production (Weggen *et al.* 2001), decrease the proximity of PS1 N- and C-termini and PS1 to the APP C-terminus (Lleo *et al.* 2004). Nevertheless, how these proximity measures of FAD mutants relate to conformational changes in the catalytic or substrate docking site of γ-secretase in a manner that fosters elevated Aβ42 production remains to be understood. Recent studies also suggest that the gain of function property of FAD-linked PS alleles is not limited to altering cleavage site specificity of substrates APP and Notch. For example, large-scale gene expression profiling in brains of conditional PS1 knockout and transgenic mice expressing wild-type or FAD-linked mutant PS1 suggests that the FAD-linked PS1 variant produces transcriptome changes primarily by gain of aberrant function (Mirnics *et al.* 2005). Furthermore, the expression and activity of neprilysin (NEP), an Aβ degrading enzyme, is regulated by γ-secretase activity. *In vitro* studies show that transcription from the NEP gene promoter can be activated by cytosolic domains released from APP, APLP1, or APLP2 by γ-secretase cleavage, and FAD-associated PS1 mutations increase NEP levels in brains of patients with mutant PS1 alleles (Pardossi-Piquard *et al.* 2005). Finally, since Aβ42 activates neutral sphingomyelinase and Aβ40 inhibits hydroxymethyl-CoA reductase, an indirect role for PS in maintaining cholesterol and sphingomyelin levels through Aβ40 and Aβ42 production has been proposed (Grimm *et al.* 2005). As expected from increased production of Aβ42, FAD-associated PS1 mutants specifically increase cellular cholesterol and decrease sphingomyelin levels.

In addition to modulating Aβ42 production, FAD-linked PS1 variants also appear to influence other processes that are highly relevant to neurodegeneration in AD. For example, neurofibrillary tangle-like tau pathology was observed in the hippocampus of FAD mutant (I213T) PS1 knock-in mice in the absence of Aβ deposition, suggesting that PS1 mutations contribute to the onset of AD not only by enhancing Aβ42 production but by also accelerating the formation and accumulation of filamentous tau (Tanemura *et al.* 2006). Interestingly, loss of *PSEN1* and *PSEN2* expression in mouse forebrain is also associated with tau hyperphosphorylation (Saura *et al.* 2004). PS deficiency or expression of FAD-linked PS1 mutants in cultured cells inhibits PI3K/Akt signalling and promotes glycogen synthase kinase-3 (GSK-3) dependent phosphoryla-tion of tau (Baki *et al.* 2004; Kang *et al.* 2005), suggesting a loss-of-function mechanism relating PS function to tau phosphorylation. Recent studies show that melanin synthesis is another physiological function that is impaired by FAD-linked mutation in PS1 (Wang *et al.* 2006).

8.9 Presenilin 1 regulates trafficking of select membrane proteins

In addition to its function as the catalytic subunit of γ-secretase, PS directly or indirectly regulates the trafficking of select membrane proteins. PS1 deficiency in neurons acceler-ated the secretion of α- and β-secretase cleaved APP ectodomain (Naruse *et al.* 1998). Further analysis revealed that PS1 regulates biosynthetic secretory trafficking of APP. Absence of PS1 or the expression of a loss of function PS1 variant resulted in increased budding/generation of APP-containing vesicles from both ER and TGN with a concomi-tant increase in complex glycosylation and cell surface appearance of APP (Leem *et al.* 2002a; Cai *et al.* 2003). In addition, the half-life and steady-state residence of full-length APP and APP CTFs at the cell surface were greatly increased (Kim *et al.* 2001; Kaether *et al.* 2002). Interestingly, FAD-linked PS1 variants significantly reduced budding of APP-containing vesicles from both ER and TGN, resulting in decreased delivery of APP to the cell surface (Cai *et al.* 2003). These findings raised the possibility that FAD-linked PS1 variants may influence APP processing by increasing the time of APP residing at TGN, consequently prolonging their availability for cleavage by β- and γ-secretases within the TGN. Direct evidence to support a trafficking role of PS1, independent of its function in γ-secretase activity, emerged from analysis of membrane proteins that are not substrates for γ-secretase cleavage. For example, complex oligosaccharide modifica-tion and brain-derived neurotrophic factor (BDNF) induced phosphorylation of the neurotrophin receptor TrkB was severely affected in neurons lacking PS1 expression (Naruse *et al.* 1998). Furthermore, neuron specific intercellular adhesion molecule (ICAM-5)/telencephalin accumulated intracellularly in autophagic vacuoles and dis-played an extended half-life in the absence of PS1 (Annaert *et al.* 2001). Remarkably, increased assembly and cell surface delivery of nicotinic acetylcholine receptors (a mutimeric polytopic membrane protein complex) was observed in cells expressing a loss of function PS1 mutant which exhibited accelerated cell surface delivery of APP and slower kinetics of Aβ secretion (Leem *et al.* 2002a).

While a role for PS1 in regulating protein trafficking remains attractive, the growing number of γ-secretase substrates raises the possibility of impaired turnover of CTFs derived from type I membrane proteins directly or indirectly influencing protein trafficking in cells lacking PS1 expression or γ-secretase activity. Indeed, PS deficiency results in accumulation of C-terminal fragments derived from tyrosinase, a substrate of γ-secretase, and aberrant accumulation of post-Golgi vesicles containing tyrosinase

(Wang *et al.* 2006). On the other hand, interaction of PS with protein trafficking factors suggests that PS1 can indirectly regulate protein trafficking via its binding partners. For example, PS1 interacts with the Ras-related GTP binding proteins Rab11 and Rab6, as well as with Rab GDP dissociation inhibitors that are involved in regulation of intracellular vesicle trafficking (Dumanchin *et al.* 1999; Scheper *et al.* 2000, 2004). Interestingly, expression of dominant-negative Rab6 increased α-secretase processing of APP with only a minor effect on Aβ production (McConlogue *et al.* 1996). Recently, Cai *et al.* (2006a,b) also showed that PS1 interacts with phospholipase D1 (PLD1), a phospholipid-modifying enzyme which facilitates membrane vesicle trafficking. The results demonstrated that this interaction recruited PLD1 to the Golgi/TGN network and thus possibly modulated APP trafficking, since overexpression of PLD1 promoted generation of APP-containing vesicles from the TGN. Finally, PS1 deficiency in blastocyst-derived cells leads to intracellular retention of caveolin-1 and loss of caveolae at the plasma membrane; this effect appears to be indirect as caveolin-1 was found not to interact physically with PS1. Similarly, α- and β-synuclein were found mislocalized in autophagic organelles in PS1-deficient neurons (Wilson *et al.* 2004).

8.10 **Presenilin-interacting proteins**

Over the past several years many investigators employed yeast two-hybrid assays and candidate approaches to identify proteins that interact with various domains of PS (Van Gassen *et al.* 2000) (Fig.8.2 [**Plate 3**]). As an outcome, several PS-interacting proteins have been identified, including: members of a family of armadillo-related proteins such as β-catenin; cell surface transmembrane protein E-cadherin; neuronal cell adhesion molecule telencephalin; filamin, an actin binding protein; PBP/MOCA, a protein with limited homology to Dock180; the enzyme glycogen synthase kinase-3β; microtubule-associated protein tau; calcium binding proteins such as calsenilin, calmyrin, sorcin, mu-calpain, and CALP/KChIP4; anti-apoptotic molecule Bcl-X_L; Rab11 and Rab6, small GTPases involved in regulation of vesicular transport; RabGDI, a regulatory factor in vesicular transport; PLD1, a phospholipid-modifying enzyme involved in membrane trafficking events; syntaxin 1A, a *t*-SNARE present in synaptic plasma membrane; syntaxin 5, a *t*-SNARE that regulates vesicular trafficking in the ER and Golgi; adaptor proteins X11a and X11b; brain G-protein G_o; Ubiquilin, a protein containing ubiquitin-related domains; HC5 and ZETA subunits of the catalytic 20S proteasome; TPIP, a tetratricopeptide-repeat-containing protein; PSAP, a PDZ-like protein, QM/Jif-1, a negative regulator of c-Jun; DRAL, a LIM-domain protein; proliferation-associated gene product, a protein of the thioredoxin peroxidase family; β-secretase, BACE1; PAMP and PARL, two novel putative metalloproteases; a novel putative methyl-transferase (Metl); mitochondrial immunophilin FKBP38; a splice variant of glial fibril-lary acidic protein, etc. The *C.elegans* PS homologue SEL-12 was recently reported to interact with SEL-10, a Cdc4p-related protein. In many instances the physiological role of the identified interaction between the putative protein with PS1 or PS2 is not clearly defined. For example, the PS interaction with the anti-apoptotic Bcl2 family member Bcl-X_L offers a potential mechanism by which PSs might regulate apoptosis (Passer *et al.* 1999).

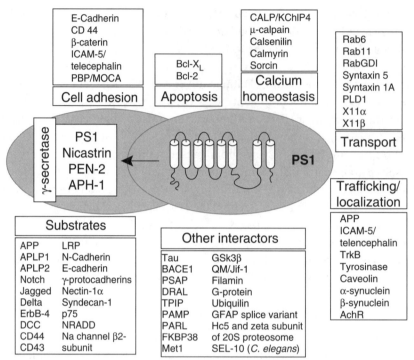

Fig. 8.2 Multifaceted functions of presenilin 1 mediated through the interacting proteins. The figure represents the multiple roles of PS1 in cellular functions. PS1 associates with nicastrin, APH-1, and PEN-2 to form γ-secretase, which cleaves a set of type I membrane protein substrates including APP and Notch. In addition, PS1 interacts with several proteins that are involved in important physiological functions such as calcium homeostasis, vesicular transport, and cell adhesion. In addition, PS1 deficiency, inhibition of γ-secretase activity, or expression of FAD-linked PS1 mutant, influences membrane protein trafficking.

However, it is unclear at present whether an increased apoptotic response associated with the expression of FAD-linked PS variants noted in several studies can be attributed to differential interaction of mutant PS1 or PS2 with Bcl-X$_L$. Similarly, the absence of PS1 or the expression of FAD-linked PS1 mutant results in increased glycogen synthase kinase-3β activity, enhanced kinesin light chain phosphorylation, and concomitant reduction in kinesin-based axonal transport (Pigino *et al.* 2003). However, it remains unclear whether PS1–glycogen synthase kinase-3β interaction plays a direct role in regulating kinesin-based axonal transport. Finally, it is somewhat puzzling that neither regulated metabolism of PS nor the enhanced production of Aβ42 by FAD mutants appears to be influenced by any of the reported PS-interacting proteins.

8.11 Conclusions

Mutations in *PSEN1* and *PSEN2* genes co-segregate with early-onset FAD cases, and genetic ablation of these genes eliminates Aβ production in transgenic mice. Biochemical and pharmacological evidence strongly supports the notion that γ-secretase is a transmembrane aspartyl protease, and PS1 (or PS2) functions as the catalytic subunit of γ-secretase. The mechanism of FAD-associated PS variants in elevating levels of Aβ42 is still being explored. Several pharmacological inhibitors that selectively target γ-secretase and abolish Aβ production have been developed as potential therapeutic strategies for reducing Aβ burden in AD. However, it is now clear that several functionally important type I membrane proteins such as the Notch receptor also undergo γ-secretase cleavage. Thus generalized inhibition of γ-secretase could potentially result in severe consequences by interfering with normal physiological processes carried out by diverse γ-secretase substrates. Nevertheless, γ-secretase processing of APP and other substrates appears to differ with respect to the relative position within the transmembrane segment, as well as subcellular site(s) of cleavage. Therefore, through logical design and high-throughput screening it might be possible to develop inhibitors that specifically target γ-secretase processing of APP but not other substrates. In this regard, a subset of non-steroidal anti-inflammatory drugs have been shown to be effective in specifically reducing the levels of Aβ42 peptides without affecting Aβ40 production or Notch processing. Since the first clues to the association of PS with FAD emerged a decade ago, we have gained tremendous insights into the biological functions of these novel proteins under normal conditions and during AD pathogenesis. With the information we have gained on the biology of presenilins and γ-secretase, developing a viable therapeutic strategy to cure the devastating AD is still promising.

Acknowledgments

The authors are supported by grants from the National Institutes of Health, the Alzheimer's Association, and the American Health Assistance Foundation.

References

Annaert WG, Esselens C, Baert V, *et al.* (2001). Interaction with Telencephalin and the amyloid precursor protein predicts a ring structure for presenilins. *Neuron* **32**, 579–89.

Baki L, Shioi J, Wen P, *et al.* (2004). PS1 activates PI3K thus inhibiting GSK-3 activity and tau overphosphorylation: effects of FAD mutations. *EMBO Journal* **23**, 2586–96.

Beher D, Wrigley JD, Nadin A, *et al.* (2001). Pharmacological knock-down of the presenilin 1 heterodimer by a novel gamma-secretase inhibitor: implications for presenilin biology. *Journal of Biological Chemistry* **276**, 45394–402.

Beher D, Fricker M, Nadin A, *et al.* (2003). *In vitro* characterization of the presenilin-dependent gamma-secretase complex using a novel affinity ligand. *Biochemistry* **42**, 8133–42.

Berezovska O, Ramdya P, Skoch J, Wolfe MS, Bacskai BJ, and Hyman BT (2003). Amyloid precursor protein associates with a nicastrin-dependent docking site on the presenilin 1-gamma-secretase complex in cells demonstrated by fluorescence lifetime imaging. *Journal of Neuroscience* **23**, 4560–6.

Berezovska O, Lleo A, Herl LD, *et al.* (2005). Familial Alzheimer's disease presenilin 1 mutations cause alterations in the conformation of presenilin and interactions with amyloid precursor protein. *Journal of Neuroscience* **25**, 3009–17.

Bertram L and Tanzi RE (2004). The current status of Alzheimer's disease genetics: what do we tell the patients? *Pharmacological Research* **50**, 385–96.

Borchelt DR, Thinakaran G, Eckman CB, *et al.* (1996). Familial Alzheimer's disease-linked presenilin 1 variants elevate Abeta1–42/1–40 ratio *in vitro* and *in vivo. Neuron* **17**, 1005–13.

Brunkan AL, Martinez M, Wang J, *et al.* (2005). Two domains within the first putative transmembrane domain of presenilin 1 differentially influence presenilinase and gamma-secretase activity. *Journal of Neurochemistry* **94**, 1315–28.

Cai D, Leem JY, Greenfield JP, *et al.* (2003). Presenilin-1 regulates intracellular trafficking and cell surface delivery of beta-amyloid precursor protein. *Journal of Biological Chemistry* **278**, 3446–54.

Cai D, Netzer WJ, Zhong M, *et al.* (2006a). Presenilin-1 uses phospholipase D1 as a negative regulator of beta-amyloid formation. *Proceedings of the National Academy of Sciences of the USA* **103**, 1941–6.

Cai D, Zhong M, Wang R, *et al.* (2006b). Phospholipase D1 corrects impaired betaAPP trafficking and neurite outgrowth in familial Alzheimer's disease-linked presenilin-1 mutant neurons. *Proceedings of the National Academy of Sciences of the USA* **103**, 1936–40.

Chyung JH, Raper DM, and Selkoe DJ (2005). Gamma -secretase exists on the plasma membrane as an intact complex that accepts substrates and effects intramembrane cleavage. *Journal of Biological Chemistry* **280**, 4383–92.

De Strooper B, Saftig P, Craessaerts K, *et al.* (1998). Deficiency of presenilin-1 inhibits the normal cleavage of amyloid precursor protein. *Nature* **391**, 387–90.

De Strooper B, Annaert W, Cupers P, *et al.* (1999). A presenilin-1-dependent gamma-secretase-like protease mediates release of Notch intracellular domain. *Nature* **398**, 518–22.

Doan A, Thinakaran G, Borchelt DR, *et al.* (1996). Protein topology of presenilin 1. *Neuron* **17**, 1023–30.

Dumanchin C, Czech C, Campion D, *et al.* (1999). Presenilins interact with rab11, a small GTPase involved in the regulation of vesicular transport. *Human Molecular Genetics* **8**, 1263–9.

Edbauer D, Winkler E, Regula JT, Pesold B, Steiner H, and Haass C (2003). Reconstitution of gamma-secretase activity. *Nature Cell Biology* **5**, 486–8.

Esler WP, Kimberly WT, Ostaszewski BL, *et al.* (2002). Activity-dependent isolation of the presenilin-gamma-secretase complex reveals nicastrin and a g substrate. *Proceedings of the National Academy of Sciences of the USA* **99**, 2720–5.

Francis R, McGrath G, Zhang J, *et al.* (2002). aph-1 and pen-2 are required for Notch pathway signaling, gamma-secretase cleavage of betaAPP, and presenilin protein accumulation. *Developmental Cell* **3**, 85–97.

Goutte C, Hepler W, Mickey KM, and Priess JR (2000). aph-2 encodes a novel extracellular protein required for GLP-1-mediated signaling. *Development*, **127**, 2481–92.

Goutte C, Tsunozaki M, Hale VA, and Priess JR (2002). APH-1 is a multipass membrane protein essential for the Notch signaling pathway in Caenorhabditis elegans embryos. *Proceedings of the National Academy of Sciences of the USA* **99**, 775–9.

Grimm MO, Grimm HS, Patzold AJ, *et al.* (2005). Regulation of cholesterol and sphingomyelin metabolism by amyloid-beta and presenilin. *Nature Cell Biology* **7**, 1118–23.

Gu Y, Misonou H, Sato T, Dohmae N, Takio K, and Ihara Y (2001). Distinct intramembrane cleavage of the beta-amyloid precursor protein family resembling gamma-secretase-like cleavage of Notch. *Journal of Biological Chemistry* **276**, 35235–8.

Iwatsubo T (2004). The gamma-secretase complex: machinery for intramembrane proteolysis. *Current Opinion in Neurobiology* **14**, 379–83.

Kaether C, Lammich S, Edbauer D, *et al.* (2002). Presenilin-1 affects trafficking and processing of betaAPP and is targeted in a complex with nicastrin to the plasma membrane. *Journal of Cell Biology* **158**, 551–61.

Kaether C, Capell A, Edbauer D, *et al.* (2004). The presenilin C-terminus is required for ER-retention, nicastrin-binding and gamma-secretase activity. *EMBO Journal* **23**, 4738–48.

Kang DE, Yoon IS, Repetto E, *et al.* (2005). Presenilins mediate phosphatidylinositol 3-kinase/AKT and ERK activation via select signaling receptors. Selectivity of PS2 in platelet-derived growth factor signaling. *Journal of Biological Chemistry* **280**, 31537–47.

Kim SH and Sisodia SS (2005). Evidence that the 'NF' motif in transmembrane domain 4 of presenilin 1 is critical for binding with PEN-2. *Journal of Biological Chemistry* **280**, 41953–66.

Kim SH, Leem JY, Lah JJ, *et al.* (2001). Multiple effects of aspartate mutant presenilin 1 on the processing and trafficking of amyloid precursor protein. *Journal of Biological Chemistry* **276**, 43343–50.

Kim SH, Ikeuchi T, Yu C, and Sisodia SS (2003). Regulated hyperaccumulation of presenilin-1 and the "gamma-secretase" complex. Evidence for differential intramembranous processing of transmembrane substrates. *Journal of Biological Chemistry* **278**, 33992–4002.

Kimberly WT, LaVoie MJ, Ostaszewski BL, Ye W, Wolfe MS, and Selkoe DJ (2003). Gamma-secretase is a membrane protein complex comprised of presenilin, nicastrin, Aph-1, and Pen-2. *Proceedings of the National Academy of Sciences of the USA* **100**, 6382–7.

Laudon H, Mathews PM, Karlstrom H, *et al.* (2004). Co-expressed presenilin 1 NTF and CTF form functional gamma-secretase complexes in cells devoid of full-length protein. *Journal of Neurochemistry* **89**, 44–53.

Laudon H, Hansson EM, Melen K, *et al.* (2005). A Nine-transmembrane Domain Topology for Presenilin 1. *Journal of Biological Chemistry* **280**, 35352–60.

LaVoie MJ, Fraering PC, Ostaszewski BL, *et al.* (2003). Assembly of the gamma-secretase complex involves early formation of an intermediate sub-complex of Aph-1 and Nicastrin. *Journal of Biological Chemistry* **278**, 37213–22.

Lazarov VK, Fraering PC, Ye W, Wolfe MS, Selkoe DJ, and Li H (2006). Electron microscopic structure of purified, active gamma-secretase reveals an aqueous intramembrane chamber and two pores. *Proceedings of the National Academy of Sciences of the USA* **103**, 6889–94.

Lee MK, Slunt HH, Martin LJ, *et al.* (1996). Expression of presenilin 1 and 2 (PS1 and PS2) in human and murine tissues. *Journal of Neuroscience* **16**, 7513–25.

Leem JY, Saura CA, Pietrzik C, *et al.* (2002a). A role for presenilin 1 in regulating the delivery of amyloid precursor protein to the cell surface. *Neurobiology of Disease* **11**, 64–82.

Leem JY, Vijayan S, Han P, *et al.* (2002b). Presenilin 1 is required for maturation and cell surface accumulation of nicastrin. *Journal of Biological Chemistry* **277**, 19236–40.

Levy-Lahad E, Wasco W, Poorkaj P, *et al.* (1995). Candidate gene for the chromosome 1 familial Alzheimer's disease locus. *Science* **269**, 973–7.

Li T, Ma G, Cai H, Price DL, and Wong PC (2003). Nicastrin is required for assembly of presenilin/gamma-secretase complexes to mediate Notch signaling and for processing and trafficking of b-amyloid precursor protein in mammals. *Journal of Neuroscience* **23**, 3272–7.

Li YM, Lai MT, Xu M, *et al.* (2000a). Presenilin 1 is linked with gamma-secretase activity in the detergent solubilized state. *Proceedings of the National Academy of Sciences of the USA* **97**, 6138–43.

Li YM, Xu M, Lai MT, *et al.* (2000b). Photoactivated gamma-secretase inhibitors directed to the active site covalently label presenilin 1. *Nature* **405**, 689–94.

Lleo A, Berezovska O, Herl L, *et al.* (2004). Nonsteroidal anti-inflammatory drugs lower Abeta42 and change presenilin 1 conformation. *Nature Medicine* **10**, 1065–6.

McConlogue L, Castellano F, deWit C, Schenk D, and Maltese WA (1996). Differential effects of a Rab6 mutant on secretory versus amyloidogenic processing of Alzheimer's beta-amyloid precursor protein. *Journal of Biological Chemistry* **271**, 1343–8.

Mastrangelo P, Mathews PM, Chishti MA, *et al.* (2005). Dissociated phenotypes in presenilin transgenic mice define functionally distinct gamma-secretases. *Proceedings of the National Academy of Sciences of the USA* **102**, 8972–7.

Mirnics K, Korade Z, Arion D, *et al.* (2005). Presenilin-1-dependent transcriptome changes. *Journal of Neuroscience* **25**, 1571–8.

Mumm JS, Schroeter EH, Saxena MT, *et al.* (2000). A ligand-induced extracellular cleavage regulates gamma-secretase-like proteolytic activation of Notch1. *Molecular Cell*, **5**, 197–206.

Naruse S, Thinakaran G, Luo J-J, *et al.* (1998). Effects of PS1 deficiency on membrane protein trafficking in neurons. *Neuron* **21**, 1213–21.

Ogura T, Mio K, Hayashi I, *et al.* (2006). Three-dimensional structure of the gamma-secretase complex. *Biochemical and Biophysical Research Commun*ications **343**, 525–34.

Pardossi-Piquard R, Petit A, Kawarai T, *et al.* (2005). Presenilin-dependent transcriptional control of the Abeta-degrading enzyme neprilysin by intracellular domains of betaAPP and APLP. *Neuron* **46**, 541–54.

Parent AT, Barnes NY, Taniguchi Y, Thinakaran G, and Sisodia SS (2005). Presenilin attenuates receptor-mediated signaling and synaptic function. *Journal of Neuroscience* **25**, 1540–9.

Passer BJ, Pellegrini L, Vito P, Ganjei JK, and D'Adamio L (1999). Interaction of Alzheimer's Presenilin-1 and Presenilin-2 with Bcl-X(L). A potential role in modulating the threshold of cell death. *Journal of Biological Chemistry* **274**, 24007–13.

Pigino G, Morfini G, Pelsman A, Mattson MP, Brady ST, and Busciglio J (2003). Alzheimer's presenilin 1 mutations impair kinesin-based axonal transport. *Journal of Neuroscience* **23**, 4499–508.

Podlisny MB, Citron M, Amarante P, *et al.* (1997). Presenilin proteins undergo heterogeneous endoproteolysis between Thr291 and Ala299 and occur as stable N- and C-terminal fragments in normal and Alzheimer brain tissue. *Neurobiology of Disease* **3**, 325–37.

Ratovitski T, Slunt HH, Thinakaran G, Price DL, Sisodia SS, and Borchelt DR (1997). Endoproteolytic processing and stabilization of wild-type and mutant presenilin. *Journal of Biological Chemistry* **272**, 24536–41.

Rechards M, Xia W, Oorschot VM, Selkoe DJ, and Klumperman J (2003). Presenilin-1 exists in both pre- and post-Golgi compartments and recycles via COPI-coated membranes. *Traffic* **4**, 553–65.

Rogaev EI, Sherrington R, Rogaeva EA, *et al.* (1995). Familial Alzheimer's disease in kindreds with missense mutations in a gene on chromosome 1 related to the Alzheimer's disease type 3 gene. *Nature* **376**, 775–8.

Sastre M, Steiner H, Fuchs K, *et al.* (2001). Presenilin-dependent gamma-secretase processing of b-amyloid precursor protein at a site corresponding to the S3 cleavage of Notch. *EMBO Reports* **23**, 23.

Saura CA, Tomita T, Davenport F, Harris CL, Iwatsubo T, and Thinakaran G (1999). Evidence that intramolecular associations between Presenilin domains are obligatory for endoproteolytic processing. *Journal of Biological Chemistry* **274**, 13818–23.

Saura CA, Tomita T, Soriano S, *et al.* (2000). The non-conserved hydrophilic loop domain of presenilin (PS) is neither required for PS endoproteolysis nor enhanced Ab42 production mediated by familial Alzheimer's disease-linked PS variants. *Journal of Biological Chemistry* **275**, 17136–42.

Saura CA, Choi SY, Beglopoulos V, *et al.* (2004). Loss of presenilin function causes impairments of memory and synaptic plasticity followed by age-dependent neurodegeneration. *Neuron* **42**, 23–36.

Scheper W, Zwart R, Sluijs P, Annaert W, Gool WA, and Baas F (2000). Alzheimer's presenilin 1 is a putative membrane receptor for rab GDP dissociation inhibitor. *Human Molecular Genetics* **9**, 303–310.

Scheper W, Zwart R, and Baas F (2004). Rab6 membrane association is dependent of presenilin 1 and cellular phosphorylation events. *Brain Research Molecular Brain Res*earch **122**, 17–23.

Scheuner D, Eckman C, Jensen M, *et al.* (1996). Secreted amyloid beta-protein similar to that in the senile plaques of Alzheimer's disease is increased *in vivo* by the presenilin 1 and 2 and APP mutations linked to familial Alzheimer's disease. *Nature Medicine* **2**, 864–70.

Schroeter EH, Ilagan MX, Brunkan AL, *et al.* (2003). A presenilin dimer at the core of the gamma-secretase enzyme: insights from parallel analysis of Notch 1 and APP proteolysis. *Proceedings of the National Academy of Sciences of the USA* **100**, 13075–80.

Shah S, Lee SF, Tabuchi K, *et al.* (2005). Nicastrin functions as a gamma-secretase-substrate receptor. *Cell* **122**, 435–47.

Sherrington R, Rogaev EI, Liang Y, *et al.* (1995). Cloning of a gene bearing missense mutations in early-onset familial Alzheimer's disease. *Nature* **375**, 754–60.

Shirotani K, Edbauer D, Prokop S, Haass C, and Steiner H (2004). Identification of distinct gamma-secretase complexes with different APH-1 variants. *Journal of Biological Chemistry* **279**, 41340–5.

Steiner H, Romig H, Pesold B, *et al.* (1999). Amyloidogenic function of the Alzheimer's disease-associated presenilin 1 in the absence of endoproteolysis. *Biochemistry* **38**, 14600–14605.

Stromberg K, Hansson EM, Laudon H, *et al.* (2005). Gamma-secretase complexes containing N- and C-terminal fragments of different presenilin origin retain normal gamma-secretase activity. *Journal of Neurochemistry* **95**, 880–90.

Struhl G and Adachi A (2000). Requirements for presenilin-dependent cleavage of Notch and other transmembrane proteins. *Molecular Cell* **6**, 625–636.

Struhl G and Greenwald I (1999). Presenilin is required for activity and nuclear access of Notch in *Drosophila*. *Nature* **398**, 522–5.

Takasugi N, Tomita T, Hayashi I, *et al.* (2003). The role of presenilin cofactors in the gamma-secretase complex. *Nature* **422**, 438–41.

Tanemura K, Chui DH, Fukuda T, *et al.* (2006). Formation of tau inclusions in knock-in mice with familial Alzheimer disease (FAD) mutation of presenilin 1 (PS1). *Journal of Biological Chemistry* **281**, 5037–41.

Thinakaran G and Parent AT (2004). Identification of the role of presenilins beyond Alzheimer's disease. *Pharmacological Research* **50**, 411–8.

Thinakaran G, Borchelt DR, Lee MK, *et al.* (1996). Endoproteolysis of presenilin 1 and accumulation of processed derivatives *in vivo*. *Neuron* **17**, 181–90.

Thinakaran G, Harris CL, Ratovitski T, *et al.* (1997). Evidence that levels of presenilins (PS1 and PS2) are coordinately regulated by competition for limiting cellular factors. *Journal of Biological Chemistry* **272**, 28415–22.

Van Gassen G, Annaert W, and Van Broeckhoven C (2000). Binding partners of Alzheimer's disease proteins: are they physiologically relevant? *Neurobiology of Disease* **7**, 135–51.

Vetrivel KS, Cheng H, Lin W, *et al.* (2004). Association of gamma-secretase with lipid rafts in post-golgi and endosome membranes. *Journal of Biological Chemistry* **279**, 44945–54.

Wang R, Tang P, Wang P, Boissy RE, and Zheng H (2006). Regulation of tyrosinase trafficking and processing by presenilins: partial loss of function by familial Alzheimer's disease mutation. *Proceedings of the National Academy of Sciences of the USA* **103**, 353–8.

Watanabe N, Tomita T, Sato C, Kitamura T, Morohashi Y, and Iwatsubo T (2005). Pen-2 is incorporated into the gamma-secretase complex through binding to transmembrane domain 4 of presenilin 1. *Journal of Biological Chemistry* **280**, 41967–75.

Weggen S, Eriksen JL, Das P, *et al.* (2001). A subset of NSAIDs lower amyloidogenic Abeta42 independently of cyclooxygenase activity. *Nature* **414**, 212–16.

Wilson CA, Murphy DD, Giasson BI, Zhang B, Trojanowski JQ, and Lee VM (2004). Degradative organelles containing mislocalized alpha-and beta-synuclein proliferate in presenilin-1 null neurons. *Journal of Cell Biology* **165**, 335–46.

Wolfe MS, Xia W, Ostaszewski BL, Diehl TS, Kimberly WT, and Selkoe DJ (1999). Two transmembrane aspartates in presenilin-1 required for presenilin endoproteolysis and gamma-secretase activity. *Nature* **398**, 513–17.

Yu C, Kim SH, Ikeuchi T, Xu H, Gasparini L, Wang R, and Sisodia SS (2001). Characterization of a presenilin-mediated amyloid precursor protein carboxyl-terminal fragment gamma. Evidence for distinct mechanisms involved in gamma -secretase processing of the APP and Notch1 trans-membrane domains. *Journal of Biological Chemistry* **276**, 43756–60.

Yu G, Nishimura M, Arawaka S, *et al.* (2000). Nicastrin modulates presenilin-mediated notch/glp-1 signal transduction and betaAPP processing. *Nature* **407**, 48–54.

Zhao G, Mao G, Tan J, *et al.* (2004). Identification of a new presenilin-dependent zeta-cleavage site within the transmembrane domain of amyloid precursor protein. *Journal of Biological Chemistry* **279**, 50647–50.

Zhou S, Zhou H, Walian PJ, and Jap BK (2005). CD147 is a regulatory subunit of the gamma-secretase complex in Alzheimer's disease amyloid beta-peptide production. *Proceedings of the National Academy of Sciences of the USA* **102**, 7499–504.

Chapter 9

Multiple transgenic mouse models for Alzheimer's disease

Dick Terwel, Ilse Dewachter, Tom Van Dooren, Tom Vandebroek, Thomas Van Helmont, David Muyllaert, Peter Borghgraef, Sophie Croes, Joris Winderickx, and Fred Van Leuven

The identification of the genetic causes of familial Alzheimer's disease (AD) and frontotemporal dementia (FTD) by clinical geneticists has yielded the necessary genes for experimental biologists to model these diseases in transgenic mice. The first models to be developed were transgenic mice that express wild-type and mutant forms of amyloid precursor protein (APP). Some of these models recapitulate surprisingly faithfully the formation of amyloid plaques in brain parenchyma as well as the accompanying cerebral amyloid angiopathy (CAA), typical of sporadic and familial AD. The mice have served as fundamental research tools and are also excellent preclinical test banks for proven or experimental drugs, diets and natural products, metals and metal chelators, and active and passive vaccination.

Furthermore, transgenic mice in general, and the APP mice in particular, provide a specified genetic background with a specified phenotype or pathology to investigate direct modulation by intrinsic or extrinsic genes *in vivo* by inactivation, mutation, or coexpression of the candidate gene. In the case of AD models, the resulting multiple transgenic mouse strains are studied comparatively with the original parental single strains with respect to amyloid and tau pathology, for cognitive and behavioural defects, and for related physiological and biochemical parameters. The APP transgenic mice have been very informative in the delineation and analysis of the proteinases that cleave APP, i.e. the secretases, and in defining their contribution to the normal function and the pathological role of APP and its metabolites, which remain largely unknown *in vivo*. We review here our contributions in this respect, based on our APP-V717I transgenic mice (Moechars *et al.* 1999; Van Dorpe *et al.* 2000).

Best known is the demonstration of presenilin-1 (PS1) as the most essential subunit of γ-secretase in the generation of amyloid peptides *in vivo* (De Strooper *et al.* 1998). Mutant PS1 particularly increases the production of the neurotoxic Aβ42 isoform, while neuron-specific inactivation of PS1 prevents the amyloid pathology completely but is not capable of rescuing the cognitive defects of the APP-V717I transgenic mice

(Dewachter *et al.* 2000, 2002). Moreover, these and parallel studies have revealed other functions of PS1, unrelated to the cleavage of APP. Proteolysis of other transmembrane proteins could be involved, although a totally different activity could be at work, unrelated to the proteolytic activity of γ-secretase, i.e. as a regulator of calcium ion homeostasis (Herms *et al.* 2003; Ris *et al.* 2003).

The contribution of the β-secretase BACE1 (β-APP site cleaving enzyme) to APP processing was similarly explored and found as expected, at least in terms of the large increase in parenchymal amyloid pathology in the brain of ageing APP-V717I × BACE1 double transgenic mice (Willem *et al.* 2004). However, the correlated finding that these brains were practically devoid of CAA was totally unexpected. The close correlation of CAA with large increased levels of the less soluble N-truncated amyloid peptide Aβ11–42, demonstrated *in vivo* that BACE1 is in tight control of the balance in amyloid pathology in brain, promoting either parenchyma or vasculature.

Human ADAM10 (a disintegrin and metalloprotease) was validated as the major α-secretase by its expression in combination with mutant APP-V717I, allowing us to demonstrate the correlated prevention of amyloid pathology and the rescue of cognitive defects in APP-V717I × ADAM10 double transgenic mice (Postina *et al.* 2004). Therefore ADAM10 was upgraded to a major and potentially very interesting therapeutic target in AD. Nevertheless, the pharmacological task remains a challenge, since selective activation of ADAM10 is needed; this is probably already being approached *in vivo* with drugs that activate protein kinase C (PKC) (Etcheberrigaray *et al.* 2004).

The strongest genetic risk factor for AD is the e4 allele of the apolipoprotein gene coding for ApoE4, concerning at least 40% of all sporadic AD cases. Overexpression of human ApoE4 in either neurons or glial cells, in combination with APP-V717I in double transgenic mice, strongly promoted an earlier onset of the amyloid pathology (Van Dooren *et al.* 2006). ApoE4 was as effective as mutant PS1 in this respect, but acted by a completely different mechanism, since it did not affect the processing of APP, even when coexpressed in the same neurons. Thus it has been proposed that the ApoE4 effect in sporadic AD is exerted downstream of the production of the amyloid peptides, i.e. by interfering with their clearance by transport or by degradation (Van Dooren *et al.* 2006). It is also significant that neuronal coexpression of ApoE4 with mutant APP and PS1 did not cause any overt signs of tau pathology, although this could have been expected based on the proven tauopathy in the parental thy1-ApoE4 transgenic mice (Tesseur *et al.* 2000a,b).

The cerebral presence of neurofibrillary tangles (NFTs) in the typical combination with amyloid remains the only pathological definition of AD, and allowing definitive post-mortem diagnosis. Therefore a major problem remains unsolved in that none of the single transgenic amyloid mouse models show any major sign of tauopathy in the form of intraneuronal tangles or neuropil treads.

In mice, the combinatorial pathology can only be obtained in APP × Tau (× PS1) double and triple transgenic mice (Lewis *et al.* 2000; D. Terwel *et al.*, unpublished data). Since these models prove that both pathologies can coexist in mouse brain, the apparent contradiction of the amyloid cascade hypothesis in murine brain needs to be approached by an alternative route, which we have attempted to do by generating multiple transgenic mice that overexpress the tau kinases glycogen synthase kinase 3β (GSK-3β) and cyclin-dependent protein kinase-5 (cdk5).

Neuronal overexpression of the wild-type human protein tau-4R isoform did not result in a tauopathy in tau-4R transgenic mice, but produced in a severe, although not lethal, motor neuron disease (Spittaels *et al.* 1999). The specified axonopathy and Wallerian degeneration which occurred was manifest mainly in motor neurons and was completely rescued by coexpression of human GSK-3β (Spittaels *et al.* 2000). Despite the pronounced increase in phosphorylation of human and mouse protein tau in the brain of tau-4R × GSK-3β double transgenic mice, no tauopathy resulted from this combination.

On the other hand, the triple combination of tau-4R × cdk5 × p35 transgenic mice even failed to produce the anticipated phosphorylation of tau-4R, indicating that cdk5/p35 does not contribute directly to the phosphorylation of protein tau (Van den Haute *et al.* 2001). Obviously, additional factors are needed to make cdk5 as effective a tau kinase *in vivo* as it is claimed to be *in vitro*.

Finally, only the expression of clinical mutants of protein tau, such as tau-P301L or R406W, which cause familial FTD have resulted in transgenic models with a robust formation of NFTs in the brain (Terwel *et al.* 2005 and references cited therein). Tau-P301L transgenic mice developed intense tau pathology not only in the forebrain, but also in the spinal cord and brainstem. The latter is probably the cause of the moribund phenotype, i.e. tau-P301L mice do not survive for more than 10–12 months (Terwel *et al.* 2005).

The APP-V717I, Tau-P301L, and GSK-3β transgenic mice are now being intercrossed in all possible combinations, and they are currently being phenotypically characterized. The preliminary data indicate that all these combinations yield novel indications for the mechanisms whereby the amyloid and tau pathology 'talk to each other' in the brain of transgenic mice.

9.1 Introduction

9.1.1 Alzheimer's disease, amyloid pathology, and mutant amyloid precursor protein and presenilin 1

AD is an age-related neurodegenerative disorder with retrograde progressive amnesia that is mentally and physically completely debilitating. The pathological hallmarks are extracellular amyloid deposits in the brain parenchyma and vasculature, combined with intraneuronal tangles of filaments of tau protein. The synaptic loss and neuronal death are associated with severe astrogliosis and microglial activation that destroy the brain in later stages. Amyloid peptides (Aβ) are derived by multiple and complex endoproteolytic processing steps from APP, generating a variety of biochemical processing products with incompletely understood or unknown functions (reviewed by Reinhard *et al.* 2005). Protein tau is a microtubule-associated protein that is normally involved in the neuronal differentiation, outgrowth, and maintenance of neurites and axons by stabilizing the microtubules as cytoskeletal elements, mainly responsible for transport in post-mitotic neurons (reviewed by Terwel *et al.* 2002).

Most cases of AD are sporadic, stemming from unknown causes. Only a small percentage, i.e. less than 1% of all cases, can be inherited in an autosomal dominant manner. These are due to mutations in the genes coding for APP, PS1,or PS2. The mutations in APP are located close to the proteolytic cleavage sites at the N- and C-termini of the amyloid sequence in APP, and lead to increased production of Aβ, particularly the Aβ42 amino acid isoform (reviewed by Hardy 1997; Esler and Wolfe 2001; Sisodia and St. George-Hyslop 2002). The combined genetic and biochemical findings suggest a central role for the amyloid peptides in the pathogenesis of AD, commonly referred to as the 'amyloid cascade hypothesis' (Hardy and Allsop 1991; Hardy and Selkoe 2002). This hypothesis is further corroborated by a variety of data, not least by the fact that amyloid peptides are derived by endoproteolytic processing from APP as the main components of the amyloid plaques. Since amyloid peptides negatively affect synaptic plasticity and cognition, these defective processes are believed to be already critical in the early phases of the disease.

Amyloid peptides have various neurotoxic properties and can induce phosphorylation and aggregation of protein tau, at least in cellular models (Gotz *et al.* 2004; Iqbal *et al.* 2005). However, these findings urgently require confirmation and/or in-depth analysis to establish the exact molecular relations and the exact nature of the amyloid peptides; their chemical and physical forms can vary from monomeric to di-, multi- and oligomeric, and from soluble to deposited, fibrillar, or diffusely amorphous. Finally, amyloid peptides are present in the various somatic, dendritic, and axonal subcellular and extracellular compartments of the brain, not only as soluble monomers or multimers, but also as membrane-associated and embedded dimers, as recently proposed (Marchesi 2005). This complex structural picture is further complicated by the physiological and pathological contributions of the other APP processing products, i.e. the secreted ectodomains (α- and β-APPs), the various C-terminal fragments (α-, β-, γ-CTFs), phosphorylated or non-phosphorylated, and finally the intracellular domain (AICD). This diversity of biochemical metabolites, the various molecular mechanisms, and the very diverse cellular and animal models in which their production and resultant effects are being studied illustrate the enormous scientific and logistic problems posed by AD.

The need to develop and study adequate *in vivo* models for clarifying the underlying pathological mechanisms, as well as the modulation of normal brain processes, led us to generate transgenic mice, in the first instance overexpressing mutant APP-V717I (reviewed by Van Leuven 2000; Dewachter and Van Leuven 2002). Subsequently we combined these amyloid model mice with other transgenic strains that express different human genes, suspected or implied, directly or indirectly, in familial and sporadic AD (e.g. PS1, BACE1, ADAM10, ApoE4, Tau). The choice of these genes and mice is evident from the genetic and clinical studies, and will be discussed further below.

9.1.2 Protein tau, tauopathy, and mutant tau

9.1.2.1 Normal and abnormal phosphorylation of tau

Protein tau is predominantly expressed in neurons in both the central and peripheral nervous system. It is encoded by a single gene with 16 exons located on the long arm of chromosome 17 (reviewed by Buée *et al.* 2000). By alternative splicing of the primary gene transcript, up to six different mRNA species in the brain can be produced, coding for six tau protein isoforms ranging from 352 to 441 amino acids. These six isoforms contain either three or four imperfect C-terminal tandem repeats and either one or two shorter N-terminal domains. The three or four C-terminal domains mediate binding to microtubules (MT) and thus are involved in the assembly, elongation, and stabilization of MT (Terwel *et al.* 2002). *In vitro*, tau promotes tubulin polymerization and stabilizes the resulting MT (Goedert and Jakes 1990). *In vivo*, tau is redundant, at least in mouse brain, with only minor disturbances of brain development in tau-deficient mice (Harada *et al.*1994; Dawson *et al.* 2001). The binding of tau and other MT-associated proteins (MAPs) to preformed MT is intuitively thought to occur on the outer MT surface, although copolymerization could also include tau inside MT (Kar *et al.* 2003). This may explain observations of 'undissociable' binding of tau to MT (Makrides *et al.* 2004), although unsaturated binding of tau to MT has also been reported (Ackmann *et al.* 2000).

Obviously, the issue of tau–MT interactions becomes even more complicated when the contribution of the dynamic phosphorylation of protein tau is considered. Tau is a naturally unfolded protein, without a predefined three-dimensional structure. In normal conditions it is always phosphorylated to some extent, although this is variable, but in pathological situations such as AD and FTD the degree of phosphorylation is dramatically increased. This condition is referred to as hyperphosphorylation, and although no exact molecular definition is possible, the detection of pathological phospho-epitopes, defined by specified monoclonal antibodies, in combination with an increased apparent molecular mass is the accepted operational definition.

Phosphorylation of tau is considered important in relation to binding to MT, whereas hyperphosphorylation is considered to contribute to the pathogenesis in tauopathies, including AD (Avila *et al.* 2004; Iqbal *et al.* 2005; Stoothoff and Johnson 2005). *In vitro*, many kinases can phosphorylate tau, but it is very difficult to establish the equivalent in brain *in vivo*, and to define exactly which kinases are responsible for the phosphorylation of tau at precise amino acid residues. Two kinases that are most likely candidates *in vivo* are GSK-3β and cdk5, and therefore studies and modelling of the functions of protein tau and of tauopathy *in vivo* need to consider these kinases, among others, as intrinsic partners in the overall disease process.

9.1.2.2 Mutant tau or excess tau-4R causes frontotemporal dementia

Of particular note in the current context is the fact that mutations in the gene coding for protein tau do not cause AD but cause a diverse group of tauopathies, named frontotemporal dementia with parkinsonism associated with chromosome 17 (FTDP-17) (Delacourte and Buée 2000; Heutink 2000; Goedert and Jakes 2005). Mutations in the tau gene are exonic, but can also be located in introns, and are then proposed to increase the splicing in of exon 10 that encodes the second microtubule binding domain in protein tau, resulting in a relative over-abundance of the tau-4R isoform. Exonic mutations evidently influence the properties of the expressed cellular product of mutant protein tau, but also appear to alter tau mRNA splicing, similar to the intronic mutations.

Protein tau accumulates as filaments inside neurons not only in the brain of FTD patients, but also in all tauopathy patients including AD patients, although with marked brain-regional differences. Moreover 'tangled tau' is invariably hyperphosphorylated, i.e. carries more phosphate groups than normal tau and/or is phosphorylated at different sites, defined by a series of monoclonal antibodies as pathological epitopes (Terwel *et al.* 2005; Vandebroek *et al.* 2005). These biochemical findings are highly suggestive of an important or even pivotal contribution of kinases, and possibly also phosphatases, to the development of any tauopathy in patients who evidently express mostly wild-type tau-4R (Buée *et al.* 2000; Ingram and Spillantini 2002).

The properties of tau-4R that are thought to increase its self-aggregation are the combination of lesser binding to the microtubule because of an increased propensity to be phosphorylated. The exact relation is not clear and the evidence varies in quality, and the eventual contribution to the ensuing tauopathy is also uncertain (reviewed by Goedert and Jakes 2005; Iqbal *et al.* 2005; Stoothoff and Johnson 2005). An innate tendency to self-aggregation of protein tau as a naturally unfolded protein is further promoted by diverse polyanions, including RNA, that are plentiful in the cell cytoplasm. This tendency to aggregate could be controlled by limited phosphorylation and, more likely, by the fact that all protein tau appears to be bound to MT in normal conditions. Conversely, mutant protein tau appears to require a lesser degree of phosphorylation to induce its self-aggregation, at least *in vitro* (Barghorn *et al.* 2000; Alonso *et al.* 2004).

The common trait of patients with FTDP-17 and sporadic tauopathies is the tau-related pathology characterized by intraneuronal twisted, straight, or paired helical filaments, depending on the particular mutation or disease. However, this common pathology and denomination hides important differences in clinical symptoms resulting from the different mutations in the tau gene, while even the same mutation in different families or in different members of the same family results in clinical heterogeneity (Ingram and Spillantini 2002). As well as the primary genetic problem, additional genetic or epigenetic factors must be invoked to modify the primary molecular defect of mutant tau or of the increased tau-4R/tau-3R concentration ratio.

Finally, and most importantly, the genetic findings that mutations in the gene coding for protein tau can themselves cause neurodegeneration in FTD patients, particularly those who overexpress tau-4R because of intronic mutations, have contributed tremendously to the experimental attention directed towards protein tau and tau pathologies, not only in FTD but in neurodegenerative diseases in general, including AD as the most frequent tauopathy in the human population.

9.2 Transgenic mice as pathological models

The identification of the genetic defects and mutations that cause AD, FTD, and other tauopathies has yielded the genes needed to generate transgenic mice as models for these devastating diseases.

Transgenic mice that overexpress mutant APP, alone or in combination with mutant PS1, were among the first to be generated and continue to be valuable models for analysis of the pathophysiological role of the amyloid peptides and other APP-processing products. Moreover, they are valuable testbanks for analysis of the potential of therapeutic strategies aimed at decreasing amyloid peptide levels in brain, although they do not recapitulate all aspects of AD. Particularly lacking is the key pathological hallmark of intraneuronal fibrillar tangles that are readily formed in neurons of transgenic mice that overexpress mutants of protein tau which cause FTDP-17 in humans.

In the long run, the need to understand the contribution of other genes, as well as to produce a more complete model for AD, motivated us to generate double and multiple transgenic mouse strains that eventually comprise both pathological hallmarks, i.e. amyloid plaques and neurofibrillary tangles. Crossing APP-V717I transgenic mice with mice that overexpress tau-P301L yields such a model that helps to understand the duality and the relation of the pathological lesions, as well as helping to define disease-modifying therapeutic strategies.

Since much of this effort is ongoing 'work in progress', in this chapter we focus on our recent progress with single, double, and triple transgenic mice that relates to AD and FTDP-17. The first part deals with amyloid pathology and its modification by co-overexpression of BACE1 (β-site APP cleaving enzyme 1), ADAM10 (a disintegrin and metallopeptidase 10), PS1, or ApoE4 (apolipoprotein E4). The second part deals with animal models for tau and FTDP-17 and modifications of the tau pathology by coexpression of kinases. Some preliminary information on different combinations of double transgenic mice that reveal synergistic actions of APP, protein tau, and GSK-3β will be discussed.

9.2.1 Transgenic models for amyloid pathology and its modulation *in vivo*

9.2.1.1 Amyloid precursor protein transgenic mice

Animal models recapitulating characteristic features of AD pathology have been generated by overexpressing clinical mutants of APP, known to cause EOFAD (early-onset familial AD). We have generated APP-V717I transgenic mice that display early phenotypic changes similar to AD, including behavioural deficits (increased aggression and neophobia) and cognitive deficits assessed in a spatial navigation task and an object recognition task. The deficits in learning and memory are reflected at the cellular level as a deficit in long-term potentiation (LTP) in CA1 of the hippocampus, a region involved in the generation of spatial memories which becomes dysfunctional early in AD patients. The APP-V717I transgenic mice exhibit a clear dissociation between early clinical deficits or symptoms and late amyloid pathology, indicating that amyloid plaques are not the cause of the early deficits (Moechars *et al.* 1999; Dewachter *et al.* 2000; Van Dorpe *et al.* 2000). The cognitive and behavioural deficits precede by many months the formation of typical brain amyloid pathology, i.e. diffuse and later neuritic amyloid plaques in the brain parenchyma at the age of 12 months, followed by amyloid deposits in the vasculature, all progressively increasing with age.

Double transgenic mice with coexpression of human APP-V717I with mutant PS1-A246E underlined the essential role of the Aβ42 peptides in the development of the amyloid pathology in the brain parenchyma and vasculature. Both types of amyloid pathology had already developed at the earlier age of 6–8 months in APP × PS1 double mutant transgenic mice, compared with 2 months in single APP transgenic mice (Borchelt *et al.* 1996, 1997; Citron *et al.* 1997; Holcomb *et al.* 1998; Moechars *et al.* 1999; Dewachter *et al.* 2000; Van Dorpe *et al.* 2000).

9.2.1.2 Contribution of presenilin as γ-secretase and more

In view of the proposed central role of the amyloid peptides in the pathogenesis of AD, by far the majority of therapeutic strategies aim at inhibiting the production or increasing the clearance of amyloid peptides. The β- and γ-secretases that generate the Aβ peptides by sequential cleavage from APP are obvious prominent targets for the development of specific inhibitors (Dewachter and Van Leuven 2002). Presenilins are polytopic transmembrane proteins which are, in combination with at least three other proteins (aph1, pen2 and nicastrin), required for an efficient γ-secretase complex and activity to generate amyloid peptides (Edbauer *et al.* 2003).

We have used LoxP/Cre-recombinase-mediated deletion to generated mice with a post-natal neuron-specific deficiency in PS1, denoted PS1(n–/–) mice, thereby circumventing the embryonic lethality of a complete deletion of PS1 (Dewachter *et al.* 2002). Adult PS1(n–/–) mice were normal in most respects, and displayed a normal brain morphology. However, the levels of endogenous brain amyloid peptides were strongly decreased, concomitantly with an accumulation of APP C-terminal fragments (CTFs). The crossing APP-V717I × PS1(n–/–) resulted in double transgenic mice in which the neuronal absence of PS1 effectively prevented amyloid pathology even at the advanced age of 16 months (Dewachter *et al.* 2002). This contrasts sharply with the APP-V717I single transgenic mice which all develop amyloid pathology at the age of 12 months. Moreover, in the APP-V717I × PS1(n–/–) mice, the LTP in CA1 was practically normalized at the end of the 2 h observation period, demonstrating the involvement of amyloid peptides in defective LTP in APP transgenic mice (Dewachter *et al.* 2002). The data corroborated evidence for inhibition of LTP and cognition by intracerebral injection of oligomeric amyloid peptides (Walsh *et al.* 2002) and of vaccinations that specifically decrease soluble amyloid peptides (Dodart *et al.* 2003).

Importantly, however, neuronal absence of PS1 failed to normalize the cognitive defect assessed by the object recognition task, which remained impaired in the APP-V717I × PS1(n–/–) mice as in the parent APP-V717I transgenic mice (Dewachter *et al.* 2002). These data point to a potentially detrimental effect of accumulating β-CTF or C99 fragments from APP, the remnants after β-secretase cleavage, which was recently confirmed independently (Saura *et al.* 2004). The implication is that when the accumulating APP β-CTFs are responsible for the impaired cognition, they must also be detrimental in the parental single APP transgenic mice, and by extrapolation also in AD patients. A pathological role for neuronally accumulating APP β-CTFs, in combination with the ever-growing list of substrates of γ-secretase with many pleiotropic effects, is shedding grave doubts on the therapeutic potential use of γ-secretase inhibitors in AD.

In addition, our data highlight once more the complex functional relation of APP and PS1 to cognition and neuronal plasticity in adult and ageing brain. The clinical mutations proved that the polytopic transmembrane PS1 protein is an essential subunit of the γ-secretase complex. The mutations all increase the formation of the longer, less soluble, and more fibrillogenic Aβ42. In addition to being essential for γ-secretase activity, PS1 is implicated in a variety of cellular processes, from assisting in the trafficking of proteins, to signal transduction, apoptosis, and regulation of calcium homeostasis.

Interestingly, patients with clinical mutations in PS1 that cause EOFAD most often present with the more severe and earlier pathology, characterized by enhanced deposition of amyloid and, for some mutations, even with an earlier onset of neurofibrillary tangle formation and neuronal loss. These do not appear to be correlated with the increased Aβ42 levels (Gomez-Isla *et al.* 1999). These data suggest that, as well as increased amyloid concentrations, other mechanisms contribute to the pathogenesis in these EOFAD patients, and eventually also in (some) sporadic AD cases. These are proposed to be related to other (patho)physiological functions of (mutant) PS1, particularly including their role in calcium homeostasis, apoptosis and synaptic dysfunction (La Ferla 2002). Compelling evidence for a role for PS1 and PS2 mutations in the (dys)regulation of calcium homeostasis has been obtained (Leissring *et al.* 2000; Yoo *et al.* 2000; Herms *et al.* 2003; Ris *et al.* 2003). In addition, mutant PS1 alters synaptic plasticity; in this case LTP in the CA1 of the hippocampus is facilitated in transgenic mice that overexpress mutant PS1 (I. Dewachter *et al.*, unpublished data). Deregulation of the equilibrium between induced LTP and LTD has been demonstrated to be deleterious for the generation of associative memory in computational models (Small 2004) and *in vivo* in transgenic mice models (Moechars *et al.* 1999; Dewachter *et al.* 2002).

The exact mechanisms underlying the deregulation of synaptic plasticity and its potential relation to altered calcium homeostasis will require further detailed analysis of PS1-overexpressing and PS1-deficient transgenic mice at the electrophysiological, physiological, and molecular levels.

9.2.1.3 Modulation by BACE1: balance between parenchymal and vascular amyloid

The type I transmembrane aspartyl proteinase β-site APP cleaving enzyme (BACE1) was identified as the major β-secretase (Cai *et al.* 2001; Luo *et al.* 2003). BACE cleaves APP at Asp1 and Glu11, whereas subsequent cleavage by γ-secretase gives rise to the Aβ(1–40/42) and Aβ(11–40/42) amyloid peptides. Deficiency of BACE1 in a double transgenic combination with the Swedish mutant APP (APPsw) rescued the early hippocampal memory deficits and correlated with dramatic reduction in Aβ levels (Ohno *et al.* 2004). Moreover, the complete absence of any overt phenotypic problems in the parental BACE1-deficient mice (Roberts 2002; Vassar 2001), in addition to the enhanced BACE1 activity correlating with Aβ production in sporadic AD (Yang *et al.* 2003; Li *et al.* 2004), makes the BACE1 endoproteinase an almost perfect drug target for disease-modifying therapy in AD.

Since no clear-cut physiological or pathological functions of BACE are known *in vivo*, human BACE1 mice were generated and studied alone and in the double transgenic combination with mutant human APP to investigate the contribution of increased BACE1 activity to the amyloid pathology (Bodendorf *et al.* 2002; Mohajeri *et al.* 2004; Willem *et al.* 2004). As expected, in our model BACE1 overexpression increased the amyloidogenic processing of APP as revealed by increased levels of the telltale APP metabolites APPsb, β-CTF, and Aβ peptides in brain of APP × BACE1 double transgenic mice (Willem *et al.* 2004). The concomitant dramatic increase in parenchymal plaque load in these double transgenic mice (Fig. 9.1 [Plate 3]) emphasized the very large remaining margin of amyloidogenic potential in mouse brain, i.e. more than 50% amyloid plaque load was measured in the subiculum of old APP-V717I × BACE1 double transgenic mice at age 22 months (Willem *et al.* 2004).

In sharp contrast, almost no CAA was observed in these same old APP-V717I × BACE1 double transgenic mice, indicating a complete shift in balance between parenchyma and vasculature. Most interestingly, the levels of the less soluble N-terminal-truncated Aβ peptides, as opposed to full-length Aβ species, were dramatically enhanced in the brain of aged APP-V717I × BACE1 double transgenic mice and were highly significantly

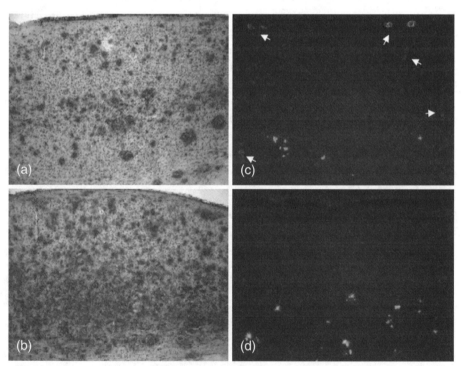

Fig. 9.1 Amyloid pathology in brain parenchyma of APP-V717I × BACE1 double transgenic mice at age 22 months (b, d) compared with age-matched APP-V717I single transgenic mice (a,c). Amyloid deposits in subiculum immunostained with antibody 3D6 (a,b) and with thioflavin S (c, d). Note the complete absence of amyloid pathology in the vasculature of the double transgenic mice (d) as opposed to the CAA in the parental APP transgenic mice (c, arrows). (from Willem *et al.* 2004). Please see the colour plate section for a colour version of this figure.

inversely correlated with the CAA (Willem *et al.* 2004). The almost complete absence of CAA is explained by the much higher rate of self-association and fibrillogenic capacity of the shorter and less soluble N-truncated Aβ11–42 peptides that form nidi for amyloid deposits in the parenchyma. These deposits close to the neurons that are their origin evidently prevents the diffusion or transport of the peptides into the perivascular spaces, reducing or eliminating CAA altogether (Willem *et al.* 2004 and references cited therein; CAA reviewed by Weller *et al.* 2004).

These data also emphasize the potential and deciding role of BACE1 in the relative balance of amyloid pathology in either the parenchyma or the vasculature by generating more or less of the 11–42 truncated amyloid peptides. In this respect it will be interesting to learn about possible correlations between the levels in CSF and/or blood plasma of N-truncated Aβ species that are being measured for their potential as additional diagnostic markers in AD (Vanderstichele *et al.* 2005).

9.2.1.4. Modulation of amyloid pathology by apolipoprotein E4 is complex: depending on the neuronal or glial origin of apolipoprotein E4, and differing for plaques and cerebral amyloid angiopathy in different brain regions

In addition to ageing as the most important, but least understood, risk factor for AD, it has been found in epidemiological investigations that the ApoE4 allele is genetically associated with sporadic AD (Corder *et al.* 1993). Lipoprotein ApoE exists in three variant isoforms, with ApoE3 having the highest abundance in the general population (e3 allele frequency of ~0.75). The ApoE4 isoform is dramatically enriched in the population of sporadic AD patients, with an e4 allele frequency of ~0.45 compared with 0.15 in the general population. This finding identified the e4 allele as the most important genetic risk factor, resulting in earlier age of onset of AD (Corder *et al.* 1993; Bales *et al.* 2002; St George-Hyslop and Petit 2005). The mechanism by which this genetic effect is translated into the neurodegenerative pathology has remained very enigmatic.

In contrast with the peripheral functions of ApoE which are well understood, its physiological and pathological actions in the central nervous system (CNS) remain largely unknown. This lack evidently also led to the difficulty of pinpointing its precise contribution to the increased risk in AD (Beffert *et al.* 2004; Poirier *et al.* 2005). The pathological contributions of ApoE to amyloid and tau pathology in AD have been studied in different types of transgenic mice, deficient in endogenous murine ApoE and/or overexpressing different human ApoE isoforms, including various combinations with mutant human APP and PS1 (reviewed by Holtzman 2004). The finding that ApoE deficiency delayed amyloid plaque deposition in mice, while overexpression of human ApoE4, and not ApoE3, by transferrin gene promoter accelerated plaque formation in transgenic mice, suggested a gain of function of ApoE4 (Bales *et al.* 1999; Carter *et al.* 2001). Nevertheless, ApoE4 could exert effects on the production, fibril formation, degradation, and clearance of amyloid peptides, and none of these options can be totally excluded.

Hyperphosphorylation of protein tau, resulting in an axonopathy very similar to that in tau-4R transgenic mice, was provoked by a direct test of this hypothesis, i.e. by neuronal but not by glial expression of human ApoE4 (Tesseur *et al.* 2000a,b). These findings support the hypothesis that increased or aberrant synthesis of ApoE4 in neurons, as opposed to its normal synthesis in glial cells, contributes to the pathology, particularly the tau pathology. Moreover, neuron-specific proteolysis of ApoE4 was recently linked to increased phosphorylation of tau in the brain of ApoE transgenic mice (Brecht *et al.* 2004), further supporting the hypothesis that neuronal synthesis of the E4 isoform promotes earlier development of AD (Roses 1998; Xu *et al.* 1999).

We continued testing this hypothesis experimentally to define the eventual contribution of either neuronal or glial ApoE4 synthesis to the parenchymal and vascular amyloid pathology, and to the eventual tauopathy in the APP-V717I amyloid mouse model of AD. Therefore we generated a unique set of double and triple transgenic mice carrying different combinations of APP-V717I, mutant PS1, and ApoE4, with the latter driven by either the mouse thy1 gene promoter or the GFAP promoter (Van Dooren *et al.* 2006). Coexpression of ApoE4 with APP-V717I in neurons increased the total amyloid plaque load in the hippocampus, cortex, and thalamus, while glial ApoE4 expression only increased amyloid plaques in the thalamus. Neither neuronal nor glial expression of ApoE4 in combination with APP-V717I and PS1-A246E in triple transgenic mice affected plaque pathology appreciably, emphasizing the already very high amyloidogenic potential of mutant PS1, which probably cannot be further increased.

Likewise, in the development of CAA differential repercussions were evident in the cortex and hippocampus compared with the thalamus. In the first instance, neuronal and glial expression of ApoE4 with APP-V717I in double transgenic mice increased the CAA to the same extent as the combination with mutant PS1, without marked additional effects in the triple transgenic lines. Remarkably, CAA in the thalamus was affected differentially by neuronal and glial expression of ApoE4, with neuronal ApoE4 expression advancing onset of plaque deposition in the thalamus to an earlier age in both double and triple transgenic mice. These findings emphasize the importance of the different mechanisms that are operating in different brain regions (Van Dooren *et al.* 2006).

Even when coexpressed with mutant APP and PS1 in neurons, ApoE4 did not change the balance of amyloidogenic to non-amyloidogenic pathways, based on the measurements of brain concentrations of all known APP metabolites. Nevertheless, the levels of Aβ40 and Aβ42 were increased by ApoE4 overexpression, indicating not that the production of amyloid peptides was increased, but that ApoE4 acted downstream of their production, i.e. slowed down the degradation and clearance of the Aβ peptides (Van Dooren *et al.* 2006).

The differential effect of neuronal and glial ApoE4 expression on amyloid plaque load could be related to the site of production of Aβ and ApoE4, i.e. either together in the region of deposition or at distant locations. In these transgenic mice, the thy1 gene promoter drives expression in most, but not all, neurons of the CNS, including the cortex and hippocampus. However, the thalamus is notoriously lacking in transgene expression, implying that some form of transport is operational. This can include axonal transport of APP and/or Aβ and also carrier-mediated diffusion of Aβ via extracellular, perivascular, and ventricular spaces. These mechanisms are needed to explain the contribution of

ApoE4 to the increased amyloid pathology in the thalamus without any local expression of APP (Van Dooren *et al.* 2006).

Another most interesting finding relative to the known effects of mutant PS1 is the huge impact of ApoE4 on CAA relative to a smaller effect on amyloid plaque load. This again emphasizes that ApoE4 does not increase the production of Aβ, but that its prime mode of action is in the pathways of clearance by physical drainage and proteolytic degradation (Fryer *et al.* 2005; Van Dooren *et al.* 2006). Despite experimental evidence for direct interactions of Aβ with ApoE lipoproteins, further investigations are required to understand their relevance not only to the physiological functions of Aβ and ApoE in brain, but also to their pathological interactions in relation to AD.

9.2.1.5 Modulation by ADAM10: validation of activated α-secretase as a major therapeutic goal

In addition to the obvious strategies aiming at inhibition of β- and γ-secretases as disease-modifying therapeutic approaches, the increase in the neuronal activity of α-secretase in the brain suggests an alternative and attractive strategy. The endoproteolysis of APP within the Aβ sequence by α-secretase can preclude the formation of any Aβ peptides (Dewachter and Van Leuven 2002; Kojro and Fahrenholz 2005). In addition, cleavage by α-secretase releases the N-terminal soluble ectodomain of APP, known as APPsα, which has been claimed to exert neurotrophic and neuroprotective properties (Mattson 1997; Meziane *et al.* 1998). Proteinases belonging to the ADAM family (a disintegrin and metalloproteinase) were the main candidates as physiologically relevant α-secretases, but early lethality of various deficient mice prevented detailed analysis *in vivo* or in neuronal cells. To overcome this restriction, we studied transgenic mice that overexpress either ADAM10 or a dominant-negative catalytically inactive ADAM10 mutant.

Moderate neuronal levels of ADAM10, coexpressed with human APP-V717I in double transgenic mice, increased the secretion of the neurotrophic APPsα concomitantly with considerable reduction in the production of Aβ peptides, preventing any deposition of amyloid plaques. The functional repercussions were evident by significant alleviation of the impaired LTP and the cognitive deficits (Postina *et al.* 2004). On the other hand, the expression of a mutant catalytically inactive ADAM10 significantly enhanced the amyloid load in the brain of double transgenic mice.

These observations have several important implications. They provide the first *in vivo* evidence that ADAM10, a proteinase of the ADAM family, acts very efficiently as the physiological α-secretase of APP. On the other hand, any lowering of the activity of ADAM10 in ageing brain would contribute to the development or the progression of sporadic AD. Moreover, the data validate the concept that the activation of ADAM10 is a promising therapeutic target.

Finally, indirect findings further corroborated that the PKC activator bryostatin reduced brain amyloid levels with a concomitant amelioration of behavioural deficits in APP × PS1-A246E mice. The parallel increase in secreted APPsα, observed in cells and *in vivo* in the brain of transgenic mice, strongly suggest that the mechanism of action of these PKC activators is to be sought in the activation of the non-amyloidogenic α-secretase pathway of APP cleavage (Etcheberrigaray *et al.* 2004).

9.2.2 Transgenic models for the tau pathology in Alzheimer's disease and frontotemporal dementia

9.2.2.1 Axonopathy in human tau-4R transgenic mice

We have generated transgenic mice that overexpress the longest wild-type human tau-4R isoform. These are considered to mimic excessive splicing in of exon10, coding for the second MT-binding domain in tau (Spittaels at al. 1999; Terwel *et al.* 2005). Like any transgenic model, this approach is not perfect because, unlike human brain which contains both tau-3R and tau-4R isoforms, only the tau-4R isoform is expressed in adult mouse brain.

Tau-4R transgenic mice did not develop any sign of tauopathy defined by NFT, but suffered from a severe axonopathy instead. Major symptoms are the progressive paralysis of the hindlimbs, extending to the forelimbs with age, accompanied by increasing impairment in the performance of tasks like beam walking and rotarod (Spittaels *et al.* 1999; Terwel *et al.* 2005). Neuropathologically, the axonopathy manifested itself by numerous large axonal dilatations (balloons or spheroids), particularly in the spinal cord, brainstem, and midbrain. Despite severe motor problems due to Wallerian degeneration and muscle wasting, the lifespan of the tau-4R transgenic mice is not markedly lower than that of non-transgenic mice.

Clearly, the tau-4R model does not recapitulate a tauopathy, as opposed to a model that expressed all six human tau isoforms from a genomic construct in mice that are otherwise deficient in mouse tau (Andorfer *et al.* 2003). NFT formation at very old age was interpreted to mean that mouse tau-4R isoforms, present in adult mouse brain, could interfere with the pathological filament formation of human tau-4R (Andorfer *et al.* 2003). While it is conceivable that the presence of tau-3R in combination with its lesser affinity for binding to MT could help tau filaments to form, the apparent or presumed interference by mouse tau-4R in this process is not understood. Mice that overexpress human tau-3R were reported to develop a different type of pathology in the hippocampus, characterized by straight tau filaments, at the advanced age of over 18 months (Ishihara *et al.* 2001).

9.2.2.2 Tauopathy in mutant tau-P301L transgenic mice

The many clinical mutations in the tau gene that are associated with FTD and cause its tauopathy, have led to a flurry of mouse models, the discussion of which surpasses the scope and space of the current review. Suffice it to state that models based on missense mutations that are present only in tau-4R isoforms or in both tau-3R and tau-4R isoforms (Lewis *et al.* 2000; Allen *et al.* 2002; Tanemura *et al.* 2002; Tatebayashi *et al.* 2002; Zhang *et al.* 2004; Terwel *et al.* 2005). Again, these are a priori imperfect, in that they rely on the expression of a single mutant tau isoform, and are driven by diverse gene promoters to sometimes very high expression levels. Evidently these do not recapitulate the human clinical disease, neither with respect to brain regional localization, nor to the type of tau filaments formed.

Overexpression of tau-P301L or tau-P301S has yielded the most robust and reliable phenotypes with a conspicuous clinical trait of a severely reduced lifespan of less than 12 months (Lewis *et al.* 2000; Allen *et al.* 2002; Terwel *et al.* 2005). The pathological appearance of intraneuronal tangles with age was demonstrated by thioflavinS, X34 and Gallyas silver staining and by tau phospho-epitopes with specified and validated monoclonal antibodies (Terwel *et al.* 2005 and references cited therein) (Fig. 9.2 [Plate 4]).

Lifespan and histopathology are critically dependent on tau mutant concentration in the transgenic mice. The heterozygous tau-P301L transgenic mice do not develop pathology and have a normal lifespan, while their homozygous counterparts do not survive beyond the age of 9–12 months because of the severe tau pathology (Terwel *et al.* 2005).

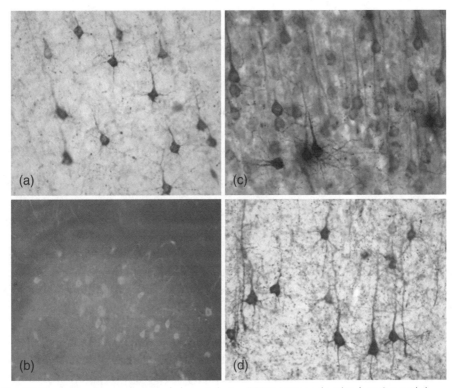

Fig. 9.2 The unique tau-pathological changes in tau-P301L transgenic mice (age 9 months) are identified by immunostaining with (a) PG5 in cerebral cortex and (b) X-34 in medial fastigial nucleus of the cerebellum. (c) Tauopathy defined in the cortex of tau-P301L transgenic mouse by AT8 immunohistochemistry; (d) similar but after 30 minutes post-mortem delay to demonstrate that most normal tau phosphorylation recognized by AT8 has disappeared, while the pathological hyperphosphorylation remains unchanged, typical for the paired helical filaments in AD brain. Please see the colour plate section for a colour version of this figure.

Remarkably, expression of wild-type or mutant P301L tau to the same overall brain levels, resulted in very different phenotypes despite the same tau-4R isoform. Motor impairment was absent or less severe and much later in onset in tau-P301L mice than in tau-4R mice, with axonal dilatations only present in the latter. Massive somatodendritic accumulation of hyperphosphorylated and conformationally altered tau was demonstrated, i.e. epitopes AT8, AT270, AT100, AD2, PG5, AP422, and MC-1, resulting in sarkosyl-insoluble tau aggregates. Nevertheless, the bulk of tau-P301L was in fact less phosphorylated than tau-4R in age-matched mice, indicating that increased phosphorylation is not the a priori status or condition of mutant tau. The conformational epitope detected with MC-1 could not be separated, either regionally or temporally, from the formation of pathological phospho-epitopes, indicating a close mechanistic relationship between hyperphosphorylation, conformational change, and NFT formation (Terwel *et al.* 2005).

Our data demonstrated that axonopathy and tauopathy exist as two separate pathological entities, at least in our mice (Terwel *et al.* 2005) as opposed to others which developed a combination (Lewis *et al.* 2000). In fact, mice heterozygous for both human tau-4R and tau-P301L develop only axonopathy (D. Terwel *et al.*, unpublished data), demonstrating that filament formation depends on more than just the total concentration of tau-4R, and corroborating that filaments in the brain of FTD-P301L patients consist almost exclusively of mutant protein tau, not wild-type tau-4R (Goedert and Jakes 2005).

9.2.2.3 GSK-3β but not cdk5/p35 phosphorylates tau-4R in brain *in vivo*

GSK-3β and cdk5 were identified as important tau kinases, and overexpression alone and in combination with wild-type and mutant tau have been attempted to model their contribution to tauopathy. In addition, cdk5 and its activating subunit p35 or its N-truncated p25 proteolytic remnant have been inactivated or expressed in transgenic mouse brain, with various degrees of success.

Neuronal overexpression of GSK-3β by itself reduced the brain size of transgenic mice, without any phenotypic repercussions or development of tauopathy despite increased phosphorylation of endogenous mouse tau (Spittaels *et al.* 2000; 2002). The benign microcephaly could be related to a direct effect of GSK-3β on the neuronal cytoskeleton, including tau and MT, but also to signalling pathways that determine size and morphology of neurons (Spittaels *et al.* 2002 and references cited therein). Moreover, both aspects could be opposing faces of the same coin.

Surprisingly, in the tau-4R × GSK-3β double transgenic combination, the axonopathy was practically completely rescued, i.e. elimination of axonal dilatations in all regions of the CNS concomitant with normalization of motoric disabilities (Spittaels *et al.* 2000). Increasing the phosphorylation of tau by the extra GSK-3β activity must effectively reduce binding of protein tau to MT and thereby reduce the initial interference by excess MT-bound tau-4R of the axonal transport by kinesin motor proteins.

The exact mechanisms are further explored in cells and *in vitro*, although it is clear that even hyperphosphorylation in tau-4R × GSK-3β mice failed to induce tau filaments, leading to the conclusion that hyperphosphorylation of protein tau does not cause tauopathy *per se*. Moreover, it is a clear signal that we are in need of a molecular definition of hyperphosphorylation to replace the current operational one.

In our hands, the expression of cdk5 with p35 and tau-4R in triple transgenic mice has yielded no additional new insight in the problem at hand. Despite evidence for increased phosphorylation of microfilaments and MT-associated proteins, no marked contribution to the phosphorylation of endogenous mouse or transgenic human tau was noted, nor any pathological consequences since tau-4R × cdk5 × p35 triple transgenic mice behaved normally (Van den Haute *et al.* 2001). Nevertheless, silver impregnation and immunohistochemistry demonstrated that neurofilament proteins became redistributed into apical dendrites of cortical neurons, pointing to a cytoskeletal effect of the increased cdk5 activity. Overall, however, the claim that cdk5/p35 is a major protein tau-kinase in mouse brain was not directly supported (Van den Haute *et al.* 2001). This was recently emphasized by the findings of increased phosphorylation in the brain of p35–/– mice (Hallows *et al.* 2003), which we have corroborated in a totally different system of yeast (Vandebroek *et al.* 2005) (see also section 3 below).

The problems surrounding the eventual precise role of p25 in normal and pathological conditions are even more controversial. While claimed as a pathological marker in AD brain, this N-truncated version of p35 generated by calpain cleavage was claimed to contribute to neurodegeneration by exacerbating the tauopathy via cdk5 mediated phosphorylation (Ahlijanian *et al.* 2000). We have generated inducible p25 mice controlled by tetracycline, which indeed demonstrated the dramatic neurodegeneration that can be induced by p25. In fact, a 30% decrease in brain weight was evident in a 3 month observation period after the induction of p25 at the age of 6 weeks (D. Muyllaert *et al.*, unpublished data).

In contrast, a minor increase in the levels of p25 appeared to be beneficial and improved learning and memory (Angelo *et al.* 2003). Both aspects were most recently proposed to be two faces of the same coin (Fischer *et al.* 2005), which makes straightforward interpretation of the role of cdk5/p35/p25 in AD and in related tauopathies nearly impossible.

9.3 Humanized yeast: linking phosphorylation, conformation, and aggregation of tau

Although this review is focused on transgenic mice as valuable experimental models, it is clear that definition of underlying molecular mechanisms also requires cell-biological and *in vitro* models (Ko *et al.* 2005; Stoothoff and Johnson 2005). We have developed a cellular model of bakers yeast, and derived from it *in vitro* assays to model tauopathy. A brief overview is fitting here, because the yeast model complements our animal models rather well, and fills in some important gaps in our understanding of the structure–function relation of protein tau.

The need for a 'less complex' cellular model for tauopathies led us to express human tau isoforms and mutants in *Saccharomyces cerevisiae*, apparently without very pronounced negative consequences. Nevertheless, human tau readily became phosphorylated at most of the sites that are also targets in the brain of tau transgenic mice and even in the brain of AD and FTD patients (Vandebroek *et al.* 2005). One reason that the pathological epitopes were formed is thought to be because human protein tau does not bind effectively to yeast MT. Consequently, a pool of soluble protein tau that is readily available for cytosolic kinases is provided in the yeast cell cytoplasm, without negative or confounding actions on the microtubular network.

The phosphorylation state of human tau is readily modified by inactivating yeast kinase mds1, the yeast orthologue of human GSK-3β and of pho85, the equivalent of mammalian cdk5. Lack of mds1 reduced the phosphorylation of tau, which was completely restored by expression of human GSK-3β. Unexpectedly, the inactivation of pho85 dramatically increased the phosphorylation, as well as the conformational change and the aggregation, of protein tau-4R (Vandebroek *et al.* 2005). This surprising outcome lends support to *in vivo* findings in transgenic mouse brain (Hallows *et al.* 2003) and can be combined and explained if cdk5 operates as a negative regulator of GSK-3β and other kinases such as JNK3.

Recombinant differentially phosphorylated human protein tau-4R and tau-P301L were isolated from wild-type and kinase-deficient yeast strains and used for *in vitro* assays. In particular, the phosphorylated mutant tau-P301L aggregated very readily *in vitro* under near-physiological conditions, forming straight and twisted filaments that have been characterized by electron and atomic force microscopy (Vandebroek *et al.* 2005). Interestingly, as opposed to a bacterial expression system, the yeast cells allowed us to isolate specifically phosphorylated isoforms of tau, and even a hyperphosphorylated form in the MC1 conformation, designated hP-tau/MC1 (Vandebroek *et al.* 2005). The biochemical and functional analysis demonstrated this hP-tau/MC1form to be unable to bind to MT, and to be most active in seeding tau filament formation *in vitro*. Thus this form, that has never been isolated before, is the most prone intermediate for aggregation and pivotal in tau filament formation (Vandebroek *et al.* 2005 and unpublished data) (Fig. 9.3).

Further structural and functional characterization is ongoing and yielding insight not only into its precise phosphorylation status, but also into other types of modifications.

9.4 Models under construction: combinations of APP-V717I, tau-P301L, and GSK-3β transgenic mice

Our efforts to develop transgenic mice to study the pathology in AD has yielded two robust models: for amyloid in the APP-V717I transgenic mice, and for tauopathy by way of the tau-P301L transgenic strain. Under extensive phenotypic characterization is the obvious double transgenic combination of APP-V717I × tau-P301L. We breed these double transgenic mice so that they are heterozygous for the mutant APP-V717I and homozygous for the mutant tau-P301L, because we want to maintain the same genetic configuration as their respective parental single transgenic strains. The data demonstrate the strongly synergistic interaction between both pathologies, with a marked increase in amyloid pathology in the subiculum, hippocampus, and cortex, i.e. the same regions that

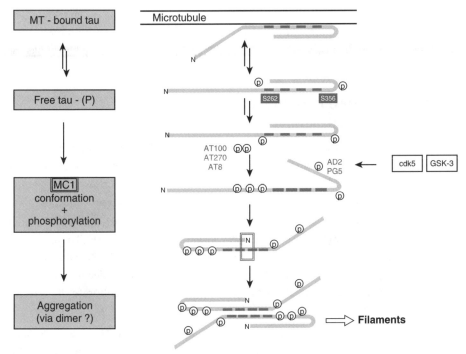

Fig. 9.3 Schematic overview of the proposed sequential mechanism of tau hyperphosphorylation, conformational change, and aggregation into filaments, based on data obtained in transgenic mice and humanized yeast models (Terwel *et al.* 2005; Vandebroek *et al.* 2005). Some of the most important pathological epitopes are depicted, with cdk5 and GSK-3β included as the proposed responsible kinases, although other kinases and phosphatases may be participating as well.

are affected in the parental APP-V717I mice (Moechars *et al.* 1999; Van Dorpe *et al.* 2000). Most interesting, however, is the pronounced exacerbation of the tauopathy, which becomes evident in the same brain regions as the amyloid pathology, as opposed to the more confined regional tauopathy in more caudal brain regions of the parental tau-P301L mice (Terwel *et al.* 2005). Notably, the intense intracellular amyloid pathology in pyramidal neurons in CA1 and in other regions of the hippocampal formation is accompanied by pre-tangle pathology not observed in either of the parental strains. In addition, we generated and are developing the APP-V717I × GSK-3β as well as the tau-P301L × GSK-3β double transgenic strains. GSK-3β, the primary tau-kinase, is most likely responsible for or considerably contributes to the formation of filamentous tau and NFT in the limbic brain structures.

The in-depth phenotyping of these double transgenic mouse models, comparatively to their parental single transgenic predecessors, is already yielding important clues to the mechanistic relations of amyloid and tau pathology. This relation remains enigmatic and needs to be elucidated because it is the signature combination of pathology in the brain of all AD patients, familial or sporadic, with early or late onset. We believe that the recruitment of these and other multiple transgenic mice to come is pivotal to eluci-dating the aetiology and pathogenic processes in AD, in addition to providing insight, identifying novel candidates, validating therapeutic targets, and providing test-banks for therapeutic approaches.

9.5 Proposed relation of the amyloid and tau pathology in Alzheimer's disease

Although a comprehensive scenario of the pathogenesis of AD cannot be drafted yet, the data from the transgenic mice, combined with data from the yeast model, allows a framework to be proposed.

1. Neuronal activity releases Aβ from synapses after transport of APP and proteolytic processing in axons. Initial deposition of Aβ in the hippocampus is activated through the perforant pathway (Buxbaum *et al.* 1998; Lazarov *et al.* 2002). A similar 'projection mechanism' must underlie plaque formation in the thalamus, where local expression of APP is negligible and deposited Aβ is probably released from projecting axons (Van Dooren *et al.* 2006). The localization of ADAM10 would corroborate the relative importance of synaptically released Aβ (Postina *et al.* 2004).

2. Concomitant with Aβ, the APP intracellular domain (AICD) is produced and either or both alter the activities of GSK-3β and cdk5, although by different mechanisms (Kim *et al.* 2003; Ryan *et al.* 2005). The resulting increase in tau phosphorylation destabilizes MT and axonal transport, further impairing synaptic activity and structure.

3. Intracellular oligomers of Aβ accumulate in endosomes, multivesicular bodies, and acidic vesicular structures (Wirths *et al.* 2001; Langui *et al.* 2004; Gouras *et al.* 2005; D. Terwel *et al.*, unpublished data) following impaired secretion or reuptake at synapses and by (auto)phagocytosis.

4. Released or secreted Aβ oligomers further disturb synaptic structure and function (Walsh *et al.* 2002; Lacor *et al.* 2004), and this is followed by formation of extracellular plaques, first diffuse and evolving into dense core plaques, as a byproduct due to the overflow of the drainage (perivascular, CSF) and degradation pathways of Aβ (lysosome, neprilysin (NEP), insulin degrading enzyme (IDE), etc.).

5. Neurons form new, or extend existing, neurites, resulting in aberrant sprouting dystrophic and swollen dendrites and axons in and around plaques (Su *et al.* 1998; Phinney *et al.* 1999; D. Terwel *et al.*, unpublished observations in APP-V717I × tau-P301L transgenic mice). Amyloid plaques and non-functional amyloid- and tangle-laden neurons constitute the detrimental environment for inflammatory reactions of surrounding microglial and astroglial cells as the final stages of the AD pathology.

These steps do not necessarily occur in all AD patients and are not strictly sequential, since most will take place in different cells or brain regions at different time-points, resulting in the typically 'blurred' snapshot views of the brain of human patients showing different stages or aspects of the overall pathology. Importantly, however, all these steps are observed in the experimental models, although not all are equally well studied. Therefore they are accessible for further study with the emphasis on first combining the amyloid and tau pathology in the multiple transgenic models described, and then 'deconvoluting' them by comparison with the single transgenic parental strains to define the important brain-regional, spatial, and temporal details of the mechanisms at work.

Acknowledgements

We thank P. Davies, A. Delacourte, M. Willem, C. Haass, F. Fahrenholtz, R. Postina, H. Hirokawa, and W. Klunk for collabaration, generous gifts of reagents, and advice. This work was supported by the Fonds voor Wetenschappelijk Onderzoek-Vlaanderen (FWO-Vlaanderen), the EEC Fifth and Sixth Framework Programs, the Rooms-fund, the KULeuven Research Fund and KULeuven-R&D, the Instituut voor Aanmoediging van Wetenschappelijk en Technisch Onderzoek (IWT), the European Space Agency (ESA), and an EEC-Marie Curie Training Grant (EURON PhD-School for Neurosciences).

References

Ackmann M, Wiech H, and Mandelkow E (2000). Nonsaturable binding indicates clustering of tau on the microtubule surface in a paired helical filament-like conformation. *Journal of Biological Chemistry*, **275**, 30335–43.

Ahlijanian MK, Barrezueta NX, Williams RD *et al.* (2000). Hyperphosphorylated tau and neurofilament and cytoskeletal disruptions in mice overexpressing human p25, an activator of cdk5. *Proceedings of the National Academy of Sciences of the USA* **97**, 9210–5.

Allen B, Ingram E, Takao M, *et al.* (2002). Abundant tau filaments and nonapoptotic neurodegeneration in transgenic mice expressing human P301S tau protein. *Journal of Neuroscience* **22**, 9340–51.

Andorfer C, Kress Y, Espinoza M, *et al.* (2003).Hyperphosphorylation and aggregation of tau in mice expressing normal human tau isoforms. *Journal of Neurochemistry* **86**, 582–90.

Angelo M, Plattner F, Irvine EE, and Giese KP (2003). Improved reversal learning and altered fear conditioning in transgenic mice with regionally restricted p25 expression. *European Journal of Neuroscience* **18**, 423–31.

Alonso A. del C., Mederlyova A, Novak M, Grundke-Iqbal I, and Iqbal K (2004). Promotion of hyperphosphorylation by frontotemporal dementia tau mutations. *Journal of Biological Chemistry* **279**, 34873–81.

Avila J, Lucas JJ, Perez M, and Hernandez F (2004). Role of tau protein in both physiological and pathological conditions. *Physiological Reviews* **84**, 361–84.

Bales KR, Verina T, Cummins DJ, *et al.* (1999). Apolipoprotein E is essential for amyloid deposition in the APP(V717F) transgenic mouse model of Alzheimer's disease. *Proceedings of the National Academy of Sciences of the USA* **96**, 15233–8.

Bales KR, Dodart JC, DeMattos RB, Holtzman DM, and Paul SM (2002). Apolipoprotein E, amyloid, and Alzheimer disease. *Molecular Interventions* **6**, 363–75.

Barghorn S, Zheng-Fischhofer Q, Ackmann M, *et al.* (2000). Structure, microtubule interactions, and paired helical filament aggregation by tau mutants of frontotemporal dementias. *Biochemistry* **39**, 11714–24.

Beffert U, Stolt PC, and Herz J (2004). Functions of lipoprotein receptors in neurons. *Journal of Lipid Research* **45**, 403–9.

Bodendorf U, Danner S, Fischer F, *et al.* (2002). Expression of human beta-secretase in the mouse brain increases the steady-state level of beta-amyloid. *Journal of Neurochemistry* **80**, 799–806.

Borchelt DR, Thinakaran G, Eckman CB, *et al.* (1996). Familial Alzheimer's disease-linked presenilin 1 variants elevate Aβ1–42/1–40 ratio *in vitro* and *in vivo. Neuron* **17**, 1005–13.

Borchelt DR, Ratovitski T, van Lare J, *et al.* (1997). Accelerated amyloid deposition in the brains of transgenic mice coexpressing mutant presenilin 1 and amyloid precursor proteins. *Neuron* **19**, 939–45.

Brecht WJ, Harris FM, Chang S, *et al.* (2004) Neuron-specific apolipoprotein e4 proteolysis is associated with increased tau phosphorylation in brains of transgenic mice. *Journal of Neuroscience* **24**, 2527–34.

Buée L, Bussière T, Buée-Scherrer V, Delacourte A, and Hof PR (2000). Protein tau isoforms, phopshorylation and role in neurodegenerative disorders. *Brain Research. Brain Research Reviews* **33**, 95–130.

Buxbaum JD, Thinakaran G, Koliatsos V, *et al.* (1998). Alzheimer amyloid protein precursor in the rat hippocampus: transport and processing through the perforant path. *Journal of Neuroscience* **18**, 9629–37.

Cai H, Wang Y, McCarthy D, *et al.* (2001). BACE1 is the major beta-secretase for generation of Abeta peptides by neurons. *Nature Neuroscience* **4**, 233–4.

Carter DB, Dunn E, McKinley DD, *et al.* (2001). Human apolipoprotein E4 accelerates β-amyloid deposition in APPsw transgenic mouse brain. *Annals of Neurology* **50**, 468–75.

Corder EH, Saunders AM, Strittmatter WJ, *et al.* (1993) Gene dose of apolipoprotein E type 4 allele and the risk of Alzheimer's disease in late onset families. *Science* **261**, 921–3.

Citron M, Westaway D, Xia W, *et al.* (1997). Mutant presenilins of Alzheimer's disease increase production of 42-residue amyloid β-protein in both transfected cells and transgenic mice. *Nature Medicine* **3**, 28–9.

Dawson HN, Ferreira A, Eyster MV, Ghoshal N, Binder LI, and Vitek MP (2001). Inhibition of neuronal maturation in primary hippocampal neurons from tau deficient mice. *Journal of Cell Science* **114**, 1179–87.

Delacourte A and Buée L (2000). Tau pathology, a marker of neurodegenerative disorders. *Current Opinion in Neurology* **13**, 371–6.

De Strooper B, Saftig P, Craessaerts K, *et al.* (1998). Deficiency of presenilin-1 inhibits the normal cleavage of amyloid precursor protein. *Nature* **391**, 387–90.

Dewachter I and Van Leuven F (2002). Secretases as targets for the treatment of Alzheimer's disease: the prospects. *Lancet Neurology* **1**, 409–16.

Dewachter I, Van Dorpe J,Smeijers L, *et al.* (2000). Aging increased amyloid peptide and caused amyloid plaques in brain of old APP/V717I transgenic mice by a different mechanism than mutant presenilin 1. *Journal of Neuroscience* **20**, 6452–8.

Dewachter I, Reverse D, Caluwaerts N, *et al.* (2002). Neuronal deficiency of presenilin 1 inhibits amyloid plaque formation and corrects hippocampal long-term potentiation but not a cognitive defect of amyloid precursor protein [V717I] transgenic mice. *Journal of Neuroscience* **22**, 3445–53.

Dodart JC, Bales KR, and Paul SM (2003). Immunotherapy for Alzheimer's disease: will vaccination work? *Trends in Molecular Medicine* **9**, 85–7.

Edbauer D, Winkler E, Regula JT, Pesold B, Steiner H, and Haass C (2003). Reconstitution of γ-secretase activity. *Nature Cell Biology* **5**, 486–8.

Esler WP and Wolf MS (2001). A portrait of Alzheimer secretases: new features and familiar faces. *Science* **293**, 1449–54.

Etcheberrigaray R, Tan M, Dewachter I, *et al.* (2004). Therapeutic effects of PKC activators in Alzheimer's disease transgenic mice. *Proceedings of the National Academy of Sciences of the USA* **101**, 11141–6.

Fischer A, Sananbenesi F, Pang PT, Lu B, and Tsai L-H (2005). Opposing roles of transient and prolonged expression of p25 in synaptic plasticity and hippocampus-dependent memory. *Neuron* **48**, 825–38.

Fryer JD, Simmons K, Parsadanian M, *et al.* (2005). Human apolipoprotein E4 alters the amyloid-b 40, 42 ratio and promotes the formation of cerebral amyloid angiopathy in an amyloid precursor protein transgenic model. *Journal of Neuroscience* **25**, 2803–10.

Goedert M and Jakes R (1990). Expression of separate isoforms of human tau protein, correlation with the tau pattern in brain and effects on tubulin polymerization. *EMBO Journal* **9**, 4225–30.

Goedert M and Jakes R (2005). Mutations causing neurodegenerative tauopathies. *Biochimica Biophysica Acta* **1739**, 240–50.

Gomez-Isla T, Growdon WB, McNamara MJ, *et al.* (1999). The impact of different presenilin 1 and presenilin 2 mutations on amyloid deposition, neurofibrillary changes and neuronal loss in the familial Alzheimer's disease brain: evidence for other phenotype-modifying factors. *Brain* **122**, 1709–19.

Gotz J, Schild A, Hoerndli F, and Pennanen L (2004). Amyloid-induced neurofibrillary tangle formation in Alzheimer's disease: insight from transgenic mouse and tissue-culture models. *International Journal of Developmental Neuroscience* **22**, 453–65.

Gouras GK, Almeida CG, and Takahashi RH (2005). Intraneuronal Aβ accumulation and origin of plaques in Alzheimer's disease. *Neurobiology of Aging* **9**, 1235–44.

Hallows JL, Chen K, DePinho RA, and Vincent I (2003). Decreased cyclin-dependent kinase 5 (cdk5) activity is accompanied by redistribution of cdk5 and cytoskeletal proteins and increased cytoskeletal protein phosphorylation in p35 null mice. *Journal of Neuroscience* **23**, 10633–44.

Harada A, Oguchi K, Okabe S, *et al.* (1994). Altered microtubule organization in small-calibre axons of mice lacking tau protein. *Nature* **369**, 488–91.

Hardy J (1997). Amyloid, the presenilins and Alzheimer's disease. *Trends in Neuroscience* **20**, 154–9.

Hardy J and Allsop D (1991). Amyloid deposition as the central event in the aetiology of Alzheimer's disease. *Trends in Pharmacological Science* **12**, 383–8.

Hardy J and Selkoe DJ (2002). The amyloid hypothesis of Alzheimer's disease: progress and problems on the road to therapeutics. *Science* **297**, 353–6.

Herms J, Schneider I, Dewachter I, Caluwaerts N, Kretzschmar H, and Van Leuven F (2003). Capacitive calcium entry is directly attenuated by mutant presenilin-1, independent of the expression of the amyloid precursor protein. *Journal of Biological Chemistry* **278**, 2484–9.

Heutink P (2000). Untangling tau-related dementia. *Human Molecular Genetics* **9**, 979–86.

Holcomb L, Gordon MN, McGowan E, *et al.* (1998). Accelerated Alzheimer-type phenotype in transgenic mice carrying both mutant amyloid precursor protein and presenilin 1 transgenes. *Nature Medicine* **4**, 97–100.

Holtzman DM (2004). *In vivo* effects of ApoE and clusterin on amyloid-beta metabolism and neuropathology. *Journal of Molecular Neuroscience* **23**, 247–54.

Ingram EM and Spillantini MG (2002). Tau gene mutations: dissecting the pathogenesis of FTDP-17. *Trends in Molecular Medicine* **8**, 555–61.

Iqbal K, Alonso Adel C, Chen S, *et al.* (2005). Tau pathology in Alzheimer disease and other tauopathies. *Biochimica Biophysica Acta* **1739**, 198–210.

Ishihara T, Zhang B, Higuchi M, Yoshiyama Y, Trojanowski JQ, and Lee VM (2001). Age-dependent induction of congophilic neurofibrillary tau inclusions in tau transgenic mice. *American Journal of Pathology* **158**, 555–62.

Kar S, Fan J, Smith MJ, Goedert M, and Amos LA (2003). Repeat motifs of tau bind to the insides of microtubules in the absence of taxol. *EMBO Journal* **22**, 70–7.

Kim HS, Kim EM, Lee JP, *et al.* (2003). C-terminal fragments of amyloid precursor protein exert neurotoxicity by inducing glycogen synthase kinase-3β expression. *FASEB Journal* **17**, 1951–3.

Ko L-W, DeTure M, Sahara N, Chihab R, Vega IE, and Yen S-H (2005). Recent advances in experimental modelling of the assembly of tau filaments. *Biochimica Biophysica Acta* **1729**, 125–39.

Kojro E and Fahrenholz F (2005). The non-amyloidogenic pathway: structure and function of α-secretases. *Subcelullar Biochemistry* **38**, 105–27.

Lacor PN, Buniel MC, Chang L, *et al.* (2004). Synaptic targeting by Alzheimer's-related amyloid β oligomers. *Journal of Neuroscience* **24**, 10191–200.

LaFerla FM (2002). Calcium dyshomeostasis and intracellular signalling in Alzheimer's disease. *Nature Reviews Neuroscience* **3**, 862–72.

Langui D, Girardo N, El Hachimi KH, *et al.* (2004). Subcellular topography of Aβ peptide in APP × PS1 transgenic mice. *American Journal of Pathology* **165**, 1465–77.

Lazarov O, Lee M, Peterson DA, and Sisodia SS (2002). Evidence that synaptically released β-amyloid accumulates as extracellular deposits in the hippocampus of transgenic mice. *Journal of Neuroscience* **22**, 9785–93.

Leissring MA, Akbari Y, Fanger CM, Cahalan MD, Mattson MP, and LaFerla FM (2000). Capacitative calcium entry deficits and elevated luminal calcium content in mutant presenilin-1 knockin mice. *Journal of Cell Biology* **149**, 793–8.

Lewis J, McGowan E, Rockwood J, *et al.* (2005). Neurofibrillary tangles, amyotrophy and progressive motor disturbance in mice expressing mutant (P301L) tau protein. *Nature Genetics* **25**, 402–5.

Li R, Lindholm K, Yang LB, Yue X, *et al.* (2004). Amyloid beta peptide load is correlated with increased beta-secretase activity in sporadic Alzheimer's disease patients. *Proceedings of the National Academy of Sciences of the USA* **101**, 3632–7.

Luo Y, Bolon B, Damore MA, *et al.* (2003). BACE1 (beta-secretase) knockout mice do not acquire compensatory gene expression changes or develop neural lesions over time. *Neurobiology of Disease* **14**, 81–8.

Makrides V, Massie MR, Feinstein SC, and Lew J (2004). Evidence for two distinct binding sites for tau on microtubules. *Proceedings of the National Academy of Sciences of the USA* **101**, 6747–51.

Marchesi VT (2005). An alternative interpretation of the amyloid Aβ hypothesis with regard to the pathogenesis of Alzheimer's disease. *Proceedings of the National Academy of Sciences of the USA* **102**, 9093–8.

Mattson MP (1997). Cellular actions of β-amyloid precursor protein and its soluble and fibrillogenic derivatives. *Physiological Reviews* **77**, 1081–132.

Meziane H, Dodart JC, Mathis C, *et al.* (1998). Memory-enhancing effects of secreted forms of the β-amyloid precursor protein in normal and amnestic mice. *Proceedings of the National Academy of Sciences of the USA* **95**, 12683–8.

Moechars D, Dewachter I, Lorent K, *et al.* (1999). Early phenotypic changes in transgenic mice that overexpress different mutants of amyloid precursor protein in brain. *Journal of Biological Chemistry* **274**, 6483–92.

Mohajeri MH, Saini KD, and Nitsch RM (2004). Transgenic BACE expression in mouse neurons accelerates amyloid plaque pathology. *Journal of Neural Transmission* **111**, 413–25.

Ohno M, Sametsky EA, Younkin LH, *et al.* (2004). BACE1 deficiency rescues memory deficits and cholinergic dysfunction in a mouse model of Alzheimer's disease. *Neuron* **41**, 27–33.

Phinney AL, Deller T, Stalder M, *et al.* (1999). Cerebral amyloid induces aberrant axonal spouring and ectopic terminal formation in amyloid precursor protein transgenic mice. *Journal of Neuroscience* **19**, 8552–9.

Poirier J (2005). Apolipoprotein E, cholesterol transport and synthesis in sporadic Alzheimer's disease. *Neurobiology of Aging* **26**, 355–61.

Postina R, Schroeder A, Dewachter I, *et al.* (2004) A disintegrin-metalloproteinase prevents amyloid plaque formation and hippocampal defects in an Alzheimer disease mouse model. *Journal of Clinical Investigation* **113**, 1456–64.

Reinhard C, Hebert SS, and De Strooper B (2005). The amyloid-beta precursor protein: integrating structure with biological function. *EMBO Journal* **24**, 3996–4006.

Ris L, Dewachter I, Reverse D, Godaux E, and Van Leuven F (2003). Capacitative calcium entry induces hippocampal long term potentiation in the absence of presenilin-1. *Journal of Biological Chemistry* **278**, 44393–9.

Roberts SB (2002). Gamma-secretase inhibitors and Alzheimer's disease. *Advanced Drug Delivery Reviews* **54**, 1579–88.

Roses AD (1998). Apolipoprotein E and Alzheimer's disease. The tip of the susceptibility iceberg. *Annals of the NewYork Academy of Sciences* **855**, 738–43.

Ryan KA and Pimplikar SW (2005). Activation of GSK-3 and phosphorylation of CRMP2 in transgenic mice expressing APP intracellular domain. *Journal of Chemical Biology* **171**, 327–35.

Saura CA, Choi SY, Beglopoulos V, *et al.* (2004). Loss of presenilin function causes impairments of memory and synaptic plasticity followed by age-dependent neurodegeneration. *Neuron* **42**, 23–36.

Sisodia SS and St George-Hyslop PH (2002). γ-Secretase, Notch, Aβ and Alzheimer's disease: where do the presenilins fit in? *Nature Reviews Neuroscience* **3**, 281–90.

Small DH (2004). Mechanisms of synaptic homeostasis in Alzheimer's disease. *Current Alzheimer Research* **1**, 27–32.

Spittaels K, Van den Haute C, Van Dorpe J, *et al.* (1999). Prominent axonopathy in the brain and spinal cord of transgenic mice overexpressing four-repeat human tau protein. *American Journal of Pathology* **155**, 2153–65.

Spittaels K, Van den Haute C, Van Dorpe J, *et al.* (2000). Glycogen synthase kinase-3β phosphorylates protein tau and rescues the axonopathy in the central nervous system of human four-repeat tau transgenic mice. *Journal of Biological Chemistry* **275**, 41340–9.

Spittaels K, Van den Haute C, Van Dorpe J, *et al.* (2002). Neonatal neuronal overexpression of glycogen synthase kinase-3 b reduces brain size in transgenic mice. *Neuroscience* **113**, 797–808.

St George-Hyslop PH and Petit A (2005). Molecular biology and genetics of Alzheimer's disease *Comptes Rendus Biologies* **328**, 119–30

Stoothoff WH and Johnson GVW (2005). Tau phosphorylation: physiological and pathological consequences. *Biochimica Biophysica Acta* **1739**, 280–97.

Su JH, Cummings BJ, and Cotman CW (1998). Plaque biogenesis in brain aging and Alzheimer's disease. II: Progressive transformation and developmental sequence of dystrophic neurites. *Acta Neuropathologica* **96**, 463–71.

Tanemura K, Mutajama M, Akagi T, *et al.* (2002). Neurodegeneration with tau accumulation in a transgenic mouse expressing V337M human tau. *Journal of Neuroscience* **22**, 133–41.

Tatebayashi Y, Miyasaka T, Chui DH, *et al.* (2002). Tau filament formation and associative memory deficit in aged mice expressing mutant (R406W) human tau. *Proceedings of the National Academy of Sciences of the USA* **99**, 13896–901.

Terwel D, Dewachter I, and Van Leuven F (2002). Axonal transport, tau protein, and neurodegeneration in Alzheimer's disease. *Neuromolecular Medicine* **2**, 151–65.

Terwel D, Lasrado R, Snauwaert J, *et al.* (2005) Changed conformation of mutant tau-P301L underlies the moribund tauopathy, absent in progressive, nonlethal axonopathy of tau-4R/2N transgenic mice. *Journal of Biological Chemistry* **280**, 3963–73.

Tesseur I, Van Dorpe J, Spittaels K, Van den Haute C, Moechars D, and Van Leuven F (2000a). Expression of human apolipoprotein E4 in neurons causes hyperphosophorylation of protein tau in the brains of transgenic mice. *American Journal of Pathology* **156**, 951–64.

Tesseur I, Van Dorpe J, Bruynseels K, *et al.* (2000b) Prominent axonopathy and disruption of axonal transport in transgenic mice expressing human apolipoprotein E4 in neurons of brain and spinal cord. *American Journal of Pathology* **157**, 1495–10.

Vandebroek T, Vanhelmont T, Terwel D, *et al.* (2005). Identification and isolation of a hyperphosphory-lated, conformationally changed intermediate of human protein tau expressed in yeast. *Biochemistry* **44**, 11466–75.

Van den Haute C, Spittaels K, Van Dorpe J, *et al.* (2001).Coexpression of human cdk5 and its activator p35 with human protein tau in neurons in brain of triple transgenic mice. *Neurobiology of Disease* **8**, 32–44.

Vanderstichele H, De Meyer G, Andreasen N, *et al.* (2005). Amino-truncated beta-amyloid42 peptides in cerebrospinal fluid and prediction of progression of mild cognitive impairment. *Clinical Chemistry* **51**, 1650–60.

Van Dooren T, Muyllaert D, Borghgraef P, *et al.* (2006). Neuronal or glial expression of human ApoE4 affects parenchymal and vascular amyloid pathology differentially in different brain regions of double and triple transgenic mice. *American Journal of Pathology* **168**, 245–60.

Van Dorpe J, Smeijers L, Dewachter I, *et al.* (2000). Prominent cerebral amyloid angiopathy in transgenic mice overexpressing the London mutant of human APP in neurons. *American Journal of Pathology* **157**, 1283–98.

Van Leuven F. (2000). Single and multiple transgenic mice as models for Alzheimer's disease. *Progress in Neurobiology* **61**, 305–12.

Vassar R (2001). The beta-secretase, BACE: a prime drug target for Alzheimer's disease. *Journal of Molecular Neuroscience* **17**, 157–70.

Walsh DM, Klyubin I, Fadeeva JV, *et al.* (2002). Naturally secreted oligomers of amyloid b protein potently inhibit hippocampal long-term potentiation *in vivo*. *Nature* **416**, 535–9.

Weller RO, Cohen NR, and Nicoll JA (2004). Cerebrovascular disease and the pathophysiology of Alzheimer's disease. Implications for therapy. *Panminerva Medica* **46**, 239–51.

Willem M, Dewachter I, Smyth N, *et al.* (2004). β-Site amyloid precursor protein cleaving enzyme 1 increases amyloid deposition in brain parenchyma but reduces cerebrovascular amyloid angiopathy in aging BACE × APP[V717] double-transgenic mice. *American Journal of Pathology* **165**, 1621–31.

Wirths O, Multhaup G, Czech C, *et al.* (2001). Intraneuronal Abeta accumulation precedes plaque formation in beta-amyloid precursor protein and presenilin-1 double-transgenic mice. *Neuroscience Letters* **306**, 116–20.

Xu PT, Gilbert JR, Qiu HL, *et al.* (1999). Specific regional transcription of apolipoprotein E in human brain neurons. *American Journal of Pathology* **154**, 601–11.

Yang LB, Lindholm K, Yan R, *et al.* (2003). Elevated beta-secretase expression and enzymatic activity detected in sporadic Alzheimer disease. *Nature Medicine* **9**, 3–4.

Yoo AS, Cheng I, Chung S, *et al.* (2000). Presenilin-mediated modulation of capacitative calcium entry. *Neuron* **27**, 561–72.

Zhang B, Higuchi M, Yoshiyama Y, *et al.* (2004). Retarded axonal transport of R406W mutant tau in transgenic mice with a neurodegenerative tauopathy. *Journal of Neuroscience* **24**, 4657–67.

Chapter 10

Inflammation and Alzheimer's disease

Mark P. Mattson

Both brain-intrinsic and humoral inflammatory processes occur in Alzheimer's disease (AD). The contributions of such immune-response-related cellular interactions and molecular cascades to the dysfunction and death of neurons are beginning to be elucidated. Local activation of microglia and astrocytes which produce pro-inflammatory cytokines, together with activation of complement cascades, may occur early in the course of the disease in response to amyloid β peptide (Aβ) accumulation and oxidative stress. As the disease progresses the production of signalling molecules by damaged cells and activated glia may recruit leucocytes to the affected brain regions. Epidemiological findings and data from studies of animal models of AD suggest that anti-inflammatory drugs and dietary manipulations can delay disease onset and may slow the progression of AD. Stimulation of an immune response by immunization with Aβ or Aβ antibodies also holds promise for disease prevention and treatment.

10.1 Ageing, oxidative stress, and Aβ pathology: setting the stage for inflammation

AD involves the selective degeneration of vulnerable populations of neurons in brain regions involved in learning and memory and affective behaviours including the hippocampus and associated cortical and limbic structures (Yankner 1996; Selkoe 2002). The vulnerable brain regions exhibit progressive accumulation of Aβ in the form of diffuse deposits and plaques, in combination with neurons exhibiting neurofibrillary tau pathology. Age is the major risk factor for AD, with most subjects becoming symptomatic in their seventh and eighth decades of life. Among the changes that occur in the brain during normal ageing, increased oxidative stress and impaired neuronal energy metabolism are likely to contribute to the development of AD (Mattson 2004). Increased oxidative damage to proteins, DNA, and lipids occurs in neurons during normal ageing, and the same types of damage are dramatically increased in AD (Moreira et al. 2005). Several such oxidative processes are evident early in the disease process, suggesting a role for age-related oxidative stress as an important factor contributing to the pathogenesis of AD (Lim et al. 2001; Kruman et al. 2002; Markesbery et al. 2005). Factors that increase the risk of AD also enhance oxidative stress. Examples include a high calorie intake (Mattson 2003), elevated homocysteine levels and folate deficiency (Mattson and Shea 2003), sedentary and cognitively unchallenging lifestyles (Friedland et al. 2001), and apolipoprotein E4 genotype (Smith 2002).

Although rare, some cases of AD have an early age of onset (forties and fifties) and are inherited in an autosomal dominant manner. Mutations in three different genes, the amyloid precursor protein (APP), presenilin-1, and presenilin-2, can cause such 'familial' forms of AD (Hardy 1997; St George-Hyslop and Petit 2005). Transgenic mice expressing familial AD mutant APP alone or in combination with mutant presenilin-1 exhibit age-related Aβ deposition in the cerebral cortex and hippocampus and associated behavioural abnormalities (Holcomb et al. 1998). Triple-mutant mice with APP, presenilin-1 and tau mutations (3×TgAD mice) also exhibit neurofibrillary-tangle-like pathology (Oddo et al. 2003). Studies of such mouse models have shown that the APP and presenilin-1 mutations, and age-related increases in oxidative stress and metabolic impairment, alter the proteolytic processing of APP such that increased amounts of a long 42 amino acid form of Aβ (Aβ42) are produced. Oligomeric forms of Aβ may promote neuronal dysfunction and death in AD by inducing oxidative stress and disrupting cellular calcium homeostasis (Mattson 2004). Depending upon factors that are not well understood, Aβ can adopt different morphologies which can be propagated and exhibit different neuro-toxic properties (Petkova et al. 2005). Synapses may be particularly vulnerable to neurotoxic effects of Aβ because they are subjected to high levels of ion fluxes and oxidative stress.

Oxidative stress, cellular damage, and the presence of abnormal proteins are all capable of activating both innate and humoral inflammatory processes (Wyss-Coray and Mucke 2002). During ageing and in AD, increased oxidative damage to neurons manifests, in part, as post-translational modifications of proteins. These include modification by lipid peroxidation products such as 4-hydroxynonenal (Bruce-Keller et al. 1998), nitration of tyrosine residues (Castegna et al. 2003), and glycation (Yan et al. 2000). Each of the latter modifications can induce inflammatory processes (Parola et al. 1999; Basta et al. 2004; Dedon and Tennenbaum 2004). In addition to such intrinsic inflammatory mechanisms, it has been proposed that infectious agents such as herpes simplex virus type 1 and *Chlamydia pneumoniae* play an important role (Itzhaki et al. 2004).

10.2 **Cellular and molecular characteristics of inflammation in Alzheimer's disease**

There is no doubt that inflammatory processes occur in association with Aβ and neurofibrillary pathology in AD. Activated microglia and astrocytes are intimately associated with neuritic plaques (McGeer and McGeer 2002). Immunoreactivity for several different inflammatory mediators, including proteins involved in the complement cascade and cytokines, is typically intensified in the proximity of amyloid plaques and neu-rofibrillary tangles. For example, complement factors C1q, C3b, C3c, C3d, and C4, and activated complement products, are localized to amyloid plaques (Eikelenboom et al. 1989), although the cascade may not proceed beyond C3 (Veerhuis et al. 1995). The complement protein C1q is present in particularly high amounts in plaques containing fibril-lar amyloid β-peptide compared with diffuse plaques; Aβ may induce the production of C1q in neurons (Afagh et al. 1996). Levels of several acute phase proteins are elevated in the vicinity of plaques in the cerebral cortex of AD patients compared with neurologically normal control subjects (Rozemuller et al. 1990).

Levels of several different pro-inflammatory cytokines are increased in plaque-associated cells in AD including tumour necrosis factor-α (TNF-α) (Dickson et al. 1993), interleukin-1β (IL-1β) (Mrak and Griffin 2001), and transforming growth factor-β_1 (TGF-β1) (van der Wal et al. 1993). TGF-β_2 levels are increased in large tangle-bearing neurons and in astrocytes of patients with sporadic AD and those with presenilin-1 mutations (Flanders et al. 1995). In addition, IL-6 and α_2-macroglobulin levels are elevated in cerebral cortex of AD patients compared with control subjects (Bauer et al. 1991; Griffin et al. 1998). Proteases and protease inhibitors involved in inflammatory processes, such as a1-antichymotrypsin (Abraham et al. 1988) and thrombin (Akiyama et al. 1992) are also increased in association with amyloid plaques in brain tissue from AD patients. Cyclooxygenase-2 (COX-2), which catalyses the production of prostaglandins from arachidonic acid and is an important target for anti-inflammatory drugs, is increased in hippocampal pyramidal neurons of AD patients compared with age-matched controls in a plaque-density-dependent manner (Ho et al. 1999). Neurons from mice overexpressing COX-2 are more vulnerable to Aβ toxicity than are neurons from non-transgenic mice, suggesting that the increased levels of COX-2 in AD could contribute to the neurodegenerative process.

Several different changes occur in AD that result in the recruitment of peripheral leukocytes to the sites of disease in the brain. Levels of cell adhesion molecules (CAMs), such as integrins, are increased in vascular endothelial cells associated with amyloid plaques (Eikelenboom et al. 1989), and vitronectin receptor levels are elevated in reactive microglia and neurofibrillary-tangle-bearing neurons in the brains of AD patients (Akiyama et al. 1991). Similarly, levels of LFA-1 leukocyte adhesion molecules are increased in microglial cells surrounding amyloid plaques (Rozemuller et al. 1989). While there is certainly not a massive infiltration of leukocytes into affected brain regions in AD, increasing evidence suggests more subtle roles for leukocytes in the disease process. For example, leukocyte common antigen positive cells and T-cytotoxic suppressor cells were detected in brain tissues of AD patients but not of control subjects (Itagaki et al. 1988). Antibodies against Aβ have also been demonstrated associated with amyloid deposits (Eikelenboom and Stam 1982) and circulating in the blood (Gaskin et al. 1993; Nath et al. 2003), although whether such antibodies play a detrimental or beneficial role in the disease process is unknown (Mattson and Chan 2003).

10.3 Inflammation and experimental models of Alzheimer's disease

Because Aβ plaques are sites where inflammatory processes are localized, there has been much interest in determining if and how Aβ incites inflammation. Aβ can activate the classical complement pathway by interacting with a site within the collagen-like domain of C1q (Rogers et al. 1992; Jiang et al. 1994). C1q enhances the aggregation of Aβ (Webster et al. 1994), and studies of C1q-deficient APP mutant mice have demonstrated a pivotal role for C1q in amyloidogenesis (Fonseca et al. 2004). When microglia encounter Aβ they become activated and increase their production of pro-inflammatory cytokines such as TNF and interferon-γ (Meda et al. 1995). Such cyokines, together with oxyradicals and excitotoxins produced by activated microglia, may promote neuronal degeneration. APP mutant mice exhibit microglial and astrocyte activation in association with Aβ deposits (Hsiao *et al.* 1996; Chen *et al.* 1998; Matsuoka *et al.* 2001), suggesting that Aβ plays a major role in inducing glial activation and inflammatory cytokine cascades. Chronic overproduction of TGF-β_1 in transgenic mice promotes amyloidogenesis and neurodegeneration through its stimulatory effects on the deposition of extracellular matrix proteins (Finch *et al.* 1993; Wyss-Coray *et al.* 1995). The activation of microglia and production of nitric oxide and pro-inflammatory cytokines in response to lipopolysaccharide is enhanced in presenilin-1 mutant knockin mice, suggesting that presenilin-1 mutations may exert a pro-inflammatory effect on brain cells that is independent of Aβ accumulation (Lee *et al.* 2002). Oxygenases may contribute to the neurodegenerative process in AD because lipoxygenase inhibitors such as nordihydroguaiaretic acid can protect cultured hippocampal neurons against Aβ toxicity (Goodman *et al.* 1994) and indomethacin (a COX inhibitor) can suppress Aβ-induced microglial activation (Netland *et al.* 1998). The inflammatory processes associated with amyloid pathology are correlated with age-related cognitive deficits in APP/PS2 double-mutant transgenic mice (Richards *et al.* 2003). IL-1β has been shown to impair long-term potentiation of synaptic transmission in the hippocampus (Vereker *et al.* 2000), suggesting a mechanism by which inflammation could promote cognitive dysfunction.

10.4 The good side of inflammation

While many aspects of inflammatory processes are likely to be detrimental to neurons, inflammation evolved as an adaptive response to invasion of pathogens and tissue injury (Schwartz *et al.* 1999). Rat hippocampal neurons treated with TNF-α exhibit increased resistance to cell death induced by excitotoxic and oxidative insults, and exposure to amyloid β-peptide (Cheng *et al.* 1994; Barger *et al.* 1995). Mice deficient in TNF-α receptor exhibit increased vulnerability to excitotoxin-induced hippocampal damage, suggesting a neuroprotective role for TNF-α *in vivo* (Bruce *et al.* 1996). Moreover, synaptic plasticity is impaired in mice lacking TNF-α receptors, suggesting that this pro-inflammatory cytokine may play a role in learning and memory (Albensi and Mattson 2000). The signal transduction pathway that mediates the anti-apoptotic and anti-excitotoxic effects of TNF-α involves activation of the transcription factor NFκB which induces the expression of several neuroprotective proteins including manganese superoxide dismutase and Bcl-2 (Mattson *et al.* 1997; Yu *et al.* 1999; Mattson and Meffert 2005). TNF-α might also enhance synaptic transmission by a rapid induction of translocation of glutamate receptors to the cell surface (Ogoshi *et al.* 2005). TGF-β_1 may also serve an adaptive neuroprotective role in AD as pretreatment of cultured cortical neurons with TGF-β_1 confers resistance to excitotoxicity and amyloid β-peptide toxicity (Prehn *et al.* 1993; Chao *et al.* 1994). Because of these *in vitro* findings and because TGF-β_1 is localized to sites where neuronal degeneration is occurring in AD brains (see above), it is possible that TGF-β_1 serves an important neuroprotective function.

10.5 Targeting inflammatory processes to prevent and treat Alzheimer's disease

Inflammatory processes in AD include several different targets for prophylactic and therapeutic intervention. Data from epidemiological studies suggest that individuals who use glucocorticoid and/or non-steroidal anti-inflammatory drugs for extended time periods are at a reduced risk for AD (Breitner 1996). The results of a double-blind placebo-controlled trial of indomethacin 100–150 mg daily suggested that anti-inflammatory drugs can slow cognitive decline in patients with mild to moderate AD (Rogers *et al.* 1993). COX-2 inhibitors may be of benefit in the prevention and treatment of AD, but progress in their development as AD drugs has been hindered by evidence of adverse cardiovascular side effects. Antioxidants of many types might be expected to suppress inflammatory processes in AD, although their benefit in preventing and treating AD is as yet unclear. Exercise and dietary restriction can suppress ageing and injury-related inflammatory processes in the brain, suggesting that such lifestyle modifications may suppress inflammatory processes associated with AD (Fig. 10.1). Finally, recent findings suggest that it may be possible to

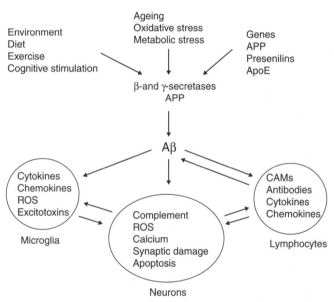

Fig. 10.1 Roles of inflammatory processes in the pathogenesis of AD. The ageing process involves oxidative and metabolic stress which sets the stage for the development of AD. Environmental factors (e.g. overeating, and sedentary and cognitively unchallenging lifestyles) and genetic factors (mutations in APP or presenilins, or apoE4 genotype) can interact with ageing to increase the production of Aβ. Aβ can form oligomers and fibrillar aggregates which induce oxidative stress in neurons and microglial cells, and activate microglia. Activated microglia produce pro-inflammatory cytokines, reactive oxygen species (ROS), and excitotoxins that can damage neurons. Factors produced by activated microglia and damaged neurons, as well as Aβ, may recruit lymphocytes to sites of amyloid and neuronal pathology. Lymphocytes can produce antibodies against Aβ; the antibodies may promote clearance of Aβ from the brain and/or may enhance damage to neurons.

activate the immune system to target Aβ directly and remove it from the brain (Mattson and Chan 2003). In mouse models of AD, immunization with Aβ or purified Aβ antibodies results in a reduction in Aβ accumulation in the brain and improved cognitive function (Janus 2003). However, in an initial clinical trial in AD patients, active immunization with Aβ resulted in a severe side effect in some patients, and it is unclear whether or not there was/will be a clinical benefit in those who tolerated the immunization (Gilman *et al.* 2005). A passive immunization trial is currently in progress. Future investigations are likely to reveal novel targets and more refined preventative and therapeutic strategies that focus on inflammatory processes in AD.

References

Abraham CR, Selkoe DJ, and Potter H (1988). Immunohistochemical identification of the serine protease inhibitor a1-antichymotrypsin in the brain amyloid deposits of Alzheimer's disease. *Cell* **52**, 487–501.

Afagh A, Cummings BJ, Cribbs DH, Cotman CW, and Tenner AJ (1996). Localization and cell association of C1q in Alzheimer's disease brain. *Experimental Neurology* **138**, 22–32.

Akiyama H, Kawamata T, Dedhar S and McGeer PL (1991). Immunohistochemical localization of vitronectin, its receptor and beta-3 integrin in Alzheimer brain tissue. *Journal of Neuroimmunology* **32**, 19–28.

Akiyama H, Ikeda K, Kondo H, and McGeer PL (1992). Thrombin accumulation in brains of patients with Alzheimer's disease. *Neuroscience Letters* **146**, 152–4.

Albensi BC and Mattson MP (2000). Evidence for the involvement of TNF and NF-kappaB in hippocampal synaptic plasticity. *Synapse* **35**, 151–9.

Barger SW, Horster D, Furukawa K, Goodman Y, Krieglstein J, and Mattson MP (1995). Tumor necrosis factors a and β protect neurons against amyloid β-peptide toxicity: evidence for involvement of a kB-binding factor and attenuation of peroxide and Ca^{2+} accumulation. *Proceedings of the National Academy of Sciences of the USA* **92**, 9328–32.

Basta G, Schmidt AM, and De Caterina R. (2004). Advanced glycation end products and vascular inflammation: implications for accelerated atherosclerosis in diabetes. *Cardiovascular Research* **63**, 582–92.

Bauer J, Strauss S, Schreiter-Gasser U, *et al.* (1991). Interleukin-6 and a2-macroglobulin indicate an acute-phase state in Alzheimer's disease cortices. *FEBS Letters* **285**, 111–14.

Breitner JC (1996). The role of anti-inflammatory drugs in the prevention and treatment of Alzheimer's disease. *Annual Reviews of Medicine* **47**, 401–11.

Bruce AJ, Boling W, Kindy MS, *et al.* (1996). Altered neuronal and microglial responses to brain injury in mice lacking TNF receptors. *Nature Medicine* **2**, 788–94.

Bruce-Keller AJ, Li YJ, Lovell MA, *et al.* (1998). 4-Hydroxynonenal, a product of lipid peroxidation, damages cholinergic neurons and impairs visuospatial memory in rats. *Journal of Neuropathology and Experimental Neurology* **57**, 257–267.

Castegna A, Thongboonkerd V, Klein JB, Lynn B, Markesbery WR, and Butterfield DA. (2003). Proteomic identification of nitrated proteins in Alzheimer's disease brain. *Journal of Neurochem*istry **85**, 1394–1401.

Chao CC, Hu S, Kravitz FH, Tsang M, Anderson WR, and Peterson PK (1994). Transforming growth factor-β protects human neurons against beta-amyloid-induced injury. *Molecular and Chemical Neuropathology* **23**, 159–78.

Chen KS, Masliah E, Grajeda H, *et al.* (1998). Neurodegenerative Alzheimer-like pathology in PDAPP 717V–>F transgenic mice. *Progress in Brain Research* **117**, 327–34.

Cheng B, Christakos S, and Mattson MP (1994). Tumor necrosis factors protect neurons against excito-toxic/metabolic insults and promote maintenance of calcium homeostasis. *Neuron* **12**, 139–153.

Dedon PC and Tannenbaum SR. (2004). Reactive nitrogen species in the chemical biology of inflammation. *Archives of Biochemistry and Biophysics* **423**, 12–22.

Dickson DW, Lee SC, Mattiace LA, Yen SH, and Brosnan C (1993). Microglia and cytokines in neurological disease, with special reference to AIDS and Alzheimer's disease. *Glia* **7**, 75–83.

Eikelenboom P and Stam FC. (1982). Immunoglobulins and complement factors in senile plaques. An immunoperoxidase study. *Acta Neuropathologica* **57**, 239–42.

Eikelenboom P, Hack CE, Rozemuller JM, and Stam FC (1989). Complement activation in amyloid plaques in Alzhiemer's dementia. *Virchows Archiv* **56**, 259–62.

Finch CE, Laping NJ, Morgan TE, Nichols NR, and Pasinetti GM (1993). TGF-β1 is an organizer of responses to neurodegeneration. *Journal of Cellular Biochemistry* **53**, 314–22.

Flanders KC, Lippa CF, Smith TW, Pollen DA, and Sporn MB (1995). Altered expression of transforming growth factor-beta in Alzheimer's disease. *Neurology* **45**, 1561–9.

Fonseca MI, Zhou J, Botto M, and Tenner AJ. (2004). Absence of C1q leads to less neuropathology in transgenic mouse models of Alzheimer's disease. *Journal of Neuroscience* **24**, 6457–65.

Friedland RP, Fritsch T, Smyth KA, *et al.* (2001). Patients with Alzheimer's disease have reduced activities in midlife compared with healthy control-group members. *Proceedings of the National Academy of Sciences of the USA* **98**, 3440–5.

Gaskin F, Finley J, Fang Q, Xu S, and Fu SM (1993). Human antibodies reactive with beta-amyloid protein in Alzheimer's disease. *Journal of Experimental Medicine* **177**, 1181–6.

Gilman S, Koller M, Black RS, *et al.* (2005). Clinical effects of Abeta immunization (AN1792). in patients with AD in an interrupted trial. *Neurology* **64**, 1553–62.

Goodman Y, Steiner MR, Steiner SM, and Mattson MP (1994). Nordihydroguaiaretic acid protects hippocampal neurons against amyloid β-peptide toxicity, and attenuates free radical and calcium accumulation. *Brain Research* **654**, 171–6.

Griffin WS, Sheng JG, Royston MC, *et al.* (1998). Glial-neuronal interactions in Alzheimer's disease: the potential role of a 'cytokine cycle' in disease progression. *Brain Pathology* **8**, 65–72.

Hardy J (1997). Amyloid, the presenilins and Alzheimer's disease. *Trends in Neuroscience* **20**, 154–9.

Ho L, Pieroni C, Winger D, Purohit DP, Aisen PS, and Pasinetti GM (1999). Regional distribution of cyclooxygenase-2 in the hippocampal formation in Alzheimer's disease. *Journal of Neuroscience Research* **57**, 295–303.

Holcomb L, Gordon MN, McGowan E, *et al.* (1998). Accelerated Alzheimer-type phenotype in transgenic mice carrying both mutant amyloid precursor protein and presenilin 1 transgenes. *Nature Medicine* **4**, 97–100.

Hsiao K, Chapman P, Nilsen S, *et al.* (1996). Correlative memory deficits, Abeta elevation, and amyloid plaques in transgenic mice. *Science* **274**, 99–102.

Itagaki S, McGeer PL, and Akiyama H (1988). Presence of T-cytotoxic suppressor and leucocyte common antigen positive cells in Alzheimer's disease brain tissue. *Neuroscience Letters* **91**, 259–64.

Itzhaki RF, Wozniak MA, Appelt DM, and Balin BJ (2004). Infiltration of the brain by pathogens causes Alzheimer's disease. *Neurobiology of Aging* **25**, 619–27.

Janus C (2003). Vaccines for Alzheimer's disease: how close are we? *CNS Drugs* **17**, 457–74.

Jiang H, Burdick D, Glabe CG, Cotman CW, and Tenner AJ (1994). β-amyloid activates complement by binding to a specific region of the collagen-like domain of C1q A chain. *Journal of Immunology* **152**, 5050–9.

Kruman II, Kumaravel TS, Lohani A, *et al.* (2002). Folic acid deficiency and homocysteine impair DNA repair in hippocampal neurons and sensitize them to amyloid toxicity in experimental models of Alzheimer's disease. *Journal of Neuroscience* **22**, 1752–62.

Lee J, Chan SL, and Mattson MP (2002). Adverse effect of a presenilin-1 mutation in microglia results in enhanced nitric oxide and inflammatory cytokine responses to immune challenge in the brain. *Neuromolecular Medicine* **2**, 29–45.

Lim GP, Chu T, Yang F, Beech W, Frautschy SA, and Cole GM (2001). The curry spice curcumin reduces oxidative damage and amyloid pathology in an Alzheimer transgenic mouse. *Journal of Neuroscience* **21**, 8370–7.

McGeer PL and McGeer EG (2002). Local neuroinflammation and the progression of Alzheimer's disease. *Journal of Neurovirology* **8**, 529–38.

Markesbery WR, Kryscio RJ, Lovell MA, and Morrow JD (2005). Lipid peroxidation is an early event in the brain in amnestic mild cognitive impairment. *Annals of Neurology* **58**, 730–5.

Matsuoka Y, Picciano M, Malester B, *et al.* (2001). Inflammatory responses to amyloidosis in a transgenic mouse model of Alzheimer's disease. *American Journal of Pathology* **158**, 1345–54.

Mattson MP (2003). Will caloric restriction and folate protect against AD and PD? *Neurology* **60**, 690–5.

Mattson MP (2004). Pathways towards and away from Alzheimer's disease. *Nature* **430**, 631–9.

Mattson MP and Chan SL (2003). Good and bad amyloid antibodies. *Science* **301**, 1847–9.

Mattson MP and Meffert M (2006). Roles for NF-kappaB in nerve cell survival, plasticity, and disease. *Cell Death and Differentiation* **13**, 852–60.

Mattson MP and Shea TB (2003). Folate and homocysteine metabolism in neural plasticity and neurodegenerative disorders. *Trends in Neuroscience* **26**, 137–46.

Mattson MP, Goodman Y, Luo H, Fu W, and Furukawa K (1997). Activation of NF-kB protects hippocampal neurons against oxidative stress-induced apoptosis: evidence for induction of Mn-SOD and suppression of peroxynitrite production and protein tyrosine nitration. *Journal of Neuroscience Research* **49**, 681–97.

Meda L, Cassatella MA, Szendrei GI, *et al.* (1995). Activation of microglial cells by β-amyloid protein and interferon-g. *Nature* **374**, 647–650.

Moreira PI, Smith MA, Zhu X, Nunomura A, Castellani JR, and Perry G (2005). Oxidative stress and neurodegeneration. *Annals of New York Academy of Sciences* **1043**, 545–552.

Mrak RE and Griffin WS (2001). Interleukin-1, neuroinflammation, and Alzheimer's disease. *Neurobiology of Aging* **22**, 903–8.

Nath A, Hall E, Tuzova M, Dobbs M, *et al.* (2003). Autoantibodies to amyloid beta-peptide (Abeta). are increased in Alzheimer's disease patients and Abeta antibodies can enhance Abeta neurotoxicity: implications for disease pathogenesis and vaccine development. *Neuromolecular Medicine* **3**, 29–39.

Netland EE, Newton JL, Majocha RE, and Tate BA (1998). Indomethacin reverses the microglial response to amyloid beta protein. *Neurobiology of Aging* **19**, 201–4.

Oddo S, Caccamo A, Shepherd JD, *et al.* (2003). Triple-transgenic model of Alzheimer's disease with plaques and tangles: intracellular Abeta and synaptic dysfunction. *Neuron* **39**, 409–21.

Ogoshi F, Yin HZ, Kuppumbatti Y, Song B, Amindari S, and Weiss JH (2005). Tumor necrosis-factor-alpha (TNF-alpha). induces rapid insertion of Ca^{2+}-permeable alpha-amino-3-hydroxyl-5-methyl-4-isoxazole-propionate (AMPA)/kainate (Ca-A/K). channels in a subset of hippocampal pyramidal neurons. *Experimental Neurology* **193**, 384–93.

Parola M, Bellomo G, Robino G, Barrera G, and Dianzani MU (1999). 4-Hydroxynonenal as a biological signal: molecular basis and pathophysiological implications. *Antioxidants and Redox Signaling* **1**, 255–84.

Petkova AT, Leapman RD, Guo Z, Yau WM, Mattson MP, and Tycko R (2005). Self-propagating, molecular-level polymorphism in Alzheimer's beta-amyloid fibrils. *Science* **307**, 262–5.

Prehn JHM, Backhaub C, and Krieglstein J (1993). Transforming growth factor-β1 prevents glutamate neurotoxicity in rat neocortical cultures and protects mouse from ischemic injury *in vivo*. *Journal of Cerebral Blood Flow and Metabolism* **13**, 521–5.

Richards JG, Higgins GA, Ouagazzal AM, *et al.* (2003). PS2APP transgenic mice, coexpressing hPS2mut and hAPPswe, show age-related cognitive deficits associated with discrete brain amyloid deposition and inflammation. *Journal of Neuroscience* **23**, 8989–9003.

Rogers J, Cooper NR, Webster S, *et al.* (1992). Complement activation by b-amyloid in Alzheimer's disease. *Proceedings of the National Academy of Sciences of the USA* **89**, 10016–11120.

Rogers J, Kirby LC, Hempleman SR, *et al.* (1993). Clinical trial of indomethacin in Alzheimer's disease. *Neurology* **43**, 1609–11.

Rozemuller JM, Eikelenboom P, Pals ST, and Stam FC (1989). Microglial cells around amyloid plaques in Alzheimer's disease express leucocyte adhesion molecules of the LFA-1 family. *Neuroscience Letters* **101**, 288–92.

Rozemuller JM, Stam FC, and Eikelenboom P (1990). Acute phase proteins are present in amorphous plaques in the cerebral but not in the cerebellar cortex in patients with Alzheimer's disease. *Neuroscience Letters* **119**, 75–8.

Schwartz M, Moalem G, Leibowitz-Amit R, and Cohen IR (1999). Innate and adaptive immune responses can be beneficial for CNS repair. *Trends in Neuroscience.* **22**, 295–9.

Selkoe DJ (2002). Alzheimer's disease is a synaptic failure. *Science* **298**, 789–91.

Smith JD (2002). Apolipoproteins and aging: emerging mechanisms. *Ageing Research Reviews*, **1**, 345–65.

St George-Hyslop PH and Petit A (2005). Molecular biology and genetics of Alzheimer's disease. *Comptes Rendus Biologies* **328**, 119–30.

van der Wal EA, Gomez-Pinilla F, and Cotman CW (1993). Transforming growth factor-β1 is in plaques in Alzheimer and Down pathologies. *Neuroreport* **4**, 69–72.

Veerhuis R, van der Valk P, Janssen I, Zhan SS, Van Nostrand WE, and Eikelenboom P (1995). Complement activation in amyloid plaques in Alzheimer's disease does not proceed further than C3. *Virchows Archiv* **426**, 603–10.

Vereker E, O'Donnel E, and Lynch MA (2000). The inhibitory effect of interleukin-1beta on long-term potentiation is coupled with increased activity of stress-activated protein kinases. *Journal of Neuroscience* **20**, 6811–19.

Webster S, O'Barr S, and Rogers J (1994). Enhanced aggregation and β structure of the amyloid β peptide after coincubation with C1q. *Journal of Neuroscience Research* **39**, 448–56.

Wyss-Coray T and Mucke L (2002). Inflammation in neurodegenerative disease: a double-edged sword. *Neuron* **35**, 419–32.

Wyss-Coray T, Feng L, Masliah E, *et al.* (1995). Increased central nervous system production of extracellular matrix components and development of hydrocephalus in transgenic mice overexpressing transforming growth factor-beta 1. *American Journal of Pathology* **147**, 53–67.

Yan SD, Roher A, Chaney M, Zlokovic B, Schmidt AM, and Stern D. (2000). Cellular cofactors potentiating induction of stress and cytotoxicity by amyloid beta-peptide. *Biochimica Biophysica Acta.* **1502**, 145–57.

Yankner BA (1996). Mechanisms of neuronal degeneration in Alzheimer's disease. *Neuron* **16**, 921–32.

Yu ZF, Zhou D, Bruce-Keller AJ, Kindy MS, and Mattson MP (1999). Lack of the p50 subunit of NF-kB increases the vulnerability of hippocampal neurons to excitotoxic injury. *Journal of Neuroscience* **19**, 8856–5.

Chapter 11

Cellular targets of amyloid β peptide: potential roles in neuronal cell stress and toxicity

Xi Chen, David Stern, and Shi Du Yan

11.1 Introduction

Amyloid plaque formation, caused by deposition of amyloid beta peptide (Aβ), is a pathological hallmark of Alzheimer's disease (AD) (Glenner and Wong 1984; Hardy and Selkoe 2002). However, the role of Aβ in causing neuronal cell stress and toxicity, thereby leading to AD, is poorly understood. Debate continues regarding the pathological properties of Aβ. Questions concerning whether the effects of Aβ are direct or indirect, and whether they reflect specific or non-specific mechanisms remain to be answered. These questions can only be addressed by identifying specific cellular targets for Aβ and assessing how interaction of Aβ with such targets impacts on cellular functions. As a molecule prone to aggregation and non-specific interactions, Aβ has been shown to interact with diverse macromolecular structures, including lipid bilayers (Terzi *et al.* 1997) and proteoglycans (Narindrasorasak *et al.* 1991; Buée *et al.* 1993; Snow *et al.* 1994, 1995; Watson *et al.* 1997), as well as a number of proteins. Proteins which interact with Aβ include those expressed by neurons, such as Aβ precursor protein (Lorenzo *et al.* 2000), the α_7-nicotinic acetylcholine receptor (Yan *et al.* 2000a), the neurotropin receptor (p75) (Yaar *et al.* 1997), and the extracellular collagen-like Alzheimer amyloid plaque component precursor (CLAC-P/collagen type XXV) (Hashimoto *et al.* 2002), and those on microglia or other cell types such as the macrophage scavenger receptor (type A) (El Khoury *et al.* 1996; Paresce *et al.* 1996), $\alpha_5\beta_1$ integrins (Matter *et al.* 1998), serpin enzyme complex receptor (Boland *et al.* 1995), low-density lipoprotein (LDL) receptor-related protein (Hammad *et al.* 1997), LRP-1 (Deane *et al.* 2004), and β_2-macroglobulin (Narita *et al.* 1997). These studies provide compelling evidence that extracellular macromolecular assemblies of Aβ can lead to neuronal/cellular stress and toxicity by interacting with specific cellular targets. In this chapter, we present *in vitro* and *in vivo* evidence that Aβ binds receptor for advance glycation endproducts (RAGE) extracellularly and Aβ binding alcohol dehydrogenase (ABAD) intracellularly. Further studies indicate that these interactions affect neurons, microglia, and the vasculature, leading to neuronal stress, activation of intracellular signal cascades, disturbance of energy metabolism, free-radical production, and potentiation of local neuro-inflammation. Taken together, these events propagate cellular injury in AD, and exacerbate pathological and behavioural defects in transgenic murine models of AD-type pathology. Our data provide support for the hypothesis that accumulation of Aβ in the central nervous system (CNS) can lead to AD pathology through interaction with specific cellular targets.

11.2 RAGE as extracellular target

RAGE is a cell surface multi-ligand receptor. As shown in Fig. 11.1, its extracellular portion contains one V-domain followed by two C-domains and bears close homology to other members of the immunoglobulin superfamily such as the neural cell adhesion molecule (NCAM) and *muc*-18. The N-terminal V-domain appears to be responsible for ligand binding. RAGE also contains a single transmembrane spanning domain and a short highly charged cytosolic tail that has a key role in mediating intracellular signalling (Yan *et al.* 2000b). RAGE was initially identified by our group for its ability to bind advanced glycation endproducts (AGE), non-enzymatically glycoxidized adducts which accumulate in environments rich in reducing sugars and oxidant stress (Miyata *et al.* 1996). Further investigations have suggested that RAGE is not likely to be a scavenger of AGEs; instead it is a signalling receptor which modulates cellular functions upon binding to its ligands (Yan *et al.* 2001).

RAGE–AGE interaction has been implicated in vascular injury associated with diabetes mellitus, renal failure, systemic amyloidoses, inflammation, oxidant stress, and ageing (Schmidt *et al.* 1994). RAGE activation by AGEs was initially found to be involved in the

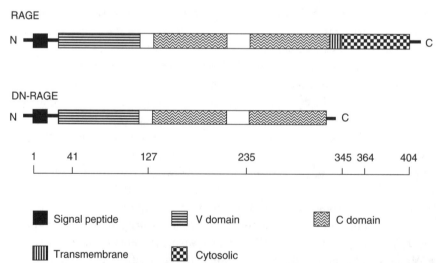

Fig. 11.1 Schematic structure of RAGE and DN-RAGE. RAGE is a 404 amino acid protein with an N-terminal V-domain followed by two C-domains, which bears close homology to other members of immunoglobulin superfamily. It also contains a single transmembrane domain and a short highly charged cytosolic C-terminal. DN-RAGE (dominant negative RAGE) is a truncated form which does not contain the transmembrane domain and cytosolic C-terminal.

oxidative process triggered by infusion of AGE in a rodent model (Yan *et al.* 1994b). The interaction of AGEs with RAGE on vascular cells (principally endothelium) induces vascular dysfunction, including hyperpermeability and cytokine generation. At the cellular level, RAGE–AGE interaction causes activation and nuclear translocation of NFκB, a transcription factor important in inflammatory responses. Such AGE–RAGE-induced NFκB activation results in increased expression of vascular cell adhesion molecule-1 (VCAM-1) in the endothelium, causing monocyte adhesion (Collins 1993, Schmidt *et al.* 1995). In addition, NFκB activation stimulates upregulation of RAGE itself, producing an autocrine feedback loop (Yan *et al.* 1994b). A possible basis for these observations has been provided by analysis of the RAGE promoter, which has two functional NFκB DNA binding sites capable of regulating the expression of receptor (Yan *et al.* 1997). RAGE–AGE interaction also modulates synthesis of a number of other proteins, including increased expression of tissue factor (TF) by both endothelium and macrophages (Ichikawa *et al.* 1998, Yamagishi *et al.* 1998), endothelin-1 expression, and reduced production of prostacyclin (Veyssier-Belot and Cacoub 1999).

The finding of Aβ binding to RAGE was made during our investigation of AGE modification of Aβ. As the optimal substrate for AGE modification would be a lysine-rich peptide with delayed turnover exposed to multiple reducing sugars (as in the cytosol), it was not surprising that PHF tau displayed AGE epitopes (Smith *et al.* 1994; Vitek *et al.* 1994; Yan *et al.* 1995, 1999a). This observation led us to explore the possibility that AGE modification of Aβ might render it a ligand for RAGE. At this point, two lines of inquiry in our laboratory converged: radioligand binding studies with glycated/non-glycated Aβ, and experiments to identify cell surface receptors for Aβ. First, we studied glycation of Aβ, in view of evidence that glycoxidized forms of the amyloid peptide were apparently present *in vivo* (Vitek *et al.* 1994). We observed that glycation of Aβ under *in vitro* conditions produced a species which easily precipitated in aqueous solution, precluding further studies with culture cells. However, when we tested the control, non-glycated Aβ, we were surprised to observe dose-dependent and sequence-specific binding to RAGE. At the same time, another line of investigation in our laboratory had just achieved the purification of a polypeptide, derived from lung, based on its capacity to bind Aβ. The N-terminal sequence of this polypeptide was identified to be RAGE. Observation there led us to explore the role of RAGE as a cell surface receptor for Aβ (Yan *et al.* 1996).

Several lines of evidence suggested that RAGE plays an important role in propagating cellular stress in pathophysiological states. Although RAGE is present in the brain at high levels during early development (virtually all cortical neurons express RAGE in the neonatal period, based on studies in rodents (Yan *et al.* 2000a)), expression of the receptor falls off during maturity, with only an occasional cortical neuron staining positively (Brett *et al.* 1993; Yan *et al.* 1996). However, concomitant with development of pathological conditions, especially those in which RAGE ligands accumulate, receptor expression is enhanced (Ritthaler *et al.* 1995; Miyata *et al.* 1996; Yan *et al.* 1996; Lue *et al.* 2001). Thus, in AD brain, RAGE is evident in neurons, microglia, and astrocytes, especially those proximal to deposits of Aβ and in neurons bearing neurofibrillary tangles. The presence of cells displaying high levels of RAGE near sites of accumulation of ligands appears to be a feature of RAGE regulation. Like AGEs, exposure of cells to Aβ has been found to cause RAGE-dependent nuclear translocation of NFκB (Yan *et al.* 1994a, 1996, 2000a). Thus Aβ binding to RAGE may initiate a positive feedback mechanism, causing increased expression of the receptor and sustained RAGE-induced cellular perturbation.

RAGE is expressed by cells of monocyte lineage, including microglia in the brain. Unlike the macrophage scavenger receptor (MSR), which is also known to be able to bind and internalize Aβ (El Khoury *et al.* 1996; Paresce *et al.* 1996), RAGE does not effectively mediate internalization and degradation of Aβ. On microglia, Aβ binding to RAGE causes cellular activation through activation of NFκB with resultant expression of cytokines, such as tumour necrosis factor-α (TNF-α) and macrophage colony-stimulating factor (M-CSF) (Yan *et al.* 2000a; Lue *et al.* 2001). The presence of M-CSF antigen in AD brain, especially in proximity to sites of Aβ deposition, was evident compared with age-matched controls. Furthermore, increased levels of M-CSF antigen were observed in cerebrospinal fluid (CSF) of patients with AD, suggesting the relevance of M-CSF as a barometer of cellular perturbation in AD (Yan *et al.* 2000a). Stimulation of microglial c-*fms* (Stanley *et al.* 1997), the M-CSF receptor, with M-CSF enhances its survival, promotes proliferation in response to cellular stress, and modulates the gene expression with upregulation of the scavenger receptor and apoE. The protective effect of M-CSF for the microglia in response to Aβ was modelled by exposing transformed murine macrophages (BV-2 cells) to Aβ; preincubation of BV-2 cells with M-CSF maintained MTT (3-(4,5-dimethylthiazol-2-yl)-2,5-diphenylterazolium bromide) reduction and intact cellular morphology in the presence of Aβ compared with BV-2 cells exposed to Aβ in the absence of M-CSF (Yan *et al.* 2000a). In contrast, neurons, which do not display c-*fms*, did not benefit from the protective effects of M-CSF and must withstand the added stress of an environment rich in activated microglia releasing potentially toxic products.

On neurons, Aβ binding to RAGE suppresses MTT reduction and activates NFκB, the latter resulting in enhanced expression of M-CSF. The cytotoxic effects of the RAGE–Aβ interaction are exemplified by results of studies in COS cells overexpressing RAGE; dose-dependent suppression of MTT reduction was exaggerated in the presence of increasing concentrations of Aβ compared with mock-transfected controls. Furthermore, a line of PC12 cells with exogenous expression of RAGE showed diminished MTT reduction in the presence of Aβ. Evidence of toxicity was abrogated by addition of soluble (s) RAGE, which intercepts the interaction of Aβ with cell surface RAGE. Consistent with these data, RAGE-bearing cells displayed enhanced NFκB activation and toxic morphological changes in the presence of Aβ, including retraction of cellular processes (Yan *et al.* 2000a). In each of these experiments, it was evident that RAGE-mediated potentiation of Aβ-induced cell stress occurred at the lower range of Aβ concentrations (10^{-9}–10^{-6} M), whereas at higher levels of amyloid peptide ($\geq 10^{-5}$ M) RAGE-independent cytotoxicity occurred. These data are consistent with the concept that RAGE-dependent effects of Aβ occurs at lower concentrations of Aβ, whereas other mechanisms predominate at higher levels. Meanwhile, M-CSF produced by neurons at the sites of cellular perturbation can recruit microglia, further exacerbating neuronal stress.

The presence of RAGE in the vascular wall, in both endothelial and smooth muscle cells, is also likely to mediate interactions with amyloid peptide deposited in the vessel wall. Using an *in vitro* system to analyse Aβ transport across the endothelial monolayer, blockade of RAGE suppressed approximately 36% of apical to basal transit of Aβ (Mackic *et al.* 1998). These data, together with other evidence demonstrating the presence of RAGE in choroid plexus vasculature, suggest that RAGE could contribute to transfer of Aβ between the blood and CSF compartments. Recent studies in a murine model have demonstrated RAGE-dependent uptake of Aβ by brain capillaries and RAGE-mediated transport of Aβ when the animals were infused with Aβ at nanomolar levels. Interaction of Aβ with RAGE within the vascular wall can also suppress cerebral blood flow by inducing production of endothelin-1, a potent vasoconstrictor (Deane *et al.* 2004).

The role of RAGE–Aβ interaction in exacerbating AD pathology was further investigated with double transgenic mice (Tg mAPP × RAGE)(Arancio *et al.* 2004). Tg mAPP × RAGE mice demonstated more significant Aβ accumulation, as shown by ELISA and immunohistochemistry for amyloid plaque load in cortex as well as hippocampus, at the age of 14–18 months (Fig. 11.2 and Fig. 11.3 (Plate 5)). More importantly, young (3–4 months of age) Tg mAPP × RAGE mice displayed functional and pathological evidence of neuronal perturbation, before accumulation of cerebral Aβ and plaque formation occurred. The behavioural abnormalities were studied in a radial-arm water maze and included impaired learning and memory. While Tg mAPP mice first showed deficits at age of 5–6 months, Tg mAPP × RAGE mice showed deficits at 3–4 months of age. Using electrophysiological techniques to record synaptic tranismission under basal conditions and during long-term potentiation (LTP), our data demonstrated that field-excitatory postsynaptic potential in the CA1 stratum radiatum was impaired at the level of basal synaptic transmission in both Tg mAPP and Tg mAPP × RAGE mice. Consistently, synaptic transmission associated with LTP was more severely affected in Tg mAPP × RAGE than in Tg mAPP mice aged 8–10 months. At 60 min after LTP induction, we observed only 140% ± 9% increase in potentiation in Tg mAPP × RAGE mice compared with increases of 202% ± 13%, 188% ± 20%, and 203% ± 17% in non-Tg, Tg RAGE, and Tg mAPP mice, respectively. This result implies that overexpression of RAGE in an Aβ-rich environment further perturbs synaptic function and synaptic plasticity. Neuropathologically, acetylcholinesterase (AChE) positive neurites and synaptophysin immunoreactivity (markers of presynaptic teminals) in the subiculum (Sb), CA1, and entorhinal cortex (EC) were all decreased at 3–4 months in Tg mAPP × RAGE mice compared to Tg mAPP mice. Consistent with the hypothesis that these effects exerted by RAGE–Aβ interaction are due to signal transduction via RAGE as a cellular receptor, similar experiments showed that mice bearing a DN-RAGE (dominant negative RAGE or RAGE with the C-terminal intracellular domain truncated as shown in Fig.11.1) transgene under the control of the same promoter were protected from the behavioural, electrophysiological, and neuropathological changes observed in Tg mAPP × RAGE mice. Intracellular signalling cascades underlying neuronal dysfunction in Tg mAPP × RAGE mice include increased activation and phosphorylation of cAMP-response element binding protein (CREB), mitogen-activated protein kinases (MAPKs), and calcium/calmodulin-dependent

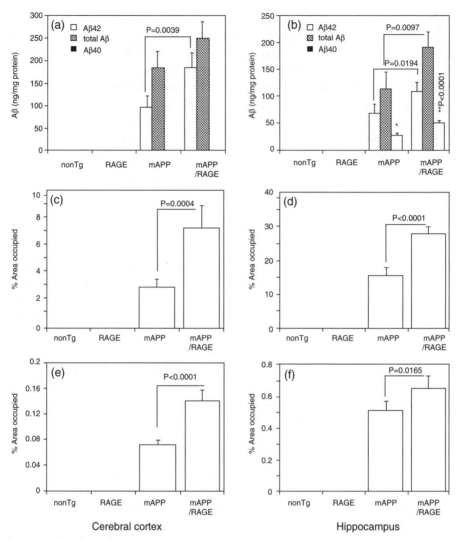

Fig. 11.2 Aβ level and amyloid plaque load in the cerebral cortex and hippocampus in 14–18-month-old transgenic mice. (a, b) The Aβ level was significantly increased in double transgenic mice (Tg mAPP × RAGE) compared with Tg mAPP mice. While Aβ42, total Aβ, and Aβ40 were all increased in the hippocampus, only Aβ42 and the total Aβ levels were increased in the cerebral cortex. Quantification of amyloid load after staining with the 3D6 antibody (an anti-Aβ antibody provided by Eli Lilly) (c, d) or thioflavin S (e, f) showed significantly increased amyloid load in Tg mAPP × RAGE mice compared with Tg mAPP mice.

protein kinase II (CAMKII) in hippocampal extracts from Tg mAPP × RAGE mice. Onyango *et al.* (2005) demonstrated that exposure of cybrids containing platelet mitochondria from AD patients or age-matched controls to Aβ induced activation of p38 and JNK pathways. In this model, anti-RAGE antibody blocked Aβ-induced activation and reduced the neurotoxic effects of Aβ.

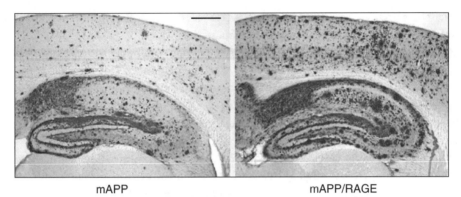

mAPP   mAPP/RAGE

Fig. 11.3 Amyloid load in the cerebral cortex and hippocampus in Tg mAPP mice and Tg mAPP × RAGE mice after staining the section with 3D6 antibody. Please see the colour plate section for a colour version of this figure.

Taken together, these data indicate that Aβ interacts with RAGE on the surface of microglia, neurons, and cells in the vasculature. These interactions can exacerbate AD-type pathology through direct induction of neuronal stress, cytokine release, and trapping of Aβ in the CNS.

11.3 ABAD as an intracellular target

Using yeast two-hybrid system, we found that ABAD, a member of the short-chain dehydrogenase–reductase family, binds Aβ in sequence-specific manner (Yan *et al.* 1997). ABAD is identical to L-3-hydroxyacyl coenzyme A dehydrogenase type II, first identified by investigators analysing fatty acid β-oxidation in their study of 3-hydroxyacyl CoA dehydrogenase activity in liver (Kobayashi *et al.* 1996; Furuta *et al.* 1997). ABAD is a mitochondrial enzyme (Furuta *et al.* 1997; He *et al.* 2001) which shares many properties with other members of the superfamily of short-chain dehydrogenase reductase or short-chain alcohol dehydrogenase, including NAD/NADH binding and catalytic sites (Krozowski 1994; Eaton *et al.* 1996). Functional inactivation of *scully*, the ABAD counterpart in *Drosophila*, resulted in a developmental lethal phenotype with multiple abnormalities (Torroja *et al.* 1998). Salient features of the ABAD-deficient flies included small and underdeveloped testes with cytoplasmic accumulation of fat-containing vesicles and scarce mitochondria in spermatocytes, consistent with a defect in fatty acid oxidation and mitochondrial dysfunction. Photoreceptor cell mitochondria were smaller and fewer, and showed swollen cristae, compared with controls.

Another line of work has led to the identification of a genetic defect in humans in the enzyme methyl-3-hydroxybutyryl-CoA dehydrogenase (MHBD) which participates in isoleucine and branched-chain fatty acid catabolism. Patients deficient in MHBD display infantile neurodegeneration and psychomotor retardation. MHBD is identical to ABAD. Biochemically, these patients show elevated levels of two urinary metabolites, 2-methyl-3-hydroxybutyrate and tiglyl glycine (Ofman *et al.* 2003), consistent with the predicted catabolic defect. These data suggest that ABAD plays an essential role in fatty acid and isoleucine metabolism under homeostatic conditions.

In vitro studies have revealed that ABAD is an enzyme with broad substrate specificity. The enzyme catalyses reversible NAD/NADH-dependent oxidation–reduction of a range of substrates, including linear alcohols, steroid substrates (such as 17β-estradiol), S-acetoacetyl-CoA and β-hydroxybutyrate (Yan *et al.* 1999b, 2000b). To analyse the possible role of ABAD in β-hydroxybutyrate metabolism, we have overexpressed the enzyme in COS cells and maintained cultures in medium devoid of glucose using β-hydroxybutyrate as the principal energy substrate (Yan *et al.* 2000b). Whereas control (vector-transfected) COS cells displayed loss of viability over 2–4 days in culture under these conditions, accompanied by a rapid fall in ATP, COS cells overexpressing ABAD better maintained their viability and ATP content. NMR analysis of cultures following addition of [$^{13}P^{13}C$]D-β-hydroxybutyrate demonstrated an increased flux of acetyl-CoA through the tricarboxylic acid/Krebs cycle in COS cells overexpressing ABAD compared with controls (Yan *et al.* 2000b). To determine the relevance of these *in vitro* data to potential cytoprotective properties of ABAD *in vivo*, experiments were performed using a murine model of stroke: transient middle cerebral artery ischaemia. First, wild-type animals were subjected to stroke and expression of ABAD was assessed. Increased ABAD antigen was observed in neurons proximal to the infarcted area, especially those in the penumbra, compared with controls. Image analysis of immunohistochemical data demonstrated approximately fivefold increased levels of ABAD antigen in cortical neurons in the ischaemic territory compared with controls. These experiments, indicating the relevance of ABAD to the setting of ischaemic stress, led us to produce transgenic (Tg) mice overexpressing ABAD under the control of platelet-derived growth factor-β (PDGF-β) chain promoter. The latter animals, termed Tg ABAD mice, displayed increased neuronal expression of ABAD in the cerebral cortex, which was present in the appropriate subcellular locations, including mitochondria. Tg ABAD animals had higher baseline levels of brain ATP and showed increased flux of acetyl-CoA through the tricarboxylic acid/Krebs cycle following administration of [$^{13}P^{13}C$]D-β-hydroxybutyrate in the fasted state. Following induction of stroke, Tg ABAD mice displayed infarcts of smaller volume and decreased neurological deficit, consistent with resistance to ischaemic stress, compared with non-Tg littermates. These data are consistent with a role for ABAD in the metabolic defence against ischaemic stress (Yan *et al.* 2000b).

In addition to its roles in metabolic homeostasis related to energy metabolism and catabolism of isoleucine and branched-chain fatty acids, ABAD was also found to bind the mitochondrial chaperone molecule cyclophilin D, a peptidylprolyl *cis–trans* isomerase whose activity and association with the inner mitochondrial membrane have been linked to opening of the mitochondrial membrane permeability transition pore (MPT) (Crompton *et al.* 1998; Woodfield *et al.* 1998, Crompton 1999). Overexpression of ABAD was also found to block hypoxia-induced cellular stress and apoptosis by inhibiting the JNK signalling pathway (Yan *et al.*, unpublished data). Furthermore, in the 1-methyl-4-phenyl-1,2,3,6-tetrahydropyridine (MPTP)-induced mouse model of murine Parkinsonism, overexpression of ABAD in neurons had a cytoprotective effect.

It appeared that increased levels of ABAD mitigated MPTP-induced inhibition of oxidative phosphorylation and ATP production (Tieu *et al.* 2004). Thus there is reason to believe that ABAD might have multiple protective effects on cellular functions.

In order to assess the possible relevance of ABAD to AD, studies in AD brain ($n = 19$) and non-demented controls ($n = 15$), matched for age, were performed using samples harvested according to the rapid autopsy method developed at Sun Health Research Institute in order to minimize post-mortem delay (Lue *et al.* 2001). Immunoblots of brain extracts were compared by laser densitometry, and each band was normalized according to the intensity of the β-actin band. The results showed that the intensity of the ABAD band was increased by approximately 28% (inferior temporal lobe gyrus) and 40% (hippocampus) in samples from AD patients compared with non-demented controls (Lustbader *et al.* 2004). These data are consistent with our previous work, demonstrating enhanced expression of ABAD in AD brain by immunoblotting with anti-ABAD antibody (Yan *et al.* 2000b). In contrast, protein extracts prepared from the cerebellum, a portion of the brain not affected by AD, showed no significant differences in ABAD expression compared with non-demented controls. These data indicate that ABAD expression increases in affected regions of AD brain. Of course, at this level it is equally possible that such elevated levels of ABAD could reflect either a protective or a deleterious response to the Aβ-rich environment.

ABAD–Aβ interaction was further demonstrated by immunoprecipitation in AD brain. Protein extracts of AD brain were subjected to immunoprecipitation with anti-Aβ antibody followed by SDS–PAGE, transfer of proteins to nitrocellulose membranes, and immunoblotting with anti-ABAD antibody (Lustbader *et al.* 2004). The results demonstrated a strong immunoreactive band in AD brain, but not in non-demented controls. Because preparation of samples for immunoprecipitation disturbs tissue architecture, morphological studies were performed. Confocal microscopy of affected regions of AD brain displayed co-localization of ABAD and Aβ antigens. In additon, ABAD antigen was co-localized with a mitochondrial marker, voltage-dependent anion channel. At the level of transmission electron microscopy, ABAD and Aβ were also co-localized to mitochondria using gold-labelled antibodies. These studies demonstrate that ABAD–Aβ interaction can occur in AD brain, and that Aβ is present in mitochondria, a previously unrecognized observation (Caspersen *et al.* 2005).

Parameters of Aβ–ABAD interaction were further characterized by the following studies. Using plasmon resonance, it was found that Aβ(1–40), Aβ(1–42), and Aβ(1–20) were able to bind to ABAD in a dose-dependent manner with half-maximal occupancy of binding sites at 40–80 nM. In contrast, Aβ(25–35), Aβ-derived peptides with a reversed sequence, amylin, and prion-derived peptide (109–141) displayed no specific binding. To characterize this interaction further, structural studies were performed in the presence of NAD. To our surprise, NAD was not bound to ABAD in the crystal structure, suggesting that Aβ resulted in a considerable distortion of ABAD. This indeed proved to be the case; native ABAD shows an NAD binding pocket and the expected catalytic triad associated with its enzymatic activity (Fig. 11.4 (Plate 6); the native enzyme is shown in cyan).

Note the prominent distortion of ABAD in the presence of Aβ, with deformation of the LB$_{DB}$ loop as well as the NAD binding pocket (Fig. 11.4 (Plate 6); Aβ-bound ABAD is shown in magenta) (Lustbader *et al.* 2004). These data suggested that the LB$_{DB}$ region of ABAD (residue 92–120) might be the actual binding site for Aβ. Studies with the ABAD (92–120) fragment (also termed ABAD decoy peptide (DP)) indicated that it is an

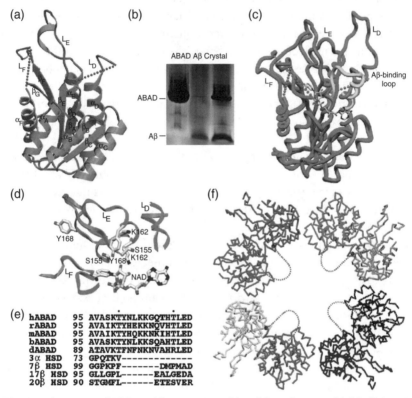

Fig. 11.4 Crystal structure of Aβ-bound human ABAD. (a) A ribbon diagram with labelled secondary structures and the LB$_{DB}$, LB$_{EB}$, and LB$_{FB}$ loops. Helices are shown in green, β strands are shown in pink, and disordered regions are shown by dotted lines. (b) SDS–PAGE of washed and dissolved crystals of human ABAD and Aβ. Lanes from left to right: ABAD standard, Aβ(1–40) standard and dissolved crystals. (c) Superposition of Aβ-bound human ABAD (pink) and rat ABAD in complex with NAD (blue). The LB$_{DB}$ loop of 3β-hydroxysteroid dehydrogenase (3β-HSD) is shown in yellow. NAD is shown as a stick model with grey for carbon atoms, red for oxygen atoms, blue for nitrogen atoms, and yellow for phosphate atoms. The proposed Aβ-binding loop is indicated. (d) Superposition of the active sites of Aβ-bound human ABAD (pink) and rat ABAD (blue), showing distortion of the NAD binding site and the catalytic triad S^{155}, K^{162}, and Y^{168}. Colours are the same as in (c). (e) Sequence alignment of the disordered part of the LB$_{DB}$ loop (residues 95–113) among human, rat, mouse, bovine, and *Drosophila* ABAD and several hydroxysteroid dehydrogenases (HSDs), showing the insertion in ABAD relative to other HSDs. (f) Section of the crystal packing interactions, showing the large solvent channels. Each ABAD molecule is shown in a different colour. The ordered ends of the LB$_{DB}$ loop, residues 94 and 114, are marked as red and blue balls, respectively, and the hypothetical loops are shown as pink dotted lines. Please see the colour plate section for a colour version of this figure.

inhibitor of the ABAD–Aβ interaction, whereas a peptide with the same amino acids but a reversed sequence (also termed ABAD reversed peptide (RP)) was inactive (Lustbader *et al.* 2004).

In vitro studies demonstrated that Aβ inhibits ABAD enzymatic activity towards all its substrates, including *S*-acetoacetyl-CoA, octanol and 17β-estradiol (Yan *et al.* 1999b). However, the concentration of Aβ necessary for half-maximal inhibition of ABAD is in the low micromolar range (1–3 μM), rather than in the nanomolar range (40–80 nM); the latter corresponds to concentrations of Aβ which bind ABAD. Our concept is that the N-terminus of the peptide serves to localize Aβ at critical sites within the cell (in this case, associated with ABAD), and the C-terminal portion of the peptide is free to interact with additional molecules of Aβ forming oligomers, potentially fibrillar structures that further deform protein structure. Thus binding of an initial molecule of Aβ to ABAD probably does not affect the enzyme's activity significantly. However, when oligomers of Aβ form, using the initial ABAD-bound Aβ as a nidus, the resulting macromolecular assembly distorts the enzyme and alters its function. In this context, it is important to note that, from a structural standpoint, there is a large cavity in the structure of the ABAD–Aβ complex which could accommodate a fibrillar assembly. Inspection of the crystal packing shows that ordered ends of the LB_{DB} loop point into interconnected solvent channels with dimensions of about 70 Å. We estimate that the ordered part of the crystal accounts for only about 30% of the total crystal volume. Therefore sufficient space is available for the disordered loops and the bound Aβ (as well as macromolecular assemblies of Aβ), which could non-specifically bind and clog the active site region to inactivate the enzyme.

In order to investigate ABAD–Aβ interaction and its impact *in vivo*, Tg mice overexpressing ABAD in an Aβ-rich environment were generated (Lustbader *et al.* 2004; Takuma *et al.* 2005). Overexpression of mAPP was achieved using mice made in D.L. Mucke's laboratory (Hsia *et al.* 1999). These mice express an alternatively spliced human APP minigene that encodes human APP695, APP751, and APP770 bearing mutations linked to familial AD (V717F, K670M, N671L) under control of the PDGF-β chain promoter (Yan *et al.* 2000b). Tg mice overexpressing ABAD, also under control of the PDGF-β chain promoter were crossed with Tg mAPP mice to generate four genotypes; Tg mAPP × ABAD (double transgenics), Tg mAPP, Tg ABAD (single transgenics) and non-Tg littermates. We first analysed cortical neurons from Tg mAPP × ABAD mice to determine whether they displayed Aβ-induced stress in cell culture. We evaluated the induction of oxidant stress, which has a counterpart in AD brain and transgenic models of AD-type pathology (Wild-Bode *et al.* 1997; Pappolla *et al.* 1998; Perry *et al.* 1998; Eckert *et al.* 2001; Tabner *et al.* 2002; Marques *et al.* 2003; Tamagno *et al.* 2003). Neurons from Tg mAPP × ABAD mice demonstrated evidence of free-radical-induced cell stress. In order to antagonize the ABAD–Aβ interaction, we produced a membrane-permeable form of ABAD decoy peptide (ABAD-DP). The latter was accomplished using the membrane transduction domain of human immunodeficience virus type 1 (Aarts *et al.* 2002). A fusion peptide, tat-ABAD-DP, was added to a culture medium of neurons from Tg mAPP × ABAD mice. In the presence of tat-ABAD-DP, induction of oxidant stress and, specifically, generation of superoxide anion and hydrogen peroxide were inhibited, although the reversed peptide was without effect (Lustbader *et al.* 2004, Takuma *et al.* 2005).

These data suggested that increased levels of ABAD in an Aβ-rich environment potentiate Aβ-induced cell stress. To probe this issue further, brains from each of the genotypes of transgenic mice were studied using electron paramagnetic resonance (EPR) spectroscopy to identify free radicals. Higher levels of radicals were observed in brains from Tg mAPP × ABAD mice compared with samples from non-Tg, Tg ABAD, or Tg mAPP mice (Lustbader *et al.* 2004). Such radicals clearly connote a high level of oxidant stress in Tg mAPP × ABAD mice which might underlie the impaired spatial learning and memory observed in these animals in the radial-arm water maze test (Bliss and Collingridge 1993, Morgan *et al.* 2000). In contrast with non-Tg littermates, Tg mAPP, and Tg ABAD mice, double transgenic Tg mAPP × ABAD mice were already failing to learn efficiently at only 5 months of age. Furthermore, the area occupied by acetylcholinesterase-positive neurites in the subiculum, entorhinal cortex, and CA-1 in brains of Tg mAPP × ABAD mice was decreased compared with the other genotypes. Further, bigenic mice (Tg mAPP and ABAD) showed impaired energy metabolism and mitochondrial dysfunction, as evidenced by decreased ATP levels and mitochondrial enzymes associated with the respiratory chain compared with the single Tg mice. These data indicate that Tg mAPP × ABAD mice display exaggerated cytotoxicity of a type associated with AD-type pathology.

Our findings in transgenic mice led us to analyse properties of cortical neurons cultured from each of the genotypes in further detail. Although these studies are ongoing, several findings have already been clearly delineated. First, neurons from Tg mAPP × ABAD mice spontaneously generate superoxide anion and hydrogen peroxide as a consequence of mitochondrial leakage of reactive oxygen species (Takuma *et al.* 2005). Production of oxygen radicals occurs at the level of complex III and is also associated with opening of the MPT. These oxygen-free radicals trigger a programmed cell death pathway, leading to activation of caspase-3, DNA fragmentation, and ultimately loss of cell viability. Thus neurons from mice overexpressing both mutant APP and ABAD display evidence of severe cellular stress. We propose that these mice are a model of accelerated and exaggerated cell stress relevant to the role of ABAD in AD. Further studies to link ABAD to the pathogenesis of cellular dysfunction in Alzheimer's type pathology in murine models are underway through generation of animals deficient in ABAD and studies using agents to inhibit ABAD–Aβ interaction *in vivo*.

11.4 **Summary**

As summarized in Fig. 11.5, extracellular Aβ can interact with RAGE on microglia, neurons, and blood vessels in the CNS. RAGE–Aβ interaction on microglia leads to NFκB activation and release of TNF-α and M-CSF through a positive feedback loop. RAGE–Aβ interaction on neurons also activates the NFκB pathway while suppressing MTT reduction and inducing toxic morphological changes on neurons. RAGE–Aβ interaction on the blood vessel facilitates transfer of Aβ across the blood–brain barrier. In a transgenic murine model, interaction between Aβ and RAGE can lead to early and exaggerated

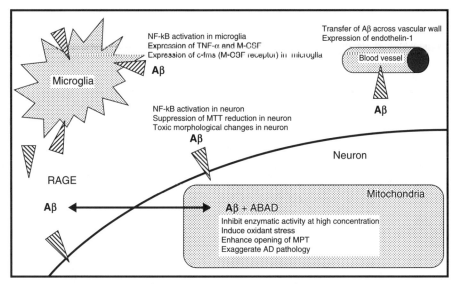

Fig. 11.5 Schematic summary of Aβ interaction with RAGE and ABAD.

impairment of spatial memory and learning. When Aβ gains access to the mitochondrial matrix, ABAD–Aβ interaction has the capacity to induce neuronal and mitochondrial oxidant stress, opening of the MPT, and exaggeration of AD pathology.

Acknowledgements

This work was supported by grants from the NIH (National Institute on Aging, AG16736, PO1 AG17490, P50 AG08702), the Michael J Fox Foundation, and the Alzheimer's Association.

References

Aarts M, Liu Y, Liu L, *et al.* (2002) Treatment of ischemic brain damage by perturbing NMDA receptor- PSD-95 protein interactions. *Science* **298**, 846–50.

Arancio O, Zhang HP, Chen X, *et al.* (2004). RAGE potentiates Abeta-induced perturbation of neuronal function in transgenic mice. *EMBO Journal* **23**, 4096–105.

Bliss TV and Collingridge GL (1993). A synaptic model of memory: long-term potentiation in the hippocampus. *Nature* **361**, 31–9.

Boland K, Manias K, and Perlmutter DH (1995). Specificity in recognition of amyloid-beta peptide by the serpin-enzyme complex receptor in hepatoma cells and neuronal cells. *Journal of Biological Chemistry* **270**, 28022–8.

Brett J, Schmidt AM, Yan SD, *et al.* (1993). Survey of the distribution of a newly characterized receptor for advanced glycation end products in tissues. *American Journal of Pathology* **143**, 1699–1712.

Buée L, Ding W, Delacourte A, and Fillit H (1993). Binding of secreted human neuroblastoma proteoglycans to the Alzheimer's amyloid A4 peptide. *Brain Research* **601**, 154–63.

Caspersen C, Wang N, Yao J, *et al.* (2005). Mitochondrial Abeta: a potential focal point for neuronal metabolic dysfunction in Alzheimer's disease. *FASEB Journal* **19**, 2040–1.

Collins T (1993). Endothelial nuclear factor-kappa B and the initiation of the atherosclerotic lesion. *Laboratory Investigation* **68**, 499–508.

Crompton M (1999). The mitochondrial permeability transition pore and its role in cell death. *Biochemical Journal* **341** (Pt 2), 233–49.

Crompton M, Virji S, and Ward JM (1998). Cyclophilin-D binds strongly to complexes of the voltage-dependent anion channel and the adenine nucleotide translocase to form the permeability transition pore. *European Journal of Biochemistry* **258**, 729–35.

Deane R, Wu Z, and Zlokovic BV (2004). RAGE (yin) versus LRP (yang) balance regulates alzheimer amyloid beta-peptide clearance through transport across the blood–brain barrier. *Stroke* **35**, 2628–31.

Eaton S, Bartlett K, and Pourfarzam M (1996). Mammalian mitochondrial beta-oxidation. *Biochemical Journal* **320**, 345–57.

Eckert A, Steiner B, Marques C, *et al.* (2001). Elevated vulnerability to oxidative stress-induced cell death and activation of caspase-3 by the Swedish amyloid precursor protein mutation. *Journal of Neuroscience Research* **64**, 183–92.

El Khoury J, Hickman SE, Thomas CA, Cao L, Silverstein SC, and Loike JD (1996). Scavenger receptor-mediated adhesion of microglia to beta-amyloid fibrils. *Nature* **382**, 716–19.

Furuta S, Kobayashi A, Miyazawa S, and Hashimoto, T. (1997). Cloning and expression of cDNA for a newly identified isozyme of bovine liver 3-hydroxyacyl-CoA dehydrogenase and its import into mitochondria. *Biochimica Biophysica Acta* **1350**, 317–24.

Glenner GG and Wong CW (1984). Alzheimer's disease: initial report of the purification and characterization of a novel cerebrovascular amyloid protein. *Biochemical and Biophysical Research Communications* **120**, 885–90.

Hammad SM, Ranganathan S, Loukinova E, Twal WO, and Argraves WS (1997). Interaction of apolipoprotein J-amyloid beta-peptide complex with low density lipoprotein receptor-related protein-2/megalin. A mechanism to prevent pathological accumulation of amyloid beta-peptide. *Journal of Biological Chemistry* **272**, 18644–9.

Hardy J and Selkoe DJ (2002). The amyloid hypothesis of Alzheimer's disease: progress and problems on the road to therapeutics. *Science* **297**, 353–6.

Hashimoto T, Wakabayashi T, Watanabe A, *et al.* (2002). CLAC: a novel Alzheimer amyloid plaque component derived from a transmembrane precursor, CLAC-P/collagen type XXV. *EMBO Journal* **21**, 1524–34.

He XY, Merz G, Yang YZ, Mehta P, Schulz H, and Yang SY (2001). Characterization and localization of human type10 17beta-hydroxysteroid dehydrogenase. *European Journal of Biochemistry* **268**, 4899–907.

Hsia AY, Masliah E, McConlogue L, *et al.* (1999). Plaque-independent disruption of neural circuits in Alzheimer's disease mouse models. *Proceedings of the National Academy of Sciences of the USA* **96**, 3228–33.

Ichikawa K, Yoshinari M, Iwase M, *et al.* (1998). Advanced glycosylation end products induced tissue factor expression in human monocyte-like U937 cells and increased tissue factor expression in monocytes from diabetic patients. *Atherosclerosis* **136**, 281–7.

Kobayashi A, Jiang LL, and Hashimoto T (1996). Two mitochondrial 3-hydroxyacyl-CoA dehydrogenases in bovine liver. *Journal of Biochemistry* **119**, 775–82.

Krozowski Z (1994). The short-chain alcohol dehydrogenase superfamily: variations on a common theme. *journal of Steroid Biochemistry and Molecular Biology* **51**, 125–30.

Lorenzo A, Yuan M, Zhang Z, *et al.* (2000). Amyloid beta interacts with the amyloid precursor protein: a potential toxic mechanism in Alzheimer's disease. *Nature Neuroscience* **3**, 460–4.

Lue LF, Walker DG, Brachova L, *et al.* (2001). Involvement of microglial receptor for advanced glycation endproducts (RAGE) in Alzheimer's disease: identification of a cellular activation mechanism. *Experimental Neurology* **171**, 29–45.

Lustbader JW, Cirilli M, Lin C, *et al.* (2004). ABAD directly links Abeta to mitochondrial toxicity in Alzheimer's disease. *Science* **304**, 448–52.

Mackic JB, Stins M, McComb JG, *et al.* (1998). Human blood–brain barrier receptors for Alzheimer's amyloid-beta 1-40. Asymmetrical binding, endocytosis, and transcytosis at the apical side of brain microvascular endothelial cell monolayer. *Journal of Clinical Investigation* **102**, 734–43.

Marques CA, Keil U, Bonert A, *et al.* (2003). Neurotoxic mechanisms caused by the Alzheimer's disease-linked Swedish amyloid precursor protein mutation: oxidative stress, caspases, and the JNK pathway. *Journal of Biological Chemistry* **278**, 28294–302.

Matter ML, Zhang Z, Nordstedt C, and Ruoslahti E. (1998). The alpha5beta1 integrin mediates elimination of amyloid-beta peptide and protects against apoptosis.*Journal of Cell Biology* **141**, 1019–30.

Miyata T, Hori O, Zhang J, *et al.* (1996). The receptor for advanced glycation end products (RAGE) is a central mediator of the interaction of AGE-beta2microglobulin with human mononuclear phagocytes via an oxidant-sensitive pathway. Implications for the pathogenesis of dialysis-related amyloidosis. *Journal of Clinical Investigation* **98**, 1088–94.

Morgan D, Diamond DM, Gottschall PE, *et al.* (2000). A beta peptide vaccination prevents memory loss in an animal model of Alzheimer's disease. *Nature* **408**, 982–5.

Narindrasorasak S, Lowery D, Gonzalez-DeWhitt P, *et al.* (1991). High affinity interactions between the Alzheimer's beta-amyloid precursor proteins and the basement membrane form of heparan sulfate proteoglycan. *Journal of Biological Chemistry* **266**, 12878–83.

Narita M, Holtzman DM, Schwartz AL, and Bu G (1997). Alpha2-macroglobulin complexes with and mediates the endocytosis of beta-amyloid peptide via cell surface low-density lipoprotein receptor-related protein. *Journal of Neurochemistry* **69**, 1904–11.

Ofman R, Ruiter JP, Feenstra M, *et al.* (2003). 2-Methyl-3-hydroxybutyryl-CoA dehydrogenase deficiency is caused by mutations in the HADH2 gene. *American Journal of Human Genetics* **72**, 1300–7.

Onyango IG, Tuttle JB, and Bennett, JP, Jr (2005). Altered intracellular signaling and reduced viability of Alzheimer's disease neuronal cybrids is reproduced by beta-amyloid peptide acting through receptor for advanced glycation end products (RAGE). *Molecular and Cellular Neurosciences* **29**, 333–43.

Pappolla MA, Chyan YJ, Omar RA, *et al.* (1998). Evidence of oxidative stress and *in vivo* neurotoxicity of beta-amyloid in a transgenic mouse model of Alzheimer's disease: a chronic oxidative paradigm for testing antioxidant therapies *in vivo*. *American Journal of Pathology* **152**, 871–7.

Paresce DM, Ghosh RN, and Maxfield FR (1996). Microglial cells internalize aggregates of the Alzheimer's disease amyloid beta-protein via a scavenger receptor. *Neuron* **17**, 553–65.

Perry G, Castellani RJ, Hirai K, and Smith MA (1998). Reactive oxygen species mediate cellular damage in Alzheimer disease. *Journal of Alzheimer's Disease* **1**, 45–55.

Ritthaler U, Deng Y, Zhang Y, *et al.* (1995). Expression of receptors for advanced glycation end products in peripheral occlusive vascular disease. *American Journal of Pathology* **146**, 688–94.

Schmidt AM, Hori O, Brett J, Yan SD, Wautier JL, and Stern D (1994). Cellular receptors for advanced glycation end products. Implications for induction of oxidant stress and cellular dysfunction in the pathogenesis of vascular lesions. *Arteriosclerosis and Thrombosis* **14**, 1521–8.

Schmidt AM, Hori O, Chen JX, *et al.* (1995). Advanced glycation endproducts interacting with their endothelial receptor induce expression of vascular cell adhesion molecule-1 (VCAM-1) in cultured human endothelial cells and in mice. A potential mechanism for the accelerated vasculopathy of diabetes. *Journal of Clinical Investigation* **96**, 1395–403.

Smith MA, Richey PL, Taneda S, *et al.* (1994). Advanced Maillard reaction end products, free radicals, and protein oxidation in Alzheimer's disease. *Annals of the New York Academy of Sciences* **738**, 447–54.

Snow AD, Sekiguchi R, Nochlin D, *et al.* (1994). An important role of heparan sulfate proteoglycan (Perlecan) in a model system for the deposition and persistence of fibrillar A beta-amyloid in rat brain. *Neuron* **12**, 219–34.

Snow AD, Kinsella MG, Parks E, *et al.* (1995). Differential binding of vascular cell-derived proteoglycans (perlecan, biglycan, decorin, and versican). to the beta-amyloid protein of Alzheimer's disease. *Archives of Biochemistry and Biophysics* **320**, 84–95.

Stanley ER, Berg KL, Einstein DB, *et al.* (1997). Biology and action of colony-stimulating factor-1. *Molecular Reproduction and Development* **46**, 4–10.

Tabner BJ, Turnbull S, El-Agnaf OM, and Allsop D (2002). Formation of hydrogen peroxide and hydroxyl radicals from A(beta). and alpha-synuclein as a possible mechanism of cell death in Alzheimer's disease and Parkinson's disease. *Free Radical Biology and Medicine* **32**, 1076–83.

Takuma K, Yao J, Huang J, *et al.* (2005). ABAD enhances Abeta-induced cell stress via mitochondrial dysfunction. *FASEB Journal* **19**, 597–8.

Tamagno E, Parola M, Guglielmotto M, *et al.* (2003). Multiple signaling events in amyloid beta-induced, oxidative stress-dependent neuronal apoptosis. *Free Radical Biology and Medicine* **35**, 45–58.

Terzi E, Holzemann G, and Seelig J. (1997). Interaction of Alzheimer beta-amyloid peptide(1–40). with lipid membranes. *Biochemistry* **36**, 14845–52.

Tieu K, Perier C, Vila M, *et al.* (2004). L-3-hydroxyacyl-CoA dehydrogenase II protects in a model of Parkinson's disease. *Annals of Neurology* **56**, 51–60.

Torroja L, Ortuno-Sahagun D, Ferrus A, Hammerle B, and Barbas JA (1998). *scully*, an essential gene of *Drosophila*, is homologous to mammalian mitochondrial type II L-3-hydroxyacyl-CoA dehydrogenase/amyloid-beta peptide-binding protein. *Journal of Cell Biology* **141**, 1009–17.

Veyssier-Belot C and Cacoub P (1999). Role of endothelial and smooth muscle cells in the physiopathology and treatment management of pulmonary hypertension. *Cardiovascular Research* **44**, 274–82.

Vitek MP, Bhattacharya K, Glendening JM, *et al.* (1994). Advanced glycation end products contribute to amyloidosis in Alzheimer disease. *Proceedings of the National Academy of Sciences of the USA* **91**, 4766–70.

Watson DJ, Lander AD, and Selkoe DJ (1997). Heparin-binding properties of the amyloidogenic peptides Abeta and amylin. Dependence on aggregation state and inhibition by Congo red. *Journal of Biological Chemistry* **272**, 31617–24.

Wild-Bode C, Yamazaki T, Capell A, *et al.* (1997). Intracellular generation and accumulation of amyloid beta-peptide terminating at amino acid 42. *Journal of Biological Chemistry* **272**, 16085–8.

Woodfield K, Ruck A, Brdiczka D, and Halestrap AP (1998). Direct demonstration of a specific interaction between cyclophilin-D and the adenine nucleotide translocase confirms their role in the mitochondrial permeability transition. *Biochemical Journal* **336**, 287–90.

Yaar M, Zhai S, Pilch PF, *et al.* (1997). Binding of beta-amyloid to the p75 neurotrophin receptor induces apoptosis: a possible mechanism for Alzheimer's disease. *Journal of Clinical Investigation* **100**, 2333–40.

Yamagishi S, Fujimori H, Yonekura H, Yamamoto Y, and Yamamoto H. (1998). Advanced glycation endproducts inhibit prostacyclin production and induce plasminogen activator inhibitor-1 in human microvascular endothelial cells. *Diabetologia* **41**, 1435–41.

Yan SD, Chen X, Schmidt AM, *et al.* (1994a). Glycated tau protein in Alzheimer disease: a mechanism for induction of oxidant stress. *Proceedings of the National Academy of Sciences of the USA* **91**, 7787–91.

Yan SD, Schmidt AM, Anderson GM, *et al.* (1994b). Enhanced cellular oxidant stress by the interaction of advanced glycation end products with their receptors/binding proteins. *Journal of Biological Chemistry* **269**, 9889–97.

Yan SD, Yan SF, Chen X, *et al.* (1995). Non-enzymatically glycated tau in Alzheimer's disease induces neuronal oxidant stress resulting in cytokine gene expression and release of amyloid beta-peptide. *Nature Medicine* **1**, 693–9.

Yan SD, Chen X, Fu J, *et al.* (1996). RAGE and amyloid-beta peptide neurotoxicity in Alzheimer's disease. *Nature* **382**, 685–91.

Yan SD, Fu J, Soto C., *et al*. (1997). An intracellular protein that binds amyloid-beta peptide and mediates neurotoxicity in Alzheimer's disease. *Nature* **389**, 689–95.

Yan SD, Roher A, Schmidt AM, and Stern DM (1999a). Cellular cofactors for amyloid beta-peptide-induced cell stress. Moving from cell culture to *in vivo*. *American Journal of Pathology* **155**, 1403–11.

Yan SD, Shi Y, Zhu A, *et al*. (1999b). Role of ERAB/L-3-hydroxyacyl-coenzyme A dehydrogenase type II activity in Abeta-induced cytotoxicity. *Journal of Biological Chemistry* **274**, 2145–56.

Yan SD, Zhu Y, Stern ED, *et al*. (2000a). Amyloid beta -peptide-binding alcohol dehydrogenase is a component of the cellular response to nutritional stress. *Journal of Biological Chemistry* **275**, 27100–9.

Yan SD, Roher A, Chaney M, Zlokovic B, Schmidt AM, and Stern D. (2000b). Cellular cofactors potentiating induction of stress and cytotoxicity by amyloid beta-peptide. *Biochimica Biophysica Acta* **1502**, 145–57.

Yan SD, Schmidt AM, and Stern D (2001). Alzheimer's disease: inside, outside, upside down. *Biochemical Society Symposium* 15–22.

Chapter 12

Alzheimer's disease as a neurotransmitter disease

Mitchell K.P. Lai, María J. Ramírez,
Shirley W.Y. Tsang, and Paul T. Francis

12.1 Introduction

Since the first reports of a profound loss of basal forebrain acetylcholine-synthesizing neurons and associated neurochemical perturbation of pre-synaptic cholinergic markers (Whitehouse *et al.* 1982; Wilcock *et al.* 1982), neurotransmitter system dysfunction and synapse loss have been recognized as hallmarks of Alzheimer's disease (AD) along with senile plaques and neurofibrillary tangles (Terry *et al.* 1991). Although it is generally accepted that dysregulation of the processing and metabolism of amyloid precursor protein (APP) play a key pathogenic role in AD and contribute towards the accumulation of β-amyloid (Aβ) fibrils and senile plaque formation (Selkoe 2001), clinical and preclinical studies point to neuronal loss and associated neurochemical alterations of several transmitter systems as the main factor underlying both cognitive and neuropsychiatric symptoms. The latter is collectively termed Behavioural and Psychological Symptoms of Dementia (BPSD) and encompasses a wide range of behaviours, including depression, psychosis, anxiety, and aberrant motor activity (IPA 1996). BPSD also occurs in other conditions, for example frontotemporal dementia (FTD) and Lewy Body dementia (LBD). It is not yet clear whether the neurochemical substrates of BPSD in these diseases are similar to AD. BPSD here refers to behavioural symptoms associated with AD.

BPSD occurs frequently over the course of AD and can be more distressing to care-givers than the cognitive decline, often leading to institutionalization of the patient (Steele *et al.* 1990; O'Donnell *et al.* 1992). However, pharmacological treatment of BPSDs is at present empirical and often causes troublesome side effects. Because blood concentrations of medications can quickly accumulate to toxic levels in AD patients, orthostatic hypotension falls and delirium develops even at low doses of many drugs. Therefore although neuroleptics such as risperidone are effective in reducing behavioural disturbances or disrupted sleep related to psychosis, they frequently cause extrapyramidal side effects as well as cardiovascular events, thus necessitating considerable caution in their use in AD (Folstein and Bylsma 1999; Ballard and Waite 2006). Similarly, although benzodiazepines are efficacious against anxiety disorders in non-demented subjects, they should be avoided in AD because they can further impair cognitive function (Costa and Guidotti 1996).

Occasionally, serendipitous effects of drugs on neuropsychiatric behaviours provide clues to novel neurochemical substrates of behaviours in AD. For example, clinical trials of cholinergic replacement therapies targeted at cognitive improvement have found unexpected improvements in certain BPSDs including psychosis (Cummings 2000; Cummings *et al.* 2000), which suggests that cholinergic alterations may be involved in both cognitive and neuropsychiatric features of AD. On the other hand, neurochemical alterations in behaviourally perturbed AD patients which parallel changes found in non-demented psychiatric subjects suggest common neurochemical substrates for the behaviours and similar pharmacotherapeutic strategies. For example, the post-mortem finding of losses of adrenergic, serotonergic, and dopaminergic neurons (D'Amato *et al.* 1987; Zweig *et al.* 1988; Zubenko *et al.* 1990, 1991) in depressed AD patients parallels findings of neurotransmitter depletion in non-demented depressed patients. This indicates that tricyclic antidepressants may be efficacious in AD patients with depression (Reifler *et al.* 1989). However, it is important to note that a pharmacological agent may act at a site that is independent of the neurochemical abnormality causing the behaviour yet produce changes in brain function that compensate for the abnormality. In addition, it is likely that BPSDs are associated with several neurochemical processes involving different neurotransmitter systems and occurring in multiple sites in the brain, thus giving further impetus for comprehensively characterizing neurochemical alterations in AD and studying how each of the neurochemical changes correlates with clinical symptoms. Over two decades of post-mortem and, more recently, neuroimaging studies with positron emission tomography (PET) or single-photon emission computed tomography (SPECT) using specific radioligands has yielded some insight into the nature of neurotransmitter disturbances in AD as well as their putative roles in the clinical and neuropathological features of the disease.

12.1.1 Transmitter systems are selectively affected in Alzheimer's disease

Results of early biopsy and post-mortem studies suggested that neurotransmitter systems are not uniformly affected in AD. Thus, while cholinergic, serotonergic, and glutamatergic deficits occur in moderate stages of AD, there is a relative sparing of catecholamine, GABAergic, and somatostatin neurons (Francis *et al.* 1993; Selden *et al.* 1994; Davies *et al.* 1998). This selective involvement of transmitter systems indicates that neurochemical changes do not reflect a generalized cortical atrophy but may follow disease-specific patterns of neurodegeneration. However, if left unchecked, the neurodegenerative process (as indicated by neurofibrillary tangle pathology) proceeds from the hippocampus and entorhinal cortex to involve increasingly the cortical association areas in a typical orderly topographical progression (Braak and Braak 1991; Delacourte *et al.* 1999), such that all major transmitter systems of the entire neocortex (with relatively minor effects on the occipital cortex) are affected. For example, more recent studies have reported variable dysfunction in the GABAergic system of AD which may in part be dependent on disease severity and BPSD (Lanctot *et al.* 2004; Garcia-Alloza *et al.* 2006). Taken together, these results suggest that neurotransmitter systems show selective vulnerability in earlier stages of AD but are universally affected in late stages of the disease.

12.1.2 Neurochemical profiling

Neurochemical changes in AD patients are not uniform, despite sharing general features like cholinergic denervation. This neurochemical variability is especially evident when considering levels of individual pre-and post-synaptic neurotransmitter receptors which are not only affected by synaptic or neuronal loss, but are also regulated by neuronal processes such as synaptic plasticity. Genetic factors such as the presence of functional polymorphisms, as well as environmental or disease factors like educational attainment or cerebrovascular disease, may also play a part in the patient's neurochemical state. Importantly, studies by our group as well as others indicate that particular neurochemical alterations are correlated with BPSD or faster cognitive decline. For example, loss of serotonin 5-HT_{1A} receptors in the temporal cortex seemed to associate with symptoms of aggression (Lai *et al.* 2003b), while reduced 5-HT_{1A} receptors in the hippocampus correlated with cognitive decline (Kepe *et al.* 2006). The finding that groups of AD patients who exhibited certain neurochemical changes are at higher risk of faster cognitive decline or developing BPSD compared to those who did not, led us and others to propose the concept of 'neurochemical profiling'. With further characterization of the neurochemical correlates of BPSD as well as the advent of safe specific ligands for *in vivo* imaging, neurochemical profiling of living patients as a tool to evaluate prognostic factors and risk of BPSD may be feasible in the future.

12.1.3 Neuropathological correlates of neurotransmitter dysfunction

It is increasingly clear that Aβ peptides adversely affect neurotransmission in certain transmitter systems. For example, Aβ has been shown to depress acetylcholine synthesis and interfere with cholinergic receptor signalling, and may impair retrograde transport of nerve growth factor to the basal nucleus, thus leading to loss of trophic support and hypofunction of the basal forebrain cholinergic system (Auld *et al.* 2002). These findings provide a pathophysiological basis linking deposition of Aβ with cholinergic deficits. However, Nitsch *et al.* (1992) reported that the activation of muscarinic receptors positively regulates APP processing and secretion in a manner which precludes Aβ formation. This non-amyloidogenic effect seems to be specific only to $G\alpha_q$-coupled muscarinic M_1 and M_3 receptors, a finding later corroborated by studies on $G\alpha_q$-coupled serotonin (5-HT_{2A} and 5-HT_{2C}) as well as metabotropic glutamate (mGluR1a) receptors (Nitsch *et al.* 1996, 1997). Cell-based and animal studies (Sadot *et al.* 1996; Genis *et al.* 1999) also found that activity on M_1 receptors decreased the phosphorylation of microtubule-associated tau, the hyperphosphorylated form of which is implicated in tangle formation and cell death in AD (Spillantini and Goedert 1998). Taken together, these studies suggest that dysfunction of neurotransmission may exacerbate the neuropathology of AD (Hellstrom-Lindahl 2000). Perry *et al.*(1978) reported that cholinergic depletion, as measured by loss of choline acetyltransferase (ChAT) activities, correlated with senile plaque counts, while Arendt *et al.* (1999) showed a strong association between cholinergic deficits and the cortical distribution of hyperphosphorylated tau in AD. Although these correlative studies cannot confirm the cause–effect relationships between cholinergic dysfunction and plaque and tangle formation, they do suggest that the cholinergic system, APP and tau interact during the AD process.

When assessing the evidence supporting a role of the disruption of particular neurotransmitter systems in BPSD and cognitive impairment, we take into consideration biological plausibility as determined by physiological functions of the transmitter system, reports of disruptions of the system in AD, and correlations with clinical features, as well as the reported and potential efficacy of specific ligands in treating BPSD or related neuropsychiatric conditions. In this chapter, we conceptualize neuronal death and neurotransmitter system dysfunction as the direct cause of clinical symptoms of AD, being influenced by genetic and environmental factors, and having reciprocal interactions with Aβ-containing senile plaques and hyperphosphorylated tau-containing neurofibrillary tangles (Fig. 12.1). Despite various limitations, this scheme represents an attempt to resolve in a biologically coherent manner the different neuropathological features of AD, as well as their translation into the clinical manifestations of the disease. Importantly, this scheme suggests that neurotransmitter system dysfunction and the plaque and tangle neuropathological hallmarks may aggravate one another in a vicious circle which eventually leads to AD occurrence or progression. In the subsequent sections we will provide a brief overview of the major neurotransmitter systems affected in AD and their involvement in the cognitive and behavioural features of the disease, as well as evidence of interactions with APP and tau biology. It should be noted that this chapter is focused on changes in the 'classical' neurotransmitter systems, where more research has been done and more is known. However, the potential relevance of more recently discovered neurotransmitters, such as neuropeptides, nitric oxide, and D-serine to the neurochemistry of AD needs to be studied further (Snyder and Ferris 2000).

12.2 The cholinergic system and Alzheimer's disease

12.2.1 Neuroanatomy and neurochemistry

Neurons which synthesize acetylcholine (ACh) in the brain are dispersed across nuclear groups in the basal forebrain cholinergic (BFC) projection system, which are arranged in a continuum on an antero-posterior axis (Záborsky *et al.* 1991). Axons arising from the BFC, including the basal nucleus and septum, form extensive connections throughout the neocortex and hippocampus. In addition, there are secondary circuits within the hippocampal formation as well as from the thalamus. Cholinergic terminals contain the enzyme choline acetyltransferase (ChAT) which catalyses the transfer of the acetyl group from acetyl coenzyme A to choline, resulting in the formation of ACh. Because of ChAT's specific localization in cholinergic terminals, measurement of brain ChAT activity has been used as an indicator of cholinergic innervation and reflects the state of cholinergic neurons. After synthesis, ACh is transported into storage vesicles in the nerve endings by the vesicular ACh transporter (VAChT). The rate of formation of ACh in cholinergic

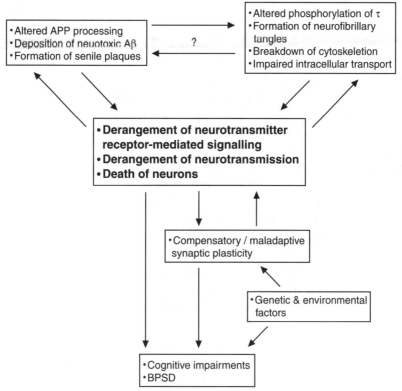

Fig. 12.1 A representative scheme for the interplay of neurochemical, neuropathological, and genetic/environmental factors in the pathogenesis of AD. The emphasis on neurotransmitter system dysfunction intentionally reflects the slant of this chapter and the generally accepted direct contribution of neurochemical abnormalities to clinical symptoms; However, the extent of interactions among neurotransmission, APP, and τ AD (evidence is particularly weak for the effects of τ phosphorylation on APP processing), and the relative importance of each in the pathogenesis of the disease, remain unclear. Genetic factors like presenilin mutations also affect APP but only in a small minority of familial AD cases and thus are not indicated here.

neurons is limited by the intracellular concentration of choline, which is determined in part by the Na$^+$-dependent high-affinity choline uptake (HACU) system (K_m = 1–5 µM). Because the brain concentration of choline is between 10 and 20 µM under physiological conditions, there should be adequate choline for sustained ACh synthesis even under conditions of high demand (Taylor and Brown 1999). In addition, ACh in the synaptic cleft is catabolized by cholinesterases, especially acetylcholinesterase (AChE). Therefore AChE plays a role in determining the intensity and duration of cholinergic neurotransmission, which has been implicated in a number of behavioural functions including arousal, sensory processing, learning. and memory (Taylor and Brown 1999).

ACh signalling in target neurons is mediated by two main classes of receptor, nicotinic and muscarinic, originally classified on the basis of the pharmacological activity of nicotine and muscarine long before the structures of these naturally occurring agonists were determined. Nicotinic receptors are heteropentameric ligand-gated cation channels which were first purified from the electric organ of *Torpedo* based on their high affinity for, and their inactivation by, snake α-bungarotoxin. Subsequently, nicotinic receptors have been isolated from neuromuscular junctions as well as from the central and peripheral nervous systems. In the brain, nicotinic receptors exist as combinations of α and β subunits, of which at least eight different sequences of α subunits and three different sequences of β subunits have been identified (Role and Berg 1996). This enables functional nicotinic receptors to have a potentially diverse range of sensitivity towards agonists.

The muscarinic receptors belong to a group of G-protein-associated seven-transmembrane-domain receptors whose signal transduction is mediated by intracellular second messengers. At present, at least five different muscarinic receptor genes have been cloned and sequenced (m_1–m_5), and the corresponding M_1–M_4 receptors have been identified pharmacologically, with M_5 being actively studied. The subtypes differ in their relative distributions in the brain as well as their ability to couple to different G proteins and hence to trigger signalling events. For example, activation of $Ga_{i/o}$-coupled M_2 and M_4 receptors leads to inhibition of adenylyl cyclase, while stimulation of $Ga_{q/11}$-coupled M_1, M_3, and M_5 receptors results in phosphoinositide hydrolysis and activation of phospholipase C (PLC) and protein kinase C (PKC) (Caulfield 1993). At present, the two subtypes which have been well studied in AD are the M_1 receptors, which predominate in the cerebral cortex and hippocampus, and the M_2 receptors, which are abundant in the central nervous system (CNS) as well as in cardiac tissues. M_1 receptors are widely thought to be post-synaptic heteroreceptors while M_2 receptors are mainly pre-synaptic autoreceptors. However, later studies using cholinergic deafferentation models found that M_2 receptors are also localized in peri- or extra-synaptic spaces, as well as on non-cholinergic terminals (Mrzljak *et al.* 1998), suggesting further regulatory roles for M2 receptors.

12.2.2 Cholinergic changes in Alzheimer's disease and associated clinical correlations

Abnormalities of the cholinergic system were among the earliest pathological findings in AD, and were initially limited to pre-synaptic components (Bowen *et al.* 1982, Davies and Maloney 1976). Therefore findings of losses of cholinergic neurons in the basal forebrain, as well as reduced ChAT, ACh, and AChE in the hippocampus and neocortex (summarized in Table 12.1), together with ample preclinical evidence of the involvement of cholinergic function in learning and behaviour, formed the basis of the 'cholinergic hypothesis of AD'(Bartus *et al.* 1982) which ascribed the clinical features of dementia to cholinergic deficits. Indeed, the cholinergic hypothesis formed the conceptual framework on which perturbations in a number of other transmitter systems are approached.

Initial findings of correlation between dementia severity and cholinergic markers (Table 12.1) supported the cholinergic hypothesis, and research into pharmacotherapeutic strategies which target the cholinergic system has led to the development of cholinesterase inhibitors (AChEIs) which increase availability of ACh in the synapse and which, until recently, have been the only class of pharmaceuticals approved for AD.

Table 12.1 Cholinergic changes in AD and their clinical correlates

	Changes in AD	Clinical correlates
Cholinergic neurons and nerve terminals	*Post-mortem* ↓ acetylcholine-producing neurons in the BFC system (Davies and Maloney 1976; Whitehouse *et al*. 1982)	*Post-mortem* Correlated with ↓ ChAT, thus indirectly with dementia severity (Wilcock *et al*. 1982)
ACh	*Biopsy* ↓ [^{14}C]ACh synthesis in FC and TC (Francis *et al*. 1985; Sims *et al*. 1983)	*Biopsy* ↓ [^{14}C]ACh synthesis correlates with cognitive impairment (Francis *et al*. 1985)
ChAT	*Post-mortem* ↓ widespread: cerebral cortex, HP, and basal nucleus (Davies 1979; Bowen *et al*. 1982; Rossor *et al*. 1984; Araujo *et al*. 1988) *Biopsy* ↓ neocortex (Sims *et al*. 1983)	*Post-mortem* ↓ ChAT correlated with dementia severity (Perry *et al*. 1978; Wilcock *et al*. 1982) but ↓ cerebral cortex in AD patients both with and without aggressive symptoms (Procter *et al*. 1992)[1]
AChE	*Post-mortem* ↓ (Perry *et al*. 1978; Davies 1979; Garcia-Alloza *et al*. 2005) *Biopsy* ↓ neocortex (Sims *et al*. 1983)	*Post-mortem* AChE/serotonin ratios correlated with pre-death cognitive scores (Garcia-Alloza *et al*. 2005)
HACU	*Post-mortem* ↓ FC and HP; ↔ caudate-putamen and cerebellum (Pascual *et al*. 1991)	Not reported
Muscarinic receptors		
M$_1$	*Post-mortem* ↔ FC and TC (Lai *et al*. 2001); ↔ cerebral cortex(Mash *et al*. 1985); ↔ cortical, subcortical regions, but modestly ↑ in HP and ST(Araujo *et al*. 1988); uncoupled from G-proteins (Flynn *et al*. 1991)	*Post-mortem* M$_1$–G-protein uncoupling correlated with dementia severity (Tsang *et al*. 2005)[2]
M$_2$	*Post-mortem* ↓ in cerebral cortex . (Mash *et al* 1985); ↓ all cortical areas and HP, but ↔ in subcortical areas (Araujo *et al*. 1988)	*Post-mortem* Delusion and hallucinations correlated with ↑ in FC and TC (Lai *et al*. 2001)[2]
M$_3$	Not reported	Not reported
M$_4$	*Post-mortem* ↓ in HP (Mulugeta *et al*. 2003)	Not reported
M$_5$	Not reported	Not reported
Nicotinic receptors	*Post-mortem* ↓ cortical areas and HP, but not in subcortical regions (Araujo et al. 1988)3	Not reported
(nAChR)	*Post-mortem* specific ↓ α4β2 populations in TC (Warpman and Nordberg 1995); ↓ α4, but ↔α3 and α7-containing nAChR in TC (Martin-Ruiz et al. 1999)	

↓, decreased; ↑, increased; ↔, no change; ACh, acetylcholine; AChE, acetylcholinesterase; ChAT, choline acetyltransferase; FC, frontal cortex; HACU, high-affinity choline uptake; HP, hippocampus; ST, striatum; TC, temporal cortex.

[1] Results may suggest that 5-HT deficits are more closely related than ACh to aggression in AD (Procter *et al*. 1992).

[2] Correlations with longitudinally assessed clinical data.

[3] Use of [^{3}H]carbamylcholine which has limited specificity for nAChR subtypes.

However, clinical observations of generally modest cognitive recovery, as well as unexpected improvements in non-cognitive behavioural aspects with AChEI treatment (Mega *et al.* 1999), led to a re-examination of the cholinergic hypothesis of AD, and the current thinking is that the cholinergic system may play a less direct role in cognitive processes, perhaps via the regulation of arousal and attention; instead, other transmitter systems (e.g. serotonergic and glutamatergic) may be involved (Francis *et al.* 1999) (see subsequent sections). However, cholinergic alterations may be important in a number of BPSDs (Cummings and Kaufer 1996). For example, alterations in ChAT activities and M_2 receptor levels have been associated with overactivity and psychosis, respectively (Minger *et al.* 2000; Lai *et al.* 2001). Nevertheless, the role of other muscarinic and nicotinic receptor subtypes in BPSD is not well studied, partly because of the lack of specific ligands.

As mentioned above, initial findings of cholinergic deficits were restricted to pre-synaptic components. Thus the selective reductions of M_2 receptors with relative preservation of putative post-synaptic M_1 receptors was believed to conform to this pre-synaptic pattern of cholinergic loss (Mash *et al.* 1985; Araujo *et al.* 1988). However, this postulate has now been challenged on two fronts. Firstly, the finding of M_2 heteroreceptors suggest that at least a proportion of M_2 may be lost through neurodegeneration of M_2-positive neurons rather than being due entirely to the loss of autoreceptors (Mrzljak *et al.* 1998). Secondly, although binding to M_1 receptors appeared intact, it was subsequently reported that M_1-mediated signalling may be impaired in AD since they were found to be uncoupled from G-proteins (Flynn *et al.* 1991). More recently, we found that the extent of M_1–G-protein uncoupling correlated with dementia severity in AD (Tsang *et al.* 2005). These data suggest that the post-synaptic component of the cholinergic system is abnormal in AD, and the breakdown in M_1-mediated intracortical signalling may underlie the limited efficacy of AChEI in general, and of M_1 agonists in particular (Mouradian *et al.* 1988; Greenlee *et al.* 2001), since the ability of ACh or agonists to ameliorate deficits in cholinergic neurotransmission would have been severely limited. Instead, rational therapeutic strategies point to targets downstream of the receptor, of which the successful use of the non-tumorigenic PKC activator, bryostatin 1, in an AD animal model is but one example (Etcheberrigaray *et al.* 2004).

12.2.3 The cholinergic system and Alzheimer's disease pathology

As mentioned above, one of the earliest reports of an interaction between cholinergic neurotransmission and Aβ came from Nitsch *et al.* (1992) (see also Fig. 12.2), who found that activation of $Ga_{q./11}$-coupled M_1 and M_3 receptors led to increased non-amyloidogenic secretion of APP. These cell-based studies have since been confirmed in hippocampal and cortical slices (Nitsch *et al.* 1993; Pittel *et al.* 1996), as well as *in vivo* using AD transgenic models (Caccamo *et al.* 2006). Furthermore, activation of PKC, a downstream effector of M_1 signalling, was shown to inhibit the production of Aβ (Hung *et al.* 1993), while in post-mortem studies losses of ChAT and AChE correlated with amyloid plaque counts (Perry *et al.* 1978). Interestingly, α_7-containing nicotinic receptors also seemed to mediate soluble APP release in a Ca^{2+}-influx-dependent manner, but without reducing Aβ

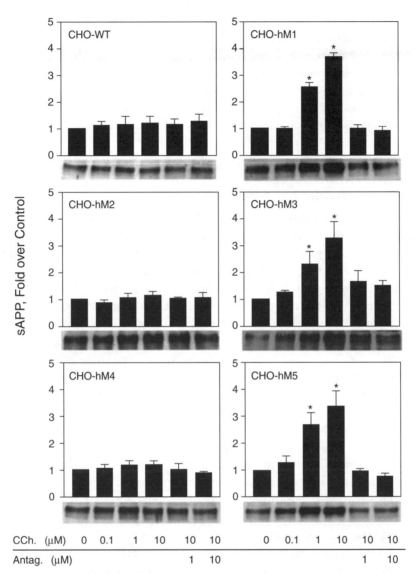

Fig. 12.2 Regulation of APP secretion by $Ga_{q/11}$-coupled muscarinic receptors. Bar graphs and representative immunoblots of endogenous APP secretion in wild-type Chinese hamster ovary (CHO) cells as well as in CHO cells stably transfected with human M_1, M_2, M_3, M_4, and M_5 receptors. After incubation with the agonist carbachol (CCh) in the presence or absence of a competing antagonist (Antag.) atropine sulphate, the media were collected and lyophilized, and the protein content was normalized with protein in the cell layer before immunoblotting with a polyclonal antibody directed against the N-terminal of APP. Values are expressed as fold change over control (no CCh) and represent the mean ± SD of three independent experiments. *Significantly different from control, $P < 0.05$.

production (Kim *et al.* 1997). Studies in transgenic mice also indicate that treatment with nicotine significantly reduces Aβ deposition (Hellstrom-Lindahl *et al.* 2004). In addition to affecting APP processing, stimulation of M_1 receptors also results in reductions of hyperphosphorylated tau in cultured cells and animal models (Genis *et al.* 1999; Sadot *et al.* 1996; Caccamo *et al.* 2006). These results suggested that restoring cholinergic neurotransmission in AD may also ameliorate senile plaque and neurofibrillary tangle formation. However, this postulate has not yet been borne out in clinical studies.

12.3 The serotonergic system and Alzheimer's disease

12.3.1 Neuroanatomy and neurochemistry

The serotonergic system is widespread throughout the brain, with most of the cell bodies of serotonergic neurons located in the dorsal and median raphe nuclei of the caudal midbrain (Palacios *et al.* 1990; Jacobs and Azmitia 1992). The neurons of these nuclei project widely over the thalamus, hypothalamus, basal ganglia, basal forebrain, and the entire neocortex. Serotonin (5-hydroxytryptamine, 5-HT) is synthesized from its precursor L-tryptophan in a process catalysed by the enzyme tryptophan hydroxylase. Once it has been released, 5-HT-mediated neurotransmission is terminated by its removal from the synaptic cleft through an active 5-HT transporter (5-HTT, or 5-HT reuptake sites), or subject to catalytic deactivation by the mitochondrial monoamine oxidase (MAO) to 5-hydroxyindoleacetic acid (5-HIAA), the main 5-HT metabolite in brain.

There are at least seven families of serotonergic receptors (Barnes and Sharp 1999), designated 5-HT_{1-7}. At present these have been subdivided into approximately 14 subtypes, and their number continues to grow. With the exception of the ionotropic 5-HT_3 receptors, 5-HT receptors are heptahelical G-protein-coupled receptors. The $5\text{-HT}_{1B/D}$ receptors are found largely pre-synaptically, while 5-HT_{1A} receptor exists in both a pre- and post-synaptic form. The remaining receptor subtypes are predominantly expressed post-synaptically. Converging evidence suggests that the administration of $5\text{-HT}_{2A/2C}$ and 5-HT_4 receptor agonists, or 5-HT_{1A}, 5-HT_3, and $5\text{-HT}_{1B/D}$ receptor antagonists prevents memory impairment and facilitates learning (Meneses 1999; Buhot *et al.* 2000). Additionally, these and other 5-HT receptors are known to be involved in a wide range of non-cognitive behaviours, including appetite, sexual behaviour, sleep–wake cycle, aggression, and mood (Lucki 1998; Barnes and Sharp 1999). Dysregulation of 5-HT function is believed to be involved in the pathogenesis of neuropsychiatric conditions such as obsessive–compulsive disorders, depression and suicidal behaviour, bipolar disorder, and schizophrenia, some of which are treated with serotonergic medications.

12.3.2 Serotonergic changes in Alzheimer's disease

Extensive serotonergic denervation has been demonstrated in AD, although its clinical signification has been only partially defined (Bowen *et al.* 1983; Chen *et al.* 2000). Reduction of 5-HT as well as its metabolite levels have been reported in post-mortem AD brains (Palmer *et al.* 1987a; Nazarali and Reynolds 1992; Garcia-Alloza *et al.* 2005).

In the temporal cortex, 5-HT reuptake sites are depleted to a similar extent as 5-HT and metabolites in patients with early AD, further suggesting a loss of serotonergic nerve terminals. As the raphe nucleus is a preferential site for neurofibrillary tangle formation and neuronal loss in AD (Curcio and Kemper 1984), 5-HT reductions in the cortex probably reflect loss of projections from the raphe nuclei. Interestingly, depletions of 5-HT and 5-HIAA were more severe in early-onset than late-onset AD brains (Arai *et al.* 1992). Similarly, there is an inverse relationship between age and [^{3}H]5-HT binding in AD patients, with similar levels in controls and AD subjects reached at approximately 85 years of age (Bowen *et al.* 1983). These data suggest that serotonergic dysfunction may be most prominent in AD patients with an earlier onset of disease, and that serotonergic changes occurring in both ageing and AD are degenerative and potentially additive.

5-HT receptors are selectively affected in AD, with the 5-HT$_1$ and 5-HT$_2$ receptor families generally showing deficits (although some studies report no changes), while 5-HT$_3$ and 5-HT$_4$ receptors seem to be unaffected (Table 12.2). Not much work has been done on the other 5-HT receptors, except for a recent study showing reduced 5-HT$_6$ receptors in the frontal and temporal cortex of AD patients (Garcia-Alloza *et al.* 2004). It has been argued that serotonergic changes are secondary to cholinergic changes; as 5-HT has an inhibitory effect on cholinergic neurons, decreases in the serotonergic tone may help maintain the cholinergic input in deficient cholinoceptive target areas. However, the pattern of preferential serotonergic depletion in the temporal and parietal cortex and hippocampus favours a disease pathogenesis model of retrograde degeneration of ascending 5-HT pathways (Palmer *et al.* 1987b).

12.3.3 Clinical correlations of serotonergic changes in Alzheimer's disease

Data on the clinical correlates of serotonergic alterations are summarized in Table 12.2. Alterations of specific 5-HT receptors seem to be correlated with BPSDs as well as cognitive decline. Interestingly, neuronal losses in the raphe nuclei do not by themselves correlate with behavioural or cognitive changes (Chen *et al.* 2000), suggesting a role of plasticity in the expression (both depletion and upregulation) of 5-HT receptors in the manifestation of clinical symptoms. For example, although deficits in both 5-HT$_{2A}$ receptors and 5-HTT have been reported in AD (Cross *et al.* 1984a; Jansen *et al.* 1990; Chen *et al.* 1996), the rate of cognitive decline correlated with 5-HT$_{2A}$ receptor loss (Lai *et al.* 2005), while anxiety correlated with a subgroup of patients whose 5-HTT densities were similar to those of controls (Tsang *et al.* 2003), suggesting that preserved levels of 5-HTT may act concomitant to pre-synaptic serotonergic deficits to exacerbate synaptic 5-HT depletion. It should be noted that, as with studies on the cholinergic system, much of the work done on the serotonergic system are post-mortem in nature, although there are an increasing number of reports on *in vivo* imaging in AD whose results generally concur with the post-mortem studies (see Table 12.2). Importantly, with the exception of 5-HT$_6$ receptors (Garcia-Alloza *et al.* 2004), little is known about the status of the more recently discovered 5-HT receptors (5-HT$_{5-7}$) in AD.

Table 12.2 Serotonergic changes in AD and their clinical correlates

	Changes in AD	Clinical correlates
Serotonergic neurons and nerve terminals	*Post-mortem* ↓ neurons in RN (Wilcock et al. 1988; Zweig et al. 1988; Aletrino et al. 1992; Chen et al. 2000)	*Post-mortem* No correlation with behavioural change or cognitive decline (Chen et al. 2000); delusions and hallucinations correlated with ↓ neurons in dorsal RN (Forstl et al. 1994)
5-HT	*Post-mortem* ↓ TC, HP, and ST (Middlemiss et al. 1986; Cross 1990; Reinikainen et al. 1990)	*Post-mortem* Rate of cognitive decline correlated with ↓ in FC (Lai et al. 2002); ↓ levels of 5-HT in cortical and subcortical regions in AD patients with psychotic behaviours compared with non-psychotic AD patients (Zubenko et al. 1991); ↔ TC in AD patients with psychosis (Lawlor et al. 1995); ↓ cortical levels of 5-HT in patients with aggression (Palmer et al. 1988)
5-HIAA	*Post-mortem* ↓ TC (Cross 1990)	*Post-mortem* Anxiety correlated with ↓ in cortex, thalamus and putamen (Reinikainen et al. 1990)
5-HTT	*Post-mortem* ↓ neocortex (Cross 1990, Chen et al. 1996)	*Post-mortem* Depression correlated with ↓ in FC and TC (Chen et al. 1996); ↓ in TC in non-anxious AD patients only, anxiety correlated with LL genotype of the 5-HTTLPR functional polymorphism (Tsang et al. 2003)[1]
5-HT₁ receptors		
1A	*Post-mortem* ↓ FC, TC, HP, and AMG (Bowen et al. 1983; Cross et al. 1984a; Crow et al. 1984; Middlemiss et al. 1986); ↔ FC (Sparks 1989)[2]	*Post-mortem* ↓ TC with moderate/severe aggression (Lai et al. 2003b); ↓ 5-HT to 5-HT₁A ratio in FC with higher rate of cognitive decline (Lai et al. 2002) *Neuroimaging* ↓ in HP and RN correlated with faster cognitive decline and ↓ in glucose utilization (Kepe et al. 2006)
1B,1D	*Post-mortem* ↓ FC and TC (Garcia-Alloza et al. 2004)	*Post-mortem* Cognitive decline correlated with ↓ in FC only (Garcia-Alloza et al. 2004)[3]
1E, 1F	Not reported	Not reported
5-HT2 receptors		
2A	*Post-mortem* ↓ neocortex (particularly TC), HP and AMG (Cross et al. 1984a; Crow et al. 1984; Jansen et al. 1990); ↔ in FC and HP(Dewar et al. 1990) *Neuroimaging* ↓ [¹⁸F]setoperone binding in FC,	*Post-mortem* Cognitive decline correlated with ↓ in TC (Lai et al. 2005); a functional polymorphism in 5-HT₂A receptor (102-T/C) correlated with the presence of visual and auditory hallucinations in AD patients (Holmes et al. 1998); ↔ in cerebral cortex in a group of patients without aggressive symptoms (Procter et al. 1992)

	TC, PC, OC in moderately to severely demented patients (Blin et al. 1993)[4]	
2B	Not reported	Not reported
2C	Not reported	Post-mortem A functional polymorphism in 5-HT$_{2C}$ receptor (Cys23Ser) correlated with visual hallucinations in AD patients (Holmes et al. 1998)
5-HT3 receptors	Post-mortem ↔ AMG and HP (Barnes et al. 1990)	Not reported[5]
5-HT4 receptors	Post-mortem ↔ FC and TC	None (Lai et al. 2003a)
5-HT5 receptors	Not reported	Not reported
5-HT6 receptors	Post-mortem ↓ FC and TC (Garcia-Alloza et al. 2004)	Post-mortem Overactivity correlated with ↓ in TC; aggression correlated with 5-HT$_6$:ChAT ratio in both FC and TC (Garcia-Alloza et al. 2004)[1]
5-HT7 receptors	Not reported	Not reported

↓, decreased; ↑, increased; ↔, no change; 5-HIAA, 5-hydroxyindoleacetic acid; 5-HT, 5-hydroxytryptamine; 5-HTT, 5-HT transporter; 5-HTTLPR, 5-HTT gene promoter region; AMG, amygdala; FC, frontal cortex; HP, hippocampus; OC, occipital cortex; PC, parietal cortex; RN, raphe nucleus; ST, striatum; TC, temporal cortex;

[1] Correlations with longitudinally assessed clinical data.

[2] May be due to cross-reactivity of [3H]5-HT for other receptor subtypes (Sparks 1989).

[3] ↓ 5-HT$_{1B/D}$ receptors could reflect serotonergic neuronal loss in AD, or possibly plasticity responses to a deteriorated cholinergic system as 5-HT$_{1B/D}$ receptors act as inhibitors of ACh release (Garcia-Alloza et al. 2004).

[4] Potential confounding by cross-reactivity to dopamine receptors (Blin et al. 1993).

[5] 5-HT$_3$ antagonist ondansetron also failed to improve cognition in AD in a controlled trial (Dysken et al. 2002).

12.3.4 Clinical efficacy of serotonergic compounds in Alzheimer's disease

Observations from clinical studies suggest that serotonergic dysfunction is probably of particular relevance to the neuropsychiatric aspects of AD, with important implications for therapeutic intervention of BPSD. Selective serotonin reuptake inhibitors (SSRIs) such as alaproclate, fluoxetine, citalopram, paroxetine, or sertraline have been tested. Treatment trials of AD patients with major depression using fluoxetine showed improvements in both depressive symptoms and cognitive function (Taragano *et al.* 1997). Similarly, citalopram administered to AD patients resulted in significant improvement in a range of behavioural symptoms, including confusion, anxiety, irritability, fear, depression, and motor restlessness (Nyth and Gottfries 1990). Interestingly, increased depression was reported by Lawlor *et al.* (1991) after administration of the 5-HT agonist meta-chlorophenylpiperazine (m-CPP) to affected AD patients. On the other hand, most of the new atypical antipsychotics have strong affinities for $5-HT_2$ and $5-HT_6$ receptors in addition to dopamine D_2 antagonism. Several studies have reported the efficacy of risperidone, clozapine, olanzapine, and quetiapine in treating agitation and psychosis in demented patients, although it is unclear whether these results are secondary to serotonergic or dopaminergic antagonism, since typical antipsychotics (e.g. haloperidol) have also been demonstrated to have significant benefit. Therefore there is some clinical evidence to support the SSRIs and other serotonergic agents in the management of BPSD. However, the possibility of cognitive effects of $5-HT_2$ antagonism from atypical antipsychotics needs to be studied in view of our finding that patients with decreased $5-HT_{2A}$ binding showed faster cognitive decline (Lai *et al.* 2005).

12.3.5 The serotonergic system and Alzheimer's disease pathology

Initial *in vitro* evidence of serotonergic regulation of APP processing was reported by Nitsch *et al.* (1996), who showed that 5-HT dose dependently increased APP release three- to fourfold in 3T3 cells stably overexpressing $5-HT_{2A}$ receptors or $5-HT_{2C}$ receptors. Pharmacological activation of $5-HT_{2C}$ receptors can also stimulate APP secretion in CSF and reduce $A\beta$ production *in vivo* (Arjona *et al.* 2002). Importantly, tricyclic antidepressants (TCAs) and SSRIs also increased APP secretion in primary rat neuronal cultures by more than threefold, the latter via a PKC-independent pathway (Pakaski *et al.* 2005). Given that depression is often comorbid with dementia, and indeed may be the presenting symptom in the clinical setting before the onset of obvious cognitive changes (Petry *et al.* 1988), it is interesting to speculate whether serotonergic deficits play a role in both depression and the AD process, and whether administration of TCA or SSRI will show both behavioural- and disease-modifying efficacies.

Serotonergic stimulation of non-amyloidogenic APP processing seems to be mediated via PKC/PLC-activating G_q-coupled receptors, similar to the cholinergic system (see above). However, different 5-HT receptors seem to regulate APP processing via different mechanisms, underlined by the involvement of phospholipase A2 (PLA2), but not PKC, in the regulation of sAPP by $5-HT_{2A}$ receptors; while both PLA2 and PKC are necessary for 5-HT$_{2C}$-receptor-mediated APP secretion (Nitsch *et al.* 1996). Additionally, activation of 5-HT$_4$ receptors which are coupled to G$_s$ proteins and stimulate adenylyl cyclase activity has also been shown to enhance APP secretion *in vitro* (Lezoualc'h and Robert 2003). Given that 5-HT$_4$ receptor levels seem to be preserved in AD (Lai *et al.* 2003a), these, together with 5-HT$_{2A}$ and 5-HT$_{2C}$ receptors, may be potential pharmacotherapeutic targets for AD. In post-mortem studies, reductions of 5-HT in the prosubiculum has been correlated with both increased density of senile plaques and psychosis (Zubenko *et al.* 1991). However, there is as yet little data on the relationship between alterations of 5-HT receptors and neuropathological hallmarks, and more work also needs to be done on assessing the long-term clinical effects of serotonergic ligands in AD.

12.4 The catecholaminergic system and Alzheimer's disease

12.4.1 Neuroanatomy and neurochemistry

Dopamine (DA) and noradrenaline (NA) are biogenic amine neurotransmitters of the catecholaminergic system which play critical functions in both the CNS and peripheral nervous system. Receptors for DA and NA are also found in other organs, including the kidney, liver, and spleen. Although DA and NA have closely related structures (with an additional –OH group in NA) and are found along the same biosynthetic pathway, neurons which synthesize DA and NA are located in disparate areas in the brain and are functionally separate interacting systems. A chief source of dopaminergic neurons is the midbrain substantia nigra with a major tract to the corpus striatum (the nigrostriatal tract). Discrete regions in the frontal cortex, nucleus accumbens, and anterior cingulate cortex also receive dopaminergic innervation. On the other hand, the locus coeruleus, which is anatomically part of the reticular formation, constitutes the source of noradrenergic neurons, with axons coursing anteriorly through the medial forebrain bundle to reach the entire cerebral cortex and hippocampus.

The rate-limiting enzyme in the biosynthesis of catecholamines is tyrosine hydroxylase (TH) which is found in both dopaminergic and noradrenergic neurons and catalyses the oxidation of L-tyrosine to 3,4-dihydroxy-L-phenylalanine (L-DOPA). L-DOPA is further decarboxylated by DOPA decarboxylase (also known as aromatic amino acid decarboxylase) to form DA which is stored in the vesicles of dopaminergic terminals. Neurons that synthesize NA contain another enzyme, dopamine β-hydroxylase, which then converts DA to NA. Within the terminals, catecholamines not taken up into the storage vesicles are inactivated by monoamine oxidase (MAO), which deaminates catecholamines to their respective aldehydes. After release into synapses, the action of catecholamines is terminated by diffusion or reuptake into pre-synaptic terminals via a family of 11–13-transmembrane-domained neuronal transporters: NA transporter (NET), DA transporter (DAT), and vesicular membrane transporter VMAT-2 (Amara and Kuhar 1993).

The actions of DA and NA are mediated by multiple DA and NA receptors, respectively. DA receptors are septahelical G-protein-coupled receptors which can be classified as D$_1$-like (D1 and D$_5$), which positively regulate adenylyl cyclase, and D$_2$-like (D$_2$, D$_3$, and D$_4$), which inhibit adenylyl cyclase and regulate the Ca^{2+} and K^+ channels. Three classes of adrenergic receptors exist (α_1, α_2, β), each of which has three receptor subtypes. Members of the α_1-adrenergic receptors (α_{1A}, α_{1B}, α_{1D}) are coupled to effector mechanisms regulating Ca^{2+} channels or inositol triphosphate (IP3), while the α_2 receptors (α_{2A}, α_{2B}, α_{2C}) all inhibit adenylyl cyclase. In contrast, the β-adrenergic receptors (β_1, β_2, β_3) all stimulate adenylyl cyclase.

Central catecholaminergic neurons are believed to have important roles in normal CNS functions and in neuropsychiatric conditions, including learning and memory, sleep–wake cycle regulation, reinforcement, psychosis, depression, and aggressive behaviours. Various psychotropic medications effective in treating these neuropsychiatric conditions have affinity for DA and NDA receptors (see below).

12.4.2 Catecholaminergic changes in Alzheimer's disease and associated clinical correlations

Disruptions of the dopaminergic and noradrenergic systems in AD as well as their cognitive or behavioural correlates are summarized in Tables 12.3 and 12.4, respectively. Generally, the dopaminergic neurons seem to be intact, but may be dysfunctional (Palmer and DeKosky 1993), while loss of TH-positive neurons in the locus coeruleus is evident. Patterns of receptor alterations are more complicated, with selective decreases and increases of DA and NA receptor subtypes in different regions of the brain, and may result from compensatory plasticity in response to pre-synaptic neurotransmitter deficits, or may be maladaptive and associated with BPSD (Tables 12.3 and 12.4). Conflicting results also occur, probably because of different methodologies and patient selection. Interestingly, certain polymorphisms of D_1 and D_3 receptors also render carriers with increased risk of having troublesome BPSD, while not affecting risk of developing AD itself (Holmes *et al.* 2001), suggesting that the disease process may overcome compensatory mechanisms for, or exacerbate, the altered receptor physiology resulting from such polymorphisms.

12.4.3 Clinical efficacy of catecholaminergic compounds in Alzheimer's disease

Psychotic symptoms and agitation in AD can be treated with typical antipsychotics which have antidopaminergic properties in addition to anticholinergic, antihistaminergic, and antinoradrenergic actions. However, the high risk of developing extrapyramidal side effects (including parkinsonism, dystonia, and akathisia) and tardive dyskinesia, especially with high-potency compounds such as haloperidol, limits their use in the geriatric population. In addition, the antinoradrenergic actions of typical antipsychotics can lead to orthostatic hypotension and falls, a side effect of particular concern for the elderly. Similarly, TCAs which inhibit reuptake of DA and NA are poorly tolerated by older people with organic brain diseases, including AD (Dubovsky and Buzan 2000). BPSDs such as depression, psychosis, and agitation have been successfully treated with alternative medications such as atypical antipsychotics and SSRIs (see above). However, some catecholaminergic compounds are still used routinely for dementia and other geriatric patients.

Table 12.3 Dopaminergic changes in AD and their clinical correlates

	Changes in AD	**Clinical correlates**
Dopaminergic neurons	*Post-mortem* dopaminergic neurons intact in AD (Palmer and DeKosky 1993)	*Post-mortem* Preservation of SN neurons was associated with aggression (Victoroff *et al.* 1996)
DA	*Post-mortem* ↓ caudate region (Nazarali and Reynolds 1992)	*Post-mortem* Psychotic behaviours and depression not correlated with DA when controlled for level of cognitive impairment (Zubenko *et al.* 1991; Bierer *et al.* 1993); ChAT:DA ratio in TC negatively correlated with aggression (Minger *et al.* 2000) *Neuroimaging* ↓ DA metabolism, measured with [^{18}F]-fluorodopa (FDOPA) influx predicted for lower MMSE scores (Itoh *et al.* 1994)
DAT	*Post-mortem* ↔ in ST (Piggott *et al.* 1999) *Neuroimaging* ↔ in ST of AD with parkinsonism (Ceravolo *et al.* 2004) and AD (O'Brien *et al.* 2004)	Not reported
DA receptors		
D1	*Post-mortem* ↓ in HP and putamen (Cortes *et al.* 1988); or ↔ in putamen (Cross *et al.* 1984b); ↔ in FC and TC (Matthews *et al.* 2002)	*Post-mortem* Dopamine DRD1 B2 allele associated with aggression and hallucinations (Holmes *et al.* 2001); ChAT: D1 ratio in TC negatively correlated with aggression (Minger *et al.* 2000)[1]
D2	*Post-mortem* ↓ in ST of AD-P but not in pure AD (Joyce *et al.* 1998); ↓ in putamen (Cross *et al.* 1984b) *Neuroimaging* ↓ in ST (Pizzolato *et al.* 1996); ↓ in HP (Kemppainen *et al.* 2003)	*Neuroimaging* Binding in HP correlated with verbal memory (Kemppainen *et al.* 2003)[2]; ↓ in ST correlated with more severe BPSD[3] (Tanaka *et al.* 2003)
D3	Unknown	Dopamine DRD3 1/1 genotype associated with delusions (Holmes *et al.* 2001)
D4	Unknown	Unknown

↓, decreased; ↑, increased; ↔, no change; AD-P, AD with parkinsonism; ChAT, choline acetyltransferase activity; DA, dopamine; DAT, DA transporter; FC, frontal cortex; HP, hippocampus; MMSE, Mini-Mental State Examination (Folstein *et al.* 1975); ST, striatum; TC, temporal cortex.

[1] Correlations with longitudinally assessed clinical data.

[2] The results suggest a role for temporal lobe D2 receptors in the memory and naming performance in AD.

[3] Patients with AD who manifest severe BPSD may have some dysfunction of striatal dopamine metabolism compared with those without BPSD.

Table 12.4 Noradrenergic changes in AD and their clinical correlates

	Changes in AD	Clinical correlates
Noradrenergic neurons	*Post-mortem* ↓ LC neurons (Hoogendijk *et al.* 1999; Matthews *et al.* 2002); ↓ LC neurons with ↑ TH mRNA in remaining neurons indicating compensatory changes (Szot *et al.* 2006)	*Post-mortem* Loss of LC neurons correlated with depression (Zubenko and Moossy 1988; Zweig *et al.* 1988; Chan-Palay and Asan 1989) or not correlated with depression (Hoogendijk *et al.* 1999); loss of rostral LC neurons correlated with aggressive behaviours (Matthews *et al.* 2002)[1]
NA	*Post-mortem* ↓ cerebral cortex (Rossor *et al.* 1984); ↓ TC (Nazarali and Reynolds 1992); ↓ in many brain areas innervated by LC (Arai *et al.* 1984; Francis *et al.* 1985; Nazarali and Reynolds 1992)	*Post-mortem* ↓ cortical NA correlated with depression (Zubenko *et al.* 1990); preserved NA in the SN correlated with psychosis (Zubenko *et al.* 1991); ↓ NA in FC and TC correlated with cognitive impairment (Matthews 2002) *et al.*
NA receptors[2]		
α$_1$	*Post-mortem* ↓ in PFC but ↔ in HP, CB, putamen (Kalaria 1989); or ↓ HP, CB (Shimohama *et al.* 1986); ↔ α1A mRNA and ↓ α1D mRNA in HP (Szot *et al.* 2006)	Not reported
α$_2$	*Post-mortem* ↔ in FC and TC (Matthews *et al.* 2002); ↓ in PFC but ↔ in HP, TC, CB, putamen (Kalaria and Andorn 1991); ↓ NB (Shimohama *et al.* 1986); ↔ α2A mRNA and ↓ α2C mRNA in HP (Szot *et al.* 2006)	*Post-mortem* No correlations with BPSD (Matthews *et al.* 2002)[1]; ↑ ↑ (70%) in CB of aggressive patients compared with non-aggressive patients (Russo-Neustadt and Cotman 1997)
β$_1$	*Post-mortem* ↓ in PFC, ↑ in HP, ↓ in putamen, ↔ in CB (Kalaria *et al.* 1989); disruption of β$_1$ receptor–G-protein coupling in the TC (Cowburn *et al.* 1993)	*Post-mortem* ↑ in the CB of aggressive patients compared with non-aggressive patients (Russo-Neustadt and Cotman 1997)[3]
β$_2$	*Post-mortem* ↑ in PFC, ↑ in HP, ↔ in putamen, ↔ in CB (Kalaria *et al.* 1989)	*Post-mortem* ↑ in CB of aggressive patients compared with non-aggressive patients (Russo-Neustadt and Cotman 1997)[3]

↓, decreased; ↑, increased; ↔, no change; CB, cerebellum; FC, frontal cortex; HP, hippocampus; LC, locus coeruleus; NA, noradrenaline; NB, nucleus basalis; OC, occipital cortex; PFC, prefrontal cortex; SN, substantia nigra; TC, temporal cortex; TH, tyrosine hydroxylase.

[1] Correlations with longitudinally assessed clinical data.

[2] NA receptors: α$_1$ is the major post-synaptic subtype, while α$_2$ is found pre- and post-synaptically.

[3] Magnitude of increase for β$_1$ and β$_2$ receptors was less than that reported for α$_2$ receptors (Russo-Neustadt and Cotman 1997).

For example, monoamine oxidase inhibitors have been shown to be as effective as TCAs in treating geriatric depression (Georgotas *et al.* 1986).

12.4.4 The catecholaminergic system and Alzheimer's disease pathology

At present, there are no reports of *in vitro* or *in vivo* regulation of APP processing or tau phosphorylation by dopaminergic or noradrenergic receptors.

12.5 The glutamatergic system and Alzheimer's disease

12.5.1 Neuroanatomy and neurochemistry

Glutamate is the primary excitatory neurotransmitter of the central nervous system, and is used by approximately two-thirds of synapses in the neocortex and hippocampus while ACh is found in perhaps 5% (Fonnum 1984). Therefore glutamate is involved in all aspects of cognition and higher mental function. In particular, normal (physiological) glutamate stimulation of N-methyl-D-aspartate (NMDA) receptors is essential to learning and memory processes, mainly through long-term potentiation (LTP), a form of synaptic strengthening with repeated use. NMDA receptors are $Ca^{2+}/K^+/Na^+$ channels formed from tetrameric complexes of NR1 and one of NR2A/2B/2C/2D subunits (Ishii *et al.* 1993). Apart from the agonist binding site, NMDA receptors contain allosteric glycine and polyamine recognition sites which act to potentiate NMDA receptor currents (Scatton 1993). In addition, two other classes of ionotropic glutamate receptors, kainate (KA) and α-amino-3-hydroxy-5-methylisoxazole-4-propionic acid (AMPA) receptors function in conjunction with NMDA receptors to mediate and regulate fast excitatory neurotransmission. There are also eight subtypes of G-protein-coupled metabotropic glutamate receptors ($mGluR_{1-8}$) divided into three groups, with the group I receptors ($mGluR_1$, $mGluR_5$) positively coupled to PKC and PLC, while group II receptors ($mGluR_{2-3}$) and group III receptors ($mGluR_4$, $mGluR_{6-8}$) inhibit adenylyl cyclase and downregulate intracellular cAMP levels. Instead of direct modulation of channel currents, mGluRs induce long term changes in glutamatergic neurotransmission via the activation of second messengers and structural changes or plasticity at synapses. In addition to dementia, mGluRs are implicated in a range of psychiatric or neurological conditions such as pain, anxiety, Parkinson's disease, and schizophrenia (Shipe *et al.* 2005).

12.5.2 Glutamatergic changes in Alzheimer's disease

Pyramidal neuron loss is a feature of AD (Francis *et al.* 1993; Morrison and Hof 1997), and this can be seen reflected in loss of wet weight and total protein of cortical regions (Najlerahim and Bowen 1988a,b). Glutamatergic pyramidal neurons account for many of the neurons lost in the cerebral cortex and hippocampus in AD (Greenamyre *et al.* 1988; Morrison and Hof 1997). Similarly, a large decrease in glutamate receptors has been observed in the hippocampus (Greenamyre *et al.* 1987; Procter *et al.* 1989) and cortex (Greenamyre *et al.* 1985) (Fig. 12.3) of the brains of AD patients presumably due in part to the accompanying neuronal loss described above. Subsequent reports of subunit-specific

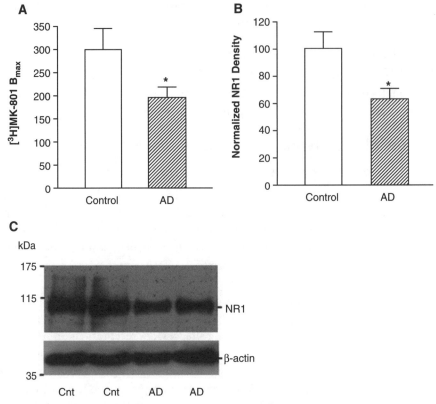

Fig. 12.3 Loss of NMDA receptors in AD. Bar graphs of (a) [³H]MK-801 binding densities B_{max} (fmol/mg protein) and (b) optical densities from immunoblots of the NR1 subunit of NMDA receptors (normalized as percentage of control values) for a cohort of aged controls ($n = 12$) and AD subjects ($n = 18$). Error bars indicate SEM. (c) Representative NR1 immunoblot (top lane) from brain membrane homogenates was stripped and reblotted with antibody against β-actin (bottom lane) for controls (Cnt) and AD. Molecular weight markers in kilodaltons (kDa) are as indicated. *Significantly different from control, $P < 0.05$ (Student's t-test).

abnormalities in the expression and pharmacological properties of NMDA receptors (Procter *et al.* 1989; Hynd *et al.* 2001, 2004) and mGluR (Lee *et al.* 2004; Albasanz *et al.* 2005) suggest that these changes do not simply reflect generalized cortical atrophy, but may indicate disease-specific processes. Abnormalities in AMPA receptor expression have also been observed in the AD entorhinal cortex and dentate gyrus (Armstrong *et al.* 1994; Wakabayashi *et al.* 1999), while KA receptors seem to be unaffected (Cowburn *et al.* 1989). However, since all biochemical data are expressed relative to either wet weight or total protein, unexpected findings can occur later in the disease process in severely affected regions. Markers of extrinsic neurons, such as ChAT activity for cholinergic neurons, are reduced (as they represent only a small proportion of the neuropil) in the temporal and other cortical regions in pathologically severe cases; markers of pyramidal neurons and their synapses do not always follow this pattern, perhaps because such structures contribute significantly to total protein. This appears to be the case in our own studies of β-tubulin and synaptophysin since relative to total protein, neither was reduced in temporal cortex from AD patients (Kirvell *et al.* 2006). Similarly, the glial marker GFAP was also unaltered in this region, although it was increased in a region (parietal cortex) which is less severely affected, even at the endstage of the disease.

In addition to the consequences of glutamatergic cell loss, there is evidence for dysfunction in remaining neurons. For example, although the relative concentration of the major glial glutamate transporter (EAAT2 equivalent to rat GLT-1, responsible for 90–95% of glutamate uptake) was unaffected in any of the regions studied (Beckstrom *et al.* 1999), the ability of glial cells to remove glutamate from the synaptic cleft (the main method of neurotransmitter inactivation following release) was impaired in several brain regions including the temporal cortex (Procter *et al.* 1988). This impairment of functional activity may be due to the action of free radicals on GLT-1 (Keller *et al.* 1997). Similarly there was a reduction in the vesicular glutamate transporter (VGLUT1) in the parietal but not the temporal cortex (Kirvell *et al.* 2006) and the activity of this protein was lower in AD temporal cortex than in controls (Westphalen *et al.* 2003). Again, this suggests an alteration in the activity of the protein by an as yet unknown mechanism.

12.5.3 Clinical correlations of glutamatergic changes in Alzheimer's disease

As a consequence of these alterations in proteins or protein function we propose (Francis 2003) that in AD there is inadequate removal of glutamate in the synaptic cleft between the pre- and post-synaptic neuron, creating an excessive level of background 'noise' at glutamate receptors within the synapse and adversely affecting the ability of the NMDA receptor to generate LTP (Danysz *et al.* 2000). Under normal resting conditions the NMDA receptor channel is blocked by Mg^{2+} ions in a voltage-dependent manner and only becomes responsive to glutamate binding when the membrane is depolarized. As a consequence of higher background concentrations of glutamate in the synaptic cleft, the membrane tends to be more, or more frequently, depolarized, with the result that the Mg^{2+} blockade is less efficient and the role of the NMDA as a coincidence detector, capable of generating LTP, is impaired. It is proposed that this disruption of glutamatergic neurotransmission may contribute to cognitive impairment in AD (Francis 2003). Furthermore, when taken to the extreme, excessive (pathological) glutamate acting at all classes of glutamate receptors can cause neuronal death and is likely to be a major factor in stroke (Greenamyre *et al.* 1988; Danysz *et al.* 2000). It is likely that such excitotoxicity, while not the major cause of cell death, is a contributory factor in AD. However, although atrophy of cortical grey matter (which contains a preponderance of glutamatergic neurons) correlated with the rate of cognitive decline (Mungas *et al.* 2002), there is as yet no report of a direct correlation between neurochemical alterations of glutamatergic markers and cognitive decline in AD. The possible involvement of NMDA or other glutamate receptor changes in BPSD is also not well studied. However, recent clinical studies suggest that memantine reduces agitation in people with dementia (Areosa *et al.* 2005; Gauthier *et al.* 2005). Therefore further studies should be conducted in this area.

12.5.4 Clinical efficacy of glutamatergic compounds in Alzheimer's disease

Several approaches to correcting glutamatergic dysfunction in AD have been attempted, including positive modulation of both AMPA and NMDA receptors. AMPAKines, which are considered to work by increasing the sensitivity of these receptors, are currently in clinical trial for mild cognitive impairment (Johnson and Simmon 2002). Modulation of the NMDA receptor has been attempted via the glycine co-agonist site with clear indications in preclinical studies that the partial agonist D-cycloserine improved learning and memory (Myhrer and Paulsen 1997). Clinical studies have suggested some benefit, but full-scale trials have not been initiated (Schwartz *et al.* 1996). There is currently no evidence that these drugs enhance excitotoxicity.

Perhaps the most surprising development is the success of the non-competitive NMDA antagonist memantine in clinical trials in moderate and severe AD (Reisberg *et al.* 2003). One would normally consider that such an approach, blockade of a receptor that would normally be activated in learning and memory, would be counter-intuitive. However, there is evidence that this molecule acts like magnesium ions, able to prevent background activation of the NMDA receptor ('noise') whilst allowing activation of this receptor for LTP formation (Chen *et al.* 1992; Danysz *et al.* 2000; Francis 2003). Since AChE inhibitors are likely to act in part by increasing glutamate release ('signal'), any benefit from their use in conjunction with memantine may be hypothesized to come from the combination of a reduction in glutamate 'noise' by memantine and an increase in discrete glutamate signals by donepezil (Francis 2003).

12.5.5 The glutamatergic system and Alzheimer's disease pathology

Excitotoxic cell death involves excess activation of receptors, leading to raised intracellular Ca^{2+} and consequent activation of a cascade of enzymes, resulting in cell death by necrosis or apoptosis (Lipton 1999). During the 1980s it was also suggested that endogenous glutamate could accumulate and become excitotoxic, perhaps as a result of impaired clearance (as a consequence of disrupted transporter function or indirectly in conditions of reduced energy availability). There is some evidence that energy levels may be reduced in AD because of perturbed mitochondrial function (Francis *et al.* 1993), and considerable evidence for oxidative damage of proteins including the glutamate transporter (Keller *et al.* 1997). Others have cautioned that there is no simple relationship between raised extracellular glutamate concentrations and cell death *in vivo*. It remains possible that changes in numbers of glutamate receptors or changes in ion selectivity may lead over time to cell death. For instance, the large numbers of calcium-permeable AMPA receptors present on basal forebrain cholinergic neurones may be linked to their loss in AD (Ikonomovic and Armstrong 1996).

A further possible contribution links reduced glutamatergic neurotransmission to cell death. Activation of receptors linked to phospholipase C (such as some $Ga_{q/11}$-coupled metabotropic glutamate receptors) has been shown to increase the secretion of neuroprotective forms of APP and decrease Aβ while at the same time reducing the phosphorylation state of tau (Nitsch *et al.* 1997). If glutamate neurotransmission is reduced as a consequence of tangle formation, one may hypothesize that Aβ production may increase and tau become more hyperphosphorylated in neurons innervated by the affected neuron (Francis *et al.* 1999). A recent study also suggests that exposure to Aβ in cells can lead to internalization of NMDA receptors and this may provide another mechanism for observed reductions in binding to NMDA receptors in AD (Snyder *et al.* 2005). Taken together, these data suggest that, as with the other transmitter systems, glutamatergic changes may interact with Aβ and tau in a pathophysiologically relevant manner during the AD process.

12.6 The GABAergic system and Alzheimer's disease

12.6.1 Neuroanatomy and neurochemistry

γ-Aminobutyric acid (GABA) is a major inhibitory neurotransmitter in the CNS, with GABAergic interneurons exerting their inhibitory effect on cholinergic, dopaminergic, and serotonergic neurons via the activation of $GABA_A$, $GABA_B$, and $GABA_C$ receptors, with the last of these receptor classes being found only in the retina (Decker and McGaugh 1991; Zorumski and Isenberg 1991; Bormann 2000). GABA is synthesized from glutamate by glutamic acid decarboxylase (GAD) and is metabolized by GABA-transaminase. $GABA_A$ receptors are ligand-gated channels which mediate Cl^- influx, leading to membrane hyperpolarization and inhibition of neuronal excitation (Zorumski and Isenberg 1991). $GABA_A$ receptors can be assembled from a pentameric combination of six α, four β, and three γ subunits, in addition to δ, ε, and θ subunits (Mehta and Ticku 1999). Therefore many permutations of $GABA_A$ receptor assemblies can theoretically be derived from the subunit combinations, but previous studies have indicated that functional receptor assemblies seem to be composed primarily of α, β, and one of γ, d, or ε subunits (Mehta and Ticku 1999). The specific composition of subunits determines the presence of regulatory sites for benzodiazepines, barbiturates, neurosteroids, and ethanol, which in turn affect the functions and pharmacological characteristics of particular receptor populations (Zorumski and Isenberg 1991). For example, benzodiazepine (BZ) binding sites are located at the interface of α and γ subunits (Zezula *et al.* 1996), with $α_1$-containing receptors mediating the sedative and amnesic properties of diazepam (Low *et al.* 2000; McKernan *et al.* 2000), while $α_2$- or $α_3$-containing receptors mediate diazepam's anxiolytic and alcohol-potentiating effects (Rudolph *et al.* 1999; Low *et al.* 2000), and $α_4$ mediates the withdrawal properties of the neurosteroid allopregnanolone (Smith *et al.* 1998). Because of these pharmacological properties, $GABA_A$ receptors and their regulatory sites are targeted by pharmaceuticals to treat anxiety disorder, seizure, and alcohol withdrawal.

In contrast to $GABA_A$ receptors, which are mainly post-synaptic and are widely distributed in the brain (Zorumski and Isenberg 1991), $GABA_B$ receptors are found in relatively low levels both pre- and post-synaptically, and mediate slow inhibitory potentials via signalling to adenylyl cyclase as well as Ca^{2+} and K^+ channels. Interestingly, functional $GABA_B$ receptors are obligate heterodimers and may play a role in cognitive processes as well as in depression and anxiety (Marshall 2005).

12.6.2 GABAergic changes in Alzheimer's disease and associated clinical correlations

Both post-mortem and neuroimaging studies show that GABAergic dysfunction occurs in AD, with reported changes in neurotransmitter levels, $GABA_A$ receptors, and associated binding sites, although the nature and extent of dysfunction may be quite variable (Table 12.5). However, although certain GABAergic compounds have demonstrable efficacy in BPSD (see below), few studies other than that of Garcia-Alloza *et al.* (2006) have correlated GABAergic alterations to clinical features. Such studies may be especially important for their potential utility in identifying patients who are likely to benefit from GABAergic pharmacotherapies, since the variability of GABAergic alterations also implies large variations in the efficacy, tolerability, and severity of side effects in response to GABAergic medications (Lanctot *et al.* 2004). At present, the status of BZ sites does not seem to correlate with cognitive functions in AD (Ohyama *et al.* 1999). Furthermore, neuroimaging studies suggest that BZ sites are generally preserved in the frontal and temporal cortices, areas often shown to be affected in post-mortem studies (Table 12.5). Instead, BZ sites are more consistently reduced in the parietal cortex in neuroimaging studies. Since these studies are usually conducted on subjects with less advanced disease, such differences may reflect disease-stage-specific GABAergic changes (Lanctot *et al.* 2004).

12.6.3 Clinical efficacy of GABAergic compounds in Alzheimer's disease

A variety of GABAergic agents are used routinely as anxiolytics, anticonvulsants, and sedatives, but only a few studies have evaluated their efficacy in treating BPSD. Many clinically useful agents target various binding sites on GABA receptors, including those for GABA, BZ, barbiturates, and neurosteroids, each of which may mediate inhibitory effects of GABAergic neurons on other neurotransmitter systems. A major concern with using compounds such as benzodiazepines on psychogeriatric patients is the rapid accumulation of these compounds in the blood because of altered pharmacokinetic and pharmacodynamic characteristics, resulting in troublesome side effects as well as having an adverse impact on cognitive function (Costa and Guidotti 1996). Therefore more clinical studies to evaluate the efficacy of novel shorter-acting compounds in treating BPSD are clearly indicated. Table 12.6 provides a brief overview on the current utility of various GABAergic agents in treating BPSD.

12.6.4 The GABAergic system and Alzheimer's disease pathology

There is at present no evidence of a direct correlation between GABAergic markers and neuropathological features in AD. However, it is possible that GABAergic neuronal function may indirectly affect APP processing and tau phosphorylation via inhibitory effects on cholinergic and serotonergic neurotransmission, systems which are known to be related to amyloid plaques and neurofibrillary tangles (see above). More research is clearly needed in this area.

Table 12.5 GABAergic changes in AD and their clinical correlates

	Changes in AD	Clinical correlates
GABAergic neurons	*Post-mortem* ↔ FC and HP (Yew *et al.* 1999); ↔ putamen (Cross *et al.* 1984b)	Not reported
GAD	*Post-mortem* ↔ in AD brain (Reinikainen *et al.* 1988); or ↓ only in MB and HP (Davies 1979)	Not reported
GABA	*Post-mortem* ↓ neocortex, HP, CG (Rossor *et al.* 1984; Perry *et al.* 1987; Lowe *et al.* 1988; Perry *et al.* 1987); or ↔ HP(Tarbit *et al.* 1980; Ellison *et al.* 1986)	*Post-mortem* ↓ in FC correlated with depression (Garcia-Alloza *et al.* 2006)
GABA$_A$ receptors		
GABA binding site	*Post-mortem* ↔ in neocortex (Garcia-Alloza *et al.* 2006)[1]; or ↓ FC and HP (Chu *et al.* 1987a,b)[2]	*Post-mortem* ↑ [^{3}H]muscimol binding to GABA$_A$ correlated with depression (Garcia-Alloza *et al.* 2006)[3]
BZ binding site	*Post-mortem* ↓[^{3}H]flunitrazepam binding in FC, TC, and HP (Shimohama *et al.* 1988; Carlson *et al.* 1993); or ↔ TC (Garcia-Alloza *et al.* 2006)[1] and OC (Carlson *et al.* 1993) *Neuroimaging* ↔ [^{11}C]flumazenil binding in FC, PC, and TC (Meyer *et al.* 1995, Ohyama *et al.* 1999)[4]; ↓[^{123}I]iomazenil binding in PC, ↔ FC, TC, and OC (Fukuchi *et al.* 1997)	*Post-mortem* ↑ [^{3}H]flunitrazepam binding to BZ sites correlated with depression (Garcia-Alloza *et al.* 2006)[3] *Neuroimaging* No correlations of BZ sites with MMSE scores (Ohyama *et al.* 1999)
GABA$_B$ receptors	*Post-mortem* ↔ neocortex (Garcia-Alloza *et al.* 2006)[5]; or ↓ FC and HP (Chu *et al.* 1987a,b)[2]	None (Garcia-Alloza *et al.* 2006)

↓, decreased; ↑, increased; ↔, no change; GABA, γ-aminobutyric acid; BZ, benzodiazepine; CG, cingulate gyrus; FC, frontal cortex; GAD, glutamic acid decarboxylase; HP, hippocampus; MB, midbrain; MMSE, Mini-Mental State Examination (Folstein *et al.* 1975); OC, occipital cortex; PC, parietal cortex; TC, temporal cortex.

[1] Lack of reduction in receptor densities may be explained by a high proportion of AD subjects with depression (Garcia-Alloza *et al.* 2006).

[2] Measured by [^{3}H]GABA binding under specific cold blockers.

[3] Correlations with longitudinally assessed clinical data.

[4] Reduced blood to brain transport (K_1) evident in regions under study (Meyer *et al.* 1995; Ohyama *et al.* 1999)

[5] Measured by [^{3}H]CGP54626 binding.

12.7 **Concluding remarks**

In this chapter we have demonstrated that deficit and dysfunction of neurotransmitter systems are hallmarks of AD, and probably play important roles in the cognitive and behavioural features of the disease. This assertion comes from three lines of evidence: the involvement of various neurotransmitter systems in cognitive and behavioural processes as gleaned from preclinical studies and work on non-dementia conditions, post-mortem and neuroimaging findings of neurochemical abnormalities in AD, and reversal or amelioration of certain neurochemical abnormalities by pharmacological means which also improves some of the cognitive and behavioural symptoms. Furthermore, there is some

Table 12.6 Clinical efficacy of GABAergic compounds in dementia

	Cognitive function	**BPSD**
GABA receptor ligands		
THIP	*CS* No effect on cognition using six neuropsychological tests (Mohr *et al*. 1986)	Not reported
BZ site ligands		
Diazepam	Not reported	*RCT* ↓ agitation and non-psychotic behaviours but efficacy < thioridazine; high risk of developing tolerance (Kirven and Montero 1973; Covington 1975)[1]
Lorazepam	Not reported	*RCT* ↓ agitation with efficacy ↔ alprazolam but with more side effects (Ancill *et al*. 1991)[1]
Alprazolam[2]	Not reported	*RCT* ↓ agitation with efficacy ↔ lorazepam but with fewer side effects (Ancill *et al*. 1991)[1] *RCT* ↓ BPSD with efficacy ↔ haloperidol (Christensen and Benfield 1998)[1]
Non-BZ hypnotics Zolpidem	Not reported	*CS* ↓ sleep disturbances and night-time wandering (Shelton and Hocking 1997)[1]
Anticonvulsants Gabapentin	*CS* Improved cognition (Hawkins *et al*. 2000)[1,3]	*CS* ↓ agitation and aggression for 21/22 patients with few side effects over 24 months (Hawkins *et al*. 2000)[1] *CS* ↓ BPSD and caregiver stress with few side effects (Moretti *et al*. 2003)

↓, decreased; ↑, increased; ↔, similar; THIP, 4,5,6,7-tetrahydroisoxazolo [5,4,-c]pyridin-3-ol; BZ, benzodiazepine; CS, case study or case series; RCT, randomized controlled trial.

[1] Subjects had dementia but no formal diagnosis of AD had been made.

[2] Short- to intermediate-acting BZ agonist.

[3] Likely to be related to behavioural improvement (Hawkins *et al*. 2000).

indication that neurochemical abnormalities may exacerbate disease processes concerning aberrant APP processing or tau phosphorylation, although strong evidence for this is not yet available. However, this simplified assertion is complicated by the fact that not all systems degenerate uniformly, ranging from extensive losses of basal forebrain cholinergic neurons and pre-synaptic cholinergic markers, to relative preservation of GABAergic neurons and minimal changes to GABAergic markers. Why do certain neurotransmitter systems, but not others, show particular vulnerability in AD? Future work attempting to connect neurotransmitter system pathology with other pathological features of AD is likely to yield further insights into the disease mechanisms. In addition, at present it is unclear why certain patients exhibit particular neurochemical abnormalities but others do not. Are such variations a reflection of biological differences (e.g. genetic polymorphisms), environmental factors (e.g. years of education), or disease heterogeneity (AD subtypes)? Would successful prediction of therapeutic response or adverse side effects, or prognosis based on individual patterns of neurochemical abnormalities detected *in vivo* be feasible in the future? Clearly, more work is needed to elucidate further the nature and effects of neurotransmitter system dysfunctions, a feature which occupies a prominent position in landscape of AD pathogenesis.

Acknowledgements

Studies by the authors described in this chapter were supported by the Wellcome Trust in the UK, the Secretaría de Estado de Educación y Universidades and the Fondo Social Europeo in Spain, and the Department of Clinical Research, Singapore General Hospital, the National Medical Research Council, and the Biomedical Research Council in Singapore. MKPL would like to thank H. H. Ching for helpful discussions.

References

Albasanz JL, Dalfo E, Ferrer I, and Martin M (2005). Impaired metabotropic glutamate receptor/phospholipase C signaling pathway in the cerebral cortex in Alzheimer's disease and dementia with Lewy bodies correlates with stage of Alzheimer's-disease-related changes. *Neurobiology of Disease* 20, 685–93.

Aletrino MA, Vogels OJ, Van Domburg PH, and Ten Donkelaar HJ (1992). Cell loss in the nucleus raphes dorsalis in Alzheimer's disease. *Neurobiology of Aging* 13, 461–8.

Amara SG and Kuhar MJ (1993). Neurotransmitter transporters: recent progress. *Annual Review of Neuroscience* 16, 73–93.

Ancill RJ, Carlyle WW, Liang RA, and Holliday SG (1991). Agitation in the demented elderly: a role for benzodiazepines? *International Clinical Psychopharmacology* 6, 141–6.

Arai H, Kosaka K, and Iizuka R (1984). Changes of biogenic amines and their metabolites in postmortem brains from patients with Alzheimer-type dementia. *Journal of Neurochemistry* 43, 388–93.

Arai H, Ichimiya Y, Kosaka K, Moroji T, and Iizuka R (1992). Neurotransmitter changes in early- and late-onset Alzheimer-type dementia. *Progress in Neuro-Psychopharmacology and Biological Psychiatry* 16, 883–90.

Araujo,D.M., Lapchak,P.A., Robitaille,Y., Gauthier,S., and Quirion,R. (1988) Differential alteration of various cholinergic markers in cortical and subcortical regions of human brain in Alzheimer's disease. *Journal of Neurochemistry* 50, 1914–1923.

Arendt T, Holzer M, Gertz HJ, and Bruckner MK (1999). Cortical load of PHF-tau in Alzheimer's disease is correlated to cholinergic dysfunction. *Journal of Neural Transmission* **106**, 513–23.

Areosa SA, Sherriff F, and McShane R (2005). Memantine for dementia. *Cochrane Database of Systematic Reviews* CD003154.

Arjona AA, Pooler AM, Lee RK, and Wurtman RJ (2002). Effect of a 5-HT$_{2C}$ serotonin agonist, dexnor-fenfluramine, on amyloid precursor protein metabolism in guinea pigs. *Brain Research* **951**, 135–40.

Armstrong DM, Ikonomovic MD, Sheffield R, and Wenthold RJ (1994). AMPA-selective glutamate receptor subtype immunoreactivity in the entorhinal cortex of non-demented elderly and patients with Alzheimer's disease. *Brain Research* **639**, 207–16.

Auld DS, Kornecook TJ, Bastianetto S, and Quirion R (2002). Alzheimer's disease and the basal forebrain cholinergic system: relations to β-amyloid peptides, cognition, and treatment strategies. *Progress in Neurobiology* **68**, 209–45.

Ballard C and Waite J (2006). The effectiveness of atypical antipsychotics for the treatment of aggression and psychosis in Alzheimer's disease. *Cochrane Database of Systematic Reviews* CD003476.

Barnes NM and Sharp T (1999). A review of central 5-HT receptors and their function. *Neuropharmacology* **38**, 1083–152.

Barnes NM, Costall B, Naylor RJ, Williams TJ, and Wischik CM (1990). Normal densities of 5-HT$_3$ receptor recognition sites in Alzheimer's disease. *Neuroreport* **1**, 253–4.

Bartus RT, Dean RL III, Beer B, and Lippa AS (1982). The cholinergic hypothesis of geriatric memory dysfunction. *Science* **217**, 408–14.

Beckstrom H, Julsrud L, Haugeto O, *et al.* (1999) Interindividual differences in the levels of the glutamate transporters GLAST and GLT, but no clear correlation with Alzheimer's disease. *Journal of Neuroscience Research* **55**, 218–29.

Bierer LM, Knott PJ, Schmeidler JM, *et al.* (1993). Post-mortem examination of dopaminergic parameters in Alzheimer's disease: relationship to noncognitive symptoms. *Psychiatry Research* **49**, 211–17.

Blin J, Baron JC, Dubois B, *et al.* (1993). Loss of brain 5-HT$_2$ receptors in Alzheimer's disease. *In vivo* assessment with positron emission tomography and [^{18}F]setoperone. *Brain* **116**, 497–510.

Bormann J (2000). The 'ABC' of GABA receptors. *Trends in Pharmacological Sciences* **21**, 16–19.

Bowen DM, Benton JS, Spillane JA, Smith CC, and Allen SJ (1982). Choline acetyltransferase activity and histopathology of frontal neocortex from biopsies of demented patients. *Journal of the Neurological Sciences* **57**, 191–202.

Bowen DM, Allen SJ, Benton JS, *et al.* (1983). Biochemical assessment of serotonergic and cholinergic dysfunction and cerebral atrophy in Alzheimer's disease. *Journal of Neurochemistry* **41**, 266–72.

Braak H and Braak E (1991). Neuropathological stageing of Alzheimer-related changes. *Acta Neuropathologica* **82**, 239–59.

Buhot MC, Martin S, and Segu L (2000). Role of serotonin in memory impairment. *Annals of Medicine* **32**, 210–21.

Caccamo A, Oddo S, Billings LM *et al.* (2006). M$_1$ receptors play a central role in modulating AD-like pathology in transgenic mice. *Neuron* **49**, 671–82.

Carlson MD, Penney JB, Jr, and Young AB (1993). NMDA, AMPA, and benzodiazepine binding site changes in Alzheimer's disease visual cortex. *Neurobiology of Aging* **14**, 343–52.

Caulfield MD (1993). Muscarinic receptors: characterization, coupling and function. *Pharmacology and Therapeutics* **58**, 319–379.

Ceravolo R, Volterrani D, Gambaccini G, *et al.* (2004). Presynaptic nigro-striatal function in a group of Alzheimer's disease patients with parkinsonism: evidence from a dopamine transporter imaging study. *Journal of Neural Transmission* **111**, 1065–1073.

Chan-Palay V and Asan E (1989). Alterations in catecholamine neurons of the locus coeruleus in senile dementia of the Alzheimer type and in Parkinson's disease with and without dementia and depression. *Journal of Comparative Neurology* **287**, 373–92.

Chen CP, Alder JT, Bowen DM, *et al.* (1996). Presynaptic serotonergic markers in community-acquired cases of Alzheimer's disease: correlations with depression and neuroleptic medication. *Journal of Neurochemistry* **66**, 1592–8.

Chen CP, Eastwood SL, Hope T, McDonald B, Francis PT, and Esiri MM (2000). Immunocytochemical study of the dorsal and median raphe nuclei in patients with Alzheimer's disease prospectively assessed for behavioural changes. *Neuropathology and Applied Neurobiology* **26**, 347–55.

Chen HS, Pellegrini JW, Aggarwal SK, *et al.* (1992). Open-channel block of N-methyl-D-aspartate (NMDA) responses by memantine: therapeutic advantage against NMDA receptor-mediated neurotoxicity. *Journal of Neuroscience* **12**, 4427–36.

Christensen DB and Benfield WR (1998). Alprazolam as an alternative to low-dose haloperidol in older, cognitively impaired nursing facility patients. *Journal of the American Geriatrics Society* **46**, 620–5.

Chu DC, Penney JB, Jr, and Young AB (1987a). Cortical $GABA_B$ and $GABA_A$ receptors in Alzheimer's disease: a quantitative autoradiographic study. *Neurology* **37**, 1454–9.

Chu DC, Penney JB, Jr, and Young AB (1987b). Quantitative autoradiography of hippocampal $GABA_B$ and $GABA_A$ receptor changes in Alzheimer's disease. *Neuroscience Letters* **82**, 246–52.

Cortes R, Probst A, and Palacios JM (1988). Decreased densities of dopamine D1 receptors in the putamen and hippocampus in senile dementia of the Alzheimer type. *Brain Research* **475**, 164–7.

Costa E and Guidotti A (1996). Benzodiazepines on trial: a research strategy for their rehabilitation. *Trends in Pharmacological Sciences* **17**, 192–200.

Covington JS (1975). Alleviating agitation, apprehension, and related symptoms in geriatric patients: A double-blind comparison of a phenothiazine and a benzodiazepien. *Southern Medical Journal* **68**, 719–24.

Cowburn RF, Hardy JA, Briggs RS, and Roberts PJ (1989). Characterisation, density, and distribution of kainate receptors in normal and Alzheimer's diseased human brain. *Journal of Neurochemistry* **52**, 140–7.

Cowburn RF, Vestling M, Fowler CJ, Ravid R, Winblad B, and O'Neill C (1993). Disrupted β_1-adrenoceptor-G protein coupling in the temporal cortex of patients with Alzheimer's disease. *Neuroscience Letters* **155**, 163–6.

Cross AJ (1990). Serotonin in Alzheimer-type dementia and other dementing illnesses. *Annals of the New York Academy of Sciences* **600**, 405–15.

Cross AJ, Crow TJ, Ferrier IN, Johnson JA, Bloom SR, and Corsellis JA (1984a). Serotonin receptor changes in dementia of the Alzheimer type. *Journal of Neurochemistry* **43**, 1574–81.

Cross AJ, Crow TJ, Ferrier IN, Johnson JA, and Markakis D (1984b). Striatal dopamine receptors in Alzheimer-type dementia. *Neuroscience Letters* **52**, 1–6.

Crow TJ, Cross AJ, Cooper SJ, *et al.* (1984). Neurotransmitter receptors and monoamine metabolites in the brains of patients with Alzheimer-type dementia and depression, and suicides. *Neuropharmacology* **23**, 1561–9.

Cummings JL (2000). Cholinesterase inhibitors: a new class of psychotropic compounds. *American Journal of Psychiatry* **157**, 4–15.

Cummings JL and Kaufer D (1996). Neuropsychiatric aspects of Alzheimer's disease: the cholinergic hypothesis revisited. *Neurology* **47**, 876–83.

Cummings JL, Donohue JA, and Brooks RL (2000). The relationship between donepezil and behavioral disturbances in patients with Alzheimer's disease. *American Journal of Geriatric Psychiatry* **8**, 134–40.

Curcio CA and Kemper T (1984). Nucleus raphe dorsalis in dementia of the Alzheimer type: neurofibrillary changes and neuronal packing density. *Journal of Neuropathology and Experimental Neurology* **43**, 359–68.

D'Amato RJ, Zweig RM, Whitehouse PJ, *et al.* (1987). Aminergic systems in Alzheimer's disease and Parkinson's disease. *Annals of Neurology* **22**, 229–36.

Danysz W, Parsons CG, and Quack G (2000). NMDA channel blockers: memantine and amino-aklylcyclohexanes: *in vivo* characterization. *Amino Acids* **19**, 167–72.

Davies P (1979). Neurotransmitter-related enzymes in senile dementia of the Alzheimer type. *Brain Research* **171**, 319–27.

Davies P and Maloney AJ (1976). Selective loss of central cholinergic neurons in Alzheimer's disease. *Lancet* **ii**, 1403.

Davies P, Anderton B, Kirsch J, Konnerth A, Nitsch R, and Sheetz M (1998). First one in, last one out: the role of gabaergic transmission in generation and degeneration. *Progress in Neurobiology* **55**, 651–8.

Decker MW and McGaugh JL (1991). The role of interactions between the cholinergic system and other neuromodulatory systems in learning and memory. *Synapse* **7**, 151–68.

Delacourte A, David JP, Sergeant N, *et al.* (1999). The biochemical pathway of neurofibrillary degeneration in aging and Alzheimer's disease. *Neurology* **52**, 1158–65.

Dewar D, Graham DI, and McCulloch J (1990). 5 HT$_2$ receptors in dementia of Alzheimer type: a quantitative autoradiographic study of frontal cortex and hippocampus. *Journal of Neural Transmission. Parkinson's Disease and Dementia Section* **2**, 129–37.

Dubovsky SL and Buzan R (2000). Psychopharmacology. In Coffey CE and Cummings JL (eds), *Textbook of Geriatric Neuropsychiatry*, pp. 779–828 (American Psychiatric Press, Washington, DC).

Dysken M, Kuskowski M, and Love S (2002). Ondansetron in the treatment of cognitive decline in Alzheimer dementia. *American Journal of Geriatric Psychiatry* **10**, 212–15.

Ellison DW, Beal MF, Mazurek MF, Bird ED, and Martin JB (1986). A postmortem study of amino acid neurotransmitters in Alzheimer's disease. *Annals of Neurology* **20**, 616–21.

Etcheberrigaray R, Tan M, Dewachter I, *et al.* (2004). Therapeutic effects of PKC activators in Alzheimer's disease transgenic mice. *Proceedings of the National Academy of Sciences of the USA* **101**, 11141–6.

Flynn DD, Weinstein DA, and Mash DC (1991). Loss of high-affinity agonist binding to M_1 muscarinic receptors in Alzheimer's disease: implications for the failure of cholinergic replacement therapies. *Annals of Neurology* **29**, 256–62.

Folstein MF and Bylsma FW (1999). Noncognitive symptoms of Alzheimer disease. In Terry RD, Katzman R, Bick KL, and Sisodia SS (eds), *Alzheimer Disease*, pp. 25–37 (Lippincott–Williams & Wilkins, Philadelphia, PA).

Folstein MF, Folstein SE, and McHugh PR (1975). 'Mini-mental state': a practical method for grading the cognitive state of patients for the clinician. *Journal of Psychiatric Research* **12**, 189–98.

Fonnum F (1984). Glutamate: a neurotransmitter in mammalian brain. *Journal of Neurochemistry* **42**, 1–11.

Forstl H, Burns A, Levy R, and Cairns N (1994). Neuropathological correlates of psychotic phenomena in confirmed Alzheimer's disease. *British Journal of Psychiatry* **165**, 53–9.

Francis PT (2003). Glutamatergic systems in Alzheimer's disease. *International Journal of Geriatric Psychiatry* **18**, S15–21.

Francis PT, Palmer AM, Sims NR, *et al.* (1985). Neurochemical studies of early-onset Alzheimer's disease. Possible influence on treatment. *New England Journal of Medicine* **313**, 7–11.

Francis PT, Sims NR, Procter AW, and Bowen DM (1993). Cortical pyramidal neurone loss may cause glutamatergic hypoactivity and cognitive impairment in Alzheimer's disease: investigative and therapeutic perspectives. *Journal of Neurochemistry* **60**, 1589–1604.

Francis PT, Palmer AM, Snape M, and Wilcock GK (1999). The cholinergic hypothesis of Alzheimer's disease: a review of progress. *Journal of Neurology, Neurosurgery, and Psychiatry* **66**, 137–47.

Fukuchi K, Hashikawa K, Seike Y, *et al.* (1997). Comparison of iodine-123-iomazenil SPECT and technetium-99m-HMPAO-SPECT in Alzheimer's disease. *Journal of Nuclear Medicine* **38**, 467–470.

Garcia-Alloza M, Hirst WD, Chen CP, Lasheras B, Francis PT, and Ramirez MJ (2004). Differential involvement of 5-HT$_{1B/1D}$ and 5-HT$_6$ receptors in cognitive and non-cognitive symptoms in Alzheimer's disease. *Neuropsychopharmacology* **29**, 410–16.

Garcia-Alloza M, Gil-Bea FJ, ez-Ariza M, *et al.* (2005). Cholinergic–serotonergic imbalance contributes to cognitive and behavioral symptoms in Alzheimer's disease. *Neuropsychologia* **43**, 442–9.

Garcia-Alloza M, Tsang SW, Gil-Bea FJ, *et al.* (2006). Involvement of the GABAergic system in depressive symptoms of Alzheimer's disease. *Neurobiology of Aging* **27**, 1110-17.

Gauthier S, Wirth Y, and Mobius HJ (2005). Effects of memantine on behavioural symptoms in Alzheimer's disease patients: an analysis of the Neuropsychiatric Inventory (NPI) data of two randomised, controlled studies. *International Journal of Geriatrics Psychiatry* **20**, 459–64.

Genis I, Fisher A, and Michaelson DM (1999). Site-specific dephosphorylation of t of apolipoprotein E-deficient and control mice by M$_1$ muscarinic agonist treatment. *Journal of Neurochemistry* **72**, 206–13.

Georgotas A, McCue RE, Hapworth W, *et al.* (1986). Comparative efficacy and safety of MAOIs versus TCAs in treating depression in the elderly. *Biological Psychiatry* **21**, 1155–66.

Greenamyre JT, Penney JB, Young AB, D'Amato CJ, Hicks SP, and Shoulson I (1985). Alterations in L-glutamate binding in Alzheimer's and Huntington's diseases. *Science* **227**, 1496–9.

Greenamyre JT, Penney JB, D'Amato CJ, and Young AB (1987). Dementia of the Alzheimer's type: changes in hippocampal L-[^{3}H]glutamate binding. *Journal of Neurochemistry* **48**, 543–51.

Greenamyre JT, Maragos WF, Albin RL, Penney JB, and Young AB (1988). Glutamate transmission and toxicity in Alzheimer's disease. *Progress in Neuro-Psychopharmacology and Biological Psychiatry* **12**, 421–30.

Greenlee W, Clader J, Asberom T, *et al.* (2001). Muscarinic agonists and antagonists in the treatment of Alzheimer's disease. *Farmaco* **56**, 247–50.

Hawkins JW, Tinklenberg JR, Sheikh JI, Peyser CE, and Yesavage JA (2000). A retrospective chart review of gabapentin for the treatment of aggressive and agitated behavior in patients with dementias. *American Journal of Geriatric Psychiatry* **8**, 221–5.

Hellstrom-Lindahl E (2000). Modulation of β-amyloid precursor protein processing and tau phosphorylation by acetylcholine receptors. *European Journal of Pharmacology* **393**, 255–63.

Hellstrom-Lindahl E, Court J, Keverne J, *et al.* (2004). Nicotine reduces Aβ in the brain and cerebral vessels of APPsw mice. *European Journal of Neuroscience* **19**, 2703–10.

Holmes C, Arranz MJ, Powell JF, Collier DA, and Lovestone S (1998). 5-HT$_{2A}$ and 5-HT$_{2C}$ receptor polymorphisms and psychopathology in late onset Alzheimer's disease. *Human Molecular Genetics* **7**, 1507–1509.

Holmes C, Smith H, Ganderton R, *et al.* (2001). Psychosis and aggression in Alzheimer's disease: the effect of dopamine receptor gene variation. *Journal of Neurology, Neurosurgery, and Psychiatry* **71**, 777–9.

Hoogendijk WJ, Sommer IE, Pool CW, *et al.* (1999). Lack of association between depression and loss of neurons in the locus coeruleus in Alzheimer disease. *Archives of General Psychiatry* **56**, 45–51.

Hung AY, Haass C, Nitsch RM, *et al.* (1993). Activation of protein kinase C inhibits cellular production of the amyloid β-protein. *Journal of Biological Chemistry* **268**, 22959–62.

Hynd MR, Scott HL, and Dodd PR (2001). Glutamate(NMDA) receptor NR1 subunit mRNA expression in Alzheimer's disease. *Journal of Neurochemistry* **78**, 175–82.

Hynd MR, Scott HL, and Dodd PR (2004). Differential expression of N-methyl-D-aspartate receptor NR2 isoforms in Alzheimer's disease. *Journal of Neurochemistry* **90**, 913–19.

Ikonomovic MD and Armstrong DM (1996). Distribution of AMPA receptor subunits in the nucleus basalis of Meynert in aged humans: implications for selective neuronal degeneration. *Brain Research* **716**, 229–32.

IPA (1996). Behavioral and Psychological Signs and Symptoms of Dementia: Implications for Research and Treatment. Proceedings of an International Consensus Conference, Lansdowne, Virginia. *International Psychogeriatrics*, **8** (Suppl 3), 215–552.

Ishii T, Moriyoshi K, Sugihara H, *et al.* (1993). Molecular characterization of the family of the N-methyl-D-aspartate receptor subunits. *Journal of Biological Chemistry* **268**, 2836–43.

Itoh M, Meguro K, Fujiwara T, *et al.* (1994). Assessment of dopamine metabolism in brain of patients with dementia by means of [^{18}F]-fluorodopa and PET. *Annals of Nuclear Medicine* **8**, 245–251.

Jacobs BL and Azmitia EC (1992). Structure and function of the brain serotonin system. *Physiological Reviews* **72**, 165–229.

Jansen KL, Faull RL, Dragunow M, and Synek BL (1990). Alzheimer's disease: changes in hippocampal N-methyl-D-aspartate, quisqualate, neurotensin, adenosine, benzodiazepine, serotonin and opioid receptors: an autoradiographic study. *Neuroscience* **39**, 613–627.

Johnson SA and Simmon VF (2002) Randomized, double-blind, placebo-controlled international clinical trial of the Ampakine CX516 in elderly participants with mild cognitive impairment: a progress report. *Journal of Molecular Neuroscience* **19**, 197–200.

Joyce JN, Murray AM, Hurtig HI, Gottlieb GL, and Trojanowski JQ (1998). Loss of dopamine D2 receptors in Alzheimer's disease with parkinsonism but not Parkinson's or Alzheimer's disease. *Neuropsychopharmacology* **19**, 472–80.

Kalaria RN (1989). Characterization of [^{125}I]HEAT binding to α_1-receptors in human brain: assessment in aging and Alzheimer's disease. *Brain Research* **501**, 287–94.

Kalaria RN and Andorn AC (1991). Adrenergic receptors in aging and Alzheimer's disease: decreased a2-receptors demonstrated by [^{3}H]p-aminoclonidine binding in prefrontal cortex. *Neurobiology of Aging* **12**, 131–6.

Kalaria RN, Andorn AC, Tabaton M, Whitehouse PJ, Harik SI, and Unnerstall JR (1989). Adrenergic receptors in aging and Alzheimer's disease: increased β_2-receptors in prefrontal cortex and hippocampus. *Journal of Neurochemistry* **53**, 1772–81.

Keller JN, Mark RJ, Bruce AJ, *et al.* (1997). 4-Hydroxynonenal, an aldehydic product of membrane lipid peroxidation, impairs glutamate transport and mitochondrial function in synaptosomes. *Neuroscience* **80**, 685–96.

Kemppainen N, Laine M, Laakso MP, *et al.* (2003). Hippocampal dopamine D_2 receptors correlate with memory functions in Alzheimer's disease. *European Journal of Neuroscience* **18**, 149–54.

Kepe V, Barrio JR, Huang SC, *et al.* (2006). Serotonin 1A receptors in the living brain of Alzheimer's disease patients. *Proceedings of the National Academy of Sciences of the USA* **103**, 702–7.

Kim SH, Kim YK, Jeong SJ, Haass C, Kim YH, and Suh YH (1997). Enhanced release of secreted form of Alzheimer's amyloid precursor protein from PC12 cells by nicotine. *Molecular Pharmacology* **52**, 430–436.

Kirvell SL, Esiri MM, and Francis PT (2006). Down regulation of vesicular glutamate transporters precedes cell loss and pathology in Alzheimer's disease. *Journal of Neurochemistry*, **98**, 939–50.

Kirven LE and Montero EF (1973). Comparison of thioridazine and diazepam in the control of nonpsychotic symptoms associated with senility: double-blind study. *Journal of the American Geriatrics Society* **21**, 546–51.

Lai MK, Lai OF, Keene J, *et al.* (2001). Psychosis of Alzheimer's disease is associated with elevated muscarinic M_2 binding in the cortex. *Neurology* **57**, 805–11.

Lai MK, Tsang SW, Francis PT, *et al.* (2002). Postmortem serotoninergic correlates of cognitive decline in Alzheimer's disease. *Neuroreport* **13**, 1175–8.

Lai MK, Tsang SW, Francis PT, *et al.* (2003a). [³H]GR113808 binding to serotonin 5-HT₄ receptors in the postmortem neocortex of Alzheimer disease: a clinicopathological study. *Journal of Neural Transmission* **110**, 779–88.

Lai MK, Tsang SW, Francis PT, *et al.* (2003b). Reduced serotonin 5-HT$_{1A}$ receptor binding in the temporal cortex correlates with aggressive behavior in Alzheimer disease. *Brain Research* **974**, 82–7.

Lai MK, Tsang SW, Alder JT, *et al.* (2005) Loss of serotonin 5-HT$_{2A}$ receptors in the postmortem temporal cortex correlates with rate of cognitive decline in Alzheimer's disease. *Psychopharmacology* **179**, 673–7.

Lanctot KL, Herrmann N, Mazzotta P, Khan LR, and Ingber N (2004). GABAergic function in Alzheimer's disease: evidence for dysfunction and potential as a therapeutic target for the treatment of behavioural and psychological symptoms of dementia. *Canadian Journal of Psychiatry* **49**, 439–453.

Lawlor BA, Sunderland T, Mellow AM, Molchan SE, Martinez R, and Murphy DL (1991). A pilot placebo-controlled study of chronic m-CPP administration in Alzheimer's disease. *Biological Psychiatry* **30**, 140–4.

Lawlor BA, Ryan TM, Bierer LM, *et al.* (1995). Lack of association between clinical symptoms and postmortem indices of brain serotonin function in Alzheimer's disease. *Biological Psychiatry* **37**, 895–6.

Lee HG, Ogawa O, Zhu X, *et al.* (2004). Aberrant expression of metabotropic glutamate receptor 2 in the vulnerable neurons of Alzheimer's disease. *Acta Neuropathologica* **107**, 365–71.

Lezoualc'h F and Robert SJ (2003). The serotonin 5-HT₄ receptor and the amyloid precursor protein processing. *Experimental Gerontology* **38**, 159–66.

Lipton P (1999). Ischemic cell death in brain neurons. *Physiological Reviews* **79**, 1431–1568.

Low K, Crestani F, Keist R, *et al.* (2000). Molecular and neuronal substrate for the selective attenuation of anxiety. *Science* **290**, 131–4.

Lowe SL, Francis PT, Procter AW, Palmer AM, Davison AN, and Bowen DM (1988). γ-aminobutyric acid concentration in brain tissue at two stages of Alzheimer's disease. *Brain* **111**, 785–99.

Lucki I (1998). The spectrum of behaviors influenced by serotonin. *Biological Psychiatry* **44**, 151–62.

McKernan RM, Rosahl TW, Reynolds DS, *et al.* (2000). Sedative but not anxiolytic properties of benzodiazepines are mediated by the GABA$_A$ receptor α$_1$ subtype. *Nature Neuroscience* **3**, 587–92.

Marshall FH (2005). Is the GABA B heterodimer a good drug target? *Journal of Molecular Neuroscience* **26**, 169–76.

Martin-Ruiz CM, Court JA, Molnar E, *et al.* (1999). α$_4$ but not α$_3$ and α$_7$ nicotinic acetylcholine receptor subunits are lost from the temporal cortex in Alzheimer's disease. *Journal of Neurochemistry* **73**, 1635–40.

Mash DC, Flynn DD, and Potter LT (1985). Loss of M$_2$ muscarine receptors in the cerebral cortex in Alzheimer's disease and experimental cholinergic denervation. *Science* **228**, 1115–17.

Matthews KL, Chen CP, Esiri MM, Keene J, Minger SL, and Francis PT (2002). Noradrenergic changes, aggressive behavior, and cognition in patients with dementia. *Biological Psychiatry* **51**, 407–16.

Mega MS, Masterman DM, O'Connor SM, Barclay TR, and Cummings JL (1999). The spectrum of behavioral responses to cholinesterase inhibitor therapy in Alzheimer disease. *Archives of Neurology* **56**, 1388–93.

Mehta AK and Ticku MK (1999). An update on GABA$_A$ receptors. *Brain Research Reviews* **29**, 196–217.

Meneses A (1999). 5-HT system and cognition. *Neuroscience and Biobehavioral Reviews* **23**, 1111–25.

Meyer M, Koeppe RA, Frey KA, Foster NL, and Kuhl DE (1995). Positron emission tomography measures of benzodiazepine binding in Alzheimer's disease. *Archives of Neurology* **52**, 314–17.

Middlemiss DN, Palmer AM, Edel N, and Bowen DM (1986). Binding of the novel serotonin agonist 8-hydroxy-2-(di-n-propylamino) tetralin in normal and Alzheimer brain. *Journal of Neurochemistry* **46**, 993–6.

Minger SL, Esiri MM, McDonald B, *et al.* (2000). Cholinergic deficits contribute to behavioral disturbance in patients with dementia. *Neurology* **55**, 1460–7.

Mohr E, Bruno G, Foster N, *et al.* (1986). GABA-agonist therapy for Alzheimer's disease. *Clinical Neuropharmacology* **9**, 257–63.

Moretti R, Torre P, Antonello RM, Cazzato G, and Bava A (2003). Gabapentin for the treatment of behavioural alterations in dementia: preliminary 15-month investigation. *Drugs and Aging* **20**, 1035–40.

Morrison JH and Hof PR (1997). Life and death of neurons in the aging brain. *Science* **278**, 412–19.

Mouradian MM, Mohr E, Williams JA, and Chase TN (1988). No response to high-dose muscarinic agonist therapy in Alzheimer's disease. *Neurology* **38**, 606–8.

Mrzljak L, Levey AI, Belcher S, and Goldman-Rakic PS (1998). Localization of the m2 muscarinic acetylcholine receptor protein and mRNA in cortical neurons of the normal and cholinergically deafferented rhesus monkey. *Journal of Comparative Neurology* **390**, 112–32.

Mulugeta E, Karlsson E, Islam A, *et al.* (2003). Loss of muscarinic M_4 receptors in hippocampus of Alzheimer patients. *Brain Research* **960**, 259–62.

Mungas D, Reed BR, Jagust WJ, *et al.* (2002). Volumetric MRI predicts rate of cognitive decline related to AD and cerebrovascular disease. *Neurology* **59**, 867–73.

Myhrer T and Paulsen RE (1997). Infusion of D-cycloserine into temporal-hippocampal areas and restoration of mnemonic function in rats with disrupted glutamatergic temporal systems. *European Journal of Pharmacology* **328**, 1–7.

Najlerahim A and Bowen DM (1988a). Biochemical measurements in Alzheimer's disease reveal a necessity for improved neuroimaging techniques to study metabolism. *Biochemical Journal* **251**, 305–8.

Najlerahim A and Bowen DM (1988b). Regional weight loss of the cerebral cortex and some subcortical nuclei in senile dementia of the Alzheimer type. *Acta Neuropathologica* **75**, 509–12.

Nazarali AJ and Reynolds GP (1992). Monoamine neurotransmitters and their metabolites in brain regions in Alzheimer's disease: a postmortem study. *Cellular and Molecular Neurobiology* **12**, 581–7.

Nitsch RM, Slack BE, Wurtman RJ, and Growdon JH (1992). Release of Alzheimer amyloid precursor derivatives stimulated by activation of muscarinic acetylcholine receptors. *Science* **258**, 304–7.

Nitsch RM, Farber SA, Growdon JH, Wurtman RJ (1993). Release of amyloid β-protein precursor derivatives by electrical depolarization of rat hippocampal slices. *Proceedings of the National Academy of Sciences of the USA* **90**, 5191–3.

Nitsch RM, Deng M, Growdon JH, and Wurtman RJ (1996). Serotonin 5-HT_{2a} and 5-HT_{2c} receptors stimulate amyloid precursor protein ectodomain secretion. *Journal of Biological Chemistry* **271**, 4188–94.

Nitsch RM, Deng A, Wurtman RJ, and Growdon JH (1997). Metabotropic glutamate receptor subtype mGluR1alpha stimulates the secretion of the amyloid β-protein precursor ectodomain. *Journal of Neurochemistry* **69**, 704–12.

Nyth AL and Gottfries CG (1990). The clinical efficacy of citalopram in treatment of emotional disturbances in dementia disorders. A Nordic multicentre study. *British Journal of Psychiatry* **157**, 894–901.

O'Brien JT, Colloby S, Fenwick J, *et al.* (2004). Dopamine transporter loss visualized with FP-CIT SPECT in the differential diagnosis of dementia with Lewy bodies. *Archives of Neurology* **61**, 919–25.

O'Donnell BF, Drachman DA, Barnes HJ, Peterson KE, Swearer JM, and Lew RA (1992). Incontinence and troublesome behaviors predict institutionalization in dementia. *Journal of Geriatric Psychiatry and Neurology* **5**, 45–52.

Ohyama M, Senda M, Ishiwata K, *et al.* (1999). Preserved benzodiazepine receptors in Alzheimer's disease measured with C-11 flumazenil PET and I-123 iomazenil SPECT in comparison with CBF. *Annals of Nuclear Medicine* **13**, 309–15.

Pakaski M, Bjelik A, Hugyecz M, Kasa P, Janka Z, and Kalman J (2005). Imipramine and citalopram facilitate amyloid precursor protein secretion *in vitro*. *Neurochemistry International* **47**, 190–5.

Palacios JM, Waeber C, Hoyer D, and Mengod G (1990). Distribution of serotonin receptors. *Annals of the New York Academy of Sciences* **600**, 36–52.

Palmer AM and DeKosky ST (1993) Monoamine neurons in aging and Alzheimer's disease. *Journal of Neural Transmission* **91**, 135–59.

Palmer AM, Francis PT, Benton JS, *et al.* (1987a). Presynaptic serotonergic dysfunction in patients with Alzheimer's disease. *Journal of Neurochemistry* **48**, 8–15.

Palmer AM, Middlemiss DN, and Bowen DM (1987b). [^{3}H]8-OH-DPAT binding in Alzheimer's disease. An index of pyramidal cell loss? In Dourish C, Ahlenius A, and Hutson P (eds), *Brain 5-HT$_{1A}$ Receptors*, pp. 286–99 (Ellis Howood, Chichester).

Palmer AM, Stratmann GC, Procter AW, and Bowen DM (1988). Possible neurotransmitter basis of behavioral changes in Alzheimer's disease. *Annals of Neurology* **23**, 616–20.

Pascual J, Fontan A, Zarranz JJ, Berciano J, Florez J, and Pazos A (1991). High affinity choline uptake carrier in Alzheimer's disease: implications for cholinergic hypothesis of dementia. *Brain Research* **552**, 170–4.

Perry EK, Tomlinson BE, Blessed G, Bergmann K, Gibson PH, and Perry RH (1978). Correlation of cholinergic abnormalities with senile plaques and mental test scores in senile dementia. *British Medical Journal* **ii**, 1457–9.

Perry TL, Yong VW, Bergeron C, Hansen S, and Jones K (1987). Amino acids, glutathione, and glutathione transferase activity in the brains of patients with Alzheimer's disease. *Annals of Neurology* **21**, 331–6.

Petry S, Cummings JL, Hill MA, and Shapira J (1988). Personality alterations in dementia of the Alzheimer type. *Archives of Neurology* **45**, 1187–90.

Piggott MA, Marshall EF, Thomas N, *et al.* (1999). Striatal dopaminergic markers in dementia with Lewy bodies, Alzheimer's and Parkinson's diseases: rostrocaudal distribution. *Brain* **122**, 1449–68.

Pittel Z, Heldman E, Barg J, Haring R, and Fisher A (1996). Muscarinic control of amyloid precursor protein secretion in rat cerebral cortex and cerebellum. *Brain Research* **742**, 299–304.

Pizzolato G, Chierichetti F, Fabbri M, *et al.* (1996). Reduced striatal dopamine receptors in Alzheimer's disease: single photon emission tomography study with the D$_2$ tracer [^{123}I]-IBZM. *Neurology* **47**, 1065–8.

Procter AW, Palmer AM, Francis PT, *et al.* (1988). Evidence of glutamatergic denervation and possible abnormal metabolism in Alzheimer's disease. *Journal of Neurochemistry* **50**, 790–802.

Procter AW, Wong EH, Stratmann GC, Lowe SL, and Bowen DM (1989). Reduced glycine stimulation of [^{3}H]MK-801 binding in Alzheimer's disease. *Journal of Neurochemistry* **53**, 698–704.

Procter AW, Francis PT, Stratmann GC, and Bowen DM (1992). Serotonergic pathology is not widespread in Alzheimer patients without prominent aggressive symptoms. *Neurochemical Research*, **17**, 917–22.

Reifler BV, Teri L, Raskind M, *et al.* (1989) Double-blind trial of imipramine in Alzheimer's disease patients with and without depression. *American Journal of Psychiatry* **146**, 45–9.

Reinikainen KJ, Paljarvi L, Huuskonen M, Soininen H, Laakso M, and Riekkinen PJ (1988). A post-mortem study of noradrenergic, serotonergic and GABAergic neurons in Alzheimer's disease. *Journal of the Neurological Sciences* **84**, 101–16.

Reinikainen KJ, Soininen H, and Riekkinen PJ (1990). Neurotransmitter changes in Alzheimer's disease: implications to diagnostics and therapy. *Journal of Neuroscience Research* **27**, 576–86.

Reisberg B, Doody R, Stoffler A, Schmitt F, Ferris S, and Mobius HJ (2003). Memantine in moderate-to-severe Alzheimer's disease. *New England Journal of Medicine* **348**, 1333–41.

Role LW and Berg DK (1996). Nicotinic receptors in the development and modulation of CNS synapses. *Neuron* **16**, 1077–85.

Rossor MN, Iversen LL, Reynolds GP, Mountjoy CQ, and Roth M (1984). Neurochemical characteristics of early and late onset types of Alzheimer's disease. *British Medical Journal* **288**, 961–4.

Rudolph U, Crestani F, Benke D, *et al.* (1999) Benzodiazepine actions mediated by specific γ-aminobutyric acid$_A$ receptor subtypes. *Nature* **401**, 796–800.

Russo-Neustadt A and Cotman CW (1997). Adrenergic receptors in Alzheimer's disease brain: selective increases in the cerebella of aggressive patients. *Journal of Neuroscience* **17**, 5573–80.

Sadot E, Gurwitz D, Barg J, Behar L, Ginzburg I, and Fisher A (1996). Activation of m1 muscarinic acetylcholine receptor regulates tau phosphorylation in transfected PC12 cells. *Journal of Neurochemistry* **66**, 877–80.

Scatton B (1993). The NMDA receptor complex. *Fundamental and Clinical Pharmacology* **7**, 389–400.

Schwartz BL, Hashtroudi S, Herting RL, Schwartz P, and Deutsch SI (1996). D-Cycloserine enhances implicit memory in Alzheimer patients. *Neurology* **46**, 420–4.

Selden N, Geula C, Hersh L, and Mesulam MM (1994). Human striatum: chemoarchitecture of the caudate nucleus, putamen and ventral striatum in health and Alzheimer's disease. *Neuroscience* **60**, 621–36.

Selkoe DJ (2001). Alzheimer's disease: genes, proteins, and therapy. *Physiological Reviews* **81**, 741–66.

Shelton PS and Hocking LB (1997). Zolpidem for dementia-related insomnia and nighttime wandering. *Annals of Pharmacotherapy* **31**, 319–22.

Shimohama S, Taniguchi T, Fujiwara M, and Kameyama M (1986). Biochemical characterization of α-adrenergic receptors in human brain and changes in Alzheimer-type dementia. *Journal of Neurochemistry* **47**, 1295–1301.

Shimohama S, Taniguchi T, Fujiwara M, and Kameyama M (1988). Changes in benzodiazepine receptors in Alzheimer-type dementia. *Annals of Neurology* **23**, 404–6.

Shipe WD, Wolkenberg SE, Williams DL Jr, and Lindsley CW (2005). Recent advances in positive allosteric modulators of metabotropic glutamate receptors. *Current Opinion in Drug Discovery and Development* **8**, 449–57.

Sims NR, Bowen DM, Allen SJ, *et al.* (1983). Presynaptic cholinergic dysfunction in patients with dementia. *Journal of Neurochemistry* **40**, 503–9.

Smith SS, Gong QH, Hsu FC, Markowitz RS, ffrench-Mullen JM, and Li X (1998). GABA$_A$ receptor a4 subunit suppression prevents withdrawal properties of an endogenous steroid. *Nature* **392**, 926–30.

Snyder SH and Ferris CD (2000). Novel neurotransmitters and their neuropsychiatric relevance. *American Journal of Psychiatry* **157**, 1738–51.

Snyder EM, Nong Y, Almeida CG, *et al.* (2005). Regulation of NMDA receptor trafficking by amyloid-β. *Nature Neuroscience* **8**, 1051–8.

Sparks DL (1989). Aging and Alzheimer's disease: altered cortical serotonergic binding. *Archives of Neurology* **46**, 138–40.

Spillantini MG and Goedert M (1998). Tau protein pathology in neurodegenerative diseases. *Trends in Neuroscience* **21**, 428–33.

Steele C, Rovner B, Chase GA, and Folstein M (1990). Psychiatric symptoms and nursing home placement of patients with Alzheimer's disease. *American Journal of Psychiatry* **147**, 1049–51.

Szot P, White SS, Greenup JL, Leverenz JB, Peskind ER, and Raskind MA (2006). Compensatory changes in the noradrenergic nervous system in the locus ceruleus and hippocampus of postmortem subjects with Alzheimer's disease and dementia with Lewy bodies. *Journal of Neuroscience* **26**, 467–78.

Tanaka Y, Meguro K, Yamaguchi S, *et al* (2003) Decreased striatal D2 receptor density associated with severe behavioral abnormality in Alzheimer's disease. *Annals of Nuclear Medicine* **17**, 567–73.

Taragano FE, Lyketsos CG, Mangone CA, Allegri RF, and Comesana-Diaz E (1997). A double-blind, randomized, fixed-dose trial of fluoxetine vs. amitriptyline in the treatment of major depression complicating Alzheimer's disease. *Psychosomatics* **38**, 246–52.

Tarbit I, Perry EK, Perry RH, Blessed G, and Tomlinson BE (1980). Hippocampal free amino acids in Alzheimer's disease. *Journal of Neurochemistry* **35**, 1246–9.

Taylor P and Brown JH (1999). Acetylcholine. In Siegel GJ, Agranoff BW, Albers RW, Fisher SK, and Uhler MD (eds), *Basic Neurochemistry. Molecular, Cellular and Medical Aspects*, pp. 213–42 (Lippincott–Raven, Philadelphia, PA).

Terry RD, Masliah E, Salmon DP, *et al* (1991). Physical basis of cognitive alterations in Alzheimer's disease: synapse loss is the major correlate of cognitive impairment. *Annals of Neurology* **30**, 572–80.

Tsang SW, Lai MK, Francis PT, *et al* (2003). Serotonin transporters are preserved in the neocortex of anxious Alzheimer's disease patients. *Neuroreport* **14**, 1297–1300.

Tsang SW, Lai MK, Kirvell S, *et al* (2005). Impaired coupling of muscarinic M_1 receptors to G-proteins in the neocortex is associated with severity of dementia in Alzheimer's disease. *Neurobiology of Aging* **27**, 1216–23.

Victoroff J, Zarow C, Mack WJ, Hsu E, and Chui HC (1996). Physical aggression is associated with preservation of substantia nigra pars compacta in Alzheimer disease. *Archives of Neurology* **53**, 428–34.

Wakabayashi K, Narisawa-Saito M, Iwakura Y, *et al* (1999). Phenotypic down-regulation of glutamate receptor subunit GluR1 in Alzheimer's disease. *Neurobiology of Aging* **20**, 287–95.

Warpman U and Nordberg A (1995). Epibatidine and ABT 418 reveal selective losses of a4b2 nicotinic receptors in Alzheimer brains. *Neuroreport* **6**, 2419–23.

Westphalen RI, Scott HL, and Dodd PR (2003). Synaptic vesicle transport and synaptic membrane transporter sites in excitatory amino acid nerve terminals in Alzheimer disease. *Journal of Neural Transmission* **110**, 1013–27.

Whitehouse PJ, PriceDL, Struble RG, Clark AW, Coyle JT, and Delon MR (1982). Alzheimer's disease and senile dementia: loss of neurons in the basal forebrain. *Science* **215**, 1237–9.

Wilcock GK, Esiri MM, Bowen DM, and Smith CC (1982). Alzheimer's disease. Correlation of cortical choline acetyltransferase activity with the severity of dementia and histological abnormalities. *Journal of the Neurological Sciences* **57**, 407–17.

Wilcock GK, Esiri MM, Bowen DM, and Hughes AO (1988). The differential involvement of subcortical nuclei in senile dementia of Alzheimer's type. *Journal of Neurology, Neurosurgery, and Psychiatry* **51**, 842–9.

Yew DT, Li WP, Webb SE, Lai HW, and Zhang L (1999). Neurotransmitters, peptides, and neural cell adhesion molecules in the cortices of normal elderly humans and Alzheimer patients: a comparison. *Experimental Gerontology* **34**, 117–33.

Záborszky L, Cullinan WE, and Braun A (1991). Afferents to basal forebrain cholinergic projection neurons: an update. In Napier CT, Kalivas PW, and Hanin I (eds), *Basal Forebrain: Anatomy to Function*, pp. 43–100 (Plenum Press, New York).

Zezula J, Slany A, and Sieghart W (1996). Interaction of allosteric ligands with GABA-A receptors containing one, two, or three different subunits. *European Journal of Pharmacology* **301**, 207–14.

Zorumski CF and Isenberg KE (1991). Insights into the structure and function of GABA-benzodiazepine receptors: ion channels and psychiatry. *American Journal of Psychiatry* **148**, 162–73.

Zubenko GS and Moossy J (1988). Major depression in primary dementia. Clinical and neuropathologic correlates. *Archives of Neurology* **45**, 1182–6.

Zubenko GS, Moossy J, and Kopp U (1990). Neurochemical correlates of major depression in primary dementia. *Archives of Neurology* **47**, 209–14.

Zubenko GS, Moossy J, Martinez AJ, *et al.* (1991). Neuropathologic and neurochemical correlates of psychosis in primary dementia. *Archives of Neurology* **48**, 619–24.

Zweig RM, Ross CA, Hedreen JC, *et al.* (1988). The neuropathology of aminergic nuclei in Alzheimer's disease. *Annals of Neurology* **24**, 233–42.

Chapter 13

NGF family of neurotrophins and their receptors: early involvement in the progression of Alzheimer's disease

Elliott J. Mufson, Scott E. Counts, Margaret Fahnestock, and Stephen D. Ginsberg

Cholinergic basal forebrain (CBF) neurons of the nucleus basalis (NB) provide the major cholinergic innervation to the cortical mantle, are selectively vulnerable in late-stage Alzheimer's disease (AD), and require the neurotrophin nerve growth factor (NGF) for survival. The neurobiological events underlying the eventual demise of NB neurons during the progression of AD remain unknown. Our group investigated the molecular events underlying the degeneration of cholinotrophic basal forebrain neurons in tissue harvested from subjects participating in a longitudinal clinical pathological study of ageing and AD. The number of choline acetyltransferase (ChAT) positive neurons was unchanged in people with mild cognitive impairment (MCI) and mild AD, while the number of the neurons containing high affinity TrkA or low affinity p75^{NTR} NGF receptors, which co-localize with ChAT neurons, was significantly reduced compared with normal aged controls. These findings suggest a phenotypic down regulation rather than frank CBF neuronal degeneration in MCI. Gene profiling of single CBF neurons reveal that trkA but not p75^{NTR} mRNA is reduced in MCI. Furthermore, cortical Western blot assays revealed that TrkA but not p75^{NTR} protein is reduced in early-stage AD. These data suggest that decreased neurotrophin responsiveness may be an early biomarker for the onset of AD. Unlike endstage/severe AD, cortical NGF protein synthesis remains stable throughout the disease process, whereas the NGF precursor molecule, proNGF, which predominately binds with p75^{NTR}, a putative cell death molecule, is increased in MCI and AD. The accumulation of cortical proNGF in the presence of reduced TrkA may shift the balance between CBF neuron survival and death in prodromal AD. These perturbations coincide with a decrease in brain-derived neurotrophic factor (BDNF) and its precursor molecule, proBDNF, in MCI cortex, effectively removing additional CBF trophic support. In MCI, CBF neurons also display a shift in the ratio of three repeat tau (3Rtau) to four repeat tau (4Rtau), potentially affecting neurofibrillary tangle pathogenesis in these cells. These studies suggest that CBF neurons undergo multiple molecular and biochemical shifts during and/or prior to frank onset of AD, which lead to cell dysfunction and eventual death over the course of the disease.

13.1 **Introduction**

Neurotrophins represent a family of proteins that play a pivotal role in the mechanisms underlying neuronal survival, differentiation, growth and apoptosis (Kaplan and Miller 2000; Lad *et al.* 2003; Teng and Hempstead 2004). The potential applications for neurotrophins in the treatment of several neurological diseases for which there are currently no treatment regimens have generated great interest (Tuszynski and Blesch 2004). Before these potentially therapeutic proteins are used for the treatment of neurodegenerative diseases, it is imperative to understand the biology underlying their function(s). Since the initial discovery of the prototypic neurotrophin, nerve growth factor (NGF), 50 years ago by Levi-Montalcini (Cohen and Levi-Montalcini 1957; Levi-Montalcini 1964), several additional NGF family members have been discovered and characterized. NGF is the founding member of the neurotrophin family, which includes brain-derived neurotrophic factor (BDNF), neurotrophin-3 (NT-3), neurotrophin-4 (NT-4) and neurotrophin-5 (NT-5) (interspecies homologues referred to as NT-4 for simplicity) (Kaplan and Miller 2000; Teng and Hempstead 2004), and neurotrophin-6 (NT-6) and neurotrophin-7 (NT-7) (Dethleffsen *et al.* 2003). All members of the neurotrophin family are structurally similar, sharing ~50% sequence homology (Ibanez 1994; Timm *et al.* 1994; Dethleffsen *et al.* 2003). The members of the neurotrophin family transduce their biological effects by interacting with two types of cell surface receptors, the Trk family of receptor tyrosine kinases and the p75 pan-neurotrophin receptor (p75^{NTR}) (Kaplan and Miller 2000; Teng and Hempstead 2004) (Fig. 13.1).

NGF is produced by neurons in the hippocampus and neocortex, binds receptors on cholinergic axon terminals, and is retrogradely transported to cholinergic neuron cell bodies located within the basal forebrain (Sofroniew *et al.* 2001, Dawbarn and Allen 2003). Understanding how NGF regulates such diverse processes as cell survival, axonal and dendritic outgrowth, synapse and bouton formation, and cell proliferation have been long-standing questions in the field of neurotrophin biology. Animal models of

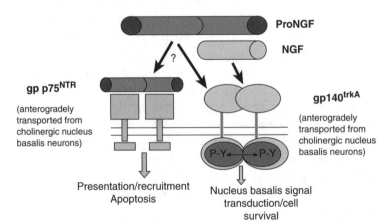

Fig. 13.1 Schematic diagram showing binding of the proNGF–NGF dimers with the low (p75^{NTR}) and high (trkA) affinity receptors and their putative functional consequences. The ? indicates uncertainty about the precise binding of proNGF with these receptors.

aging and AD have shown that NGF delivery may ameliorate cholinergic cell loss and reverse cholinergic atrophy (Hefti 1986; Sofroniew *et al.* 1990; Conner *et al.* 2001). A related neurotrophin, BDNF, is widely distributed in the brain but is found in particularly high amounts in regions affected by AD (Hofer *et al.* 1990; Phillips *et al.* 1990). BDNF is transported retrogradely to cholinergic cell bodies but can also be transported both antero- and retrogradely in hippocampus and entorhinal cortex, and can function as an autocrine or paracrine survival factor in hippocampus (Altar *et al.* 1997; Sobreviela *et al.* 1996). In this chapter, we provide an overview of the role that two members of the NGF family (NGF and BDNF) and their receptors play in CBF cell degeneration and their relation to the development of AD.

CBF neurons within the NB and the septal diagonal band provide the major source of cholinergic innervation to the cerebral cortex and hippocampus, respectively (Mesulam *et al.* 1983; Mesulam and Geula 1988), and play a key role in memory and attention function (Coyle *et al.* 1983; Baxter and Chiba 1999; Sarter *et al.* 1999; Bartus 2000; Sarter and Bruno 2004) (Fig. 13.2a) [Plate 7]. NB cortical projection neurons undergo extensive degeneration in late-stage AD (Fig. 13.2f-h) which correlates with clinical severity and disease duration (Whitehouse *et al.* 1981; Bartus *et al.* 1982; Allen *et al.* 1988; Mufson *et al.* 1989; Vogels *et al.* 1990; DeKosky *et al.* 1992). CBF neurons require NGF and BDNF for their survival and biological activity (Kaplan and Miller 1997; Sofroniew *et al.* 2001; Lad *et al.* 2003). BDNF maintains survival and function not only of CBF neurons, but also of

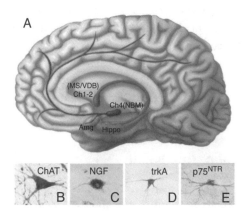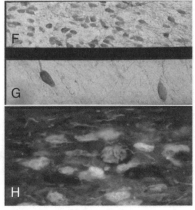

Fig. 13.2 (a). Diagram showing the cholinergic projection neurons from Ch4 (NB) (blue) to the entire cortical mantle and the cholinergic projections from Ch2 (medial septal diagonal band complex (MS–VDB)) (red) to the hippocampus in a sagittal view of the human brain. Note that the Ch4 neurons also innervate the amygdala (Amg). The lower panels show that the cholinergic neurons are immunopositive for (b) ChAT, (c) NGF, (d) trkA, and (e) p75^{NTR} (f) Low-power photomicrograph showing numerous ChAT-immunopositive cholinergic neurons within the NB from an aged control subject. (g) Higher-magnification photomicrograph showing the extensive reduction in ChAT-positive neurons within the NB from a patient with end stage AD (h) Section dual stained for p75^{NTR} (dark blue) and thioflavin S (yellow) showing both tangle-bearing (yellow) and non-tangle-bearing (dark blue) neurons in endstage AD. Please see the colour plate section for a colour version of this figure.

several other neuronal populations and their connections that are compromised in AD, including neurons of the entorhinal cortex, hippocampus, and cortex (Ghosh *et al.* 1994; Lindholm *et al.* 1996; Ando *et al.* 2002). BDNF is also a trophic factor for neural progenitor cells (Shetty and Turner 1998; Barnabe-Heider and Miller 2003).

Cellular responses to NGF are elicited through binding and activation of its cognate receptors TrkA and p75^{NTR} (Bothwell 1995; Chao and Hempstead 1995; Kaplan and Miller 1997) which are produced within CBF neurons and anterogradely transported to the cortex and hippocampus (Sobreviela *et al.* 1994) (Fig. 13.1). Similarly, BDNF biological activity is generated through binding and activation of TrkB and p75^{NTR} (Barbacid 1995). Approximately 20 years ago it was hypothesized that degeneration of CBF neurons was due to a loss of neurotrophic support from target regions that produce NGF in AD (Appel 1981; Hefti *et al.* 1989; Kordower and Mufson 1989). These findings and subsequent studies were based solely upon investigations which examined autopsy material harvested from late-stage AD patients. To gain a greater understanding of the neurobiological mechanism(s) that underlie the onset of this cholinotrophic deficit, its relation to cognitive impairment, and whether the changes seen in endstage AD occur during the prodromal stages of this disease, we examined post-mortem human brain tissue harvested from the Religious Orders Study (ROS), a longitudinal clinical pathological study of ageing and dementia in retired Catholic clergy (Gilmor *et al.* 1999; Mufson *et al.* 2000; Bennett *et al.* 2002; Counts *et al.* 2004; Peng *et al.* 2004). Each ROS participant received an annual detailed clinical evaluation, including a battery of tests for function in five cognitive domains (orientation, attention, memory, language, and perception), and agreed to brain autopsy and neuropathological evaluation (Bennett *et al.* 2002). These subjects were categorized as having no cognitive impairment (NCI), mild cognitive impairment (MCI), considered a prodromal stage of AD (Petersen 2004; Winblad *et al.* 2004), or AD. In this chapter, we describe the findings generated from our clinical pathological investigations of the cholinotrophic basal forebrain system. These studies have provided unique insights into the role that NGF, BDNF, and their receptors play in CBF neuronal dysfunction during the prodromal stages of AD.

13.2 Nerve growth factor and cholinergic basal forebrain neuron survival

A growing body of data support a role for NGF in the maintenance and survival of CBF neurons. For example, *in vitro* studies using dissociated rat CBF cultures (Hartikka and Hefti 1988; Hatanaka *et al.* 1988) or organotypic CBF slices (Humpel and Weis 2002) showed that NGF treatment prevents the degeneration of cholinergic neurons observed in untreated preparations. Similarly, the atrophy and loss of CBF neurons *in vivo* following neocortical infarction in monkeys (Burgos *et al.* 1995) or excitotoxic neocortical lesions in rats (Charles *et al.* 1996) was reversed with exogenous NGF treatment. In addition, infusion of NGF can prevent loss of cholinergic phenotypic markers as well as septal cholinergic neuron death following septohippocampal axotomy via fimbria–fornix transection (Hefti 1986; Williams *et al.* 1986; Tuszynski and Gage 1995). Finally, transgenic mice that express anti-NGF antibodies in adulthood display an age-dependent loss of CBF neurons (Capsoni *et al.* 2000; Ruberti *et al.* 2000).

CBF neurons express two NGF receptors, the NGF-specific receptor tyrosine kinase TrkA and the pan-neurotrophin receptor p75^{NTR} (Fig. 13.1). These receptors mediate the binding, internalization, and retrograde transport of target-derived NGF from the cortex to cholinergic NB neurons (Bothwell 1995; Kaplan and Miller 1997; Mufson and Kordower 1999; Sobreviela *et al.* 1994). p75^{NTR} is a member of the tumour necrosis factor receptor family and binds NGF with relatively low affinity ($K_d = 10^{-9}$) (Chao *et al.* 1986; Mahadeo *et al.* 1994). Although this pan-neurotrophin receptor lacks intrinsic catalytic activity, it interacts with cytosolic adaptor proteins to activate downstream signalling molecules including c-*jun* N-terminal kinase (JNK) and nuclear factor-κB (NF-κB) (Carter *et al.* 1996; Casaccia-Bonnefil *et al.* 1996; Roux and Barker 2002). Activation of these pathways through the p75^{NTR} receptor *in vitro* elicits either a pro- or an anti-apoptotic signal which probably depends on the cellular milieu (e.g. cell type, presence or absence of specific binding partners) (Mamidipudi and Wooten 2002; Roux and Barker 2002). The binding partners of p75^{NTR} include sortilin, Nogo receptor, and lingo-1, in addition to TrkA (Bronfman and Fainzilber 2004). TrkA is a receptor tyrosine kinase that binds NGF with relatively high affinity ($K_d = 10^{-11}$) when coexpressed with p75^{NTR} (Hempstead *et al.* 1991; Klein *et al.* 1991; Mahadeo *et al.* 1994). Following binding of NGF to TrkA, the receptor activates signal transduction cascades, including mitogen-activated protein kinases (MAPKs), phosphatidylinositol-3 kinase (PI3K), and phospholipase C-γ (PLCγ) pathways that promote survival and sustain the cholinergic phenotype of CBF neurons by activation of transcription factors that alter gene expression (Kaplan and Stephens 1994; Klesse and Parada 1999; Lad *et al.* 2003; Sofroniew *et al.* 2001).

It is well established that NGF is retrogradely transported from its site of synthesis in the cortex and hippocampus to the CBF consumer neurons (Seiler and Schwab 1984; Mufson *et al.* 1999). The mechanism by which NGF retrograde transport proceeds and culminates in cell survival remains to be answered, but a series of observations indicate that NGF-bound TrkA complexes are internalized as specialized 'signalling endosomes' (Grimes *et al.* 1996; Delcroix *et al.* 2003; Ye *et al.* 2003) by a process involving dynamin (Zhang *et al.* 2000) and PI3K (Reynolds *et al.* 1998; York *et al.* 2000). These signalling endosomes are transported in the negative direction along microtubules, as dynein inhibition or microtubule disassembly blocks the retrograde transport of both NGF and Trk receptors (Reynolds *et al.* 1998; Watson *et al.* 1999). Interestingly, NGF application to distal axons in compartmentalized sympathetic neurons induced cyclic AMP response element binding protein (CREB) phosphorylation in an endocytosis-dependent manner (Riccio *et al.* 1997). CREB-mediated gene expression was both necessary and sufficient for NGF-dependent survival (Riccio *et al.* 1999), indicating that the retrograde transport of signalling endosomes is required for NGF-induced nuclear responses and is important for cholinergic cell survival. However, there is also evidence that NGF retrograde signalling can occur in the absence of signalling endosomes, suggesting the existence of alternative and possibly parallel mechanisms of retrograde signalling (MacInnis and Campenot 2002; Campenot and MacInnis 2004).

13.3 Nerve growth factor receptors in mild cognitive impairment and early Alzheimer's disease

The cholinotrophic phenotype of NB neurons is altered during the prodromal and earliest stages of AD. Unbiased estimation using stereology has demonstrated that the number of NB perikarya expressing either ChAT, the synthetic enzyme for acetylcholine (ACh), or the vesicular ACh transporter (VAChT) remained unchanged in subjects clinically diagnosed with MCI (prodromal stage of AD) or mild AD compared with NCI (Gilmor *et al.* 1999) (Fig. 13.3a,b,i). Cortical ChAT activity was stable in the early stages of AD or increased in the superior frontal cortex of MCI cases (DeKosky *et al.* 2002). These observations suggest that basocortical cholinergic tone is preserved in MCI and early AD, supporting the notion that cholinergic NB neurotransmission dysfunction is a late-stage event in AD (Davies and Maloney 1976; Bartus *et al.* 1982; Mufson *et al.* 1989). In contrast, stereology revealed that the number of NB perikarya expressing either p75^{NTR} (Fig. 13.3c–e, i) or TrkA (Fig. 13.3f–h, i) was reduced by ~50% in MCI and mild AD compared with NCI (Mufson *et al.* 2000, 2002b). Intriguingly, there was a significant positive correlation between the number of TrkA-immunoreactive NB neurons and performance on the following measures of cognitive function: the Boston Naming Test and a composite score of 19 cognitive tests termed the Global Cognitive Score (GCS) (Mufson *et al.* 2000). Similarly, the number of p75^{NTR}-immunoreactive NB neurons was significantly correlated with performance on the Mini-Mental State Examination (MMSE) and the GCS (Mufson *et al.* 2002b). These chemo-anatomical studies suggest that a subpopulation of cholinergic NB neurons undergo a phenotypic silencing of NGF receptor protein expression in the absence of frank cell loss or cholinergic deficits during the early stages of AD, which correlates with cognitive impairment. Thus, pronounced early defects in NGF receptor expression may be an early pathological event that precedes the eventual NB cell loss seen in endstage AD.

Since NGF receptors are synthesized within NB perikarya and anterogradely transported to the cortex, quantitative immunoblotting experiments were performed to determine whether TrkA and p75^{NTR} protein levels in five NB cortical projection sites (anterior cingulate, superior frontal, superior temporal, inferior parietal, and visual cortex) are decreased during the progression of AD (Fig. 13.4) (Counts *et al.* 2004). Interestingly, whereas p75^{NTR} protein cortical levels were stable across the diagnostic groups (Fig. 13.4a), cortical TrkA protein was reduced by ~50% in mild AD compared with NCI and MCI subjects (Fig. 13.4a,b). Moreover, cortical TrkA levels were positively correlated with MMSE performance (Fig. 13.4c) (Counts *et al.* 2004). Because TrkA signalling is associated with neuronal survival (Kaplan and Stephens 1994; Klesse and Parada 1999; Sofroniew *et al.* 2001; Lad *et al.* 2003), this specific reduction of cortical TrkA receptor protein may play a pivotal role in the frank loss of cholinergic NB neurons seen in later stages of AD (Whitehouse *et al.* 1981; Mufson *et al.* 1989). It is important to bear in mind that the catalytic form of TrkB is also reduced in AD in both CBF neurons (Salehi *et al.* 1996; Mufson *et al.* 2002a) and cortical projection sites (Allen *et al.* 1999; Ferrer *et al.* 1999). TrkB and TrkC are also localized to CBF neurons, albeit at lower levels

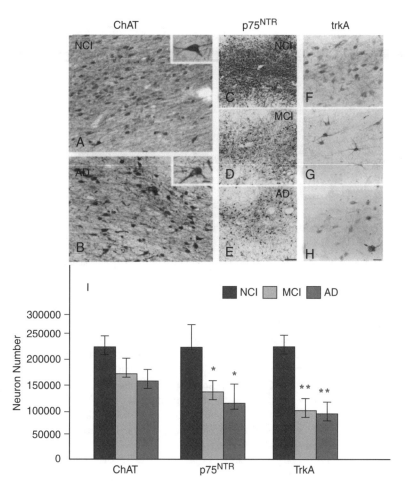

Fig. 13.3 Photomicrographs showing phenotypic differences in the number of NB neurons expressing (a, b). ChAT, (c–e). p75[NTR], or (f–h). TrkA in subjects clinically diagnosed as NCI, MCI, or mild AD. (i). Histogram showing a significant ~50% reduction in the number of NB neurons expressing NGF receptor protein despite stable expression of ChAT. **$P < 0.001$, *$P < 0.01$ via one-way ANOVA with post hoc Tukey's studentized range test for multiple comparisons.

than TrkA (Salehi *et al*. 1996; Mufson *et al*. 2002a). Whether reductions occur for TrkB and TrkC protein early in the progression of AD has not been determined.

13.4 Mechanisms of nerve growth factor receptor alterations in mild cognitive impairment and Alzheimer's disease

The mechanism(s) underlying TrkA protein reduction in NB perikarya in MCI and its projection sites in early AD is unclear. TrkA gene expression is under positive feedback from NGF signalling (Holtzman *et al*. 1992; Li *et al*. 1995) and this pathway is probably disrupted in AD by diminished retrograde transport of cortical NGF to CBF consumer

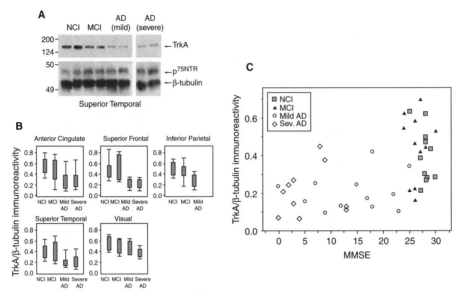

Fig. 13.4 Cortical TrkA but not p75[NTR] levels are reduced in subjects clinically diagnosed with mild–moderate or severe AD. (a). Representative immunoblot of detergent lysates from superior temporal cortex were separated by electrophoresis and immunoblotted for TrkA, p75[NTR], and β-tubulin. (b). Cortical TrkA was reduced by ~50% in anterior cingulate, superior frontal, superior temporal, and inferior parietal cortex in AD subjects compared with subjects diagnosed with NCI or MCI ($P < 0.01$ via one-way ANOVA with post hoc Tukey's studentized range test for multiple comparisons). Densitometry was performed by normalizing TrkA or p75[NTR] immunoreactive signals to β-tubulin signals on the same blots. (c). Scattergram showing that reductions in TrkA protein levels in the superior temporal cortex of the subjects examined correlate with reductions in ante-mortem MMSE scores ($r = 0.55$; $p = 0.0007$).

neurons (Mufson *et al.* 1995; Scott *et al.* 1995). This hypothesis is supported by the observation that NGF mRNA levels are not altered in AD (Goedert *et al.* 1986; Jette *et al.* 1994; Fahnestock *et al.* 1996), whereas NGF protein levels are stable (Allen *et al.* 1991; Mufson *et al.* 2003) or increased (Crutcher *et al.* 1993; Scott *et al.* 1995; Fahnestock *et al.* 1996) in the cortex and decreased in cholinergic NB neurons in late-stage AD (Mufson *et al.* 1995; Scott *et al.* 1995). Notably, defective retrograde transport of NGF within cholinergic projection neurons occurs in a segmental trisomic mouse model of Down's syndrome, a congenital neurodegenerative disorder that mimics the neuropathology of AD (Cooper *et al.* 2001). In aged rats, CBF neurons exhibited a pronounced reduction in NGF retrograde transport, reduced TrkA protein expression, and severe cellular atrophy (Cooper *et al.* 1994; De Lacalle *et al.* 1996). It is possible that defective NGF retrograde transport underlies the reductions in expression of NB TrkA mRNA (Chu *et al.* 2001; Mufson *et al.* 2002a) and protein (Mufson *et al.* 1997, 2000) observed in single NB neurons in MCI and early AD. This defect may ultimately lead to reduced TrkA protein in cortical projection sites (Mufson *et al.* 1997; Counts *et al.* 2004) and further perturbations in NGF signalling within NB neurons. This putative 'off trk' cycle of deficient NGF receptor signalling

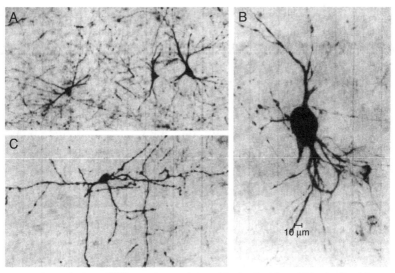

Fig. 13.5 Ectopic cortical p75^{NTR} immunoreactive multipolar (a, b) and bipolar-like (c) neurons in AD.

(Mufson 1997; Sarter and Bruno 2004) may contribute to the selective vulnerability of cholinergic NB neurons (Whitehouse *et al.* 1981; Bartus *et al.* 1982; Allen *et al.* 1988; Mufson *et al.* 1989, 1999; Vogels *et al.* 1990; Bartus 2000) and deficits in cortical cholinergic tone observed in late-stage AD (DeKosky *et al.* 1992).

In contrast, the stability of cortical p75^{NTR} protein expression levels across clinical diagnosis (NCI→MCI→AD) was surprising in light of stereological cell count data showing that NB perikarya expressing p75^{NTR} were reduced in both MCI (38%) and mild AD (43%) relative to NCI (Fig. 13.3). Moreover, cortical p75^{NTR} levels remain stable even in severe AD despite a dramatic reduction of p75^{NTR}-immunopositive neurons and fibres in the NB in late-stage AD (Mufson *et al.* 1989; Salehi *et al.* 2000a). However, gene expression profiling studies of single cholinergic NB neurons revealed that p75^{NTR} mRNA levels were unchanged in end stage AD compared with age-matched controls (Mufson *et al.* 2002a; Ginsberg *et al.* 2006a), and therefore the stability of cortical p75^{NTR} levels is not due to compensatory up regulation of p75^{NTR} by NB neurons. On the other hand, the stability of cortical p75^{NTR} may be related to the *de novo* expression of cortical p75^{NTR}-immunoreactive neurons in AD (Mufson and Kordower 1992) (Fig. 13.5). Alternatively, these data could reflect a redistribution of p75^{NTR} from the CBF soma to the nerve terminals and receptor accumulation at the terminals due to failed retrograde transport in AD.

13.5 Consequences of nerve growth factor receptor alterations in mild cognitive impairment and Alzheimer's disease

TrkA receptor reductions in basocortical projection systems may result in a loss of TrkA-mediated NGF signalling via downstream effector molecules such as MAPK and PI3K that are related to NB cell survival (Kaplan and Stephens 1994; Klesse and Parada 1999;

Sofroniew et al. 2001; Lad et al. 2003). TrkA activation of the MAP kinases, extracellular signal-related kinases 1 and 2 (ERK1/2), results in the translocation of these proteins to the nucleus (Kaplan and Stephens 1994; Klesse and Parada 1999; Sofroniew et al. 2001; Lad et al. 2003). A major downstream nuclear target of ERK1/2 is the transcription factor CREB. CREB-mediated transcription plays an important role in neuronal survival (Ginty et al. 1994; Riccio et al. 1997, 1999; Bonni et al. 1999), probably via an anti-apoptotic signal involving the upregulation of the bcl-2 gene (Liu et al. 1999; Riccio et al. 1999). Disruption of TrkA kinase activity following NGF treatment resulted in defects in nuclear CREB activation in sympathetic neurons (Riccio et al. 1997). Moreover, studies using compartmentalized dorsal root sensory neurons have shown that another MAP kinase member, ERK5, activates CREB and promotes neuronal survival following NGF application to distal axons (Watson et al. 2001).

The role of PI3K in TrkA-mediated neuronal survival is supported by observations that the PI3K inhibitors wortmannin and LY294002 induce apoptosis in NGF-protected PC12 cells and sympathetic neurons (Yao and Cooper 1995; Spear et al. 1997; Crowder and Freeman 1998). The primary downstream mediator of PI3K is phosphorylated Akt kinase. This protein promotes cell survival through a phosphorylation cascade that prevents the insertion of pro-apoptotic Bax proteins into the mitochondrial membrane, an action that breaks down the membrane potential and promotes the release of cytochrome c and apoptotic signalling (Xiang et al. 1996; Zha et al. 1996; Dudek et al. 1997; Gross et al. 1998; Putcha et al. 1999). In addition, in vitro and in vivo activation of PI3K after NGF treatment is required for TrkA internalization and retrograde transport (Reynolds et al. 1998; Kuruvilla et al. 2000; York et al. 2000). These findings suggest a potential mechanistic link between reduced cortical TrkA and diminished pro-survival signalling in CBF neurons.

The preservation of cortical p75[NTR] protein despite a reduction of cortical TrkA protein in AD could result in a relative increase in p75[NTR]-mediated NGF signalling in cholinergic NB neurons. Traditionally, p75[NTR] is thought to facilitate NGF signalling by producing high-affinity NGF binding sites and increased TrkA responsiveness to NGF (Hempstead et al. 1991; Barker and Shooter 1994; Mahadeo et al. 1994). This concept is being revisited since p75[NTR] possesses complex autonomous signalling capabilities in response to NGF which, paradoxically, regulate both cell death and cell survival under different experimental paradigms. NGF activation of p75[NTR] in the absence of TrkA induces apoptosis in neurons (von Bartheld et al. 1994; Frade et al. 1996; Bamji et al. 1998; Davey and Davies 1998; Friedman 2000; Salehi et al. 2000b) and mature oligodendrocytes (Casaccia-Bonnefil et al. 1996; Yoon et al. 1998). Moreover, retroviral infection of mature p75[NTR]-positive oligodendrocytes with TrkA cDNA reverses NGF-induced apoptosis (Yoon et al. 1998), suggesting that the coexpression of TrkA suppresses p75[NTR]-mediated pro-apoptotic signalling. The mechanism for p75[NTR]-mediated cell death is also currently being investigated. Recent data indicate that activation of the JNK pathway is a primary carrier of the p75[NTR] pro-apoptotic signal (Casaccia-Bonnefil et al. 1996; Bamji et al. 1998; Yoon et al. 1998; Friedman 2000; Roux et al. 2001). NGF binding to p75[NTR] results in the phosphorylation and activation of JNK, which translocates to the nucleus to phosphorylate the transcription factor c-Jun (Derijard *et al.* 1994). Phosphorylated c-Jun-mediated transcription activates a pro-apoptotic cascade involving the release of cytochrome *c* and caspase activation (Gu *et al.* 1999; Roux and Barker 2002; Troy *et al.* 2002; Bhakar *et al.* 2003).

Evidence is accumulating that p75^{NTR}-mediated apoptosis involves the activation of cell cycle regulatory molecules. For example, in post-mitotic p75^{NTR}-containing retinal neurons, NGF-mediated apoptosis is preceded by an increase in mitotic figures as well as an upregulation of cyclin B2 (Frade 2000). Cyclin B2 is typically active at the G2–M cell cycle transition, and apoptosis can be prevented in a dose-dependent manner by the delivery of CDK1 cyclin B2-dependent kinase inhibitor roscovitine (Frade 2000). A connection between the aberrant cell cycle re-entry of post-mitotic neurons and apoptosis exists (Freeman *et al.* 1994; Herrup and Busser 1995; Farinelli and Greene 1996; Park *et al.* 1997), and the AD brain is characterized by neuronal expression of cell cycle regulatory proteins (Vincent *et al.* 1996, 1997; Busser *et al.* 1998) and evidence for DNA replication (Yang *et al.* 2001). More recently, our group has shown the presence of the cell cycle proteins PCNA, cyclin D1, and cyclin B1 in NB neurons in MCI and mild AD (Yang *et al.* 2003). It is possible that increases in p75^{NTR}-mediated activity in the face of reduced TrkA signalling may contribute to NB neurodegeneration in AD by promoting unscheduled cell cycle re-entry and apoptosis. Taken together, these data suggest that the ratio of TrkA to p75^{NTR} receptors may determine whether neurons survive or die when exposed to NGF. Thus the ~50% reduction in cortical TrkA in the absence of significant changes in p75^{NTR} levels at the onset of AD may signify a relative increase in pro-apoptotic p75^{NTR} signalling in the cholinergic NB projection system.

13.6 Pro-nerve growth factor and nucleus basalis neuron survival

The physiological consequences of TrkA and p75^{NTR} signalling may depend upon their interactions with the NGF precursor protein proNGF. Quantitative western blotting studies demonstrated that proNGF is the predominant form of NGF present in the cortex of aged intact humans (Fahnestock *et al.* 2001). ProNGF levels increase by ~40%–60% in the inferior parietal cortex of subjects diagnosed with MCI or mild AD compared with NCI (Fig.13. 6a) (Peng *et al.* 2004). This finding was intriguing given the previous demonstration that levels of total NGF, as detected by enzyme-linked immunosorbent assay (ELISA), were unchanged in the cortex (Mufson *et al.* 2003). This discrepancy may be due to technical differences in the studies (e.g. immunoblotting versus ELISA). However, these data are consistent with ELISA assays using a different antibody which demonstrated increased total NGF in AD (Fahnestock *et al.* 1996).

The biological consequences of proNGF accumulation in the cortex during the prodromal stages of AD have yet to be determined. An emerging body of literature suggests that recombinant proNGF binds TrkA and promotes neuronal survival and neurite outgrowth similar to mature NGF, but is approximately fivefold less active than mature NGF (Rattenholl *et al.* 2001; Fahnestock *et al.* 2004). Although TrkA-mediated proNGF retrograde transport

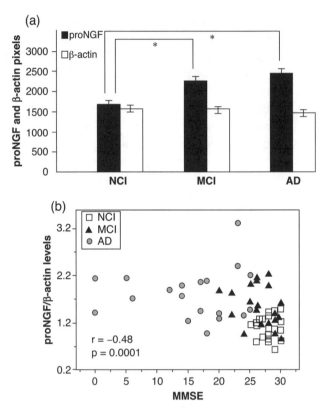

Fig. 13.6 (a). Histogram showing that proNGF protein expression increases by ~40% in the inferior parietal cortex in MCI and by ~60% in AD. Densitometry was performed by normalizing proNGF immunoreactive signals to β-actin signals on the same western blots. *$P < 0.001$ via one-way ANOVA with post hoc Tukey's studentized range test for multiple comparisons. (b). Scattergram showing that increases in proNGF protein levels in the inferior parietal cortex correlate with reduced ante mortem MMSE scores.

has not been demonstrated, proNGF accumulation in NB cortical target sites may be due to reduced cortical TrkA levels and/or retrograde transport to cholinotrophic basal forebrain perikarya. Thus, both poor utilization of proNGF and signalling of proNGF via p75^{NTR} in combination with reduced TrkA may result in CBF neurodegeneration. Significantly, while TrkA levels in the cortex were positively associated with MMSE scores (Fig. 13.4c) (Counts *et al.* 2004), cortical proNGF levels were negatively correlated with MMSE performance (Fig. 13.6b) (Peng *et al.* 2004). Thus, the concomitant reduction of TrkA and accumulation of proNGF in the cortex may be an early pathobiological marker for the onset of AD (Fig. 13.7). In fact, significantly increased CSF levels of NGF-immunoreactive material are detectable in AD patients (Blasko *et al.* 2005), demonstrating the potential utility of NGF as a diagnostic marker.

In contrast, other studies indicate that increases in cortical proNGF may result in pro-apoptotic signalling through binding to the p75^{NTR} receptor. In support of this, a different

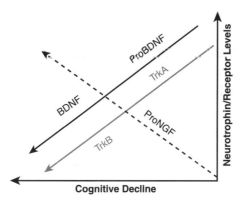

Fig. 13.7 Schematic diagram showing the relationship between cortical levels of proNGF, TrkA, proBDNF, mature BDNF, and TrkB and cognitive decline in the elderly. Note that progressive cognitive impairment is associated with increases in proNGF and decreases in TrkA, proBDNF, mature BDNF, and TrkB protein levels in the cortex.

form of recombinant proNGF was shown to bind p75^{NTR} with high affinity and promote neuronal apoptosis (Lee *et al.* 2001). More recent data indicate that the pro-apoptotic effect(s) of p75^{NTR}-mediated proNGF signalling is dependent on interactions between p75^{NTR} and the neurotensin receptor sortilin (Nykjaer *et al.* 2004). Sortilin binds the 'pro' region of proNGF, and blocking this binding event precludes high-affinity binding of proNGF to p75^{NTR}. Moreover, sortilin expression is required for p75^{NTR}-mediated apoptosis following proNGF treatments (Nykjaer *et al.* 2004), suggesting that sortilin is a p75^{NTR} binding partner associated with its role in the initiation of cell death (Mamidipudi and Wooten 2002; Roux and Barker 2002). Thus, the outcome of p75^{NTR} signalling in response to proneurotrophin may depend on the identity of the bound co-receptor. In fact, p75^{NTR} may have other receptor chaperones and binding partners which assist in its role as a 'multi-tasking' neurotrophin receptor (Bronfman and Fainzilber 2004). For example, if p75^{NTR} and the co-receptor NgR-lingo 1 interact, inhibition of axonal growth occurs, as opposed to cell death if sortilin acts as a second receptor (Nykjaer *et al.* 2004). It remains to be determined whether sortilin–p75^{NTR}–proNGF complexes with motor proteins to form a retrogradely transported ensemble on microtubules, leading to intracellular signalling of a cell death response, or whether it is internalized into the somotodendritic compartment and anterogradely transported for recycling to the cell surface (Bronfman and Fainzilber 2004). Figure 13.8 [Plate 8] shows a diagram of a potential scenario for pro-survival (Fig. 13.8a) or pro-apoptotic (Fig. 13.8b) signalling in cholinergic NB neurons as TrkA levels decline, p75^{NTR} levels remain unchanged, and proNGF levels rise in cortical projection sites during the progression of AD. Alterations in cortical sortilin levels in prodromal AD have not been evaluated to date. Moreover, Bruno and Cuello (2006) have provided novel evidence that the conversion of proNGF to NGF in the central nervous systems is regulated by an activity induced protease cascade and its endogenous regulators (e.g. matrix metalloproteinase 9). Alterations in these events are currently being investigated in brain tissue derived from people with MCI to determine whether alterations in matrix metalloproteinases occur during progression of AD.

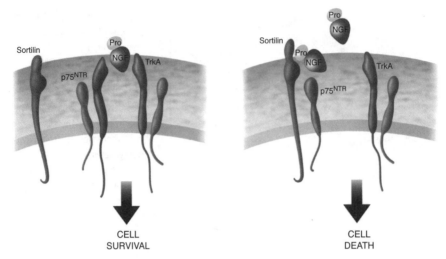

Fig. 13.8 Schematic model illustrating putative scenarios for proNGF–receptor binding during the progression of AD. (a). ProNGF complexes with TrkA to activate cell survival mechanisms in the aged human healthy brain. This binding event is facilitated by the coexpression of p75^{NTR} on the cell surface. (b). Elevated levels of cortical proNGF in the face of reduced cortical TrkA result in increased binding of proNGF to p75^{NTR}–sortilin complexes, enhancing the activation of pro-apoptotic mechanisms during the transition from MCI to AD. (Reproduced with permission from Counts and Mufson 2005.) Please see the colour plate section for a colour version of this figure.

13.7 Nerve growth factor in plaque and tangle pathology in Alzheimer's disease

The impact that a loss in NGF plays in the development of the pathological features of AD has been investigated using a phenotypic knockout of NGF activity in transgenic anti-NGF mice (AD11), which results in a progressive neurodegenerative phenotype similar to AD (Capsoni *et al.* 2000; Ruberti *et al.* 2000). Specifically, aged AD11 mice display a behavioural deficit linked to cholinergic atrophy, neuronal loss, tau hyperphosphorylation, and insolubility similar to the neuronal cytoskeletal defects seen in neurofibrillary tangles (NFTs) as well as the amyloid-plaque-like deposits seen in AD. Moreover, these impairments can be largely reversed by NGF delivery through an intranasal route (Capsoni *et al.* 2002, 2004; De Rosa *et al.* 2005). These observations suggest that defects in NGF, and perhaps in its receptors, play a role not only in the development of cholinergic cell loss, but also in the onset of cytoskeletal defects reminiscent of NFTs. It is interesting to note that these authors present data showing that the hyperphosphorylation of tau in the cortex of the AD11 mouse appears to be linked to NGF deprivation as opposed to a cholinergic deficit, since NGF, but not the AChE inhibitor galantamine, rescues this defect. It is possible that alterations in the NGF–TrkA–p75^{NTR} signalling pathways might disrupt the phosphorylation balance/ratio of tau, initiating downstream cytoskeletal neuronal effects (Capsoni *et al.* 2002, 2004; De Rosa *et al.* 2005).

The low molecular weight microtubule-associated protein tau is the major component of NFTs in select neuronal populations in AD (Kosik *et al.* 1986; Lee *et al.* 1991). The

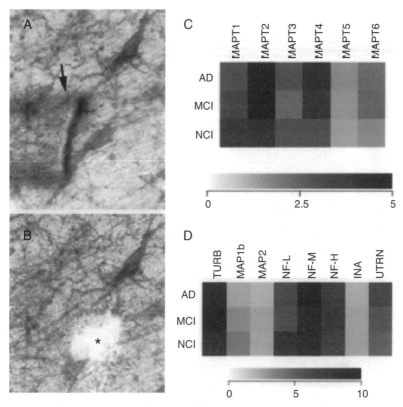

Fig. 13.9 Single cholinergic NB neuron stained for p75^{NTR} (a) before and (b) after (asterisk). Microaspiration for gene array expression analysis. The arrow in (a) indicates the position of the glass pipette prior to aspiration. (c) Dendrogram illustrating the stability of relative tau gene expression levels across NCI, MCI, and AD. In contrast, significant differences were found when a 3Rtau/4Rtau ratio was calculated (see Table 13.1). (d) Dendrogram illustrating a non-significant difference in relative expression levels for β-tubulin (TuBB), microtubule-associated proteins MAP1b and MAP2, neurofilament subunits NF-L (light molecular weight). NF-M (medium), and NF-H (heavy), alpha internexin (INA), and utrophin (UTRN) across clinical groups. Please see the colour plate section for a colour version of this figure.

adult human brain contains six tau isoforms ranging from 48 to 67 kDa, which are expressed through alternative splicing of a single tau gene on chromosome 17 that contains 16 exons (Goedert *et al.* 1989a,b). Brain contains three isoforms with three tandem repeats (3Rtau; Unigene annotation MAPT1, MAPT3, and MAPT5) and three isoforms with four tandem repeats (4Rtau; MAPT2, MAPT4, and MAPT6) of 31 or 32 amino acids in the carboxyterminus of the molecule. To evaluate whether alterations in tau isoform expression levels occur in the face of an imbalance of the cholinotrophic basal cortical system, single-cell gene array technology was applied to individual cholinergic neurons of the NB during the progression of AD (Ginsberg *et al.* 2006) (Fig. 13.9a,b) [Plate 9]. The results of our single-cell gene array studies revealed that the global expression levels of the six tau transcripts (Fig. 13.9c) and several other cytoskeletal transcripts (Fig. 13.9d) within individual CBF neurons did not vary among AD, MCI, and aged control

Table 13.1 3Rtau/4Rtau ratio in single-cell populations

Cell type	Condition	3R/4R ratio 0 insert	3R/4R ratio 1 insert	3R/4R ratio 2 insert
CBF	AD	0.66 ± 0.09*	0.69 ± 0.11*	0.71 ± 0.13**
CBF	MCI	0.62 ± 0.12*	0.61 ± 0.16*	0.71 ± 0.13**
CBF	NCI	0.93 ± 0.05	0.99 ± 0.07	0.89 ± 0.09
CA1	AD	0.63 ± 0.17***	NA	NA
CA1	CTR	1.08 ± 0.12	NA	NA
CA1	Ageing CTR	1.12 ± 0.07	NA	NA
Stellate	Ageing CTR	1.04 ± 0.11	NA	NA

The ratio 3Rtau/4Rtau was calculated and the data are presented for each pair of tau transcripts (mean ± standard deviation). relative to the amino-terminal inserts (0, 1, or 2 inserts). A ratio of approximately 1 is observed in normal CBF, CA1 pyramidal, and entorhinal cortex stellate neurons (ageing CTR), indicating a relative similarity in the expression of 3Rtau and 4Rtau transcripts. A significant shift in the 3Rtau/4Rtau ratio was observed in AD (CBF and CA1 neurons). and MCI (CBF neurons). relative to NCI and cognitively normal controls (CTR).

*$P < 0.001$; **$P < 0.005$; ***$P < 0.002$. Reprinted with permission from Ginsberg et al. 2006b.

subjects, consistent with previous reports. Surprisingly, when a 3Rtau/4Rtau ratio was calculated, a significant shift in the ratio was observed with a decrement in 3Rtau in relation to 4Rtau levels for all tau transcripts examined (Fig. 13.9c and Table 13.1). These data suggest a subtle, yet pervasive, shift in the dosage of 3Rtau and 4Rtau within vulnerable CBF neurons in MCI and AD (Ginsberg et al. 2006a; 2006b). Interestingly, a similar shift does not occur during normal ageing (Ginsberg et al. 2006b) (Table 13.1). Shifts in the ratio of tau genes may be a fundamental mechanism whereby normal tau expression is dysregulated, leading to NFT formation. The precise role, if any, that shifts in the balance between NGF and its receptors play in the development of NFTs in AD remains to be determined. Additional studies are under way to differentiate tau gene isoform differences between non-NFT-and NFT-containing CBF neurons.

13.8 Nerve growth factor and β-amyloid expression

During the last several years, NGF delivery to the brain has gained momentum as a potential therapeutic approach for the treatment of AD. For NGF to have clinical efficacy as a treatment of AD, it must protect CBF neurons without inducing harmful effects that outweigh its potential benefits. A concern regarding NGF therapy is the induction or exacerbation of the deposition of β-amyloid, a major component of plaque formation in AD. Studies in rodents demonstrate that injections of NGF increase the expression of amyloid-β precursor protein (APP) in the developing hamster brain (Mobley et al. 1988). As Aβ protein is not normally expressed at high levels in most non-primate brains, it is difficult in this and other studies to determine whether the upregulation of APP results in the cleavage of this molecule into the more toxic mature Aβ peptide. There are few non-primate animal models that naturally display β-amyloid-containing plaques, making it difficult to study the *in vivo* effects of trophic factors such as NGF on amyloid expression and deposition. Transgenic animals have been generated exhibiting both β-amyloid-containing plaques (Games et al. 1995; Borchelt et al. 1997; Hsiao 1998) and cognitive dysfunction (Holcomb *et al.* 1999; Gordon *et al.* 2001; Puolivali *et al.* 2002) which resemble some of the changes observed in AD. Recently, a triple transgenic mouse has been generated that displays amyloid plaques and tangle-like neurons, and is cognitively impaired (Oddo *et al.* 2003). These animals exhibit cholinergic deficits (Yamasaki *et al.* 2005) but have not yet been examined for defects in the NGF–TrkA–p75^{NTR} system. As discussed above, the aged AD11 NGF knockout mouse displays both amyloid-like plaques and tangle-like structures in an age-dependent manner reminiscent of that seen in AD (Capsoni *et al.* 2000; Ruberti *et al.* 2000). The introduction of NGF via an intranasal route can rescue CBF cell loss and reverse the cortical tau hyperphosphorylation reported in this animal model of AD, further indicating the importance of an intact basal forebrain neurotrophin signalling and receptor system in the disease process (Capsoni *et al.* 2000; Ruberti *et al.* 2000). Interestingly, the AChE inhibitor, galantamine, rescues cholinergic deficits but not NGF-induced APP deposition in the aged AD11 mouse (Capsoni *et al.* 2002; De Rosa *et al.* 2005), suggesting that a combinatorial pharmaceutical delivery approach may be optimal for the treatment of AD. In general, these observations strengthen the previously suggested interconnection between NGF and AD (Appel 1981; Hefti 1983) and present evidence that reducing the availability of NGF to its targets may play a role in the cascade of events leading to sporadic AD.

In contrast, certain non-human primate brains contains β-amyloid-containing plaques that are similar in structure to those seen in aged humans and in patients with AD as a normal consequence of ageing (Mufson *et al.* 1994; Walker 1997). The non-human primate brain provides an opportunity to determine whether grafts of human recombinant NGF (hNGF) secreting cells influence β-amyloid expression in a species which normally expresses this protein as a consequence of ageing. To determine whether NGF accelerates β-amyloid deposition in the brain, amyloid plaque formation was quantified in aged monkey brain following grafts of cells genetically modified to produce and secrete NGF. NGF-secreting cells were injected intracerebroventricularly and plaque density was evaluated 2 weeks post-treatment (Kordower *et al.* 1997), or cells were grafted intraparenchymally and total plaque numbers were quantified 3 months later (Tuszynski *et al.* 1998) using a specific antibody directed against amino acids 1–40 of the β-amyloid protein (Haass *et al.* 1992). Findings were compared with subjects that received control grafts and with unoperated adult (but not aged) monkeys. The adult non-aged monkey brain does not display Aβ-immunoreactive plaques. In aged monkeys, the density of Aβ-containing plaques was increased compared with young monkeys. However, plaque density was not increased in NGF-treated aged subjects compared with aged controls (Kordower *et al.* 1997). Similarly, aged monkeys that received intraparenchymal NGF-secreting fibroblast implants displayed Aβ plaques but did not show an increase in plaque number relative to control aged subjects that received uninfected fibroblast grafts (Tuszynski *et al.* 1998). These data suggest that hNGF delivery to the adult and aged brain for up to 3 months does not upregulate expression of Aβ in aged primates.

13.9 Cholinotrophic therapy for Alzheimer's disease

The reciprocal correlations of reduced cortical TrkA and elevated proNGF levels with MMSE scores suggest that cholinotrophic abnormalities play a key role in cognitive impairment and may underlie the eventual demise of NB neurons and the extensive cholinergic deficits seen in the late stages of AD. Previous studies in non-human primates have shown that recombinant hNGF reverses both age-related and lesion-induced cholinergic neuronal degeneration and promotes cholinergic neurite sprouting (Koliatsos *et al.* 1991; Tuszynski *et al.* 1991; Emerich *et al.* 1994; Kordower *et al.* 1994; Burgos *et al.* 1995; Conner *et al.* 1998; Smith *et al.* 1999). In addition, exogenous NGF rescues age-related and cholinergic lesion-induced spatial memory deficits in rodents (Backman *et al.* 1996; Fischer *et al.* 1987). Interestingly, the age-related reduction in cortical cholinergic fibre innervation can be ameliorated by cellular delivery of hNGF into cholinergic somata in the basal forebrain in non-human primates (Conner *et al.* 2001). Thus restoration of NGF signalling may prove efficacious for the prevention of cognitive deficits resulting from NB dysfunction in AD. However, secondary consequences of the restoration of NGF in human neurodegenerative disease are still not well understood. An interesting result of delivering NGF to the human brain may be to increase the expression of other neurotrophic-like substances. For example, the neuropeptide galanin (GAL), a 29 amino acid neuroactive peptide (30 amino acids in humans), is over-expressed by remaining cholinergic neurons only in advanced AD (Counts *et al.* 2006). This neuropeptide may play a role in cholinergic cell survival and the regulation of ACh production within these neurons (Counts *et al.* 2003). Since GAL is upregulated following intraventricular infusion of NGF (Planas *et al.* 1997), restoring NGF levels during the prodromal stages of AD may produce a secondary benefit by enhancing the production of galanin, resulting in increased cholinergic cell signalling, survival, and/or ACh production. Recently, GAL was shown to attenuate β-amyloid toxicity in rat CBF neurons (Ding *et al.* 2005), suggesting yet another beneficial side effect to augmenting NGF levels in the AD brain.

13.10 Means of restoring nerve growth factor signalling in brain

Several methods of delivering NGF or augmenting NGF function in the brain are being developed to deliver neurotrophins to the brain in a fashion that is less invasive, albeit less targeted, than gene therapy. Small peptide analogues of NGF which contain or mimic the putative active region of NGF should be considered. These molecules possess the potential advantage of being able to cross the blood–brain barrier, because of their low molecular weight, thus eliminating the need for invasive brain delivery via surgical procedures. *In vitro* peptide analogues have been demonstrated to mimic certain NGF actions such as elicitation of neurite outgrowth from PC-12 cells (Massa *et al.* 2003). Longo and coworkers (1997) have investigated the development of small molecule mimetics of neurotrophins (e.g. NGF), which potentially could overcome these limitations. These researchers have established the proof of principle that mimetics of specific NGF domains could prevent neuronal death. Peptidomimetics of the NGF loop 1 domain prevent death via p75^{NTR}-dependent signalling and peptidomimetics of the NGF loop 4 domains reduce death via Trk-related signalling (Longo *et al.* 1997; Massa *et al.* 2003) (Fig. 13.10) [Plate 10]. More recent work by this group has been directed at

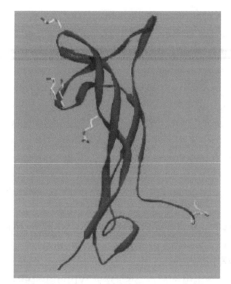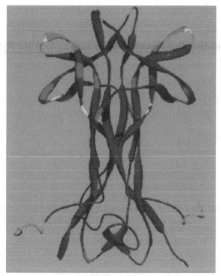

Fig. 13.10 Structure of NGF. (a). NGF monomer structure, with the variable β-turn loops 1, 2, and 4 labelled. Further denoted are residues of particular importance for binding and/or biological activity to p75^{NTR} (lysines 32, 34, and 95). or TrkA (arginine 103, and residues 2–9 of the N-terminus of the mature form, shown in yellow). (b). NGF dimer structure with mapping of residues interacting with p75^{NTR} (red: W21, D30, I31, K32, K34, E35, K74, H75, K88, K95, R100, and R103), TrkA (green: S2, S3, H4, P5, I6, F7, H8, R9, I31, N45, G94, K95, Q96, A97, A98, and R103), or both (yellow: I31, K95, and R103). Note the overlap of the binding sites for the two receptors. (Reproduced with permission from Longo *et al.* 2005.) Please see the colour plate section for a colour version of this figure.

designing pharmacophore queries corresponding to loop domains 1 or 4 of NGF which incorporate characteristics of the NGF crystal structure along with features derived from peptidomimetic structure–activity-relationships (Massa *et al.* 2003). Screening of *in silico* databases containing small non-peptide molecules has identified several candidate NGF domain mimetics. Future studies will be needed to assess whether these compounds prevent neuronal death and whether they act via the trkA or p75^{NTR} receptors. However, to date, the *in vivo* performance of peptide analogues in rescuing degenerating CBF neurons has been less than stellar. The peptide analogue approach is still confounded by the potential hypothetical caveat that it is a non-targeted approach, and thus might elicit growth responses from NGF-responsive non-targeted systems in AD, including Schwann cells, sensory axons, and sympathetic axons.

Alternatively, the ongoing development of TrkA-specific NGF peptidomimetics has provided the groundwork for another potential approach to improving pro-survival signalling in NB neurons during the early stages of AD (LeSauteur *et al.* 1995; Debeir *et al.* 1999; Maliartchouk *et al.* 2000). Most recently, the *in vivo* efficacy of a selective partial TrkA agonist has been demonstrated in cognitively impaired aged rats (Bruno *et al.* 2004). Intraventricular infusion of this ligand resulted in long-lasting cognitive improvement in spatial memory tasks, significantly higher density of cortical VAChT-positive fibre staining, and significantly increased cell size of VAChT-positive soma in the NB (Bruno *et al.* 2004). These data suggest that partial activation of TrkA can upregulate the cholinergic phenotype of the basocortical projection system and improve cognition in the absence of p75^{NTR} activation.

In the early 1990s a novel system of delivering NGF to the CNS was reported in which NGF is attached to molecules that are actively transported into the CNS by being linked to an antibody to the transferrin receptor (Friden *et al.* 1993). This technique is based upon the fact that iron enters the brain following its conversion to transferrin by an active mechanism that requires binding to the transferrin receptor. In fact, the retrograde atrophy of NB neurons that follows cortical excitotoxic damage was prevented by systemic administration of NGF that was conjugated to an antibody against the transferrin receptor (Charles *et al.* 1996). The specificity of this effect was confirmed since a non-conjugated mixture of NGF and the antibody to the transferrin receptor was ineffective. This novel approach eliminates the need for invasive CNS procedures to administer NGF, but suffers from the disadvantage that general delivery of NGF could produce adverse effects from non-targeted structures (Gaffan and Harrison 1989; Williams 1991). Recently, a phase 1 clinical trial was completed in which mild AD subjects receive grafts of autologous fibroblasts genetically modified to secrete human NGF (Tuszynski and Blesch 2004). Cells are implanted directly in the NB, so that this therapeutic strategy may bypass putative defects in the retrograde transport of NGF and may even restore TrkA expression in AD. To date these patients have not shown any adverse effects from these NGF intraparenchymal injections, and indications are that NGF therapy is efficacious and may even slow the course of the disease (Tuszinski *et al.* 2005).

13.11 Brain-derived neurotrophic factor in the progression of Alzheimer's disease

The human *BDNF* gene is complex, consisting of multiple upstream exons governed by individual promoters and differentially spliced to one downstream exon containing the coding sequence for the preproBDNF protein (Aoyama *et al.* 2001; Garzon *et al.* 2002; Liu *et al.* 2005) (Fig. 13.11a). This complex gene structure produces seven transcripts and splice variants account for three more, giving a total of 10 transcripts. Little is known about the regulation of the human *BDNF* transcripts, although in rat they are differentially regulated in a tissue-specific, developmentally specific, and insult-specific manner (Metsis *et al.* 1993; Kokaia *et al.* 1994; Timmusk and Metsis 1994; Lauterborn *et al.* 1998; Oliff *et al.* 1998). Recent evidence suggests that the antisense strand may also be transcribed, which creates the capability of producing inhibitory mRNA (Liu *et al.* 2005). Several studies have shown that the mRNA for *BDNF* is dramatically decreased in cortex, hippocampus, and basal forebrain in end stage AD (Phillips *et al.* 1991; Murray *et al.* 1994; Holsinger *et al.* 2000; Fahnestock *et al.* 2002). Current findings from our group indicate that, although seven different transcripts of the *BDNF* gene are expressed in human cortex, only three are down regulated in end stage AD (Garzon *et al.* 2002), providing clues to possible regulatory mechanisms (Fig. 13.11b). We have recently

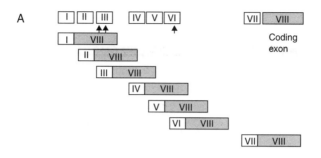

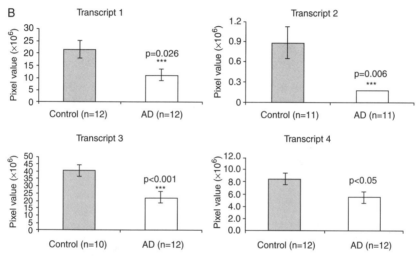

Fig. 13.11 (a) Multiple BDNF transcripts in human brain tissue. Each upstream exon (I–VII) is individually spliced to exon VIII, which codes for the 35 kDa preproBDNF protein. Arrowheads indicate alternative splice variants. Numbers to the right of each transcript represent the relative number of copies determined by qPCR of human neuroblastoma cell line SH-SY5Y; the relative copy number is similar in human hippocampal tissue. (b) Relative levels of mRNA for the four of seven BDNF transcripts in control versus AD samples. The *y* axes show pixel intensity values determined by phosphorimage analysis. Error bars represent standard error of the mean. Statistically significant *P*-values are shown. Note that transcripts 3 and 4 (Garzon *et al.* 2002) are now known as transcripts 4 and 5.

demonstrated that oligomeric, but not fibrillar, Aβ downregulates the most highly expressed of these three *BDNF* transcripts (Garzon and Fahnestock 2005).

13.12 **ProBDNF in prodromal and early Alzheimer's disease**

Both NGF and BDNF proteins are synthesized as precursors, proNGF and proBDNF, which are secreted molecules (Fahnestock *et al.* 2001; Mowla *et al.* 2001; Egan *et al.* 2003;

Pang *et al.* 2004). Whereas proNGF remains as a precursor in the CNS (Fahnestock *et al.* 2001), both proBDNF and the mature forms are present in human brain (Fahnestock *et al.* 2002; Michalski and Fahnestock 2003). Whether proBDNF is neurotrophic or apoptotic is controversial (Mowla *et al.* 2001; Fayard *et al.* 2005; Teng *et al.* 2005). ProBDNF can be converted to mature BDNF intracellularly by furin and related proprotein convertases (Seidah *et al.* 1996; Mowla *et al.* 2001) or extracellularly by plasmin (Pang *et al.* 2004). The importance of both forms of BDNF is evident, since proBDNF and mature BDNF are implicated in synaptic plasticity. In models of learning and memory, proBDNF supports long-term depression (LTD) via binding to p75^{NTR}, whereas mature BDNF enhances long-term potentiation (LTP) by binding to TrkB (Woo *et al.* 2005). Furthermore, a polymorphism in the pro-domain of proBDNF, Val66Met, affects hippocampal activation and memory in human subjects, with Met/Met variants exhibiting altered proBDNF secretion and defects in episodic memory (Egan *et al.* 2003). Lastly, there is one report of a genetic association between this pro-domain polymorphism and sporadic AD (Ventriglia *et al.* 2002).

During the progression of AD, reduced BDNF mRNA is reflected in decreased cortical proBDNF (Michalski and Fahnestock 2003; Peng *et al.* 2005) (Fig. 13.12a) and mature

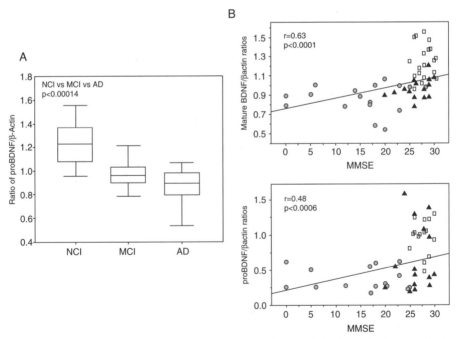

Fig. 13.12 (a) Box plot of proBDNF protein levels normalized to β-actin in the parietal cortex of subjects clinically diagnosed with NCI, MCI, or AD. The Kruskal–Wallis test was used for comparison. ProBDNF levels were significantly reduced in MCI and AD compared with NCI cases. (b) Scattergrams showing that reductions in levels of mature BDNF (top) and proBDNF (bottom) correlate with reductions in ante mortem MMSE scores. Symbols: square, NCI; triangle, MCI; circle, AD.

BDNF protein (Ferrer *et al.* 1999; Peng *et al.* 2005). The reduction in cortical proBDNF and BDNF protein in both early- and endstage AD is comparable with the decrease in total BDNF found in transgenic mice exhibiting synaptic transmission and memory defects (Korte *et al.* 1995; Patterson *et al.* 1996; Saarelainen *et al.* 2000). Furthermore, the decrease in cortical proBDNF and BDNF correlates with cognitive impairment during the progression of AD (Peng *et al.* 2005) (Fig. 13.12b). These data suggest that a reduction in proBDNF and BDNF occurs early in the disease process and plays a major role in the synaptic loss and associated cognitive dysfunction seen even in the earliest stages of AD.

In MCI subjects, the amount of proBDNF decreased by 21% compared with controls, whereas mature BDNF was reduced by 34% (Peng *et al.* 2005) (Fig. 13.12a). Thus, the decrease in proBDNF and mature BDNF precedes the decline in ChAT activity, which occurs later in AD (DeKosky *et al.* 1992, 2002; Davis *et al.* 1999). In AD subjects, proBDNF decreased by 30% while mature BDNF reduced by 62% compared with age-matched controls (Fig. 13.12a). At both early and later stages of the disease, the decrease in BDNF was more severe than for proBDNF. If cognitive function in AD depends upon an appropriate balance between LTP and LTD, then the more dramatic decrease in mature BDNF compared with proBDNF may increase the ratio of proBDNF to mature BDNF, favouring inhibition in the MCI and AD brain resulting in a disruption of cognitive function.

13.13 BDNF therapy for Alzheimer's disease

Down regulation of cortical and hippocampal BDNF and the loss of CBF and cortical TrkB (see above) in AD (Fig. 13.7) are expected to result in a loss of pro-survival signalling in the CBF similar to the loss of TrkA. In addition, reduced cortical and hippocampal proBDNF and BDNF may result in a loss of signalling within target tissues, a problem not found with NGF. The loss of BDNF/TrkB signalling within target tissues is expected to compromise neuronal survival and function within these regions (Ghosh *et al.* 1994; Lindholm *et al.* 1996).

Because of its survival activity for cortical, hippocampal, and CBF neurons and their connections (Alderson *et al.* 1990; Knusel *et al.* 1992; Ghosh *et al.* 1994; Lindholm *et al.* 1996; Ando *et al.* 2002), its regulation of synaptic transmission and excitatory properties of hippocampal and cortical neurons (Kang and Schuman 1995; Osehobo *et al.* 1999; Scharfman 1997), its effects on learning and memory and synaptic plasticity (McAllister *et al.* 1999; Aicardi *et al.* 2004; Pang *et al.* 2004; Barco *et al.* 2005; Bramham and Messaoudi 2005), and its prevalence in relatively high amounts in regions of the brain affected by AD, the dramatic decrease in the mRNA for BDNF in cortex, hippocampus, and basal forebrain is expected to have severe consequences for the AD brain. In fact, a polymorphism in the *BDNF* gene that compromises the processing of proBDNF to mature BDNF and affects secretion of mature BDNF results in poor memory in human subjects (Hariri *et al.* 2003), demonstrating that decreased BDNF in AD could ultimately lead to impaired memory and cognitive dysfunction. Furthermore, transgenic mice with reduced BDNF levels exhibit defects in synaptic transmission, LTP, and memory tests which are rescued by recombinant BDNF (Korte *et al.* 1995; Patterson *et al.* 1996; Olofsdotter *et al.* 2000; Saarelainen *et al.* 2000), suggesting that BDNF therapy may be effective in AD. BDNF has also been demonstrated to increase in the brain following cognitive performance such as environmental enrichment and physical activity such as repeated exercise sessions (Molteni *et al.* 2002; Vaynman *et al.* 2004; Gomez-Pinilla and Vaynman 2005; Vaynman and Gomez-Pinilla 2005). Exercise also restores levels of neurotrophins and synaptic plasticity following injury in animal models (Adlard and Cotman 2004; Molteni *et al.* 2004; Vaynman and Gomez-Pinilla 2005).

The poor blood–brain barrier (BBB) permeability of BDNF has led to development of several approaches for CNS delivery of BDNF. Anti-transferrin receptor antibody has been used as a BBB delivery vector for chimeric BDNF conjugated to the antibody via avidin/biotin. This BDNF chimeric peptide was neuroprotective in rats subjected to transient forebrain ischaemia, permanent focal ischaemia, or transient focal ischaemia (Pardridge 2002). Other approaches include transplantation of BDNF-expressing cells in the brain and viral delivery of BDNF (Rubio *et al.* 1999; Kells *et al.* 2004). However, two major issues remain to be resolved. First, chronic CNS administration of BDNF downregulates its receptor, and bolus administration can cause unwanted changes in excitability (Frank *et al.* 1997; Xu *et al.* 2004), making the mode of BDNF administration problematic. Secondly, targeted parenchymal administration rather than intracerebroventricular administration is necessitated by the large amount of truncated TrkB and TrkB on the ependymal cells lining the cerebral ventricles (Yan *et al.* 1994), which effectively prevents BDNF penetration from the ventricles, and by unwanted side effects such as appetite suppression and weight loss caused by penetration of BDNF into areas of the brain such as the hypothalamus (Pelleymounter *et al.* 1995). On the other hand, targeted parenchymal administration makes it difficult to combat BDNF decreases in multiple systems (e.g. entorhinal cortex, hippocampus, cortex, and basal forebrain). Combinational therapies using two neurotrophins may be desirable to achieve additive or synergistic effects on multiple cell systems.

13.14 **Concluding comments**

Figure 13.13 [Plate 11] summarizes current findings dealing with the pathobiology of the cholinotrophic basocortical projection system during the progression of AD. The preservation of ChAT/VAChT-containing NB neurons in the face of reduced numbers of TrkA- and p75[NTR]-positive NB neurons in MCI and mild AD indicates that there is not a frank loss of cholinergic perikarya *per se* but a phenotypic down regulation of TrkA protein in these neurons that may compromise neurotrophic support early in the disease process. Interestingly, the reduction of TrkA receptors in NB neurons seen in MCI also occurs in the presence of normal or elevated cortical ChAT activity, suggesting that compensatory cellular mechanisms are able to maintain the cholinergic function of these cells during the prodromal stages of AD (DeKosky *et al.* 2002). The stability of cortical TrkA and p75^{NTR} protein levels in MCI subjects, despite the loss of these receptors within the NB, may be due to a defect in anterograde transport or target-derived NGF signalling. Thus the selective loss of cortical TrkA protein observed during the transition from MCI to AD may signify the onset of neurotrophic signalling dysfunction that ultimately leads to the demise of these cholinergic neurons. We hypothesize that NB neuronal dysfunction is related to an imbalance between TrkA-mediated survival signalling and p75^{NTR}-mediated pro-apoptotic signalling. The prevalence of proNGF rather than NGF in the normal human cortex suggests that, in the presence of normal levels of TrkA, proNGF is neurotrophic for CBF. However, in AD, decreased levels of TrkA in the presence of normal or increased levels of p75^{NTR}, coupled with an increase in cortical proNGF due to failed retrograde transport, combine to cause CBF degeneration and, later in the disease, CBF cell loss (Counts and Mufson 2005). The use of NGF replacement therapy (Tuszynski and Blesch 2004; Tuszinski *et al.* 2005) in combination with TrkA-selective agonists (Bruno *et al.* 2004) may prove to be a useful pharmacological approach for the treatment of cognitive loss due to degeneration of the cholinergic NB neurons in AD. Finally, if NGF is 'off-trk' (Mufson 1997) during the early stages of AD, then this phenomenon may be a novel diagnostic marker for the prodromal stages of AD. With respect to BDNF, the early and significant downregulation of BDNF mRNA and the concomitant reduction in both proBDNF and mature BDNF protein, coupled with the loss of TrkB receptors by CBF neurons, most likely contribute to the synaptic degeneration, atrophy, and cognitive dysfunction seen in the AD brain. CBF neurons rely on both NGF and BDNF for their survival and function, and so the loss of this neurotrophic support provides a double hit. Although the administration of NGF as a therapeutic modality in AD has been the subject of intense translational study and is being investigated in clinical trials, administration of BDNF in AD has been less well explored. Finally, the extensive overlap of the cholinotrophic neurodegenerative phenotype observed between MCI and early AD suggests that MCI is probably prodromal AD. However, since the MCI population examined in the present series of studies contained both amnestic and non-amnestic/multidomain cases (Bennett *et al.* 2002; Petersen 2004), it remains to be determined whether the cholinotrophic phenotype is similar for each subtype of MCI. If there are differences, then neurotrophin therapy may be more appropriate for a specific MCI subtype early in the disease process.

Acknowledgements

These studies were supported by the National Institute of Aging grants AG000257, AG14449, AG16668, AG09446; AG10161, and AG026032, and by a grant from the Scottish Rite Charitable Foundation of Canada. We are indebted to the altruism and support of the hundreds of nuns, priests and brothers participating in the Religious Orders Study. A list of participating groups can be found at http://www.rush.edu/rumc/page-R12394.html. We would like to thank all our colleagues who participated in the numerous investigations that formed the basis of this review.

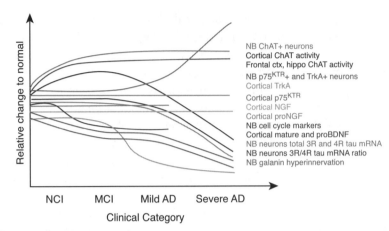

Fig. 13.13 Summary diagram of cholinotrophic alterations found in the basocortical projection system during the progression of AD. Note that the number of ChAT-positive NB neurons is preserved while the number of NB neurons expressing the NGF receptor proteins TrkA and p75^{NTR} is reduced in MCI and early-stage AD. The frank loss of ChAT-positive NB neurons is associated with late-stage AD. In contrast, cortical levels of TrkA protein are stable in MCI but reduced in AD, while p75^{NTR} levels in the cortex are stable throughout the disease progression. ProNGF protein levels increase during MCI and AD. Cortical levels of both proBDNF and mature BDNF decrease during AD progression and may also impact cholinergic cell survival as well as many other neuronal populations subserving cognitive function. In addition to these cholinotrophic alterations, there is a *de novo* expression of cell cycle markers in NB neurons in MCI and early AD that may precede apoptotic signalling attributable to increased p75^{NTR} signalling. Moreover, the ratio of 3R to 4R tau transcripts is reduced in MCI and AD in NB neurons, potentially impacting neurofibrillary pathogenesis. Finally, levels of galanin hyperinnervation within the NB are increased only during the advanced stages of the disease but may have a beneficial impact on remaining neurons. The lines do not represent actual numerical values but a general trend for each factor depicted. Please see the colour plate section for a colour version of this figure.

References

Adlard PA and Cotman CW (2004). Voluntary exercise protects against stress-induced decreases in brain-derived neurotrophic factor protein expression. *Neuroscience* **124**, 985–92.

Aicardi G, Argilli E, Cappello S, *et al* (2004). Induction of long-term potentiation and depression is reflected by corresponding changes in secretion of endogenous brain-derived neurotrophic factor. *Proceedings of the National Academy of Sciences of the USA* **101**, 15788–92.

Alderson RF, Alterman AL, Barde YA, and Lindsay RM (1990). Brain-derived neurotrophic factor increases survival and differentiated functions of rat septal cholinergic neurons in culture. *Neuron* **5**, 297–306.

Allen SJ, Dawbarn D, and Wilcock GK (1988). Morphometric immunochemical analysis of neurons in the nucleus basalis of Meynert in Alzheimer's disease. *Brain Research* **454**, 275–81.

Allen SJ, MacGowan SH, Treanor JJ, Feeney R, Wilcock GK, and Dawbarn D (1991). Normal beta-NGF content in Alzheimer's disease cerebral cortex and hippocampus. *Neuroscience Letters* **131**, 135–9.

Allen SJ, Wilcock GK, and Dawbarn D (1999). Profound and selective loss of catalytic TrkB immunoreactivity in Alzheimer's disease. *Biochemical and Biophysical Research Communications* **264**, 648–51.

Altar CA, Cai N, Bliven T, *et al.* (1997). Anterograde transport of brain-derived neurotrophic factor and its role in the brain. *Nature* **389**, 856–60.

Ando S, Kobayashi S, Waki H, *et al.* (2002). Animal model of dementia induced by entorhinal synaptic damage and partial restoration of cognitive deficits by BDNF and carnitine. *Journal of Neuroscience Research* **70**, 519–27.

Aoyama M, Asai K, Shishikura T, *et al.* (2001). Human neuroblastomas with unfavorable biologies express high levels of brain-derived neurotrophic factor mRNA and a variety of its variants. *Cancer Letters* **164**, 51–60.

Appel SH (1981). A unifying hypothesis for the cause of amyotrophic lateral sclerosis, parkinsonism, and Alzheimer disease. *Annals of Neurology* **10**, 499–505.

Backman C, Rose GM, Hoffer BJ, *et al.* (1996). Systemic administration of a nerve growth factor conjugate reverses age-related cognitive dysfunction and prevents cholinergic neuron atrophy. *Journal of Neuroscience* **16**, 5437–42.

Bamji SX, Majdan M, Pozniak CD, *et al.* (1998). The p75 neurotrophin receptor mediates neuronal apoptosis and is essential for naturally occurring sympathetic neuron death. *Journal of Cell Biology* **140**, 911–23.

Barbacid M (1995). Structural and functional properties of the TRK family of neurotrophin receptors. *Annals of the New York Academy of Sciences* **766**, 442–58.

Barco A, Patterson S, Alarcon JM, *et al.* (2005). Gene expression profiling of facilitated L-LTP in VP16-CREB mice reveals that BDNF is critical for the maintenance of LTP and its synaptic capture. *Neuron* **48**, 123–37.

Barker PA and Shooter EM (1994). Disruption of NGF binding to the low affinity neurotrophin receptor p75LNTR reduces NGF binding to TrkA on PC12 cells. *Neuron* **13**, 203–15.

Barnabe-Heider F and Miller FD (2003). Endogenously produced neurotrophins regulate survival and differentiation of cortical progenitors via distinct signaling pathways. *Journal of Neuroscience* **23**, 5149–60.

Bartus RT (2000). On neurodegenerative diseases, models, and treatment strategies: lessons learned and lessons forgotten a generation following the cholinergic hypothesis. *Experimental Neurology* **163**, 495–529.

Bartus RT, Dean RL, 3rd, Beer B, and Lippa AS (1982). The cholinergic hypothesis of geriatric memory dysfunction. *Science* **217**, 408–14.

Baxter MG and Chiba AA (1999). Cognitive functions of the basal forebrain. *Current Opinion in Neurobiology* **9**, 178–83.

Bennett DA, Wilson RS, Schneider JA, *et al.* (2002). Natural history of mild cognitive impairment in older persons. *Neurology* **59**, 198–205.

Bhakar AL, Howell JL, Paul CE, *et al.* (2003). Apoptosis induced by p75NTR overexpression requires Jun kinase-dependent phosphorylation of Bad. *Journal of Neuroscience* **23**, 11373–81.

Blasko I, Lederer W, Oberbauer H, *et al.* (2005). Measurement of thirteen biological markers in CSF of patients with Alzheimer's disease and other dementias. *Dementia and Geriatric Cognitive Disorders* **21**, 9–15.

Bonni A, Brunet A, West AE, Datta SR, Takasu MA, and Greenberg ME (1999). Cell survival promoted by the RAS-MAPK signaling pathway by transcription-dependent and -independent mechanisms. *Science* **286**, 1358–62.

Borchelt DR, Ratovitski T, van Lare J, *et al.* (1997). Accelerated amyloid deposition in the brains of transgenic mice coexpressing mutant presenilin 1 and amyloid precursor proteins. *Neuron* **19**, 939–45.

Bothwell M (1995). Functional interactions of neurotrophins and neurotrophin receptors. *Annual Reviews of Neuroscience* **18**, 223–53.

Bramham CR and Messaoudi E. (2005). BDNF function in adult synaptic plasticity: the synaptic consolidation hypothesis. *Progress in Neurobiology* **76**, 99–125.

Bronfman FC and Fainzilber M (2004). Multi-tasking by the p75 neurotrophin receptor: sortilin things out? *EMBO Reports* **5**, 867–71.

Bruno MA., Clarke PB, Seltzer A, *et al.* (2004). Long-lasting rescue of age-associated deficits in cognition and the CNS cholinergic phenotype by a partial agonist peptidomimetic ligand of TrkA. *Journal of Neuroscience* **24**, 8009–18.

Bruno MA and Cuello AC (2006). Activity-dependent release of precursor nerve growth factor, conversion to mature nerve growth factor, and its degradation by a protease cascade. *Proceedings of the National Academy of Sciences of the USA* **25**, 6735–40

Burgos I, Cuello AC, Liberini P, Pioro E, and Masliah E (1995). NGF-mediated synaptic sprouting in the cerebral cortex of lesioned primate brain. *Brain Research* **692**, 154–60.

Busser J, Geldmacher DS, and Herrup K (1998). Ectopic cell cycle proteins predict the sites of neuronal cell death in Alzheimer's disease brain. *Journal of Neuroscience* **18**, 2801–7.

Campenot RB and MacInnis BL (2004). Retrograde transport of neurotrophins: fact and function. *Journal of Neurobiology* **58**, 217–29.

Capsoni S, Ugolini G, Comparini A, Ruberti F, Berardi N, and Cattaneo A. (2000). Alzheimer-like neurodegeneration in aged antinerve growth factor transgenic mice. *Proceedings of the National Academy of Sciences of the USA* **97**, 6826–31.

Capsoni S, Giannotta S, and Cattaneo A (2002). Nerve growth factor and galantamine ameliorate early signs of neurodegeneration in anti-nerve growth factor mice. *Proceedings of the National Academy of Sciences of the USA* **99**, 12432–7.

Capsoni S, Giannotta S, Stebel M, *et al.* (2004). Ganstigmine and donepezil improve neurodegeneration in AD11 antinerve growth factor transgenic mice. *American Journal of Alzheimer's Disease and Other Dementias* **19**, 153–60.

Carter BD, Kaltschmidt C, Kaltschmidt B, *et al.* (1996). Selective activation of NF-kappa B by nerve growth factor through the neurotrophin receptor p75. *Science* **272**, 542–5.

Casaccia-Bonnefil P, Carter BD, Dobrowsky RT, and Chao MV (1996). Death of oligodendrocytes mediated by the interaction of nerve growth factor with its receptor p75. *Nature* **383**, 716–19.

Chao MV and Hempstead BL (1995). p75 and Trk: a two-receptor system. *Trends in Neuroscience* **18**, 321–6.

Chao MV, Bothwell MA, Ross AH, *et al.* (1986). Gene transfer and molecular cloning of the human NGF receptor. *Science* **232**, 518–21.

Charles V, Mufson EJ, Friden PM, Bartus RT, and Kordower JH (1996). Atrophy of cholinergic basal forebrain neurons following excitotoxic cortical lesions is reversed by intravenous administration of an NGF conjugate. *Brain Research* **728**, 193–203.

Chu Y, Cochran EJ, Bennett DA, Mufson EJ, and Kordower, JH (2001). Down-regulation of trkA mRNA within nucleus basalis neurons in individuals with mild cognitive impairment and Alzheimer's disease. *Journal of Comparative Neurology* **437**, 296–307.

Cohen S and Levi-Montalcini R (1957). Purification and properties of a nerve growth-promoting factor isolated from mouse sarcoma 180. *Cancer Research* **17**, 15–20.

Conner JM, Lauterborn JC, and Gall CM (1998). Anterograde transport of neurotrophin proteins in the CNS: a reassessment of the neurotrophic hypothesis. *Reviews in the Neurosciences* **9**, 91–103.

Conner JM, Darracq MA, Roberts J, and Tuszynski MH (2001). Nontropic actions of neurotrophins: subcortical nerve growth factor gene delivery reverses age-related degeneration of primate cortical cholinergic innervation. *Proceedings of the National Academy of Sciences of the USA* **98**, 1941–6.

Cooper JD, Lindholm D, and Sofroniew MV (1994). Reduced transport of [125I]nerve growth factor by cholinergic neurons and down-regulated TrkA expression in the medial septum of aged rats *Neuroscience* **62**, 625–9.

Cooper JD, Salehi A, Delcroix JD, *et al.* (2001). Failed retrograde transport of NGF in a mouse model of Down's syndrome: reversal of cholinergic neurodegenerative phenotypes following NGF infusion. *Proceedings of the National Academy of Sciences of the USA* **98**, 10439–44.

Counts SE and Mufson EJ (2005). The role of nerve growth factor receptors in cholinergic basal forebrain degeneration in prodromal Alzheimer disease. *Journal of Neuropathology and Experimental Neurology* **64**, 263–72.

Counts SE, Perez SE, Ginsberg SD, De Lacalle S, and Mufson EJ (2003). Galanin in Alzheimer disease. *Molecular Interventions* **3**, 137–56.

Counts SE, Nadeem M, Wuu J, Ginsberg SD, Saragovi HU, and Mufson EJ (2004). Reduction of cortical TrkA but not p75(NTR) protein in early-stage Alzheimer's disease. *Annals of Neurology* **56**, 520–31.

Counts SE, Chen EY, Che S, *et al.* (2006). Galanin fiber hypertrophy within the cholinergic nucleus basalis during the progression of Alzheimer's disease. *Dementia and Geriatric Cognitive Disorders* **21**, 205–14.

Coyle JT, Price DL, and DeLong MR (1983). Alzheimer's disease: a disorder of cortical cholinergic innervation. *Science* **219**, 1184–90.

Crowder RJ and Freeman RS (1998). Phosphatidylinositol 3-kinase and Akt protein kinase are necessary and sufficient for the survival of nerve growth factor-dependent sympathetic neurons. *Journal of Neuroscience* **18**, 2933–43.

Crutcher KA, Scott SA, Liang S, Everson WV, and Weingartner J (1993). Detection of NGF-like activity in human brain tissue: increased levels in Alzheimer's disease. *Journal of Neuroscience* **13**, 2540–50.

Davey F and Davies AM (1998). TrkB signalling inhibits p75-mediated apoptosis induced by nerve growth factor in embryonic proprioceptive neurons. *Current Biology* **8**, 915–18.

Davies P and Maloney AJ (1976). Selective loss of central cholinergic neurons in Alzheimer's disease *Lancet* **ii**, 1403.

Davis KL, Mohs RC, Marin D *et al.* (1999). Cholinergic markers in elderly patients with early signs of Alzheimer disease. *Journal of the American Medical Association* **281**, 1401–6.

Dawbarn D and Allen SJ (2003). Neurotrophins and neurodegeneration. *Neuropathology and Applied Neurobiology* **29**, 211–30

Debeir T, Saragovi HU, and Cuello AC (1999). A nerve growth factor mimetic TrkA antagonist causes withdrawal of cortical cholinergic boutons in the adult rat. *Proceedings of the National Academy of Sciences of the USA* **96**, 4067–72.

DeKosky ST, Harbaugh RE, Schmitt FA, *et al.* (1992). Cortical biopsy in Alzheimer's disease: diagnostic accuracy and neurochemical, neuropathological, and cognitive correlations. Intraventricular Bethanecol Study Group. *Annals of Neurology* **32**, 625–32.

DeKosky ST, Ikonomovic MD, Styren SD, *et al.* (2002). Upregulation of choline acetyltransferase activity in hippocampus and frontal cortex of elderly subjects with mild cognitive impairment. *Annals of Neurology* **51**, 145–55.

De Lacalle S, Cooper JD, Svendsen CN, Dunnett SB, and Sofroniew MV (1996). Reduced retrograde labelling with fluorescent tracer accompanies neuronal atrophy of basal forebrain cholinergic neurons in aged rats *Neuroscience* **75**, 19–27.

Delcroix JD, Valletta JS, Wu C, Hunt SJ, Kowal AS, and Mobley WC (2003). NGF signaling in sensory neurons: evidence that early endosomes carry NGF retrograde signals. *Neuron* **39**, 69–84.

Derijard B, Hibi M, Wu IH, *et al.* (1994). JNK1: a protein kinase stimulated by UV light and Ha-Ras that binds and phosphorylates the c-Jun activation domain. *Cell* **76**, 1025–37.

De Rosa R, Garcia AA, Braschi C, *et al.* (2005). Intranasal administration of nerve growth factor (NGF). rescues recognition memory deficits in AD11 anti-NGF transgenic mice. *Proceedings of the National Academy of Sciences of the USA* **102**, 3811–16.

Dethleffsen K, Heinrich G, Lauth M, Knapik EW, and Meyer M. (2003). Insert-containing neurotrophins in teleost fish and their relationship to nerve growth factor. *Molecular and Cellular Neurosciences* **24**, 380–94.

Ding X, Mactavish D, Kar S, and Jhamandas JH (2005). Galanin attenuates beta-amyloid (Abeta). toxicity in rat cholinergic basal forebrain neurons. *Neurobiology of Disease* **21**, 413–20.

Dudek H, Datta SR, Franke TF, *et al.* (1997). Regulation of neuronal survival by the serine-threonine protein kinase Akt. *Science* **275**, 661–5.

Egan MF, Kojima M, Callicott JH *et al.* (2003). The BDNF val66met polymorphism affects activity-dependent secretion of BDNF and human memory and hippocampal function. *Cell* **112**, 257–69.

Emerich DF, Winn SR, Harper J, Hammang JP, Baetge EE, and Kordower JH (1994). Implants of polymer-encapsulated human NGF-secreting cells in the nonhuman primate: rescue and sprouting of degenerating cholinergic basal forebrain neurons. *Journal of Comparative Neurology* **349**, 148–64.

Fahnestock M, Scott SA, Jette N, Weingartner JA, and Crutcher KA (1996). Nerve growth factor mRNA and protein levels measured in the same tissue from normal and Alzheimer's disease parietal cortex. *Brain Research: Molecular Brain Research* **42**, 175–8.

Fahnestock M, Michalski B, Xu B, and Coughlin MD (2001). The precursor pro-nerve growth factor is the predominant form of nerve growth factor in brain and is increased in Alzheimer's disease. *Molecular and Cellular Neurosciences* **18**, 210–20.

Fahnestock M, Garzon D, Holsinger RM, and Michalski B (2002). Neurotrophic factors and Alzheimer's disease: are we focusing on the wrong molecule? *Journal of Neural Transmission. Supplementum*, 241–52.

Fahnestock M, Yu G, Michalski B, *et al.* (2004). The nerve growth factor precursor proNGF exhibits neurotrophic activity but is less active than mature nerve growth factor. *Journal of Neurochemistry* **89**, 581–92.

Farinelli SE and Greene LA (1996). Nitric oxide delays the death of trophic factor-deprived PC12 cells and sympathetic neurons by a cGMP-mediated mechanism. *Journal of Neuroscience* **16**, 1150–62.

Fayard B, Loeffler S, Weis J, Vogelin E, and Kruttgen, A. (2005). The secreted brain-derived neurotrophic factor precursor pro-BDNF binds to TrkB and p75NTR but not to TrkA or TrkC. *Journal of Neuroscience Research* **80**, 18–28.

Ferrer I, Marin C, Rey MJ, *et al.* (1999). BDNF and full-length and truncated TrkB expression in Alzheimer disease. Implications in therapeutic strategies. *Journal of Neuropathology and Experimental Neurology* **58**, 729–39.

Fischer W, Wictorin K, Bjorklund A, Williams LR, Varon S, and Gage FH (1987). Amelioration of cholinergic neuron atrophy and spatial memory impairment in aged rats by nerve growth factor. *Nature* **329**, 65–8.

Frade JM (2000). Unscheduled re-entry into the cell cycle induced by NGF precedes cell death in nascent retinal neurones. *Journal of Cell Science* **113**, 1139–48.

Frade JM, Rodriguez-Tebar A, and Barde YA (1996). Induction of cell death by endogenous nerve growth factor through its p75 receptor. *Nature* **383**, 166–8.

Frank L, Wiegand SJ, Siuciak JA, Lindsay RM, and Rudge JS (1997). Effects of BDNF infusion on the regulation of TrkB protein and message in adult rat brain. *Experimental Neurology* **145**, 62–70.

Freeman RS, Estus S, and Johnson EM, Jr (1994). Analysis of cell cycle-related gene expression in postmitotic neurons: selective induction of cyclin D1 during programmed cell death. *Neuron* **12**, 343–55.

Friden PM, Walus LR, Watson P, *et al.* (1993). Blood–brain barrier penetration and *in vivo* activity of an NGF conjugate. *Science* **259**, 373–7.

Friedman WJ (2000). Neurotrophins induce death of hippocampal neurons via the p75 receptor. *Journal of Neuroscience* **20**, 6340–6.

Gaffan D and Harrison S (1989). A comparison of the effects of fornix transection and sulcus principalis ablation upon spatial learning by monkeys *Behavioural Brain Research* **31**, 207–20.

Games D, Adams D, Alessandrini R, *et al.* (1995). Alzheimer-type neuropathology in transgenic mice overexpressing V717F beta-amyloid precursor protein. *Nature* **373**, 523–7.

Garzon DJ and Fahnestock M (2005). Transcriptional down-regulation of BDNF by oligomeric amyloid-β application to human neuroblastoma cells. Presented at Society for Neuroscience, Washington, DC. Available online at: http://sfn.scholarone.com/itin2005/index.html

Garzon D, Yu G, and Fahnestock M. (2002). A new brain-derived neurotrophic factor transcript and decrease in brain-derived neurotrophic factor transcripts 1, 2 and 3 in Alzheimer's disease parietal cortex. *Journal of Neurochemistry* **82**, 1058–64.

Ghosh A, Carnahan J, and Greenberg ME (1994). Requirement for BDNF in activity-dependent survival of cortical neurons. *Science* **263**, 1618–23.

Gilmor ML, Erickson JD, Varoqui H, *et al.* (1999). Preservation of nucleus basalis neurons containing choline acetyltransferase and the vesicular acetylcholine transporter in the elderly with mild cognitive impairment and early Alzheimer's disease. *Journal of Comparative Neurology* **411**, 693–704.

Ginsberg SD, Che S, Counts SE, and Mufson EJ (2006a). Down regulation of trk but not p75NTR gene expression in single cholinergic basal forebrain neurons mark the progression of Alzheimer's disease. *Journal of Neurochemistry* **97**, 475–87.

Ginsberg SD, Che S, Counts SE and Mufson EJ (2006b). Shift in the 3Rtau and 4Rtau mRNA ratio during the progression of Alzheimer's disease. *Journal of Neurochemistry* **96**,1401–8.

Ginty DD, Bonni A, and Greenberg ME (1994). Nerve growth factor activates a Ras-dependent protein kinase that stimulates c-fos transcription via phosphorylation of CREB. *Cell* **77**, 713–25.

Goedert M, Fine A, Hunt SP, and Ullrich A (1986). Nerve growth factor mRNA in peripheral and central rat tissues and in the human central nervous system: lesion effects in the rat brain and levels in Alzheimer's disease. *Brain Research* **387**, 85–92.

Goedert M, Spillantini MG, Jakes R, Rutherford D, and Crowther RA (1989a). Multiple isoforms of human microtubule-associated protein tau: sequences and localization in neurofibrillary tangles of Alzheimer's disease. *Neuron* **3**, 519–26.

Goedert M, Spillantini MG, Potier MC, Ulrich J, and Crowther RA (1989b). Cloning and sequencing of the cDNA encoding an isoform of microtubule-associated protein tau containing four tandem repeats: differential expression of tau protein mRNAs in human brain. *EMBO Journal* **8**, 393–9.

Gomez-Pinilla F. and Vaynman S. (2005). License to run: exercise impacts functional plasticity in the intact and injured central nervous system by using neurotrophins. *Experimental Neurology* **192**, 235–43.

Gordon MN, King DL, Diamond DM, *et al.* (2001). Correlation between cognitive deficits and Abeta deposits in transgenic APP+PS1 mice. *Neurobiology of Aging* **22**, 377–85.

Grimes ML, Zhou J, Beattie EC, *et al.* (1996). Endocytosis of activated TrkA: evidence that nerve growth factor induces formation of signaling endosomes. *Journal of Neuroscience* **16**, 7950–64.

Gross A, Jockel J, Wei MC, and Korsmeyer SJ (1998). Enforced dimerization of BAX results in its translocation, mitochondrial dysfunction and apoptosis. *EMBO Journal* **17**, 3878–85.

Gu C, Casaccia-Bonnefil P, Srinivasan A, and Chao MV (1999). Oligodendrocyte apoptosis mediated by caspase activation. *Journal of Neuroscience* **19**, 3043–9.

Haass C, Schlossmacher MG, Hung AY, *et al.* (1992). Amyloid beta-peptide is produced by cultured cells during normal metabolism. *Nature* **359**, 322–5.

Hariri AR, Goldberg TE, Mattay VS, *et al.* (2003). Brain-derived neurotrophic factor va166met polymorphism affects human memory-related hippocampal activity and predicts memory performance. *Journal of Neuroscience* **23**, 6690–4.

Hartikka J and Hefti F (1988). Comparison of nerve growth factor's effects on development of septum, striatum, and nucleus basalis cholinergic neurons *in vitro*. *Journal of Neuroscience Research* **21**, 352–64.

Hatanaka H, Nihonmatsu I, and Tsukui H (1988). Nerve growth factor promotes survival of cultured magnocellular cholinergic neurons from nucleus basalis of Meynert in postnatal rats. *Neuroscience Letters* **90**, 63–8.

Hefti F (1983). Is Alzheimer's disease caused by a lack of nerve growth factor? *Annals of Neurology* **1**, 109–10.

Hefti F (1986). Nerve growth factor promotes survival of septal cholinergic neurons after fimbrial transections. *Journal of Neuroscience* **6**, 2155–62.

Hefti F, Hartikka J, and Knusel B (1989). Function of neurotrophic factors in the adult and aging brain and their possible use in the treatment of neurodegenerative diseases. *Neurobiology of Aging* **10**, 515–33.

Hempstead BL, Martin-Zanca D, Kaplan DR, Parada LF, and Chao MV (1991). High-affinity NGF binding requires coexpression of the trk proto-oncogene and the low-affinity NGF receptor. *Nature* **350**, 678–83.

Herrup K and Busser JC (1995). The induction of multiple cell cycle events precedes target-related neuronal death *Development* **121**, 2385–95.

Hofer M, Pagliusi SR, Hohn A, Leibrock J, and Barde YA (1990). Regional distribution of brain-derived neurotrophic factor mRNA in the adult mouse brain. *EMBO Journal* **9**, 2459–64.

Holcomb LA, Gordon MN, Jantzen P, Hsiao K, Duff K, and Morgan D (1999). Behavioral changes in transgenic mice expressing both amyloid precursor protein and presenilin-1 mutations: lack of association with amyloid deposits. *Behavior Genetics* **29**, 177–85.

Holsinger RM, Schnarr J, Henry P, Castelo VT, and Fahnestock M (2000). Quantitation of BDNF mRNA in human parietal cortex by competitive reverse transcription-polymerase chain reaction: decreased levels in Alzheimer's disease. *Brain Research: Molecular Brain Research* **76**, 347–54.

Holtzman DM, Li Y, Parada LF, *et al.* (1992). p140trk mRNA marks NGF-responsive forebrain neurons: evidence that trk gene expression is induced by NGF. *Neuron* **9**, 465–78.

Hsiao K (1998). Transgenic mice expressing Alzheimer amyloid precursor proteins. *Experimental Gerontology* **33**, 883–9.

Humpel C and Weis C (2002). Nerve growth factor and cholinergic CNS neurons studied in organotypic brain slices. Implication in Alzheimer's disease?. *Journal of Neural Transmission. Supplementum* 253–63.

Ibanez CF (1994). Structure–function relationships in the neurotrophin family. *Journal of Neurobiology* **25**, 1349–61.

Jette N, Cole MS, and Fahnestock M (1994). NGF mRNA is not decreased in frontal cortex from Alzheimer's disease patients. *Brain Research: Molecular Brain Research* **25**, 242–50.

Kang H and Schuman EM (1995). Long-lasting neurotrophin-induced enhancement of synaptic transmission in the adult hippocampus. *Science* **267**, 1658–62.

Kaplan DR and Miller FD (1997). Signal transduction by the neurotrophin receptors. *Current Opinion in Cell Biology* **9**, 213–21.

Kaplan DR and Miller FD (2000). Neurotrophin signal transduction in the nervous system. *Current Opinion in Neurobiology* **10**, 381–91.

Kaplan DR and Stephens RM (1994). Neurotrophin signal transduction by the Trk receptor. *Journal of Neurobiology* **25**, 1404–17.

Kells AP, Fong DM, Dragunow M, During MJ, Young D, and Connor B (2004). AAV-mediated gene delivery of BDNF or GDNF is neuroprotective in a model of Huntington disease. *Molecular Therapy* **9**, 682–8.

Klein R, Jing SQ, Nanduri V, O'Rourke E, and Barbacid M (1991). The trk proto-oncogene encodes a receptor for nerve growth factor. *Cell* **65**, 189–97.

Klesse LJ and Parada LF (1999). Trks: signal transduction and intracellular pathways. *Microscopy Research and Technique* **45**, 210–16.

Knusel B, Beck KD, Winslow JW, *et al*. (1992). Brain-derived neurotrophic factor administration protects basal forebrain cholinergic but not nigral dopaminergic neurons from degenerative changes after axotomy in the adult rat brain. *Journal of Neuroscience* **12**, 4391–402.

Kokaia Z, Metsis M, Kokaia M, *et al*. (1994). Brain insults in rats induce increased expression of the BDNF gene through differential use of multiple promoters. *European Journal of Neuroscience* **6**, 587–96.

Koliatsos VE, Clatterbuck RE, Nauta HJ, *et al*. (1991). Human nerve growth factor prevents degeneration of basal forebrain cholinergic neurons in primates. *Annals of Neurology* **30**, 831–40.

Kordower JH and Mufson EJ (1989). Nerve growth factor receptor and choline acetyltransferase remain colocalized in the nucleus basalis (Ch4). of Alzheimer's patients. *Neurobiology of Aging* **10**, 543–4, 552–3.

Kordower JH, Winn SR, Liu YT *et al*. (1994). The aged monkey basal forebrain: rescue and sprouting of axotomized basal forebrain neurons after grafts of encapsulated cells secreting human nerve growth factor. *Proceedings of the National Academy of Sciences of the USA* **91**, 10898–902.

Kordower JH, Mufson EJ, Fox N, Martel L, and Emerich DF (1997). Cellular delivery of NGF does not alter the expression of beta-amyloid immunoreactivity in young or aged nonhuman primates. *Experimental Neurology* **145**, 586–91.

Korte M, Carroll P, Wolf E, Brem G, Thoenen H, and Bonhoeffer T (1995). Hippocampal long-term potentiation is impaired in mice lacking brain-derived neurotrophic factor. *Proceedings of the National Academy of Sciences of the USA* **92**, 8856–60.

Kosik KS, Joachim CL, and Selkoe DJ (1986). Microtubule-associated protein tau (t) is a major antigenic component of paired helical filaments in Alzheimer disease. *Proceedings of the National Academy of Sciences of the USA* **83**, 4044–8.

Kuruvilla R, Ye H, and Ginty DD (2000). Spatially and functionally distinct roles of the PI3-K effector pathway during NGF signaling in sympathetic neurons. *Neuron* **27**, 499–512.

Lad SP, Neet KE, and Mufson EJ (2003). Nerve growth factor: structure, function and therapeutic implications for Alzheimer's disease. *Current Drug Targets. CNS and Neurological Disorders* **2**, 315–34.

Lauterborn JC, Poulsen FR, Stinis CT, Isackson PJ, and Gall CM (1998). Transcript-specific effects of adrenalectomy on seizure-induced BDNF expression in rat hippocampus. *Brain Research: Molecular Brain Research* **55**, 81–91.

Lee R, Kermani P, Teng KK, and Hempstead BL (2001). Regulation of cell survival by secreted proneurotrophins. *Science* **294**, 1945–8.

Lee VM, Balin BJ, Otvos L, Jr, and Trojanowski JQ (1991). A68: a major subunit of paired helical filaments and derivatized forms of normal tau. *Science* **251**, 675–8.

LeSauteur L, Wei L, Gibbs BF, and Saragovi HU (1995). Small peptide mimics of nerve growth factor bind TrkA receptors and affect biological responses. *Journal of Biological Chemistry* **270**, 6564–9.

Levi-Montalcini R (1964). The nerve growth factor. *Annals of the New York Academy of Sciences* **118**, 149–70.

Li Y, Holtzman DM, Kromer LF, *et al*. (1995). Regulation of TrkA and ChAT expression in developing rat basal forebrain: evidence that both exogenous and endogenous NGF regulate differentiation of cholinergic neurons. *Journal of Neuroscience* **15**, 2888–905.

Lindholm D, Carroll P, Tzimagiogis G, and Thoenen H (1996). Autocrine-paracrine regulation of hippocampal neuron survival by IGF-1 and the neurotrophins BDNF, NT-3 and NT-4 *European Journal of Neuroscience* **8**, 1452–60.

Liu QR, Walther D, Drgon T, *et al*. (2005). Human brain derived neurotrophic factor (BDNF). genes, splicing patterns, and assessments of associations with substance abuse and Parkinson's disease. *American Journal of Medical Genetics. Part B, Neuropsychiatric Genetics* **134**, 93–103.

Liu YZ, Boxer LM, and Latchman DS (1999). Activation of the Bcl-2 promoter by nerve growth factor is mediated by the p42/p44 MAPK cascade. *Nucleic Acids Research* **27**, 2086–90.

Longo FM, Manthorpe M, Xie YM, and Varon S (1997). Synthetic NGF peptide derivatives prevent neuronal death via a p75 receptor-dependent mechanism. *Journal of Neuroscience Research* **48**, 1–17.

McAllister AK, Katz LC, and Lo DC (1999). Neurotrophins and synaptic plasticity. *Annual Reviews of Neuroscience* **22**, 295–318.

MacInnis BL and Campenot RB (2002). Retrograde support of neuronal survival without retrograde transport of nerve growth factor. *Science* **295**, 1536–9.

Mahadeo D, Kaplan L, Chao MV, and Hempstead BL (1994). High affinity nerve growth factor binding displays a faster rate of association than p140trk binding. Implications for multi-subunit polypeptide receptors. *Journal of Biological Chemistry* **269**, 6884–91.

Maliartchouk S, Feng Y, Ivanisevic L, *et al.* (2000). A designed peptidomimetic agonistic ligand of TrkA nerve growth factor receptors. *Molecular Pharmacology* **57**, 385–91.

Mamidipudi V and Wooten MW (2002). Dual role for p75(NTR). signaling in survival and cell death: can intracellular mediators provide an explanation?. *Journal of Neuroscience Research* **68**, 373–84.

Massa SM, Xie Y, and Longo FM (2003). Alzheimer's therapeutics: neurotrophin domain small molecule mimetics. *Journal of Molecular Neuroscience* **20**, 323–6.

Mesulam MM and Geula C (1988). Nucleus basalis (Ch4). and cortical cholinergic innervation in the human brain: observations based on the distribution of acetylcholinesterase and choline acetyltransferase. *Journal of Comparative Neurology* **275**, 216–40.

Mesulam MM, Mufson EJ, Levey AI, and Wainer BH (1983). Cholinergic innervation of cortex by the basal forebrain: cytochemistry and cortical connections of the septal area, diagonal band nuclei, nucleus basalis (substantia innominata), and hypothalamus in the rhesus monkey. *Journal of Comparative Neurology* **214**, 170–97.

Metsis M, Timmusk T, Arenas E, and Persson H (1993). Differential usage of multiple brain-derived neurotrophic factor promoters in the rat brain following neuronal activation. *Proceedings of the National Academy of Sciences of the USA* **90**, 8802–6.

Michalski B and Fahnestock M (2003). Pro-brain-derived neurotrophic factor is decreased in parietal cortex in Alzheimer's disease. *Brain Research: Molecular Brain Research* **111**, 148–54.

Mobley WC, Neve RL, Prusiner SB, and McKinley MP (1988). Nerve growth factor increases mRNA levels for the prion protein and the beta-amyloid protein precursor in developing hamster brain. *Proceedings of the National Academy of Sciences of the USA* **85**, 9811–15.

Molteni R, Ying Z, and Gomez-Pinilla F (2002). Differential effects of acute and chronic exercise on plasticity-related genes in the rat hippocampus revealed by microarray *European Journal of Neuroscience* **16**, 1107–16.

Molteni R, Wu A, Vaynman S, Ying Z, Barnard RJ, and Gomez-Pinilla F (2004). Exercise reverses the harmful effects of consumption of a high-fat diet on synaptic and behavioral plasticity associated to the action of brain-derived neurotrophic factor *Neuroscience* **123**, 429–40.

Mowla SJ, Farhadi HF, Pareek S, *et al.* (2001). Biosynthesis and post-translational processing of the precursor to brain-derived neurotrophic factor. *Journal of Biological Chemistry* **276**, 12660–6.

Mufson EJ (1997). NGF, p75[NTR] and trkA in Alzheimer's disease: Neural Notes, *Promega*, **3**, pp. 16–19.

Mufson EJ and Kordower JH (1992). Nerve growth factor in Alzheimer's disease. In: Cerebral Cortex. AA Peters and JH Morrison (Eds.) Kluwer Academic/Plenum Press, New York, vol. **14**, pp.681–731.

Mufson EJ and Kordower JH (1999). Nerve growth factor in Alzheimer's disease. In: Cerebral Cortex. Peters AA and Morrison JH (eds), Kluwer Academic/Plenum Press, New York, vol. 14, pp. 681–731.

Mufson EJ, Bothwell M, and Kordower JH (1989). Loss of nerve growth factor receptor-containing neurons in Alzheimer's disease: a quantitative analysis across subregions of the basal forebrain. *Experimental Neurology* **105**, 221–32.

Mufson EJ, Benzing WC, Cole GM, *et al.* (1994). Apolipoprotein E-immunoreactivity in aged rhesus monkey cortex: colocalization with amyloid plaques. *Neurobiology of Aging* **15**, 621–7.

Mufson EJ, Conner JM, and Kordower JH (1995). Nerve growth factor in Alzheimer's disease: defective retrograde transport to nucleus basalis. *Neuroreport* **6**, 1063–6.

Mufson EJ, Lavine N, Jaffar S, Kordower JH, Quirion R, and Saragovi HU (1997). Reduction in p140-TrkA receptor protein within the nucleus basalis and cortex in Alzheimer's disease. *Experimental Neurology* **146**, 91–103.

Mufson EJ, Kroin JS, Sendera TJ, and Sobreviela T (1999). Distribution and retrograde transport of trophic factors in the central nervous system: functional implications for the treatment of neurodegenerative diseases. *Progress in Neurobiology* **57**, 451–84.

Mufson EJ, Ma SY, Cochran EJ, *et al.* (2000). Loss of nucleus basalis neurons containing trkA immunoreactivity in individuals with mild cognitive impairment and early Alzheimer's disease. *Journal of Comparative Neurology* **427**, 19–30.

Mufson EJ, Counts SE, and Ginsberg SD (2002a). Gene expression profiles of cholinergic nucleus basalis neurons in Alzheimer's disease. *Neurochemical Research* **27**, 1035–48.

Mufson EJ, Ma SY, Dills J, *et al.* (2002b). Loss of basal forebrain P75(NTR). immunoreactivity in subjects with mild cognitive impairment and Alzheimer's disease. *Journal of Comparative Neurology* **443**, 136–53.

Mufson EJ, Ikonomovic MD, Styren SD, *et al.* (2003). Preservation of brain nerve growth factor in mild cognitive impairment and Alzheimer disease. *Archives of Neurology* **60**, 1143–8.

Murray KD, Gall CM, Jones EG, and Isackson PJ (1994). Differential regulation of brain-derived neurotrophic factor and type II calcium/calmodulin-dependent protein kinase messenger RNA expression in Alzheimer's disease *Neuroscience* **60**, 37–48.

Nykjaer A, Lee R, Teng KK, Jansen P, *et al.* (2004). Sortilin is essential for proNGF-induced neuronal cell death. *Nature* **427**, 843–8.

Oddo S, Caccamo A, Shepherd JD, *et al.* (2003). Triple-transgenic model of Alzheimer's disease with plaques and tangles: intracellular Abeta and synaptic dysfunction. *Neuron* **39**, 409–21.

Oliff HS, Berchtold NC, Isackson P, and Cotman CW (1998). Exercise-induced regulation of brain-derived neurotrophic factor (BDNF). transcripts in the rat hippocampus. *Brain Research: Molecular Brain Research* **61**, 147–53.

Olofsdotter K, Lindvall O, and Asztely F (2000). Increased synaptic inhibition in dentate gyrus of mice with reduced levels of endogenous brain-derived neurotrophic factor. *Neuroscience* **101**, 531–9.

Osehobo P, Adams B, Sazgar M, Xu Y, Racine RJ, and Fahnestock M (1999). Brain-derived neurotrophic factor infusion delays amygdala and perforant path kindling without affecting paired-pulse measures of neuronal inhibition in adult rats. *Neuroscience* **92**, 1367–75.

Pang PT, Teng HK, Zaitsev E, *et al.* (2004). Cleavage of proBDNF by tPA/plasmin is essential for long-term hippocampal plasticity. *Science* **306**, 487–91.

Pardridge WM (2002). Blood–brain barrier drug targeting enables neuroprotection in brain ischemia following delayed intravenous administration of neurotrophins. *Advances in Experimental Medicine and Biology* **513**, 397–430.

Park DS, Levine B, Ferrari G and Greene LA (1997). Cyclin dependent kinase inhibitors and dominant negative cyclin dependent kinase 4 and 6 promote survival of NGF-deprived sympathetic neurons. *Journal of Neuroscience* **17**, 8975–83.

Patterson SL, Abel T, Deuel TA, Martin KC, Rose JC, and Kandel ER (1996). Recombinant BDNF rescues deficits in basal synaptic transmission and hippocampal LTP in BDNF knockout mice. *Neuron* **16**, 1137–45.

Pelleymounter MA, Cullen MJ, and Wellman CL (1995). Characteristics of BDNF-induced weight loss. *Experimental Neurology* **131**, 229–38.

Peng S, Wuu J, Mufson EJ, and Fahnestock M (2004). Increased proNGF levels in subjects with mild cognitive impairment and mild Alzheimer disease. *Journal of Neuropathology and Experimental Neurology* **63**, 641–9.

Peng S, Wuu J, Mufson EJ, and Fahnestock M (2005). Precursor form of brain-derived neurotrophic factor and mature brain-derived neurotrophic factor are decreased in the pre-clinical stages of Alzheimer's disease. *Journal of Neurochemistry* **93**, 1412–21.

Petersen RC (2004). Mild cognitive impairment as a diagnostic entity. *Journal of Internal Medicine* **256**, 183–94.

Phillips HS, Hains JM, Laramee GR, Rosenthal A, and Winslow JW (1990). Widespread expression of BDNF but not NT3 by target areas of basal forebrain cholinergic neurons. *Science* **250**, 290–4.

Phillips HS, Hains JM, Armanini M, Laramee GR, Johnson SA, and Winslow JW (1991). BDNF mRNA is decreased in the hippocampus of individuals with Alzheimer's disease. *Neuron* **7**, 695–702.

Planas B, Kolb PE, Raskind MA, and Miller MA (1997). Nerve growth factor induces galanin gene expression in the rat basal forebrain: implications for the treatment of cholinergic dysfunction. *Journal of Comparative Neurology* **379**, 563–70.

Puolivali J, Wang J, Heikkinen T, *et al.* (2002). Hippocampal A beta 42 levels correlate with spatial memory deficit in APP and PS1 double transgenic mice. *Neurobiology of Disease* **9**, 339–47.

Putcha GV, Deshmukh M, and Johnson EM, Jr (1999). BAX translocation is a critical event in neuronal apoptosis: regulation by neuroprotectants, BCL-2, and caspases. *Journal of Neuroscience* **19**, 7476–85.

Rattenholl A, Lilie H, Grossmann A, Stern A, Schwarz E, and Rudolph R (2001). The pro-sequence facilitates folding of human nerve growth factor from *Escherichia coli* inclusion bodies. *European Journal of Biochemistry* **268**, 3296–303.

Reynolds AJ, Bartlett SE, and Hendry IA (1998). Signalling events regulating the retrograde axonal transport of 125I-beta nerve growth factor *in vivo*. *Brain Research* **798**, 67–74.

Riccio A, Pierchala BA, Ciarallo CL, and Ginty DD (1997). An NGF-TrkA-mediated retrograde signal to transcription factor CREB in sympathetic neurons. *Science* **277**, 1097–100.

Riccio A, Ahn S, Davenport CM, Blendy JA, and Ginty DD (1999). Mediation by a CREB family transcription factor of NGF-dependent survival of sympathetic neurons. *Science* **286**, 2358–61.

Roux PP and Barker PA (2002). Neurotrophin signaling through the p75 neurotrophin receptor. *Progress in Neurobiology* **67**, 203–33.

Roux PP, Bhakar AL, Kennedy TE, and Barker PA (2001). K252a and CEP1347 are neuroprotective compounds that inhibit mixed-lineage kinase-3 and induce activation of Akt and ERK. *Journal of Biological Chemistry* **276**, 23097–104.

Ruberti F, Capsoni S, Comparini A, *et al.* (2000). Phenotypic knockout of nerve growth factor in adult transgenic mice reveals severe deficits in basal forebrain cholinergic neurons, cell death in the spleen, and skeletal muscle dystrophy. *Journal of Neuroscience* **20**, 2589–601.

Rubio F, Kokaia Z, Arco A, *et al.* (1999). BDNF gene transfer to the mammalian brain using CNS-derived neural precursors. *Gene Therapy* **6**, 1851–66.

Saarelainen T, Pussinen R, Koponen E, *et al.* (2000). Transgenic mice overexpressing truncated trkB neurotrophin receptors in neurons have impaired long-term spatial memory but normal hippocampal LTP. *Synapse* **38**, 102–4.

Salehi A, Verhaagen J, Dijkhuizen PA, and Swaab DF (1996). Co-localization of high-affinity neurotrophin receptors in nucleus basalis of Meynert neurons and their differential reduction in Alzheimer's disease *Neuroscience* **75**, 373–87.

Salehi A, Ocampo M, Verhaagen J, and Swaab DF (2000a). P75 neurotrophin receptor in the nucleus basalis of meynert in relation to age, sex, and Alzheimer's disease. *Experimental Neurology* **161**, 245–58.

Salehi AH, Roux PP, Kubu CJ, *et al.* (2000b). NRAGE, a novel MAGE protein, interacts with the p75 neurotrophin receptor and facilitates nerve growth factor-dependent apoptosis. *Neuron* **27**, 279–88.

Sarter M and Bruno JP (2004). Developmental origins of the age-related decline in cortical cholinergic function and associated cognitive abilities. *Neurobiology of Aging* **25**, 1127–39.

Sarter M, Bruno JP, and Turchi J (1999). Basal forebrain afferent projections modulating cortical acetylcholine, attention, and implications for neuropsychiatric disorders. *Annals of the New York Academy of Sciences* **877**, 368–82.

Scharfman HE (1997). Hyperexcitability in combined entorhinal/hippocampal slices of adult rat after exposure to brain-derived neurotrophic factor. *Journal of Neurophysiology* **78**, 1082–95.

Scott SA, Mufson EJ, Weingartner JA, Skau KA, and Crutcher KA (1995). Nerve growth factor in Alzheimer's disease: increased levels throughout the brain coupled with declines in nucleus basalis. *Journal of Neuroscience* **15**, 6213–21.

Seidah NG, Benjannet S, Pareek S, Chretien M, and Murphy RA (1996). Cellular processing of the neurotrophin precursors of NT3 and BDNF by the mammalian proprotein convertases. *FEBS Letters* **379**, 247–50.

Seiler M and Schwab ME (1984). Specific retrograde transport of nerve growth factor (NGF). from neocortex to nucleus basalis in the rat. *Brain Research* **300**, 33–9.

Shetty AK and Turner DA (1998). *In vitro* survival and differentiation of neurons derived from epidermal growth factor-responsive postnatal hippocampal stem cells: inducing effects of brain-derived neurotrophic factor. *Journal of Neurobiology* **35**, 395–425.

Smith DE, Roberts J, Gage FH, and Tuszynski MH (1999). Age-associated neuronal atrophy occurs in the primate brain and is reversible by growth factor gene therapy. *Proceedings of the National Academy of Sciences of the USA* **96**, 10893–8.

Sobreviela T, Clary DO, Reichardt LF, Brandabur MM, Kordower JH, and Mufson EJ (1994). TrkA-immunoreactive profiles in the central nervous system: colocalization with neurons containing p75 nerve growth factor receptor, choline acetyltransferase, and serotonin. *Journal of Comparative Neurology* **350**, 587–611.

Sobreviela T, Pagcatipunan M, Kroin JS, and Mufson EJ (1996). Retrograde transport of brain-derived neurotrophic factor (BDNF) following infusion in neo- and limbic cortex in rat: relationship to BDNF mRNA expressing neurons. *Journal of Comparative Neurology* **375**, 417–44.

Sofroniew MV, Galletly NP, Isacson O, and Svendsen CN (1990). Survival of adult basal forebrain cholinergic neurons after loss of target neurons. *Science* **247**, 338–42.

Sofroniew MV, Howe CL, and Mobley WC (2001). Nerve growth factor signaling, neuroprotection, and neural repair. *Annual Reviews of Neuroscience* **24**, 1217–81.

Spear N, Estevez AG, Barbeito L, Beckman JS, and Johnson GV (1997). Nerve growth factor protects PC12 cells against peroxynitrite-induced apoptosis via a mechanism dependent on phosphatidylinositol 3-kinase. *Journal of Neurochemistry* **69**, 53–9.

Teng HK, Teng KK, Lee R, *et al.* (2005). ProBDNF induces neuronal apoptosis via activation of a receptor complex of p75NTR and sortilin. *Journal of Neuroscience* **25**, 5455–63.

Teng KK and Hempstead BL (2004). Neurotrophins and their receptors: signaling trios in complex biological systems. *Cellular and Molecular Life Sciences* **61**, 35–48.

Timm DE, Ross AH, and Neet KE (1994). Circular dichroism and crosslinking studies of the interaction between four neurotrophins and the extracellular domain of the low-affinity neurotrophin receptor *Protein Science* **3**, 451–8.

Timmusk T. and Metsis M. (1994). Regulation of BDNF promoters in the rat hippocampus. *Neurochemistry Internatinsal* **25**, 11–15.

Troy CM, Friedman JE, and Friedman WJ (2002). Mechanisms of p75-mediated death of hippocampal neurons. role of caspases. *Journal of Biological Chemistry* **277**, 34295–302.

Tuszynski MH and Blesch A (2004). Nerve growth factor: from animal models of cholinergic neuronal degeneration to gene therapy in Alzheimer's disease *Progress in Brain Research* **146**, 441–9.

Tuszynski MH and Gage FH (1995). Bridging grafts and transient nerve growth factor infusions promote long-term central nervous system neuronal rescue and partial functional recovery. *Proceedings of the National Academy of Sciences of the USA* **92**, 4621–5.

Tuszynski MH, Sang H, Yoshida K, and Gage FH (1991). Recombinant human nerve growth factor infusions prevent cholinergic neuronal degeneration in the adult primate brain. *Annals of Neurology* **30**, 625–36.

Tuszynski MH, Smith DE, Roberts J, McKay H, and Mufson E (1998). Targeted intraparenchymal delivery of human NGF by gene transfer to the primate basal forebrain for 3 months does not accelerate beta-amyloid plaque deposition. *Experimental Neurology* **154**, 573–82.

Tuszynski MT, Thal L, Pay M *et al.* (2005). A phase I clinical trial of nerve growth factor gene therapy for Alzheimer's disease, *Nature Medicine* **11**, 551–5.

Vaynman S and Gomez-Pinilla F (2005). License to run: exercise impacts functional plasticity in the intact and injured central nervous system by using neurotrophins. *Neurorehabilitation and Neural Repair* **19**, 283–95.

Vaynman S, Ying Z, and Gomez-Pinilla F (2004). Hippocampal BDNF mediates the efficacy of exercise on synaptic plasticity and cognition *European Journal of Neuroscience* **20**, 2580–90.

Ventriglia M, Bocchio Chiavetto L, Benussi L, *et al.* (2002). Association between the BDNF 196 A/G polymorphism and sporadic Alzheimer's disease. *Molecular Psychiatry* **7**, 136–7.

Vincent I, Rosado M and Davies P (1996). Mitotic mechanisms in Alzheimer's disease?. *Journal of Cell Biology* **132**, 413–25.

Vincent I, Jicha G, Rosado M, and Dickson DW (1997). Aberrant expression of mitotic cdc2/cyclin B1 kinase in degenerating neurons of Alzheimer's disease brain. *Journal of Neuroscience* **17**, 3588–98.

Vogels OJ, Broere CA, ter Laak HJ, *et al* (1990). Cell loss and shrinkage in the nucleus basalis Meynert complex in Alzheimer's disease. *Neurobiology of Aging* **11**, 3–13.

von Bartheld CS, Kinoshita Y, Prevette D, Yin QW, Oppenheim RW, and Bothwell M (1994). Positive and negative effects of neurotrophins on the isthmo-optic nucleus in chick embryos. *Neuron* **12**, 639–54.

Walker LC (1997). Animal models of cerebral beta-amyloid angiopathy. *Brain Research: Brain Research Reviews* **25**, 70–84.

Watson FL, Heerssen HM, Moheban DB, *et al.* (1999). Rapid nuclear responses to target-derived neurotrophins require retrograde transport of ligand–receptor complex. *Journal of Neuroscience* **19**, 7889–900.

Watson FL, Heerssen HM, Bhattacharyya A, Klesse L, Lin MZ, and Segal RA (2001). Neurotrophins use the Erk5 pathway to mediate a retrograde survival response. *Nature Neuroscience* **4**, 981–8.

Whitehouse PJ, Price DL, Clark AW, Coyle JT, and DeLong MR (1981). Alzheimer disease: evidence for selective loss of cholinergic neurons in the nucleus basalis. *Annals of Neurology* **10**, 122–6.

Williams LR (1991). Hypophagia is induced by intracerebroventricular administration of nerve growth. *Experimental Neurology* **113**, 31–7.

Williams LR, Varon S, Peterson GM, *et al.* (1986). Continuous infusion of nerve growth factor prevents basal forebrain neuronal death after fimbria fornix transection. *Proceedings of the National Academy of Sciences of the USA* **83**, 9231–5.

Winblad B, Palmer K, Kivipelto M, *et al.* (2004). Mild cognitive impairment–beyond controversies, towards a consensus: report of the International Working Group on Mild Cognitive Impairment. *Journal of Internal Medicine* **256**, 240–6.

Woo NH, Teng HK, Siao CJ, *et al.* (2005). Activation of p75NTR by proBDNF facilitates hippocampal long-term depression *Nature Neuroscience* **8**, 1069–77.

Xiang J, Chao DT, and Korsmeyer SJ (1996). BAX-induced cell death may not require interleukin 1 beta-converting enzyme-like proteases. *Proceedings of the National Academy of Sciences of the USA* **93**, 14559–63.

Xu B, Michalski B, Racine RJ, and Fahnestock M. (2004). The effects of brain-derived neurotrophic factor (BDNF). administration on kindling induction, Trk expression and seizure-related morphological changes *Neuroscience* **126**, 521–31.

Yamasaki TR, Caccamo A, Oddo S, *et al.* (2005). Modulation of the cholinergic system in the 3XTG-AD. Presented at Society for Neuroscience, Washington, DC. Available online at: http://sfn.scholarone.com/itin2005/index.html

Yan Q, Matheson C, Sun J, Radeke MJ, Feinstein SC, and Miller JA (1994). Distribution of intracerebral ventricularly administered neurotrophins in rat brain and its correlation with trk receptor expression. *Experimental Neurology* **127**, 23–36.

Yang Y, Geldmacher DS, and Herrup K (2001). DNA replication precedes neuronal cell death in Alzheimer's disease. *Journal of Neuroscience* **21**, 2661–8.

Yang Y, Mufson EJ, and Herrup K (2003). Neuronal cell death is preceded by cell cycle events at all stages of Alzheimer's disease. *Journal of Neuroscience* **23**, 2557–63.

Yao R and Cooper GM (1995). Requirement for phosphatidylinositol-3 kinase in the prevention of apoptosis by nerve growth factor. *Science* **267**, 2003–6.

Ye H, Kuruvilla R, Zweifel LS, and Ginty DD (2003). Evidence in support of signaling endosome-based retrograde survival of sympathetic neurons. *Neuron* **39**, 57–68.

Yoon SO, Casaccia-Bonnefil P, Carter B, and Chao MV (1998). Competitive signaling between TrkA and p75 nerve growth factor receptors determines cell survival. *Journal of Neuroscience* **18**, 3273–81.

York RD, Molliver DC, Grewal SS, Stenberg PE, McCleskey EW, and Stork PJ (2000). Role of phosphoinositide 3-kinase and endocytosis in nerve growth factor-induced extracellular signal-regulated kinase activation via Ras and Rap1. *Molecular and Cellular Biology* **20**, 8069–83.

Zha J, Harada H, Yang E, Jockel J, and Korsmeyer SJ (1996). Serine phosphorylation of death agonist BAD in response to survival factor results in binding to 14–3–3 not BCL-X(L). *Cell* **87**, 619–28.

Zhang Y, Moheban DB, Conway BR, Bhattacharyya A, and Segal RA (2000). Cell surface Trk receptors mediate NGF-induced survival while internalized receptors regulate NGF-induced differentiation. *Journal of Neuroscience* **20**, 5671–8.

Chapter 14

Clinical assessment of Alzheimer's disease

Sarmishtha Bhattacharyya, Ross Overshott, and Alistair Burns

In the past the symptoms of dementia were often dismissed as an inevitable accompaniment of advanced age. With raised public awareness of dementia, and the development of memory clinics, patients and doctors are now less likely to accept cognitive impairment as 'just old age'. Patients and their carers are increasingly requesting assessment in the hope that they can receive a diagnosis and treatment.

14.1 Aims of assessment

The aim of assessment is first to determine whether the patient with cognitive impairment has dementia, rather than delirium, depression, a physical illness, or normal ageing. Once a diagnosis of dementia has been made it is important to establish the underlying aetiology: Alzheimer's disease (AD), vascular dementia, Lewy body dementia (LBD), frontal lobe dementia, or the many rarer causes. This is vital as some treatments are only licensed and available for specific forms of dementia; for example, cholinesterase inhibitors are only licensed for AD. The type of dementia can also influence the management of troublesome cognitive and non-cognitive symptoms which have been elicited during the assessment.

The clinical syndrome of dementia consists of the following triad:

- neuropsychological symptoms
- neuropsychiatric components
- inability to perform activities of daily living.

The neuropsychological element consists mainly of:

- amnesia (loss of memory)
- aphasia (receptive or expressive)
- apraxia (inability to carry out tasks despite an intact sensory and motor nervous system)
- agnosia (inability to recognize things e.g. does not recognize one's own mirror reflection, or inability to recognize a family member).

The neuropsychiatric component consists of associated psychiatric symptoms and behavioural disturbances which are often present in a substantial proportion of patients.

Behavioural problems eventually occur in nearly all patients with the disease and are a major cause of carer distress. The most common are as follows:

- depression (up to 66% at some point during the dementia)
- paranoid ideation (30%)
- misidentification (20%)
- hallucinations (commonly auditory, suspect either delirium or LBD; persistent visual hallucinations are present in 15% of cases)
- aggression (20%)
- wandering (20%).

The third component is the inability to perform activities of daily living. In the initial stages, assessment of daily function is critical to the understanding of the degree of the patient's disability and dependence on the carer. Objective evidence of memory deficits is often associated with decreased performance in housework and self-care. This may lead patients to neglect their diet, causing weight loss, or fail to wash regularly and dress to their usual standard. In the later stages, impairment may be so profound that the patient needs assistance with feeding, dressing, and toileting.

This triad is common to all types of dementia, and differentiation between the various types of dementia is based on clinical presentation and presence of other features.

14.2 Why is it important to diagnose dementias?

Dementia is a significant cause of morbidity and mortality in elderly people, with an impact on the health of the population similar to that of lung cancer and stroke (Whitthaus *et al.* 1999). Dementia affects 10% of people over 65 years and 20% of those over 80 years old. The perceived benefits of assessment by caregivers include the following (Wackerbarth and Johnson 2002):

- confirmation of medical condition
- access to treatment
- help in preparing for caregiver role.

Reaching a diagnosis provides caregivers and relatives with an explanation for the patient's difficulties and behaviours. Increasing their understanding of dementia can enhance their communication and interactions with the patient, which in turn can reduce caregiver stress. As already mentioned, diagnosis can also provide access to suitable treatments.

Assessment and diagnosis can identify potential hazards (e.g. leaving a gas cooker on, driving) which can be minimized or eliminated; this may allow the patient to continue living independently. It is also important to determine whether there is a degree of reversibility to the cognitive impairment so that appropriate intervention can be commenced. The development of dementia has legal and social implications regarding capacity to consent to future treatments and management of financial affairs. Therefore early diagnosis and assessment are crucial to allow patients to make choices regarding the future, and to preserve their autonomy even when they have become incapable of making decisions for themselves. However, making a diagnosis is often difficult, particularly in the early stages, and unfortunately a confirmed diagnosis is not possible until a post-mortem examination is made.

14.3 Clinical presentation of dementias

Lishman (1978) defined dementia as 'an acquired global impairment of intellect, memory and personality but without impairment of consciousness'.

The main clinical presentation of AD is short-term memory loss, with impairment for recent events and reduced ability to learn new information. It is insidious in onset. This may be manifest by repetitive questioning, which caregivers may find irritating, as well as forgetting significant events, such as birthdays and social appointments. The person may need reminding of the day, the date, or the names of people that they have recently met. The person may have adapted to their impairments and use a notepad or calendar to prompt themselves. Memory for more distant events is usually preserved in the early stages.

Disturbances in executive function, such as planning and regulation of complex goal-directed behaviour, are common. This causes difficulties with dealing with tasks such as cooking a meal, dealing with finances, or driving. This is most prominently seen in frontal lobe dementia and vascular dementia. However, difficulties with activities of daily living are also a core feature of AD, where impairments of praxis, visuospatial perception, and executive function are influential. The first problem to be noted may be difficulty in handling new technology around the house, such as a microwave or a video recorder. The patient may become so frustrated he or she avoids using the appliances, and newly purchased gadgets may remain in their boxes. Language problems can also occur, affecting word fluency, naming, and finding the right words, and may be present in the early stages of the disease.

Subtle changes in personality lead to reduction in confidence and range of interests, but these are often apparent in retrospect unless marked as in the case of frontal lobe dementia. Personality changes include increased rigidity, increased egocentricity, impairment of regard for the feelings of others, coarsening of affect, impairment of emotional control, laughing in inappropriate situations, diminished emotional responsiveness, sexual misdemeanour, relinquishing of hobbies, diminished initiative or growing apathy, and purposeless hyperactivity (Blessed et al. 1968). In the later stages emotional and neuropsychiatric manifestations, such as hallucinations, delusions, and mood disturbances, may also be present. Oppenheim (1994) found that a psychiatric presentation was the primary difficulty in a third of patients. AD is often accompanied by symptoms of agitation, wandering, aggression, and incontinence.

Symptoms of AD evolve over a relatively long period of time and it can be up to 4 years before the patient actually comes for help (Brodaty et al. 1993). In vascular dementia, the presentation is usually more sudden and the time of onset can be predicted more accurately than AD. The progression often follows a stepwise course, with periods of weeks or even months of no obvious decline followed by a sudden deterioration which can often be presumed to coincide with a cerebral vascular event. Aphasia may be more prominent in vascular dementia than in AD and personality changes are less marked. Depression is more apparent, with catastrophic reaction and emotional lability.

Patients with frontal lobe dementia characteristically present with personality changes, changes in social conduct, depression or euphoria, and repetitive behaviour. As there is relatively little memory impairment or perceptual problems, frontal lobe dementia can usually be easily distinguished from AD (Carlesimo *et al.* 2002). Patients with LBD present with acute or subacute confusional states, falls, florid visual hallucinations, and persecutory ideas, with parkinsonian features and a sensitivity to neuroleptic medication.

It is often difficult to separate these defined syndromes clinically and frequently there is a considerable overlap between the various types of dementia described. For example, patients with vascular dementia may have prominent frontal lobe impairment and patients with AD may have prominent aphasia.

14.4 **Clinical diagnostic criteria for dementia**

The gold standard for diagnosis of AD is regarded by many as neuropathological examination, and the success of clinical criteria to predict pathological changes is taken as a marker of their validity. Nevertheless there is significant overlap in the histological changes seen in dementia (Holmes *et al.* 1999), and increasing understanding of molecular biology has probably diminished the importance of classical descriptive neuropathological examination in research terms. However, it still remains the gold standard in terms of pathological diagnosis.

The most commonly used clinical diagnostic criteria in practice are the North American based *Diagnostic and Statistical Manual*, currently in its fourth edition (DSM-IV) (American Psychiatric Association 1994) and the European based *International Classification of Disease*, currently in the10th revision (ICD-10) (World Health Organization 1992). The main criteria are summarized in Tables 14.1 and 14.2.

Both sets of criteria require the memory problems to be evident in absence of delirium and exclusion of other diseases which may cause memory problems. In DSM-IV a person has to have multiple cognitive deficits rather than simply a decline in memory and thinking. Both emphasize that the deficit should interfere with activities of daily living and social functioning, but only ICD-10 puts a time estimate of 6 months on the duration of the illness.

Erkinjuntti *et al.* (1997) compared the DSM criteria (DSM-III (1980), DSM-IIIR (1987), and DSM-IV (1994)) and the ICD criteria (ICD-9 (1977) and ICD-10 (1992)) with the CAMDEX criteria (Roth *et al.* 1986). The percentage of people receiving a diagnosis of dementia decreased with successive revisions of DSM: 29.1% with DSM-III, 17.3% with DSM-IIIR, and 13.7% with DSM-IV. The rates with ICD-9 and ICD-10 were 5% and 3.1%, respectively. Subtle changes in criteria over time are apparent, i.e. in the earliest DSM criteria either short- or long-term memory loss was required for the diagnosis but later both were required. The comparatively low rates for ICD reflect the fact that more factors are required for diagnosis.

Table 14.1 DSM IV criteria

Diagnostic criteria for dementia of the Alzheimer's type

A The development of multiple cognitive deficits manifested by both memory impairment (impaired ability to learn new information or to recall previously learned information one (or more) of the following cognitive disturbances
  Aphasia (language disturbance)
  Apraxia (impaired ability to carry out motor activities despite intact motor function)
  Agnosia (failure to recognize or identify objects despite intact sensory function)
  Disturbance in executive functioning (ie planning, organizing, sequencing, abstracting)

B The cognitive deficits in Criteria A1 and A2 each cause significant impairment in social or occupational functioning and represent a significant decline from a previous level of functioning.

C The course is characterized by gradual onset and continuing cognitive decline.

D The cognitive deficits in Criteria A1 and A2 are not due to any of the following
  Other central nervous system conditions that cause progressive deficits in memory and cognition (e.g. cerebrovascular disease, Parkinson's disease, Huntington's disease, subdural haematoma, normal-pressure hydrocephalus, brain tumour)
  Systemic conditions that are known to cause dementia (e.g. hypothyroidism, vitamin b$_{12}$ or folic acid deficiency, niacin deficiency, hypercalcaemia, neurosyphilis, HIV infection)
  Substance-induced conditions

E The deficits do not occur exclusively during the course of a delirium

F The disturbance is not better accounted for by another Axis 1 disorder (e.g. major depressive disorder, schizophrenia).

Table 14.2 ICD-10 criteria

Diagnostic guidelines

The primary requirement for diagnosis is evidence of decline in both memory and thinking which is sufficient to impair personal activities of daily living
Symptoms and impairments should be evident *for at least 6 months* for a confident clinical diagnosis of dementia to be made
The following features are essential for a definite diagnosis:

(a) Presence of a dementia as described above

(b) Insidious onset with slow deterioration. While the onset usually seems difficult to pinpoint in time, realization by others that the defects exist may come suddenly. An apparent plateau may Occur in the progression

(c) Absence of clinical evidence, or findings from special investigations, to suggest that the mental state may be due to other systemic or brain disease which can induce a dementia (e.g. hypothyroidism, hypercalcaemia, vitamin B$_{12}$ deficiency, niacin deficiency, neurosyphilis, normal-pressure hydrocephalus, or subdural haematoma)

(d) Absence of a sudden apoplectic onset, or of neurological signs of focal damage such as hemiparesis, sensory loss, visual field defects, and incoordination occurring early in the illness (although these phenomena may be superimposed later)

Only a minority of clinicians use diagnostic criteria, 25% and 11%, respectively, in the studies by Somerfield *et al.* (1991) and Smith *et al.* (1992). Garcia *et al.* (1981) noted that the use of diagnostic criteria improved diagnostic accuracy, with the main reasons for incorrect diagnosis being related to the following:

- failure to recognize depression.
- equating cerebral atrophy as seen on CT with clinical dementia (avoided with the NINCDS/ADRDA criteria as cerebral atrophy is supportive rather than diagnostic)
- regarding focal impairment as synonymous with global impairment.

The following instruments have been developed to provide a structured diagnosis of dementia.

- The Geriatric Mental State Schedule (GMSS) (Copeland *et al.* 1976) relies on features of mental state but, using the computer-assisted algorithm (AGECAT), can provide an accurate diagnosis.
- The Cambridge Examination for Mental Disorders of the Elderly (CAMDEX) (Roth *et al.* 1986) is divided into a number of sections: a structured psychiatric interview, cognitive evaluation (often referred to separately as CAMCOG), mental state examination, informant interview, physical examination, physical investigations, and a list of medications.
- The Canberra Interview for the Elderly (CIE) (Henderson *et al.* 1992) was designed for lay interviewers. It has computer algorithms and is based on DSM and ICD criteria.
- The Structured Interview for the Diagnosis of Dementia of the Alzheimer type, Multi-infarct Dementia and Other Dementias (SIDAM) (Zaudig *et al.* 1991) is a semistructured diagnostic interview, which allows for diagnosis in accordance with ICD-10 and DSM-IIIR and includes standardized cognitive testing, taking into account medical history and severity of dementia.

14.5 Clinical diagnostic criteria for Alzheimer's disease

AD was first described by Alois Alzheimer in 1907 (Alzheimer 1907). The most prominent symptoms were behavioural disturbance (screaming, dragging bedding around the hospital ward) and psychiatric symptoms (delusional ideas directed against the patient's husband and against her physician Alzheimer), together with rapid onset of memory loss (Berrios 2000). In the late 1960s Blessed *et al.* (1968) showed that AD pathology was the most common determinant of dementia in older people, and the severity of the clinical picture correlated with the extent of pathological change. There is now an appreciation that AD can be diagnosed positively (Reisberg *et al.* 1997) following a straightforward and simple clinical approach based on history, examinations, and investigations.

The most widely used set of diagnostic criteria for AD are those of the National Institute of Neurological and Communicative Disorders and Stroke and Alzheimer's Disease and Related Disorders Association (NINCDS/ADRDA) (McKhann *et al.* 1984). The criteria are divided into probable, possible, and definite (requiring a histological diagnosis alongside a clinical picture). It requires a documentation of dementia using a scale such as the Mini-Mental State Examination (Folstein *et al.* 1975) or the Blessed scale (Blessed *et al.* 1968). Two or more cognitive deficits are required and the age range is 40–90 years, making it a diagnostic uncertainty outside this range. Physical examination and CT are recommended to exclude secondary causes (Table 14.3).

Tierney *et al.* (1988) validated these criteria against neuropathological features of ageing. Their overall accuracy was 81%–88% with a sensitivity of 64%–86% (i.e. those patients diagnosed as having the disease clinically who had the pathological features) and a specificity of 89%–91% (i.e. those patients who did not have the disease clinically who did not have pathological changes). Rates of accuracy of 78%–88% (using possible and probable categories) have been described (Burns *et al.* 1990). Gearing *et al.* (1995) and Morris *et al.* (1988) found 100% accuracy of clinical diagnosis compared with histological changes. In other words, the criteria are useful for defining a sample of patients who are highly likely to have AD.

14.6 Clinical diagnostic criteria for other dementias

The other major dementias for which there are diagnostic criteria are vascular dementia, LBD, and dementia of frontotemporal degeneration. The two main criteria for vascular dementia are ADDTC (Chui *et al.* 1992) and NINCDS/AIREN (Roman *et al.* 1993). The NINCDS/AIREN criteria are summarized in Table 14.4.

The criteria for diagnosis of LBD are given in Table 14.5 (after McKeith *et al.* 1992), and features of frontotemporal lobe dementia are given in Table 14.6 (Neary *et al.* 1998).

14.7 Diagnosis of dementia

14.7.1 Differential diagnosis

In the past the prevalence of reversible dementias was thought to be approximately 10%. For an outpatient population probably only 1% of dementias are actually fully reversible (Walstra *et al.* 1997), although a higher proportion may partially reverse. The most common causes of reversible or partially reversible dementia include the following (Clarfield 1988):

- drugs
- depression
- metabolic
  - hypothyroidism
  - vitamin B_{12}
  - calcium
  - hepatic
- normal-pressure hydrocephalus
- subdural hematoma
- neoplasm.

Table 14.3 NINCDS/ADRDA criteria

Diagnostic criteria[1]

A Alzheimer's disease is characterized by progressive decline and ultimately loss of multiple cognitive functions, including both:
   - Memory impairment (impaired ability to learn new information or to recall previously learned information)

   - And at least one of the following
     - Loss of word comprehension ability, e.g. inability to respond to 'Your daughter is on the phone' (aphasia)

     - Loss of ability to perform complex tasks involving muscle coordination, e.g. bathing or dressing (apraxia)
       Loss of ability to recognize and use familiar objects, e.g. clothing (agnosisa)
       Loss of ability to plan, organize, and execute normal activities, e.g. going shopping

B The problems in A represent a substantial decline from previous abilities and cause significant Problems in everyday functioning

C The problems in A begin slowly and gradually become more severe

D The problems in A are not due to:
   - Other conditions that cause progressive cognitive decline, including stroke, Parkinson's disease, Huntington's chorea, brain tumour, etc.

   - Other conditions that cause dementia, including hypothyroidism, HIV infection, syphilis, and deficiencies in niacin, vitamin B12, and folic acid

E The problems in A are not caused by episodes of delirium

F The problems in A are not caused by another mental illness: depression, schizophrenia, etc.

Criteria for diagnosis of probable Alzheimer's disease
   - Dementia established by clinical examination, and documented by a standard test of cognitive function (e.g. Mini-Mental State Examination, Blessed Dementia Scale, etc.) and confirmed by neuropsychological tests

   - Significant deficiencies in two or more areas of cognition, e.g. word comprehension and task-completion ability

   - Progressive deterioration of memory and other cognitive functions

   - No loss of consciousness

   - Onset from age 40–90, typically after age 65

   - No other diseases or disorders that could account for the loss of memory and cognition

A diagnosis of probable Alzheimer's disease is supported by:
   - Progressive deterioration of specific cognitive functions: language (aphasia), motor skills (apraxia), and perception (agnosia)

   - Impaired activities of daily living and altered patterns of behaviour

   - A family history of similar problems, particularly if confirmed by neurological testing

   - The following laboratory results:
     normal cerebrospinal fluid (lumbar puncture test)
     normal electroencephalogram (EEG) test of brain activity
     evidence of cerebral atrophy in a series of CT scans

Table 14.3 NINCDS/ADRDA criteria—cont'd

Other features consistent with Alzheimer's disease

- Plateaus in the course of illness progression
- CT findings normal for the person's age
- Associated symptoms, including depression, insomnia, incontinence, delusions, hallucinations, weight loss, sexual problems, and significant verbal, emotional, and physical outbursts
- Other neurological abnormalities, especially in advanced disease, including increased muscle Tone and a shuffling gait

Features that decrease the likelihood of Alzheimer's disease

- Sudden onset
- Such early symptoms as seizures, gait problems, and loss of vision and coordination

[1] Adapted from G. McKhann *et al.* (1984) Clinical diagnosis of Alzheimer's disease: Report of the NINCDS/ADRDA Work Group, Dept of HHS Task Force on Alzheimer's Disease, *Neurology* **34**, 939.

Table 14.4 NINDS/AIREN criteria for the diagnosis of probable vascular dementia

1 **Dementia**
   Impairment of memory
   Impairment of memory and $=$ 2 cognitive domains
   Orientation
   Attention
   Language
   Visiospatial functions
   Executive functions, motor control and praxis

2 **Cerebrovascular disease**
   Focal signs on neurological examination (hemiparesis, lower facial weakness, Babinski's sign, sensory deficit, hemianopia, and dysarthria)
   Evidence of relevant cerebrovascular disease by brain imaging (CT)
   Large-vessel infarcts
   Single strategically placed infarct
   Multiple basal ganglia and white matter lacunae
   Extensive periventricular white matter lesions
   Combinations thereof

3 **A relationship between the above disorders manifested or inferred by the presence of the following**
   Onset of dementia within 3 months after a recognized stroke
   Abrupt deterioration in cognitive functions
   Fluctuating, stepwise progression of cognitive deficits

4 **Clinical features consistent with the diagnosis of probable vascular dementia**
   Early presence of a gait disturbance
   History of unsteadiness or frequent unprovoked falls
   Early urinary incontinence
   Pseudobulbar palsy
   Personality and mood changes

5 **Features that make the diagnosis of vascular dementia uncertain**
   Early onset of memory deficit and progressive worsening of memory and other cognitive functions in the absence of focal neurological signs and cerebrovascular lesions on CT or MRI

Table 14.5 Clinical diagnosis criteria for FTD

I Core diagnostic features (all features must be present)
Insidious onset and gradual progression
Early decline in social interpersonal conduct
Early impairment in regulation of personal conduct
Early emotional blunting
Early loss of insight

II Supportive diagnostic features
Behavioural disorder
1. Decline in personal hygiene and grooming
2. Mental rigidity and inflexibility
3. Distractibility and impersistence
4. Hyperorality and dietary changes
5. Perseverative and stereotyped behaviour
6. Utilization behaviour

Speech and language
1. Altered speech output
   Aspontaneity and economy of speech
   Press of speed
2. Stereotype of speech
3. Echolalia
4. Perseveration
5. Mutism

Physical signs
1. Primitive reflexes
2. Incontinence
3. Akinesia, rigidity, and tremor
4. Low and labile blood pressure

III Supportive features
Onset before 65 years
Positive family history of similar disorder in first-degree relatives
Bulbar palsy, muscular weakness and fasting, fasciculations (associated motor neuron disease present in a minority of patients)

IV Diagnostic exclusion features (all features must be absent)
1. Abrupt onset with ictal events
2. Head trauma related to onset
3. Early severe amnesia
4. Spatial disorientation
5. Logoclonic festinant speech with loss of train of thought
6. Myoclonus
7. Corticospinal weakness
8. Cerebellar ataxia
9. Choreoathetosis

V Relative diagnostic exclusion features
Typical history of chronic alcoholism
Sustained hypertension
History of vascular disease (e.g. angina, claudication)

Table 14.6 Diagnostic criteria for Lewy body dementia

- Fluctuating cognitive impairment affecting both memory and higher cortical functions. The fluctuation is marked with the occurrence of both episodic confusion and lucid intervals, as in delirium, and is evident either on repeated tests of cognitive function or by variable performance in daily skills
- At least one of the following:
  - visual or auditory hallucinations which are usually accompanied by delusions
  - mild spontaneous extrapyramidal features or neuroleptic sensitivity syndrome
  - repeated unexplained falls or transient clouding or loss of consciousness
- Despite the fluctuating patterns, the clinical features persist over a long period (weeks or months), unlike delirium
- Exclusion of any underlying physical illness adequate to account for the fluctuating cognitive state by appropriate examination and investigation
- Exclusion of past history of confirmed stroke or evidence of cerebral ischaemic damage on structural brain imaging

After McKeith *et al*. 1992; 1996.

It is also vital that the initial assessment should include distinguishing dementia from other conditions with may mimic dementia, of which the main ones are delirium and depression.

Delirium is defined as a transient organic mental syndrome of acute onset associated with a physical cause, characterized by marked abnormalities of attention as well as impairment of global cognitive function, perceptual disturbances, disordered sleep–wake cycle, and increased or decreased psychomotor activity. There are marked fluctuations in clinical symptoms over a 24-hour period, although this can also be seen in most types of dementia, particularly LBD. It can be very difficult to distinguish the symptoms of delirium and dementia as delirium is commonly superimposed on dementia and, if untreated, can subsequently lead to the development of dementia. Compared with AD, attention is impaired more in delirium and the patient is more likely to suffer from visual hallucinations and have impaired conscious level. Onset is also much quicker for delirium and one would expect to find evidence of an underlying physical cause. Instruments such as the Confusion Assessment Method (Inouye *et al*. 1990), which has excellent sensitivity and specificity for delirium, can aid diagnosis.

The other main differential diagnosis that should be considered before assessing for dementia is depression. Patients with depression, unlike those with dementia, frequently complain of poor memory and concentration. Depression is generally shorter in duration, and the onset of memory loss can be pinpointed more accurately. A past history of affective illness is often elicited. Biological features of depression are present and, on cognitive testing, attention is often impaired and performance on memory and executive tasks may be inconsistent, with responses often being 'don't know'. Although depression is seen as a reversible cause of cognitive impairment, in some cases it persists after the depressive symptoms have resolved (Alexopoulos *et al*. 1993).

14.7.2 **History and physical examination**

Making a diagnosis of dementia is often like doing a jigsaw puzzle; the picture is not clear until all the pieces are in place. The largest, and most important, 'piece' is the history. This reveals the most information and is the central piece which all the other pieces (physical examination, investigations, etc.) are connected to and enhanced by. Without a good history it is difficult to obtain a clear picture of what is going on. It is becoming increasingly common for patients to present themselves with complaints of memory problems. As they are likely to have cognitive impairment, they may not be able to give a full history. The diagnosis of AD is often first suspected from the history from an informant. A collateral history from a relative, a carer, or a health care professional is essential but it is crucial to ask the patient's permission before seeing the informant. Ideally, the patient and informant should be seen separately so that the patient can feel more relaxed and the informant can give information more freely. Despite the potential bias of retrospective reporting, informants can reliably recall the onset of dementia (Friske *et al.* 2005). Informants can also accurately judge current cognitive functioning of people with dementia as measured against testing (Kemp *et al.* 2002; Cacchione *et al.* 2003). The referrer (usually the general practioner) is generally able to give valuable information about current drug treatment and other medical details, as well as family and social situation.

Ideally, the assessment should take place in the patient's home as it allows him or her to be in safe and known surroundings and might avoid a tiring journey to the hospital. An additional benefit is that it allows the patient to be assessed in the home environment which gives valuable insight into his or her orientation and competency in activities of daily living, as well as any risks which might be present in the home.

It is very important to establish the timing and the nature of the initial problem as this may give a clue to the diagnosis; AD has a gradual onset and a progressive course, whereas a sudden onset is more indicative of vascular disease.

Important details in the history would be elicited from the following points.

14.7.2.1 **History of presenting complaint**

Often the patient presents with memory difficulties which might be precipitated by a change in circumstances. Family members may be unaware of this or may have been concerned for some time, having noticed the patient forgetting appointments or repeating questions. It is important to verify what symptoms were first noticed, when they were noticed, how they started, and how they have progressed. It is helpful to have examples of how the patient's memory has changed and to have details of their current functional abilities. It is also important to ask about other areas of cognitive function such as dyspraxia, disorientation, language impairment, and reduction in interests and activities.

It is important to enquire about non-cognitive symptoms such as visual hallucinations, which may suggest LBD or delirium, as well as anxiety, agitation, aggression, sleep disturbance, and mood symptoms. Changes in personality and eating preference, along with disinhibition and features of poor social awareness, might help differentiate between a frontotemporal dementia and AD.

A risk assessment is essential as the patient might be at risk of harm, self-neglect, or exploitation due to wandering or driving a car or allowing unknown people into the house.

14.7.2.2 Medical history

Information regarding past medical problems is important. Vascular causes of dementia maybe elicited by a history of cerebral vascular events, transient ischaemic attacks (TIAs), hypertension, or ischaemic heart disease. A history of head injury is also important as it increases the risk of AD as well as normal-pressure hydrocephalus. A history of other medical functions, such as thyroid disease or diabetes, also needs to be closely monitored if the patient has developed dementia.

14.7.2.3 Psychiatric history

A history of depression raises the possibility that any cognitive impairment present is secondary to either a primary or a comorbid mood disorder. A recurrence of a psychotic illness may explain any current delusions or hallucinations, as well as any recent changes in cognition.

14.7.2.4 Drug history

Obtaining a list of medications from the general practitioner before seeing the patient is often useful as it also helps to check compliance. Inadequate compliance can be caused by cognitive impairments including memory difficulties, dyspraxia, and visuospatial problems.

14.7.2.5 Social, family and personal history

Social history provides information about both formal and informal support as well as how the patient copes with the activities of daily living. A full alcohol history also needs to be taken. Family history is important and needs to be precise, with details of any memory problems in the family. Personal history from the viewpoint of patient's education and occupational history is important to assess the patient's premorbid abilities.

14.7.3 Mental state examination

Mental state examination plays an important role in the assessment of dementia. Evidence of self-neglect and poor personal hygiene may indicate either depression or dementia. Frontal lobe impairment may be demonstrated by disinhibition on the part of the patient. Paranoia might present with guarded or hostile behaviour.

14.7.3.1 Speech and language difficulties

Language difficulties such as dysphasia, i.e difficulties with articulation of speech, can be caused by cerebrovascular disease. Patients with expressive dysphasia have significant difficulties in finding words and in motor control of the muscles of articulation. This is caused by posterior frontal lobe dysfunction.

Receptive dysphasia is caused by lesions to the temporal lobe where the patient's comprehension is severely impaired with no impairment of word fluency. Frontotemporal dementia may present with decreased rate of pressure of speech. Perseveration may occur in frontal lobe impairment. Temporal lobe impairment may be suggested by paraphasia where the patient uses slightly incorrect words.

14.7.3.2 Mood disorders

Diagnosing mood disorders in more severe cognitive impairment is difficult and is reliant on biological symptoms and a collateral history from an informant. Suicidal ideation should always be assessed. Vascular dementia classically shows lability of mood with sudden onset of laughing or crying.

14.7.3.3 Thought content

The thought content of people with dementia is often impoverished. Delusions and psychotic experiences may be present and are likely to be a consequence of the patient trying to fill in the gaps in their memory and give some meaning to their current situation.

14.7.3.4 Perception

Hallucinations, most commonly auditory or visual, can occur including detailed visual hallucinations of people as animals, which suggest a diagnosis of delirium or LBD.

14.7.3.5 Cognition

Cognitive assessment takes place throughout the assessment: by listening to the patient's speech and noting their language, evaluating their attention and memory when asked questions, etc. A more formal evaluation using one or more of the wide range of structured bedside cognitive assessment tools is also needed to quantify any impairment found and to give a comprehensive assessment.

14.7.3.6 Physical examination

Physical examination is important and is relevant for excluding other causes of dementia and identifying any comorbid medical conditions which may have an effect on cognitive function. A vascular aetiology may be suggested by the presence of hypertension, atrial fibrillation, signs of peripheral vascular disease, or focal neurological signs. Parkinsonian symptoms, such as shuffling gait, bradykinesia, and tremor, may indicate a diagnosis of LBD, depending on the history.

14.7.4 Investigations for dementia

This includes baseline blood screening tests: full blood count, urea, electrolytes, glucose, liver function, erythrocyte sedimentation rate (ESR), calcium and phosphate, thyroid function, cholesterol, serum B_{12}, and folate. These tests are not part of the diagnosis of AD but are used to identify comorbid conditions which could be treated and partially reverse the cognitive impairment present.

14.7.5 Further investigations

Further tests are indicated by the patient's history, past medical history and findings on physical examination. These could include urine culture, blood culture, VDRL and

TPHA (tests for syphilis), electroencephalograph, HIV testing, serum copper and caeruloplasmin, and lumbar puncture.

Radiological investigations, such as CT and MRI scans in the clinical setting, are more likely to be used to exclude other causes of dementia rather than to diagnose AD *per se*. Single-photon emission CT (SPECT) and particularly positron emission tomography (PET) have aroused recent interest as more specific investigations for AD which help differentiate it from other types of dementia. PET is relatively non-invasive and measures the glucose metabolic rates in the brain. Patients with AD have shown reduced use of cerebral glucose in the parietal and temporal areas, which is a distinctive pattern generally not found in vascular dementia, LBD, or frontal lobe dementia.

Neuropsychological testing can also assist in differentiating between AD and other types of dementia in patients who are difficult to diagnose. Patients with LBD have substantially greater impairment of attention, working memory, and visuospatial abilities than patients with AD (Calderon *et al.* 2001). Semantic memory seems equally affected in LBD and AD, unlike episodic memory which is worse in AD. Neuropsychological testing is also useful in telling the difference between frontal lobe dementia and AD. Patients with frontal lobe dementia have deficits predominately in executive function and less impairment of memory. The opposite is true for patients with AD (Walker *et al.* 2005).

14.8 Assessment instruments for dementia

A plethora of instruments are available for the assessment of dementia and only some of the main ones will be mentioned here. It should be remembered that these instruments can never replace good history notes and mental state examination.

14.8.1 Cognitive function

- The Mini- Mental State Examination (MMSE) (Folstein *et al.*1975) is the most commonly used instrument and is scored out of 30 points. It tests the following domains: orientation, registration of information, attention and calculation, recall language, and praxis. It has a high inter-rater and test–retest reliability and takes only a few minutes to administer. It is not suitable for assessing frontal lobe function and has been criticized for having a ceiling effect, resulting in inability to differentiate moderate from high functioning. Education and language abilities also affect performance.

- Mini-Cog (Borson *et al.* 2003) is a 3-minute cognitive screen originally designed for use in primary care. It is reported to detect cognitive impairment as well as or better than the MMSE and is less biased by cultural differences, educational level, and literacy (Borson *et al.* 2005).

- The Standardized Mini-Mental State Examination (Molloy *et al.* 1991) is a standardized version of the MMSE which comes with complete rating instructions, leading to slightly improved validity.

- The Abbreviated Mental Test Score (AMTS) (Hodkinson 1972) is a much briefer screening tool (scored out of 10) used to test only orientation and memory.

- The Alzheimer's Disease Assessment Scale (ADAS) Rosen *et al.* 1984) is now used as a standard, especially in drug trials. It assesses cognitive as well as non-cognitive symptoms.

- The clock drawing test (Brodaty and Moore1997) is a quick screening test for cognitive dysfunction secondary to dementia, delirium, or a range of neurological or psychiatric illness. It tests a wide range of skills including comprehension, visual memory, visuospatial abilities, numerical thinking, and executive functioning. The clock drawing test may complement other quick screening tests, including the MMSE, and is a component of the 'Seven Minute Neurocognitive Screening Battery' (Solomon *et al.* 1998). The scoring system has varied, but published studies demonstrate a mean sensitivity and specificity of 85% (Sunderland 1989; Shulman 2000).

- The Informant Questionnaire on Cognitive Decline in the Elderly (IQCODE) (Jorm and Jacomb 1989) is a questionnaire administered to an informant about the changes in the everyday cognitive function of an elderly person and aims to assess cognitive decline independent of premorbid ability.

14.8.2 Global measures

- The Clinical Dementia Rating (Hughes *et al.* 1983; Berg 1984; Morris 1993) is the most widely used global scale. It gives an overall severity rating in dementia, ranging from 0 (healthy), 0.5 (questionable) through (1) mild and (2) moderate to (3) severe dementia. Each patient is rated in six domains: memory, orientation, judgement and problem solving, community affairs, home and hobbies, and personal care.

- The Global Deterioration Scale (Reisberg *et al.* 1982) consists of the seven stages of dementia. Stage 1 is normal and stage 7 is where all verbal ability is lost. It is used extensively and has been validated by post-mortem findings.

14.8.3 Psychopathology

- The Cornell Scale for Depression and Dementia (Alexopoulos *et al.* 1998) is a 19-item scale, specifically assessing depression in dementia in five main domains: mood-related signs, behavioural disturbance, physical signs, cyclic functions, and ideation disturbance. Both informant and observational ratings make it ideal for assessment in patients with cognitive impairment where such information may not always be available from the patient. Significant depression is suggested by a score ≥8.

- BEHAV-AD (Reisberg *et al.* 1987) is a 25-item scale measuring many of the psychiatric symptoms and behavioural disturbances associated with dementia. The rating is on a three-point scale and a global rating of the severity of symptoms is also done. BEHAV-AD is particularly useful in assessment of patients in drug trials.

- The Neuropsychiatric Inventory (NPI) (Cummings *et al.* 1994) assesses 12 behavioural areas: delusions, hallucinations, agitation, depression, anxiety, euphoria, apathy, disinhibition, irritability, aberrant motor behaviour, night-time behaviours, and appetite/eating disorders. Each of these is rated on a four-point scale of frequency and a three-point scale of severity. Distress in carers can also be measured.

- MOUSEPAD (Allen *et al.* 1996) is based on the longer Present Behavioural Examination (PBE) and measures a number of behaviour and psychiatric symptoms, with particular reference to their development over the life of the dementia syndrome.

14.8.4 **Functional ability**

- The Bristol Activities of Daily Living Scale (Bucks *et al.* 1996) is a 20-item rating on a five-point scale of severity looking at basic activities of daily living (e.g. feeding, eating, and toileting) and also looking at instrumental activities of daily living (performance of more complex tasks such as shopping, travelling, answering the telephone, and handling finances).

- The Interview for Deterioration in Daily Living Activities in Dementia (IDDD) (Teunisse *et al.* 1991) has been used extensively to assess the effect of drugs and contains a 33-item self-care rating on a four-point scale.

- The Disability Assessment for Dementia (DAD) (Gelinas *et al.* 1999) is a rating of activities of daily living in dementia, separating them into different areas of initiation, planning, organization, and active performance. ADL and Instrumental ADL are also assessed.

- The Alzheimer's Disease Functional Assessment and Change Scale (ADFACS) (Galasko *et al.* 1997) is an 18-item scale looking at a wide range of basic activities and instrumental activities of daily living.

A more extensive summary of all the assessment instruments for dementia is given by Burns *et al.* (2004).

14.9 **Conclusion**

With the advent of cholinesterase inhibitors and other treatments there is a need for rapid accurate diagnosis of AD. With sound clinical skills, and especially thorough history taking, this is not as difficult as it was once thought to be. It is important to adhere to validated diagnostic criteria as this improves the diagnostic accuracy. Standardized scales have a role in evaluating the level of cognitive impairment and monitoring its progression, and also identifying and quantifying behavioural and psychiatric symptoms of dementia. Detailed investigations and additional assessments should only be required where the presentation or clinical course is atypical.

Acknowledgements

We are grateful to Barbara Dignan for secretarial help.

References

Alexopoulos GS, Meyes B, Young RC, *et al.* (1993) The course of geriatric depression with 'reversible dementia': a controlled study. *American Journal of Psychiatry* **150**, 1693–9.

Alexopoulos GS, Abrams R, Young R, and Shamoian C (1998). Cornell Scale for Depression and Dementia. *Biological Psychiatry* **23**, 271–84.

Allen NHP, Gordon S, Hope T, and Burns A (1996). Manchester and Oxford Universities Scale for the Psychopathological Assessment of Dementia (MOUSEPAD). *British Journal of Psychiatry* **169**, 293–307.

Alzheimer, A. (1907). Über eine eigenartige Erkrankung der Hirnrinde [Concerning a novel disease of the cortex]. *Allgemeine Zeitschrift für Psychiatrie Psychisch-Gerichtlich Medizine* **64**, 146–48.

American Psychiatric Association (1994). *Diagnostic and Statistical Manual of Mental Disorders* (4th edn). (American Psychiatric Association, Washington, DC).

Berg L (1984). Clinical dementia rating. *British Journal of Psychiatry* **145**, 339.

Berrios G (2000) Dementia: historical overview. In O'Brien J, Ames D, Burns A (eds), *Dementia*, pp. 3–13 (Hodder Arnold: London).

Blessed G, Tomlinson B, and Roth M (1968). The association between quantitative measures of dementia and of senile change in the cerebral grey matter of elderly subjects. *British Journal of Psychiatry* **114**, 797–811.

Borson S, Scanlan JM, Chen P, *et al.* (2003). The Mini-Cog as a screen for dementia: validation in a population-based sample. *Journal of the American Geriatrics Society* **51**, 1451–4.

Borson S, Scanlan JM, Watanabe J, *et al.* (2005). Simplifying detection of cognitive impairment: comparison of the Mini-Cog and Mini-Mental State Examination in a multiethnic sample. *Journal of the American Geriatrics Society* **53**, 871–4.

Brodaty H and Moore CM (1997). The clock drawing test for dementia of the Alzheimer's type: a comparison of three scoring methods in a memory disorders clinic. *International Journal of Geriatric Psychiatry* **12**, 619–27.

Brodaty H, McGilchrist C, Harris L, *et al.* (1993). Time until institutionalization and death in patients with dementia: role of caregiver training and risk factors. *Archives of Neurology* **50**, 643–50.

Bucks RS, Ashworth DL, Wilcock GK, and Siegfried K (1996). Assessment of activites of daily living in dementia: development of the Bristol Activities of Daily Living Scales. *Age and Ageing* **25**, 113–20.

Burns A, Luthert P, Levy R, Jacoby R, and Lantos P (1990). Accuracy of clinical diagnosis of Alzheimer's disease. *British Medical Journal* **301**, 1026.

Burns A, Lawlor B, and Martin, CS (eds) (2004). *Assessment Scales in Old Age Psychiatry* (2nd edn) (Dunitz, London).

Cacchione PZ, Powlishta KK, Grant EA, *et al.* (2003). Accuracy of collateral source reports in very mild to mild dementia of the Alzheimer type. *Journal of the American Geriatrics Society* **51**, 819–23.

Calderon J, Perry RJ, Erzinclioglu SW, *et al.* (2001). Perception, attention, and working memory are disproportionately impaired in dementia with Lewy bodies compared with Alzheimer's disease. *Journal of Neurology, Neurosurgery, and Psychiatry* **70**, 157–64.

Carlesimo GA, Bussione I, Fadda L, *et al.* (2002). Standardizzazione di due test di memoria per uso clinico: breve racconto e figura di rey. *Nuova Rivista di Neurologica* **12**, 1–13.

Chui HC, Victoroff JI, Margolin D, Jagust W, Shankle R, and Katzman R (1992). Criteria for the diagnosis of ischemic vascular dementia proposed by the State of California Alzheimer's Disease Society Diagnostic and Treatment Centers. *Neurology* **42**, 473–80.

Clarfield AM (1988). The reversible dementias: do they reverse. *Annals of Internal Medicine* **109**, 476–86.

Copeland J, Kelleher M, Kellett J, *et al.* (1976). The Geriatric Mental State Schedule. *Psychological Medicine* **6**, 439–49.

Cummings JL, Mega M, Gray K, Rosenberg-Thompson S, Carusi DA, and Gornbein J (1994). The Neuropsychiatric Inventory: comprehensive assessment of psychopathology in dementia. *Neurology* **44**, 2308–14.

Erkinjuntti T, Ostbye T, Steenhuis R, and Hachinski V (1997). The effect of different diagnostic criteria on the prevalence of dementia. *New England Journal of Medicine* **337**, 1667–74.

Folstein ME, Folstein SE, and McHugh PR (1975). Mini-Mental State: a practical method for grading the cognitive state of patients for the clinician. *Journal of Psychiatric Research* **12**, 189–98.

Friske A, Gatz M, Aadnoy B, and Pedersen NL (2005). Assessing age of dementia onset: validity of informant reports. *Alzheimer Disease and Associated Disorders* **19**, 128–34.

Galasko D, Bennett D, Sano M, *et al.* (1997). An inventory to assess activities of daily living for clinical trials in Alzheimer's disease. *Alzheimer Disease and Associated Disorders* **11**(Suppl 2), S33–9.

Gelinas I, Gauthier I, McIntyre M, and Gauthier S (1999). Devlopment of a functional measure for persons with Alzheimer's disease: the Disability Assessment for Dementia. *American Journal of Occupational Therapy* **53**, 471–81.

Garcia CA, Reding MJ, and Blass JP (1981). Overdiagnosis of dementia. *Journal of the American Geriatrics Society* **29**, 407–10.

Gearing M, Mirra SS, Hedreeen JC, Sumi SM, Hansen LA, and Heyman A (1995). The Consortium to Establish a Registry for Alzheimer's Disease (CERAD). Part X: Neuropathy confirmation of the clinical diagnosis of Alzheimer's disease. *Neurology* **45**, 461–6.

Henderson AS, *et al.* (1992). The Canberra Interview for Elderly: a new field instrument for the diagnosis of dementia and depression by ICD-10 AND DSM-III-R. *Acta Psychiatrica Scandinavica* **85**, 105–13.

Hodkinson H (1972). Evaluation of a mental test score for assessment of mental impairment in the elderly. *Age and Ageing* **1**, 233–8.

Holmes C, Cairns N, Lantos P *et al.* (1999). Validity of current clinical criteria for Alzheimer's disease, vascular dementia and dementia with Lewy bodies. *British Journal of Psychiatry* **174**, 45–50.

Hughes C, Berg L, Dantziger W, Coban L, and Martin R (1982). A new clinical scale for the staging of dementia. *British Journal of Psychiatry* **140**, 566–72.

Inouye SK, van Dyck CH, Alessi CA, *et al.* (1990). Clarifying confusion: the Confusion Assessment Method. *Annals of Internal Medicine* **113**, 941–8.

Jorm AF and Jacomb PA (1989). An informant questionnaire on cognitive decline in the elderly (IQCODE): socio-demographic correlates, reliability, validity and some norms. *Psychological Medicine* **19**, 1015–22.

Kemp NM, Brodaty H, Pond D, *et al.* (2002). Diagnosing dementia in primary care: the accuracy of informant reports. *Alzheimer Disease and Associated Disorders* **16**, 171–6.

Lishman WA (1978). *Organic Psychiatry* (Blackwell Scientific, Oxford).

McKeith I, Perry R, Fairbairn A, *et al.* (1992). Operational criteria for senile dementia of Lewy body type. *Psychological Medicine* **22**, 911–22.

McKeith I G, Galasko D, Kosaka K, *et al.* (1996). Consensus guidelines for the clinical and pathologic diagnosis of dementia with Lewy bodies (DLB). *Neurology* **47**, 1113–24.

McKhann G, Drachman D, Folstein M, Katzman R, Price D, and Stadlan EM (1984). Clinical diagnosis of Alzheimer's disease: Report of the NINCDS-ADRDA Work Group under the auspices of Department of Health and Human Services Task Force on Alzheimer's disease. *Neurology* **34**, 939–44.

Molloy D, Alemayehu E, and Roberts R (1991). Reliability of a standardised Mini Mental State Examination. *American Journal of Psychiatry* **148**, 102–5.

Morris J (1993). The CDR: current version and scoring rules. *Neurology* **43**, 2412–13.

Morris JC, Mckeel DW jr, Fulling K, Torack RM, Berg L (1988). Validation of Clinical diagnostic criteria for Alzheimer's disease. *Annals of Neurology* **24**, 17–22.

Neary D, Snowden JS, Gustafson L *et al.* (1998). Frontotemporal Lobe Degeneration: a consensus on clinical diagnostic criteria. *Neurology* **51**, 1546–54.

Oppenheim G (1994). The earliest signs of Alzheimer's disease. *Journal of Geriatric Psychiatry and Neurology* **7**, 118–22.

Reisberg B, Ferris S, DeLeon M, and Crook T (1982). The Global Deterioration Scale (GDS) for the assessment of primary degenerative dementia. *American Journal of Psychiatry* **139**, 1136–9.

Reisberg B, Borenstein J, Salob S, *et al.* (1987). Behavioural symptoms in Alzheimer's disease: phenomenology and treatment. *Journal of Clinical Psychiatry* **48**(Suppl 5), 9–15.

Reisberg, B, Burns A, Brodaty H, *et al.* (1997). Diagnosis of Alzheimer's disease. Report of an International Psychogeriatric Association Special Meeting Work Group under the cosponsorship of Alzheimer's Disease International, the European Federation of Neurological Societies, the World Health Organization, and the World Psychiatric Association. *International Psychogeriatrics* **9** (Suppl 1), 11–38.

Roman G, Tatemechi T, Erkinjuntti T, *et al.* (1993). Vascular dementia: diagnostic criteria for research studies. *Neurology* **43**, 250–60.

Rosen W, Mohs R, and Davis KL (1984). A new rating scale for Alzheimer's disease. *American Journal of Psychiatry* **141**, 1356–64.

Roth M, Tyme E, Mountjoy CQ, *et al.* (1986). CAMDEX: a standardised instrument for the diagnosis of mental disorder in the elderly with special reference to the early detection of dementia. *British Journal of Psychiatry* **149**, 698–709.

Shulman K (2000). Clock-drawing: is it the ideal cognitive screening test? *International Journal of Geriatric Psychiatry* **15**, 548–61.

Smith CW, Byrne EJ, Arie T, and Lilley JM (1992). Diagnosis of dementia. II: Diagnostic Methods: a survey of current consultant practice and review of the literature. *International Journal of Geriatric Psychiatry* **7**, 323–9.

Somerfield MR, Wiesman CS, Ury W, Chase GA, and Folstein MF (1991). Physician practices in the diagnosis of dementing disorders. *Journal of the American Geriatrics Society* **39**, 172–5.

Sunderland T, Hill J, Mellow A *et al.* (1989). Clock drawing in Alzheimer's disease. *Journal of the American Geriatrics Society* **37**, 725–9.

Teunisse S, Derix MMA, and van Crevel H (1991). Assessing the severity of dementia: patient and caregiver. *Archives of Neurology* **48**, 274–7.

Tierney MC, Fisher RH, Lewis AJ, *et al.* (1998). The NINCDS/ADRDA work group outcome for the clinical diagnosis of probable Alzheimer's disease: a clinicopathologic study of 57 cases. *Neurology* **38**, 359–64.

Wackerbarth SB and Johnson MMS (2002). The carrot and the stick: benefits and barriers in getting a diagnosis. *Alzheimer Disease and Associated Disorders* **16**, 213–20.

Walker AJ, Meares S, Sachdev PS, and Brodaty H (2005). The differentiation of mild frontotemporal dementia from Alzheimer's disease and healthy aging by neuropsychological tests. *International Psychogeriatrics* **17**, 57–68.

Walstra GJM, Teunisse S, van Gool WA, and van Crevel H (1997). Reversible dementia in elderly patients referred to a memory clinic. *Journal of Neurology* **244**, 17–22.

Whitthaus E., Ott A, Barendregt JJ, Breteler M, and Bonneux L (1999). Burden of mortality and morbidity from dementia. *Alzheimer Disease and Associated Disorders* **13**, 176–81.

World Health Organization (1992). *International Classification of Mental and Behavioural Disorders, 10th Revision* (World Health Organization, Geneva).

Zaudig M, Mittelhammer J, Hiller W, *et al.* (1991). SIDAM: a structural interview for the diagnosis of dementia of the Alzheimer type, multi-infarct dementia and dementia of other aetiology according to ICD-10 and DSM-IIIR. *Psychological Medicine* **21**, 225–36.

Chapter 15

Molecular and biological markers for Alzheimer's disease

Henrik Zetterberg and Kaj Blennow

In this chapter we review the biochemistry and clinical usefulness of the most established cerebrospinal fluid (CSF) protein markers for Alzheimer's disease (AD) and how to interpret their results with emphasis on early and differential diagnosis. Scientific benefits of using biomarkers as additional inclusion and exclusion criteria in the design of clinical trials and as surrogate markers for drug effects are considered. Further, the ability of CSF biomarkers to reveal AD in its preclinical stage, when novel disease-modifying drugs are likely to be the most effective, is discussed. Finally, we present an updated list of novel possible biological markers which may prove useful in the diagnostic work-up of patients with subtle memory disturbances in the clinical routine.

15.1 Introduction

Just a few decades ago AD was considered an incurable syndrome related to the 'normal' ageing process in certain individuals. After exclusion of vitamin deficiencies, anaemia, hypothyreosis, and a few other treatable causes of dementia, the syndrome was left essentially untreated. The discovery that acetylcholine esterase (AChE) inhibitors can be used for symptomatic treatment of AD has changed all this. The awareness in the population of the availability of drug treatment has also made patients seek medical advice at an earlier stage of the disease, resulting in an increased percentage of individuals with very mild or only subjective memory disturbances attending memory clinics. At present there is no clinical method to determine whether an individual with mild cognitive dysfunction will develop AD with dementia, remain stable, or progress to other dementias. Hence new diagnostic tools to aid in the diagnosis of early or incipient AD would be of fundamental importance in public health. Such diagnostic markers would be of even greater significance if new drugs with promise of disease-arresting effects, such as β-sheet breakers, β- and γ-secretase inhibitors, and anti-amyloid vaccination regimes, prove to be clinically effective. These types of drugs are likely to have the best effects in the early or even preclinical stage of the disease when the synaptic and neuronal loss has not become too widespread.

15.2 The biological rationale for Alzheimer's disease biomarkers

A biological marker for a disease should mirror central pathogenic events in the disease process. Optimally, it should also reflect the progress and intensity of the disease.

The primary cause of AD is not yet established, although a wealth of evidence points to an important and probably pathogenic role of the 42 amino acid (aa) fragment of β-amyloid (Aβ42) in both familial and sporadic AD. The neuropathological hallmarks of AD include neuronal loss and synaptic and axonal degeneration, senile plaques composed of aggregates of Aβ, and neurofibrillary tangles composed of hyperphosphorylated and poly-ubiquitinated tau. All pathogenic events in AD appear to occur selectively in the brain and are rarely seen in other more accessible compartments of the body. One possible exception is the lens of the eye. Recent data indicate that AD patients display aggregates of Aβ in the supranuclear type of cataract (Goldstein *et al.* 2003). These aggregates can be visualized using biomicroscopy in conjunction with amyloid tracing techniques. However, sensitivity and specificity figures for this possible biomarker are not available. Efforts so far to find reliable biomarkers for AD in peripheral blood have been fruitless. It is not possible to measure tau in plasma, serum, or whole blood. There are detectable fragments of Aβ in peripheral blood, including Aβ40 and Aβ42, but they do not enable AD patients to be distinguished from non-demented controls in a reproducible manner and are unlikely to reflect Aβ processing in the brain (Vanderstichele *et al.* 2000; Fukumoto *et al.* 2003). Much of the research activity has focused on the CSF, which is in direct contact with the brain interstitial fluid without any barriers. Hence changes in neuronal function, metabolism, and survival, and host responses to neuronal death, are likely to be reflected in the metabolome and proteome of CSF. It should also be noted that CSF is a fairly accessible biological fluid. It can easily be obtained through lumbar puncture, which has been used routinely and safely in the practice of neurology for decades. Two large studies performed as part of an evaluation of possible AD biomarkers corroborate the safety of the method and show that it can be applied broadly and is well tolerated in the elderly population (Blennow *et al.* 1993; Andreasen *et al.* 2001). The only recorded complication was post-lumbar puncture headache. However, when a thin needle (0.7 mm in diameter) was used, the rate of mild headache (duration less than a day, not affecting daily life) was less than 4%, and the rate of moderate or severe headache (duration more than a day and/or affecting daily life) was less than 1%.

15.3 Established CSF biomarkers for Alzheimer's disease

15.3.1 Tau proteins

Tau is a microtubule-associated protein, primarily located in the neuronal axons. It is undetectable in peripheral blood but measurable in a reproducible way in CSF. Because of alternative splicing of tau mRNA, six different isoforms of the protein exist (Goedert *et al.* 1989). By binding to tubulin in the axonal microtubules, tau promotes microtubule assembly and stability and, hence, axonal transport (Buée *et al.* 2000), which is important for axonal function.

Four different enzyme-linked immunosorbent assay (ELISA) methods for quantification of total tau (T-tau), i.e. all tau isoforms irrespective of phosphorylation status, have been published (Vandermeeren *et al.* 1993; Blennow *et al.* 1995; Mori *et al.* 1995; Vigo-Pelfrey *et al.* 1995). The 20 largest studies, including more than 2000 AD patients and

1000 controls, and evaluating the most commonly used ELISA method for T-tau in CSF, have recently been reviewed (Blennow and Hampel 2003; Blennow 2004). Taken together, these studies report an overall sensitivity to discriminate sporadic AD from non-demented age-matched controls of 81% at a specificity level of 91%. Most of the studies are cross-sectional and there is a lack of longitudinal studies of reasonable size that investigate how tau levels reflect the disease intensity in AD. Studies of this type are urgently needed in order to establish tau as a surrogate endpoint in clinical trials of disease-modifying drugs against AD. Furthermore, few studies have examined CSF T-tau in pathologically confirmed AD cases. However, in a study of 131 clinically diagnosed AD cases a subgroup was autopsy verified, and the performance of CSF T-tau and Aβ42 was found to be similar in both the clinically diagnosed and autopsy-verified groups (Sunderland et al. 2003).

Since CSF T-tau reflects neuronal and axonal degeneration, it is not specific for AD. High levels can be found in all CNS disorders with significant neuronal degeneration or damage. The highest levels are detected in acute stroke (Hesse et al. 2000) and Creutzfeldt–Jakob disease (CJD) (Otto et al. 1997, 2002; Riemenschneider et al. 2003; Van Everbroeck et al. 2003). The mean levels of CSF T-tau in CJD are 10–50 times higher than those in controls, resulting in a sensitivity close to 100% and a specificity above 90% against other dementias such as AD. This diagnostic performance is similar to that found in some of the studies that established the CSF 14–3–3 protein as a golden standard bio-marker for CJD (Hsich et al. 1996; Beaudry et al. 1999; Otto et al. 2002; Van Everbroeck et al. 2003). However, the clinical usefulness of 14–3–3 has recently been challenged. Geschwind et al. (2003) found that only 17 of 32 patients with autopsy-proven CJD were positive for the 14–3–3 test, and we recently obtained very similar results (Zetterberg and Blennow 2004; Blennow et al. 2005). Furthermore, since most of the available 14–3–3 tests are based on qualitative immunoblot, the quantitative ELISA methods for T-tau may be preferable in the clinical laboratory.

High CSF T-tau has been found in some studies of vascular dementia (VAD) (Blennow et al. 1995; Andreasen et al. 1999a; Sjogren et al. 2000), but not all (Vigo-Pelfrey et al. 1995; Shoji et al. 1998; Sjogren et al. 2001b). The reason for this discrepancy is unknown at present but may involve differences in patient characteristics and the diagnostic crite-ria used. High T-tau in clinically diagnosed VAD cases may be caused by concomitant AD pathology, which is a frequent finding at autopsy but difficult to identify clinically (Jellinger 1996; Kosunen et al. 1996). A study of longitudinal magnetic resonance tomog-raphy scans also found that VAD cases with progressive changes in white matter have normal CSF T-tau (Andreasen et al. 1999a). Furthermore, CSF T-tau is normal in patients with non-acute cerebrovascular disease without dementia (Arai et al. 1997; Nishimura et al. 1998). One way of interpreting these findings is that a high CSF-tau level in a patient with clinical and brain-imaging findings indicative of VAD suggests that this patient may have mixed AD/VAD dementia.

Most studies have found normal to mildly increased CSF T-tau levels in other demen-tias, such as frontotemporal dementia (FTD) and Lewy body dementia (LBD) (Blennow 2004). Other than in aged non-demented individuals, normal CSF T-tau levels are found in depression, alcohol dementia, other diseases with secondary cognitive dysfunction such as anaemia, vitamin B deficiency, and hypothyrosis, and in chronic neurological disorders such as Parkinson's disease (PD) and progressive supranuclear palsy (PSP) (Blennow 2004). Thus CSF T-tau has a clear diagnostic value in the differentiation between AD and these important and sometimes difficult alternative diagnoses.

Tau is a phosphoprotein, with more than 30 potential phosphorylation sites (Buée et al. 2000; Iqbal et al. 2002). The tangles in AD are made up of a hyperphosphorylated form of tau (Grundke-Iqbal et al. 1986). Because of the hyperphosphorylation, tau loses its ability to bind to the microtubules and support their assembly, which is essential for axonal transport and function (Iqbal et al. 2000; Mandelkow et al. 2003). Whether the hyperphosphorylation is a cause or a consequence of axonal dysfunction or degeneration in AD is unknown. The AD research field has for some time been divided into two main groups depending on the preferred hypothesis on the primary cause of AD, the 'tauoists' and the 'baptists', believing in either tau dysfunction or altered Aβ metabolism as a primary cause of AD; one or the other would then represent a secondary event. Interpretation of the literature to date suggests that a primary alteration of tau is not a very likely explanation for the axonopathy and neuronal loss seen in AD for at least three reasons. First, missense and splicing mutations in the TAU gene are associated with frontotemporal dementia, PSP, and Pick's disease, but not with AD (Selkoe 2005). Secondly, tau dysfunction can lead to severe neuronal degeneration in the absence of Aβ deposition. Studies in double-transgenic mice suggest that the presence of Aβ-elevating APP mutations augments the formation of tau-bearing neurofibrillary tangles, rather than the opposite (Lewis et al. 2001). Thirdly, Aβ induces tau hyperphosphorylation and loss of cholinergic neurons in rat primary septal cultures (Zheng et al. 2002). Nevertheless, the relevance of some of these models for sporadic AD in humans is questionable and, again, longitudinal and naturalistic studies of tau and Aβ metabolism in elderly patients are needed to resolve this controversy.

Six different ELISA methods have been developed for quantification of tau phosphorylated at different epitopes (Blennow 2004). A moderate to marked increase in CSF P-tau has been found using all these different methods, and the mean sensitivity to discriminate AD from non-demented age-matched individuals is 81% at a specificity level of 91% (Blennow 2004). The CSF level of P-tau probably reflects the phosphorylation state of tau. This view is based on indirect evidence such as the finding that there is no change in CSF P-tau after acute stroke (Hesse et al. 2000), although there is a marked increase in T-tau. Furthermore, CSF P-tau levels are normal or only mildly increased in CJD despite a very marked increase in T-tau (Riemenschneider et al. 2003). These data suggest that P-tau in CSF is not a marker for neuronal degeneration or damage, but that it specifically reflects the phosphorylation state of tau and thus possibly the formation of tangles in AD brains. Accordingly, the specificity of CSF P-tau to differentiate AD from other dementias seems to be higher than for T-tau and Aβ42. Normal CSF levels of P-tau are found in psychiatric disorders such as depression, chronic neurological disorders such as amyotrophic lateral sclerosis, PD, and PSP, and in most cases with other dementia disorders such as VAD, FTD, and LBD (Blennow 2004). Furthermore, although there is a very distinct increase in CSF T-tau in patients with CJD, most have normal or only mildly elevated CSF P-tau. In a large set of patients with CJD and other dementias, the ratio of P-tau to T-tau in CSF was found to discriminate CJD from other neurodegenerative disorders without any overlap (Riemenschneider *et al.* 2003).

15.3.2 β-Amyloid fragments

The main component of senile plaques is β-amyloid (Aβ) (Masters *et al.* 1985). Aβ is generated from amyloid precursor protein (APP), a type I integral membrane protein with one transmembrane domain, by enzymatic digestion involving β- and γ-secretase activities (Selkoe 2001). Cleavage of APP by β-secretase generates an approximately 100 kDa soluble N-terminal fragment and a 12 kDa C-terminal fragment (C99), which can be further cleaved by γ-secretase to yield the APP intracellular domain (AICD) and 40 and 42 aa long Aβ βpeptides (Aβ40 and Aβ42), the latter of which appears the most prone to aggregate (the amyloidogenic pathway) (Lorenzo and Yankner 1994). γ-Secretase is a membrane-bound protease complex consisting of at least four essential components: the homologous presenilins 1 and 2 (PS1 and PS2), nicastrin, anterior pharynx defective (Aph-1), and presenilin enhancer-2 (Pen-2) (De Strooper 2003). Most of the β-secretase activity originates from two membrane-anchored aspartyl proteases encoded by the β-site APP-cleaving enzyme 1 and 2 (*BACE1* and *BACE2*) genes, respectively (Sinha *et al.* 1999; Vassar *et al.* 1999; Farzan *et al.* 2000). The amyloid hypothesis predicts that a gradual elevation of Aβ42 levels in brain interstitial fluid leads to formation of synaptotoxic Aβ42 oligomers and eventually to deposition of Aβ42 into senile plaques, and that these events are pathogenic in AD (Selkoe 2002). At least three lines of evidence support this view.

(i) Aβ42 accumulation can be initiated by mutations in the genes that encode APP and PS, the latter harbouring the active site of β-secretase, and these mutations cause rare autosomal dominant forms of familial AD.

(ii) Trisomy 21 (Down's syndrome) leads to APP overexpression because of increased dosage of the *APP* gene on chromosome 21 and early onset of AD with classical AD neuropathology in the form of Aβ42 deposition.

(iii) The strongest susceptibility gene for sporadic AD is the apolipoprotein E (*APOE*) ε4 gene variant that encodes the E4 isotype of the protein. The AD-promoting effect of inheriting one or two ε4 alleles may involve an enhanced aggregation or decreased clearance of Aβ42 conferred by the E4 isotype of apoE (Prince *et al.* 2004). Transgenic mice that express E4 display higher density of Aβ plaques and vascular deposits than E3-expressing mice (Holtzman *et al.* 2000; Fagan *et al.* 2002).

After it was found that Aβ is generated as a more or less soluble protein during normal cellular metabolism and is secreted to CSF (Seubert *et al.* 1992), studies examining CSF Aβ as a candidate biomarker for AD were published. However, these initial reports used ELISA methods for 'total Aβ' that did not discriminate between different Aβ isoforms. Although some studies found a slight decrease in the CSF level of total Aβ in AD, there was a large overlap between AD patients and controls, and in some studies no change in

CSF total Aβ was found. It was soon discovered that there are several N- and C-terminally truncated forms of Aβ. The two major C-terminal variants of Aβ consist of a shorter form ending at Val-40 (Aβ40) and a longer form ending at Ala-42 (Aβ42). Aβ42 was found to aggregate more rapidly than Aβ40 (Jarrett *et al.* 1993), and to be the initial and predominant form of Aβ deposited in diffuse plaques (Iwatsubo *et al.* 1994, Miller *et al.* 1993, Tamaoka *et al.* 1994). These data made it logical to develop immunoassays specific for Aβ42.

To date, 11 different methods have been developed for quantification of Aβ42 in CSF (Blennow 2004). A moderate to marked decrease in CSF Aβ42 to about 50% of control levels in AD patients has been found using most of these methods. These studies include more than 650 AD patients and 500 controls. The mean sensitivity to discriminate between AD and normal ageing is 86% at a specificity level of 89% (Blennow 2004). Other than in non-demented aged individuals, normal CSF Aβ42 is found in psychiatric disorders such as depression and in chronic neurological disorders such as PD and PSP (Blennow 2004). Thus CSF Aβ42 helps in the clinical differentiation between AD and these important and often difficult differential diagnoses. However, data on the performance of CSF Aβ42 in the discrimination between AD and other dementias and neurological disorders are relatively limited. A mild to moderate decrease in CSF Aβ42 is found in a percentage of patients with FTD and VAD (Hulstaert *et al.* 1999; Sjogren *et al.* 2000, 2002; Riemenschneider *et al.* 2002b).

The reduced CSF level of Aβ42 in AD is often hypothesized to be caused by deposition of Aβ42 in senile plaques, with lower levels diffusing to CSF. Accordingly, a recent autopsy study found a strong correlation between low Aβ42 in ventricular CSF and high numbers of plaques in the neocortex and hippocampus (Strozyk *et al.* 2003). However, some studies have also found a marked reduction in CSF Aβ42 in disorders without Aβ plaques, such as CJD (Otto *et al.* 2000), amyotrophic lateral sclerosis (Sjogren *et al.* 2002), and multiple system atrophy (Holmberg *et al.* 2003). There is also a strong correlation between inheritance of the *APOE* ε4 allele and lower Aβ42 not only in AD patients but also in non-demented individuals (Prince *et al.* 2004). These findings make a direct link between low CSF Aβ42 and deposition of Aβ in plaques questionable. Alternative explanations include apoE4-induced formation of Aβ42 oligomers that escape ELISA detection, binding of Aβ42 to apoE4 or other chaperone-like proteins that block the antibody recognition of Aβ42, and sequestering of Aβ42 in the plasma membrane with lower levels diffusing to CSF.

There is no decrease in CSF Aβ40 in AD (Kanai *et al.* 1998; Shoji *et al.* 1998; Fukuyama *et al.* 2000; Mehta *et al.* 2000). As a consequence, a decrease in the ratio Aβ42/Aβ40 (or an increase in the ratio Aβ40/Aβ42) in CSF in AD patients has been reported in several papers (Kanai *et al.* 1998; Shoji *et al.* 1998; Fukuyama *et al.* 2000). This decrease seems more pronounced than the reduction of CSF Aβ42 alone, and hence further studies are warranted to determine if the CSF Aβ42/Aβ40 ratio has a higher diagnostic potential than CSF Aβ42 alone.

Using mass spectrometry, it has been found that there is a quite heterogeneous set of Aβ peptides in CSF (Vigo-Pelfrey *et al.* 1993). Also, by using urea-based sodium dodecyl sulphate polyacrylamide gel electrophoresis (SDS–PAGE) and immunoblot (Wiltfang *et al.* 2002) it is possible to separate several C-terminally truncated Aβ peptides in CSF, including Aβ1–37, Aβ1–38, Aβ1–39, Aβ1–40, and Aβ1–42. A recent finding is that the second most abundant Aβ peptide after Aβ1–40 is Aβ1–38 (Wiltfang *et al.* 2002). Increased CSF levels of both Aβ1–40 and Aβ1–38 are found in AD, along with a decrease in Aβ1–42. Similar data have been obtained using surface-enhanced laser desorption/ionization (SELDI) time-of-flight (TOF) mass spectrometry (Lewczuk *et al.* 2003). Further studies of large patient and control series are needed to determine the diagnostic potential of these Aβ variants. Other promising findings include those of different N-terminally truncated Aβ species present in protein extracts from AD brains (Sergeant *et al.* 2003). These fragments are also detectable in CSF and seem to be of diagnostic usefulness in early AD (Vanderstichele *et al.* 2005).

15.4 Combination of CSF biomarkers for Alzheimer's disease

The diagnostic potential of the combination of CSF T-tau and Aβ42 has been evaluated in several studies. For the most commonly used ELISA methods for T-tau and Aβ42, sensitivity and specificity figures are available from 12 studies (Blennow 2004). The sensitivity and specificity for the combination of CSF T-tau and Aβ42 are slightly higher (89% and 90%, respectively) than for the respective markers alone. Other combinations of CSF markers also tend to result in slightly better diagnostic performance than the use of single markers. In a study on the combination of CSF P-tau181 and Aβ42, the sensitivity for AD was 86% at a specificity of 97% (Maddalena *et al.* 2003), and in another study the combination of CSF T-tau and P-tau396/404 resulted in a sensitivity of 96% at a specificity of 100% (Hu *et al.* 2002).

15.5 CSF biomarkers in early Alzheimer's disease

The performance of CSF T-tau, P-tau, and Aβ42 in AD cases with Mini-Mental State Examination (MMSE) scores above 23–25, i.e. representing mild or early AD, has been examined in some studies. In this early phase of the disease, the sensitivity figures are similar to those found in more advanced AD cases (Blennow 2004). Several studies have evaluated the performance of CSF markers at baseline in patients who fulfilled the diagnostic criteria for mild cognitive impairment (MCI) (DeCarli 2003) and who developed AD during a clinical follow-up period of 1–2 years, and sensitivity figures similar to those in AD cases with clinical dementia have been found (Andreasen *et al.* 1999b, 2003; Arai *et al.* 2000; Gottfries *et al.* 2001; Lautenschlager *et al.* 2001; Maruyama *et al.* 2001; Buerger *et al.* 2002; Riemenschneider *et al.* 2002a; Zetterberg *et al.* 2003). In a study which included 78 MCI patients who were followed for more than 4 years it was concluded that a combination of CSF Aβ42 and P-tau analyses at baseline could predict development of AD with a sensitivity of 60%–67% and a specificity of 85%–90% depending on the MCI subtype (Herukka *et al.* 2005). A similar study, including 137 MCI patients who were followed for 4–6 years, reported sensitivity and specificity figures of 95% and 83–87%, respectively (Hansson *et al.* 2006). A population-based study also showed that reduced CSF Aβ42 is present in asymptomatic elderly who developed dementia during a 3-year follow-up period (Skoog *et al.* 2003). Furthermore, an association between lower CSF Aβ42 concentrations and the ε4 allele of the *APOE* gene in non-demented controls has recently been reported (Prince *et al.* 2004). One interpretation of this result is that Aβ42 concentrations in CSF may decline before the onset of clinical cognitive change and thus Aβ42 should be further explored as a potential antecedent biomarker that may predict development of dementia due to AD. Again, longitudinal studies incorporating sequential CSF sampling in the study protocol are warranted.

15.6 Use of CSF biomarkers in clinical practice

The diagnostic performance of AD biomarkers in clinical practice has been evaluated in two studies (Andreasen *et al.* 1999a, 2001). The CSF markers were evaluated on prospective patient samples from clinical practice and ELISA assays were run each week in clinical neurochemical routine. The diagnostic performance of CSF T-tau and the combination of CSF T-tau and Aβ42 was similar to that found in other studies, with a high ability to differentiate AD from normal ageing, depression, and PD, but lower specificity against other dementias like VAD and LBD.

Taken together, the high sensitivity and specificity figures reported in most investigations of CSF T-tau, P-tau, and Aβ42 as markers of AD suggest that these are ready for use in the clinical routine. Table 15.1 summarizes the typical results of the three-marker panel in AD and the most important differential diagnoses. The diagnostic performance of the CSF markers seems to be highest in the differentiation between AD and normal ageing, depression, secondary dementia caused by vitamin B deficiency, anaemia or hypothyrosis, alcohol dementia, PD, and PSP. Another useful clinical application is the identification of CJD in cases with rapidly progressive dementia, in which the combination of a very marked increase in CSF T-tau concentration and normal or only mildly increased P-tau has high diagnostic value. Lastly, these CSF markers may be useful in identifying mixed AD/VAD dementia. The lower specificity against other dementias, such as LBD and FTD, may hamper the clinical utility of the currently available CSF markers. However, in the clinic, dementias with differing history, symptoms, and findings on brain imaging (e.g. FTD, VAD) can often be identified by means of the medical history, clinical examination, and auxiliary investigations (e.g. blood tests, single-photon emission computed tomography (SPECT), CT, or MRI). A major diagnostic challenge is whether or not a patient with mild cognitive dysfunction will progress to AD with dementia. Although larger longitudinal studies are needed, the three-marker test seems very useful for this application.

15.7 The importance of biomarkers in clinical trials

Biomarkers can be applied to development of drugs against AD in a number of ways. First, they can be applied as additional diagnostic measures in a population clinically identified as having AD and hence provide additional inclusion or exclusion criteria. This would hypothetically ensure the recruitment of 'pure' AD cases to studies of anti-AD drugs. For example, in a recent study of MCI patients who were followed clinically for

Table 15.1 Typical three-marker test results in AD and important differential diagnoses.

Diagnosis	CSF biomarker		
	T-tau	P-tau	Aβ42
AD	Increase	Increase	Decrease
Normal ageing	Normal	Normal	Normal
Depression	Normal	Normal	Normal
Vitamin B deficiency	Normal	Normal	Normal
Anaemia	Normal	Normal	Normal
Hypothyreosis	Normal	Normal	Normal
Alcohol dementia	Normal	Normal	Normal
PD	Normal	Normal	Normal
PSP	Normal	Normal	Normal
CJD	Very marked increase	Normal	Normal to marked decrease
FTD	Normal to mild increase	Normal	Normal to moderate decrease
LBD	Normal to mild increase	Normal to mild increase	Moderate decrease
VAD	Normal to mild increase	Normal	Mild to moderate decrease
Acute stroke	Mild to very marked increase	Normal	Normal
Non-acute CVD without dementia	Normal	Not examined	Normal

AD, Alzheimer's disease; PD, Parkinson's disease; PSP, progressive supranuclear palsy; CJD, Creutzfeldt–Jakob disease; FTD, frontotemporal dementia; LBD, Lewy body dementia; VAD, vascular dementia; CVD, cerebrovascular disease.

2 years, CSF biomarkers at baseline could correctly predict 20 out of 21 stable MCI patients and 10 out of 10 MCI patients with progression to other dementias (Zetterberg *et al.* 2003). It would be extremely difficult to interpret a clinical trial of an anti-AD drug if these patients were to be included in the study, and the likelihood of a result in favour of the null hypothesis would be very high. To date, CSF analysis is the only clinical method of identifying these patients. Secondly, biomarkers may offer an indirect measure of disease intensity and progression. A number of points should be established for such use: the marker must have a scientific rationale, it should change with disease progression in longitudinal observational studies, and it must be measurable and reproducible. Unlike typical diagnostic measures, when biomarkers are used for this purpose, high specificity is not required. Particularly in mid-phase trials, biomarkers can be used to identify appropriate dosage, improve safety assessments, demonstrate pharmacological activity, and identify preliminary evidence of efficacy. As previously mentioned, longitudinal studies of T-tau, P-tau, and Aβ42 in CSF from AD patients are needed to prove their possible usefulness for these purposes.

15.8 Novel candidate CSF biomarkers for Alzheimer's disease

There are several promising candidate CSF biomarkers that to date have only been evaluated in a few studies. These include ubiquitin (Wang *et al.* 1991; Blennow *et al.* 1994), neurofilament proteins (Rosengren *et al.* 1999; Sjogren *et al.* 2001a), GAP43 (neuromodulin) (Sjogren *et al.* 2000), different markers for oxidative damage and defective lipid peroxidation, such as isoprostanes (Pratico 2005) and 12/15 lipoxygenase products (Yao *et al.* 2005), and several cytokines involved in microglia and astrocyte activation (Mrak and Griffin 2005). Even more biomarkers are being discovered in the exploratory studies that are employing proteomics technology on samples from different well-defined dementia disorders (Davidsson and Sjogren 2005; Ruetschi *et al.* 2005). Further studies of the diagnostic potential of all of these markers are warranted. Furthermore, the time seems ripe for larger longitudinal studies combining CSF biomarkers with recently developed and very promising neuroimaging techniques, such as 2-[^{18}F]fluoro-2-deoxy-D-glucose positron emission tomography (FDG-PET) (Mosconi *et al.* 2005) and amyloid visualization using molecular tracers in combination with PET or SPECT (Mathis *et al.* 2004). Possibly, we may end up with a CSF multimarker profile that can be used in conjunction with clinical examination, neuropsychological testing, and neuroimaging techniques for early diagnosis of AD, preclinical diagnosis of the disease in certain risk groups of patients, and drug selection and monitoring of drug effects for optimal treatment using upcoming disease-modifying drugs.

References

Andreasen N, Minthon L, Clarberg A, *et al.* (1999a). Sensitivity, specificity, and stability of CSF-tau in AD in a community-based patient sample. *Neurology* 53, 1488–94.

Andreasen N, Minthon L, Vanmechelen E, *et al.* (1999b). Cerebrospinal fluid tau and Abeta42 as predictors of development of Alzheimer's disease in patients with mild cognitive impairment. *Neuroscience Letters* 273, 5–8.

Andreasen N, Minthon L, Davidsson P, *et al.* (2001). Evaluation of CSF-tau and CSF-Abeta42 as diagnostic markers for Alzheimer disease in clinical practice. *Archives of Neurology* 58, 373–9.

Andreasen N, Vanmechelen E, Vanderstichele H, Davidsson P, and Blennow K (2003). Cerebrospinal fluid levels of total-tau, phospho-tau and A beta 42 predicts development of Alzheimer's disease in patients with mild cognitive impairment. *Acta Neurologica Scandinavica Supplementum* 179, 47–51.

Arai H, Morikawa Y, Higuchi M, *et al.* (1997). Cerebrospinal fluid tau levels in neurodegenerative diseases with distinct tau-related pathology. *Biochemical and Biophysical Research Communications* 236, 262–4.

Arai H, Ishiguro K, Ohno H, *et al.* (2000). CSF phosphorylated tau protein and mild cognitive impairment: a prospective study. *Experimental Neurology* 166, 201–3.

Beaudry P, Cohen P, Brandel JP, *et al.* (1999). 14–3–3 protein, neuron-specific enolase, and S-100 protein in cerebrospinal fluid of patients with Creutzfeldt–Jakob disease. *Dementia, Geriatric and Cognitive Disorders* 10, 40–6.

Blennow K (2004). Cerebrospinal fluid protein biomarkers for Alzheimer's disease. *NeuroRx* 1, 213–25.

Blennow K and Hampel H (2003). CSF markers for incipient Alzheimer's disease. *Lancet Neurology* 2, 605–13.

Blennow K, Wallin A, and Hager O (1993). Low frequency of post-lumbar puncture headache in demented patients. *Acta Neurologica Scandinavica* 88, 221–3.

Blennow K, Davidsson P, Wallin A, Gottfries CG, and Svennerholm L (1994). Ubiquitin in cerebrospinal fluid in Alzheimer's disease and vascular dementia. *International Psychogeriatrics* **6**, 13–22, 59–60.

Blennow K, Wallin A, Agren H, Spenger C, Siegfried J, and Vanmechelen E (1995). Tau protein in cerebrospinal fluid: a biochemical marker for axonal degeneration in Alzheimer disease? *Molecular and Chemical Neuropathology* **26**, 231–45.

Blennow K, Johansson A, and Zetterberg H (2005). Diagnostic value of 14-3-3beta immunoblot and T-tau/P-tau ratio in clinically suspected Creutzfeldt-Jakob disease. *International Journal of Molecular Medicine* **16**, 1147–9.

Buée L, Bussiere T, Buée-Scherrer V, Delacourte A, and Hof PR (2000). Tau protein isoforms, phosphorylation and role in neurodegenerative disorders. *Brain Research and Brain Research Reviews* **33**, 95–130.

Buerger K, Teipel SJ, Zinkowski R, *et al.* (2002). CSF tau protein phosphorylated at threonine 231 correlates with cognitive decline in MCI subjects. *Neurology* **59**, 627–9.

Davidsson P and Sjogren M (2005). The use of proteomics in biomarker discovery in neurodegenerative diseases. *Disease Markers* **21**, 81–92.

De Strooper B (2003). Aph-1, Pen-2, and nicastrin with presenilin generate an active gamma-secretase complex. *Neuron* **38**, 9–12.

DeCarli C (2003). Mild cognitive impairment: prevalence, prognosis, aetiology, and treatment. *Lancet Neurology* **2**, 15–21.

Fagan AM, Watson M, Parsadanian M, Bales KR, Paul SM, and Holtzman DM (2002). Human and murine ApoE markedly alters A beta metabolism before and after plaque formation in a mouse model of Alzheimer's disease. *Neurobiology of Disease* **9**, 305–18.

Farzan M, Schnitzler CE, Vasilieva N, Leung D, and Choe H (2000). BACE2, a beta -secretase homolog, cleaves at the beta site and within the amyloid-beta region of the amyloid-beta precursor protein. *Proceedings of the National Academy of Sciences of the USA* **97**, 9712–17.

Fukumoto H, Tennis M, Locascio JJ, Hyman BT, Growdon JH, and Irizarry MC (2003). Age but not diagnosis is the main predictor of plasma amyloid beta-protein levels. *Archives of Neurology* **60**, 958–64.

Fukuyama R, Mizuno T, Mori S, Nakajima K, Fushiki S, and Yanagisawa K (2000). Age-dependent change in the levels of Abeta40 and Abeta42 in cerebrospinal fluid from control subjects, and a decrease in the ratio of Abeta42 to Abeta40 level in cerebrospinal fluid from Alzheimer's disease patients. *European Neurology* **43**, 155–60.

Geschwind MD, Martindale J, Miller D, *et al.* (2003). Challenging the clinical utility of the 14–3–3 protein for the diagnosis of sporadic Creutzfeldt–Jakob disease. *Archives of Neurology* **60**, 813–16.

Goedert M, Spillantini MG, Potier MC, Ulrich J, and Crowther RA (1989). Cloning and sequencing of the cDNA encoding an isoform of microtubule-associated protein tau containing four tandem repeats: differential expression of tau protein mRNAs in human brain. *EMBO Journal*, **8**, 393–9.

Goldstein LE, Muffat JA, Cherny RA, *et al.* (2003). Cytosolic beta-amyloid deposition and supranuclear cataracts in lenses from people with Alzheimer's disease. *Lancet* **361**, 1258–65.

Gottfries J, Blennow K, Lehmann MW, Regland B, and Gottfries CG (2001). One-carbon metabolism and other biochemical correlates of cognitive impairment as visualized by principal component analysis. *Journal of Geriatric Psychiatry and Neurology* **14**, 109–14.

Grundke-Iqbal I, Iqbal K, Tung YC, Quinlan M, Wisniewski HM, and Binder LI (1986). Abnormal phosphorylation of the microtubule-associated protein tau (τ) in Alzheimer cytoskeletal pathology. *Proceedings of the National Academy of Sciences of the USA* **83**, 4913–17.

Hansson O, Zetterberg H, Buchhave P, Londos E, Blennow K, and Minthon L (2006). Association between CSF biomarkers and incipient Alzheimer's disease in patients with mild cognitive impairment: a follow-up study. *Lancet Neurology* **5**, 228–34.

Herukka SK, Hallikainen M, Soininen H, and Pirttila T (2005). CSF Abeta42 and tau or phosphorylated tau and prediction of progressive mild cognitive impairment. *Neurology* **64**, 1294–7.

Hesse C, Rosengren L, Vanmechelen E, *et al.* (2000). Cerebrospinal fluid markers for Alzheimer's disease evaluated after acute ischemic stroke. *Journal of Alzheimer's Disease* **2**, 199–206.

Holmberg B, Johnels B, Blennow K, and Rosengren L (2003). Cerebrospinal fluid Abeta42 is reduced in multiple system atrophy but normal in Parkinson's disease and progressive supranuclear palsy. *Movement Disorders* **18**, 186–90.

Holtzman DM, Bales KR, Tenkova T, *et al.* (2000). Apolipoprotein E isoform-dependent amyloid deposition and neuritic degeneration in a mouse model of Alzheimer's disease. *Proceedings of the National Academy of Sciences of the USA* **97**, 2892–7.

Hsich G, Kenney K, Gibbs CJ, Lee KH, and Harrington MG (1996). The 14–3–3 brain protein in cerebrospinal fluid as a marker for transmissible spongiform encephalopathies. *New England Journal of Medicine* **335**, 924–30.

Hu YY, He SS, Wang X, *et al.* (2002). Levels of nonphosphorylated and phosphorylated tau in cerebrospinal fluid of Alzheimer's disease patients: an ultrasensitive bienzyme-substrate-recycle enzyme-linked immunosorbent assay. *American Journal of Pathology* **160**, 1269–78.

Hulstaert F, Blennow K, Ivanoiu A, *et al.* (1999). Improved discrimination of AD patients using beta-amyloid(1–42) and tau levels in CSF. *Neurology* **52**, 1555–62.

Iqbal K, Alonso AD, Gondal JA, *et al.* (2000). Mechanism of neurofibrillary degeneration and pharmacologic therapeutic approach. *Journal of Neural Transmission Supplementum* **59**, 213–22.

Iqbal K, Alonso Adel C, El-Akkad E, *et al.* (2002). Pharmacological targets to inhibit Alzheimer neurofibrillary degeneration. *Journal of Neural Transmission Supplementum*, 309–19.

Iwatsubo T, Odaka A, Suzuki N, Mizusawa H, Nukina N, and Ihara Y (1994). Visualization of A beta 42(43) and A beta 40 in senile plaques with end-specific A beta monoclonals: evidence that an initially deposited species is A beta 42(43). *Neuron* **13**, 45–53.

Jarrett JT, Berger EP, and Lansbury PT, Jr (1993). The carboxy terminus of the beta amyloid protein is critical for the seeding of amyloid formation: implications for the pathogenesis of Alzheimer's disease. *Biochemistry* **32**, 4693–7.

Jellinger KA (1996). Diagnostic accuracy of Alzheimer's disease: a clinicopathological study. *Acta Neuropathologica* **91**, 219–20.

Kanai M, Matsubara E, Isoe K, *et al.* (1998). Longitudinal study of cerebrospinal fluid levels of tau, A beta1–40, and A beta1–42(43) in Alzheimer's disease: a study in Japan. *Annals of Neurology* **44**, 17–26.

Kosunen O, Soininen H, Paljarvi L, Heinonen O, Talasniemi S, and Riekkinen PJ Sr (1996). Diagnostic accuracy of Alzheimer's disease: a neuropathological study. *Acta Neuropathologica* **91**, 185–93.

Lautenschlager NT, Riemenschneider M, Drzezga A, and Kurz AF (2001). Primary degenerative mild cognitive impairment: study population, clinical, brain imaging and biochemical findings. *Dementia, Geriatric and Cognitive Disorders* **12**, 379–86.

Lewczuk P, Esselmann H, Meyer M, *et al.* (2003). The amyloid-beta (Abeta) peptide pattern in cerebrospinal fluid in Alzheimer's disease: evidence of a novel carboxyterminally elongated Abeta peptide. *Rapid Communications in Mass Spectrometry* **17**, 1291–6.

Lewis J, Dickson DW, Lin WL, *et al.* (2001). Enhanced neurofibrillary degeneration in transgenic mice expressing mutant tau and APP. *Science* **293**, 1487–91.

Lorenzo A and Yankner BA (1994). Beta-amyloid neurotoxicity requires fibril formation and is inhibited by congo red. *Proceedings of the National Academy of Sciences USA*, **91**, 12243–7.

Maddalena A, Papassotiropoulos A, Muller-Tillmanns B, *et al.* (2003). Biochemical diagnosis of Alzheimer disease by measuring the cerebrospinal fluid ratio of phosphorylated tau protein to beta-amyloid peptide42. *Archives of Neurology* **60**, 1202–6.

Mandelkow EM, Stamer K, Vogel R, Thies E, and Mandelkow E (2003). Clogging of axons by tau, inhibition of axonal traffic and starvation of synapses. *Neurobiology of Aging* **24**, 1079–85.

Maruyama M, Arai H, Sugita M, *et al.* (2001). Cerebrospinal fluid amyloid beta(1–42) levels in the mild cognitive impairment stage of Alzheimer's disease. *Experimental Neurology* **172**, 433–6.

Masters CL, Simms G, Weinman NA, Multhaup G, McDonald BL, and Beyreuther K (1985). Amyloid plaque core protein in Alzheimer disease and Down syndrome. *Proceedings of the National Academy of Sciences of the USA* **82**, 4245–9.

Mathis C.A, Wang Y, and Klunk WE (2004). Imaging beta-amyloid plaques and neurofibrillary tangles in the aging human brain. *Current Pharmaceutical Design* **10**, 1469–92.

Mehta PD, Pirttila T, Mehta SP, Sersen EA, Aisen PS, and Wisniewski HM (2000). Plasma and cerebrospinal fluid levels of amyloid beta proteins 1–40 and 1–42 in Alzheimer disease. *Archives of Neurology* **57**, 100–5.

Miller DL, Papayannopoulos IA, Styles J, *et al.* (1993). Peptide compositions of the cerebrovascular and senile plaque core amyloid deposits of Alzheimer's disease. *Archives of Biochemistry and Biophysics* **301**, 41–52.

Mori H, Hosoda K, Matsubara E, *et al.* (1995). Tau in cerebrospinal fluids: establishment of the sandwich ELISA with antibody specific to the repeat sequence in tau. *Neuroscience Letters* **186**, 181–3.

Mosconi L, Tsui WH, De Santi S, *et al.* (2005). Reduced hippocampal metabolism in MCI and AD: automated FDG-PET image analysis. *Neurology* **64**, 1860–7.

Mrak RE and Griffin WS (2005). Glia and their cytokines in progression of neurodegeneration. *Neurobiology of Aging* **26**, 349–54.

Nishimura T, Takeda M, Nakamura, Y *et al.* (1998). Basic and clinical studies on the measurement of tau protein in cerebrospinal fluid as a biological marker for Alzheimer's disease and related disorders: multicenter study in Japan. *Methods and Findings in Experimental and Clinical Pharmacology* **20**, 227–35.

Otto M, Wiltfang J, Tumani H, *et al.* (1997). Elevated levels of tau-protein in cerebrospinal fluid of patients with Creutzfeldt–Jakob disease. *Neuroscience Letters* **225**, 210–12.

Otto M, Esselmann H, Schulz-Shaeffer W, *et al.* (2000). Decreased beta-amyloid1–42 in cerebrospinal fluid of patients with Creutzfeldt–Jakob disease. *Neurology* **54**, 1099–102.

Otto M, Wiltfang J, Cepek L, *et al.* (2002). Tau protein and 14–3–3 protein in the differential diagnosis of Creutzfeldt–Jakob disease. *Neurology* **58**, 192–7.

Pratico D (2005). Peripheral biomarkers of oxidative damage in Alzheimer's disease: the road ahead. *Neurobiology of Aging* **26**, 581–3.

Prince JA, Zetterberg H, Andreasen N, Marcusson J, and Blennow K (2004). *APOE* epsilon4 allele is associated with reduced cerebrospinal fluid levels of Abeta42. *Neurology* **62**, 2116–18.

Riemenschneider M, Lautenschlager N, Wagenpfeil S, Diehl J, Drzezga A, and Kurz A (2002a). Cerebrospinal fluid tau and beta-amyloid 42 proteins identify Alzheimer disease in subjects with mild cognitive impairment. *Archives of Neurology* **59**, 1729–34.

Riemenschneider M, Wagenpfeil S, Diehl J, *et al.* (2002b). Tau and Abeta42 protein in CSF of patients with frontotemporal degeneration. *Neurology* **58**, 1622–8.

Riemenschneider M, Wagenpfeil S, Vanderstichele H, *et al.* (2003). Phospho-tau/total tau ratio in cerebrospinal fluid discriminates Creutzfeldt–Jakob disease from other dementias. *Molecular Psychiatry* **8**, 343–7.

Rosengren LE, Karlsson JE, Sjogren M, Blennow K, and Wallin A (1999). Neurofilament protein levels in CSF are increased in dementia. *Neurology* **52**, 1090–3.

Ruetschi U, Zetterberg H, Podust VN, *et al.* (2005). Identification of CSF biomarkers for frontotemporal dementia using SELDI-TOF. *Experimental Neurology* **196**, 273–81.

Selkoe DJ (2001). Alzheimer's disease: genes, proteins, and therapy. *Physiological Reviews* **81**, 741–66.

Selkoe DJ (2002). Alzheimer's disease is a synaptic failure. *Science* **298**, 789–91.

Selkoe DJ (2005). Defining molecular targets to prevent Alzheimer disease. *Archives of Neurology* **62**, 192–5.

Sergeant N, Bombois S, Ghestem A, *et al.* (2003). Truncated beta-amyloid peptide species in pre-clinical Alzheimer's disease as new targets for the vaccination approach. *Journal of Neurochemistry* **85**, 1581–91.

Seubert P, Vigo-Pelfrey C, Esch F, *et al.* (1992). Isolation and quantification of soluble Alzheimer's beta-peptide from biological fluids. *Nature* **359**, 325–7.

Shoji M, Matsubara E, Kanai M, *et al.* (1998). Combination assay of CSF tau, A beta 1–40 and A beta 1–42(43) as a biochemical marker of Alzheimer's disease. *Journal of Neurological Sciences* **158**, 134–40.

Sinha S, Anderson JP, Barbour R, *et al.* (1999). Purification and cloning of amyloid precursor protein beta-secretase from human brain. *Nature* **402**, 537–40.

Sjogren M, Minthon L, Davidsson P, *et al.* (2000). CSF levels of tau, beta-amyloid(1–42) and GAP-43 in frontotemporal dementia, other types of dementia and normal aging. *Journal of Neural Transmission* **107**, 563–79.

Sjogren M, Blomberg M, Jonsson M, *et al.* (2001a). Neurofilament protein in cerebrospinal fluid: a marker of white matter changes. *Journal of Neuroscience Research* **66**, 510–16.

Sjogren M, Davidsson P, Gottfries J, *et al.* (2001b). The cerebrospinal fluid levels of tau, growth-associated protein-43 and soluble amyloid precursor protein correlate in Alzheimer's disease, reflecting a common pathophysiological process. *Dementia, Geriatric and Cognitive Disorders* **12**, 257–64.

Sjogren M, Davidsson P, Wallin A, *et al.* (2002). Decreased CSF-beta-amyloid 42 in Alzheimer's disease and amyotrophic lateral sclerosis may reflect mismetabolism of beta-amyloid induced by disparate mechanisms. *Dementia, Geriatric and Cognitive Disorders* **13**, 112–18.

Skoog I, Davidsson P, Aevarsson O, Vanderstichele H, Vanmechelen E, and Blennow K (2003). Cerebrospinal fluid beta-amyloid 42 is reduced before the onset of sporadic dementia: a population-based study in 85-year-olds. *Dementia, Geriatric and Cognitive Disorders* **15**, 169–76.

Strozyk D, Blennow K, White LR, and Launer LJ (2003). CSF Abeta 42 levels correlate with amyloid-neuropathology in a population-based autopsy study. *Neurology* **60**, 652–6.

Sunderland T, Linker G, Mirza N, *et al.* (2003). Decreased beta-amyloid1–42 and increased tau levels in cerebrospinal fluid of patients with Alzheimer disease. *Journal of the American Medical Association* **289**, 2094–103.

Tamaoka A, Kondo T, Odaka A, *et al.* (1994). Biochemical evidence for the long-tail form (A beta 1–42/43) of amyloid beta protein as a seed molecule in cerebral deposits of Alzheimer's disease. *Biochemical and Biophysical Research Communications* **205**, 834–42.

Van Everbroeck B, Quoilin S, Boons J, Martin JJ, and Cras P (2003). A prospective study of CSF markers in 250 patients with possible Creutzfeldt–Jakob disease. *Journal of Neurology Neurosurgery and Psychiatry* **74**, 1210–14.

Vandermeeren M, Mercken M, Vanmechelen E, *et al.* (1993). Detection of tau proteins in normal and Alzheimer's disease cerebrospinal fluid with a sensitive sandwich enzyme-linked immunosorbent assay. *Journal of Neurochemistry* **61**, 1828–34.

Vanderstichele H, Van Kerschaver E, Hesse C, *et al.* (2000). Standardization of measurement of beta-amyloid(1–42) in cerebrospinal fluid and plasma. *Amyloid* **7**, 245–58.

Vanderstichele H, De Meyer G, Andreasen N, *et al.* (2005). Amino-truncated β-Amyloid42 peptides in cerebrospinal fluid and prediction of progression of mild cognitive impairment. *Clinical Chemistry* **51**, 1650–60.

Wang GP, Khatoon S, Iqbal K, and Grundke-Iqbal I (1991). Brain ubiquitin is markedly elevated in Alzheimer disease. *Brain Research* **566**, 146–51.

Vassar R, Bennett BD, Babu-Khan S, *et al.* (1999). Beta-secretase cleavage of Alzheimer's amyloid precursor protein by the transmembrane aspartic protease BACE. *Science* **286**, 735–41.

Vigo-Pelfrey C, Lee D, Keim P, Lieberburg I, and Schenk DB (1993). Characterization of beta-amyloid peptide from human cerebrospinal fluid. *Journal of Neurochemistry* **61**, 1965–8.

Vigo-Pelfrey C, Seubert P, Barbour R, *et al.* (1995). Elevation of microtubule-associated protein tau in the cerebrospinal fluid of patients with Alzheimer's disease. *Neurology* **45**, 788–93.

Wiltfang J, Esselmann H, Bibl M, *et al.* (2002). Highly conserved and disease-specific patterns of carboxyterminally truncated Abeta peptides 1–37/38/39 in addition to 1–40/42 in Alzheimer's disease and in patients with chronic neuroinflammation. *Journal of Neurochemistry* **81**, 481–96.

Yao Y, Clark CM, Trojanowski JQ, Lee VM, and Pratico D (2005). Elevation of 12/15 lipoxygenase products in AD and mild cognitive impairment. *Annals of Neurology* **58**, 623–6.

Zetterberg H and Blennow K (2004). Elevated total tau/phospho-tau ratio in autopsy-proven Creutzfeldt–Jakob disease with negative 14–3–3 test results. *Neurological Sciences* **25**, 301–2.

Zetterberg H, Wahlund LO, and Blennow K (2003). Cerebrospinal fluid markers for prediction of Alzheimer's disease. *Neuroscience Letters* **352**, 67–9.

Zheng WH, Bastianetto S, Mennicken F, Ma W, and Kar S (2002). Amyloid beta peptide induces tau phosphorylation and loss of cholinergic neurons in rat primary septal cultures. *Neuroscience* **115**, 201–11.

Chapter 16

Current pharmacological approaches to treating Alzheimer's disease

Gordon K. Wilcock and David Dawbarn

16.1 Introduction

Pharmacological approaches to the treatment of Alzheimer's disease (AD) can be broadly grouped into those that aim for symptomatic treatment, those that attempt to modify the disease process, and those that may lower the risk or delay the onset of AD. The distinction between these approaches is not clear cut, as will become apparent.

Symptomatic treatments have mainly centred on neurotransmitter systems. The well-established deficits in the cholinergic system in AD provided an obvious focus for pharmacological research, and led to the development of the first drugs to be licensed for the treatment of AD. However, numerous other neurotransmitter systems are affected in AD, including the noradrenergic, serotonergic, glutaminergic, and dopaminergic systems. These systems have physiological roles in normal cognitive functioning, and multiple neurotransmitter deficits probably contribute to the cognitive decline in AD.

The behavioural and psychiatric complications of AD are common, and may be related to monoamine dysfunction. Although they can have an important impact on both patients and carers, they have received considerably less attention than the cognitive symptoms. A small number of studies have investigated various medications for specific behavioural or psychiatric disorders, and measures of behavioural and psychiatric disturbance are beginning to be included in clinical trials of general treatments for AD.

The mechanisms which lead to neuronal death are complex and undoubtedly multifactorial. Many different agents show promise as disease-modifying treatments, although their actions may not be specific for AD. Combination therapy with cognitive enhancers may be an avenue for future research. Some neuroprotective approaches may lower the risk for developing AD, although the evidence for this remains inconclusive.

16.2 Issues in clinical drug research

Several issues need to be considered when interpreting the results of clinical trials in patients with AD. A definite diagnosis of AD can only be made with histopathological evidence. Ante-mortem diagnosis of probable AD relies on the use of standardized diagnostic criteria. Even using the most stringent diagnostic criteria, the error rate is still of the order of 10%–15%. Additionally, AD is probably a syndrome rather than a single disease entity. This may account for some of the inter-individual variation in treatment responses in many drug trials. Group treatment effects can appear modest because the marked response of a subgroup of patients is masked by the non-response of others. No distinguishing characteristics have been identified to help predict which patients will benefit, and in clinical practice a trial of treatment is often required.

Outcome measures in drug trials have to be clinically meaningful. Changes in cognitive test scores may be statistically significant but meaningless in terms of the day-to-day life of patients and carers. Most drug trials now include instruments which at least attempt to measure 'global function'. Scales which measure the ability to perform activities of daily living (ADL) and neuropsychiatric function arguably provide the most relevant clinical information. However, the impact of therapy on biomarkers (e.g. levels of tau or β-amyloid peptides in cerebrospinal fluid (CSF)) and neuroimaging is increasingly being explored. To date, however, there is no convincing evidence that such indices are sufficiently reliable to act as worthwhile surrogates for assessing drug efficacy.

Ideally, assessment schedules should include standard instruments. Common measures of cognitive function include the cognitive component of the Alzheimer's Disease Assessment Scale (ADAS-Cog) and the Mini-Mental State Examination (MMSE). Scales measuring non-cognitive items have been more varied, making it harder to compare results.

A common criticism levelled at drug treatments for AD is the lack of evidence for improvement in the quality of life (QOL) for sufferers or their carers. QOL is neither easy to define nor easy to quantify. Some studies have included QOL scales for patients, although these mainly reflect functional ability. Investigating well-being in carers is equally complex but probably more amenable to future evaluation.

16.3 **Cholinergic strategies**

One of the earliest neurochemical changes to be seen in AD is the loss of choline acetyltransferase (ChAT) activity in the cerebral cortex and hippocampus seen post-mortem (Bowen *et al.* 1976; Davies and Maloney 1976) and in biopsy specimens (Bowen *et al.* 1982; Sims *et al.* 1983). The majority of the ChAT in these regions is located in nerve terminals; the cell bodies of these fibres are found in the medial septal nucleus, the diagonal band of Broca (DBB), and the nucleus basalis of Meynert (NBM) (Mesulam *et al.* 1983, 1986; Mesulam and Geula 1988). Many lines of evidence indicate that this cholinergic system is involved in cognitive processes in rats, monkeys, and humans. Lesions of the NBM and DBB in rats result in a range of behavioural impairments including retention of passive avoidance (Dunnett *et al.* 1987), matching to position tasks (Dunnett *et al.* 1989), maze performance (Wenk *et al.* 1989; Connor *et al.* 1991; Turner *et al.* 1992), and attention to brief stimuli (Robbins *et al.* 1989). Similarly, cognitive impairment has been observed following lesions to the NBM and DBB in primates (Ridley *et al.* 1985, 1986).

In addition to lesion studies, impairment of the central cholinergic system with drugs (scopolamine, hemicholinium-3) has also been shown to result in impaired acquisition of new information in primates (Ridley *et al.* 1984a,b). Furthermore, the cholinergic antagonist scopolamine has been shown to cause learning impairments in human volunteers (Crow and Grove-White 1973). Since both cognitive deficits and cholinergic deficits appear early in AD patients, it has been widely assumed that these events are linked and that it is the loss of cholinergic function in the cerebral cortex and hippocampus that is responsible for the observed memory deficits. In addition, as previously discussed, the available drug treatments nearly all work by augmenting cholinergic synaptic transmission. If ways could be found to stop the degeneration of the basal forebrain cholinergic neurons it would be likely that disease progression would be slowed.

16.3.1 Acetylcholine precursors

Choline is a precursor for acetylcholine (ACh). Animal studies reported that choline and lecithin (phosphatidyl choline) increased the production of the brain ACh. Therefore it was logical to attempt to treat the cholinergic deficit in AD with ACh precursors. This strategy proved unsuccessful (Higgins and Flicker 1998), possibly because once the disease process in AD had become symptomatic, there were too few presynaptic cholinergic neurons for precursor loading to have any clinical effect.

16.3.2 Cholinesterase inhibitors

Cholinesterase inhibitors (CEIs) have been the most widely investigated and so far the most successful therapeutic agents for the symptomatic treatment of AD. ACh is hydrolysed by acetylcholinesterase (AChE) and butyrylcholinesterase. Cholinesterase inhibition increases the availability of ACh for synaptic transmission. Cholinesterases are widely distributed throughout the central nervous system (CNS), the peripheral nervous system (PNS), plasma, and liver. AChE is the predominant enzyme in the CNS.

Three cholinesterase inhibitors have been licensed for the treatment of mild to moderate AD. In the UK they were originally approved for funding by the National Institute for Clinical Excellence (NICE) for patients with mild to moderate AD, but this decision has recently been reviewed, and on grounds of cost-effectiveness they are now only funded by the NHS for use in moderate AD. This decision is presently subject to legal challenge.

As one would imagine, each of the licensed compounds has been vying with the others for market supremacy, but most clinicians believe that they are probably equally effective, albeit modestly, in the ~50% of AD sufferers who respond. There have been few head-to-head comparative studies, but taken overall they support this conclusion (Wilcock *et al.* 2003).

16.3.2.1 Physostigmine

Physostigmine is a non-selective reversible CEI in the carbamate class. It is reasonably well absorbed after oral administration and crosses the blood–brain barrier satisfactorily. However, physostigmine has a plasma half-life of approximately 30 minutes, necessitating the use of parenteral administration, high-frequency oral dosing schedules, or sustained release preparations.

Early double-blind trials reported a significant improvement in memory function (Davis *et al.* 1979; Christie *et al.* 1981), but these findings were not universal. More recent studies have also yielded conflicting results, possibly reflecting differences in study design. Asthana *et al.* (1995) found that continuous intravenous infusion of physostigmine improved memory in five of nine AD patients. Marin *et al.* (1995) reported no significant improvement in cognitive functioning in AD patients receiving intermittent physostigmine infusions when compared with controls receiving placebo. Supporting evidence for the beneficial effects of physostigmine is provided by studies which have reported increased cerebral blood flow, particularly in those areas of the brain most affected in AD (Geaney *et al.* 1990; Wilson *et al.* 1991; Gustafson 1993).

The short half-life of physostigmine has limited its clinical applications, but attempts have been made to overcome this. Fairly stable plasma concentrations were achieved by a continuous transdermal delivery system (Levy *et al.* 1994) in a single-blind study of 12 AD patients. Four of these patients improved on all four cognitive tests employed, and two patients on two tests. No major adverse effects were reported. Thal *et al.* (1996) evaluated a controlled release preparation of physostigmine in a double-blind placebo-controlled trial in 1111 patients with mild or moderate AD by. A modest improvement in cognitive and global function was observed in a subset of patients after 6 weeks active treatment. Common adverse effects included nausea and vomiting, diarrhoea, and anorexia. These findings helped stimulate the search for more effective and practical cholinergic strategies.

16.3.2.2 Tacrine

Tacrine or tetrahydroaminoacridine (THA) is a centrally active reversible CEI in the acridine class. It is reasonably well absorbed after oral administration but has a relatively short half life of approximately 3.5 hours in the elderly. Interest in cholinergic therapies had flagged in the 1980s but was rekindled by promising results reported for tacrine treatment in 17 AD patients (Summers *et al.* 1986). Early optimism was lost when a series of studies failed to replicate these finding (Chatellier and Lacomblez 1990; Fritten *et al.* 1990). The first convincing study after Summers' paper was that of Eagger and colleagues (1991). They were able to demonstrate an improvement in cognitive functioning equivalent to the decline which would be expected with disease progression over 6–12 months.

This was followed by further studies reporting modest benefit in about 50% of patients, but its side-effect profile led to tacrine falling out of favour as other drugs became available. However, it does remain licensed in a number of countries. It was the first drug to be approved for the symptomatic treatment of AD and as such represented a landmark development.

16.3.2.3 Donepezil

Donepezil (Aricept), a second-generation CEI, is a piperidine-based agent which was developed specifically for the treatment of AD and is now licensed for clinical use in several countries. It is a non-competitive reversible antagonist of AChE, and is highly selective for AChE as opposed to butyrylcholinesterase. It is well absorbed after oral administration and peak plasma concentration is reached in 3–4 hours. Donepezil is metabolized by the hepatic isoenzymes CYP2D6 and CYP3A4 and undergoes glucuronidation. The plasma half-life is approximately 70 hours, thus allowing once-daily dosing.

The first phase II clinical trial to be published indicated that donepezil improved cognitive and global function in patients with mild to moderate AD (Rogers *et al.* 1996). These promising findings have now been replicated in three large multicentre studies. The shortest of these was a 12 week double-blind placebo-controlled parallel-group trial (Rogers *et al.* 1998a) in which 468 patients with mild to moderate AD were randomized to receive placebo, donepezil 5 mg/day, or donepezil 10 mg/day (5 mg/day during the first week). Intention to treat (ITT) analysis, which included 97% of the patients who were enrolled, revealed a statistically significant improvement in cognitive and global function. Clinically significant improvements in cognitive function were found for 48%–57% of patients who received donepezil and 30% of patients who received placebo. A trend towards a dose-related effect was apparent but did not reach statistical significance. Plasma concentrations of donepezil were significantly correlated with red cell membrane AChE inhibition (thought to be indicative of cerebral AChE inhibition), and red cell AChE inhibition was correlated with changes in the cognitive and global function test scores. In general, donepezil was well tolerated and the only adverse effects, which occurred more commonly than in the placebo group, were nausea, insomnia, and diarrhoea. Many of these events were mild and resolved within 1–2 days with continued treatment. A decrease in group mean heart rate was observed (2.7 beats/min in the 5 mg/day group, and 2.3 beats/min in the 10 mg/day group) but this was not clinically significant. No evidence of hepatotoxicity was found.

The results of a 24 week phase III trial (Rogers *et al.* 1998b) provided further evidence of donepezil's clinical efficacy and good tolerability. In this trial 473 patients with mild to moderate AD were randomly assigned to treatment with placebo, donepezil 5 mg/day, or donepezil 10 mg/day. Clinically significant improvements in cognitive test scores were achieved in 38% of patients receiving 5 mg/day and 54% of patients receiving 10 mg/day compared with 27% of patients in the placebo group. Lack of deterioration in cognitive function was observed in over 80% of the donepezil-treated patients and in 58% of those who received placebo. Measures of global function also improved in the donepezil treatment groups. At the end of the 6 week placebo washout which followed the treatment phase, there were no significant differences on cognitive or global function measures between any of the three groups. The only adverse effects which occurred significantly more frequently in the active treatment groups were fatigue, diarrhoea, nausea, vomiting, and muscle cramps.

The first analysis (at 98 weeks) of the effects of long-term treatment with donepezil in an open-label study of 133 patients was published (Rogers and Friedhoff 1998). Improvements from baseline levels in cognitive and global function were still evident after 38 weeks and 26 weeks respectively. Thereafter the deterioration was slower than would be expected in untreated AD patients. There was no evidence of hepatotoxicity.

The evidence to date indicates that donepezil is generally well tolerated and can improve cognitive symptoms and functional ability in a subgroup of AD patients. It is also being increasingly recognized that the cholinesterase inhibitors as a class, including donepezil (Holmes *et al.* 2004) may be useful in ameliorating difficult behavioural traits in some patients.

Long-term use may slow the rate of cognitive deterioration. Whether or not it is disease modifying, i.e. is neuroprotective, is controversial; however, a recent study has suggested that donepezil slows the rate of hippocampal atrophy (Hashimoto *et al.* 2005) in line with pre-clinical evidence suggesting that cholinergic modulation may affect β-amyloid production.

Finally, the use of donepezil has recently been extended in the USA for treating people with severe AD, and evidence in support of this indication is now becoming available for the other cholinesterases. Thus the range of indications is widening.

16.3.2.4 Rivastigmine

Rivastigmine (ENA 713, Exelon), also a second generation CEI, is a centrally selective carbamate CEI that is already licensed as a symptomatic treatment for mild to moderate AD. It binds to the esteratic site of the AChE enzyme but dissociates more slowly than ACh and so has pseudo-irreversible activity. Thus cholinesterase inhibition lasts approximately 10 hours although the plasma half-life is only about 2 hours (Anand and Gharabawi 1996). Rivastigmine is inactivated mainly by the interaction with AChE, and its cleavage product is excreted by the kidneys. This reduces the potential for interaction with drugs which are metabolized by the hepatic cytochrome *P*-450 system. Rivastigmine is well absorbed and peak plasma levels are reached 1–2 hours after ingestion. It inhibits both acetylcholinesterase and butyrylcholinesterase, whereas donepezil and galantamine are specific for the former.

The first large multicentre clinical trial recruited 699 patients with mild to moderate AD (Corey-Bloom *et al.* 1998). Patients with concomitant illnesses were eligible for inclusion unless the illness was severe or unstable. Concurrent medication was allowed with the exception of anticholinergic drugs, any potential memory enhancers, and all psychotropics apart from chloral hydrate. Patients were randomly allocated to receive placebo, low-dose rivastigmine (1–4 mg daily), or high-dose rivastigmine (6–12 mg daily). The study consisted of an initial 7 week fixed-dose–titration phase, followed by a flexible-dose phase up to week 26. Overall, 78% of patients completed the study. At 26 weeks 56% of the patients who had received high-dose rivastigmine had either improved or not deteriorated on tests of cognitive function compared with 27% of those who received placebo. The improvement in cognitive function relative to the patients receiving placebo represented a delay in cognitive deterioration of about 6 months. As a group the patients who received active treatment deteriorated less on global function measures than those on placebo. A quarter of the patients in the high-dose rivastigmine group had shown clinically significant improvements in their ability to perform activities of daily living compared with 15% patients in the placebo group. The most common adverse effects were dose related and cholinergic in origin, mainly nausea, vomiting, diarrhoea, and anorexia. Most were mild or moderate in severity, transient, and occurred on initiation or dose escalation. Weight loss also appeared to be dose related, and was ≥7% body weight in 21% of the high-dose treatment group, 6% of the low-dose treatment group, and 2% of the placebo group. There was no evidence of any organ toxicity.

Several subsequent clinical trials have shown that rivastigmine has beneficial effects on cognitive and global function, may delay the worsening in ability to perform activities of daily living, and may have beneficial effects on behavioural abnormalities. It has also been shown to be effective in Lewy body dementia (McKeith *et al.* 2000), and there are claims for neuroprotection from one small clinical study using neuroimaging (Venneri *et al.* 2004).

16.3.2.5 Galantamine

Galantamine (Reminyl) is a tertiary alkaloid derived from snowdrop and narcissus bulbs. Apart from its action as a specific reversible inhibitor of brain AChE, it appears to modulate nicotinic receptors allosterically. Early open studies (Wilcock *et al.* 1993) indicated that galantamine showed promise as a symptomatic treatment for AD. Phase II trials confirmed the efficacy and good tolerability of galantamine (Wilcock and Wilkinson 1997). Like donepezil and rivastigmine, galantamine has been evaluated in a number of pivotal phase III studies (e.g. Wilcock *et al.* 2000). Patients who received galantamine showed a significant improvement on measures of cognition, global function, and behavioural disturbance compared with those who received placebo. The patients who received treatment with galantamine 24 mg daily over 12 months showed no deterioration from their baseline level of functioning. Cholinergic side effects mainly affected the gastrointestinal system, and their incidence was influenced by the titration rate and dose.

It has been suggested, as for the other cholinesterase inhibitors, that galantamine may actually modify the disease process in AD, and if so it is possible that the dual mode of action may contribute to this.

16.3.3 Muscarinic agonists

The degeneration of cholinergic neuronal systems in AD primarily affects the projections from the basal forebrain to the hippocampus and cerebral cortex. Presynaptic muscarinic ACh receptors are reduced in number but neocortical post-synaptic muscarinic receptors are relatively spared. Direct action on muscarinic receptors bypasses the dysfunctional presynaptic terminals, and theoretically agonists could offer therapeutic advantages over CEIs (Avery *et al.* 1997).

The results of early studies with muscarinic agonists such as arecoline, oxotremorine, bethanechol, and RS86 were disappointing (Spiegel 1991). Compounds had a narrow therapeutic window, and activity on peripheral muscarinic receptors led to high rates of side effects. These included cardiovascular problems and hyperidrosis (increased sweating).

The discovery of subtypes of muscarinic receptor which were differently distributed was a major development. M_1 receptors (mainly post-synaptic) were found to predominate in the frontal cortex and hippocampus, and M_2 (mainly pre-synaptic) and M_3 subtypes predominated peripherally.

Xanomeline, an M_1/M_4 agonist, has been evaluated in a multicentre trial of 343 patients with AD (Bodick *et al.* 1997). This study used a randomized parallel-group double-blind placebo-controlled design. Patients received placebo or xanomeline in doses of 25, 50, or 75 mg three times daily for 6 months. Significant treatment effects on cognitive and global function were demonstrated for high-dose xanomeline, and there were significant dose-dependent reductions in behavioural disturbance and psychotic symptomatology. However, adverse effects, particularly gastrointestinal, were troublesome and led to the withdrawal of 52% patients in the high-dose group. Syncope occurred in 12.6%, 13.3%, and 3.5% of patients receiving high-, medium-, and low-dose xanomeline, respectively, and in 4.6% of patients on placebo. Mild to moderate elevations in transaminase levels occurred in 13% of patients who received xanomeline. In view of the poor tolerability and side-effect profile of the oral preparation, a transdermal formulation was developed, but this drug has not been licensed.

However, it has been suggested that earlier trials with M_1 agonists were not successful because of their lack of specificity. Recently, a specific M1 agonist, AF267B (NGX267), entered its second phase I trial. The animal data from this compound suggest that it is able to affect Aβ and tau pathology, by increasing α-secretase activity and reducing β-secretase and GSK-3β activity (Caccamo *et al* 2006). A first phase I trial, completed at the end of 2005, reported the safety of single doses of NGX267 in healthy adult volunteers. The second phase I trial will measure the safety and tolerability of single and multiple doses in 65 healthy elderly subjects.

16.3.4 Nicotinic agonists

Some cognitive processes involve nicotinic receptors, and the density of these is reduced in AD. Initial case–control studies found a negative association between smoking and AD, and this helped to increase research interest in the possible therapeutic role of nicotinic agonists. However, the negative association between smoking and AD was not confirmed in later studies, and a population-based cohort study of 6870 elderly people found that smoking actually doubled the risk of dementia and AD (Ott *et al.* 1998). Despite the lack of convincing evidence that smoking protects against the development of AD, nicotine may have neuroprotective properties (Shimohama *et al.* 1996). *In vitro* studies suggest that nicotine may exert an effect on amyloid processing. Nicotine is able to inhibit formation of the amyloid β-peptide (Aβ) (Salomon *et al.* 1996) and stimulates the release of amyloid precursor protein (APP) (Kim *et al.* 1997). Additional support for this hypothesis comes from a post-mortem study of brain tissue from 301 people over the age of 65 (Ulrich *et al.* 1997). Smoking appeared to protect against senile plaque formation but correlated with neurofibrillary changes.

There are numerous subtypes of nicotinic receptors which probably have differing effects on neurotransmitter systems. Stimulation of the presynaptic nicotinic receptors increases ACh release. Nicotine is an unselective agonist for nicotinic receptors. In addition to increasing the release of ACh, it may also increase the release of noradrenaline, serotonin, and dopamine. A small pilot study in patients with AD found that nicotine patches improved learning (Wilson *et al.* 1995), but a placebo-controlled double-blind cross-over study in 18 patients with AD failed to demonstrate any benefits on cognitive function (Snaedal *et al.* 1996).

As our knowledge about the function of different nicotinic receptors and their relationship with AD increases, it may be possible to target particular receptor subtypes selectively. At present we cannot judge what the clinical applications will be.

16.3.5 **Acetylcholine releasers**

Linopirdine is a phenylindolinone which enhances the depolarization-induced release of ACh, dopamine, serotonin, and glutamate, possibly via potassium-channel blockade. Basal neurotransmitter release is not increased, and so theoretically it should enhance normal brain activity. However, the results of clinical trials have been disappointing. In a double-blind parallel-group study, 375 patients with mild to moderate AD were randomized to receive either placebo or linopirdine 30 mg three times daily (Rockwood *et al.* 1997). A small difference in cognitive function in favour of linopirdine was observed, but this was not clinically significant. More potent anthracenone analogues of linopirdine have now been identified and may enter clinical trials.

16.3.6 **Neurotrophins and acetylcholine**

16.3.6.1 Structure and function

The development and maintenance of the nervous system is controlled by a series of neurotrophic factors which direct cell targeting, proliferation, differentiation, and phenotypic expression of cell markers (Purves 1989). Particularly pertinent here are the neurotrophins, which include nerve growth factor (NGF) (Thoenen and Barde 1980; Scott *et al.* 1983; Ullrich *et al.* 1983; Thoenen *et al.* 1987; Thoenen 1991), brain-derived neurotrophic factor (BDNF) (Barde *et al.* 1982, 1987; Leibrock *et al.* 1989), neurotrophin-3 (NT-3) (Hohn *et al.* 1990; Maisonpierre *et al.* 1990a), and neurotrophin-4 (NT-4) (Berkemeir *et al.* 1991; Hallbook *et al.* 1991; Ip *et al.* 1992). These proteins are structurally similar and share approximately 50% sequence homology at the amino acid level. They are all basic proteins with pI > 9.0 and are processed from pre-prohormones to give monomers of approximately 120 amino acids (Maisonpierre *et al.* 1990b). All the neurotrophins exist as homodimers and their crystal structures have all been determined (McDonald *et al.* 1991; Holland *et al.* 1994; Robinson *et al.* 1995, 1996; Butte *et al.* 1998). They are part of the cystine knot superfamily which includes transforming growth factor-β (TGF-β), platelet-derived growth factor (PDGF), and human choriogonadotrophin (McDonald and Hendrickson 1993). Each neurotrophin monomer consists of four antiparallel β-strands, with the fourth β-strand twisted around the third, a feature conserved across all the cystine knot growth factors.

16.3.6.2 Neurotrophin receptors

All the neurotrophins bind to the p75[NTR] receptor with a K_d of ~10^{-9}M (Rodrigues-Tebar *et al.* 1992). Specificity is defined through their interaction with tyrosine receptor kinases (Trk) where the K_d is ~10^{-11}M. NGF binds to Trk A (Kaplan *et al.* 1991), BDNF and NT-4 bind to TrkB (Kleine *et al.* 1992; Ip *et al.* 1993), and NT-3 binds to TrkC (Lamballe *et al.* 1991) and an alternatively spliced version of TrkA (Clary and Reichardt 1994). Following ligand binding, the Trk receptors dimerize (Jing *et al.* 1992) and intracellular tyrosines are phosphorylated, which leads to signal transduction through a complex cascade involving Cg1, PI-3 kinase, ERK, Raf1, Ras, and SHC (Loeb *et al.* 1994; Stephens *et al.* 1994). The extracellular region of the Trk receptors can be subdivided into different domains using protein structure prediction programmes (Schneider and Schweiger 1991). Thus at the N-terminus there are three leucine-rich motifs which are flanked by two cysteine-rich regions. These are followed by two immunoglobulin-like domains and a proline-rich region adjacent to the membrane. It has been shown using mutagenesis studies that if the immunoglobulin like domains of TrkA are swapped with those from TrkB ligand specificity is altered, indicating that these domains represent the binding domain (Perez *et al.* 1995). It has been shown that if all the extracellular domains except for the second Ig domain of TrkA are deleted, addition of NGF leads to phosphorylation and cell signalling (Urfer *et al.* 1995). Furthermore, recombinant Ig domains made in *Escherichia coli* are able to bind NGF with a similar affinity to the wild-type membrane-bound receptor (Holden *et al.* 1997) with all the binding activity resides in the Ig domain nearest to the membrane (Robertson *et al.* 2001). Subsequent crystallization and structure determination of this domain shows it to belong to the I-set of Ig-like domains with an unusual exposed disulphide bridge linking two neighbouring strands in the same β-sheet (Ultsch *et al.* 1999).

It has recently been shown that a large proportion of NGF in the brain exists as proNGF (Fahnestock *et al* 2001). This proNGF has been shown to bind to p75^{NTR}, TrkA, and the sortilin receptor (Lee *et al.* 2001; Harrington *et al.* 2004; Nykjaer *et al.* 2004). When mature NGF binds to p75^{NTR} and TrkA a cell survival cascade is elicited. When proNGF binds to p75^{NTR} and sortilin an apoptotic event occurs. Therefore the balance between the amount of proNGF and mature NGF may decide the fate of cells with p75^{NTR}, TrkA, and sortilin receptors.

This may have implications for therapy. A pharmacophore generated using the crystal structure of NGF bound to p75^{NTR} (He and Garcia 2004) was designed to select for compounds which bind at the p75^{NTR} receptor. This pharmacophore was used for *in silico* screening of small-molecule non-peptide libraries. Small molecules were subsequently identified that bind to the p75^{NTR} and prevented proneurotrophin-mediated apoptosis (Massa *et al.* 2006).

16.3.6.3 Distribution of NGF

The levels of NGF mRNA in the brain have been shown to correlate with the degree of basal forebrain cholinergic innervation (Korsching *et al.* 1985). Thus the highest levels of NGF and NGF mRNA in the brain are found within the cerebral cortex and hippocampus. Injection of [^{125}I]NGF into the hippocampus or cerebral cortex reveals that it is taken up by the terminals of basal forebrain cholinergic cells and retrogradely transported back to the cell bodies (Seiler and Schwab 1984). Immunohistochemical and *in situ* analysis of the distribution of p75^{NTR} in rat and human brain show that it is almost exclusively located on basal forebrain neurons (Riopelle *et al.* 1987; Dawbarn *et al.* 1988a; Kordower *et al.* 1988; Schatteman *et al.* 1988; Allen *et al.* 1989). Co-localization studies with antibodies to ChAT and p75^{NTR} demonstrate that nearly all the basal forebrain cholinergic neurons express p75^{NTR} (Dawbarn *et al.* 1988b; Kordower *et al.* 1988). Examination of the distribution of TrkA protein and mRNA in rat and human brain shows that TrkA is also expressed on basal forebrain cholinergic neurons (Steininger *et al.* 1993; Sobreviela *et al.* 1994).

Administration of NGF to basal forebrain cholinergic neurons *in vitro* results in increased survival and upregulation of ChAT (Gnahn *et al.* 1983; Hefti *et al.* 1985; Martinez *et al.* 1987; Hartikka and Hefti 1988a;b; Hatanaka *et al.* 1988).

Knockout mice have been generated to investigate the function of both NGF and TrkA in the brain (Crowley *et al.* 1994; Smeyne *et al.* 1994). The phenotypes show dramatic cell loss, similar to that seen after administration of neutralizing antibodies to NGF *in utero*. Both phenotypes survive, but show weight loss compared with unaffected littermates and usually die by the end of the third month. In the PNS there is cell loss in the superior cervical, dorsal root, and trigeminal ganglia. Surprisingly, in the CNS there is no cell loss in the cholinergic basal forebrain cells in either the NGF or TrkA knockout mice. However, there are marked reductions in basal forebrain ChAT staining and axonal cholinesterase activity in both the hippocampal and cortical target fields. Mice which are heterozygous for the NGF deletion exhibit aberrations in spatial memory function in the Morris water maze (Chen *et al.* 1993, 1994). These behavioural deficits correlate with losses in both the size and number of basal forebrain cholinergic neurons.

16.3.6.4 Rescue of cholinergic function by NGF

Since the memory loss seen in AD patients is thought to be a reflection of basal forebrain cholinergic cell loss, several attempts have been made to produce such lesions in rodents and primates. Lesion of the fimbria-fornix in rats severs the septal cholinergic fibres such that the terminals are not able to retrogradely transport NGF back to the cell bodies, resulting in cell loss. Intracerebroventricular (ICV) administration of NGF has been shown to completely stop this degeneration (Haroutunian *et al.* 1986; Hefti 1986; Kromer 1987; Williams *et al.* 1987; Gage *et al.* 1988) and the cognitive deficits arising from such lesions (Will and Hefti 1985; Williams *et al.* 1987; Mandel *et al.* 1989). Infusion of NGF in a subpopulation of aged rats which have been shown to perform poorly in the Morris water maze increases the performance of these animals to within the normal range (Fischer *et al.* 1987, 1991, 1994). It appears that NGF can protect cells from degenerating after lesions and has direct effects on cognition by acting specifically on basal forebrain cholinergic neurons.

Because of this early loss of cholinergic function in AD it has been suggested that infusion of NGF into the brains of patients should increase ChAT levels, prevent further degeneration of cholinergic cells, and thereby prevent further decline in memory function. However, if NGF is to be efficacious, it will be necessary for the basal forebrain cells still to be present to some extent and also to possess NGF receptors. Some considerable effort has been made to address these issues.

16.3.6.5 NGF receptors in AD

Immunohistochemical staining for $p75^{NTR}$ in basal forebrain cholinergic neurons has revealed a range of results. A large decrease in neuronal number has been reported (Hefti and Mash 1989; Mufson *et al.* 1989), although others (Allen *et al.* 1990) have shown that even severely demented patients may have very little cell loss, most of which has occurred in the posterior region of the NBM. The main differences in these studies are the age of onset of the disease, and it may be that an earlier onset of AD gives rise to much greater basal forebrain cell loss. Analysis of the mRNA for p75[NTR] in normal and AD basal forebrain by northern blotting showed no loss of p75[NTR] mRNA in AD (Goedert *et al.* 1989). In support of this, determination of p75[NTR] receptors by ligand binding showed that there was no difference between normal and AD brain (Treanor *et al.* 1991). Although studies examining the number of TrkA positively staining cells in AD basal forebrain have generally shown a reduction (Mufson *et al.* 1997; Boissiere *et al.* 1997), in all cases there are still a significant number of viable cells remaining. It has been suggested that the observed degeneration of these cholinergic cells in AD may be due to reduced trophic support, i.e. a lack of production of NGF in the cerebral cortex and hippocampus. Measurements of the content of NGF protein in AD brain have shown non-significant or significant increases in cerebral cortex in AD (Allen *et al.* 1991; Crutcher *et al.* 1993) and no change in hippocampus (Allen *et al.* 1991). Similarly, no change in NGF mRNA level was found in AD cerebral cortex and hippocampus compared with normal brain (Goedert *et al.* 1986). One possible explanation of the degeneration of basal forebrain cholinergic neurons in AD is disruption of the ability to retrogradely transport NGF from the target tissues to the cell bodies. Support for this hypothesis comes from a study which shows that NGF content, whilst normal in the cerebral cortex and hippocampus, is reduced in the basal forebrain (Mufson *et al.* 1995).

16.3.6.6 NGF, cholinergic neurons, and AD neuropathology

A transgenic mouse has been developed which makes antibodies to NGF (Capsoni *et al.* 2000; 2001). These mice (AD11) produce low levels of antibodies during gestation and show an age-dependent accumulation such that NGF sequestration only occurs during adulthood. They show an expected cholinergic deficit in the basal forebrain; however, they also show enlarged cerebral ventricles, cortical cell loss, amyloid plaques, and hyperphosphorylated tau in hippocampal and cortical neurons. In addition, they exhibit cognitive deficits. Intranasal administration of NGF restores cholinergic function and reduces the number of amyloid plaques, as also does intraperitoneal injection of the cholinesterase inhibitor galantamine. Both these therapies restore the number of ChAT immunopositive neurons to normal and both reduce the numbers of amyloid plaques. Furthermore, intranasal application of NGF (although not galantamine administration) prevented deposition of hyperphosphorylated tau. The same group reported that the antibodies produced in these mice are able to sequester mature NGF but not the proform of NGF (Capsoni and Cattaneo, 2006). This would inevitably lead to a change in the ratio of proNGF to NGF, which may be responsible for the ensuing neuropathological changes. In order to examine this further, AD11 mice were bred with a p75[NTR] knockout mouse. These produce a mutant p75[NTR] receptor lacking the extracellular domain. These mice (AD12), like their AD11 counterparts, lose cholinergic function at about 2 months of age; however, this is then gradually restored. Much less amyloid is deposited, but they seem to have increased hyperphosphorylated tau in the hippocampus and cortex at an early age. The explanation for this is not clear.

16.3.6.7 **Clinical trials with NGF**

It seems likely that if recombinant human NGF were to be given to AD patients, basal forebrain cholinergic function would be increased.

NGF administered intracerebroventricularly In a Swedish study in the1990s three AD patients were given mouse NGF ICV (Olson *et al.* 1992; Jonhagen *et al.* 1998). A total of 6.6 mg of NGF was infused continuously for 3 months in the first two patients. However, in Patient Three 0.21 mg of NGF was infused for 2 weeks, then the infusion was stopped, and after 5 weeks NGF was restarted with a total of 0.24 mg NGF over a period of 10 weeks. A third treatment period was started 8 months later with a total NGF dose of 0.11 mg over 10 weeks.

All the patients were diagnosed using Diagnostic and Statistical Manual (DSM-III-R) and National Institute of Neurological and Communicative Disorders and Stroke and Alzheimer's Disease and Related Disorders Association (NINCDS/ADRDA) criteria, and all were evaluated using a cognitive test battery, electroencephalography (EEG), magnetic resonance imaging (MRI), and positron emission tomography (PET). All the patients were tested for cognitive function using MMSE before NGF administration, and had scores of 18, 23, and 23. These scores did not stabilize over the time of treatment but declined with time. However Patient One showed a slight improvement in the selective reminding test and delayed recognition of word list. Patient Two showed a slight improvement in delayed recognition of faces and word list. Patient Three did not show any improvement in any subtest. Patient One was tested again 10 weeks after the NGF infusion had been stopped, and the improved results in selective reminding and delayed recall of word list found during treatment were no longer seen.

Patient One showed a marked increase in cerebral blood flow in the frontal and temporal cortices that outlasted the NGF treatment by several months. The other two patients did not show such improved cerebral blood flow. A general improvement of glucose metabolism was observed in Patient Two after the end of the NGF treatment. No marked effect on glucose metabolism was observed in Patient Three. However, the glucose metabolism in the cortical brain regions remained constant during the whole period of NGF treatments compared with the decrease in glucose metabolism that had been observed during the year prior to the initiation of the NGF treatment.

Using [^{11}C]nicotine binding (a measure of cholinergic synapses) an increase was seen in frontal and temporal cortices in Patient One after 3 months of NGF treatment and also at 3 months after infusion compared with the pre-infusion value. Similarly, increased binding was observed in the frontal, temporal, and parietal cortices as well as in the hippocampus in Patient Two after 3 months of NGF treatment and 3 months after the end of treatment. In Patient Three nicotine binding was generally decreased in the frontal, temporal, and parietal cortices during and after the end of the different periods of NGF treatment. However, increased binding was observed in the hippocampus of Patient Three 7 months after the end of the second period of NGF treatment.

Neurophysiological evaluation showed a reduction in slow-wave activity expressed as an increased alpha/theta ratio in Patients One and Two. In Patient Three there were no significant changes in EEG activity before, during, or after the three treatment periods.

Two types of side effects occurred during NGF treatment. All three patients reported the development of pain. Patient One complained of muscle pain in the lower back that appeared 11–14 days after the start of NGF infusion. Similar symptoms were also reported by Patient Two 11 days after the start of NGF infusion. When the NGF infusion was stopped the pain disappeared within 2 days. Patient Three, who received a lower dose of NGF (16 mg/24 hours), reported onstant muscle pain in the lumbar region. The NGF infusion was stopped after 2 weeks and the pain disappeared in 2 days. When the NGF dose was lowered to 3.4 mg/24 hours and restarted 5 weeks later the pain returned after four days but was reported to be less intense and could be alleviated with dextropropoxyphene. When the NGF infusion was stopped 2 months later the pain disappeared. Eight months later the patient received a lower dose of NGF (1.6 mg/24 hours) for 10 weeks. Again the pain returned but was reported to be of low intensity and also stopped completely following cessation of NGF treatment.

In the first two patients NGF infusion was also associated with a loss of weight and appetite: 6.7 kg (12%) in 3 months (Patient One) and 6.5 kg (15%) in 3 months (Patient Two). In the third patient there was a clear reduction in weight over time but no correlation with the treatment periods and no marked weight gain after the treatment periods.

It is uncertain what caused the side effects in the clinical trials. However, weight loss has been reported following infusion of NGF into the cerebral ventricles of rats (Day-Lollini *et al.* 1997) and primates (Fernandez *et al.* 1998). It is worth noting that in the rodent study it was only observed at the higher dose (300 ng/24 hour). It is difficult to project from rodent to human in terms of dosage, but it is likely that the CSF concentration obtained here would be higher than those achieved above. In the primate study the dose used was 2.5 mg/kg/24 hours which is substantially higher than that used in AD patients. Other studies where lower doses of NGF have been infused into primates have not resulted in weight loss. Furthermore, the weight loss reported in primates was reversible following cessation of NGF infusion, similarly to that seen in patients.

Two studies have shown that NGF infusion into rats and primates also results in Schwann cell hyperplasia. In one study (Day-Lollini *et al.* 1997) recombinant human NGF (5 mg/24 hours) was administered ICV to rats for 12 weeks, after which the rats were histologically examined immediately and 2, 8, 12, and 52 weeks after termination of the NGF infusion. Examination of tissue sections showed that the cervical cord was surrounded by atypical hyperplastic tissue from the caudal part of the medulla, at the hypoglossal nerve, to the C3 level. The hyperplastic tissue was predominantly attached to the dorsal and lateral surface of the medulla and the spinal cord underneath the pia arachnoid. The tissue was primarily composed of densely packed spindle-shaped cells, up to 30 layers thick. No mitotic figures or malignant transformed cells were observed.

Immunohistochemical analysis of the cells showed that they stained for p75[NTR] but not for TrkA. In addition, a strong cellular S-100 immunoreactivity was observed together with vimentin- and GFAP-positive fibres, suggesting that the hyperplasia consisted of proliferating Schwann cells. Associated neuronal fibres were also observed which were immunopositive for either tyrosine hydroxlase or calcitionin gene related protein, indicating the presence of sympathetic and sensory fibres. After stopping the NGF infusion, the hyperplasia, measured as the mean thickness of the pia arachnoid at the most affected place, was shown to be completely reversible. Thus the hyperplasia was significantly reduced after 8 weeks and completely removed after 52 weeks.

In the second study (Winkler *et al.* 1997) recombinant human NGF was infused into both rat and primate cerebral ventricles. The doses used were 0, 6, 60, or 300 ng per day for rats and 0, 0.6, 6, or 60 mg per day for primates; both groups were infused for 6 months. Similar hyperplasia was reported except that the effects were shown to be dose related such that at the lowest dose used for rodents the hyperplasia was completely absent.

Use of genetically engineered fibroblasts secreting NGF In a recent study (Tuszynski *et al.* 2005) fibroblasts from AD patients were cultured and transfected with the retroviral vector (modified from Maloney leukaemia virus) to enable the cells to secrete recombinant human NGF. NGF expression was measured by ELISA; levels were 25–75 ng NGF/10^6 cells/day. Production of proNGF constituted less than 1% of total NGF.

Eight patients with early-stage AD were recruited to the trial. The first two subjects received five injections of cells unilaterally over the basal forebrain. These subjects were operated on without general anaesthetic; unfortunately, movement during the procedure resulted in subcortical haemorrage. One patient subsequently died from pulmonary embolism and cardiac arrest. The other subject suffered hemiparesis of the right side. The following six subjects received bilateral injections under general anaesthetic. Two patients received 50 ml cells bilaterally, and four patients received 100 ml bilaterally. No adverse effects were noticed for up to 42 months post-treatment (Tuszynski *et al.* 2006). Cognition was assessed in the six subjects transplanted bilaterally using the MMSE. Before treatment the annual rate of decline was approximately 6 points per year, whereas after treatment it was reduced to 3 points/year. Cognition was also measured using the ADAS-Cog. With this scale, an average reduction of 6 points/year is usually recorded in the moderate to severe AD range. Following implantation, an amelioration of 5% in the rate of decline was seen at 6–18 months compared with 1–12 months after treatment.

PET analysis of four subjects showed significant increase of fluorodeoxyglucose uptake 6–8 months after implantation of cells. Interestingly, only the areas receiving cholinergic input from the basal forebrain showed this increase. This initial study, while inconclusive, paves the way for larger phase I/phase II clinical trials with NGF. One such trial has currently been approved in the USA which will be conducted at the Rush Medical Centre, Chicago, and is expected to start late 2006/early 2007. Rather than implantation of modified cells it will involve a double-blind study involving administration of an adeno-associated viral vector modified to secrete NGF.

16.3.6.8 Conclusions

It seems clear that ICV administration of NGF is associated with side effects. Transplantation of NGF-secreting fibroblasts may show proof of the concept that NGF therapy is beneficial but will clearly not be widely available as a general procedure.

The administration of an adenoviral vector secreting NGF may remove some problems but may also provide new ones.

An alternative and novel strategy may be to give NGF via an intranasal route. Following injection of [^{125}I]NGF into the olfactory bulb of rats, radiolabelled NGF was found to be specifically retrogradely transported to basal forebrain cholinergic nuclei (Altar and Bakhit 1991). Similarly, it has been shown that radiolabelled NGF can be transported into the brain following administration as nose drops (Frey *et al.* 1997; Chen *et al.* 1998)

Additionally, since ICV-administrated NGF elicits side effects via the p75^{NTR} receptor, the design of specific small-molecule TrkA agonists may be beneficial for the treatment of AD and significant advances have recently been made towards this end (Dawbarn *et al.* 2006).

16.4 Other neurotransmitter approaches

16.4.1 Monoamines

Although the cholinergic system has been investigated the most extensively, other neurotransmitter systems are involved in both cognitive processing and AD. Dopaminergic activity influences working memory and the noradrenergic systems are involved in maintaining attention and concentration. Monoamine oxidases (MAOs) occur in two forms, MAO-A and MAO-B. Noradrenaline and serotonin (5-HT) are deaminated by MAO-A, and dopamine is deaminated by MAO-B. MAO-B accounts for about 80% of human brain MAO, and the highest concentrations are found in the hippocampus. Patients with AD have higher MAO-B activity than healthy elderly people (Alexopoulos *et al.* 1984), and so strategies aimed at increasing monoamine availability have a logical rationale.

16.4.1.1 Selegiline

Selegiline (L-deprenyl) is an MAO inhibitor. At low doses (10 mg/day) it acts as an irreversible inhibitor of MAO-B. At higher doses (40 mg/day) it loses its selectivity for MAO-B and also inhibits MAO-A. The possible beneficial effects of selegiline may not rely solely on its ability to enhance monoamine activity (Tolbert and Fuller 1996). Selegiline is metabolized in the liver to desmethylselegiline and metamphetamine, which are subsequently metabolized to amphetamine. The stimulant effects of these metabolites may exert some effect on mood and arousal. Additionally, selegiline may have neuroprotective attributes as MAO-B inhibition helps to decrease the production of oxidative free radicals (Cohen and Spina 1989). However, a meta-analysis of relevant trials of selegiline concluded that there is no overall evidence of meaningful clinical benefit (Wilcock *et al.* 2002).

16.4.2 Glutamate

L-Glutamate is the major excitatory neurotransmitter in the CNS. The ionotropic glutamate receptors can be distinguished by the binding of three agonists: *N*-methyl-D-aspartate (NMDA), α-amino-3-hydroxy-5-methyl-4-isoxazole propionic acid (AMPA), and kainate. The metabotropic glutamate receptors, of which there are several subtypes, act via a G-protein on secondary messenger systems.

Glutamate is involved in both learning and memory, but is also a powerful neurotoxin and has been implicated in the pathogenesis of AD. *In vitro* studies have demonstrated that NMDA antagonists such as amantadine and memantine have neuroprotective properties, and there is now an extensive clinical trial database for memantine. This has been licensed for use in moderate to severe AD, and is notable for its relatively benign side-effect profile. It has been shown to have beneficial effects on cognition and other clinically relevant parameters such as functional ability and behaviour (Reisberg *et al.* 2003). It is available in many countries, but although it is licensed in the UK, NICE has not recommended that it be funded by the NHS.

Combination treatment with memantine and a cholinesterase inhibitor may have some benefit over monotherapy; however, the evidence to date is scanty. The most convincing trial involved a little over 400 patients with moderate to severe AD, already on stable doses of donepezil. Memantine or a placebo was added to the donepezil for 6 months, and in those taking the two drugs there were significantly better outcomes on measures of cognition, activities of daily living, global outcome, and behaviour than in those taking only donepezil and placebo. The combination was well tolerated (Tariot *et al.* 2004).

16.5 **Neuroprotective strategies**

16.5.1 **Oestrogen replacement therapy**

Indirect evidence supporting the potential benefits of ERT comes from a number of observational studies and can be inferred from the physiological actions of oestrogen and the results of animal experiments.

Animal studies indicate that oestrogens have a direct effect on neuronal development and neurotransmitter systems. Oestrogen receptors are found in the equivalent areas of rat brain to those that are preferentially damaged in humans suffering from AD. The poor performance on tests of memory and learning of ovariectomized rats is associated with reduced cholinergic activity, but can be prevented by oestrogen replacement (Singh *et al.* 1994). In rats 17β-oestradiol modulates CA1 hippocampal dendritic spine density (Woolley and McEwen 1994). Oestrogen may exert some of its effects through its actions on NGF. Rodent mRNA for NGF declines after ovariectomy but is restored by replacement therapy (Singh *et al.* 1995).

The effects of oestrogen on processes thought to be relevant in the pathogenesis of AD give further credence to the oestrogen hypothesis (Henderson 1997). Regional vascular insufficiency may contribute to the disease process in AD (Birge 1997), and oestrogen improves cerebral blood flow in post-menopausal women. In cell cultures with oestrogen receptors, 17β -oestradiol promotes the metabolism of APP to the soluble fragments, thus reducing the accumulation as the neurotoxic Aβ amyloid fragment (Jaffe *et al.* 1994). In addition, oestrogens have potent anti-oxidant properties.

Clinic-based case–control studies of the effects of ERT on the development of AD have yielded inconsistent results. Some of these studies were open to criticism because different methods were used to determine ERT usage in patients and controls. Moreover, prescribing practices may have been different for female AD patients compared with controls.

The population-based studies have been more consistent in their findings that ERT has beneficial effects on the development of AD, but not universally so. Brenner *et al.* (1994) compared ERT usage as determined by computerized pharmacy data in 107 female AD sufferers and 120 age-matched controls, and found no significant differences between the two groups. Paganini-Hill and Henderson (1996) conducted a case–control study nested within a prospective cohort study of 8877 women. The participants were sent a health questionnaire at the beginning of the study period; 3760 of the female cohort died in the following 14 years. AD, or other terms likely to indicate AD, was recorded on the death certificates of 248 women. Each case was matched with five women for year of birth and death. The risk of AD was reduced in those who had received ERT (odds ratio 0.65, 95% CI 0.49–0.88). Increased doses and duration of therapy with oral conjugated equine oestrogen were associated with increased risk reduction. The main limitation of this study was the use of death certificates to define the cases.

In the study by Tang *et al.* (1996), ERT use in 1124 non-demented elderly women was ascertained by personal interview at the beginning of the study. The women were then followed up annually for between 1 and 5 years. The diagnosis of AD was made by an independent group of clinicians based on information from the assessments, medical records, and imaging studies. The relative risk (RR) of developing AD was reduced in ERT users (RR 0.40, 95% CI 0.22–0.85). The RR was reduced most for women who had taken ERT for more than a year. ERT use was also documented prospectively in the Baltimore Longitudinal Study of Aging (Kawas *et al.* 1997) in which 472 post- or peri-menopausal women were assessed every 2 years for up to 16 years. The RR for AD among women who had ever used ERT (oral or transdermal oestrogen) was 0.46 (95% CI 0.209–0.997). The authors did not find any correlation with duration of ERT use, but this had been recorded in broad time-period categories.

The earliest treatment study in AD patients was an open-label uncontrolled trial of unopposed conjugated oestrogens (Fillit *et al.* 1986). Three of seven women improved on cognitive test scores. Ohkura *et al.* (1995) reported on the effects of long-term (5–45 months) low-dose ERT given to seven women with mild to moderate AD. Four women improved in areas of cognition and ability to perform activities of daily living.

More recently, a rather negative outcome has been reported from an analysis of a trial based upon the Women's Health Initiative (WHI) in the USA. Oestrogen plus progestin therapy increased the risk for probable dementia in postmenopausal women aged 65 years or older. In addition, oestrogen plus progestin therapy did not prevent mild cognitive impairment, which in many cases is preclinical AD, in these women. These findings, coupled with previously reported WHI data, support the conclusion that the risks of oestrogen plus progestin outweigh the benefits (Rapp *et al.* 2003; Shumaker *et al.* 2003).

16.5.2 **Anti-inflammatory drugs**

Inflammatory and immune mechanisms are implicated in the pathological processes which lead to nerve cell death in AD (Aisen and Davis 1994). Early reports of a decreased incidence of AD among rheumatoid arthritis sufferers led to the hypothesis that anti-inflammatory drugs might be protective for AD. A number of studies examining arthritis, non-steroidal anti-inflammatory drugs (NSAIDs), and steroid use as risk factors for AD followed. A meta-analysis of case–control studies (McGeer *et al* 1996) found that the chance of developing AD was lower than in the general population for individuals who had suffered from arthritis (odds ratio 0.56) or who had taken NSAIDs (odds ratio 0.5) or steroids (odds ratio 0.66).

The findings of population-based studies have been harder to interpret because of differences in methodologies. Andersen *et al.* (1995) examined the cross-sectional relation between NSAID use and the risk for AD in a large population-based study of elderly people. The relative risk for AD in NSAID users ($n = 365$) compared with non-users ($n = 5893$) was 0.38. The Baltimore Longitudinal Study of Aging also found an inverse association between aspirin and NSAID use and the onset of AD (Stewart *et al.* 1997). Among NSAID users, the relative risk of AD was influenced by the duration of drug use. Treatment with NSAIDs for more than 2 years was associated with a lower relative risk for AD than treatment of less than 2 years duration.

Prince *et al.* (1998) examined the relationship between NSAID use and cognitive function in 2651 people aged 65–74, who were enrolled in the UK MRC treatment trial of hypertension. NSAID use was ascertained by a nurse interviewer every 3 months for up to 8 years. NSAID users showed less decline on the Paired Associate Learning Test than non-users, with younger patients benefiting most. Other longitudinal observational studies have found no evidence that NSAIDs protect against the development of dementia or cognitive decline (Fourrier *et al.* 1996; Henderson *et al.* 1997).

Despite this promise, trials of anti-inflammatory drugs have been disappointing (Aisen *et al.* 2000, 2003). However, the explanation for these negative results seems to lie in the choice of drug. It transpires that, despite the abundant evidence for an inflammatory component to the pathology of AD, some members of one class of the anti-inflammatory agents, the NSAIDs, have an effect that is independent of their anti-inflammatory properties. This is an ability to modulate β-amyloid production and is not a class effect, i.e. this property is only vested in some members of the class, and also is not a property of steroids such as prednisone (Weggen *et al.* 2001, 2003; Eriksen *et al.* 2003). The results of a large phase II study of R-flurbiprofen, which is based upon but is not itself an NSAID, has recently been reported (Wilcock *et al.* 2006). This was of 12 months duration with a further 12 months semi-blinded extension phase, and in those subjects achieving adeqate plasma levels there was evidence of statistically significant slowing of disease progression in terms of functional ability, and a non-statistically significant trend in cognition in favour of active treatment. This early promise is now under review in two large phase III studies of 1600 subjects. It will also provide further evidence to help decide if the benefits, should they be maintained, are really the result of neuroprotection or are symptomatic, as in the case of the cholinesterase inhibitors (Gasparini *et al* 2005).

There are a number of other amyloid-based neuroprotective strategies, including those targeted at the secretases and the removal of amyloid using antibody strategies.

16.5.3 Antioxidants

Brain tissue has low levels of endogenous anti-oxidants and therefore is particularly vulnerable to free-radical damage (Lethem and Orrell 1997). Oxidative stress has been implicated in the pathogenesis of AD (Smith *et al.* 1996), and this has inevitably aroused interest in the possible therapeutic benefits of anti-oxidants. A number of medications with actions which include anti-oxidant activity have been investigated but so far there is no convincing evidence that any of these strategies are of value.

16.5.3.1 *Ginkgo biloba*

Ginkgo biloba extract (Egb 761) is a plant extract that has been approved as a treatment for dementia in Germany. Several studies have reported on the beneficial effects of *Ginkgo biloba* but few have used standard assessments of cognitive function or behaviour.

Maurer *et al.* (1997) conducted a randomized double-blind placebo-controlled trial in 20 patients with AD over 3 months. The active treatment group received 240 mg/day of *Ginkgo biloba* extract. A significant treatment effect was found on tests of attention and memory, but the sample size was too small to demonstrate any other treatment effects. A recent placebo-controlled double-blind randomized trial of *Ginkgo biloba*, which used standard measures of cognition and behaviour, included patients with AD and multi-infarct dementia (Le Bars *et al.* 1997). A total of 327 patients with mild to moderate dementia (251 with AD) were enrolled in the 52 week study; 166 patients were allocated to receive Egb 120 mg/day, and 161 patients were allocated to the placebo group. In the ITT analysis, the active treatment group deteriorated less than the placebo group on measures of cognitive function, activities of daily living, and social behaviour. Similar results were demonstrated for the subgroup of AD patients. A clinically significant improvement in cognitive function was achieved in 27% of patients treated with Egb compared with 14% of patients who received placebo. Adverse events were generally mild to moderate and, with the exception of gastrointestinal symptoms, equally distributed between the two groups.

The availability of *Ginkgo biloba* extract in over-the-counter herbal supplements has raised concern because of the potential risk of adverse events (Vale 1998). The bioavailability of the active compound varies in different preparations, and in addition to its anti-oxidant properties, *Ginkgo biloba* inhibits platelet-activating factor. Long-term use has been associated with an increased bleeding time and spontaneous haemorrhage (Rowin and Lewis 1996). The efficacy of *Ginkgo biloba* appears to be comparable to that of tacrine, but it should not be used without clinical supervision. A Cochrane Database review concluded that further evaluation in a large trial is required (Birks *et al.* 2002).

16.5.3.2 Alpha-tocopherol

Alpha-tocopherol (vitamin E) is a lipid-soluble vitamin with potent anti-oxidant activity. Animal and *in vitro* studies have suggested that α-tocopherol has neuro-protective properties. It has been evaluated in a double-blind placebo-controlled trial in patients with moderate AD (Sano *et al.* 1997). In this trial 341 patients were randomly assigned to receive selegiline 5 mg twice daily, α-tocopherol 1000 IU twice daily, selegiline and α-tocopherol, or placebo. The primary outcome measure was the time to the occurrence of one endpoint: death, institutionalization, or loss of ability to perform two of three activities of daily living (i.e. eating, grooming, using the toilet). After adjusting for baseline functioning, all three active treatments significantly delayed the primary outcome (risk ratio for α-tocopherol 0.47; risk ratio for selegiline 0.57; risk ratio for combination treatment 0.69). Only α-tocopherol delayed institutionalization (risk ratio 0.42) when individual endpoints were examined. Falls and syncope were more common in the treatment groups but did not lead to discontinuation of therapy. However, more recent studies have been more disappointing, especially a large 3 year trial of donepezil and vitamin E, or a combination of these, in people who had a diagnosis of mild cognitive impairment (MCI). This is diagnosed when an individual has mild memory problems that are abnormal for him or her, but insufficient for a diagnosis of dementia. Many such individuals have pre-clinical AD, and around 15% annually convert from MCI to AD, as happened in this trial population. Although donepezil was of some benefit in slowing down the conversion rate, especially in the first 12 months, vitamin E had no effect. This implies that it had no neuroprotective benefit (Petersen *et al.* 2005).

If anti-oxidants have a potential role as neuroprotective agents further investigation is required to clarify their method of action, efficacy, and long-term safety.

16.5.4 Glial cell modulation

Immune mechanisms contribute to neuronal death in both AD and ischaemia, and ischaemia can enhance the β-amyloidogenic reaction. Propentofylline, a xanthine derivative, is an adenosine re-uptake/phosphodiesterase inhibitor with a broad range of effects. It appears to inhibit the cytotoxic functions of activated microglia and reinforce the neuroprotective properties of astrocytes by stimulating NGF secretion. It may have additional beneficial activity as an anti-oxidant.

Animal studies indicate that propentofylline may have benefits on some aspects of cognitive functioning. Randomized double-blind placebo-controlled studies including 901 patients with mild to moderate AD have shown significant treatment effects for propentofylline in tests of global and cognitive function and activities of daily living (Kittner *et al.* 1997). Treatment effects were comparable in patients with vascular dementia. Adverse effects, most commonly gastrointestinal disturbance, headaches, and dizziness, were mainly mild. Despite this evidence, most clinicians do not consider propentofylline to have a place in the routine treatment of AD and, in the UK at least, it does not have a licence for use.

16.5.5 Calcium-channel blockers

Calcium dysregulation may play a role in neuronal cell degeneration. High intracellular calcium concentrations are neurotoxic, and the enhancement of glutamate toxicity by Aβ may be calcium dependent. Two randomized trials in 1648 patients with AD failed to demonstrate any significant benefits for nimodipine, a calcium-channel blocker, over placebo on primary outcome measures (Morich *et al.* 1996).

16.6 Immunotherapy for AD

The seminal work of Schenk and colleagues (Schenk *et al.* 1999) showed that inoculation of transgenic mice, producing human cerebral amyloid Aβ, with Aβ and an adjuvant, resulted in high anti-Aβ antibody levels and attenuated brain amyloid deposition, even in older mice. Administration of exogenous mouse monoclonal antibodies against Aβ produced a similar outcome (Bard *et al.* 2000). The most likely mechanism was enhanced clearance of amyloid by the microglial/macrophage system. However, it has also been suggested that little antibody actually penetrates the blood brain barrier, and that the amyloid lowering properties of the antibodies lie more in trapping the Aβ in the blood-stream (DeMattos *et al.* 2001), called the 'peripheral sink' effect. Another possibility is that binding of the antibody to the Aβ reduces or prevents formation of amyloid fibrils (Solomon 2001). Since amyloid deposition is probably a balance between production and removal of amyloid, and not just overproduction of amyloid, all three mechanisms and probably others as yet un-identified may play a part. This led to the Elan and Wyeth trial of an amyloid vaccine in 2001, which included 375 subjects. Unfortunately about 6% of the subjects developed an aseptic meningoencephalitis and the trial had to stop prematurely. Autopsy on one of the subjects who died confirmed that their brain had much less amyloid than expected, compared to brains from control subjects at a comparable stage of the disease, thus mirroring the transgenic mouse data (Nicoll *et al.* 2003). Analysis of the clinical data from one centre suggested that 20 subjects who generated an antibody response experienced some decline in the rate of progression of their symptoms (Hock *et al.* 2003). Other studies are now underway which will eventually clarify the place of this approach to treating people with AD.

16.7 **Summary and conclusions**

So far, CEIs and memantine have been the most successful of the symptomatic treatments for AD. However, most clinical trials of the former have been confined to patients with mild or moderately severe dementia, although efficacy in more severely affected patients is beginning to become available, and conversely further studies of memantine in less severe AD are also in progress. The most severely impaired patients and those with minor degrees of cognitive impairment, possibly representing subclinical AD, have been largely excluded. The efficacy of individual CEIs appears broadly similar.

Advances in our understanding of the neuropathological processes involved in AD have paved the way for new and exciting opportunities to alter the disease progression, or even delay the onset of AD. The people who may benefit most from neuroprotective treatments are those who are asymptomatic but at risk of developing the illness. One of our most important challenges at the moment is to find a way of identifying people at the subclinical stage of AD in order to maximize their benefits from disease-modifying, i.e. neuroprotective, strategies.

References

Aisen PS and Davis KL (1994). Inflammatory mechanisms in Alzheimer's disease: implications for therapy. *American Journal of Psychiatry* **151**, 1105–13.

Aisen PS, Davis KL, Berg JD, *et al.* (2000). A randomized controlled trial of prednisone in Alzheimer's disease. Alzheimer's Disease Cooperative Study. *Neurology* **54**, 588–93.

Aisen PS, Schafer KA, Grundman M, *et al*. (2003). Effects of rofecoxib or naproxen vs placebo on Alzheimer disease progression: a randomized controlled trial. *JAMA*, vol. 289, no. 21, pp. 2819–2826.

Alexopoulos GS, Lieberman KW, and Young RC (1984). Platelet MAO activity in primary degenerative dementia. *American Journal of Psychiatry* **141**, 97–9.

Allen SJ, Dawbarn D, Spillantini MG, *et al*. (1989). The distribution of β-nerve growth factor receptors in the human basal forebrain. *Journal of Comparative Neurology* **289**, 626–40.

Allen SJ, Dawbarn D, MacGowan SH, Wilcock GK, Treanor JJS, and Moss TH (1990). A quantitative morphometric analysis of basal forebrain neurons expressing β-NGF receptors in normal and Alzheimer's disease brains. *Dementia* **1**, 125–37.

Allen SJ, MacGowan SH, Treanor JJS, Feeney R, Wilcock GK, and Dawbarn D (1991). Normal NGF content in Alzheimer's disease cerebral cortex and hippocampus. *Neuroscience Letters* **131**, 135–9.

Altar C and Bakhit C (1991). Receptor mediated transport of human recombinant nerve growth factor from olfactory bulb to basal forebrain cholinergic nuclei. *Brain Research* **541**, 82–8.

Anand R and Gharabawi G (1996). Clinical development of Exelon (ENA-713): the ADENA programme. *Journal of Drug Development and Clinical Practice* **8**, 117–22.

Andersen K, Launer LJ, Ott A, Hoes AW, Bretler MM, and Hofman A (1995). Do nonsteroidal anti-inflammatory drugs decrease the risk for Alzheimer's disease? The Rotterdam Study. *Neurology* **45**, 1441–5.

Asthana S, Raffaele KC, Berardi A, *et al*. (1995) Treatment of Alzheimer disease by continuous intravenous infusion of physostigmine. *Alzheimer Disease and Associated Disorders* **9**, 223–32.

Avery EE, Baker LD, and Asthana S (1997). Potential role of muscarinic agonists in Alzheimer's disease. *Drugs and Aging* **11**, 450–9.

Bard F, Cannon C, Barbour R *et al*. (2000). Peripherally administered antibodies against amyloid beta-peptide enter the central nervous system and reduce pathology in a mouse model of Alzheimer disease. *Nature Medicine* **6**, 916-9.

Barde Y-A, Davies AM, Johnson JE, Lindsay RM, and Thoenen H (1982). Brain derived neurotrophic factor. *Progress in Brain Research* **71**, 185–9.

Barde Y-A, Edgar D, and Thoenen H (1987). Purification of a new neurotrophic factor from mammalian brain. *EMBO Journal* **1**, 549–53.

Berkemeir LR, Winslow JM, Kaplan DR, Nikolics K, Goeddel DV, and Rosenthal A (1991). Neurotrophin-5: a novel neurotrophic factor that activates trk and trkB. *Neuron* **7**, 857–66.

Birge SJ (1997). The role of estrogen in the treatment of Alzheimer's disease. *Neurology* **48** (Suppl 7), 36–41.

Birks J, Grimley EV, and Van Dongen M (2002). *Ginkgo biloba* for cognitive impairment and dementia. *Cochrane Database System Review* **4**, CD003120.

Bodick NC, Offen WW, Shannon HE, *et al*. (1997). The selective muscarinic agonists xanomeline improves both the cognitive deficits and behavioural symptoms of Alzheimer's disease. *Alzheimer Disease and Associated Disorders* **11** (Suppl 4), S16–22.

Boissiere F, Faucheux B, Ruberg M, Agid Y, and Hirsch EC (1997). Decreased TrkA gene expression in cholinergic neurons of the striatum and basal forebrain of patients with Alzheimer's disease. *Experimental Neurology* **145**, 245–52.

Bowen DM, Smith CB, White P, and Davison AN (1976). Neurotransmitter related enzymes and indices of hypoxia in senile dementia and other abiotrophies. *Brain* **99**, 459–96.

Bowen DM, Benton JS, Spillane JA, Smith CCT, and Allen SJ (1982). Choline acetyltransferase activity and histopathology of frontal neocortex from biopsies of demented patients. *Journal of Neurological Science* **57**, 191–202.

Brenner DE, Kukull WA, Stergachis A, *et al*. (1994). Postmenopausal estrogen replacement therapy and the risk of Alzheimer's disease: a population-based case–control study. *American Journal of Epidemiology* **140**, 262–7.

Butte MJ, Hwang PK, Mobley WC, and Fletterick RJ (1998). Crystal structure of neurotrophin-3 homodimer shows distinct regions are used to bind its receptors. *Biochemistry* **37**, 16846–52.

Caccamo A, Oddo S, Billings LM, *et al.* (2006). M1 receptors play a central role in modulating AD-like pathology in transgenic mice *Neuron* **49**, 671–82.

Capsoni S, Ugolini G, Comparini A, Ruberti F, Berardi N, and Cattaneo A (2000). Alzheimer-like neurodegeneration in aged antinerve growth factor transgenic mice. *Proceedings of the National Academy of Sciences of the USA* **97**, 6826–31.

Capsoni S, Giannotta S, Cattaneo A. (2002). Nerve growth factor and galantamine ameliorate early signs of neurodegeneration in anti-nerve growth factor mice. *Proceedings of the National Academy of Sciences of the USA* **99**, 12432–7.

Capsoni S and Cattaneo A (2006). On the Molecular Basis Linking Nerve Growth Factor (NGF) to Alzheimer's Disease. *Cellular and Molecular Neurobiology* **26**, 619–33.

Chatellier G and Lacomblez L (1990). Tacrine (tetrahydroaminoacridine; THA) and lecithin in senile dementia of the Alzheimer type: a multicentre trial. Groupe Francais d'Etude de la Tetrahydroaminoacridine. *British Medical Journal* **300**, 495–9.

Chen KS, Nishimura MC, Broz S, Crowley C, and Phillips HS (1993). Assessment of learning and memory performance in mice heterozygous for nerve growth factor deletion. In *Society for Neuroscience Abstracts*, 19.1963 (Society for Neuroscience, Washington, DC).

Chen KS, Nishimura MC, Spencer S, Crowley C, and Phillips HS (1994). Learning memory and the basal cholinergic system in aged mice heterozygous for nerve growth factor deletion. In *Society for Neuroscience Abstracts*, 28.856 (Society for Neuroscience, Washington, DC).

Chen XQ, Fawcett JR, Ala TA, Rahman YE, and Frey WH (1998). Olfactory route: a new pathway to deliver nerve growth factor to the brain. In *6th International Conference on Alzheimer's Disease, Amsterdam*, Abstract 1086.

Christie JE, Shering A, Ferguson J, and Glen AI (1981). Physostigmine and arecoline: effects of intravenous infusions in Alzheimer presenile dementia. *British Journal of Psychiatry* **138**, 46–50.

Clary DO and Reichardt LF (1994). An alternatively spliced form of the nerve growth factor receptor trkA confers an enhanced response to neurotrophin-3. *Proceedings of the National Academy of Sciences of the USA* **91**, 11133–7.

Cohen G and Spina MB (1989). Deprenyl suppresses the oxidant stress associated with increased dopamine turnover. *Annals of Neurology* **26**, 689–90.

Connor DJ, Langlais PL, and Thal LJ (1991). Behavioural impairments after lesions of the nucleus basalis by ibotenic acid and quisqualic acid. *Brain Research* **555**, 84–90.

Corey-Bloom J, Anand R, and Veach J, for the ENA 713 B352 Study Group (1998). A randomized trial evaluating the efficacy and safety of ENA 713 (rivastigmine tartrate), a new acetylcholinesterase inhibitor, in patients with mild to moderately severe Alzheimer's disease. *International Journal of Geriatric Psychopharmacology* **1**, 55–65.

Crow TJ and Grove-White IG (1973). An analysis of the learning deficit following hyoscine administration to man. *British Journal of Pharmacology* **49**, 322–7.

Crowley C, Spencer SD, Nishimura MC, *et al.* (1994). Mice lacking nerve growth factor display perinatal loss of sensory and sympathetic neurons yet develop basal forebrain cholinergic neurons. *Cell* **76**, 1001–11.

Crutcher KA, Scott SA, Liang S, Everson WV, and Weingartner J (1993). Detection of NGF-like activity in human brain tissue: Increased levels in Alzheimer's disease. *Journal of Neuroscience* **13**, 2540–50.

Davies P and Malony AJF (1976). Selective loss of central cholinergic neurons in Alzheimer's disease. *Lancet* **ii**, 1403.

Davis KL, Mohs RC, and Tinlenberg JR (1979). Enhancement of memory by physostigmine. *New England Journal of Medicine* **301**, 946.

Dawbarn D, Allen SJ, and Semenenko FM (1988a). Immunohistochemical localisation of β-nerve growth factor receptors in the forebrain of the rat. *Brain Research* **440**, 185–9.

Dawbarn D, Allen SJ, and Semenenko FM (1988b). Coexistence of choline acetyltransferase and β-nerve growth factor receptors in the forebrain of the rat. *Neuroscience Letters* **94**, 138–44.

Dawbarn D, Fahey M, Watson J, *et al.* (2006).NGF receptor TrkAd5: therapeutic agent and drug design target. *Biochemical Transactions* **34**, 587–90.

Day-Lollini PA, Stewart GR, Taylor MJ, Johnson RM, and Chellman GJ (1997). Hyperplastic changes within the leptomeninges of the rat and monkey in response to chronic intracerebroventricular infusion of nerve growth factor. *Experimental Neurology* **145**, 24–37.

DeMattos RB, Bales KR, Cummins DJ, Dodart JC, Paul SM and Holtzman DM (2001). Peripheral anti-A beta antibody alters CNS and plasma Abeta clearance and decreases brain Abeta burden in a mouse model of Alzheimer's disease. *Proceedings of the National Academy of Sciences of the USA* **98**, 8850–5.

Dunnett SB, Whishaw IQ, Jones GH, and Bunch ST (1987). Behavioural, biochemical and histochemical effects of different neurotoxic amino acids injected into nucleus basalis magnocellularis of rats. *Neuroscience* **20**, 653–69.

Dunnett SB, Rogers DC, and Jones GH (1989). Effects of nucleus basalis magnocellularis lesions in rats on delayed matching and non-matching to position tasks. *European Journal of Neuroscience* **1**, 395–406.

Eagger SA, Levy R, and Sahakian BJ (1991). Tacrine in Alzheimer's disease. *Lancet* **337**, 989–92.

Eriksen JL, Sagi SA, Smith TE, *et al.* (2003), NSAIDs and enantiomers of flurbiprofen target gamma-secretase and lower Abeta 42 *in vivo*. *Journal of Clinical Investigation* **112**, 440–9.

Fahnestock M, Michalski B, Xu B, and Coughlin MD (2001) The precursor pronerve growth factor is the predominant form of nerve growth factor in brain and is increased in Alzheimer's disease. *Molecular and Cellular Neurosciences* **18**, 210–20.

Fernandez CI, Gonzalez O, Diaz E, Alverez L, Antunez I, and Molina I (1998). Systemic and side effects of intracerebroventricular infusion of nerve growth factor in aged rodents and monkeys: cautionary evidence for its therapeutic use. *Alzheimer's Disease Review* **3**, 95–9.

Fillit H, Weinreb H, Cholst I, *et al.* (1986). Observations in a preliminary open trial of estradiol therapy for senile dementia (Alzheimer's type). *Psychoneuroendocrinology* **11**, 337–45.

Fischer W, Wictorin K, Bjorklund A, Williams LR, Varon S, and Gage FH (1987). Amelioration of cholinergic neuron atrophy and spatial memory impairment in aged rats by nerve growth factor. *Nature* **328**, 65–8.

Fischer W, Bjorklund A, Chen K, and Gage FH (1991). NGF improves spatial memory in aged rodents as a function of age. *Journal of Neuroscience* **11**, 1889–1906.

Fischer W, Sirevaag A, Wiegand SJ, Lindsay RM, and Bjorklund A (1994). Reversal of spatial memory impairments in aged rats by nerve growth factor and neurotrophins 3 and 4/5 but not by brain-derived neurotrophic factor. *Proceedings of the National Academy of Sciences of the USA* **91**, 8607–11.

Fourrier A, Letenneur L, Begaud B, and Dartigues JF (1996). Nonsteroidal antiinflammatory drug use and cogntive function in the elderly: inconclusive results from a population-based cohort study. *Journal of Clinical Epidemiology* **49**, 1201.

Frey WH, II, Liu J, Chen X, *et al.* (1997). Delivery of 125I-NGF to the brain via the olfactory route. *Drug Delivery* **4**, 87–92.

Fritten LJ, Perryman KM, Gross PM, Fine H, Cummins J, and Marshall C (1990). Treatment of Alzheimer's disease with short- and long-term oral THA and lecithin: a double-blind study. *American Journal of Psychiatry* **2**, 239–42.

Gage FH, Armstrong DM, Williams LR, and Varon S (1988). Morphological response of axotomized septal neurons to nerve growth factor. *Journal of Comparative Neurology* **269**, 147–55.

Gasparini L, Ongini E, Wilcock D, and Morgan D (2005). Activity of flurbiprofen and chemically related anti-inflammatory drugs in models of Alzheimer's disease. *Brain Research: Brain Research Review* **48**, 400–8.

Geaney DP, Soper N, Shepstone BJ, and Cowan PJ (1990). Effect of cholinergic stimulation on regional cerebral blood flow in Alzheimer's disease. *Lancet* **335**, 1484–7.

Gnahn H, Hefti F, Heumann R, Schwabb ME, and Thoenen H (1983). NGF-mediated increase of choline acetyltransferase (ChAT) in the neonatal rat forebrain: evidence for a physiological role of NGF in the brain. *Brain Research* **285**, 45–52.

Goedert M, Fine A, and Ullrich A (1986). Nerve growth factor RNA in peripheral and central rat tissues and in the human central nervous system: lesion effects in the rat brain and levels in Alzheimer's disease. *Brain Research: Molecular Brain Research* **1**, 85–92.

Goedert M, Fine A, Dawbarn D, Wilcock GK, and Chao MV (1989). Nerve growth factor receptor mRNA distribution in human brain: normal levels in basal forebrain in Alzheimer's disease. *Brain Research: Molecular Brain Research* **5**, 1–7.

Gustafson L (1993). Physostigmine and tetrahydroaminoacridine treatment of Alzheimer's disease. *Acta Neurologica Scandinavica Supplementum* **149**, 39–41.

Hallbook F, Ibanez CF, and Persson H (1991). Evolutionary studies of the nerve growth factor family reveal a novel member abundantly expressed in *Xenopus* ovary. *Neuron* **6**, 845–58.

Haroutunian V, Kanof PD, and Davis KL (1986). Partial reversal of lesion-induced deficits in cortical cholinergic markers by nerve growth factor. *Brain Research* **386**, 397–9.

Harrington AW, Leiner B, Blechschmitt C, *et al.* (2004) Secreted proNGF is a pathophysiological death-inducing ligand after adult CNS injury. *Proceedings of the National Academy of Sciences of the USA* **101**, 6226–30.

Hartikka J and Hefti F (1988a). Development of septal cholinergic neurons in culture: plating density and glial cells modulate effects of NGF on survival, fiber growth and expression of transmitter specific enzymes. *Journal of Neuroscience* **8**, 2967–85.

Hartikka J and Hefti F (1988b). Comparison of nerve growth factor effects on development of septum, striatum, and nucleus basalis cholinergic neurons *in vitro*. *Journal of Neuroscience Research* **21**, 352–64.

Hashimoto M, Kazui H, Matsumoto K, Nakano Y, Yasuda M, and Mori E (2005). Does donepezil treatment slow the progression of hippocampal atrophy in patients with Alzheimer's disease? *American Journal of Psychiatry* **162**, 676–82.

Hatanaka H, Tsukui H, and Nihonmatsu I (1988). Developmental change in nerve growth factor action from induction of choline acetyltransferase to promotion of cell survival in cultured basal forebrain cholinergic neurons from postnatal rats. *Brain Research* **467**, 85–95.

He XL and Garcia KC (2004). Structure of nerve growth factor complexed with the shared neurotrophin receptor p75. *Science* **304**, 870–5.

Hefti F (1986). Nerve growth factor promotes survival of septal cholinergic neurons after fimbrial transections. *Journal of Neuroscience* **6**, 2155–62.

Hefti F and Mash DC (1989). Localisation of nerve growth factor receptors in the normal human brain and in Alzheimer's disease. *Neurobiology of Ageing* **10**, 75–87.

Hefti F, Hartikka J, Eckenstein F, Gnahn H, Heumann R, and Schwab M (1985). Nerve growth factor increases choline acetyltransferase but not survival or fiber outgrowth of cultured fetal septal cholinergic neurons. *Neuroscience* **14**, 55–68.

Henderson AS, Jorm AF, Christensen H, Jacomb PA, and Korten AE (1997). Aspirin, anti-inflammatory drugs and risk of dementia. *International Journal of Geriatric Psychiatry* **12**, 926–30.

Henderson VW (1997). The epidemiology of estrogen replacement therapy and Alzheimer's disease. *Neurology* **48** (Suppl 7), S27–35.

Higgins JPT and Flicker L (1998). The efficacy of lecithin in the treatment of dementia and cognitive impairment. Cochrane Review. In *Cochrane Library*, Issue 3 (Update Software, Oxford).

Hock C, Konietzko U, Streffer JR, *et al.* (2003). Antibodies against beta-amyloid slow cognitive decline in Alzheimer's disease. *Neuron* **38**, 547–54.

Hohn A, Leibrock J, Bailey K, and Barde Y-A (1990). Identification and characterisation of a novel member of the nerve growth factor/brain derived-neurotrophic factor family. *Nature* **344**, 339–41.

Holden PH, Asopa V, Robertson AGS, *et al* (1997). Immunoglobulin-like domains define the nerve growth factor binding site of the TrkA receptor. *Nature Biotechnology* **15**, 668–72.

Holland DR, Cousens LS, Meng W, and Matthews B (1994). Nerve growth factor in different crystal forms displays structural flexibility and reveals zinc binding sites. *Journal of Molecular Biology* **239**, 385–400.

Holmes C, Wilkinson D, Dean C, *et al.* (2004). The efficacy of donepezil in the treatment of neuropsychiatric symptoms in Alzheimer disease. *Neurology* **63**, 214–19

Ip NY, Ibanez CF, Nye SH, *et al.* (1992). Mammalian neurotrophin-4: structure, chromosomal localisation, tissue distribution, and receptor specificity. *Proceedings of the National Academy of Sciences of the USA* **89**, 3060–4.

Ip NY, Stitt TN, Tapley P, *et al.* (1993). Similarities and differences in the way neurotrophins interact with trk receptors in neuronal and non-neuronal cells. *Neuron* **10**, 137–49.

Jaffe AB, Toran-Allerand CD, Greengard P, and Gandy SE (1994). Estrogen regulates metabolism of Alzheimer amyloid beta-precursor protein. *Journal of Biological Chemistry* **269**, 13065–8.

Jing S, Tapley P, and Barbacid M (1992) Nerve growth factor mediates signal transduction through trk homodimer receptors. *Neuron* **9**, 1067–79.

Jonhagen ME, Nordberg A, Amberla K, *et al.* (1998). Intracerbroventricular infusion of nerve growth factor in three patients with Alzheimer's disease. *Dementia and Geriatric Cognitive Disorders* **9**, 246–57.

Kaplan DR, Hempstead BL, Martinzanca D, Chao MV, and Parada LF (1991). The *trk* protooncogene product-a signal transducing receptor for nerve growth factor. *Science* **252**, 554–8.

Kawas C, Resnick S, Morrison A, *et al.* (1997). A prospective study of estrogen replacement therpy and the risk of developing Alzheimer's disease. The Baltimore Longitudinal Study of Aging. *Neurology* **48**, 1517–21.

Kim SH, Kim YK, Jeong SJ, Haas C, Kim YH, and Suh YH (1997). Enhanced release of secreted form of Alzheimer's amyloid precursor protein from PC12 cells by nicotine. *Molecular Pharmacology* **52**, 430–6.

Kittner B, Rossner M, and Rother M (1997). Clinical trials in dementia with propentofylline. *Annals of the New York Academy of Sciences* **826**, 307–16.

Kleine R, Lamballe F, Bryant S, and Barbacid M (1992). The trkB tyrosine protein kinase is a receptor for neurotrophin-4. *Neuron* **8**, 947–56.

Kordower JH, Bartus RT, Bothwell M, Schatteman G, and Dash DM (1988). Nerve growth factor receptor immunoreactivity in the non-human primate *Cebus appella*: distribution, morphology and colocalisation with cholinergic enzymes. *Journal of Comparative Neurology* **277**, 465–86.

Korsching S, Auburger G, Heumann R, Scott J, and Thoenen H (1985). Levels of nerve growth factor and its mRNA in the central nervous system of the rat correlate with cholinergic innervation. *EMBO Journal* **4**, 1389–93.

Kromer LF (1987). Nerve growth factor treatment after brain injury prevents neuronal death. *Science* **235**, 214–16.

Lamballe F, Klein R, and Barbacid M (1991). TrkC a new mender of the trk family of tyrosine protein-kinases is a receptor for neurotrophin-3. *Cell* **66**, 947–56.

Le Bars PL, Katz MM, Berman N, Itil TM, Freedman AM, and Schatzberg AF, for the North American Egb Study Group (1997). A placebo-controlled, double-blind, randomized trial of an extract of an extract of *Ginkgo biloba* for dementia. *Journal of the American Medical Association* **278**, 1327–32.

Lee R, Kermani P, Teng KK, and Hempstead BL (2001). Regulation of cell survival by secreted proneurotrophins. *Science* **294**,1945–8.

Leibrock J, Lottspeich F, Hohn A, *et al.* (1989). Molecular cloning and expression of brain-derived neurotrophic factor. *Nature* **341**, 149–52.

Lethem R and Orrell M (1997). Antioxidants and dementia. *Lancet* **349**, 1189.

Levy A, Brandeis R, Treves TA, *et al.* (1994) Transdermal physostigmine in the treatment of Alzheimer's disease. *Alzheimer Disease and Associated Disorders* **8**, 15–21.

Loeb DM, Stephens RM, Copeland T, Kaplan DR, and Greene LA (1994). A trk nerve growth factor (NGF) receptor point mutation affecting interaction with phosphlipase C-gamma I abolishes NGF- promoted peripherin induction and not neurite outgrowth. *Journal of Biological Chemistry* **269**, 8901–10.

McDonald NQ and Hendrickson WA (1993). A structural superfamily of growth factors containing a cysteine knot motif. *Cell* **73**, 421–4.

McDonald NQ, Lapatto R, Murray-Rust J, Gunning J, Wlodawer A, and Blundell TL (1991) New protein fold revealed by a 2,3-A resolution crystal structure of nerve growth factor. *Nature* **354**, 411–14.

McGeer PL, Schulzer M, and McGeer EG (1996). Arthritis and anti-inflammatory agents as possible protective factors for Alzheimer's disease: a review of 17 epidemiological studies. *Neurology* **47**, 425–32.

McKeith I, Del Ser T, Spano P, *et al.* (2000), Efficacy of rivastigmine in dementia with Lewy bodies: a randomised, double-blind, placebo-controlled international study, *Lancet* **356**, 2031–6.

Maisonpierre PC, Belluscio L, Squinto S, *et al.* (1990a). Neurotrophin-3: a neurotrophic factor related to NGF and BDNF. *Science* **247**, 1446–51.

Maisonpierre PC, Le Beau MM, Espinosa R, *et al.* (1990b). Human and rat brain-derived neurotrophic factor and neurotrophin-3: gene structures, distributions and chromosomal localisations. *Genomics* **10**, 558–68.

Mandel RJ, Gage FH, and Thal LJ (1989) Spatial learning in rats: correlation with cortical choline acetyltransferase and improvement with NGF following NBM damage. *Experimental Neurology* **104**, 208–17.

Marin DB, Bierer LP, Lawlor BA, *et al.* (1995). L-Deprenyl and physostigmine for the treatment of Alzheimer's disease. *Psychiatry Research* **58**, 181–9.

Martinez HJ, Dreyfus CF, Jonkait GM, and Black IB (1987). Nerve growth factor selectively increase cholinergic markers but not neuropeptides in rat basal forebrain culture. *Brain Research* **412**, 295–301.

Massa SM, Xie Y, Yang T, *et al.* (2006). Small, nonpeptide p75^{NTR} ligands induce survival signaling and inhibit proNGF-induced death. *Journal of Neuroscience* **26**, 5288–300.

Maurer K, Ihl R, Dierks T, and Frolich L (1997). Clinical efficacy of *Ginkgo biloba* special extract Egb 761 in dementia of the Alzheimer type. *Journal of Psychiatric Research* **31**, 645–55.

Mesulam MM and Geula C (1988). Nucleus basalis (CH4) and cortical cholinergic innervation in the human brain: observations based on the distribution of acetylcholinesterase and choline acetyltransferase. *Journal of Comparative Neurology* **275**, 216–40.

Mesulam MM, Mufson EJ, Levey AI, and Wainer BH (1983). Cytochemistry and cortical connections of the septal area, diagonal band, nucleus basalis (substantia innominata), and hypothalamus in the rheusus monkey. *Journal of Comparative Neurology* **214**, 170–97.

Mesulam MM, Mufson EJ, and Wainer BH (1986). Three-dimensional representation and cortical projection topography of the nucleus basalis (CH4) in the macaque: concurrent demonstration of choline acetylttransferase and retrograde transport with a stabilised tetramethybenzidine method for horseradish peroxidase. *Brain Research* **367**, 301–8.

Morich FJ, Bieber F, Lewis JM, *et al.* (1996). Nimodipine in the treatment of probable Alzheimer's disease. Results of two multicentre trials. *Clinical Drug Investigation* **11**, 185–95.

Mufson EJ, Bothwell M, and Kordower JH (1989). Loss of nerve growth factor receptor-containing neurons in Alzheimer's disease: a quantitative analysis across sub-regions of the basal forebrain. *Experimental Neurology* **105**, 221–32.

Mufson EJ, Conner JM, and Kordower JH (1995). Nerve growth factor in Alzheimer's disease: defective retrograde transport to nucleus basalis. *Neuroreport* **6**, 1063–6.

Mufson EJ, Lavine N, Jaffar S, Kordower JH, Quirion R, and Saragovi U (1997). Reduction in p140-trkA receptor protein within the nucleus basalis and cortex in Alzheimer's disease. *Experimental Neurology* **146**, 91–103.

Nicol JA, Wilkinson D, Holmes C, Steart P, Markham H, and Weller RO (2003). Neuropathology of human Alzheimer disease after immunization with amyloid-beta peptide: a case report. *Nature Medicine* **9**, 448–52.

Nykjaer A, Lee R, Teng KK, *et al.* (2004) Sortilin is essential for proNGF-induced neuronal cell death. *Nature* **427**, 843–8.

Ohkura T, Isse K, Akazawa K, Hamamoto M, Yaoi Y, and Hahino N (1995). Long-term estrogen replacement therapy in female patients with dementia of the Alzheimer-type: 7 case reports. *Dementia* **6**, 99–107.

Olson L, Nordberg A, von Holst H, *et al.* (1992). Nerve growth factor affects ^{11}C-nicotine binding, blood flow, EEG, and verbal episodic memory in an Alzheimer patient (case report). *Journal of Neural Transmission* **4**, 79–95.

Ott A, Slooter AJC, Hofman A, *et al.* (1998). Smoking and risk of dementia and Alzheimer's disease in a population-based cohort study: the Rotterdam Study. *Lancet* **351**, 1840–3.

Paganini-Hill A and Henderson VW (1996). Estrogen replacement therapy and risk of Alzheimer's disease. *Archives of Internal Medicine* **156**, 19, 2213–17.

Perez P, Coll PM, Hempstead BL, Martin-Zanca D, and Chao MV (1995). NGF binding to the trk tyrosine kinase receptor requires the extracellular immunoglobulin-like domains. *Molecular and Cellular Neuroscience* **6**, 97–105.

Petersen RC, Thomas RG, Grundman M, *et al.* (2005), Vitamin E and donepezil for the treatment of mild cognitive impairment. *New England Journal of Medicine* **352**, 2379–88.

Prince M, Rabe-Hesketh S, and Brennan P (1998). Do antiarthritic drugs decrease the risk for cognitive decline? An analysis based on data from the MRC treatment trial of hypertension in older adults. *Neurology* **50**, 374–9.

Purves D (1989). *Body and Brain: A Theory of Neural Connectivity* (Oxford University Press, Oxford).

Rapp SR, Espeland MA, Shumaker SA, *et al.* (2003). Effect of estrogen plus progestin on global cognitive function in postmenopausal women: the Women's Health Initiative Memory Study: a randomized controlled trial. *Journal of the American Medical Association* **289**, 2663–72.

Reisberg B, Doody R, Stoffler A, Schmitt F, Ferris S, and Mobius HJ (2003). Memantine in moderate-to-severe Alzheimer's disease, *New England Journal of Medicine* **348**, 1333–41.

Ridley RM, Barratt NG, and Baker HF (1984a). Cholinergic learning deficits in the marmoset produced by scopolamine and ICV hemicholinium. *Psychopharmacology* **83**, 340–5.

Ridley RM, Bowes PM, Baker HF, and Crow TJ (1984b). An involvement of acetylcholine in object discrimination learning and memory in the marmoset. *Neuropsychologia* **22**, 253–63.

Ridley RM, Baker HF, Drewett B, and Johnson JA (1985). Effects of ibotenic acid lesions of the basal forebrain on serial reversal learning in marmosets. *Psychopharmacology* **86**, 438–43.

Ridley RM, Murray TK, Johnson JA, and Baker HF (1986). Learning impairment following lesion of the basal nucleus of Meynert in the marmoset: modification by cholinergic drugs. *Brain Research* **376**, 108–16.

Riopelle JM, Richardson PM, and Verge VM (1987). Distribution and characteristics of nerve growth factor binding on cholinergic neurons of rat and monkey forebrain. *Neurochemical Research* **12**, 923–8.

Robbins TW, Everitt BJ, Marston HM, Wilkinson J, Jones GH, and Page KL (1989). Comparative effects of ibotenic acid and quisqualate acid induced lesions of the substantia innominata on attentional function in the rat: further implications for the role of the cholinergic neurons of the nucleus basalis in cognitive processes. *Behavioural Brain Research* **35**, 221–4.

Robertson AGS, Banfield MJ, Allen SJ, *et al.* (2001) Identification and structure of the nerve growth factor binding site on TrkA. *Biochemical and Biophysical Research Communications* **282**, 131–41.

Robinson RC, Radziejewski C, Stuart DI, and Jones EY (1995). Structure of the brain-derived neurotrophic factor/neurotrophin 3 heterodimer. *Biochemistry* **34**, 4139–46.

Robinson RC, Radziejewski C, Stuart DI, and Jones EY (1996). Crystals of the neurotrophins. *Protein Science* **15**, 973–7.

Rockwood K, Beattie BL, Eastwood MR, *et al.* (1997). A randomised, controlled trial of linopirdine in the treatment of Alzheimer's disease. *Canadian Journal of Neurological Sciences* **24**, 140–5.

Rodrigues-Tebar A, Dechant G, Gotz R, and Barde Y-A (1992). Binding of neurotrophin-3 to its neuronal receptors and interactions with nerve growth factor and brain-derived growth factor. *EMBO Journal* **11**, 917–22.

Rogers SL and Friedhoff LT (1998). Long-term efficacy and safety of donepezil in the treatment of Alzheimer's disease: an interim analysis of the results of a US multicentre open label extension study. *European Neuropsychopharmacology* **8**, 67–75.

Rogers SL, Friedhoff LT, and the Donepezil Study Group (1996). The efficacy and safety of donepezil in patients with Alzheimer's disease: results of a US multicenter, randomized, double-blind, placebo-controlled trial. *Dementia* **7**, 293–303.

Rogers SL, Doody RS, Mohs RC, Friedhoff LT, and the Donepezil Study Group (1998a). Donepezil improves cognition and global function in Alzheimer's disease: a 15-week, double-blind, placebo-controlled study. *Archives of Internal Medicine* **158**, 1021–31.

Rogers SL, Farlow MR, Doody RS, Mohs R, Friedhoff LT, and the Donepezil Study Group (1998b). A 24-week, double-blind, placebo-controlled trial of donepezil in patients with Alzheimer's disease. *Neurology* **50**, 136–45.

Rowin J and Lewis SL (1996). Spontaneous bilateral subdural haematomas associated with chronic *Ginkgo biloba* ingestion. *Neurology* **46**, 1775–6.

Salomon AR, Marcinowski KJ, Friedland RP, and Zagorski MG (1996). Nicotine inhibits amyloid formation by the beta-peptide. *Biochemistry* **35**, 13568–78.

Sano M, Ernesto C, Thomas RG, *et al.* (1997). A controlled trial of selegiline, alpha-tocopherol, or both as treatment for Alzheimer's disease. *New England Journal of Medicine* **336**, 1216–22.

Schatteman GC, Gibbs L, Lanahan AA, Claude P, and Bothwell M (1988). Expression of NGF receptors in the developing and adult primate central nervous system. *Journal of Neuroscience* **8**, 860–73.

Schenk D, Barbour R, Dunn W, *et al.* (1999). Immunization with amyloid-β attenuates Alzheimer-disease-like pathology in the PDAPP mouse. *Nature* **400**, 173–7.

Schneider R and Schweiger M (1991). A novel modular mosaic of cell adhesion motifs in the extracellular domains of the neurogenic trk and trkB tyrosine kinase receptors. *Oncogene* **6**, 1807–11.

Scott J, Selby M, Urdea M, Quironga M, Bell GI, and Rutter WJ (1983). Isolation and nucleotide sequence of a cDNA encoding the sequence of mouse nerve growth factor. *Nature* **302**, 538–40.

Seiler M and Schwab ME (1984). Specific retrograde transport of nerve growth factor (NGF) from neocortex to nucleus basalis in the rat. *Brain Research* **300**, 33–9.

Shimohama S, Akaike A, and Kimura J (1996). Nicotine-induced protection against glutamate cytotoxicity: nicotinic cholinergic receptor-mediated inhibition of nitric oxide formation. *Annals of the New York Academy of Sciences* **777**, 356–61.

Shumaker SA, Legault C, Rapp SR, *et al.* (2003), Estrogen plus progestin and the incidence of dementia and mild cognitive impairment in postmenopausal women: the Women's Health Initiative Memory Study: a randomized controlled trial, *Journal of the American Medical Association* **289**, 2651–62.

Sims NR, Bowen DM, Allen SJ, *et al.* (1983). Presynaptic cholinergic dysfunction in patients with dementia. *Journal of Neurochemistry* **40**, 503–9.

Singh M, Meyer EM, Millard WJ, and Simpkins JW (1994). Ovarian steroid deprovation results in a reversible learning impairment and compromised cholinergic function in female Sprague–Dawley rats. *Brain Research* **644**, 305–12.

Singh M, Meyer EM, and Simpkins JW (1995). The effect of ovariectomy and estradiol replacement on brain-derived neurotrophic factor messenger ribonucleic acid expression in cortical and hippocampal brain regions of female Sprague–Dawley rats. *Endocrinology* **136**, 2320–4.

Smeyne RJ, Klein R, Schnapp A, *et al.* (1994). Severe sensory and sympathetic neuropathies in mice carrying a disrupted Trk/NGF receptor gene. *Nature* **368**, 246–9.

Smith MA, Perry G, Richey PL, *et al.* (1996). Oxidative damage in Alzheimer's. *Nature* **382**, 120–1.

Snaedal J, Johannesson T, Jonsson JE, and Gylfadottir G (1996). The effects of nicotine in dermal plaster on cogntive functions in patients with Alzheimer's disease. *Dementia* **7**, 47–52.

Sobreviela T, Clary DO, Reichardt LF, Brandabur MM, Kordower JH, and Mufson EJ (1994). TrkA-immunoreactive profiles in the central nervous system: colocalisation with neurons containing p75 nerve growth factor receptor, choline acetyltransferase, and serotonin. *Journal of Comparative Neurology* **350**, 587–611.

Solomon, B (2001). Immunotherapeutic strategies for prevention and treatment of Alzheimer's disease. *DNA Cell Biology* **20**, 697–703.

Spiegel R (1991). Cholinergic drugs, affective disorders and dementia: problems of clinical research. *Acta Psychiatrica Scandinavica Supplementum* **366**, 47–51.

Steininger TL, Wainer BH, Klein R, Barbacid M, and Palfrey HC (1993). High affinity nerve growth factor receptor (trk0 is localised in cholinergic neurons of the basal forebrain and striatum in the adult rat brain. *Brain Research* **612**, 330–5.

Stephens RM, Loeb DM, Copeland TD, Pawson T, Greene LA, and Kaplan DR (1994). Trk receptors use redundant signal transduction pathways involving SHC and PLC-gamma 1 to mediate NGF responses. *Neuron* **12**, 691–705.

Stewart WF, Kawas C, Corrada M, and Metter EJ (1997). Risk of Alzheimer's disease and duration of NSAID use. *Neurology* **48**, 626–32.

Summers WK, Majovski LV, Marsh GM, Tachiki K, and Kling A (1986). Oral tetrahydroaminoacridine in long-term treatment of senile dementia, Alzheimer type. *New England Journal of Medicine* **315**, 1241–5.

Tang M-X, Jacobs D, Stern Y, *et al.* (1996). Effect of oestrogen during menopause on risk and age at onset of Alzheimer's disease. *Lancet* **348**, 429–32.

Tariot PN, Farlow MR, Grossberg GT, Graham SM, McDonald S, and Gergel I (2004). Memantine treatment in patients with moderate to severe Alzheimer disease already receiving donepezil: a randomized controlled trial, *Journal of the American Medical Association* **291**, 317–24.

Thal LJ, Schwartz G, Sano M, *et al.* (1996). A multicenter double-blind study of controlled-release physostigmine for the treatment of symptoms secondary to Alzheimer's disease. *Neurology* **47**, 1389–95.

Thoenen H (1991). The changing scene of neurotrophic factors. *Trends in Neuroscience* **14**, 165–70.

Thoenen H and Barde Y-A (1980). Physiology of nerve growth factor. *Physiological Reviews* **60**, 1284–1335.

Thoenen H, Bandtlow C, and Heumann R (1987). The physiological function of nerve growth factor in the central nervous system: comparison with the periphery. *Reviews in Physiology, Biochemistry and Pharmacology* **109**, 145–78.

Tolbert SR and Fuller MA (1996). Selegiline in treatment of behavioural and cognitive symptoms of Alzheimer's disease. *Annals of Pharmacotherapy* **30**, 1122–9.

Treanor JJS, Dawbarn D, Allen SJ, MacGowan SH, and Wilcock GK (1991). Nerve growth factor receptor binding in normal and Alzheimer's disease basal forebrain. *Neuroscience Letters* **121**, 73–6.

Turner JJ, Hodges H, Sinden JG, and Grey JJ (1992). Comparison of radial maze performance of rats after ibotenate and quisqualate lesions of the forebrain cholinergic projection system: effects of pharmacological challenge and changes in training regime. *Behavioural Pharmacology* **3**, 765–9.

Tuszynski MH, Thal L, Pay M, *et al.* (2005). A phase 1 clinical trial of nerve growth factor gene therapy for Alzheimer disease. *Nature Medicine* **11**, 551–5.

Tuszynski MH, Thal L, Pay M, *et al.* (2006). NGF gene therapy. Presented at 10th International conference on Alzheimer's disease and related disorders, Madrid 53-04-02.

Ullrich A, Grey A, Berman C, and Dull TJ (1983). Human β-nerve growth factor sequence highly homologous to that of mouse. *Nature* **303**, 821–5.

Ulrich J, Johannson-Locher G, Seiler WO, and Stahelin HB (1997). Does smoking protect from Alzheimer's disease? Alzheimer-type changes in 301 unselected brains from patients with known smoking history. *Acta Neuropathologica* **94**, 450–4.

Ultsch MH, Wiesmann C, Simmons LC, *et al.* (1999). Crystal structure of the neurotrophin-binding domain of TrkA, TrkB and TrkC. *Journal of Molecular Biology* **290**, 149–59.

Urfer R, Tsoulfas P, Occonnel L, Shelton DL, Parada LF, and Presta LG (1995). An immunoglobulin-like domain determines the specificity of neurotrophin receptors. *EMBO Journal*, **14**, 2795–805.

Vale S (1998). Subarachnoid haemorrhage associated with *Ginkgo biloba*. *Lancet* **352**, 36.

Venneri A, McGeown WJ, and Shanks MF (2004). Empirical evidence of neuroprotection by dual cholinesterase inhibition in Alzheimer's disease. *Neuroreport* **16**, 107–10.

Weggen S, Eriksen JL, Das P, *et al.* (2001). A subset of NSAIDs lower amyloidogenic Abeta42 independently of cyclooxygenase activity, *Nature* **414**, 212–16.

Weggen S, Eriksen JL, Sagi SA, *et al.* (2003). Evidence that nonsteroidal anti-inflammatory drugs decrease amyloid beta 42 production by direct modulation of gamma-secretase activity, *Journal of Biological Chemistry* **278**, 31831–7.

Wenk GL, Markowska AL, and Olton DS (1989). Basal forebrain lesions and memory: alteration in neurotensin not acetylcholine may cause amnesia. *Behavioural Neuroscience* **103**, 765–9.

Wilcock GK and Wilkinson D (1997). Galanthamine hydrobromide: Interim results of a group comparative, placebo-controlled study of efficacy and safety in patients with a diagnosis of senile dementia of the Alzheimer type. In Iqbal K, Winblad B, Nishimura T, Takeda M, and Wisniewski HM (eds), *Alzheimer's Disease: Biology, Diagnosis and Therapeutics*, pp. 661–4 (John Wiley, Chichester).

Wilcock GK, Scott M, Pearsall T, Neubauer K, Boyle M, and Razay G (1993). Galanthamine and the treatment of Alzheimer's disease. *International Journal of Geriatric Psychiatry* **8**, 781–2.

Wilcock GK, Lilienfeld S, and Gaens E (2000). Efficacy and safety of galantamine in patients with mild to moderate Alzheimer's disease: multicentre randomised controlled trial. Galantamine International-1 Study Group. *British Medical Journal* **321**, 1445–9.

Wilcock GK, Birks J, Whitehead A, and Evans, SJ (2002). The effect of selegiline in the treatment of people with Alzheimer's disease: a meta-analysis of published trials, *International Journal of Geriatric Psychiatry* **17**, 175–83.

Wilcock G, Howe I, Coles H, *et al.* (2003). A long-term comparison of galantamine and donepezil in the treatment of Alzheimer's disease. *Drugs and Aging* **20**, 777–789.

Wilcock GK, Black S, Haworth J, *et al.* (2006). Efficacy and safety of MPC-7869 (R-flurbiprofen), a selective Aβ42-lowering agent, in Alzheimer's Disease (AD): Results of a 12-month phase 2 trial and 1-year follow-on study. *Alzheimer's and Dementia* **2**, Supplement 1; S81.

Will B and Hefti F (1985). Behavioural and neurochemical effects of chronic intraventricular injections of nerve growth factor in adult rats with fimbria lesions. *Behavioural Brain Research* **17**, 17–24.

Williams LR, Varon S, Peterson GM, *et al.* (1987). Continuous infusion of nerve growth factor prevents basal forebrain neuronal death after fimbria fornix transection. *Proceedings of the National Academy of Sciences of the USA* **83**, 9231–5.

Wilson AL, Langley LK, Monley J, *et al.* (1995). Nicotine patches in Alzheimer's disease: pilot study on learning, memory, and safety. *Pharmacology, Biochemistry, and Behavior* **51**, 509–14.

Wilson K, Bowen D, Francis P, and Tyrell P (1991). Effect of central cholinergic stimulation on regional cerebral blood flow in Alzheimer's disease. *British Journal of Psychiatry* **158**, 558–62.

Winkler J, Ramirez GA, Kuhn HG, *et al.* (1997). Reversible Schwann cell hyperplasia and sprouting of sensory and sympathetic neurites after intraventricular administration of nerve growth factor. *Annals of Neurology* **41**, 82–93.

Woolley CS and McEwen BS (1994). Estradiol regulates hippocampal dendritic spine density via an N-methyl-D-aspartate receptor-dependent mechanism. *Journal of Neuroscience* **14**, 7680–7.

Chapter 17

Amyloid-based therapies

Weiming Xia

Alzheimer's disease (AD) is characterized by the progressive accumulation of extracellular neuritic plaques and intracellular neurofibrillary tangles. Neuritic plaques are composed of the amyloid β protein (Aβ), which is derived from the β-amyloid precursor protein (APP) after β- and γ-secretase cleavages. Because the neuritic amyloid plaque is an early and invariant feature of AD, reducing amyloid deposition is a rational approach to intervening in the neuropathological progression of disease. The current search for amyloid-based therapies is mainly focused on blocking Aβ generation, enhancing Aβ degradation, clearing Aβ from brain by vaccine, and interfering with Aβ oligomerization. Successful identification of novel molecular targets and Aβ-related pathways involved in AD pathogenesis has been the foundation for these approaches, and it will continue to be the driving force for the creation of novel amyloid-based therapies.

17.1 Introduction

Senile amyloid plaques and neurofibrillary tangles are two neuropathological hallmarks of AD. Neurofibrillary tangles are composed of aggregated hyperphosphorylated tau protein. Senile plaques consist of largely insoluble deposits of Aβ, which is derived from the β-amyloid precursor protein (APP). Processing APP to yield Aβ plays an important role in the pathogenesis of AD. Sequential cleavages of APP by β- and γ-secretases yield two major species of Aβ ending at residue 40 (Aβ40) or 42 (Aβ42). Genetic studies show that familial AD-linked missense mutations in APP occur either within the Aβ peptide sequence or immediately flanking this sequence. Certain missense mutations in APP specifically increase the production of Aβ42, the longer 42 residue of Aβ species initially deposited in AD brains. Mutations in presenilin 1 (PS1) and its homologue presenilin 2 (PS2) account for roughly 50% of early onset familial AD, and the increase in Aβ42 and/or decrease in Aβ40 are the characteristic phenotype of cultured cells, animals, and AD patients carrying missense mutations in PS1 or PS2. A major risk factor for late-onset AD is associated with inheritance of one or two copies of the apolipoprotein E4 allele, and brains of patients carrying *ApoE4* alleles were found to have increased density of Aβ deposits.

Soluble Aβ molecules tend to form oligomers, Aβ-derived diffusible ligands (ADDLs), and protofibrils, which have been suggested to be responsible for neuronal dysfunction in the brains of AD patients. These Aβ aggregates may cause neuronal injury directly by acting on synapses or indirectly by activating microglia and astrocytes.

Pharmacological interventions have been developed to target the sequential events originating from Aβ synthesis, mainly by inhibiting β- or γ-secretase activity. Studies also suggest that activating α-secretase is feasible in reducing Aβ plaques in an animal model. Degradation of soluble and oligomeric Aβ is achievable by modulating several amyloid-degrading enzymes, and an Aβ vaccine can effectively remove Aβ from brain. Compounds preventing Aβ oligomerization and aggregation may directly reduce the Aβ-induced neuronal toxicity.

17.2 Blocking Aβ generation

17.2.1 Amyloid precursor protein processing and Aβ generation

APP occurs in three alternatively spliced forms: 695-, 751-, and 770-residue polypeptides. Cleavage of APP starts with β-secretase to generate an ~100 kDa soluble N-terminal fragment and a 12 kDa C-terminal stub of APP (C99), followed by PS1-mediated γ-secretase cleavage to yield Aβ. Alternatively, α-secretase cleaves APP roughly in the middle of the Aβ region (residues 16 and 17 of Aβ), thereby precluding Aβ production. The α-secretase cleavage product, an ~10 kDa C-terminal stub (C83), can subsequently be cleaved by γ-secretase to yield p3.

APP is required for kinesin-mediated axonal transport of β-secretase/PS1-containing membrane vesicles, and axonal transport deficit has been pathologically characterized in brains of AD patients (Stokin *et al.* 2005). In cultured cells, APP is transported and post-translationally modified through the central vacuolar pathway to the cell surface, where it can either be cleaved by α-secretase to release APP$_s$-α or undergo re-internalization (endocytosis) as a holoprotein. Early studies indicate that extracellular Aβ can be generated in both the secretory pathway and endosomes during the endocytic recycling of APP. Most of the intracellular pools of Aβ40 and Aβ42 are generated in the secretory pathway. Because the proteasome might be an important site for the Aβ immediate precursor (C99) degradation, it could block the non-specific catabolism of C99. Studies have shown that treating cells with proteasome inhibitors can increase insoluble intracellular Aβ42 in endoplasmic reticulum (ER) and intermediate compartment (IC) vesicles. Treatment of cells with brefeldin A or introducing ER retention signals into APP also leads to APP processing and Aβ generation within ER and IC vesicles (reviewed by Xia 2001).

17.2.2 Inhibiting β-secretase

β-Secretase (BACE1) is a 501 amino acid (aa) type 1 transmembrane protein (Sinha *et al.* 1999; Vassar *et al.* 1999; Yan *et al.* 1999; Lin *et al.* 2000). BACE1 is expressed at high levels across all brain regions, but at very low levels in peripheral tissues except in the pancreas. Before the cloning of the *BACE1* gene, β-secretase activity was extensively characterized, mainly in cultured cells. The BACE1 protein identified bears all the properties of β-secretase. It is an aspartyl protease, with optimal pH ~4.5–5.0. Cleavage of APP by BACE1 occurs only at the known β-secretase cleavage sites, Asp 1 and Glu 11 (Vassar *et al.* 1999). Introducing Swedish mutations into APP (mutations at two residues before Asp1 of Aβ, KM/NL) renders it a much better substrate for BACE1 cleavage at the Asp1 of Aβ region.

Kinetic analysis of BACE1 cleavage of Swedish mutant APP shows that K_m for wild-type APP is 15-fold higher than that for APP carrying Swedish mutations (Lin *et al.* 2000). *In vitro*, overexpression of BACE1 results in increased β-secretase cleavage products; knocking down BACE1 reduces the generation of these products (Vassar *et al.* 1999; Yan *et al.* 1999). Specific cleavage of synthetic peptide substrates can be achieved *in vitro* by purified soluble BACE1 from 293T cells (Vassar *et al.* 1999) or by truncated BACE1 (without the C-terminal region) from *Escherichia coli* (Lin *et al.* 2000).

BACE2, the homologue of BACE1, has been identified; it expresses at relatively high levels in heart, kidney, and placenta, but at low levels in brain. It is not clear whether BACE2 is involved in the degradation of Aβ by cleavage within the Aβ region. BACE2 cleaves APP at Asp1 of Aβ as well as at Phe19 and Phe20 within the Aβ region (Farzan *et al.* 2000). The latter cleavage sites (Phe19 and Phe20) are very close to the Flemish mutation at residue 21 (A692G), which significantly increases the Aβ production generated by BACE2 but not by BACE1. Swedish mutations around Asp1 increase Aβ production by both BACE1 and BACE2.

BACE1 knockout mice have the expected depletion of Aβ; therefore BACE1 represents the major β-secretase activity for Aβ generation, and BACE2 does not play significant role in APP processing and Aβ generation (Cai *et al.* 2001; Luo *et al.* 2001). BACE1 has a 21 aa signal peptide, and the pro-BACE1 domain spans amino acids 22–45 (Vassar *et al.* 1999). Mutation of Arg 45 prevents the cleavage of pro-BACE. Since pro-BACE1 is not processed in furin-deficient cells, and *in vitro* incubation of recombinant pro-BACE1 and furin results in the cleavage of the pro-peptide, a furin-like convertase is responsible for the cleavage of pro-BACE1. Upon exiting ER, post-translational modification of BACE1 occurs. Treatment of cells with tunicamycin or *N*-glycosidase F abolishes the N-glycosylation of BACE1. The cytoplasmic tail of BACE1 is required for maturation and trafficking through the Golgi, while a soluble BACE1 molecule without the transmembrane domain and the cytoplasmic tail matures at an enhanced rate compared with full-length BACE1. BACE1 is localized to the Golgi, the trans-Golgi network (TGN), the secretory vesicles, and the cell surface. The cytoplasmic dileucine motif also renders BACE1 to endosomes. The half-life of BACE1 is over 16 hours, and there is no co-localization of BACE1 with lysosomal markers (reviewed by Xia 2001).

BACE1 not only clears APP but also its homologous amyloid precursor-like proteins APLP1 and APLP2. While the cleavage sequences in APP and APLP1/2 are different, a short sequence spanning seven residues from APP, once inserted into a non-BACE1 substrate, is sufficient to convert it to a BACE1 substrate. In addition, the low-density lipoprotein receptor-related protein (LRP) which interacts with and internalizes APP at the cell surface is also shown to be cleaved by BACE1. Similar to APP C99, interaction of BACE1 and LRP leads to an increase in LRP C-terminal fragment, followed by subsequent γ-secretase cleavage (see below) to release the LRP intracellular domain. However, a lack of cleavage of these substrates in BACE1 knockout mice does not cause developmental abnormality (Cai *et al.* 2001; Luo *et al.* 2001).

Development of a variety of BACE1 inhibitors provides new insight into the structure/function of BACE1. Two inhibitors that block BACE1 in the nanomolar range can inhibit cathepsin D and E, and pepsin at even lower concentrations (Gruninger-Leitch *et al.* 2002). The crystal structure of BACE1 complexed with an inhibitor has been elucidated (Hong *et al.* 2000). The large catalytic site (Asp32 and Asp228) has been a challenge in the identification of small molecular weight compounds which can cross the blood–brain barrier. Therefore low molecular weight compounds with greater space-filling capability would be ideal for blocking BACE1 activity. For example, high-throughput screening of compound libraries reveals a series of acyl guanidines which can selectively inhibit BACE1. Guanidine forms direct hydrogen bonds with catalytic Asp residues of BACE1, and they reduce Aβ and C99 without changing the levels of APP and APPs-α. Certain acyl guanidine compounds with nanomolar potency have high specificity for BACE1, compared with cathepsin D and pepsin. However, they do not discriminate human and murine BACE1, and they show same potency towards wild-type and Swedish mutant APP substrates (Cole *et al.* 2006). In conclusion, much progress has been made in the development of potent BACE1 inhibitors, and blocking BACE1 is considered to be one of the most promising therapeutic approaches for Aβ-based therapies. This is largely attributed to the observation that BACE1 knockout mice do not exhibit noticeable phenotypes, and therefore inhibiting BACE1 does not have foreseeable side effects (Cai *et al.* 2001; Luo *et al.* 2001).

17.2.3 Inhibiting γ-secretase

17.2.3.1 Identity of γ-secretase

The γ-secretase complex is composed of PS together with Nicastrin (Yu *et al.* 2000), anterior pharynx defective-1 (Aph-1), and presenilin enhancer-2 (Pen-2) (Francis *et al.* 2002). PS1 and PS2 are 467 and 448 aa polypeptides with ~60% homology. A significant increase of Aβ42 has been detected in the brain tissue and plasma of FAD mutant PS gene carriers. All FAD mutations in PS1 and PS2 studied to date selectively enhance the production of Aβ42 in both transfected cells and the brains of transgenic mice (Xia 2000). Conversely, PS is required for γ-secretase cleavage of APP for Aβ generation, as *in vivo* studies have shown that decreased Aβ production in neurons derived from PS knockout embryos (De Strooper *et al.* 1998). Importantly, cell culture studies have shown that mutations in either of two aspartates in TM domains 6 and 7 of PS1 abolish γ-secretase cleavage of APP (Wolfe *et al.* 1999). The analogous aspartate residues in PS2 are also required for APP processing by γ-secretase. Although PS1 lacks the classical D(T/S)G motif as the active site for proteases, database search reveals that G384/D385 is part of a motif conserved in a bacterial aspartyl protease family called type 4 prepilin peptidases. The conservation of this glycine 384 from bacteria to humans suggests that PS1 is a part of the γ-secretase complex (Steiner *et al.* 2000).

Co-immunoprecipitation studies demonstrate that PS1 and C83/C99 can form complexes in cells expressing endogenous levels of PS1. PS1 and substrates form complexes in Golgi/TGN-type vesicles, the same vesicles in which a major portion of intracellular steady-state Aβ is found and where *de novo* Aβ generation can occur (Xia *et al.* 2000). In addition to PS, which is the main component in the high molecular weight complex,

Nicastrin has been identified as a second important component in the complex (Yu *et al.* 2000). Nicastrin is a type I integral membrane glycoprotein which associates with the CTF fragments of APP and Notch and is required for γ-secretase activity. N-linked glycosylation of Nicastrin in the Golgi apparatus is associated with its entry into the active γ-secretase complex, and this mature form interacts preferentially with the functional PS1 heterodimers. Both PS and Nicastrin are capable of binding to an immobilized transition state analogue γ-secretase inhibitor, and the complex formation is closely associated with γ-secretase activity.

Two additional components of the γ-secretase complex have been revealed by genetic screens in *Caenorhabditis elegans*. The *aph-1* gene encodes a new protein, APH-1, which is predicted to have seven TM domains (Goutte *et al.* 2002), and the *pen-2* gene encodes a small protein, PEN-2, consisting of 101 aa with two TM domains (Francis *et al.* 2002). Downregulation of Aph1 or Pen-2 by RNAi is associated with reduced levels of PS NTF/CTF heterodimers and deficient γ-secretase function. Reducing APH-1 stabilizes the full-length PS1, whereas reducing PEN-2 decreases endoproteolytic processing of PS1. Overexpression of PS, Nicastrin, APH-1, and PEN-2 results in increased PS endoproteolysis and γ-secretase activity in both mammalian cells (Takasugi *et al.* 2003) and yeast cells (Edbauer *et al.* 2003).

17.2.3.2 Essential function of four γ-secretase components

Although the γ-secretase complex contains at least four different components, the odds of identifying a novel target within γ-secretase complex do not seem to be quadruplicated. Knockout of PS1 is embryonic lethal (Shen *et al.* 1997; Wong *et al.* 1997) as PS is required for neurogenesis in mouse embryos. In the adult mouse brain, PS1 is required for hippocampal memory and synaptic plasticity (Saura *et al.* 2004). Similar to PS1/PS2 double-knockout mice, knockout of Nicastrin in mice leads to embryonic lethality, severe patterning defects, and abnormal somite segmentation (J. Li *et al.* 2003; T. Li *et al.* 2003). Knockout of each of the three Aph-1 genes in mice shows different phenotypes (Serneels *et al.* 2005). Knockout of Aph-1A in mice cause a similar phenotype to Notch1, Nicastrin or PS null embryos. It causes defects in angiogenesis, neural tube formation, and somitogenesis. Knockout of Aph-1B or Aph-1C causes very mild phenotypes, and the mice survive into adulthood. Currently there is no knockout model for PEN-2, and knockdown of zebrafish PEN-2 (Pen-2) results in enhanced p53-dependent apoptosis throughout the whole animal (Campbell *et al.* 2006); the loop region of Pen-2 is involved in protecting zebrafish from caspase activation and apoptosis (Zetterberg *et al.* 2006). Therefore blocking the function of each γ-secretase component in animals leads to alteration of critical metabolic pathways, making the γ-secretase complex a difficult target for therapeutic intervention.

17.2.3.3 Multiple substrates for γ-secretase cleavage

Most phenotypes in animals deficient in individual γ-secretase components are associated with the dysfunction of γ-secretase substrates. There are over two dozen γ-secretase substrates, including APP, Notch, ErbB-4, E- and N-cadherin, CD44, LRP, nectin1α,

Delta and Jagged, Glutamate Receptor Subunit 3, APLP1 and APLP2, DCC, p75 neurotrophin receptor, Syndecan3, and Colony Stimulating factor-1 (reviewed by Xia and Wolfe 2003). All the known γ-secretase substrates are type I membrane proteins and are cleaved in their TM domains in a PS-dependent manner. Cleavage of these substrates can be blocked by several well-characterized γ-secretase inhibitors. For APP and Notch, these substrates are cut in the middle of their TM domains and at a residue close to the interface of the membrane and cytoplasm. Cleavage at these two sites by γ-secretase releases the Notch-1-β peptide (Nβ), which is similar to Aβ of APP, and Notch intracellular domain (NICD), which translocates to the nucleus where it regulates gene expression.

Proper Notch signalling is critical to a wide variety of cell fate determinations during embryonic development and adulthood. Currently, most potent γ-secretase inhibitors block Aβ production but also inhibit essential Notch signalling and have severe side effects in therapeutic applications. Although great effort has been devoted to identifying γ-secretase inhibitors which cleave only APP and not Notch, the toxicity of γ-secretase inhibitors (including the induction of apoptosis) is still the most significant problem for therapeutic applications.

17.2.3.4 γ-Secretase inhibitors

A variety of γ-secretase inhibitors have been used to treat cultured cells and animals. When a peptidomimetic γ-secretase inhibitor was used to block Aβ production in Chinese hamster ovary (CHO) cells, the efficacy of this inhibitor was significantly reduced when the cells overexpressed familial AD-linked mutant PS1, suggesting that PS1 is likely to be in the same γ-secretase cleavage complex with the inhibitor (Xia *et al.* 2000). In addition, transition-state analogue affinity reagents for γ-secretase prevent Aβ generation and bind directly to PS1 and PS2 (Esler *et al.* 2000; Li *et al.* 2000). Earlier studies have reported three different γ-secretase inhibitors that bind to PS1 directly. L-685,458, an aspartyl protease transition state mimic, has an IC_{50} of 17 nM for inhibiting Aβ synthesis. Photoactivated L-685,458 can covalently label the N- and C-terminal fragments of PS1. Since the full-length PS1 was not photolabelled, PS1 holoprotein may function as an aspartyl protease zymogen and have to undergo endoproteolysis to be functional (Li *et al.* 2000). Next, a biotinylated and bromoacetylated transition-state analogue affinity compound for γ-secretase is also found to bind to PS1, supporting the proposal that PS1 contains the active site of γ-secretase (Esler *et al.* 2000). Thirdly, a benzophenone analogue of non-transition-state mimic γ-secretase inhibitor was shown to specifically photo-crosslink several major membrane polypeptides, including N- and C-terminal fragments of PS1. This is further confirmed by a reduction of the binding of γ-secretase inhibitors to membranes derived from PS1 knockout embryos (Seiffert *et al.* 2000).

Because γ-secretase cleaves over two dozen substrates, it has been a challenge to develop non-toxic γ-secretase inhibitors. Treating mice with a potent γ-secretase inhibitor for 15 days resulted in the expected reduction in Aβ production, however, lymphocyte development was impaired, and the goblet cell number in intestine was increased with dramatic alteration of tissue morphology (Wong *et al.* 2004). Despite concerns regarding the toxicity of γ-secretase inhibitors, results from clinical trials of γ-secretase inhibitor LY450139 dihydrate seem to be encouraging. After 5 weeks of treatment, LY450139 was well tolerated in 70 AD patients, with a concomitant reduction of plasma Aβ (Siemers *et al.* 2006).

The recent discovery of Gleevec (imatinib mesylate) for inhibiting Aβ production represents a new class of inhibitors that specifically block Aβ production but not Notch cleavage. Gleevec has been shown to reduce Aβ production in both cultured cells and a cell-free system. This reduction was also observed in rat primary neuronal cultures and guinea pig brain. Importantly, Gleevec does not inhibit the γ-secretase-mediated S3 cleavage of Notch-1. Although Gleevec is known to target the ATP-binding site of Abl and several other tyrosine kinases, no difference in Aβ reduction was detected in fibroblasts cultured from wild-type versus $Abl^{-/-}$ mice (Netzer *et al.* 2003). It is believed that the metabolite(s) of Gleevec (instead of Gleevec itself) specifically blocks the γ-secretase cleavage of APP, but not Notch, by an unknown mechanism.

Treating cultured neurons with γ-secretase inhibitor causes an increase in cell surface *N*-methyl-D-aspartate (NMDA) receptors. It is known that Aβ inhibits the induction of NMDA-receptor-dependent long-term potentiation (LTP) via multiple pathways involving activation of microglia and stimulation of nitric oxide synthase and superoxide (Wang *et al.* 2004). In addition, Aβ is able to promote endocytosis of NMDA receptors. Neurons with high levels of Aβ have reduced amounts of cell surface NMDA receptors, and treating cortical neurons with Aβ causes immediate depression of NMDA-evoked currents. Treating neurons with a γ-secretase inhibitor leads to a reduction in Aβ and an increase in surface NMDA receptors (Snyder *et al.* 2005). The FDA approved drug Memantine is a fast voltage-dependent NMDA receptor antagonist. Currently, it is not clear whether the attenuation of NMDA receptor function by Memantine is related to Aβ-induced endocytosis of NMDA receptor.

γ-Secretase can also be inhibited by endogenous modulators such as phospholipase D1 (PLD1). PLD1 catalyses the hydrolysis of phosphatidylcholine to generate choline and phosphatidic acid. This enzymatic activity has been shown to regulate membrane trafficking. Overexpressing PLD1 in FAD-linked mutant PS1 cells rescues impaired neurite outgrowth and APP trafficking from the TGN to the plasma membrane. PLD1 is not only able to promote APP trafficking, but can also inhibit γ-secretase activity (Cai *et al.* 2006). PLD1 was shown to interact with the loop region of PS1, which facilitates the distribution of PLD1 to Golgi/TGN. Interaction between PS1 and PLD1 is specific, as PLD1 does not interact with other γ-secretase components. Overexpression of PLD1 in cultured cells leads to dissociation of the functional γ-secretase complex, resulting in a reduction of γ-secretase cleavage of both APP and Notch. Downregulation of PLD1 by small hairpin RNA increases the association of γ-secretase components and enhances Aβ production. Because PLD1-mediated conversion of phosphatidylcholine to choline affects the levels of choline for acetylcholine synthesis, exploration of PLD1 as a therapeutic target may not only regulate the γ-secretase activity and Aβ production but also contribute to the survival of cholinergic neurons (Cai *et al.* 2006).

17.2.3.5 Activating α-secretase

Historically, activating a protease has not been a favourite choice for therapeutic intervention, and activating α-secretase is not an exception. Studies have indicated that members of the α-disintegrin and metalloprotease (ADAM) family might be responsible for α-secretase activity to initiate a non-amyloidogenic APP processing pathway (Lammich *et al.* 1999). Overexpression of ADAM10 leads to a several-fold increase in α-secretase activity, and a dominant negative form of ADAM10 inhibits endogenous α-secretase activity. The pro-enzyme of ADAM10 is located in the Golgi while the activated form is located at the cell surface. Similarly to APP, α-secretase-like cleavage of Notch is mediated by a member of ADAM family proteins, TNFα-converting enzyme (TACE) (Brou *et al.* 2000). Studies using TACE$^{-/-}$ bone-marrow-derived monocytic precursor cells have shown that TACE-mediated cleavage of Notch is required for subsequent γ-secretase-like proteolytic activation of Notch1 (Mumm *et al.* 2000).

Most intracellular TACE and ADAM-10 activities can be regulated by protein kinase C (PKC), which stimulates α-secretase and shifts more APP undergoing α-secretase cleavage other than β-secretase cleavage. In TGN, TACE/ADAM10 is proteolytically active and can compete with β-secretase for the substrate APP, and PKC promotes α-secretase-mediated cleavage of APP to generate C83. The pituitary adenylate cyclase-activating polypeptide (PACAP) is a neurotrophic peptide and can stimulate the G-protein-coupled receptor PAC1, resulting in an increase of APPs-α secretion. Overexpression of PAC1 receptor also increases α-secretase cleavage of APP. Activation of muscarinic acetylcholine receptors (M_1 and M_3 receptor subtypes) with carbachol increases the basal release of APPs from cultured cells (Nitsch *et al.* 1992). Total Aβ levels are reduced in the cerebrospinal fluid (CSF) from AD patients treated with the muscarinic M_1 agonist AF102B (Nitsch *et al.* 2000). Nevertheless, a clinical trial of AF102B for AD treatment was discontinued because of unwanted side effects. Another muscarinic M_1 agonist, AF267B, has entered clinical trial. It has been shown to rescue the cognitive deficits in a triple-transgenic model of AD (3 × Tg-AD), which was generated by introducing transgenes encoding human Swedish mutant APPsw and Tau$_{P301L}$ into single-cell embryos harvested from homozygous mutant PS1$_{M146V}$ knockin mice. The reduction of plaques and tangles in the AF267B-treated mice was caused by selective activation of α-secretase and suppression of GSK3β, respectively (Caccamo *et al.* 2006). Therefore low molecular weight compounds like AF267B can be peripherally administered to activate non-amyloidogenic pathway and alleviate amyloid burden in the brains of AD patients.

17.2.3.6 Other modulators of Aβ generation

Cholesterol In addition to the three secretases as potential targets to modulate Aβ generation, a number of drugs have been reported to modulate Aβ production, including statins, oestrogen, and non-steroidal anti-inflammatory drugs (NSAIDs). Reduced prevalence of AD was found in patients taking statins. Statins can efficiently reduce levels of Aβ in CSF and brain in guinea pigs (Fassbender *et al.* 2001), and low levels of cholesterol might be able to stimulate α-secretase and promote the non-amyloidogenic pathway (Kojro *et al.* 2001). Furthermore, studies show that levels of cholesteryl ester, but not free cholesterol, are directly correlated to Aβ generation. Inhibitors of acyl-coenzyme A: cholesterol acyltransferase (ACAT), the enzyme responsible for the formation of cholesteryl esters, have been shown to decrease Aβ generation. Therefore changes in cholesterol compartmentalization may modulate Aβ synthesis (Puglielli *et al.* 2001). Further tests of ACAT inhibitor were carried in APP transgenic mice. While brain cholesteryl esters of transgenic mice are decreased as expected, Aβ levels and Aβ plaques are significantly reduced, concomitant with a slight improvement in spatial learning tests (Hutter-Paier *et al.* 2004).

Liver X receptors (LXR) indirectly control cellular cholesterol efflux and membrane composition and are involved in Aβ generation. Treating cells with LXR ligand activators elevates levels of the ATP-binding cassette transporter A1 (ABCA1) and reduces Aβ generation, and overexpression of ABCA1 itself is sufficient to decrease Aβ levels (Sun *et al.* 2003). Specifically, ligand-treated human fibroblasts and mouse primary neurons have shown a dose-dependent decrease in Aβ generation (Koldamova *et al.* 2005). Fibroblasts lacking ABCA1 produce more Aβ than wild-type fibroblasts, supporting a role for ABCA1 in APP processing and Aβ production. Treating APP transgenic mice with LXR ligand for 6 days results in an increase in ABCA1 expression and a significant reduction in the levels of soluble Aβ40 and Aβ42 in brain (Koldamova *et al.* 2005).

Depletion of cholesterol from neurons occurs when it is oxidized by cholesterol 24- and 27-hydroxylase to generate oxysterols. In AD brains, 24-hydroxylase (Cyp46) is expressed around amyloid plaques and in astrocytes, but 27-hydroxylase is present at lower levels around amyloid plaques. Both 24(S)-hydroxycholesterol and 27-hydroxycholesterol inhibit Aβ production in neurons, with the former showing higher potency (over 1000-fold with $IC_{50}=1$ nM). ABCA1 expression can be induced by both 24(S)- and 27-hydroxycholesterol, similar to LXR ligand activators that are shown to induce ABCA1 expression and reduce Aβ secretion (Brown *et al.* 2004). On the other hand, exposure of hippocampal neurons to Aβ leads to the accumulation of ceramide species and cholesterol; treating neurons with α-tocopherol (vitamin E) or an inhibitor of sphingomyelin synthesis blocks accumulation of ceramides and cholesterol, preventing cell death induced by Aβ. α-Tocopherol is under clinical-trial to test its protection of neurons from Aβ-induced membrane oxidative stress and cell death.

NSAIDs Retrospective epidemiological studies have demonstrated a low incidence of AD in people taking NSAIDs. However, rofecoxib and naproxen do not slow cognitive decline in mild to moderate AD (Aisen *et al.* 2003), suggesting that only a subset of NSAIDs may be effective.

NSAIDs inhibit cyclo-oxygenase (COX) and may slow AD progression by reducing the inflammatory response in patient brains. Both ibuprofen and indomethacin are believed to have significant suppressive effects on Aβ-induced inflammatory responses, thus reducing the risk and delaying the onset of AD. NCX-2216, a nitric oxide (NO) releasing derivative of the NSAID flurbiprofen, has an even stronger effect than ibuprofen in decreasing Aβ deposits in brains of transgenic mice, as ibuprofen may only reduce non-fibrillar Aβ deposits but not fibril plaques. Compared with non-NO-releasing

NSAIDs that do not activate microglia, NCX-2216 increases activation of microglia, which might be necessary for Aβ clearance.

Independent of their COX inhibitory effects, a subset of NSAIDS, including ibuprofen, indomethacin, and sulindac sulphide, specifically decrease Aβ42 in cultured cells, with a concomitant increase of Aβ38 (Weggen *et al.* 2001). The mechanism for inhibition of Aβ42 production by NSAIDs is not completely understood. NSAIDs are believed to modulate γ-secretase directly. In a cell-free system non-competitive inhibition of γ-secretase activity was achieved, as a radiolabelled transition-state γ-secretase inhibitor was displaced by NSAIDs. Because γ-secretase substrates are believed to enter the γ-secretase complex first at the docking site followed by proteolysis at the catalytic cleavage site in PS1 (presumably by two aspartate residues at TM 6 and 7 of PS1), NSAIDs may bind to a site other than the docking and active sites of γ-secretase, i.e. the modulating site. Therefore, at least in cultured cells, NSAIDs do not seem to rely on COX to reduce Aβ42 production; instead, they may act directly on the γ-secretase complex. Consistent with these findings, alteration in lipoxygenases, peroxisome-proliferator-activated receptor (PPARg), inhibitor κB kinase β, or nuclear factor κB does not affect Aβ42 production. APP-overexpressing mice treated with pioglitazone, a potent agonist of PPARγ, only showed a slightly reduction in Aβ levels with no effect on amyloid plaque burden or microglia activation, consistent with the findings that PPARγ activation is not involved in the Aβ-lowering effect of NSAIDs (Yan *et al.* 2003). Therefore these non-COX targets do not seem to be involved in NSAID-mediated reduction of Aβ generation. On the other hand, overexpression of human COX in cultured cells results in an increase of Aβ40 and Aβ42, along with the AICD; this effect can be suppressed by the inhibition of non-selective COX inhibitor ibuprofen or γ-secretase inhibitor.

Another mechanism by which NSAIDs might selectively reduce Aβ42 could be through the inhibition of the small GTPase Rho (Zhou *et al.* 2003). Among the NSAIDs examined, only those that are capable of inhibiting Rho are shown to reduce Aβ42. One inhibitor of Rho-associated kinase (ROCK) clearly decreased the levels of Aβ42 in brains of transgenic mice overexpressing APP (Zhou *et al.* 2003). However, in cultured cells, two ROCK inhibitors reduced total Aβ secretion without selective reduction of Aβ42 levels. In contrast with certain NSAIDs that reduce Aβ42 production with a concomitant increase in Aβ38 generation, ROCK inhibitors do not increase Aβ38 levels in cultured cells. These ROCK inhibitors also fail to reduce Aβ production in an *in vitro* γ-secretase activity assay (Leuchtenberger *et al.* 2006). While the ROCK inhibitor has been shown to reduce Aβ42 *in vivo* (Zhou *et al.* 2003), it is not clear whether the ROCK pathway is the main mechanism for its modulation in Aβ generation.

NSAIDs may also reduce Aβ production through prostaglandins. NSAIDs inhibit the enzymatic activity of COX-1 and inducible COX-2, which catalyse the first step in the synthesis of prostaglandins. The prostaglandin E_2 (PGE$_2$) E prostanoid subtype 2 (EP2) receptor is involved in the development of the innate immune response in brain. In primary neuronal cultures, EP2 and EP4 agonists were neuroprotective against Aβ42 toxicity by increasing intracellular cAMP concentration. Stimulation of EP2 and EP4 receptors attenuates the increase of reactive oxygen species induced by Aβ (Echeverria *et al.* 2005).

Deletion of PGE_2 EP2 receptor in APP/PS transgenic mice results in a significant reduction of lipid peroxidation. A marked decrease in levels of Aβ40 and Aβ42 and amyloid deposition was found, which is associated with the reduction of oxidative stress. Lower levels of APP CTF were found in EP2-receptor-deficient mice, suggesting that PGE_2 signalling through the EP2 receptor increases Aβ generation via enhancement of BACE1 activity. Therefore the EP2 receptor can serve as a target to modulate Aβ production (Liang *et al*. 2005).

Treatment of transgenic mice overexpressing APP with the NSAID ibuprofen over a period of 16 weeks resulted in significant reduction of SDS-soluble Aβ42 with less effect on Aβ40. A 60% reduction of amyloid plaque load was observed in the cortex of the treated animals. In addition, measurement of CD45 and CD11b has shown that microglial activation is suppressed upon the treatment of ibuprofen. Therefore chronic NSAID treatment of a transgenic animal can efficiently reduce Aβ levels, amyloid plaque burden, and microglial activation in brains (Yan *et al*. 2003). Furthermore, among a number of NSAIDs tested in APP transgenic mice, purified R enantiomer of flurbiprofen (R-flurbiprofen) significantly reduces Aβ42 (Eriksen *et al*. 2003). Currently, both ibuprofen and R-flurbiprofen have entered phase III clinical trials for the treatment of AD.

Oestrogen Oestrogen has been shown to reduce the generation of Aβ in cultured rat, mouse, and human embryonic neurons (Xu *et al*. 1998). In both neuroblastoma cells and primary neurons, oestrogen was found to stimulate formation of vesicles from the TGN by recruiting soluble trafficking factors like Rab11. Enhanced formation of APP-containing vesicles from TGN and facilitation of APP trafficking through the late secretory pathway minimize the amount of Aβ generated in the TGN. Oestrogen is also involved in Aβ clearance, as it may enhance the internalization of Aβ by microglial cells, which can be inhibited by certain oestrogen receptor antagonists. Through the oestrogen receptor, oestrogen can function as a hormone-activated transcription complex and directly increase gonadotrophin-releasing hormone-1 (GnRH-1) neuronal activity in brain, where its release is essential for reproduction (Temple *et al*. 2004). Because of the complex function of oestrogen, clinical trials have generated conflicting results. A 5 year follow-up study among healthy postmenopausal women showed that overall health risks exceeded benefits from hormone replacement therapy (HRT). For cognitive improvement, oestrogen and HRT have not proved successful in the clinic (Craig *et al*. 2005), and latest clinical studies of specific cognitive functions continue to show both possible beneficial and detrimental effects of HRT (Resnick *et al*. 2006).

Lithium Lithium, which has been used for treatment of bipolar disorder, has shown an effective reduction of Aβ production and accumulation in cultured cells and brains of mice overexpressing APP. Among many activities, Li^+ targets glycogen synthase kinase-3α (GSK3α) and blocks Aβ production without affecting the cleavage of Notch at low millimolar concentrations. Another GSK-3 inhibitor, kenpaullone, which is structurally unrelated to lithium, also reduces both Aβ40 and Aβ42. Since kenpaullone also inhibits cyclin-dependent kinases (cdks), this class of compounds is apparently not ideal for therapeutic application. On the other hand, depletion of GSK3α by RNAi efficiently reduces

Aβ levels, while overexpressing GSK3α causes a dose-dependent increase in Aβ levels, indicating that GSK3α is the regulator for Aβ production (Phiel *et al.* 2003). Although serum lithium levels from mice (~1 mM) are within the therapeutic range for bipolar disorder patients under the treatment of lithium, side effects of lithium treatment are wide-ranging. This is not unexpected as GSK-3 has a broad spectrum of substrates, and blocking GSK-3 activity will disrupt normal metabolism at different levels.

17.3 Enhancing Aβ degradation

17.3.1 Insulin-degrading enzyme and neutral endopeptidase

A number of Aβ-degrading enzymes have been identified, and *in vitro* and *in vivo* studies have shown that these proteases play important roles in degrading Aβ. In differentiated pheochromocytoma cells (PC12) and primary rat cortical neurons, Aβ was found to be degraded by the protease insulin-degrading enzyme (IDE). Both cell-associated and released forms of IDE can degrade Aβ from cultured cells, and purified IDE can degrade Aβ peptide *in vitro*. A point mutation at the catalytic site of IDE (E111Q) abolishes its Aβ-degrading activity. Crossing APP transgenic mice with transgenic mice overexpressing IDE shows a clear decrease of Aβ levels and plaque numbers in adult mice, providing *in vivo* evidence that IDE degrades Aβ and prevents plaque formation (Leissring *et al.* 2003).

A second protease was identified by examining candidate proteases and tracing Aβ-degradation fragments upon injecting radiolabelled Aβ peptide into rat hippocampus. Neutral endopeptidase, also known as neprilysin (NEP), was identified as the principal protease degrading Aβ42 in rat. Degradation of Aβ42 could be prevented by infusing an NEP-specific inhibitor, resulting in Aβ42 deposition in these brains (Iwata *et al.* 2000). Similar to IDE, crossing NEP transgenic mice with APP transgenic mice significantly decreases Aβ load in brains of adult mice (Leissring *et al.* 2003).

17.3.2 Plasmin, angiotensin-converting enzyme, and HrtA1

In addition to IDE and NEP, a number of proteins have been associated with Aβ degradation, including plasmin, angiotensin-converting enzyme (ACE), and HrtA1. Early studies have shown that Aβ-like peptide can bind tissue-type plasminogen activator (tPA), which cleaves plasminogen to generate plasmin. Plasmin is a serine protease which cleaves Aβ monomers, oligomers, and fibrils *in vitro*. Aβ40 and oligomeric and fibrillar Aβ42 can stimulate tPA-mediated plasminogen activation, inducing cell-associated plasmin generation. Plasminogen activator inhibitor-1 (PAI-1) can inhibit tPA to generate plasmin. In mice, elevated Aβ in brain correlates with the upregulation of PAI-1. Injection of Aβ into hippocampus of tPA or plasminogen-deficient mice induces PAI-1 expression. The injected Aβ is stable in knockout mice, compared with the short half life of injected Aβ in wild-type mice. These results suggest that tPA-plasmin is involved and assists at certain levels in Aβ clearance *in vivo* (Melchor *et al.* 2003), and inhibitors of PAI-1 may enhance proteolytic clearance of Aβ.

Expression of human ACE leads to degradation of secreted Aβ40 and Aβ42 from cultured cells (Hemming and Selkoe 2005). While *in vitro* analysis suggests that the

N-terminal domain of ACE may degrade Aβ, introduction of individual mutations at the active sites at N- and C-domains within ACE fails to degrade Aβ. Inhibition of ACE with a prescribed drug, captopril, results in an accumulation of Aβ in the media of cultured cells. Another protease, HrtA1, belongs to a widely conserved family of serine proteases involved in various aspects of protein quality control and cell fate. HrtA1 co-localizes with Aβ deposits in human brain, and inhibition of HrtA1 leads to an accumulation of Aβ in cultured astrocyte cells (Grau et al. 2005).

In addition to amyloid-degrading proteases, brain microglia cells are known to be involved in Aβ clearance. Recent studies have shown that some originate from the bone marrow. These cells can migrate towards Aβ plaques, mainly because of attraction by Aβ42. These microglia cells are able to eliminate Aβ by phagocytosis, which provides a completely novel therapeutic opportunity for bone marrow stem cells to remove Aβ deposit in brains of AD patients (Simard et al. 2006).

17.4 Clearing Aβ from brain by vaccine

Active and passive vaccines have been explored in preclinical tests and clinical trials. The groundbreaking discovery was made with a vaccination of APP transgenic mice using a synthetic aggregated form of Aβ42 (AN-1792). Immunization of mice with Aβ42 effectively prevented Aβ plaque formation, neuritic dystrophy, and astrogliosis in adult brains (Schenk et al. 1999), and later studies have shown that memory loss in those APP transgenic mice vaccinated with Aβ was clearly reduced (Janus et al. 2000; Morgan et al. 2000). However, the phase II trial of AN-1792 was discontinued because several patients developed aseptic meningoencephalitis. Following reports of meningoencephalitis (overall 18/300 (6%)), immunization was stopped after one injection (two patients), two injections (274 patients), or three injections (24 patients). Over 20% of these Aβ42-immunized patients developed antibody response; additional injection increased the percentage of antibody responders to close to 60%. CSF samples from this group of patients showed lower tau levels ($n = 11$) than those from placebo subjects ($n = 10$). Antibodies produced in these patients recognize Aβ plaques, diffuse deposits, and vascular Aβ deposit in blood vessels, but not soluble Aβ42 or full-length APP. Antibodies in the CSF from one of the patients who developed aseptic meningoencephalitis showed similar specificity, indicating that the symptom was unlikely to be related to an antibody response against cellular brain structures other than Aβ deposits (Hock et al. 2002). The cause of the meningoencephalitis in the patients who received the vaccine might be due to a T-cell-mediated autoimmune response; if interferon γ is expressed in mouse brain, Aβ immunization can induce temporary meningoencephalitis (Monsonego et al. 2006). A newly designed vaccine contains the immunodominant B-cell epitope of Aβ with the pan-HLADR-binding peptide (PADRE). This PADRE-Aβ1–15 does not have the T-cell epitope of Aβ, and immunization of mice with PADRE-Aβ1–15 produced high titres of anti-Aβ antibodies (Agadjanyan et al. 2005).

Furthermore, antibodies generated in mice through immunization with aggregated Aβ42 recognize residues 4–10 of Aβ, and these antibodies can inhibit Aβ fibril formation without causing inflammatory response (McLaurin *et al.* 2002). Studies have also shown that the Aβ vaccine can stimulate IgG receptor (FcR) mediated phagocytic clearance of Aβ by microglial phagocytosis, However, pro-inflammatory cytokines suppress this process through NF-κB activation, which can be counteracted by ibuprofen (Koenigsknecht-Talboo and Landreth 2005).

An alternative approach using peripheral administration of antibodies against Aβ has been shown to reduce Aβ plaque formation in several lines of APP transgenic mice (Bard *et al.* 2000; DeMattos *et al.* 2001). Two possible mechanisms are implicated. First, antibodies may trigger microglial cells to clear plaques through Fc-receptor-mediated phagocytosis, as supported by an *ex vivo* assay with sections of brain tissue (Bard *et al.* 2000). Secondly, brain-soluble Aβ might be drained to plasma because of disequilibrium between the CNS and plasma Aβ levels. A rapid increase in plasma Aβ was found after peripheral administration of Aβ antibody in mice (DeMattos *et al.* 2001). Currently, the humanized monoclonal antibody against Aβ has entered clinical trials. In addition, a novel Aβ immuno-conjugate with an Aβ fragment attached to a carrier protein intended to cross the blood–brain barrier has entered clinical trials. It is designed to induce an antibody response against Aβ directly in the CNS. Studies have shown that single-chain variable fragments (scFvs) can be used as an alternative to Aβ antibody vaccine to decrease Aβ toxicity. An scFv binding Aβ17–28 can inhibit aggregation of Aβ *in vitro*, in contrast with an scFv binding Aβ29–40 that failed to inhibit Aβ aggregation. Using human neuroblastoma cells SY5Y, scFv against Aβ17–28 eliminates the toxic effects of aggregated Aβ (Liu *et al.* 2004).

While vaccine holds great hope as a promising AD therapy, it is still essential to immunize patients at early stage. Immunization of the triple transgenic model of AD ($3 \times$ Tg-AD) with vaccine has demonstrated an efficient removal of Aβ plaques and early tau pathology, however, Aβ antibody vaccine fails to clear late-stage hyperphosphorylated Tau and NFT formation (Oddo *et al.* 2004–2006). Furthermore, passive immunization may not be suitable for AD patients with cerebral amyloid angiopathy, because in a particular line of APP transgenic mice modelled for spontaneous hemorrhagic stroke, passive immunization caused an increase in small haemorrhages in brain areas with amyloid deposits in blood vessels (Pfeifer *et al.* 2002).

17.5 Interfering with Aβ oligomerization

17.5.1 Aβ oligomerization

Although the major component of neuritic plaques in brains of AD patients is fibrillar Aβ, an increasing number of studies have demonstrated that low molecular weight Aβ oligomers, instead of high molecular weight fibril Aβ, have the most damaging effect on neurons.

After Aβ is generated, both intracellular and extracellular (secreted) Aβ can undergo oligomerization to form low molecular weight oligomers. Studies using synthetic Aβ peptides show an intermediate Aβ assembly form, protofibrils, which share a similar secondary structure to the mature amyloid fibrils. The protofibril has been found to acutely alter the electrical activity of neurons and gradually cause neuronal loss. Like protofibrils, Aβ-derived diffusible ligands (ADDLs), which have higher molecular weight than oligomers, have also been shown to cause toxicity in neurons. Levels of ADDLs in AD brains are 70-fold higher than control brains, and they attach to cultured hippocampal neurons (Gong *et al.* 2003).

It is possible that the majority of Aβ oligomers derive from initial intracellular Aβ dimerization. Nanomolar concentration of Aβ oligomers (primarily dimers) can be detected in the conditioned media of CHO cells stably overexpressing APP, and incubation of this conditioned media at 37°C does not generate new Aβ oligomers. Dimers can also be detected in conditioned media of CHO cells expressing mutant APP or PS gene, which contain higher levels of Aβ42. Intracellularly, small amounts of Aβ dimers can be detected in both non-neuronal and neuronal cell lines and in primary cultured human neurons. Dimers are also detectable in some CSF samples from both control and AD patients. In 3 × Tg-AD mice, oligomerization of Aβ seems to occur intraneuronally first, and Aβ oligomers gradually accumulate during aging (Oddo *et al.* 2006). Antibodies against Aβ oligomers stain neuronal cell bodies, and the binding sites co-localize with the synaptic marker post-synaptic density protein 95 (PSD-95). Aβ oligomers have been shown to inhibit hippocampal LTP once injected into animals (Walsh *et al.* 2002), and studies in APP transgenic mice have demonstrated that Aβ oligomers are responsible for memory loss (Lesne *et al.* 2006).

17.5.2 Inhibitors for Aβ oligomerization

Designing compounds to inhibit Aβ oligomerization and aggregation is a challenge. There are many unknown factors that play important roles in accelerating Aβ oligomerization from monomers. Such factors have been detected in the supernatants of autopsy-derived neocortical homogenates from AD patients. Aβ-immunoreactive senile plaques and vascular deposits are found in the hippocampus and neocortex of APP transgenic mice which are infused *in vivo* with supernatants prepared from AD cortex; mice infused with supernatants derived from control subject do not exhibit immunoreactive senile plaques. It is not clear which factor(s) from brains of AD patients accelerate Aβ deposition in these mouse brains. One candidate is Aβ42, which may have a 'seeding' effect for plaque formation. Separate studies have indicated that ApoE might be another contributing factor. First, mutant APP V717F transgenic mice expressing mouse or human *ApoE* develop Aβ deposition and neuritic degeneration, while APP transgenic mice lacking *ApoE* show less Aβ deposition and neuritic dystrophy. Secondly, mice carrying *ApoE*4 have more fibrillar Aβ deposits than mice carrying *ApoE*3, indicating an allele-specific effect of *ApoE* on Aβ deposition and fibril formation (Holtzman *et al.* 2000).

A number of compounds which block the process of Aβ oligomerization and aggregation have been identified. A low molecular weight hydroxyanaline derivative was found to block oligomerization of synthetic Aβ *in vitro* and naturally secreted Aβ from living cells. In contrast with Aβ-oligomer-containing media from control cells that inhibit LTP (Walsh *et al.* 2002), media from compound-treated cells no longer inhibit LTP because there is a lack of Aβ oligomers (Walsh *et al.* 2005). Compounds containing a disrupting element (three or more lysines) and residue Aβ 16–10 (KLVFF) disrupt Aβ aggregation and protect neurons from toxic Aβ aggregates (Lowe *et al.* 2001). A peptide containing *N*-methyl amino acid in alternating, but not consecutive, positions of Aβ sequence (Aβ16–22m) probably binds to the seeding site for Aβ aggregation and inhibits fibrillogenesis of Aβ (Gordon *et al.* 2001). On the basis of findings on certain 'plant extracts' capable of inhibiting Aβ fibrillogenesis, a basement membrane protein, laminin, was found to bind Aβ1–40 and prevent Aβ fibril formation, as well as partially disassembling a preformed Aβ fibril. The globular repeats on the laminin A chain might be the potential Aβ binding site to prevent Aβ fibril formation (Castillo *et al.* 2000). However, it is not clear whether compounds shown to disassemble the existing fibrils will result in increased neuronal toxicity because of the conversion of fibril Aβ to more toxic oligomeric Aβ.

Small molecules have been synthesized for binding to chaperones and interacting with Aβ, and these inhibitors were tested for Aβ aggregation (Gestwicki *et al.* 2004). One peptide chaperone, mini-αA-crystallin, contains an Aβ-interacting domain and a complex solubilizing domain. Mini-αA-crystallin was shown to stop Aβ fibril formation and reduce the toxic effect of Aβ on PC12 cells (Santhoshkumar and Sharma 2004). Another small molecule drug, CPHPC, was developed to interact with serum amyloid P component (SAP), which is a non-fibrillar plasma glycoprotein present in amyloid deposits. *In vitro* assay shows that SAP binds to and protects amyloid from degradation by proteolysis or phagocytic cells (Pepys *et al.* 2002). CPHPC, once injected into patients with systemic amyloidosis, can interact with SAP, resulting in a depletion of circulating SAP which can no longer protect amyloid from degradation.

In vitro, curcumin was found to inhibit aggregation of Aβ and prevent dose-dependent Aβ fibril formation detected by electron microscope. Incubation of curcumin with mouse brain sections reveals labelling of amyloid plaques. *In vivo*, peripherally injected curcumin could cross the blood–brain barrier and bind plaques in brains of APP transgenic mice. Furthermore, curcumin, when fed to APP transgenic mice, could label plaques and reduce Aβ levels and plaque burdens. These data suggest that curcumin can not only prevent Aβ oligomer and fibril formation but also disaggregate Aβ fibril (Yang *et al.* 2005).

Earlier studies have shown that Aβ aggregation requires copper and zinc, and metal-protein attenuating compounds (MPAC) have been explored to remove metal and prevent Aβ aggregation. Treatment of APP transgenic mice with clioquinol (a copper and zinc chelator) was shown to reduce Aβ accumulation in brains (Cherny *et al.* 2001). Clioquinol is an oral antibiotic which was withdrawn in the early 1970s because of overdose-associated side effects. Although vitamin B_{12} supplementation can suppress the side effects occurring in clioquinol-treated animals, modifications are needed to improve identification of clioquinol-related compounds. Although a clinical trial of clioquinol was recently stopped, a more potent derivative has entered clinical trial for the treatment of AD. Another compound, tramiprosate (3-amino-1-propanesulphonic acid; Alzhemed), has entered phase III clinical trials. It prevents Aβ aggregation *in vitro* and reduces plaque deposition in animals (Aisen 2005). The drug is taken orally and was shown to be safe in phase II trials. Most people with mild AD who received the drug for up to 16 months showed stable cognitive function tests. Apparently, results from a number of clinical trials will emerge in the near future, which will reveal the status of our endeavours to discover amyloid-based therapies for AD.

17.6 Conclusion

Many therapeutic approaches based on understanding of amyloid metabolism and secretase properties are being explored. Current scientific evidence continues to support the hypothesis that the accumulation of amyloid is the primary cause of AD pathogenesis, and increased Aβ production leads to a series of sequential events, including enhanced Aβ oligomerization, inflammatory responses, and neuronal damage. Blocking Aβ generation has been the first choice for amyloid-based therapies. Elucidating the crystal structure of BACE1 was a critical step for obtaining optimal inhibitors. Inhibitors designed to block this rate-limiting enzyme for Aβ production will undoubtedly reduce the Aβ burden in brains with minimum side effects, as mice deficient in BACE1 seem to be normal. Blocking PS1-associated γ-secretase activity raises concerns about interference with the signalling by Notch and other critical γ-secretase substrates. Since potent γ-secretase inhibitors are available to eliminate Aβ production, screening for non-toxic γ-secretase inhibitors is a natural step for drug development. Although several key proteases have been identified for their involvement in Aβ degradation, enhancing protease activity to eliminate Aβ seems to be more challenging than inhibiting activity *in vivo*, as they may have unwanted effects on the catabolism of other proteins. Despite the setback in clinical trials, the initial success of Aβ vaccine to prevent Aβ plaque formation, neuritic dystrophy, and astrogliosis in APP transgenic mice provides a complete new route for interfering with amyloid deposition and clearance. Tremendous efforts have been devoted to searching for a modified safer vaccine, and both active and passive immunization have been explored to reduce amyloid deposition. Compounds blocking Aβ oligomerization have clearly prevented neurotoxicity in cultured cells and animals, as increasing evidence suggests that Aβ oligomers and protofibrils/ADDLs might be the toxic Aβ species that cause neuronal dysfunction. However, it may not be desirable to develop drugs that disassemble fibrillar Aβ to form oligomers and protofibril/ADDLs. Advances in understanding molecular mechanisms for pathogenesis of AD will continue to expand the array of amyloid-based therapies that constitute the most promising interventions to prevent and slow the progression of AD.

Acknowledgements

This work was supported in part by a National Institute of Health Grant (AG015379), the author would like to thank Dr W.J. Ray (Merck Research Laboratories) for helpful discussions.

References

Agadjanyan MG, Ghochikyan A, Petrushina I, *et al.* (2005). Prototype Alzheimer's disease vaccine using the immunodominant B cell epitope from beta-amyloid and promiscuous T cell epitope pan HLA DR-binding peptide. *Journal of Immunology* **174**, 1580–6.

Aisen PS.(2005). The development of anti-amyloid therapy for Alzheimer's disease: from secretase modulators to polymerisation inhibitors. *CNS Drugs* **19**, 989–96.

Aisen PS, Schafer, KA, Grundman, M, *et al.* (2003). Effects of rofecoxib or naproxen vs placebo on Alzheimer disease progression: a randomized controlled trial. *Journal of the American Medical Association* **289**, 2819–26.

Bard F, Cannon C, Barbour R, *et al.* (2000). Peripherally administered antibodies against amyloid beta-peptide enter the central nervous system and reduce pathology in a mouse model of Alzheimer disease. *Nature Medicine* **6**, 916–19.

Brou C, Logeat F, Gupta N, *et al.* (2000). A novel proteolytic cleavage involved in Notch signaling: the role of the disintegrin-metalloprotease TACE. *Molecular Cell* **5**, 207–16.

Brown J, 3rd, Theisler C, Silberman S, *et al.* (2004). Differential expression of cholesterol hydroxylases in Alzheimer's disease. *Journal of Biological Chemistry* **279**, 34674–81.

Caccamo A, Oddo S, Billings LM, *et al.* (2006). M_1 receptors play a central role in modulating AD-like pathology in transgenic mice. *Neuron* **49**, 671–82.

Cai H, Wang Y, McCarthy D, *et al.* (2001). BACE1 is the major beta-secretase for generation of Abeta peptides by neurons. *Nature Neuroscience* **4**, 233–4.

Cai D, Netzer W, Zhong M, *et al.* (2006). Presenilin-1 uses phospholipase D1 as a negative regulator of beta-amyloid formation. *Proceedings of the National Academy of Sciences of the USA* **103**, 1941–6.

Campbell WA, Yang HW, Zetterberg H *et al.* (2006). Zebrafish lacking Alzheimer Presenilin Enhancer 2 (Pen-2) demonstrate excessive p53 dependent apoptosis and neuronal loss. *Journal of Neurochemistry* **96**, 1423–1440.

Castillo GM, Lukito W, Peskind E, *et al.* (2000). Laminin inhibition of beta-amyloid protein (Abeta) fibrillogenesis and identification of an Abeta binding site localized to the globular domain repeats on the laminin a chain. *Journal of Neuroscience Research* **62**, 451–62.

Cherny RA, Atwood CS, Xilinas ME, *et al.* (2001). Treatment with a copper–zinc chelator markedly and rapidly inhibits beta-amyloid accumulation in Alzheimer's disease transgenic mice. *Neuron* **30**, 665–76.

Cole D, Manas E, Stock J, *et al.* (2006). Acylguanidines as small-molecule β-secretase inhibitors. *Journal of Medicinal Chemistry* **49**, 6158–61.

Craig MC, Maki PM, and Murphy DG (2005). The Women's Health Initiative Memory Study: findings and implications for treatment. *Lancet Neurology* **4**, 190–4.

DeMattos RB, Bales KR, Cummins DJ, Dodart JC, Paul SM., and Holtzman DM (2001). Peripheral anti-Aβ antibody alters CNS and plasma A beta clearance and decreases brain A beta burden in a mouse model of Alzheimer's disease. *Proceedings of the National Academy of Sciences of the USA* **98**, 8850–5.

De Strooper B, Saftig P, Craessaerts K, *et al.* (1998). Deficiency of presenilin-1 inhibits the normal cleavage of amyloid precursor protein. *Nature* **391**, 387–90.

Echeverria V, Clerman A, and Dore S (2005). Stimulation of PGE receptors EP2 and EP4 protects cultured neurons against oxidative stress and cell death following beta-amyloid exposure. *European Journal of Neuroscience* **22**, 2199–206.

Edbauer D, Winkler E, Regula JT, Pesold B, Steiner H, and Haass C (2003). Reconstitution of gamma-secretase activity. *Nature Cell Biology* **5**, 486–488.

Eriksen JL, Sagi SA, Smith TE, *et al.* (2003). NSAIDs and enantiomers of flurbiprofen target gamma-secretase and lower Abeta 42 *in vivo*. *Journal of Clinical Investigation* **112**, 440–9.

Esler WP, Kimberly WT, Ostaszewski BL, *et al.* (2000). Transition-state analogue inhibitors of gamma-secretase bind directly to presenilin-1. *Nature Cell Biology* **2**, 428–34.

Farzan M, Schnitzler CE, Vasilieva N, Leung D, and Choe H (2000). BACE2, a beta-secretase homolog, cleaves at the beta site and within the amyloid-beta region of the amyloid-beta precursor protein. *Proceedings of the National Academy of Sciences of the USA* **97**, 9712–17.

Fassbender K, Simons M, Bergmann C, *et al.* (2001). Simvastatin strongly reduces levels of Alzheimer's disease beta -amyloid peptides Abeta 42 and Abeta 40 *in vitro* and *in vivo*. *Proceedings of the National Academy of Sciences of the USA* **98**, 5371–3.

Francis R, McGrath G, Zhang J, *et al.* (2002). *aph-1* and *pen-2* are required for Notch pathway signaling, γ-secretase cleavage of bAPP and presenilin protein accumulation. *Developmental Cell* **3**, 85–97.

Gestwicki JE, Crabtree GR, and Graef IA (2004). Harnessing chaperones to generate small-molecule inhibitors of amyloid beta aggregation. *Science* **306**, 865–9.

Gong Y, Chang L, Viola KL, *et al.* (2003). Alzheimer's disease-affected brain: presence of oligomeric A beta ligands (ADDLs) suggests a molecular basis for reversible memory loss. *Proceedings of the National Academy of Sciences of the USA* **100**, 10417–22.

Gordon DJ, Sciarretta KL, and Meredith SC (2001). Inhibition of beta-amyloid(40) fibrillogenesis and disassembly of beta-amyloid(40) fibrils by short beta-amyloid congeners containing N-methyl amino acids at alternate residues. *Biochemistry* **40**, 8237–45.

Goutte C, Tsunozaki M, Hale VA, and Priess JR (2002). APH-1 is a multipass membrane protein essential for the Notch signaling pathway in *Caenorhabditis elegans* embryos. *Proceedings of the National Academy of Sciences of the USA* **99**, 775–9.

Grau S, Baldi A, Bussani R, Tian X, *et al.* (2005). Implications of the serine protease HtrA1 in amyloid precursor protein processing. *Proceedings of the National Academy of Sciences of the USA* **102**, 6021–6.

Gruninger-Leitch F, Schlatter D, Kung E, Nelbock P, and Dobeli H (2002). Substrate and inhibitor profile of BACE (β-secretase) and comparison with other mammalian aspartic proteases. *Journal of Biological Chemistry* **277**, 4687–93.

Hemming ML and Selkoe DJ (2005). Amyloid beta-protein is degraded by cellular angiotensin-converting enzyme (ACE) and elevated by an ACE inhibitor. *Journal of Biological Chemistry* **280**, 37644–50.

Hock C, Konietzko U, Papassotiropoulos A, *et al.* (2002). Generation of antibodies specific for b-amyloid by accination of patients with Alzheimer disease. *Nature Medicine* **8**, 1270–5.

Holtzman DM, Bales KR, Tenkova T, *et al.* (2000). Apolipoprotein E isoform-dependent amyloid deposition and neuritic degeneration in a mouse model of Alzheimer's disease. *Proceedings of the National Academy of Sciences of the USA* **97**, 2892–7.

Hong L, Koelsch G, Lin X, *et al.* (2000). Structure of the protease domain of memapsin 2 (beta-secretase) complexed with inhibitor. *Science* **290**, 150–3.

Hutter-Paier B, Huttunen HJ, Puglielli L, *et al.* (2004). The ACAT inhibitor CP-113,818 markedly reduces amyloid pathology in a mouse model of Alzheimer's disease. *Neuron* **44**, 227–38.

Iwata N, Tsubuki S, Takaki Y, *et al.* (2000). Identification of the major Abeta1–42-degrading catabolic pathway in brain parenchyma: suppression leads to biochemical and pathological deposition. *Nature Medicine* **6**, 143–50.

Janus C, Pearson J, McLaurin J, *et al.* (2000). A beta peptide immunization reduces behavioural impairment and plaques in a model of Alzheimer's disease. *Nature* **408**, 979–82.

Koenigsknecht-Talboo J, and Landreth GE (2005). Microglial phagocytosis induced by fibrillar beta-amyloid and IgGs are differentially regulated by proinflammatory cytokines. *Journal of Neuroscience* **25**, 8240–9.

Kojro E, Gimpl G, Lammich S, Marz W, and Fahrenholz F (2001). Low cholesterol stimulates the nonamyloidogenic pathway by its effect on the alpha -secretase ADAM 10. *Proceedings of the National Academy of Sciences of the USA* **98**, 5815–20.

Koldamova RP, Lefterov IM, Staufenbiel M, *et al.* (2005). The liver X receptor ligand T0901317 decreases amyloid beta production *in vitro* and in a mouse model of Alzheimer's disease. *Journal of Biological Chemistry* **280**, 4079–88.

Lammich S, Kojro E, Postina R, *et al.* (1999). Constitutive and regulated alpha-secretase cleavage of Alzheimer's amyloid precursor protein by a disintegrin metalloprotease. *Proceedings of the National Academy of Sciences of the USA* **96**, 3922–7.

Leissring MA, Farris W, Chang AY, *et al.* (2003). Enhanced proteolysis of beta-amyloid in APP transgenic mice prevents plaque formation, secondary pathology, and premature death. *Neuron* **40**, 1087–93.

Lesne S, Koh MT, Kotilinek L, *et al.* (2006). A specific amyloid-beta protein assembly in the brain impairs memory. *Nature* **440**, 352–7.

Leuchtenberger S, Kummer MP, Kukar T, *et al.* (2006). Inhibitors of Rho-kinase modulate amyloid-beta (Abeta) secretion but lack selectivity for Abeta42. *Journal of Neurochemistry* **96**, 355–65.

Li J, Fici GJ, Mao CA, *et al.* (2003). Positive and negative regulation of the gamma-secretase activity by nicastrin in a murine model. *Journal of Biological Chemistry* **278**, 33445–9.

Li T, Ma G, Cai H, Price DL, and Wong PC (2003). Nicastrin is required for assembly of presenilin/gamma-secretase complexes to mediate Notch signaling and for processing and trafficking of beta-amyloid precursor protein in mammals. *Journal of Neuroscience* **23**, 3272–7.

Li Y-M, Xu M, Lai M-T, *et al.* (2000). Photoactivated γ-secretase inhibitors directed to the active site covalently label presenilin 1. *Nature* **405**, 689–94.

Liang X, Wang Q, Hand T, *et al.* (2005). Deletion of the prostaglandin E2 EP2 receptor reduces oxidative damage and amyloid burden in a model of Alzheimer's disease. *Journal of Neuroscience* **25**, 10180–7.

Lin X, Koelsch G, Wu S, Downs D, Dashti A, and Tang J (2000). Human aspartic protease memapsin 2 cleaves the beta-secretase site of beta-amyloid precursor protein. *Proceedings of the National Academy of Sciences of the USA* **97**, 1456–60.

Liu R, Yuan B, Emadi S, *et al.* (2004). Single chain variable fragments against beta-amyloid (Abeta) can inhibit Abeta aggregation and prevent abeta-induced neurotoxicity. *Biochemistry* **43**, 6959–67.

Lowe TL, Strzelec A, Kiessling LL, and Murphy RM (2001). Structure–function relationships for inhibitors of beta-amyloid toxicity containing the recognition sequence KLVFF. *Biochemistry* **40**, 7882–9.

Luo Y, Bolon B, Kahn S, *et al.* (2001). Mice deficient in BACE1, the Alzheimer's beta-secretase, have normal phenotype and abolished beta-amyloid generation. *Nature Neuroscience* **4**, 231–2.

McLaurin J, Cecal R, Kierstead ME, *et al.* (2002). Therapeutically effective antibodies against amyloid-beta peptide target amyloid-beta residues 4–10 and inhibit cytotoxicity and fibrillogenesis. *Nature Medicine* **8**, 1263–9.

Melchor JP, Pawlak R, and Strickland S (2003). The tissue plasminogen activator-plasminogen proteolytic cascade accelerates amyloid-beta (Abeta) degradation and inhibits Abeta-induced neurodegeneration. *Journal of Neuroscience* **23**, 8867–71.

Monsonego A, Imitola J, Petrovic S, *et al.* (2006). Aβ-induced meningoencephalitis is IFN-γ-dependent and is associated with T cell-dependent clearance of Aβ in a mouse model of Alzheimer's disease. *Proceedings of the National Academy of Sciences of the USA* **103**, 5048–53.

Morgan D, Diamond DM, Gottschall PE, *et al.* (2000). A beta peptide vaccination prevents memory loss in an animal model of Alzheimer's disease. *Nature* **408**, 982–5.

Mumm JS, Schroeter EH, Saxena MT, *et al.* (2000). A ligand-induced extracellular cleavage regulates gamma-secretase-like proteolytic activation of Notch1. *Molecular Cell* **5**, 197–206.

Netzer WJ, Dou F, Cai D, *et al.* (2003). Gleevec inhibits beta-amyloid production but not Notch cleavage. *Proceedings of the National Academy of Sciences of the USA* **100**, 12444–9.

Nitsch RM, Slack BE, Wurtman RJ, and Growdon JH (1992). Release of Alzheimer amyloid precursor derivatives stimulated by activation of muscarinic acetylcholine receptors. *Science* **258**, 304–7.

Nitsch RM, Deng M, Tennis M, Schoenfeld D, and Growdon JH (2000). The selective muscarinic M₁ agonist AF102B decreases levels of total Abeta in cerebrospinal fluid of patients with Alzheimer's disease. *Annals of Neurology* **48**, 913–18.

Oddo S, Billings L, Kesslak JP, Cribbs DH, and LaFerla FM (2004). Abeta immunotherapy leads to clearance of early, but not late, hyperphosphorylated tau aggregates via the proteasome. *Neuron* **43**, 321–32.

Oddo S, Caccamo A, Tran L, *et al.* (2006). Temporal profile of amyloid-beta (Abeta) oligomerization in an *in vivo* model of Alzheimer disease: a link between Abeta and tau pathology. *Journal of Biological Chemistry* **281**, 1599–1604.

Pepys MB, Herbert J, Hutchinson WL, *et al.* (2002). Targeted pharmacological depletion of serum amyloid P component for treatment of human amyloidosis. *Nature* **417**, 254–9.

Pfeifer M, Boncristiano S, Bondolfi L, *et al.* (2002). Cerebral hemorrhage after passive anti-abeta immunotherapy. *Science* **298**, 1379.

Phiel CJ, WC, Lee VM, and Klein PS (2003). GSK-3alpha regulates production of Alzheimer's disease amyloid-beta peptides. *Nature* **423**, 435–9.

Puglielli L, Konopka G, Pack-Chung E, *et al.* (2001). Acyl-coenzyme A: cholesterol acyltransferase modulates the generation of the amyloid beta-peptide. *Nature Cell Biology* **3**, 905–12.

Resnick SM, Maki PM, Rapp SR, *et al.* (2006). Effects of combination estrogen plus progestin hormone treatment on cognition and affect. *Journal of Clinical Endocrinology and Metabolism* **91**, 1802–10.

Santhoshkumar P and Sharma KK (2004). Inhibition of amyloid fibrillogenesis and toxicity by a peptide chaperone. *Molecular and Cellular Biochemistry* **267**, 147–55.

Saura CA, Choi SY, Beglopoulos V, *et al.* (2004). Loss of presenilin function causes impairments of memory and synaptic plasticity followed by age-dependent neurodegeneration. *Neuron* **42**, 23–36.

Schenk D, Barbour R, Dunn W, *et al.* (1999). Immunization with amyloid-β attenuates Alzheimer-disease-like pathology in the PDAPP mouse. *Nature* **400**, 173–7.

Seiffert D, Bradley JD, Rominger CM, *et al.* (2000). Presenilin-1 and -2 are molecular targets for gamma-secretase inhibitors. *Journal of Biological Chemistry* **275**, 34086–91.

Serneels L, Dejaegere T, Craessaerts K, *et al.* (2005). Differential contribution of the three Aph1 genes to γ-secretase activity *in vivo*. *Proceedings of the National Academy of Sciences of the USA* **102**, 1719–24.

Shen J, Bronson RT, Chen DF, Xia W, Selkoe DJ, and Tonegawa S (1997). Skeletal and CNS defects in presenilin-1 deficient mice. *Cell* **89**, 629–39.

Siemers ER, Quinn JF, Kaye J, *et al.* (2006). Effects of a gamma-secretase inhibitor in a randomized study of patients with Alzheimer disease. *Neurology* **66**, 602–4.

Simard AR, Soulet D, Gowing G, Julien JP, and Rivest S (2006). Bone marrow-derived microglia play a critical role in restricting senile plaque formation in Alzheimer's disease. *Neuron* **49**, 489–502.

Sinha S, Anderson JP, Barbour R, *et al.* (1999). Purification and cloning of amyloid precursor protein beta-secretase from human brain. *Nature* **402**, 537–40.

Snyder EM, Nong Y, Almeida CG, *et al.* (2005). Regulation of NMDA receptor trafficking by amyloid-beta. *Nature Neuroscience* **8**, 1051–8.

Steiner H, Kostka M, Romig H, *et al.* (2000). Glycine 384 is required for presenilin-1 function and is conserved in bacterial polytopic aspartyl proteases. *Nature Cell Biology* **2**, 848–51.

Stokin GB, Lillo C, Falzone TL, *et al.* (2005). Axonopathy and transport deficits early in the pathogenesis of Alzheimer's disease. *Science* **307**, 1282–8.

Sun Y, Yao J, Kim TW, and Tall AR (2003). Expression of liver X receptor target genes decreases cellular amyloid beta peptide secretion. *Journal of Biological Chemistry* **278**, 27688–94.

Takasugi N, Tomita T, Hayashi I, *et al.* (2003). The role of presenilin cofactors in the gamma-secretase complex. *Nature* **422**, 438–41.

Temple JL, Laing E, Sunder A, and Wray S (2004). Direct action of estradiol on gonadotropin-releasing hormone-1 neuronal activity via a transcription-dependent mechanism. *Journal of Neuroscience* **24**, 6326–33.

Vassar R, Bennett BD, Babu-Khan S, *et al.* (1999). Beta-secretase cleavage of Alzheimer's amyloid precursor protein by the transmembrane aspartic protease BACE. *Science* **286**, 735–41.

Walsh DM, Klyubin I, Fadeeva JV, *et al.* (2002). Naturally secreted oligomers of amyloid beta protein potently inhibit hippocampal long-term potentiation *in vivo*. *Nature* **416**, 535–9.

Walsh DM, Townsend M, Podlisny MB, *et al.* (2005). Certain inhibitors of synthetic amyloid beta-peptide (Abeta) fibrillogenesis block oligomerization of natural Abeta and thereby rescue long-term potentiation. *Journal of Neuroscience* **25**, 2455–62.

Wang Q, Rowan MJ, and Anwyl R (2004). Beta-amyloid-mediated inhibition of NMDA receptor-dependent long-term potentiation induction involves activation of microglia and stimulation of inducible nitric oxide synthase and superoxide. *Journal of Neuroscience* **24**, 6049–56.

Weggen S, Eriksen JL, Das P, *et al.* (2001). A subset of NSAIDs lower amyloidogenic Abeta42 independently of cyclooxygenase activity. *Nature* **414**, 212–16.

Wolfe MS, Xia W, Ostaszewski BL, Diehl TS, Kimberly WT, and Selkoe DJ (1999). Two transmembrane aspartates in presenilin-1 required for presenilin endoproteolysis and γ-secretase activity. *Nature* **398**, 513–17.

Wong GT, Manfra D, Poulet FM, *et al.* (2004). Chronic treatment with the gamma-secretase inhibitor LY-411,575 inhibits beta-amyloid peptide production and alters lymphopoiesis and intestinal cell differentiation. *Journal of Biological Chemistry* **279**, 12876–82.

Wong PC, Zheng H, Chen H, *et al.* (1997). Presenilin 1 is required for Notch1 and DII1 expression in the paraxial mesoderm. *Nature* **387**, 288–92.

Xia W (2000). Role of presenilin in gamma-secretase cleavage of amyloid precursor protein. *Experimental Gerontology* **35**, 453–460.

Xia W. (2001). Amyloid metabolism and secretases in Alzheimer's disease. *Current Neurology and Neuroscience Reports* **1**, 422–7.

Xia W and Wolfe M (2003). Intramembrane proteolysis by presenilin and presenilin-like proteases. *Journal of Cell Science* **116**, 2839–44.

Xia W, Ostaszewski BL, Kimberly WT, *et al.* (2000). FAD mutations in presenilin-1 or amyloid precursor protein decrease the efficacy of a γ-secretase inhibitor: a direct involvement of PS1 in the γ-secretase cleavage complex. *Neurobiology of Disease* **7**, 673–81.

Xu H, Gouras GK, Greenfield JP, *et al.* (1998). Estrogen reduces neuronal generation of Alzheimer β amyloid peptides. *Nature Medicine* **4**, 447–51.

Yan Q, Zhang J, Liu H, *et al.* (2003). Anti-inflammatory drug therapy alters beta-amyloid processing and deposition in an animal model of Alzheimer's disease. *Journal of Neuroscience* **23**, 7504–9.

Yan R, Bienkowski MJ, Shuck ME, *et al.* (1999). Membrane-anchored aspartyl protease with Alzheimer's disease beta-secretase activity. *Nature* **402**, 533–7.

Yang F, Lim GP, Begum AN, *et al.* (2005). Curcumin inhibits formation of amyloid beta oligomers and fibrils, binds plaques, and reduces amyloid *in vivo*. *Journal of Biological Chemistry* **280**, 5892–901.

Yu G, Nishimura M, Arawaka S, *et al.* (2000). Nicastrin modulates presenilin-mediated notch/glp-1 signal transduction and betaAPP processing. *Nature* **407**, 48–54.

Zetterberg H, Campbell WA, Yang HW, and Xia W (2006). The cytosolic loop of the gamma-secretase component presenilin enhancer 2 (Pen-2) protects zebrafish embryos from apoptosis. *Journal of Biological Chemistry* **281**, 11933–9.

Zhou Y, Su Y, Li B, *et al.* (2003). Nonsteroidal anti-inflammatory drugs can lower amyloidogenic Abeta42 by inhibiting Rho. *Science* **302**, 1215–17.

Chapter 18

Future directions: the Aβ amyloid pathway as the target for diagnosing, preventing, or treating Alzheimer's disease*

Colin L. Masters

18.1 Introduction

Future research into Alzheimer's disease (AD) will continue to be dominated by efforts to develop rational therapeutic strategies based on our progressive understanding of the molecular basis of synaptic degeneration in AD. In this chapter, the current status of discovering targets and developing drugs for use in the amyloid precursor protein/amyloid beta (APP/Aβ) pathway (Fig. 18.1) is reviewed. The 2005–2006 literature is surveyed, which includes several useful reviews (Benson 2005; Bloom *et al.* 2005; Golde 2005; Higuchi *et al.* 2005; Pangalos *et al.* 2005; Selkoe 2005; Hilbush *et al.* 2005; Wisniewski and Frangione 2005). Also reviewed are two public data bases of clinical trials which enumerate the current status of clinical activity in this area (Novartis International, maintained by Herrling and colleagues (Kwon and Herrling 2005) and the Clinical Trials registry sponsored by the NIH (www.clinicaltrials.gov)).

Why is there so much interest in the APP/Aβ pathway? The theory which underlies this pathway as the principal and proximal causal mechanism in AD is pinned to two critical series of observations: first, mutations in the gene encoding APP (and the presenilin (PS) genes as components of the γ-secretase machinery) are causally linked to early-onset familial AD; secondly, genetically engineered mice with these mutations recapitulate the human disease. More recently, a very tight association between the mean age at onset of pedigrees with PS-mutation-related familial AD and the ratio of secreted $A\beta_{40}$ to $A\beta_{42}$ has emerged (Duering *et al.* 2005). This, together with the development of a robust Aβ-neuroimaging ligand (a thioflovin T analogue), which as a biomarker clearly differentiates AD and mild cognitive impairment (MCI) from normal controls and other neurological diseases (Mathis *et al.* 2005; Rowe *et al.*, submitted for publication), adds much more strength to the Aβ theory. However, the single most important challenge to test the theory remains the demonstration that a drug targeting the APP/Aβ pathway actually modifies the

---

*This chapter is adapted and updated from Masters and Beyreuther (2006).

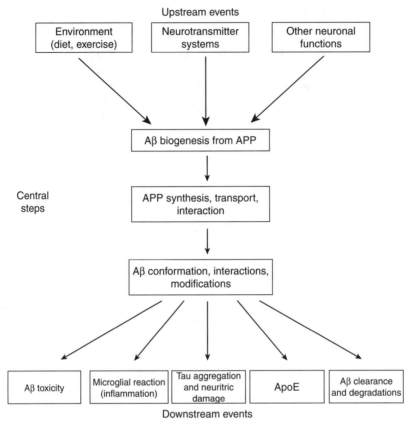

Fig. 18.1 Schematic outline of the upstream and downstream events which surround the central APP/Aβ pathway.

natural history of the disease. To this end, the criteria set out by Cummings (2006), and listed in Table 18.1, have clarified the standards to be met when we come to assessing this test of the Aβ theory of AD. The first criterion (a plausible mechanism of action in a validated model) has been achieved by many of the therapeutic strategies reviewed below. However, no drug has yet met any of the other four criteria, although we remain optimistic that the current pace of activity will deliver a result in the not too distant future.

Table 18.1 Criteria for AD modification by drug intervention

1. Plausible mechanism of action in a validated model
2. Clinical trial evidence based on the Lerber staggered start design
3. Difference in survival to a meaningful clinical outcome
4. Change in rate (slope) of decline
5. Demonstrable drug-placebo difference on an accepted biomarker of disease progression

After Cummings 2006.

18.2 Upstream events in the APP/Aβ pathway

The targets derived from the APP/Aβ pathway outlined in Fig. 18.1 are listed in more detail in Table 18.2. While it is not a comprehensive or exhaustive listing, it does present a novel and logical way of classifying the wide range of current research activity being undertaken in this area.

18.2.1 Age and environmental factors

Of all the external variables which determine risk of getting AD, age and the environment stand out as factors which demand explanations. Yet for all their obviousness, no reasonable explanations have been forthcoming. While many of the biochemical events listed in the APP/Aβ pathway are known to be developmentally regulated, very little information is yet available on what happens under normal ageing conditions. Partial loss of function of a critical biochemical reaction would seem to be a good starting point for investigation, either as an upstream or downstream event, or a 'double-hit' phenomenon could be invoked, as seen in the early development of ideas on oncogenesis. Whichever, the incontrovertible link between ageing and AD remains obscure in mechanistic terms.

Similarly, the interactions between the environment and the risk for AD have attracted many epidemiological studies. Diet and exercise remain as the two most interesting variables. General caloric restriction has often been associated with longevity in rodent models of ageing, and recent studies in transgenic AD models (Patel *et al.* 2005; J. Wang *et al.* 2005) and normal rodents (Tang 2005) suggest an effect on Aβ plaque load or α-secretase processing of APP. The effects of exercise (Adlard *et al.* 2005) and environmental enrichment (Jankowsky *et al.* 2005a; Lazarov *et al.* 2005) have also been examined in transgenic AD models with encouraging results. One group (Lazarov *et al.* 2005) found a change in a downstream event—an increase in the enzymatic activity of neprilysin, an Aβ-degrading protease, in response to environmental enrichment. These downstream events are discussed in section 18.4.6.

Specific dietary intakes, especially naturally occurring anti-oxidants (see section 18.4.1) or metal ions (section 18.3.4.1) remain largely under-investigated as AD risk factors. As methods for diagnosis and population-based screening improve (using plasma biomarkers or specific ligands of Aβ for neuroimaging), it will become more feasible to examine the dietary risk profiles of discrete populations analytically, overcoming current limitations on sensitivity and specificity of case-ascertainment. A surprising study (Youm *et al.* 2005) has already pre-empted dietary modulation of AD through the consumption of transgenic Aβ-expressing potatoes! The proposed mechanism involves low-level immune-mediated clearance of Aβ deposits (section 18.4.5).

18.2.2 The effect of modulation of neurotransmitter systems on amyloid precursor protein processing

Acetylcholinesterase (AChE) was discovered to be present in AD amyloid plaques 40 years ago, and the activity of choline acetyl transferase (ChAT) was found to be decreased

Table 18.2 AD therapeutic targets derived from the APP/Aβ pathway

Pathway step/Event	Target/Drug
A. *Upstream events*	
Ageing	
Environment	Exercise, diet
Neurotransmitter systems modulation	Cholinergic: AChE/BuChE inhibition
	Glutamatergic: NMDA antagonism
	Serotonergic: antidepressants
Other cerebral or general systemic factors	
B. *Central steps: Aβ biogenesis from APP*	
APP gene target	
APP interactions, transport	X11 gene target, growth factors, metal homeostasis, ZnT3, oestrogen, cholesterol
APP proteolytic processing	
• γ-Secretase (PS, Aph1, PEN2, Nct) and associated ε and ζ cleavages	Inhibitor, modulator (NSAID), gene target
• β-Secretase (BACE 1)	Inhibitor, modulator/interaction (PAR 4), immunomodulations of β-cleavage site, gene target
• α-Secretase	Stimulation (PKC activation)
Aβ and its varied conformations	
• Monomers/dimer/trimer (A_4, A_8, A_{12})	
▪ Metal binding sites	Metal-protein attenuating compounds (MPACs)
▪ GAG binding sites	
▪ Oxidative modifications	Di-tyrosine, methionine oxidation
▪ Aβ-lipid interactions	Lipid–protein attenuating compounds (LPACs)
▪ Aβ-protein interactions	Protein–protein attenuating compounds (PPACs)
• β-oligomers/protofibrils	Anti-aggregants/disaggregants
• Polymers/fibrils	Anti-fibrillogenics/defibrillants
C. *Downstream events*	
Aβ-induced 'toxicity' through oxidative damage (protein, mitochondria, lipid, sterols, nucleic acid, etc)	Anti-oxidants, natural product screens, oestrogen
Anti-inflammatory (anti-microglia)	Sigma 1 receptor, PPARγ, PGE_2, NSAID, iNOS inhibition
Aβ-tau direct/indirect interactions	Microtubule stabilizers, kinase, inhibitors, anti-aggregants, etc.
Aβ-ApoE interactions	Statins/cholesterol
Aβ-clearance/neutralization	Immunization, immunomodulation
Aβ-degradation	IDE, NPE, ACE

in the AD brain 30 years ago. From these observations the cholinergic hypothesis/ theory of AD arose, which led to the development of AChE inhibitors (AChE-I) as a therapeutic strategy, with apparent success, despite the lack of any plausible explanation for the presence of AChE in plaques and the underlying loss of ChAT. A paradox then emerged: subjects treated with AChE-I responded with a compensatory increase in

AChE levels, and subjects with a clinical response showed the largest increases (Davidsson *et al.* 2001). This might have been expected to negate the intended effect of the AChE-I on the availability of ACh for cholinergic transmission. At the same time, clinical trials of AChE-Is and their meta-analyses continued to show favourable, albeit mild, effects on cognitive parameters, at least during the first 6–12 months of treatment. Against this background, basic and clinical investigators have recently turned their attention towards other possible mechanisms of action of the AChE-Is, especially on the APP/Aβ pathway, and have begun to ask whether these drugs might have any disease-modifying effects (Caccamo *et al.* 2006).

Various aspects of AChE-I actions on the upstream and downstream APP/Aβ pathway have been reported: attenuating the effects of Aβ-induced neuronal cytoxicity (Kimura *et al.* 2005a), promoting α-secretase or decreasing β-secretase activity (Zimmermann *et al.* 2005; Caccamo *et al.* 2006), inhibiting Aβ aggregation (Belluti *et al.* 2005; Bolognesi *et al.* 2005), or inhibiting GSK 3β activity and tau phosphorylation (Caccamo *et al.* 2006). One group found no effect on Aβ amyloid plaque load while still improving behavioural deficits in a transgenic mouse model (Dong *et al.* 2005), while another group found that inhibitors of butyrylcholinesterase had a lowering effect on cellular APP and Aβ and brain Aβ in transgenic mice (Greig *et al.* 2005).

The modulation of glutamatergic transmission in AD has also received increasing attention with the results of the memantine clinical trials aimed at blocking (non-competitively) the action of NMDA receptors. With the growing awareness that the toxic soluble oligomers of Aβ may inhibit long-term potentiation (LTP) at the pre-synaptic level and that Aβ promotes the endocytosis of the *N*-methyl-D-aspartate (NMDA) receptor (mediated in part through α-7 nicotinic receptor, protein phosphatase PP2B, and striatal enriched protein tyrosine phosphatase (STEP) (Snyder *et al.* 2005)), the findings that memantine has beneficial behavioural effects in both Aβ toxicity models (Yamada *et al.* 2005) and APP transgenic mouse models (Van Dam *et al.* 2005) requires further work which might tie all these observations together.

Finally, behavioural intervention with antidepressants has also been explored in relation to *in vitro* APP processing (Pákáski *et al.* 2005). It would seem less likely that the tricyclics or serotonin-reuptake inhibitors will ever be subjected to AD-modification trials, but if further preclinical studies emerge showing effects on APP processing, an argument could be made for additional clinical studies in the early phases of AD.

18.2.3 Other cerebral or general systemic factors

It is likely that there will be many other upstream factors which play into the APP/Aβ pathway, but few have been identified to date. A particularly contentious area has been the role of the vascular supply to the brain and the effects of ischaemia (atherosclerosis) and hypertension. Historically, this has deep roots, going back to the days when 'arteriosclerosis' was thought to cause all forms of dementia. Similarly, head trauma has been considered as a risk factor for AD, and APP has been identified as a sensitive marker of axonal damage following traumatic brain injury. However, neither hypoxia nor trauma has yet been shown to be a major risk factor for AD, and neither has been shown to promote the long-term amyloidogenic processing of APP.

18.3 **Central steps in the APP/Aβ pathway**

18.3.1 **Targeting the *APP* gene or genes with products interacting directly with amyloid precursor protein**

With the advent of RNA interference (RNAi) silencing, it is to be expected that attempts at direct *APP* gene regulation will emerge. As a forerunner to this, models in which the overexpressed human *APP* transgene in mice can be downregulated with doxycline provide a proof of principle that rapid control over Aβ expression and deposition can be obtained without gross adverse side effects (Jankowsky *et al.* 2005b). Unexpectedly, Aβ deposits formed before the onset of downregulation of APP seemed to be remarkably stable, indicating that any treatment of this type in isolation might have to be administered early in the natural history of AD. Using RNAi techniques in transfected cell lines (Xie *et al.* 2005b), targeting the *X11* gene (APAB) successfully increased APP C-terminal fragments and lowered Aβ levels; *X11* is a known interactor with the cytoplasmic domain of APP, and presents a novel method of possibly modulating γ-secretase cleavage.

18.3.2 **Amyloid precursor protein-interacting systems**

As a presumptive cell surface receptor, APP probably has ligands and effector mechanisms for signal transduction. Nearly 200 proteins have been reported as having direct interactions with APP. Suspected ligands in the extracellular domain include growth factors (nerve growth factor (NGF) in particular), heparin-containing extracellular matrix, metals (through the extracellular Cu–Zn binding domain) and APP itself through hetero- and homodimerization. Small compounds such as propentofyline (Chauhan *et al.* 2005) can affect NGF release, and through this modulate the amyloidogenic pathway. Other small compounds may bind directly to APP (Espeseth *et al.* 2005) and affect its processing.

A controversial area involves the effects of hormones (oestrogens and testosterone especially) and how they may affect APP metabolism. Conflicting results in experimental models have appeared, in which oestrogen deficiency exacerbates Aβ in the APP23 transgenic model (Yue *et al.* 2005) but neither oestrogen deprivation nor replacement affected Aβ deposition in the PDAPP transgenic model (Green *et al.* 2005). Clearly, further studies are required for unravelling this important area where there is an epidemiological impression that females have a higher incidence of AD than males (this impression does not appear to have ever been subjected to a prospective analytical epidemiological study). The mechanisms through which oestrogen/testosterone might act remain obscure, but include oestrogen-dependent regulation of metal homeostasis in the brain through the expression of the neuronal zinc transporter ZnT3.

Cholesterol and inhibitors of cholesterol synthesis (statins) have been shown to significantly alter APP processing *in vitro*, with a reduction in β-secretase cleavage and lessened Aβ production. While some early phase clinical trials with stains have shown encouraging results (Masse *et al.* 2005), others have not (Höglund *et al.* 2005a,b). Cholesterol-independent effects have also been noted for statins acting on isoprenyl intermediates in the cholesterol biosynthetic pathways, with a putative anti-inflammatory effect induced by reactive microglia (Cole *et al.* 2005; Cordle and Landreth 2005). This might conflict with the current theory that microglia are involved in the beneficial process of clearing Aβ deposits (sections 18.4.2 and 18.4.5). Statins have also been implicated in the toxic gain of function of Aβ interacting with α7-nicotinic AChR (Si *et al.* 2005), although the mechanism for this remains unclear.

If cholesterol eventually does prove to be a risk factor for AD, then the observations (Papassotiropoulos *et al.* 2005) of an association between AD and the expression levels and haplotypes of the 5' region of the cholesterol 25-hydroxylase (*CH25H*) gene on chromosome 10 may provide a plausible explanation—one in which cerebral cholesterol metabolism (as distinct from systemic cholesterol and its association with atherosclerosis) directly plays into the APP processing and transport pathways.

18.3.3 Amyloid precursor protein proteolytic processing

The biogenesis of Aβ has been the prime validated drug target for AD since the discovery of the proteolytic processing of APP in 1987 (providing fertile ground for nearly 20 years of intensive research). Molecular details of the C-terminal cleavage (γ-secretase) were the first to emerge, followed by the α- and β-cleavage mechanisms. Subsequent elucidation of δ-, ε-, and ξ-cleavages has added another layer of complexity. Drug discovery programmes reflect this sequence of events; many large pharmaceutical companies have γ-secretase inhibitors or modulators in clinical development, while the β-secretase inhibitors are several years behind, largely in preclinical discovery.

18.3.3.1 γ-Secretase inhibitors and modulators

During 2005, the first published reports of *in vivo* γ-secretase inhibition/modulation of Aβ$_{42}$ biogenesis appeared. One of the first known inhibitors, N-[N-(3,5-difluorophenacetyl-L-alanyl)]-S-phenylglycine t-butyl ester (DAPT), was shown to be effective in acute experiments in behavioural tests (contextual fear conditioning) in the Tg 2576 AD mouse model (Comery *et al.* 2005). Modifications to the chemical structure of DAPT have improved its delivery to the brain (Quéléver *et al.* 2005), as with other compounds (Laras *et al.* 2005), in the hope of achieving lower effective dosages, thus minimizing the risk of adverse peripheral effects. Many diverse classes of inhibitors and modulators are showing very favourable acute pharmacokinetics, with rapid lowering of plasma and CSF Aβ levels (Anderson *et al.* 2005; Barten *et al.* 2005; Best *et al.* 2005; Grimwood *et al.* 2005; Lanz *et al.* 2005; Peretto *et al.* 2005). Importantly, there is now strong evidence linking plasma and CSF Aβ levels, indicating that the brain/CSF pool of Aβ is at least in part a significant proportion of the plasma Aβ pool. There are still methodological issues in measuring Aβ, using either ELISA or western blotting techniques (which soluble oligomeric species are being measured, and what forms of Aβ: total, Aβ$_{40}$, or Aβ$_{42}$?). Nevertheless, these preliminary data offer some hope that plasma Aβ species may eventually prove to be a reliable marker of cerebral Aβ turnover. Further explorations of the properties of γ-secretase inhibitors are revealing unanticipated effects on synaptic function (Dash *et al.* 2005). New classes of γ-secretase inhibitors/modulators continue to be disclosed (Gundersen *et al.* 2005; Lewis *et al.* 2005; Ravi Keerti *et al.* 2005) as part of the effort to develop compounds devoid of side effects. The major concern is the inhibition of signalling in the Notch pathway, which affects cellular differentiation (Curry *et al.* 2005; van Es *et al.* 2005). Ironically, γ-secretase inhibitor compounds originally developed for AD are now being trialled in phase II studies of acute lymphoblastic leukaemia (Clinical Trials.gov NCT00100152) and advanced breast cancer (Clinical Trials.gov NCT00106145).

The first in-human phase I results to be published (Siemers *et al.* 2005, 2006) have shown that the Lilly compound LY450139 achieved a significant lowering of plasma Aβ, but not CSF Aβ, in normal volunteers (up to 50 mg/day for 14 days) or subjects with AD (up to 40 mg/day for 6 weeks). The drug was well tolerated. Higher dosages may be required to achieve a reduction in CSF levels. The results of phase II studies with readouts on cognitive variables are eagerly awaited. In the meantime, further research on the mechanistic operations of the γ-secretase complex (Sato *et al.* 2005) may lead to new paths of drug discovery, as might gene targeting of presenilin, PEN-2, APH-1, and nicastrin lead to selective regulation of γ-secretase activity (Saura *et al.* 2005; Xie *et al.* 2005a).

18.3.3.2 β-Secretase (BACE) inhibitors

Although approximately 5 years behind the development of the γ-secretase inhibitors, much progress has been made in the discovery and design of compounds which target the active site of BACE-1. Improved assays (Pietrak *et al.* 2005) and structural-based *in silico* designs (Ghosh *et al.* 2005; Hanessian *et al.* 2005; Huang *et al.* 2005; Polgár and Keserü 2005; Turner *et al.* 2005) have added to the existing pipeline of drugs in early pre-clinical development (Kimura *et al.* 2005b; Kornacker *et al.* 2005; Lefranc-Jullien *et al.* 2005; Pietrancosta *et al.* 2005) or early discovery programmes (Byun *et al.* 2005; H.J. Lee *et al.* 2005). Other proteins interacting with BACE-1 may become drug targets (Xie and Guo 2005), and gene targeting of BACE-1 mRNA using siRNA is also producing encouraging preliminary results (Singer *et al.* 2005). As with γ-secretase, unanticipated side effects on other BACE-1 substrates or downstream consequences of BACE-1 inhibition may prove difficult to circumvent. As a consolation, inhibitors of BACE-1 may also turn out to have anti-angiogenic and anti-neoplastic activities (Paris *et al.* 2005).

18.3.4 Drugs targeting Aβ and its varied conformations

18.3.4.1 Monomers (A_4), dimers (A_8), and trimers (A_{12})

In contrast with the inhibition of Aβ biogenesis, therapeutic strategies which directly target Aβ itself should inherently have a lower risk of throwing up unanticipated side effects as the accumulated Aβ molecule is restricted to AD. However, if the Aβ fragment (or its domain within APP) does subserve some critical normal function, targeting Aβ itself might interfere with this function and thereby lead to adverse side effects, but to date a normal function for Aβ has not been identified. APP knockout mice are viable and healthy, providing some support for this idea.

Current models of the physical state of Aβ are evolving. Whilst resident in the membrane, Aβ is assumed to be in a α-helical conformation. Following sequential β- and γ- cleavages, Aβ as a monomer (A_4), dimer (A_8), or perhaps even a trimer (A_{12}), is translocated into the extra-cytosolic space and may undergo transition there into a β-strand enriched structure. These structures may then progress towards β-oligomers/protofibrils through to polymers/fibrils of amyloid filaments.

The mechanisms through which Aβ causes damage to neurons ('the toxic gain of function') are slowly emerging. There are many theories: the two most favoured include the ability of Aβ to generate oxidative stress and the hydrophobic interaction of Aβ with lipid membranes, particularly the synaptic plasma membrane. Our current working model incorporates both theories: we have defined a metal binding domain near the N-terminus of Aβ which is capable of binding Zn^{2+} (which causes Aβ to precipitate) or redox-active Cu^{2+}. When Cu^{2+} binds Aβ, it not only causes a significant increase in insolubility, but also induces a series of electron transfers which result in histidine bridge formation, tyrosine 10 radicalization, di-tyrosine cross-linking, and oxidation of methionine 35. Ultimately, in the presence of reductants, this results in the production of H_2O_2 and hydroxyl radicals, capable of inflicting short-range oxidative damage on proteins, lipids, sterols, nucleic acids, etc. Our studies show that toxicity to neurons in culture is associated with the ability of Aβ to associate with the lipid head-group on the outer surface of the plasma membrane.

If this schema is only partially correct, then it is clear that any therapeutic strategy targeting Aβ directly might have multiple routes, many intersecting and overlapping. Thus, targeting the metal binding site of Aβ might relate to Aβ in one or more of its varied conformations (α-helix, β-strand, β-sheet; A_4, A_8, A_{16} versus higher-order oligomers versus polymerized fibril) or whilst interacting with other proteins or lipids.

In consideration of targeting the metal binding site on Aβ, we have developed the concept of a a metal–protein attenuating compound (MPAC), in distinction to the more widely known term of metal chelator. The MPAC has relatively weak binding constants for metals and is able to compete with the target site for the metal ion. As a consequence, an MPAC should not alter the general homeostasis of metal ions in the whole animal. In contrast, a metal chelator has high, effectively irreversible, binding constants for metal ions. A chelator might affect the metal binding to Aβ through deletion of the total pool of bioavailable metal, but is not necessarily expected to interact with the Aβ metal binding site itself.

The utility of MPACs in AD has been initiated with studies of clioquinol, an 8-OH quinoline, with encouraging preclinical (Cherny *et al.* 2001; Raman *et al.* 2005) and early phase II clinical (Ritchie *et al.* 2003; Ibach *et al.* 2005) results. Other groups have considered chelators (Liu *et al.* 2005a; Gaeta and Hider 2005; Sutoh *et al.* 2005) or other novel compounds (Cui *et al.* 2005; Ji and Zhang 2005; Zheng *et al.* 2005a,b). Our own studies have progressed with a new chemical entity based around the 8-OH quinoline structure. This compound, PBT2 (Prana Biotechnology), has passed phase I and will soon commence phase II clinical development.

Additional binding sites on Aβ, such as the glycosaminoglycan (GAG) site (HHQK (13–16)), have been targeted with compounds such as 3-amino-1-propanesulfonic acid (3-APS (Alzhemed), Neurochem Inc.). The results of early clinical trials have been released by the company, with some effects seen on CSF $A\beta_{42}$, but none on ADAS-Cog or MMSE. A large phase III study is under way, coupled to an open-label extension study. The double-blind study results are expected in January 2007.

We have identified other structural changes or mechanisms of toxicity for Aβ which include the oxidative modifications of Tyr10 and Met35, the interaction of Aβ with the polar head groups of the lipid bilayer, and the interaction of Aβ with other proteins. These areas remain very much in the early discovery phase and may deliver lipid–protein attenuating compounds (LPACs) or protein–protein attenuating compounds (PPACs) (Mettenburg *et al.* 2005; S.P. Yang *et al.* 2005).

18.3.4.2 β-oligomers/protofibrils/and polymers/fibrils of Aβ

For a long time the pharmaceutical industry has interrogated its libraries for compounds that are anti-aggregants and/or anti-fibrillogenic. Many hits with compounds that look similar to Congo red have never been developed. Similarly, compounds capable of disaggregating or defibrillating Aβ have been sought, but not with the intensity of the search for anti-aggregants. While many peptidyl/protein-like designs have been examined (Gibson and Murphy 2005; S. Lee *et al.* 2005; Schmuck *et al.* 2005; Schuster *et al.* 2005; Szegedi *et al.* 2005), other small molecules have been discovered which hold some promise (Cohen *et al.* 2006; Hennessy and Buchwald 2005; Kanapathipillai *et al.* 2005; K.H. Lee *et al.* 2005; Liu *et al.* 2005b; Sabaté and Estelrich 2005; Török *et al.* 2006; S.S. Wang *et al.* 2005). Most interesting, however, is the development of assays specifically designed to examine the effects of soluble β-oligomers of Aβ (possibly the trimeric form A_{12}) and to use these assays in a discovery process for small compounds capable of inhibiting β-oligomer formation (Walsh *et al.* 2005).

18.4 Targeting the downstream effects of Aβ

Many productive lines of enquiry are being applied to the downstream effects of Aβ, beginning with the direct consequences of Aβ toxicity and oxidative damage through to the promotion of Aβ clearance/degradation. Major questions remain on the role of the innate immune system and the value of targeting neurofibrillary tangle formation.

18.4.1 Ameliorating the toxic gain of function of Aβ: anti-oxidants, neuroprotectants, and other products of natural origin

Existing knowledge and screens of natural product libraries have thrown up a wide variety of anti-oxidants and 'neuroprotectants' which have an effect on the actions of Aβ in experimental assays of its toxicity. Many of these assays are difficult to control, and there is little agreement in the field as to their validity. Nevertheless, an increasing number of papers are appearing reporting efficacy of compounds derived from plants (e.g. ferulic acid (Cho *et al.* 2005; Jin *et al.* 2005; Mohmmad Abdul and Butterfield 2005; Ono *et al.* 2005; Sultana *et al.* 2005), green tea extracts (S.Y. Lee *et al.* 2005; Mandel *et al.* 2006; Rezai-Zadeh *et al.* 2005), curcumin (Ono *et al.* 2004; F. Yang *et al.* 2005), resveratol (Marambaud *et al.* 2005), fucoidan (Jhamandas *et al.* 2005), and various other plant materials (Jeong *et al.* 2005; Kim *et al.* 2005a; Lecanu *et al.* 2005; Yu *et al.* 2005)), other natural products (docosahexaenoic acid (Hashimoto *et al.* 2005; Lim *et al.* 2005), vitamin E (McDaid *et al.* 2005; Quintanilla *et al.* 2005), oestrogens (Coma *et al.* 2005; Quintanilla *et al.* 2005), glutathione (Ju *et al.* 2005; Woltjer *et al.* 2005), melatonin (Quinn *et al.* 2005), coenzyme Q10 (Moreira *et al.* 2005), gelsolin (Qiao *et al.* 2005), and insulin-like growth factor 1 (Aguado-Llera *et al.* 2005)), or a variety of small compounds (Caraci *et al.* 2005; Marrazzo *et al.* 2005; Youdim *et al.* 2005). A common theme has emerged from these investigations: a wide variety of anti-oxidants can ameliorate the toxic gain of function of Aβ. This is consistent with our argument that Aβ itself is the principal pro-oxidant in AD. Other lines of evidence are emerging which contribute to an understanding of the oxidative stress (Nathan *et al.* 2005) or form a feedforward mechanism (Tong *et al.* 2005) to account for the progressive nature of AD.

18.4.2 Suppressing brain 'inflammation'

There is considerable controversy around the concept that the AD brain is undergoing inflammation. As usually understood, inflammatory changes are not visible. What Alzheimer, Cajal, and their contemporaries recognized was that the microglia were increased in number, activated, and, together with astrocytes, were reacting to some underlying factor, possibly the amyloid within the plaque. They also recognized that the dystrophic neurites and '*drusige Entartung*' associated with the perivascular amyloid deposits could represent the reactive and regenerative response of neurons to the same injurious process. Therefore it is surprising that the idea of 'inflammation' in AD has gained such ground in recent times. In this scenario, the microglia are seen as inflammatory invaders causing damage through their release of cytokines and other powerful destructive molecules designed to respond to injury. This innate immune reaction would exacerbate the clinical expression of AD and lead to its progression towards neuronal dysfunction and death. Based on this, trials of anti-inflammatories in AD have been conducted, and considerable research efforts undertaken to examine their effects in a variety of experimental models. The anti-inflammatories investigated include the non-steroidal anti-inflammatory drugs (NSAIDs) (Farías *et al.* 2005; Heneka *et al.* 2005; Morihara *et al.* 2005), peroxisome proliferator-activated receptor γ agonists (Costello *et al.* 2005; Echeverria *et al.* 2005; Shie *et al.* 2005), cannabinoids (Ramírez *et al.* 2005), glucocorticoids (Boedker *et al.* 2005), and zingansikpoongtang, a Korean herbal medicine (Kim *et al.* 2005b). To date, no prospective clinical trial with an anti-inflammatory has shown a convincing beneficial outcome. Perhaps the underlying theory is wrong? In the light of the data emerging around the immunization/immunomodulation strategies against Aβ (see below), the counter-hypothesis that microglia are actually beneficial could prove to be correct.

18.4.3 Targeting tau aggregation in the Aβ pathway

While Aβ has captured the imagination of most AD researchers, studies of the neurofibrillary tangle and its constituent, the tau microtubule associated protein, have progressed to a point where clear therapeutic strategies are emerging. The exact form of tau which causes neuronal degeneration is now being re-examined (Duff and Planel 2005), with data emerging that the soluble aggregated species, akin to soluble β-oligomers of Aβ, might represent the best target. The binding sites on tau (Mukrasch *et al.* 2005) for a variety of interactors are potential targets. Downregulation of expression of the tau gene

(Santacruz *et al.* 2005) or altering the alternative splicing (Rodriguez-Martin *et al.* 2005) also offers some new strategies.

As the molecular basis for the accumulation of tau in the AD brain becomes clearer, so will the precise therapeutic target. If tau accumulation is closely linked to Aβ toxicity, then oxidative modifications of tau become understandable (Reynolds *et al.* 2005a,b, 2006; Santa-Maria *et al.* 2005; Zhang *et al.* 2005) and subject to anti-oxidative classes of drugs. Metal ions might also affect this pathway (Ma *et al.* 2005, 2006). Looking at the normal function and processing of tau has raised the possibility of using microtubule-stabilizing agents such as paclitaxel (Taxol) (Michaelis *et al.* 2005). Great controversy still persists on the role of normal and abnormal phosphorylation of tau in its passage from a highly soluble cytoskeletal-associated protein to an aggregated neurofibrillary tangle. If phosphorylation of specific amino acids by specific kinases such as c-Abl (Derkinderen *et al.* 2005), Cdk5 (Sakaue *et al.* 2005), glycogen synthase kinase-3 (GSK-3) (Noble *et al.* 2005), or mitogen-activated protein kinase (MAPK) (Lambourne *et al.* 2005) proves to be pathogenic, then specific kinase inhibitors, including lithium (Noble *et al.* 2005), might be developed for AD; indeed, a trial with lithium is currently in progress in the UK. However, if phosphorylation proves to be a secondary event, following aggregation and accumulation of intracellular tau, this approach would not be expected to be useful. Other post-translational modifications, including proteolytic cleavages, have been proposed (Cotman *et al.* 2005)—all amenable to therapeutic drug discoveries. As with Aβ, small compounds capable of inhibiting aggregation and fibrillization of tau are now being examined *in vitro* (Necula *et al.* 2005; Taniguchi *et al.* 2005), but much more work in animal models is required.

18.4.4 How does apolipoprotein E fit within the Aβ pathway?

As apoE status is the major (if not the sole) genetic risk factor for determining the age at onset of AD, it is surprising that we still do not have a definitive explanation for its mechanism of action. Targeting the ApoE gene directly, or aiming for the delivery of the protective ApoEs isoform (Dodart *et al.* 2005), offers some prospect of therapeutic intervention. However, understanding the precise interaction between ApoE and the processing of APP/Aβ is likely to yield more amenable therapeutic strategies.

18.4.5 Using immunization and immunomodulation of Aβ to promote clearance and inhibit toxicity (neutralization)

Since 1999, increasing evidence has accumulated to make a compelling antibody-mediated Aβ clearance/neutralization strategy. Experiments in mouse models continue to demonstrate efficacy (Bales *et al.* 2006; Banks *et al.* 2005; Brendza *et al.* 2005; Buttini *et al.* 2005; Klyubin *et al.* 2005; Rowan *et al.* 2005). The aborted clinical trial with the Elan Aβ$_{42}$ antigen (AN1792) has provided a wealth of clinical information (Bayer *et al.* 2005; Gilman *et al.* 2005; M. Lee *et al.* 2005; Masliah *et al.* 2005) which will assist further development of strategies designed to avoid autoimmune adverse events (E.B. Lee *et al.* 2005a; Racke *et al.* 2005). Chief among these will be avoidance of T-cell-mediated responses (Agadjanyan *et al.* 2005) and the development of passive immunization protocols

(Asami-Odaka *et al.* 2005; Chauhan and Siegel 2005; Gaugler *et al.* 2005; Hartman *et al.* 2005; Li *et al.* 2005; Maier *et al.* 2005). The results of the current clinical trials by Elan using passive immunization are awaited with great interest (see below). In the meantime, novel methods of antigen presentation (Bowers *et al.* 2005; Frenkel *et al.* 2005; Qu *et al.* 2006; Solomon 2005; Youm *et al.* 2005; Zurbriggen *et al.* 2005) and the use of neo-epitopes (Arbel *et al.* 2005; Paganetti *et al.* 2005; Yamamoto *et al.* 2005) are under investigation. Neo-epitopes generated post-transationally by modification of Aβ (through oxidative mechanisms, as discussed above) should have inherently less potential to generate an autoimmune adverse reaction.

A startling process of lateral thinking has emerged with the report (Alvarez *et al.* 2006) of the use of Cerebrolysin in a successful phase II study of AD in Spain and Romania. The product is a proteolytic extract of pig brain, and is administered by multiple intravenous infusions over an 8 week period. Putting aside the possibility of transmitting a porcine form of prion disease, the method raises interesting regulatory and religious issues. Should someone look at the sera of these subjects to see what forms of reactivity to Aβ have been generated?

18.4.6 Modulating the Aβ degradation pathway

The reuptake, clearance, and degradation of Aβ is still subject to considerable uncertainties. If sporadic AD is the result of a low-level shift (e.g. <10%) in the efficiency in any of these mechanisms, a therapeutic strategy aimed at restoring or by-passing this faulty mechanism could be very useful. Each of the different pools of Aβ probably has slightly different mechanisms of elimination, varying with the cellular compartment in which Aβ resides over the course of its catabolic cycle. Several pieces of evidence point towards the enzymes neprilysin (NEP) and insulin-degrading enzyme (IDE) as key players (Saido and Iwata 2006; Saito *et al.* 2005), but the highly sought evidence from gene linkage studies remains elusive (Eckman and Eckman 2005). A new candidate, angiotension-converting enzyme (ACE), has emerged (Hemming and Selkoe 2005), and it will be of great interest to find out whether ACE-inhibitors could be having an adverse influence on the natural history of AD.

18.5 The future

The clinical development of drugs directly targeting the Aβ pathway is at an early stage of evolution. Table 18.3 lists of the publicly disclosed trials that are in progress or which have completed/discontinued with drugs that have been developed specifically to target the Aβ pathway. The γ-secretase inhibitor trials are of immense theoretical interest, as they are likely to provide the most compelling support for the Aβ theory of AD. The trials around the Aβ metal binding site or the CAG binding sites also have the potential to address this aspect. Immunization/immunomodulation of Aβ holds great promise for elucidating the Aβ clearance/neutralization strategies in which there is currently a dearth of information. A variety of prospective statin-mediated approaches will also test the hypothesis that cholesterol has an important role in the biogenesis of AD. The anti-oxidant trials

Table 18.3 Drugs in clinical development directly targeting the Aβ pathway

Target	Drug	Company	Status	Clinical trials gov. identifier
γ-Secretase inhibition	R-flurbiprofen (Flurizan) MCP-7869	Myriad Genetics	Phase III (in progress)	NCT00105547
	LY-450139	Eli Lilly & Co	Phase II	NCT00244322
	Merck compound	Merck Inc	Phase II	
Aβ monomer/oligomer				
Metal binding site	PBT2(PBT1)	Prana Biotechnology	Phase II	
GAG binding site	3APS Tramiprosate (Alzhemed)	Neurochem Inc.	Phase III	NCT00217769
Aβ clearance/neutralization	AN-1792 (acute immunization synthetic Aβ₄₂)	Elan/Wyeth	Phase III (discontinued)	
	AAB-001 Bapineuzamab (passive immunotherapy)	Elan/Wyeth	Phase II	NCT00112073
	ACC-001 (immunization)	Elan/Wyeth	Phase I	
APP/Aβ processing				
Cholesterol synthesis	Statins	Merck/Pfizer	Phase 2	
Antioxidants	Curcumin	French Foundation	Phase 2	NCT00099710
	VitE/VitC/α lipoic acid; coenzyme Q	NIH	Phase 1	NCT00117403

have the disadvantage of lacking specificity for Aβ, but nonetheless will continue to provide much needed guidance for the general theory of the AD brain being under oxidative stress.

It is extremely unlikely that a single class of compound or targeting a single mechanism of action will be sufficient to treat AD. It is far more likely that, for this complex disease, a combination of drugs targeting various aspects of the greater APP/Aβ pathway will evolve into some form of rational therapy. Trials now in progress should represent the very beginning of the enlightenment required to find the right combinations—all predicated on the assumption that the APP/Aβ pathway underlies the cause of AD.

Acknowledgements and disclosures

Some of the work described in this chapter is supported by research grants from the National Health and Medical Research Council of Australia. Colin L. Masters discloses interests in Prana Biotechnology.

References

Adlard PA, Perreau VM, Pop V, *et al.* (2005). Voluntary exercise decreases amyloid load in a transgenic model of Alzheimer's disease. *Journal of Neuroscience* 25, 4217–21.

Agadjanyan MG, Ghochikyan A, Petrushina I, *et al.* (2005). Prototype Alzheimer's disease vaccine using the immunodominant b cell epitope from β-amyloid and promiscuous T cell epitope pan HLA DR-binding peptide. *Journal of Immunology* 174, 1580–6.

Aguado-Llera D, Arilla-Ferreiro E, Campos-Barros A, *et al.* (2005). Protective effects of insulin-like growth factor-I on the somatostatinergic system in the temporal cortex of β-amyloid-treated rats. *Journal of Neurochemistry* 92, 607–15.

Alvarez XA, Cacabelos R, Laredo M, *et al.* (2006). A 24-week, double-blind, placebo-controlled study of three dosages of Cerebrolysin in patients with mild to moderate Alzheimer's disease. *European Journal of Neurology* 13, 43–54.

Anderson JJ, Holtz G, Baskin PP, *et al.* (2005). Reductions in β-amyloid concentrations *in vivo* by the γ-secretase inhibitors BMS-289948 and BMS-299897. *Biochemical Pharmacology* 69, 689–98.

Arbel M, Yacoby I, and Solomon B (2005). Inhibition of amyloid precursor protein processing by β-secretase through site-directed antibodies. *Proceedings of the National Academy of Sciences of the United States of America* 102, 7718–23.

Asami-Odaka A, Obayashi-Adachi Y, Matsumoto Y, *et al.* (2005). Passive immunization of the Aβ42(43) C-terminal-specific antibody BC05 in a mouse model of Alzheimer's disease. *Neurodegenerative Diseases* 2, 36–43.

Bales KR, Tzavara ET, Wu S, *et al.* (2006). Cholinergic dysfunction in a mouse model of Alzheimer disease is reversed by an anti-Aβ antibody. *Journal of Clinical Investigation* 116, 825–32.

Banks WA, Pagliari P, Nakaoke R, *et al.* (2005). Effects of a behaviorally active antibody on the brain uptake and clearance of amyloid beta proteins. *Peptides* 26, 287–94.

Barten DM, Guss VL, Corsa JA, *et al.* (2005). Dynamics of β-amyloid reductions in brain, cerebrospinal fluid, and plasma of β-amyloid precursor protein transgenic mice treated with a γ-secretase inhibitor. *Journal of Pharmacology and Experimental Therapeutics* 312, 635–43.

Bayer AJ, Bullock R, Jones RW, *et al.* (2005). Evaluation of the safety and immunogenicity of synthetic Aβ42 (AN1792) in patients with AD. *Neurology* 64, 94–101.

Belluti F, Rampa A, Piazzi L, *et al.* (2005). Cholinesterase inhibitors: xanthostigmine derivatives blocking the acetylcholinesterase-induced β-amyloid aggregation. *Journal of Medicinal Chemistry* **48**, 4444–56.

Benson A (2005). Alzheimer's disease: a tangled issue. *Drug Discovery Today* **10**, 749–51.

Best JD, Jay MT, Otu F, *et al.* (2005). *In vivo* characterisation of Aβ(40) changes in brain and CSF using the novel γ-secretase inhibitor MRK-560 (*N*-[*cis*-4-[(4-chlorophenyl) sulfonyl]-4-(2,5-difluorophenyl)cyclohexyl]-1,1,1- trifluoromethanesulfonamide) in the rat. *Journal of Pharmacology and Experimental Therapeutics* **317**, 786–90.

Bloom FE, Reilly JF, Redwine JM, *et al.* (2005). Mouse models of human neurodegenerative disorders. Requirements for medication development. *Archives of Neurology* **62**, 185–7.

Boedker M, Boetkjaer A, Bazan NG, *et al.* (2005). Budesonide epimer R, LAU-8080 and phenyl butyl nitrone synergistically repress cyclooxygenase-2 induction in [IL-1β+Aβ42]- stressed human neural cells. *Neuroscience Letters* **380**, 176–80.

Bolognesi ML, Andrisano V, Bartolini M, *et al.* (2005). Propidium-based polyamine ligands as potent inhibitors of acetylcholinesterase and acetylcholinesterase-induced amyloid-β aggregation. *Journal of Medicinal Chemistry* **48**, 24–7.

Bowers WJ, Mastrangelo MA, Stanley HA, *et al.* (2005). HSV amplicon-mediated Aβ vaccination in Tg2576 mice: differential antigen-specific immune responses. *Neurobiology of Aging* **26**, 393–407.

Brendza RP, Bacskai BJ, Cirrito JR, *et al.* (2005). Anti-Aβ antibody treatment promotes the rapid recovery of amyloid-associated neuritic dystrophy in *PDAPP* transgenic mice. *Journal of Clinical Investigation* **115**, 428–33.

Buttini M, Masliah E, Barbour R, *et al.* (2005). β-amyloid immunotherapy prevents synaptic degeneration in a mouse model of Alzheimer's disease. *Journal of Neuroscience* **25**, 9096–101.

Byun H-G, Kim Y-T, Park P-J, *et al.* (2005). Chitooligosaccharides as a novel β-secretase inhibitor. *Carbohydrate Polymers* **61**, 198–202.

Caccamo A, Oddo S, Billings LM, *et al.* (2006). M1 receptors play a central role in modulating AD-like pathology in transgenic mice. *Neuron* **49**, 671–82.

Caraci F, Chisari M, Frasca G, *et al.* (2005). Nicergoline, a drug used for age-dependent cognitive impairment, protects cultured neurons against β-amyloid toxicity. *Brain Research* **1047**, 30–7.

Chauhan NB and Siegel GJ (2005). Efficacy of anti-Aβ antibody isotypes used for intracerebroventricular immunization in TgCRND8. *Neuroscience Letters* **375**, 143–7.

Chauhan NB, Siegel GJ, and Feinstein DL (2005). Propentofylline attenuates tau hyperphosphorylation in Alzheimer's Swedish mutant model Tg2576. *Neuropharmacology* **48**, 93–104.

Cherny RA, Atwood CS, Xilinas ME, *et al.* (2001). Treatment with a copper-zinc chelator markedly and rapidly inhibits β-amyloid accumulation in Alzheimer's disease transgenic mice. *Neuron* **30**, 665–76.

Cho JY, Kim HS, Kim DH, *et al.* (2005). Inhibitory effects of long-term administration of ferulic acid on astrocyte activation induced by intracerebroventricular injection of β-amyloid peptide (1–42) in mice. *Progress in Neuropsychopharmacology and Biological Psychiatry* **29**, 901–7.

ClinicalTrials.gov. A Notch signalling pathway inhibitor for patients with advanced breast cancer. ClinicalTrials.gov Identifier: NCT00106145. Available online at: http://www.clinicaltrials.gov/ct/show?NCT00106145?order=1

ClinicalTrials.gov. A notch signalling pathway inhibitor for patients with T-cell acute lymphoblastic leukemia/lymphoma (ALL). ClinicalTrials.gov Identifier: NCT00100152. Available online at: http://www.clinicaltrials.gov/ct/show/NCT00100152?order=2

Cohen T, Frydman-Marom A, Rechter M, *et al.* (2006). Inhibition of amyloid fibril formation and cytotoxicity by hydroxyindole derivatives. *Biochemistry* **45**, 4727–35.

Cole SL, Grudzien A, Manhart IO, *et al.* (2005). Statins cause intracellular accumulation of amyloid precursor protein, β-secretase-cleaved fragments, and amyloid β-peptide via an isoprenoid-dependent mechanism. *Journal of Biological Chemistry* **280**, 18755–70.

Coma M, Guix FX, Uribesalgo I, *et al.* (2005). Lack of oestrogen protection in amyloid-mediated endothelial damage due to protein nitrotyrosination. *Brain* **128**, 1613–21.

Comery TA, Martone RL, Aschmies S, *et al.* (2005). Acute γ-secretase inhibition improves contextual fear conditioning in the Tg2576 mouse model of Alzheimer's disease. *Journal of Neuroscience* **25**, 8898–902.

Cordle A and Landreth G (2005). 3-Hydroxy-3-methylglutaryl-coenzyme a reductase inhibitors attenuate β-amyloid-induced microglial inflammatory responses. *Journal of Neuroscience* **25**, 299–307.

Costello DA, O'Leary DM, Herron CE (2005). Agonists of peroxisome proliferator-activated receptor-γ attenuate the Aβ-mediated impairment of LTP in the hippocampus *in vitro*. *NeuroPharmacology* **49**, 359–66.

Cotman CW, Poon WW, Rissman RA, *et al.* (2005). The role of caspase cleavage of tau in Alzheimer disease neuropathology. *Journal of Neuropathology and Experimental Neurology* **64**, 104–12.

Cui Z, Lockman PR, Atwood CS, *et al.* (2005). Novel D-penicillamine carrying nanoparticles for metal chelation therapy in Alzheimer's and other CNS diseases. *European Journal of Pharmaceutics and Biopharmaceutics* **59**, 263–72.

Cummings JL (2006). What we can learn from open-label extensions of randomized clinical trials. *Archives of Neurology* **63**, 18–19.

Curry CL, Reed LL, Golde TE, *et al.* (2005). Gamma secretase inhibitor blocks Notch activation and induces apoptosis in Kaposi's sarcoma tumor cells. *Oncogene* **24**, 6333–44.

Dash PK, Moore AN, Orsi SA. (2005). Blockade of γ-secretase activity within the hippocampus enhances long-term memory. *Biochemical and Biophysical Research Communications* **338**, 777–782.

Davidsson P, Blennow K, Andreasen N, *et al.* (2001). Differential increase in cerebrospinal fluid-acetylcholinesterase after treatment with acetylcholinesterase inhibitors in patients with Alzheimer's Disease. *Neuroscience Letters* **300**, 157–160.

Derkinderen P, Scales TM, Hanger DP, *et al.* (2005). Tyrosine 394 is phosphorylated in Alzheimer's paired helical filament tau and in fetal tau with c-Abl as the candidate tyrosine kinase. *Journal of Neuroscience* **25**, 6584–93.

Dodart JC, Marr RA, Koistinaho M, *et al.* (2005) Gene delivery of human apolipoprotein E alters brain Aβ burden in a mouse model of Alzheimer's disease. *Proceedings of the National Academy of Sciences of the United States of America* **102**, 1211–16.

Dong H, Csernansky CA, Martin MV, *et al.* (2005). Acetylcholinesterase inhibitors ameliorate behavioral deficits in the Tg2576 mouse model of Alzheimer's disease. *Psychopharmacology* **181**, 145–152.

Duering M, Grimm M, Grimm HS, *et al.* (2005). Mean age of onset in familial Alzheimer's disease is determined by amyloid beta 42. *Neurobiology of Aging* **26**, 785–8.

Duff K and Planel E (2005). Untangling memory deficits. *Nature Medicine,* **11**, 826–7.

Echeverria V, Clerman A, and Doré S (2005). Stimulation of PGE$_2$ receptors EP2 and EP4 protects cultured neurons against oxidative stress and cell death following β-amyloid exposure. *European Journal of Neuroscience* **22**, 2199–206.

Eckman EA and Eckman CB (2005). Aβ-degrading enzymes: modulators of Alzheimer's disease pathogenesis and targets for therapeutic intervention. *Biochemical Society Transactions* **33**, 1101–5.

Espeseth AS, Xu M, Huang Q, *et al.* (2005). Compounds that bind APP and inhibit Aβ processing *in vitro* suggest a novel approach to Alzheimer disease therapeutics. *Journal of Biological Chemistry* **280**, 17792–7.

Farías GG, Godoy JA, Vázquez MC, *et al.* (2005). The anti-inflammatory and cholinesterase inhibitor bifunctional compound IBU-PO protects from β-amyloid neurotoxicity by acting on Wnt signaling components. *Neurobiology of Disease* **18**, 176–83.

Frenkel D, Maron R, Burt DS, *et al.* (2005). Nasal vaccination with a proteosome-based adjuvant and glatiramer acetate clears β-amyloid in a mouse model of Alzheimer disease. *Journal of Clinical Investigation* **115**, 2423–33.

Gaeta A and Hider RC (2005). The crucial role of metal ions in neurodegeneration: the basis for a promising therapeutic strategy. *British Journal of Pharmacology* **146**, 1041–59.

Gaugler MN, Tracy J, Kuhnle K, *et al.* (2005). Modulation of Alzheimer's pathology by cerebro-ventricular grafting of hybridoma cells expressing antibodies against Aβ *in vivo*. *FEBS Letters* **579**, 753–6.

Ghosh AK, Devasamudram T, Hong L, *et al.* (2005). Structure-based design of cycloamide-urethane-derived novel inhibitors of human brain memapsin 2 (β-secretase). *Bioorganic and Medicinal Chemistry Letters* **15**, 15–20.

Gibson TJ and Murphy RM (2005). Design of peptidyl compounds that affect β-amyloid aggregation: importance of surface tension and context. *Biochemistry* **44**, 8898–907.

Gilman S, Koller M, Black RS, *et al.* (2005). Clinical effects of Aβ immunization (AN1792) in patients with AD in an interrupted trial. *Neurology* **64**, 1553–62.

Golde TE (2005). The Aβ hypothesis: leading us to rationally-designed therapeutic strategies for the treatment or prevention of Alzheimer disease. *Brain Pathology* **15**, 84–7.

Green PS, Bales K, Paul S, *et al.* (2005). Estrogen therapy fails to alter amyloid deposition in the PDAPP model of Alzheimer's disease. *Endocrinology* **146**, 2774–81.

Greig NH, Utsuki T, Ingram DK, *et al.* (2005) Selective butyrylcholinesterase inhibition elevates brain acetylcholine, augments learning and lowers Alzheimer β-amyloid peptide in rodent. *Proceedings of the National Academy of Sciences of the USA* **102**, 17213–18.

Grimwood S, Hogg J, Jay MT, *et al.* (2005). Determination of guinea-pig cortical γ-secretase activity ex vivo following the systemic administration of a γ-secretase inhibitor. *Neuropharmacology* **48**, 1002–11.

Gundersen E, Fan K, Haas K, *et al.* (2005). Molecular-modeling based design, synthesis, and activity of substituted piperidines as γ-secretase inhibitors. *Bioorganic and Medicinal Chemistry Letters* **15**, 1891–4.

Hanessian S, Yun H, Hou Y, *et al.* (2005). Structure-based design, synthesis, and memapsin 2 (BACE) inhibitory activity of carbocyclic and heterocyclic peptidomimetics. *Journal of Medicinal Chemistry* **48**, 5175–90.

Hartman RE, Izumi Y, Bales KR, *et al.* (2005). Treatment with an amyloid-β antibody ameliorates plaque load, learning deficits, and hippocampal long-term potentiation in a mouse model of Alzheimer's disease. *Journal of Neuroscience* **25**, 6213–20.

Hashimoto M, Tanabe Y, Fujii Y, *et al.* (2005). Chronic administration of docosahexaenoic acid ameliorates the impairment of spatial cognition learning ability in amyloid β-infused rats. *Journal of Nutrition* **135**, 549–55.

Hemming ML and Selkoe DJ (2005). Amyloid β-protein is degraded by cellular angiotensin-converting enzyme (ACE) and elevated by an ACE inhibitor. *Journal of Biological Chemistry* **280**, 37644–50.

Heneka MT, Sastre M, Dumitrescu-Ozimek L, *et al.* (2005). Acute treatment with the PPARγ agonist pioglitazone and ibuprofen reduces glial inflammation and Aβ1–42 levels in APPV717I transgenic mice. *Brain* **128**, 1442–53.

Hennessy EJ and Buchwald SL (2005). Synthesis of 4,5-dianilinophthalimide and related analogues for potential treatment of Alzheimer's disease via palladium-catalyzed amination. *Journal of Organic Chemistry* **70**, 7371–5.

Higuchi M, Iwata N, and Saido TC (2005). Understanding molecular mechanisms of proteolysis in Alzheimer's disease: progress toward therapeutic interventions. *Biochimica et Biophysica Acta* **1751**, 60–67.

Hilbush BS, Morrison JH, Young WG, *et al.* (2005). New prospects and strategies for drug target discovery in neurodegenerative disorders. *NeuroRx* **2**, 627–97.

Höglund K, Syversen S, Lewczuk P, *et al.* (2005a). Statin treatment and a disease-specific pattern of β-amyloid peptides in Alzheimer's disease. *Experimental Brain Research* **164**, 205–14.

Höglund K, Thelen KM, Syversen S, *et al.* (2005b). The effect of simvastatin treatment on the amyloid precursor protein and brain cholesterol metabolism in patients with Alzheimer's disease. *Dementia and Geriatric Cognitive Disorders* **19**, 256–65.

Huang D, Lüthi U, Kolb P, *et al.* (2005). Discovery of cell-permeable non-peptide inhibitors of β-secretase by high-throughput docking and continuum electrostatics calculations. *Journal of Medicinal Chemistry* **48**, 5108–11.

Ibach B, Haen E, Marienhagen J, *et al.* (2005). Clioquinol treatment in familiar early onset of Alzheimer's disease. A case report. *Pharmacopsychiatry* **38**,178–9.

Jankowsky JL, Melnikova T, Fadale DJ, *et al.* (2005a). Environmental enrichment mitigates cognitive deficits in a mouse model of Alzheimer's disease. *Journal of Neuroscience* **25**, 5217–224.

Jankowsky JL, Slunt HH, Gonzales V *et al.* (2005b). Persistent amyloidosis following suppression of Aβ production in a transgenic model of Alzheimer disease. *PLoS Medicine* **2**, e355.

Jeong JC, Yoon CH, Lee WH, *et al.* (2005). Effects of *Bambusae concretio salicea* (Chunchukhwang) on amyloid β-induced cell toxicity and antioxidative enzymes in cultured rat neuronal astrocytes. *Journal of Ethnopharmacology* **98**, 259–66.

Jhamandas JH, Wie MB, Harris K, *et al.* (2005). Fucoidan inhibits cellular and neurotoxic effects of β-amyloid (Aβ) in rat cholinergic basal forebrain neurons. *European Journal of Neuroscience* **21**, 2649–59.

Ji HF and Zhang HY (2005). A new strategy to combat Alzheimer's disease: combining radical-scavenging potential with metal-protein-attenuating ability in one molecule. *Bioorganic and Medicinal Chemistry Letters* **15**, 21–4.

Jin Y, Yan EZ, Fan Y, *et al.* (2005). Sodium ferulate prevents amyloid-beta-induced neurotoxicity through suppression of p38 MAPK and upregulation of ERK-1/2 and Akt/protein kinase B in rat hippocampus. *Acta Pharmacologica Sinica* **26**, 943–51.

Ju TC, Chen SD, Liu CC, *et al.* (2005). Protective effects of *S*-nitrosoglutathione against amyloid β-peptide neurotoxicity. *Free Radical Biology and Medicine* **38**, 938–49.

Kanapathipillai M, Lentzen G, Sierks M, *et al.* (2005). Ectoine and hydroxyectoine inhibit aggregation and neurotoxicity of Alzheimer's β-amyloid. *FEBS Letters* **579**, 4775–80.

Kim H, Park BS, Lee KG, *et al.* (2005a). Effects of naturally occurring compounds on fibril formation and oxidative stress of β-amyloid. *Journal of Agricultural and Food Chemistry* **53**, 8537–41.

Kim SJ, Jeong HJ, Lee KM, *et al.* (2005b). Zingansikpoongtang modulates β-amyloid and IL-1β-induced cytokine production and cyclooxygenase-2 expression in human astrocytoma cells U373MG. *Journal of Ethnoharmacology* **96**, 279–85.

Kimura M, Akasofu S, Ogura H, *et al.* (2005a). Protective effect of donepezil against Aβ(1–40) neurotoxicity in rat septal neurons. *Brain Research* **1047**, 72–84.

Kimura T, Shuto D, Hamada Y, *et al.* (2005b). Design and synthesis of highly active Alzheimer's β-secretase (BACE1) inhibitors, KMI-420 and KMI-**429**,with enhanced chemical stability. *Bioorganic and Medicinal Chemistry Letters* **15**, 211–15.

Klyubin I, Walsh DM, Lemere CA, *et al.* (2005). Amyloid β protein immunotherapy neutralizes Aβ oligomers that disrupt synaptic plasticity *in vivo*. *Nature Medicine* **11**, 556–61.

Kornacker MG, Lai Z, Witmer M, *et al.* (2005). An inhibitor binding pocket distinct from the catalytic active site on human β-APP cleaving enzyme. *Biochemistry* **44**, 11567–73.

Kwon MO and Herrling P (2005). List of drugs in development for neurodegenerative diseases. *Neurodegenerative Disorders* **2**, 61–108.

Lambourne SL, Sellers LA, Bush TG, *et al.* (2005). Increased tau phosphorylation on mitogen-activated protein kinase consensus sites and cognitive decline in transgenic models for Alzheimer's disease and FTDP-17: evidence for distinct molecular processes underlying tau abnormalities. *Molecular and Cellular Biology* **25**, 278–93.

Lanz TA, Fici GJ, and Merchant KM (2005). Lack of specific amyloid-β(1–42) suppression by nonsteroidal anti-inflammatory drugs in young, plaque-free Tg2576 mice and in guinea pig neuronal cultures. *Journal of Pharmacology and Experimental Therapeutics* **312**, 399–406.

Laras Y, Quéléver G, Garino C, *et al.* (2005). Substituted thiazolamide coupled to a redox delivery system: a new γ-secretase inhibitor with enhanced pharmacokinetic profile. *Organic and Biomolecular Chemistry* **3**, 612–18.

Lazarov O, Robinson J, Tang YP, *et al.* (2005). Environmental enrichment reduces Aβ levels and amyloid deposition in transgenic mice. *Cell* **120**, 701–13.

Lecanu L, Yao W, Piechot A, *et al.* (2005). Identification, design, synthesis, and pharmacological activity of (4-ethyl-piperazin-1-yl)-phenylmethanone derivatives with neuroprotective properties against β-amyloid-induced toxicity. *Neuropharmacology* **49**, 86–96.

Lee EB, Leng LZ, Lee VM, *et al.* (2005). Meningoencephalitis associated with passive immunization of a transgenic murine model of Alzheimer's amyloidosis. *FEBS Letters* **579**, 2564–8.

Lee HJ, Seong YH, Bae KH, *et al.* (2005). β-secretase (BACE1) inhibitors from Sanguisorbae radix. *Archives of Pharmacal Research* **28**, 799–803.

Lee KH, Shin BH, Shin KJ, *et al.* (2005). A hybrid molecule that prohibits amyloid fibrils and alleviates neuronal toxicity induced by β-amyloid (1–42). *Biochemical and Biophysical Research Communications* **328**, 816–23.

Lee M, Bard F, Johnson-Wood K, *et al.* (2005). Aβ42 immunization in Alzheimer's disease generates Aβ N-terminal antibodies. *Annals of Neurology* **58**, 430–5.

Lee S, Carson K, Rice-Ficht A, *et al.* (2005). Hsp20, a novel a-crystallin, prevents Aβ fibril formation and toxicity. *Protein Science* **14**, 593–601.

Lee SY, Lee JW, Lee H, *et al.* (2005). Inhibitory effect of green tea extract on β-amyloid-induced PC12 cell death by inhibition of the activation of NF-γB and ERK/p38 MAP kinase pathway through antioxidant mechanisms. *Brain Research: Molecular Brain Research* **140**, 45–54.

Lefranc-Jullien S, Lisowski V, Hernandez JF, *et al.* (2005). Design and characterization of a new cell-permeant inhibitor of the β-secretase BACE1. *British Journal of Pharmacology* **145**, 228–35.

Lewis SJ, Smith AL, Neduvelil JG, *et al.* (2005). A novel series of potent γ-secretase inhibitors based on a benzobicyclo[4.2.1] nonane core. *Bioorganic and Medicinal Chemistry Letters* **15**, 373–8.

Li SB, Wang HQ, Lin X, *et al.* (2005). Specific humoral immune responses in rhesus monkeys vaccinated with the Alzheimer's disease-associated β-amyloid 1–15 peptide vaccine. *Chinese Medical Journal* **118**, 660–4.

Lim GP, Calon F, Morihara T, *et al.* (2005). A diet enriched with the omega-3 fatty acid docosahexaenoic acid reduces amyloid burden in an aged Alzheimer mouse model. *Journal of Neuroscience* **25**, 3032–40.

Liu G, Garrett MR, Men P, *et al.* (2005a). Nanoparticle and other metal chelation therapeutics in Alzheimer disease. *Biochimica et Biophysica Acta* **1741**, 246–52.

Liu R, Barkhordarian H, Emadi S, *et al.* (2005b). Trehalose differentially inhibits aggregation and neurotoxicity of beta-amyloid 40 and 42. *Neurobiology of Disease* **20**, 74–81.

McDaid DG, Kim EM, Reid RE, *et al.* (2005). Parenteral antioxidant treatment preserves temporal discrimination following intrahippocampal aggregated Aβ(1–42) injections. *Behavioural Pharmacology* **16**, 237–42.

Ma QF, Li YM, Du JT, *et al.* (2005). Binding of copper (II) ion to an Alzheimer's tau peptide as revealed by MALDI-TOF MS, CD, and NMR. *Biopolymers* **79**, 74–85.

Ma Q, Li Y, Du J, *et al.* (2006). Copper binding properties of a tau peptide associated with Alzheimer's disease studied by CD, NMR, and MALDI-TOF MS. *Peptides* **27**, 841–9.

Maier M, Seabrook TJ, and Lemere CA (2005). Modulation of the humoral and cellular immune response in Aβ immunotherapy by the adjuvants monophosphoryl lipid A (MPL), cholera toxin B subunit (CTB) and *E.coli* enterotoxin LT(R192G). *Vaccine* **23**, 5149–59.

Mandel S, Amit T, Reznichenko L, *et al.* (2006). Green tea catechins as brain-permeable, natural iron chelators-antioxidants for the treatment of neurodegenerative disorders. *Molecular Nutrition and Food Research* **50**, 229–34.

Marambaud P, Zhao H, and Davies P (2005). Resveratrol promotes clearance of Alzheimer's disease amyloid-β peptides. *Journal of Biological Chemistry* **280**, 37377–82.

Marrazzo A, Caraci F, Salinaro ET, *et al.* (2005). Neuroprotective effects of sigma-1 receptor agonists against β-amyloid-induced toxicity. *Neuroreport*, **16**, 1223–6.

Masliah E, Hansen L, Adame A, *et al.* (2005). Aβ vaccination effects on plaque pathology in the absence of encephalitis in Alzheimer disease. *Neurology* **64**, 129–31.

Masse I, Bordet R, Deplanque D, *et al.* (2005). Lipid lowering agents are associated with a slower cognitive decline in Alzheimer's disease. *Journal of Neurology, Neurosurgery and Psychiatry* **76**, 1624–9.

Masters CL and Beyreuther K (2006). Disease modifying therapeutic strategies for Alzheimer's disease derived from targets in the pathways surrounding the biogenesis of the amyloid β peptide (Aβ) and the amyloid beta (A4) precursor protein (APP). In Ritchie CW, Ames DJ, Masters CL, and Cummings J (eds), *Treatment Strategies in Alzheimer's Disease and Other Dementias*, Clinical Publishing Oxford, Chapter 8, 91–110.

Mathis CA, Klunk WE, Price JC, *et al.* (2005). Imaging technology for neurodegenerative diseases. *Archives of Neurology* **62**, 196–200.

Mettenburg JM, Arandjelovic S, and Gonias SL (2005). A chemically modified preparation of a2-macroglobulin binds β-amyloid peptide with increased affinity and inhibits Aβ cytotoxicity. *Journal of Neurochemistry* **93**, 53–62.

Michaelis ML, Ansar S, Chen Y, *et al.* (2005). β-amyloid-induced neurodegeneration and protection by structurally diverse microtubule-stabilizing agents. *Journal of Pharmacology and Experimental Therapeutics* **312**, 659–68.

Mohmmad Abdul H and Butterfield DA (2005). Protection against amyloid β-peptide (1–42)-induced loss of phospholipid asymmetry in synaptosomal membranes by tricyclodecan-9-xanthogenate (D609) and ferulic acid ethyl ester: Implications for Alzheimer's disease. *Biochimica et Biophysica Acta* **1741**, 140–8.

Moreira PI, Santos MS, Sena C, *et al.* (2005). CoQ10 therapy attenuates amyloid β-peptide toxicity in brain mitochondria isolated from aged diabetic rats. *Experimental Neurology* **196**, 112–19.

Morihara T, Teter B, Yang F, *et al.* (2005). Ibuprofen suppresses interleukin-1β induction of pro-amyloidogenic a_1-antichymotrypsin to ameliorate β-amyloid (Aβ) pathology in Alzheimer's models. *Neuropsychopharmacology* **30**, 1111–20.

Mukrasch MD, Biernat J, von Bergen M, *et al.* (2005). Sites of tau important for aggregation populate β-structure and bind to microtubules and polyanions. *Journal of Biological Chemistry* **280**, 24978–86.

Nathan C, Calingasan N, Nezezon J, *et al.* (2005). Protection from Alzheimer's-like disease in the mouse by genetic ablation of inducible nitric oxide synthase. *Journal of Experimental Medicine* **202**, 1163–9.

Necula M, Chirita CN, and Kuret J (2005). Cyanine dye N744 inhibits tau fibrillization by blocking filament extension: implications for the treatment of tauopathic neurodegenerative diseases. *Biochemistry* **44**, 10227–37.

Noble W, Planel E, Zehr C, *et al.* (2005) Inhibition of glycogen synthase kinase-3 by lithium correlates with reduced tauopathy and degeneration *in vivo*. *Proceedings of the National Academy of Sciences of the USA* **102**, 6990–5.

Ono K, Hasegawa K, Naiki H, *et al.* (2004). Curcumin has potent anti-amyoidogenic effects for Alzheimer's β-amyloid fibrils *in vitro*. *Journal of Neuroscience Research* **75**, 742–750.

Ono K, Hirohata M, and Yamada M (2005). Ferulic acid destabilizes preformed β-amyloid fibrils *in vitro*. *Biochemical and Biophysical Research Communications* **336**, 444–9.

Paganetti P, Calanca V, Galli C, *et al.* (2005). β-site specific intrabodies to decrease and prevent generation of Alzheimer's Aβ peptide. *Journal of Cell Biology* **168**, 863–8.

Pákáski M, Bjelik A, Hugyecz M, *et al.* (2005). Imipramine and citalopram facilitate amyloid precursor protein secretion *in vitro*. *Neurochemistry International* **47**, 190–195.

Pangalos MN, Jacobsen SJ, and Reinhart PH (2005). Disease modifying strategies for the treatment of Alzheimer's disease targeted at modulating levels of the β-amyloid peptide. *Biochemical Society Transactions* **33**, 553–8.

Papassotiropoulos A, Lambert JC, Wavrant-De Vrièze F, *et al.* (2005). Cholesterol 25-hydroxylase on chromosome 10q is a susceptibility gene for sporadic Alzheimer's disease. *Neurodegenerative Diseases* **2**, 233–41.

Paris D, Quadros A, Patel N, *et al.* (2005) Inhibition of angiogenesis and tumor growth by β and γ-secretase inhibitors. *European Journal of Pharmacology* **514**, 1–15.

Patel NV, Gordon MN, Connor KE, *et al.* (2005). Caloric restriction attenuates Aβ-deposition in Alzheimer transgenic models. *Neurobiology of Aging* **26**, 995–1000.

Peretto I, Radaelli S, Parini C, *et al.* (2005). Synthesis and biological activity of flurbiprofen analogues as selective inhibitors of β-amyloid$_{1-42}$ secretion. *Journal of Medicinal Chemistry* **48**, 5705–20.

Pietrak BL, Crouthamel MC, Tugusheva K, *et al.* (2005). Biochemical and cell-based assays for characterization of BACE-1 inhibitors. *Annals of Biochemistry* **342**, 144–51.

Pietrancosta N, Quéléver G, Laras Y, *et al.* (2005). Design of β-secretase inhibitors by introduction of a mandelyl moiety in DAPT analogues. *Australian Journal of Chemistry* **58**, 585–94.

Polgár T and Keserü GM (2005). Virtual screening for β-secretase (BACE1) inhibitors reveals the importance of protonation states at Asp32 and Asp228. *Journal of Medicinal Chemistry* **48**, 3749–55.

Qiao H, Koya RC, Nakagawa K, *et al.* (2005). Inhibition of Alzheimer's amyloid-β peptide-induced reduction of mitochondrial membrane potential and neurotoxicity by gelsolin. *Neurobiology of Aging* **26**, 849–55.

Qu B, Boyer PJ, Johnston SA, *et al.* (2006). Aβ$_{42}$ gene vaccination reduces brain amyloid plaque burden in transgenic mice. *Journal of the Neurological Sciences* **244**, 151–8.

Quélever G, Kachidian P, Melon C, *et al.* (2005). Enhanced delivery of γ-secretase inhibitor DAPT into the brain via an ascorbic acid mediated strategy. *Organic and Biomolecular Chemistry* **3**, 2450–1457.

Quinn J, Kulhanek D, Nowlin J, *et al.* (2005). Chronic melatonin therapy fails to alter amyloid burden or oxidative damage in old tg2576 mice: implications for clinical trials. *Brain Research* **1037**, 209–13.

Quintanilla RA, Muñoz FJ, Metcalfe MJ, *et al.* (2005). Trolox and 17 β-estradiol protect against amyloid β-peptide neurotoxicity by a mechanism that involves modulation of the Wnt signaling pathway. *Journal of Biological Chemistry* **280**, 11615–25.

Racke MM, Boone LI, Hepburn DL, *et al.* (2005). Exacerbation of cerebral amyloid angiopathy-associated microhemorrhage in amyloid precursor protein transgenic mice by immunotherapy is dependent on antibody recognition of deposited forms of amyloid β. *Journal of Neuroscience* **25**, 629–36.

Raman B, Ban T, Yamaguchi K, *et al.* (2005). Metal ion-dependent effects of clioquinol on the fibril growth of an amyloid β peptide. *Journal of Biological Chemistry* **280**, 16157–62.

Ramírez BG, Blázquez C, Gómez del Pulgar T, *et al.* (2005). Prevention of Alzheimer's disease pathology by cannabinoids: neuroprotection mediated by blockade of microglial activation. *Journal of Neuroscience* **25**, 1904–13.

Ravi Keerti A, Ashok Kumar B, Parthasarathy T, *et al.* (2005). QSAR studies—potent benzodiazepine γ-secretase inhibitors. *Bioorganic and Medicinal Chemistry* **13**, 1873–8.

Reynolds MR, Berry RW, and Binder LI (2005a). Site-specific nitration and oxidative dityrosine bridging of the t protein by peroxynitrite: implications for Alzheimer's disease. *Biochemistry* **44**, 1690–1700.

Reynolds MR, Berry RW, and Binder LI (2005b). Site-specific nitration differentially influences τ assembly *in vitro*. *Biochemistry* **44**, 13997–14009.

Reynolds MR, Lukas TJ, Berry RW, *et al.* (2006). Peroxynitrite-mediated t modifications stabilize preformed filaments and destabilize microtubules through distinct mechanisms. *Biochemistry* **45**, 4314–26.

Rezai-Zadeh K, Shytle D, Sun N, *et al.* (2005). Green tea epigallocatechin-3-gallate (EGCG) modulates amyloid precursor protein cleavage and reduces cerebral amyloidosis in Alzheimer transgenic mice. *Journal of Neuroscience* **25**, 8807–14.

Ritchie CW, Bush AI, Mackinnon A, *et al.* (2003). Metal–protein attenuation with iodochloro-droxyquin (clioquinol) targeting Aβ amyloid deposition and toxicity in Alzheimer disease: a pilot phase 2 clinical trial. *Archives of Neurology* **60**, 1685–91.

Rodriguez-Martin T, Garcia-Blanco MA, Mansfield SG, *et al.* (2005). Reprogramming of tau alternative splicing by spliceosome-mediated RNA trans-splicing: implications for tauopathies. *Proceedings of the National Academy of Sciences of the United States of America* **102**, 15659–64.

Rowan MJ, Klyubin I, Wang Q, *et al.* (2005). Synaptic plasticity disruption by amyloid β protein: modulation by potential Alzheimer's disease modifying therapies. *Biochemical Society Transactions* **33**, 563–7.

Rowe CC, Ng S, Ackermann U, *et al.* Imaging β-amyloid burden in aging and dementia. Submitted for publication.

Sabaté R and Estelrich J. (2005) Stimulatory and inhibitory effects of alkyl bromide surfactants on β-amyloid fibrillogenesis. *Langmuir* **21**, 6944–9.

Saido TC and Iwata N (2006). Metabolism of amyloid β peptide and pathogenesis of Alzheimer's disease: towards presymptomatic diagnosis, prevention and therapy. *Neuroscience Research* **54**, 235–53.

Saito T, Iwata N, Tsubuki S, *et al.* (2005). Somatostatin regulates brain amyloid β peptide Aβ$_{42}$ through modulation of proteolytic degradation. *Nature Medicine* 11, 434–9.

Sakaue F, Saito T, Sato Y, *et al.* (2005). Phosphorylation of FTDP-17 mutant tau by cyclin-dependent kinase 5 complexed with p35, p25, or p39. *Journal of Biological Chemistry* 280, 31522–9.

Santacruz K, Lewis J, Spires T, *et al.* (2005). Tau suppression in a neurodegenerative mouse model improves memory function. *Science* 309, 476–81.

Santa-Maria I, Hernandez F, Smith MA, *et al.* (2005). Neurotoxic dopamine quinone facilitates the assembly of tau into fibrillar polymers. *Molecular and Cellular Biochemistry* 278, 203–12.

Sastre M, Dewachter I, Rossner S, *et al.* (2006). Nonsteroidal anti-inflammatory drugs repress β-secretase gene promoter activity by the activation of PPARγ. *Proceedings of the National Academy of Sciences of the USA* 103, 443–8.

Sato T, Tanimura Y, Hirotani N, *et al.* (2005). Blocking the cleavage at midportion between γ- and e-sites remarkably suppresses the generation of amyloid β-protein. *FEBS Letters* 579, 2907–12.

Saura CA, Chen G, Malkani S, *et al.* (2005). Conditional inactivation of presenilin 1 prevents amyloid accumulation and temporarily rescues contextual and spatial working memory impairments in amyloid precursor protein transgenic mice. *Journal of Neuroscience* 25, 6755–64.

Schmuck C, Frey P, and Heil M (2005). Inhibition of fibril formation of Aβ by guanidiniocarbonyl pyrrole receptors. *Chembiochem* 6, 628–31.

Schuster D, Rajendran A, Hui SW, *et al.* (2005). Protective effect of colostrinin on neuroblastoma cell survival is due to reduced aggregation of β-amyloid. *Neuropeptides* 39, 419–26.

Selkoe DJ (2005). Defining molecular targets to prevent Alzheimer disease. *Archives of Neurology* 62, 192–5.

Shie FS, Montine KS, Breyer RM, *et al.* (2005). Microglial eEP2 as a new target to increase amyloid β phagocytosis and decrease amyloid β-induced damage to neurons. *Brain Pathology* 15, 134–8.

Si ML, Long C, Yang DI, *et al.* (2005). Statins prevent β-amyloid inhibition of sympathetic a7-nAChR-mediated nitrergic neurogenic dilation in porcine basilar arteries. *Journal of Cerebral Blood Flow and Metabolism* 25, 1573–1585.

Siemers E, Skinner M, Dean RA, *et al.* (2005). Safety, tolerability, and changes in amyloid β concentrations after administration of a γ-secretase inhibitor in volunteers. *Clinical Neuropharmacology* 28, 126–132.

Siemers ER, Quinn JF, Kaye J, *et al.* (2006). Effects of a γ-secretase inhibitor in a randomized study of patients with Alzheimer disease. *Neurology* 66, 602–4.

Singer O, Marr RA, Rockenstein E, *et al.* (2005). Targeting BACE1 with siRNAs ameliorates Alzheimer disease neuropathology in a transgenic model. *Nature Neuroscience*, 8, 1343–9.

Snyder EM, Nong Y, Almeida CG, *et al.* (2005). Regulation of NMDA receptor trafficking by amyloid-β. *Nature Neuroscience* 8, 1051–8.

Solomon B (2005). Generation of anti-β-amyloid antibodies via phage display technology towards Alzheimer's disease vaccination. *Vaccine* 23, 2327–30.

Sultana R, Ravagna A, Mohmmad-Abdul H, *et al.* (2005). Ferulic acid ethyl ester protects neurons against amyloid β- peptide(1–42)-induced oxidative stress and neurotoxicity: relationship to antioxidant activity. *Journal of Neurochemistry* 92, 749–58.

Sutoh Y, Nishino S, and Nishida Y (2005). Metal chelates to prevent or clear the deposits of amyloid β-peptide (1–40) induced by zinc(II) Chloride. *Chemistry Letters* 34, 140.

Szegedi V, Fülöp L, Farkas T, *et al.* (2005). Pentapeptides derived from Aβ1–42 protect neurons from the modulatory effect of Aβ fibrils—an *in vitro* and *in vivo* electrophysiological study. *Neurobiology of Disease* **18**, 499–508.

Taniguchi S, Suzuki N, Masuda M. *et al.* (2005). Inhibition of heparin-induced tau filament formation by phenothiazines, polyphenols, and porphyrins. *Journal of Biological Chemistry* **280**, 7614–23.

Tang BL (2005). Alzheimer's disease: channeling APP to non-amyloidogenic processing. *Biochemical and Biophysical Research Communications* **331**, 375–8.

Tong Y, Zhou W, Fung V, *et al.* (2005). Oxidative stress potentiates BACE1 gene expression and Aβ generation. *Journal of Neural Transmission* **112**, 455–69.

Török M, Abid M, Mhadgut SC, *et al.* (2006). Organofluorine inhibitors of amyloid fibrillogenesis. *Biochemistry* **45**, 5377–83.

Turner RT, 3rd, Hong L, Koelsch G, *et al.* (2005). Structural locations and functional roles of new subsites S_5, S_6, and S_7 in memapsin 2 (β-secretase). *Biochemistry* **44**, 105–12.

Van Dam D, Abramowski D, Staufenbiel M, *et al.* (2005). Symptomatic effect of donepezil, rivastigmine, galantamine and memantine on cognitive deficits in the APP23 model. *Psychopharmacology* **180**, 177–90.

van Es JH, van Gijn ME, Riccio O, *et al.* (2005). Notch/γ-secretase inhibition turns proliferative cells in intestinal crypts and adenomas into goblet cells. *Nature* **435**, 959–63.

Walsh DM, Townsend M, Podlisny MB, *et al.* (2005). Certain inhibitors of synthetic amyloid β-peptide (Aβ) fibrillogenesis block oligomerization of natural Aβ and thereby rescue long-term potentiation. *Journal of Neuroscience* **25**, 2455–62.

Wang J, Ho L, Qin W, *et al.* (2005a). Caloric restriction attenuates β-amyloid neuropathology in a mouse model of Alzheimer's disease. *FASEB Journal* **19**, 659–61.

Wang SS, Chen YT, and Chou SW (2005b). Inhibition of amyloid fibril formation of β-amyloid peptides via the amphiphilic surfactants. *Biochimica et Biophysica Acta* **1741**, 307–13.

Wisniewski T and Frangione B (2005). Immunological and anti-chaperone therapeutic approaches for Alzheimer disease. *Brain Pathology* **15**, 72–7.

Woltjer RL, Nghiem W, Maezawa I, *et al.* (2005). Role of glutathione in intracellular amyloid-a precursor protein/carboxy-terminal fragment aggregation and associated cytotoxicity. *Journal of Neurochemistry* **93**, 1047–56.

Xie J and Guo Q (2005). Par-4 is involved in regulation of β-secretase cleavage of the Alzheimer amyloid precursor protein. *Journal of Biological Chemistry* **280**, 13824–32.

Xie Z, Romano DM, and Tanzi RE (2005a). Effects of RNAi-mediated silencing of PEN-2, APH-1a, and Nicas wild-type vs FAD mutant forms of presenilin 1. *Journal of Molecular Neuroscience* **25**, 67–77.

Xie Z, Romano DM, and Tanzi RE (2005b). RNA interference-mediated silencing of X11a and X11β attenuates amyloid β-protein levels via differential effects on β-amyloid precursor protein processing. *Journal of Biological Chemistry* **280**, 15413–21.

Yamada K, Takayanagi M, Kamei H, *et al.* (2005). Effects of memantine and donepezil on amyloid β-induced memory impairment in a delayed-matching to position task in rats. *Behavioural Brain Research* **162**, 191–99.

Yamamoto N, Yokoseki T, Shibata M, *et al.* (2005). Suppression of Aβ deposition in brain by peripheral administration of Fab fragments of anti-seed antibody. *Biochemical and Biophysical Research Communications* **335**, 45–7.

Yang F, Lim GP, Begum AN, *et al.* (2005a). Curcumin inhibits formation of amyloid β oligomers and fibrils, binds plaques, and reduces amyloid *in vivo*. *Journal of Biological Chemistry* **280**, 5892–901.

Yang SP, Kwon BO, Gho YS, *et al.* (2005b). Specific interaction of VEGF$_{165}$ with β-amyloid, and its protective effect on β-amyloid-induced neurotoxicity. *Journal of Neurochemistry* **93**, 118–27.

Youdim MB, Maruyama W, and Naoi M (2005). Neuropharmacological, neuroprotective and amyloid precursor processing properties of selective MAO-B inhibitor antiparkinsonian drug, rasagiline. *Drugs of Today* **41**, 369–91.

Youm JW, Kim H, Han JH, *et al.* (2005). Transgenic potato expressing Aβ reduce Aβ burden in Alzheimer's disease mouse model. *FEBS Letters* **579**, 6737–44.

Yu MS, Leung SK, Lai SW, *et al.* (2005). Neuroprotective effects of anti-aging oriental medicine *Lycium barbarum* against β-amyloid peptide neurotoxicity. *Experimental Gerontology* **40**, 716–27.

Yue X, Lu M, Lancaster T, *et al.* (2005) Brain estrogen deficiency accelerates Aβ plaque formation in an Alzheimer's disease animal model. *Proceedings of the National Academy of Sciences of the USA* **102**, 19198–203.

Zhang YJ, Xu YF, Chen XQ, *et al.* (2005). Nitration and oligomerization of tau induced by peroxynitrite inhibit its microtubule-binding activity. *FEBS Letters* **579**, 2421–7.

Zheng H, Youdim MB, Weiner LM, *et al.* (2005a). Novel potential neuroprotective agents with both iron chelating and amino acid-based derivatives targeting central nervous system neurons. *Biochemical Pharmacology* **70**, 1642–52.

Zheng H, Youdim MB, Weiner LM, *et al.* (2005b). Synthesis and evaluation of peptidic metal chelators for neuroprotection in neurodegenerative diseases. *Journal of Peptide Research* **66**, 190–203.

Zimmermann M, Borroni B, Cattabeni F, *et al.* (2005). Cholinesterase inhibitors influence APP metabolism in Alzheimer disease patients. *Neurobiology of Disesease* **19**, 237–42.

Zurbriggen R, Amacker M, Kammer AR, *et al.* (2005). Virosome-based active immunization targets soluble amyloid species rather than plaques in a transgenic mouse model of Alzheimer's disease. *Journal of Molecular Neuroscience* **27**, 157–66.

AI Amyloid precursor protein (APP)

A4_HUMAN
Protein name: amyloid precursor protein
Gene Name: *APP* or *A4* or *AD1*
Swiss-Prot primary accession number: P05067
Compiled using information from the Expasy database: http://us.expasy.org/

Length: **770 AA** [unprocessed precursor]

Molecular weight: **86943 Da** [unprocessed precursor]

```
            10         20         30         40         50         60
            |          |          |          |          |          |
   MLPGLALLLL AAWTARALEV PTDGNAGLLA EPQIAMFCGR LNMHMNVQNG KWDSDPSGTK

            70         80         90        100        110        120
            |          |          |          |          |          |
   TCIDTKEGIL QYCQEVYPEL QITNVVEANQ PVTIQNWCKR GRKQCKTHPH FVIPYRCLVG

           130        140        150        160        170        180
            |          |          |          |          |          |
   EFVSDALLVP DKCKFLHQER MDVCETHLHW HTVAKETCSE KSTNLHDYGM LLPCGIDKFR

           190        200        210        220        230        240
            |          |          |          |          |          |
   GVEFVCCPLA EESDNVDSAD AEEDDSDVWW GGADTDYADG SEDKVVEVAE EEEVAEVEEE

           250        260        270        280        290        300
            |          |          |          |          |          |
   EADDDEDDED GDEVEEEAEE PYEEATERTT SIATTTTTTT ESVEEVVREV CSEQAETGPC

           310        320        330        340        350        360
            |          |          |          |          |          |
   RAMISRWYFD VTEGKCAPFF YGGCGGNRNN FDTEEYCMAV CGSAMSQSLL KTTQEPLARD

           370        380        390        400        410        420
            |          |          |          |          |          |
   PVKLPTTAAS TPDAVDKYLE TPGDENEHAH FQKAKERLEA KHRERMSQVM REWEEAERQA

           430        440        450        460        470        480
            |          |          |          |          |          |
   KNLPKADKKA VIQHFQEKVE SLEQEAANER QQLVETHMAR VEAMLNDRRR LALENYITAL

           490        500        510        520        530        540
            |          |          |          |          |          |
   QAVPPRPRHV FNMLKKYVRA EQKDRQHTLK HFEHVRMVDP KKAAQIRSQV MTHLRVIYER
```

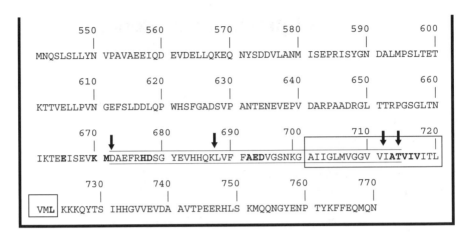

Signal peptide, putative, is underlined.

Putative transmembrane domain is boxed.

1-43 amino acid β-amyloid sequence is underlined and overlined.

Arrows show β-(M671/D672), α-(K687/L688) and γ-secretase (V711/I712 for Aβ40, A713/T714 for Aβ42) cleavage sites in order. For fuller list of cutting sites see below.

Note that this is the notation for APP770, numbers are different for other splice variants.

Sites of mutations thought to cause neurodegeneration are shown in bold

Alternative splice products of APP: Six alternative spliced variants including APP(395), APP(563), APP(695), APP(751) and APP(770). APP (751), and APP(770) contain a protease inhibitor domain belonging to the kunitz family of inhibitors.

Cleavage of APP by α-, β- (β′, β₂) and γ-secretase (sites 40, 42, ε, ζ)

α-Secretase

α-Secretase (ADAMs, e.g. TACE) cleaves at K687/L688 (i.e. between positions 16 and 17 in Aβ). APP is split to form sAPPα and C83.

β-secretase

β-Secretase (BACE1) cleaves at M671/D672 (i.e. between positions 0 and 1 in Aβ). APP is split to form sAPPβ and C99.
β-Secretase (BACE1) also cleaves at G680/Y681 (i.e. between positions 10 and 11 in Aβ). APP is split to form sAPPβ' and C89.
β-Secretase (BACE2) cleaves at F690/F691 (i.e. between positions 19 and 20 in Aβ). APP is split to form sAPPβ2 and C80. BACE2 also cleaves at M671/D672 (i.e. between positions 0 and 1 in Aβ).
(Note taken from Chapter 18. The cleavage site F690/F691 is close to the Flemish mutation at residue 21 (A692G), and the Flemish mutation significantly increases the Aβ production generated by BACE2 but not by BACE1. The Swedish mutation increases Aβ production by both BACE1 and BACE2).

γ-Secretase

γ-Secretase cleaves C83, C89, or C99 within the transmembrane domain commonly between positions 40 and 41 or between positions 42 and 43 in Aβ. γ-secretase also cleaves to form Aβ38, Aβ39 or Aβ43. The APP intracellular domain (AICD) is formed from γ-secretase cleavage.
C83 cleavage results in the production of the p3 peptide (Aβ17-40/42).
C80 cleavage results in the production of Aβ20-40/42.
C89 cleavage results in the production of Aβ11-40/42.
The ε (epsilon) cleavage site occurs between Aβ48 and 49, Aβ49 and 50 and also Aβ51 and 52 and corresponds to the γ-secretase cleavage site in Notch).
The ζ (zeta) cleavage occurs between Aβ46 and 47.
For fuller explanations of cleavages see Chapters 4, 8, and 17

Mutations in the APP gene, close to the secretase cleavage sites

Compiled using information from the following databases:
http://www.molgen.ua.ac.be/ADMutations/
http://www.alzforum.org/

- E665D   (in AD; late onset ~86 years)
- KM670/671NL   (in AD; familial; onset 44–59 years)
- A673T   *(normal)*
- H677R   (in AD; onset 55–56 years)
- D678G   (in AD; familial; onset 60 years)
- A692G   (in AD; Flemish type; large dense core plaques, more CAA than AD, onset 40–60 years)
- E693G   (in AD; arctic type; similar to idiopathic AD, onset 58 years)
- E693K   (in HCHWAD; Italian type; cerebral haemorrhage, similar to Dutch)
- E693Q   (in HCHWAD; Dutch type; CAA plus parenchymal Aβ, few NFT, onset ~50 years)
- D694N   (in HCHWAD; Iowa type; severe CAA, AD or cerebral haemorrhage; onset 69)
- A713T   (in AD; familial; onset 59 years)
- A713V   *(in one chronic schizophrenia patient; could be a polymorphism)*
- T714 I   (in AD; familial; Austrian, increased Aβ42/Aβ40 ratio)
- T714A   (in AD; familial; Iranian; onset 40–60 years)
- V715 M   (in AD; familial; French; decreased Aβ40/total Aβ; onset 40–60 years)
- V715A   (in AD; familial; German; onset 47 years)
- I716V   (in AD; familial; Florida; onset 55 years)
- I716T   (in AD; familial; onset 55 years)
- V717F   (in AD; familial; Indiana; onset 42–52 years)
- V717G   (in AD; familial; onset 45–62 years)
- V717I   (in AD; familial; London; onset 50–60 years)
- V717L   (in AD; familial; onset 38 years)
- L723P   (in AD; familial; Australian; onset 45–60 years)

All α-Secretase

α-secretase activity is known to comprise at least three enzymes of the ADAMs family (a disintegrin and metalloprotease): ADAM 9, ADAM10, and ADAM17. There is debate as to which of these contribute mainly to basal or constitutive activity and which to stimulated activity. TACE is, on balance, probably most important in stimulation-dependent activity (see Chapter 4).

TACE (ADAM 17)

ADA17_HUMAN
Protein name: ADAM 17 [precursor]
Synonyms: a disintegrin and metalloproteinase domain 17, TNF-alpha-converting enzyme, TNF-alpha convertase, snake venom-like protease, CD156b antigen
Gene name: *ADAM17* also *CSVP*, *TACE*
Swiss-Prot primary accession number: P78536
Compiled using information from the Expasy database: http://us.expasy.org/

Length: **824 aa** [unprocessed precursor]
Molecular weight: **93021 Da** [unprocessed precursor]

```
          10         20         30         40         50         60
          |          |          |          |          |          |
MRQSLLFLTS VVPFVLAPRP PDDPGFGPHQ RLEKLDSLLS DYDILSLSNI QQHSVRKRDL

          70         80         90        100        110        120
          |          |          |          |          |          |
QTSTHVETLL TFSALKRHFK LYLTSSTERF SQNFKVVVVD GKNESEYTVK WQDFFTGHVV

         130        140        150        160        170        180
          |          |          |          |          |          |
GEPDSRVLAH IRDDDVIIRI NTDGAEYNIE PLWRFVNDTK DKRMLVYKSE DIKNVSRLQS

         190        200        210        220        230        240
          |          |          |          |          |          |
PKVCGYLKVD NEELLPKGLV DREPPEELVH RVKRRADPDP MKNTCKLLVV ADHRFYRYMG

         250        260        270        280        290        300
          |          |          |          |          |          |
RGEESTTTNY LIELIDRVDD IYRNTSWDNA GFKGYGIQIE QIRILKSPQE VKPGEKHYNM

         310        320        330        340        350        360
          |          |          |          |          |          |
AKSYPNEEKD AWDVKMLLEQ FSFDIAEEAS KVCLAHLFTY QDFDMGTLGL AYVGSPRANS

         370        380        390        400        410        420
          |          |          |          |          |          |
HGGVCPKAYY SPVGKKNIYL NSGLTSTKNY GKTILTKEAD LVTTHELGHN FGAEHDPDGL

         430        440        450        460        470        480
          |          |          |          |          |          |
AECAPNEDQG GKYVMYPIAV SGDHENNKMF SNCSKQSIYK TIESKAQECF QERSNKVCGN

         490        500        510        520        530        540
          |          |          |          |          |          |
SRVDEGEECD PGIMYLNNDT CCNSDCTLKE GVQCSDRNSP CCKNCQFETA QKKCQEAINA

         550        560        570        580        590        600
          |          |          |          |          |          |
TCKGVSYCTG NSSECPPPGN AEDDTVCLDL GKCKDGKCIP FCEREQQLES CACNETDNSC

         610        620        630        640        650        660
          |          |          |          |          |          |
KVCCRDLSGR CVPYVDAEQK NLFLRKGKPC TVGFCDMNGK CEKRVQDVIE RFWDFIDQLS

         670        680        690        700        710        720
          |          |          |          |          |          |
INTFGKFLAD NIVGSVLVFS LIFWIPFSIL VHCVDKKLDK QYESLSLFHP SNVEMLSSMD

         730        740        750        760        770        780
          |          |          |          |          |          |
SASVRIIKPF PAPQTPGRLQ PAPVIPSAPA APKLDHQRMD TIQEDPSTDS HMDEDGFEKD

         790        800        810        820
          |          |          |          |
PFPNSSTAAK SFEDLTDHPV TRSEKAASFK LQRQNRVDSK ETEC
```

Signal peptide, putative, is underlined (aa 1–17).

Propeptide is in bold (aa 18–214)

Potential transmembrane domain is underlined and overlined (aa 672–692)

Extracellular region (potential) aa 215–671; cytoplasmic (potential) 693–824

AIII – β-SECRETASE

β-secretase activity comprises the action of two transmembrane aspartyl proteases named BACE1 and BACE2. BACE1 is responsible for the majority of β-secretase cleavage of APP in brain.

BACE1

Compiled using information from the Expasy database: http://us.expasy.org/
BACE1_HUMAN
Protein Name: Beta-secretase
Synonyms: beta-site APP cleaving enzyme, beta-site amyloid precursor protein cleaving enzyme; aspartyl protease 2, ASP2
Gene name: BACE
Swiss-Prot primary accession number: P56817

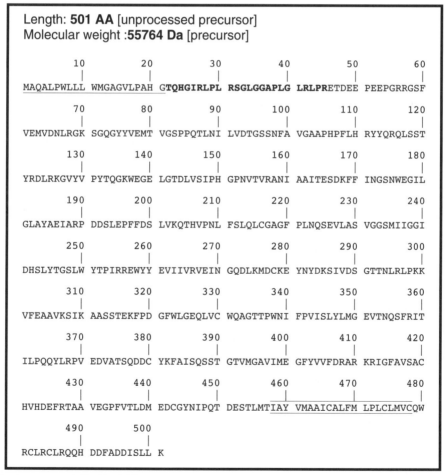

Length: **501 AA** [unprocessed precursor]
Molecular weight :**55764 Da** [precursor]

```
         10          20          30          40          50          60
          |           |           |           |           |           |
MAQALPWLLL WMGAGVLPAH GTQHGIRLPL RSGLGGAPLG LRLPRETDEE PEEPGRRGSF

         70          80          90         100         110         120
          |           |           |           |           |           |
VEMVDNLRGK SGQGYYVEMT VGSPPQTLNI LVDTGSSNFA VGAAPHPFLH RYYQRQLSST

        130         140         150         160         170         180
          |           |           |           |           |           |
YRDLRKGVYV PYTQGKWEGE LGTDLVSIPH GPNVTVRANI AAITESDKFF INGSNWEGIL

        190         200         210         220         230         240
          |           |           |           |           |           |
GLAYAEIARP DDSLEPFFDS LVKQTHVPNL FSLQLCGAGF PLNQSEVLAS VGGSMIIGGI

        250         260         270         280         290         300
          |           |           |           |           |           |
DHSLYTGSLW YTPIRREWYY EVIIVRVEIN GQDLKMDCKE YNYDKSIVDS GTTNLRLPKK

        310         320         330         340         350         360
          |           |           |           |           |           |
VFEAAVKSIK AASSTEKFPD GFWLGEQLVC WQAGTTPWNI FPVISLYLMG EVTNQSFRIT

        370         380         390         400         410         420
          |           |           |           |           |           |
ILPQQYLRPV EDVATSQDDC YKFAISQSST GTVMGAVIME GFYVVFDRAR KRIGFAVSAC

        430         440         450         460         470         480
          |           |           |           |           |           |
HVHDEFRTAA VEGPFVTLDM EDCGYNIPQT DESTLMTIAY VMAAICALFM LPLCLMVCQW

        490         500
          |           |
RCLRCLRQQH DDFADDISLL K
```

Signal sequence underlined (aa 1-21)

Pro-domain in bold (aa 22-45)

Putative transmembrane domain underlined and overlined (aa 458-478)

BACE 2

Compiled using information from the Expasy database: http://us.expasy.org/
BACE2_HUMAN
Protein Name: Beta-secretase 2
Synonym(s): Beta-site APP-cleaving enzyme 2, Aspartyl protease 1, Asp 1, ASP1
Membrane-associated aspartic protease 1, Memapsin-1, Down region aspartic protease
Gene name: BACE2
Swiss-Prot primary accession number: Q9Y5Z0

Length: **518 AA** [unprocessed precursor]
Molecular weight: **56180 Da**[unprocessed precursor]

```
        10         20         30         40         50         60
MGALARALLL PLLAQWLLRA APELAPAPFT LPLRVAAATN RVVAPTPGPG TPAERHADGL

        70         80         90        100        110        120
ALALEPALAS PAGAANFLAM VDNLQGDSGR GYYLEMLIGT PPQKLQILVD TGSSNFAVAG

       130        140        150        160        170        180
TPHSYIDTYF DTERSSTYRS KGFDVTVKYT QGSWTGFVGE DLVTIPKGFN TSFLVNIATI

       190        200        210        220        230        240
FESENFFLPG IKWNGILGLA YATLAKPSSS LETFFDSLVT QANIPNVFSM QMCGAGLPVA

       250        260        270        280        290        300
GSGTNGGSLV LGGIEPSLYK GDIWYTPIKE EWYYQIEILK LEIGGQSLNL DCREYNADKA

       310        320        330        340        350        360
IVDSGTTLLR LPQKVFDAVV EAVARASLIP EFSDGFWTGS QLACWTNSET PWSYFPKISI

       370        380        390        400        410        420
YLRDENSSRS FRITILPQLY IQPMMGAGLN YECYRFGISP STNALVIGAT VMEGFYVIFD

       430        440        450        460        470        480
RAQKRVGFAA SPCAEIAGAA VSEISGPFST EDVASNCVPA QSLSEPILWI VSYALMSVCG

       490        500        510
AILLVLIVLL LLPFRCQRRP RDPEVVNDES SLVRHRWK
```

Signal sequence, putative, underlined (aa 1-20)

Pro-domain unknown –residues 21 onwards (probably 20-30 residues long)

Potential transmembrane domain overlined and underlined (aa 474-494)

AIV γ-Secretase

γ-Secretase activity can be reconstituted by the coexpression of four components: presenilin (PS: PS1, or PS2), nicastrin, APH1 (APH1A or APH1B), and PEN2.

PRESENILIN 1 (PS1)

PSN1_HUMAN
Protein name: Human Presenilin 1 or PS-1 or S182
Gene name: *PSEN1* or *PSNL1* or *AD3* or *PS1*
Swiss-Prot primary accession number: P49768
Compiled using information from the Expasy database: http://us.expasy.org/

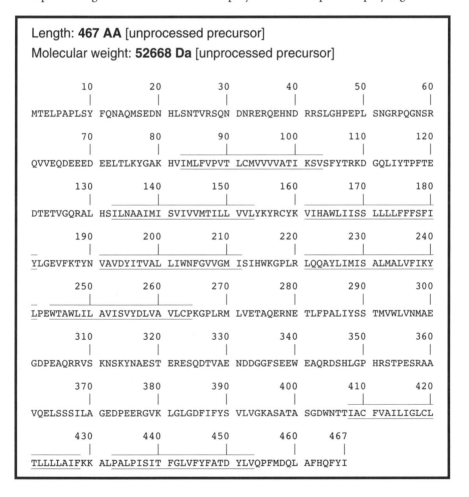

Length: **467 AA** [unprocessed precursor]
Molecular weight: **52668 Da** [unprocessed precursor]

```
          10         20         30         40         50         60
           |          |          |          |          |          |
     MTELPAPLSY FQNAQMSEDN HLSNTVRSQN DNRERQEHND RRSLGHPEPL SNGRPQGNSR

          70         80         90        100        110        120
           |          |          |          |          |          |
     QVVEQDEEED EELTLKYGAK HVIMLFVPVT LCMVVVVATI KSVSFYTRKD GQLIYTPFTE

         130        140        150        160        170        180
           |          |          |          |          |          |
     DTETVGQRAL HSILNAAIMI SVIVVMTILL VVLYKYRCYK VIHAWLIISS LLLLFFFSFI

         190        200        210        220        230        240
           |          |          |          |          |          |
     YLGEVFKTYN VAVDYITVAL LIWNFGVVGM ISIHWKGPLR LQQAYLIMIS ALMALVFIKY

         250        260        270        280        290        300
           |          |          |          |          |          |
     LPEWTAWLIL AVISVYDLVA VLCPKGPLRM LVETAQERNE TLFPALIYSS TMVWLVNMAE

         310        320        330        340        350        360
           |          |          |          |          |          |
     GDPEAQRRVS KNSKYNAEST ERESQDTVAE NDDGGFSEEW EAQRDSHLGP HRSTPESRAA

         370        380        390        400        410        420
           |          |          |          |          |          |
     VQELSSSILA GEDPEERGVK LGLGDFIFYS VLVGKASATA SGDWNTTIAC FVAILIGLCL

         430        440        450        460        467
           |          |          |          |          |
     TLLLLAIFKK ALPALPISIT FGLVFYFATD YLVQPFMDQL AFHQFYI
```

No signal peptide.

Presenilin-1 NTF subunit aa 1–298.

Presenilin-1 CTF subunit aa 299–467.

Putative transmembrane domains underlined and overlined: 83–103, 133–153, 161–181, 191–211, 221–241, 244–264, 408–428, 433–453.

Mutation D263A (in bold): reduces production of Aβ and AICD

Mutation D366A (in bold): reduces production of Aβ and AICD, and NICD in NOTCH1 processing

Mutations in PS1 known to cause early-onset Alzheimer's disease

Compiled using information from the following databases:

http://www.molgen.ua.ac.be/ADMutations/

http://www.alzforum.org/

R35Q	G206S/D/A/V/S	R358Q
A79V	G209V/E/V	S365Y
V82L	I213T/F/T	R377M
ΔI83/M84	H214Y	G378E/V
L85P	G217D	L381V
V89L	L219F/P	G384A
C92S	Q222R/H	F386S
V94M	L226R	S390I
V96F	I229F	V391F
F105I/L/Q/P	A231T/V	L392V/P
Y115C/H/D	M233T/L/V	G394V
T116N/I/S/R/L	L235V/P	N405S
E120D/K	F237I/L	A409T
E123K	L235P	C410Y
N135D/S	A246E	L418F
M139I/T/V/K	L250S/V	L424H/R
I143F/T	256S	A426P
I143N/M	A260V	A431E/V
M146I/L/V	V261F	A434C
T147I	L262F	L435F
L153V	C263R/F	P436Q/S
Y154N/C	P264L	I439V
ΔI167	P267S/T/L	ΔT440

H163R/Y	R269G/H	
W165G/C	L271V	PLUS:
L166H/P/R	V272A	Intron 4 2302ΔG
S169L/P	E273A	Five mutations of ΔE9
S170F	T274R	Two insertions
L171P	R278T/I/K	
L173W	E280A/G	
L174M/R	L282V/R	
F175S Not pathogenic	P284L	
F177L/S	A285V	
S178P	L286V	
G183V	E318G Not pathogenic	
E184D	T354I	

PRESENILIN 2 (PS2)

PSN2_HUMAN
Protein name: Human Presenilin-2 or PS-2 or STM-2
Gene name: *PSEN2* or *PSNL2* or *AD4* or *PS2* or *STM2*
Swiss-Prot primary accession number: P49810
Compiled using information from the Expasy database: http://us.expasy.org/

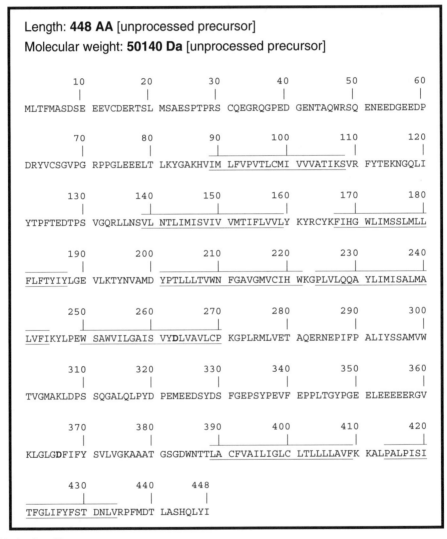

Length: **448 AA** [unprocessed precursor]

Molecular weight: **50140 Da** [unprocessed precursor]

```
          10          20          30          40          50          60
           |           |           |           |           |           |
     MLTFMASDSE  EEVCDERTSL  MSAESPTPRS  CQEGRQGPED  GENTAQWRSQ  ENEEDGEEDP

          70          80          90         100         110         120
           |           |           |           |           |           |
     DRYVCSGVPG  RPPGLEEELT  LKYGAKHVIM  LFVPVTLCMI  VVVATIKSVR  FYTEKNGQLI

         130         140         150         160         170         180
           |           |           |           |           |           |
     YTPFTEDTPS  VGQRLLNSVL  NTLIMISVIV  VMTIFLVVLY  KYRCYKFIHG  WLIMSSLMLL

         190         200         210         220         230         240
           |           |           |           |           |           |
     FLFTYIYLGE  VLKTYNVAMD  YPTLLLTVWN  FGAVGMVCIH  WKGPLVLQQA  YLIMISALMA

         250         260         270         280         290         300
           |           |           |           |           |           |
     LVFIKYLPEW  SAWVILGAIS  VYDLVAVLCP  KGPLRMLVET  AQERNEPIFP  ALIYSSAMVW

         310         320         330         340         350         360
           |           |           |           |           |           |
     TVGMAKLDPS  SQGALQLPYD  PEMEEDSYDS  FGEPSYPEVF  EPPLTGYPGE  ELEEEEERGV

         370         380         390         400         410         420
           |           |           |           |           |           |
     KLGLGDFIFY  SVLVGKAAAT  GSGDWNTTLA  CFVAILIGLC  LTLLLLAVFK  KALPALPISI

         430         440         448
           |           |           |
     TFGLIFYFST  DNLVRPFMDT  LASHQLYI
```

No signal peptide.

Presenilin-1 NTF subunit aa 1–297 .

Presenilin-1 CTF subunit aa 298–448.

Putative transmembrane domains underlined and overlined: 88–108, 139–159, 167–187, 201–221, 224–244, 250–270, 389–409, 414–434.

Mutation D263A (in bold): reduces production of Aβ and AICD.

Mutation D366A (in bold): reduces production of Aβ and AICD, and NICD in NOTCH1 processing.

Mutations in PS2 known to cause early onset Alzheimer's disease

Compiled using information from the following databases.
http://www.molgen.ua.ac.be/ADMutations/
http://www.alzforum.org/

- R62H   (in AD, mean age onset 62 years)
- T122P   (in AD, mean age onset 47.9 years)
- S130L   (in AD, mean age onset 65 years)
- N141I   (in AD; Volga German, mean age onset 56.9 years, range 57–62 years)
- V148I   (in AD; LOAD; Spanish mean age onset 71 years)
- Q228L   (in AD; mean age onset 70 years)
- M239I   (in AD; mean age onset 50.7 years)
- M239V   (in AD; Italian; mean age onset 50 years)
- P334R   *(not pathogenic, detected in 1/165 unrelated individuals.)*
- T430M   (in AD; mean age onset 56.3 years)
- D439A   (in AD; mean age onset 56 years)

Nicastrin

NICA_HUMAN
Protein name: Nicastrin
Gene name: NCSTN
Swiss-Prot primary accession number: Q92542
Compiled using information from the Expasy database: http://us.expasy.org/

Length: 709 AA [unprocessed precursor]

Molecular weight: 78411 Da [unprocessed precursor]

```
          10         20         30         40         50         60
           |          |          |          |          |          |
MATAGGGSGA DPGSRGLLRL LSFCVLLAGL CRGNSVERKI YIPLNKTAPC VRLLNATHQI

          70         80         90        100        110        120
           |          |          |          |          |          |
GCQSSISGDT GVIHVVEKEE DLQWVLTDGP NPPYMVLLES KHFTRDLMEK LKGRTSRIAG

         130        140        150        160        170        180
           |          |          |          |          |          |
LAVSLTKPSP ASGFSPSVQC PNDGFGVYSN SYGPEFAHCR EIQWNSLGNG LAYEDFSFPI

         190        200        210        220        230        240
           |          |          |          |          |          |
FLLEDENETK VIKQCYQDHN LSQNGSAPTF PLCAMQLFSH MHAVISTATC MRRSSIQSTF

         250        260        270        280        290        300
           |          |          |          |          |          |
SINPEIVCDP LSDYNVWSML KPINTTGTLK PDDRVVVAAT RLDSRSFFWN VAPGAESAVA

         310        320        330        340        350        360
           |          |          |          |          |          |
SFVTQLAAAE ALQKAPDVTT LPRNVMFVFF QGETFDYIGS SRMVYDMEKG KFPVQLENVD

         370        380        390        400        410        420
           |          |          |          |          |          |
SFVELGQVAL RTSLELWMHT DPVSQKNESV RNQVEDLLAT LEKSGAGVPA VILRRPNQSQ

         430        440        450        460        470        480
           |          |          |          |          |          |
PLPPSSLQRF LRARNISGVV LADHSGAFHN KYYQSIYDTA ENINVSYPEW LSPEEDLNFV

         490        500        510        520        530        540
           |          |          |          |          |          |
TDTAKALADV ATVLGRALYE LAGGTNFSDT VQADPQTVTR LLYGFLIKAN NSWFQSILRQ

         550        560        570        580        590        600
           |          |          |          |          |          |
DLRSYLGDGP LQHYIAVSSP TNTTYVVQYA LANLTGTVVN LTREQCQDPS KVPSENKDLY

         610        620        630        640        650        660
           |          |          |          |          |          |
EYSWVQGPLH SNETDRLPRC VRSTARLARA LSPAFELSQW SSTEYSTWTE SRWKDIRARI

         670        680        690        700
           |          |          |          |
FLIASKELEL ITLTVGFGIL IFSLIVTYCI NAKADVLFIA PREPGAVSY
```

Signal sequence, putative, underlined (aa 1–33)

Potential transmembrane domain overlined and underlined (aa 670–690)

Cytoplasmic domain (potential) 691–709

Extracellular domain (potential) 34–669

APH1A

APH1A_HUMAN
Protein name: gamma-secretase subunit APH-1A
Synonyms: APH-1a, Aph-1alpha, presenilin-stabilization factor
Gene name: *APH1A* or *PSF*
Swiss-Prot primary accession number: Q96BI3
Compiled using information from the Expasy database: http://us.expasy.org/

Length: 265 AA [unprocessed precursor]

Molecular weight: 28996 Da [unprocessed precursor]

```
         10         20         30         40         50         60
          |          |          |          |          |          |
MGAAVFFGCT FVAFGPAFAL FLITVAGDPL RVIILVAGAF FWLVSLLLAS VVWFILVHVT

         70         80         90        100        110        120
          |          |          |          |          |          |
DRSDARLQYG LLIFGAAVSV LLQEVFRFAY YKLLKKADEG LASLSEDGRS PISIRQMAYV

        130        140        150        160        170        180
          |          |          |          |          |          |
SGLSFGIISG VFSVINILAD ALGPGVVGIH GDSPYYFLTS AFLTAAIILL HTFWGVVFFD

        190        200        210        220        230        240
          |          |          |          |          |          |
ACERRRYWAL GLVVGSHLLT SGLTFLNPWY EASLLPIYAV TVSMGLWAFI TAGGSLRSIQ

        250        260
          |          |
RSLLCRRQED SRVMVYSALR IPPED
```

Putative transmembrane domains underlined and overlined: 3–23, 32–52, 69–89, 119–139, 159–179, 187–207, 214–234

APH 1B

APH1B_HUMAN
Protein name: gamma-secretase subunit APH-1B
Synonyms: APH-1b, Aph-1beta
Gene name: *APH1B*
Swiss-Prot primary accession number: Q8WW43
Compiled using information from the Expasy database: http://us.expasy.org/

```
Length: 257 AA [unprocessed precursor]
Molecular weight: 28460 Da [unprocessed precursor]

          10         20         30         40         50         60
          |          |          |          |          |          |
MTAAVFFGCA FIAFGPALAL YVFTIATEPL RIIFLIAGAF FWLVSLLISS LVWFMARVII

          70         80         90        100        110        120
          |          |          |          |          |          |
DNKDGPTQKY LLIFGAFVSV YIQEMFRFAY YKLLKKASEG LKSINPGETA PSMRLLAYVS

         130        140        150        160        170        180
          |          |          |          |          |          |
GLGFGIMSGV FSFVNTLSDS LGPGTVGIHG DSPQFFLYSA FMTLVIILLH VFWGIVFFDG

         190        200        210        220        230        240
          |          |          |          |          |          |
CEKKKWGILL IVLLTHLLVS AQTFISSYYG INLASAFIIL VLMGTWAFLA AGGSCRSLKL

         250
          |
CLLCQDKNFL LYNQRSR
```

Putative transmembrane domains underlined and overlined: 5–25, 32–52, 71–91, 115–135, 158–178, 186–206, 213–233

PEN-2

PEN2_HUMAN
Protein name: gamma-secretase subunit PEN-2
Synonym: presenilin enhancer protein 2
Gene name: *PSENEN* or *PEN2*
Swiss-Prot primary accession number: Q9NZ42
Compiled using information from the Expasy database: http://us.expasy.org/

```
Length: 101 AA [unprocessed precursor]
Molecular weight: 12029 Da [unprocessed precursor]

          10         20         30         40         50         60
          |          |          |          |          |          |
MNLERVSNEE KLNLCRKYYL GGFAFLPFLW LVNIFWFFRE AFLVPAYTEQ SQIKGYVWRS

          70         80         90        100
          |          |          |          |
AVGFLFWVIV LTSWITIFQI YRPRWGALGD YLSFTIPLGT P
```

Putative transmembrane domains underlined and overlined: 18–42, 61–81

AV Apolipoprotein E (APOE)

APOE_HUMAN
 Protein name: human apolipoprotein E or Apo-E
 Gene name: *APOE*
 Swiss-Prot primary accession number: P02649
 Compiled using information from the Expasy database: http://us.expasy.org/

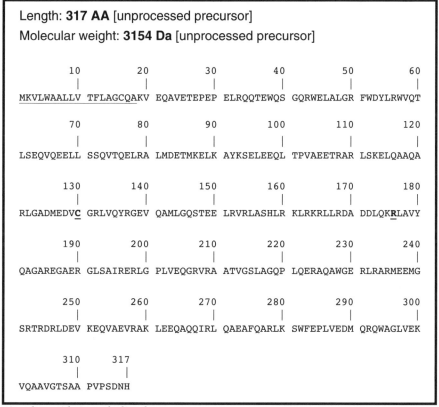

Length: **317 AA** [unprocessed precursor]

Molecular weight: **3154 Da** [unprocessed precursor]

```
            10         20         30         40         50         60
            |          |          |          |          |          |
     MKVLWAALLV TFLAGCQAKV EQAVETEPEP ELRQQTEWQS GQRWELALGR FWDYLRWVQT

            70         80         90        100        110        120
            |          |          |          |          |          |
     LSEQVQEELL SSQVTQELRA LMDETMKELK AYKSELEEQL TPVAEETRAR LSKELQAAQA

           130        140        150        160        170        180
            |          |          |          |          |          |
     RLGADMEDVC GRLVQYRGEV QAMLGQSTEE LRVRLASHLR KLRKRLLRDA DDLQKRLAVY

           190        200        210        220        230        240
            |          |          |          |          |          |
     QAGAREGAER GLSAIRERLG PLVEQGRVRA ATVGSLAGQP LQERAQAWGE RLRARMEEMG

           250        260        270        280        290        300
            |          |          |          |          |          |
     SRTRDRLDEV KEQVAEVRAK LEEQAQQIRL QAEAFQARLK SWFEPLVEDM QRQWAGLVEK

           310        317
            |          |
     VQAAVGTSAA PVPSDNH
```

Signal peptide is underlined
Two amino acids in polymorphisms shown in bold and underlined.
ε3: C112, R158 (as shown here, including 18 aa signal peptide is C130 and R176).
ε2: C112, C158.
ε4: R112, R158.

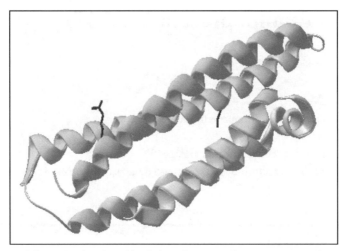

Fig. V1 Structure of Apo-E3 showing side chains of C112 and R158. Cartoon of NMR resolved structure of apoE3 showing mainly α-helix formation (Brookhaven code 1LPE).

Index

Note: Entries beginning with α, β or γ are indexed under alpha, beta or gamma respectively. Entries appear in their unabbreviated form; a list of abbreviations appears on pp. xv–xviii